Geohazards in Engineering Geology

Geological Society Engineering Geology Special Publications
Series Editor M. E. BARTON

It is recommended that reference to all or part of this book should be made in one of the following ways:

MAUND, J. G. & EDDLESTON, M. (eds) 1998. *Geohazards in Engineering Geology*. Geological Society, London, Engineering Geology Special Publications, **15**.

MOORE, R., CLARK, A. R. & LEE, E. M. Coastal cliff behaviour and management: Blackgang, Isle of White. *In:* MAUND, J. G. & EDDLESTON, M. (eds) 1998. *Geohazards in Engineering Geology*. Geological Society, London, Engineering Geology Special Publications, **15**, 49–59.

Geological Society Engineering Geology Special Publication No. 15

Geohazards in Engineering Geology

EDITED BY

Julian G. Maund
Opus International Consultants Ltd
100 Willis Street, Wellington, New Zealand

Malcolm Eddleston
Bechtel Water Technology
Chadwick House, Warrington Road, Risley
Warrington WA3 6AE, UK

1998
Published by
The Geological Society
London

THE GEOLOGICAL SOCIETY

The Society was founded in 1807 as The Geological Society of London and is the oldest geological society in the world. It received its Royal Charter in 1825 for the purpose of 'investigating the mineral structure of the Earth'. The Society is Britain's national society for geology with a membership of around 8500. It has countrywide coverage and approximately 1500 members reside overseas. The Society is responsible for all aspects of the geological sciences including professional matters. The Society has its own publishing house, which produces the Society's international journals, books and maps, and which acts as the European distributor for publications of the American Association of Petroleum Geologists, SEPM and the Geological Society of America.

Fellowship is open to those holding a recognized honours degree in geology or cognate subject and who have at least two years' relevant postgraduate experience, or who have not less than six years' relevant experience in geology or a cognate subject. A Fellow who has not less than five years' relevant postgraduate experience in the practice of geology may apply for validation and, subject to approval, may be able to use the designatory letters C Geol (Chartered Geologist).

Further information about the Society is available from the Membership Manager, The Geological Society, Burlington House, Piccadilly, London W1V 0JU, UK. The Society is a Registered Charity, No. 210161.

Published by The Geological Society from:
The Geological Society Publishing House
Unit 7 Brassmill Enterprise Centre
Brassmill Lane
Bath BA1 3JN
UK
(*Orders*: Tel. 01225 445046
Fax 01225 442836)

First published 1998

British Library Cataloguing in Publication Data
A catalogue record for this book is available from the British Library.

ISBN 1-86239-012-6
ISSN 0267-9914

Typeset by Aarontype, Bristol, UK.

Printed by The Alden Press, Osney Mead, Oxford, UK.

Distributors

USA
AAPG Bookstore
PO Box 979
Tulsa
OK 74101-0979
USA
(*Orders*: Tel. (918) 584-2555
Fax (918) 560-2632)

Australia
Australian Mineral Foundation
63 Conyngham Street
Glenside
South Australia 5065
Australia
(*Orders*: Tel. (08) 379-0444
Fax (08) 379-4634)

India
Affiliated East-West Press PVT Ltd
G-1/16 Ansari Road
New Delhi 110 002
India
(*Orders*: Tel. (11) 327-9113
Fax (11) 326-0538)

Japan
Kanda Book Trading Co.
Cityhouse Tama 204
Tsurumaki 1-3-10
Tama-Shi
Tokyo 0206-0034
Japan
(*Orders*: Tel. (0423) 57-7650
Fax (0423) 57-7651)

Contents

Preface

Geohazards was the theme of the 31st annual conference of the Engineering Group of the Geological Society, which was held at the University of Coventry from 10 to 14 September 1995. The theme was chosen as it was mid-way through the United Nations designated International Decade for Natural Disaster Reduction (IDNDR). It seemed appropriate to hold a conference on the recognition of geohazards and what measures are being used for their reduction.

The conference itself attracted a wide range of papers on geohazards from all over the world, dealing with the obvious and spectacular such as volcanic eruptions, to the more insidious 'non-spectacular', but of far greater economic importance, such as soil mineralogy. With regard to volcanic geohazards we are grateful for Dr Bill McGuire who, coming from 'outside' engineering geology, gave a presentation on volcanic hazards, which helped to provide a perspective to the conference. The scope of geohazards is large and this volume gives a good indication of this, while not claiming to be totally comprehensive. The papers focus predominantly on natural geohazards, however consideration has been given in Section 7 to geohazards associated with contaminated land. In the paper by Bell *et al.* contaminated land is natural in the sense that is the consequence of exploitation of natural resources.

To the material presented at the 1995 conference additional papers have been added from eminent academics and practitioners in respective areas of geohazards to provide an authoritative volume on the characterization and mitigation of geohazards.

As we move towards the next decade and millennium there is no doubt that geohazards will continue to exact a toll on loss of life, livelihood and property. However, research into mitigation in this volume is a small but significant contribution to help the reduction of natural hazards, by increasing our understanding of the engineering geology of the natural and built world.

The compilation of this volume has been a lengthy task, which was started by Steve Penn who convened the conference at the University of Coventry. We thank the other committee members who, largely working behind the scenes, made the conference a success. Finally we are fortunate in having David Ogden of the Geological Society Publishing House for advice and guidance and for bringing the whole volume together both literally and metaphorically.

Julian Maund
Wellington
New Zealand
June 1998

Organizing Committee

Chair	Mr Martin Culshaw	British Geological Society
Treasurer	Mr Alan Forster	British Geological Society
Secretary	Mr Stephen Penn	Coventry University
Trade Exhibition	Mr Peter Fenning	Earth Science Systems Ltd
Publicity	Dr Angull Berry	Kingston University
Technical	Dr Julian Maund	Frank Graham Cons. Engs.
Local Organization	Mr T. Davis	Coventry University
	Mr C. Hobday	Coventry University
	Mr L Moseley	Coventry University
	Mr J. Perkins	Coventry University

SECTION 1

COASTAL AND FLUVIAL GEOHAZARDS

The problem of flooding in Ladysmith, Natal, South Africa

F. G. Bell & T. R. Mason

Department of Geology and Applied Geology, University of Natal, Durban 4041, South Africa

Abstract. Ladysmith was founded in the mid-19th century alongside one of the meanders of the Klip River. The location was chosen for protection against native tribesmen, but in the last 110 years, Ladysmith has experienced 29 notable floods. The most recent floods occurred in February 1994, when the highest water levels in the last 71 years were recorded. An attempt to control flooding was made with the construction of the Windsor Dam in 1949. Unfortunately, the reservoir has been largely silted up so that its live storage capacity has been reduced to around 5%. The present flood walls in Ladysmith can be overtopped by a flood with a recurrence period of once every five years. This happens when channel flow exceeds 700 $m^3 s^{-1}$.

A number of flood alleviation schemes have been discussed over the years including the construction of artificial levees and river canalization. The scheme in vogue at present would involve the construction of another dam and reservoir. Although the reservoir would offer protection to the town, to do so it would have to remain at 10% of its storage capacity in order to retain a 100 year flood. An alternative solution, which would be less expensive, would be to relocate those properties affected most by the recent floods.

Introduction

Ladysmith is situated in the Natal Midlands. It is located on the flood-plain of the Klip River, approximately 7 km downstream of the confluence between the Klip and its tributary, the Sand River. The town was founded in 1851 by the Imperial British Government to establish greater control over that area of recently annexed Natal. The problems of flooding are directly related to its location, the original settlements being situated in a bend of the Klip River which afforded the residents protection from marauding natives. The meander that was chosen encloses that part of the town which experiences the worst flood conditions. Since then the town has grown along the meander belt of the Klip River and so is subject to periodic flooding. In addition, the meanders tend to migrate laterally and downstream, so undermining the river banks. The area worst affected by the river floods covers approximately 8 km^2.

The geology in the Ladysmith area consists of shales, mudstones and fine-grained sandstones assigned to the Ecca and Beaufort Groups of the Karoo Supergroup (late Carboniferous to early Permian). These were intruded by doleritic sills and dykes of Jurassic age. Most of the area drained by the Klip and Sand Rivers is an extensive plain of low relief, interrupted by dolerite-capped hills. Due to the generally low gradient of the river the velocity of the flood water is low relative to its volume. This means that there is usually enough time to issue evacuation warnings. The response time of a flood in the area is about 20 h, i.e. the time taken for rain to travel from where it falls to the gauging station (V1HO38-AO1) at Ladysmith. In fact there is a flood warning system in use. This involves relaying reports of heavy rainfall in the catchment area of the Klip River to Ladysmith in order to give the population a chance to move to safer ground when necessary. The system allows for the prediction of the size of the flood wave and its time of arrival.

The Klip River has its source in the Drakensberg Mountains some 40 km to the west of Ladysmith. The high rugged source area and the relatively flat area around Ladysmith play a part in the problem of flooding in that river water which has been constrained in relatively narrow, steep-sided valleys in the upper course suddenly issues onto a wide flood-plain over which it meanders. The meanders are highly pronounced to the south west of the town and over the flood-plain where their sinuosity is 1.9, increasing to over 2 further downstream. The criteria for meandering is taken to be a sinuosity of 1.5. Furthermore, river water is again somewhat constrained to the southeast of Ladysmith, restricting flow from the area. The catchment areas of the Sand River and the Klip River fall within the larger catchment area of the Tugela River (Fig. 1). The Klip River joins the Tugela River 20 km to the southeast of the town. The catchment of the Klip River and the Sand River area upstream of Ladysmith is approximately

Bell, F. G. & Mason, T. R. 1998. The problem of flooding in Ladysmith, Natal, South Africa. *In*: Maund, J. G. & Eddleston, M. (eds) *Geohazards in Engineering Geology*. Geological Society, London, Engineering Geology Special Publications, **15**, 3–10.

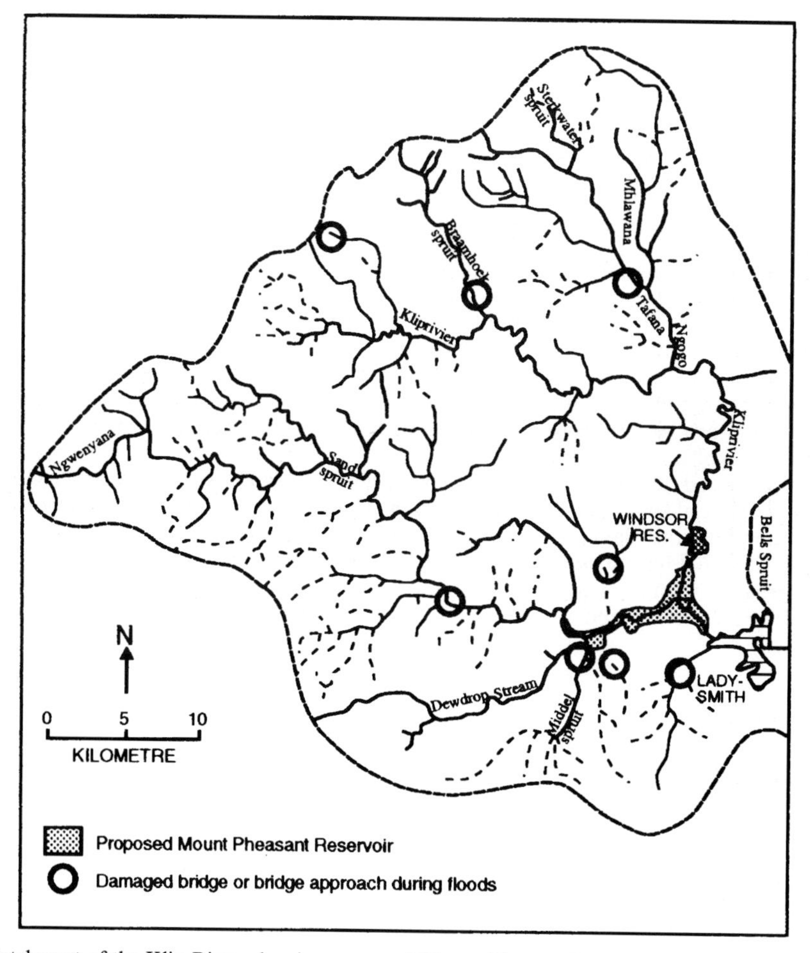

Fig. 1. Catchment of the Klip River, showing proposed Mount Pleasant Reservoir. Broken stream courses indicate non-perennial streams.

1650 km^2. Rainfall varies between 800 and 1000 mm per year over this area and falls in the spring–summer months between October and March. The mean annual temperature is 17–18°C.

Flood hydrology

Figure 2 summarizes the flood hydrology of the Klip River at Ladysmith. The Klip River has a bankfull discharge of approximately 700 m^3 s^{-1} flooding occurring once this is exceeded. This figure corresponds to a flood with a 5 to 6 year return period. As would be expected from the seasonal rainfall, the period of flood risk occurs from October to March. However, the duration of floods is relatively short (1–2 days) due to the relatively small catchment area and to the fact that rainfall occurs as thunderstorms of limited duration. In fact, Ladysmith has experienced 29 major floods in the past 110 years (Fig. 2). In other words, on average a flood occurred once every 3.9 years.

It is estimated that a flood with a 10 year return period would affect 211 families and 296 properties, the respective figures for a 20 year flood being 573 and 564. Figure 3 shows the relationship between the cost of the damage and the flood return period. It indicates that a flood with only a 10 year return period could inflict R2 000 000 worth of damage on the town. The most severely affected areas occur in the southern part of the town. The last flood occurred on 17 February 1994 (Fig. 4) and was the largest for the previous 71 years. It caused approximately R67 million in damage. The clean-up costs borne by the municipality were approximately R500 000 00. If we consider that the real costs of a flood have not changed since Ladysmith was founded, the lowest total cost of floods to the Ladysmith economy

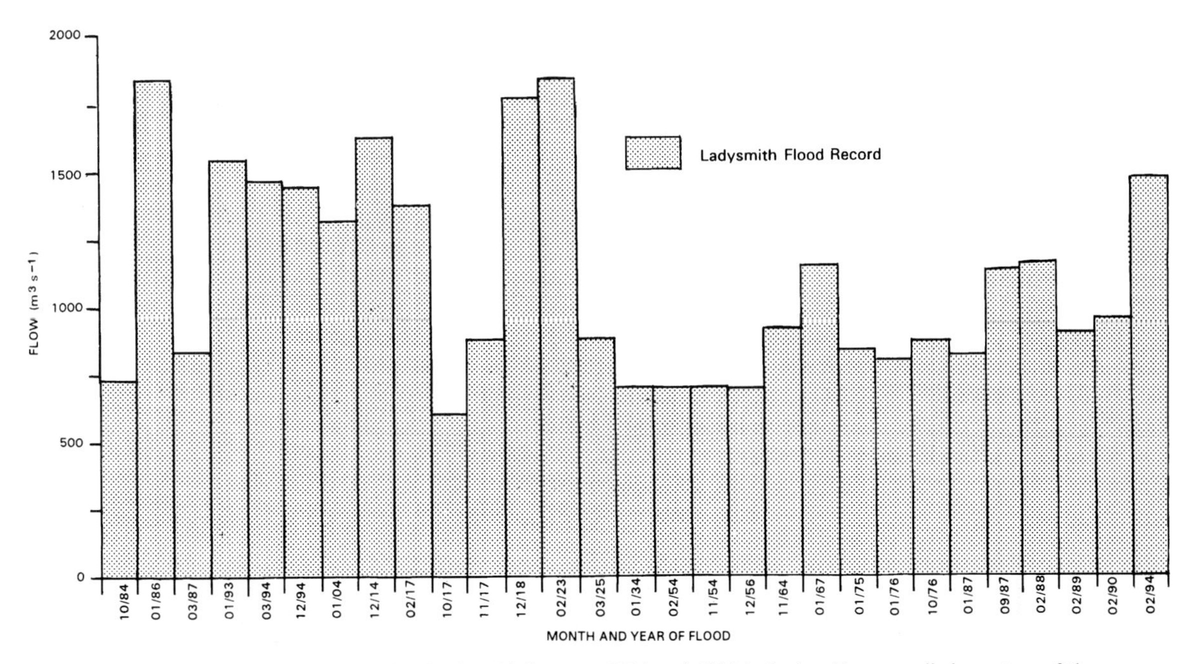

Fig. 2. Major floods experienced at Ladysmith between 1884 and 1994 inclusive. Data supplied courtesy of the Department of Water Affairs and Forestry.

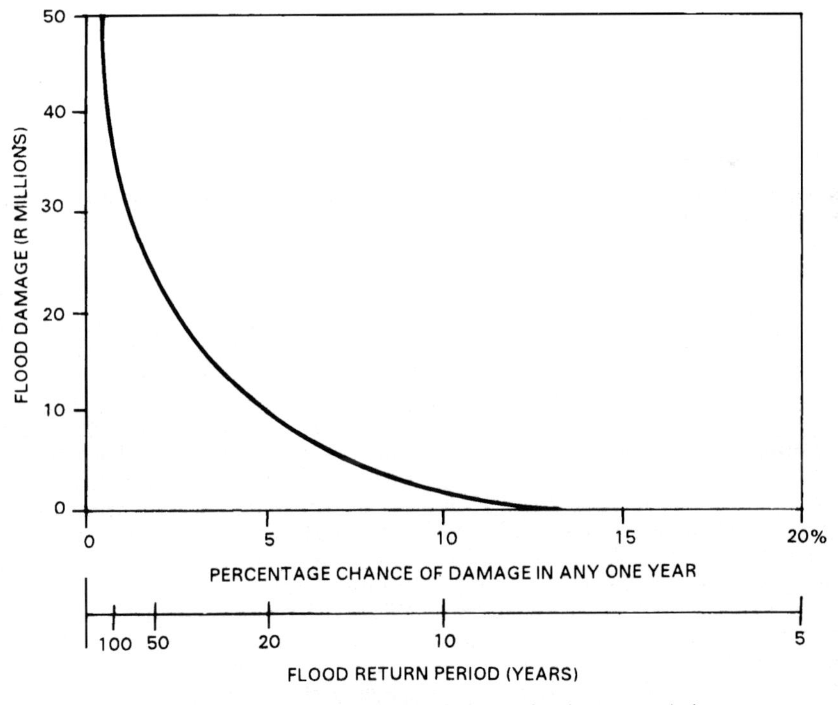

Fig. 3. Cost of flood damage in relation to flood return period.

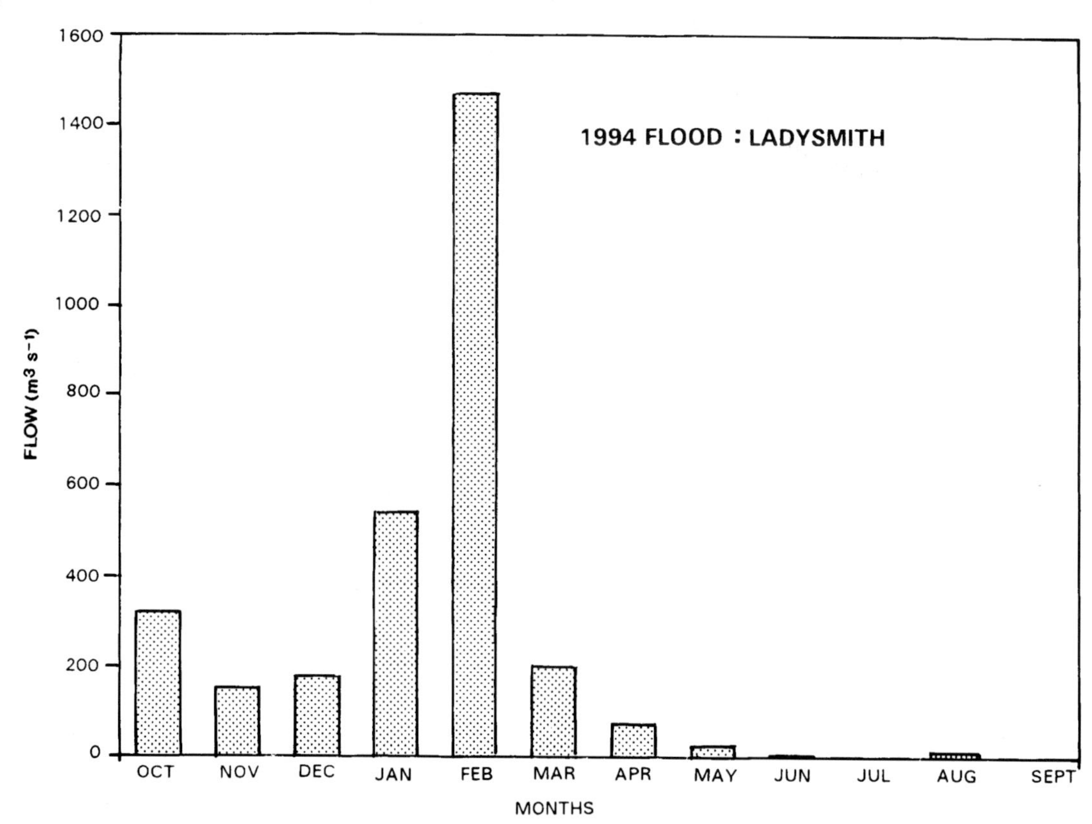

(b)

Fig. 4. (**a**) Ladysmith flood, February 1994. (**b**) Flooding in the centre of Ladysmith, February 1994.

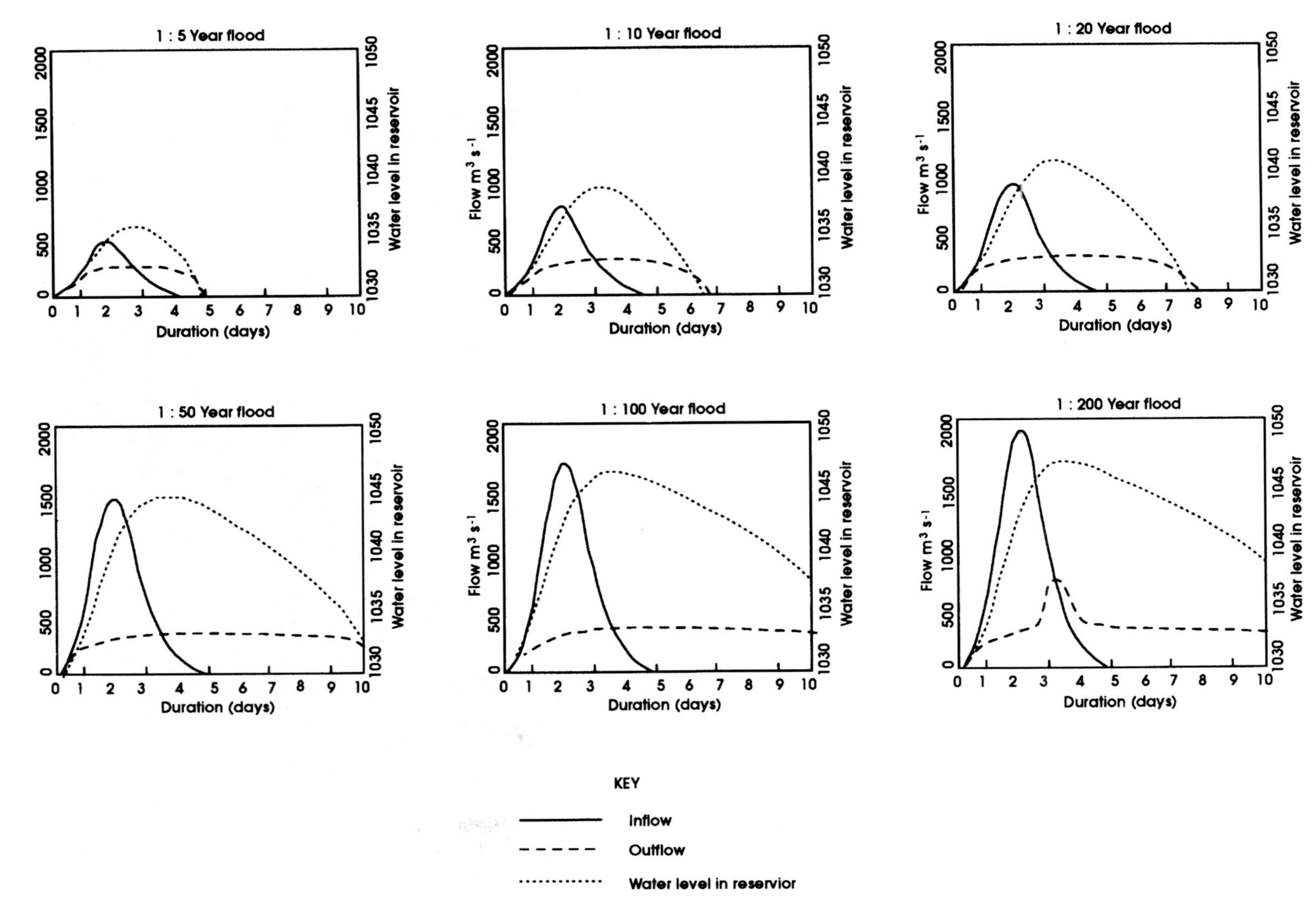

Fig. 5. Flood hydrographs for Ladysmith flood attenuation scheme, Mount Pleasant Dam. (After Claasens, G. C. D., 1988 *Verslag Oor die Voorgestelde Ladysmith Vloedbeheerskema (Mount Pleasant Dam)*. Department of Water Affairs and Forestry Pretoria.)

has been R28 million; the highest total cost, conservatively estimated, could be as much as R700 million.

Flood control measures

The most recent flooding (February 1994) of low lying areas of Ladysmith has highlighted the need for a flood control plan for the town. Several proposed schemes have been advanced over the years to mitigate the flood problem. All of the proposals are expensive and money from the central government will be needed to carry any of them out. As a rough guide, the cost of relocation is about ten times the cost of the cheapest amelioration scheme, which does not include the water storage option. Relocation, however, is likely to prove a socially less acceptable solution.

An attempt to control flooding was made in 1949 with the construction of the Windsor Reservoir. It is retained by an earth embankment dam and is located about 7.8 km NNW of Ladysmith. The capacity of the reservoir after its construction was $4.68 \times 10^6\,m^3$ with a surface area at top water level of 110 ha. The initial estimated annual rate of sedimentation in the reservoir was about 83 000 m^3. In fact this rate increased with time so that by 1989 the reservoir capacity had been reduced to 600 000 m^3. In other words, in 40 years it had been reduced to 12.8% of its original capacity and in 1995 the figure was about 5% of the original capacity (i.e. approximately 235 000 m^3).

The proposed Mount Pleasant Reservoir is the most popular of the flood control schemes and if constructed would be designed to accommodate the 100 year flood and so will reduce the risk of flooding in any year from 20% to 1%. The gross capacity of the reservoir would be $205 \times 10^6\,m^3$ with a net flood retention capacity of $183 \times 10^6\,m^3$. The minimum height of the dam above the river bed would be 30 m, which would mean that the area covered by overflow flood would be 2600 ha. The dam has been designed to withstand the force of the maximum possible flood for the area, i.e. one which has an inflow to the reservoir of $8000\,m^3\,s^{-1}$. The estimated volume of sedimentation over 50 years is $22 \times 10^6\,m^3$.

The hydrographs in Fig. 5 indicate that for up to a 100 year flood the maximum outflow from the reservoir would be $330\,m^3\,s^{-1}$ at its maximum storage capacity. This outflow is well below the bankfull discharge of the Klip River. Furthermore, Fig. 5 shows that for a 200 year flood, outflow would be $720\,m^3\,s^{-1}$ at peak water levels in the reservoir basin, which is near the flood peak for a 10 year flood (Tables 1 and 2). Levees designed to retain 10 year flood are included in the cost of the scheme. If they functioned effectively this outflow would not cause damage in the town. In fact Ladysmith has not suffered a 200 year flood (flood peak of $2090\,m^3\,s^{-1}$), and has experienced only one 100 year flood (flood peak of $1790\,m^3\,s^{-1}$), and that was in 1886.

Table 1. *Estimated flood peaks and volumes at Ladysmith*

Return period (years)	*Flood peak ($m^3\,s^{-1}$)*	*Flood volume ($\times 10^6\,m^3$)*
10	725	107
20	945	139
50	1500	229
100	1790	272
200	2090	314
RMF[a]	4000	655
PMF[b]	8000	790

[a] RMF = regional maximum floods.
[b] PMF = probable maximum floods.
Data supplied by the Department of Water Affairs and Forestry.

Table 2. *Risk of flooding*

Flood return period (years)	*Risk of flooding (exceeded at least once)*	
	In any one year	*Over a period of 20 years*
5	20%	99%
10	10%	88%
20	5%	64%
50	2%	33%
100	1%	18%

Data supplied by the Department of Water Affairs and Forestry.

Figure 6 shows the proposed site for the reservoir and the height of the water at 20 year and 100 year flood levels. It can be seen that a road and a railway line would have to be relocated to accommodate the reservoir and some flooding of access routes would occur during a 100 year flood.

Although the reservoir would protect the town from floods, a number of other factors need to be considered. First, the reservoir would have to remain at 10% capacity to maintain the net flood retention capacity. This means that the reservoir would not be of much value in terms of water supply for the surrounding area. This is a serious consideration since Ladysmith only has adequate water provision until 2010. Moreover, the reservoir would generate very little revenue from recreational activities. The cost of the reservoir in 1994 was estimated at between 200 and 250 million rands. If the reservoir is designed and constructed to enhance the water storage capability of the scheme, to allow for 50% water storage flood attenuation, then the cost escalates by R100 million.

An alternative scheme involves strengthening and upgrading of the existing artificial levees. The estimated cost of construction of artificial levees to withstand a 10 year flood is R10 million, that for the 20 year flood being R40 million However, the construction of levees can pose problems; for example, if levees are overtopped water cannot readily escape back into the river channel

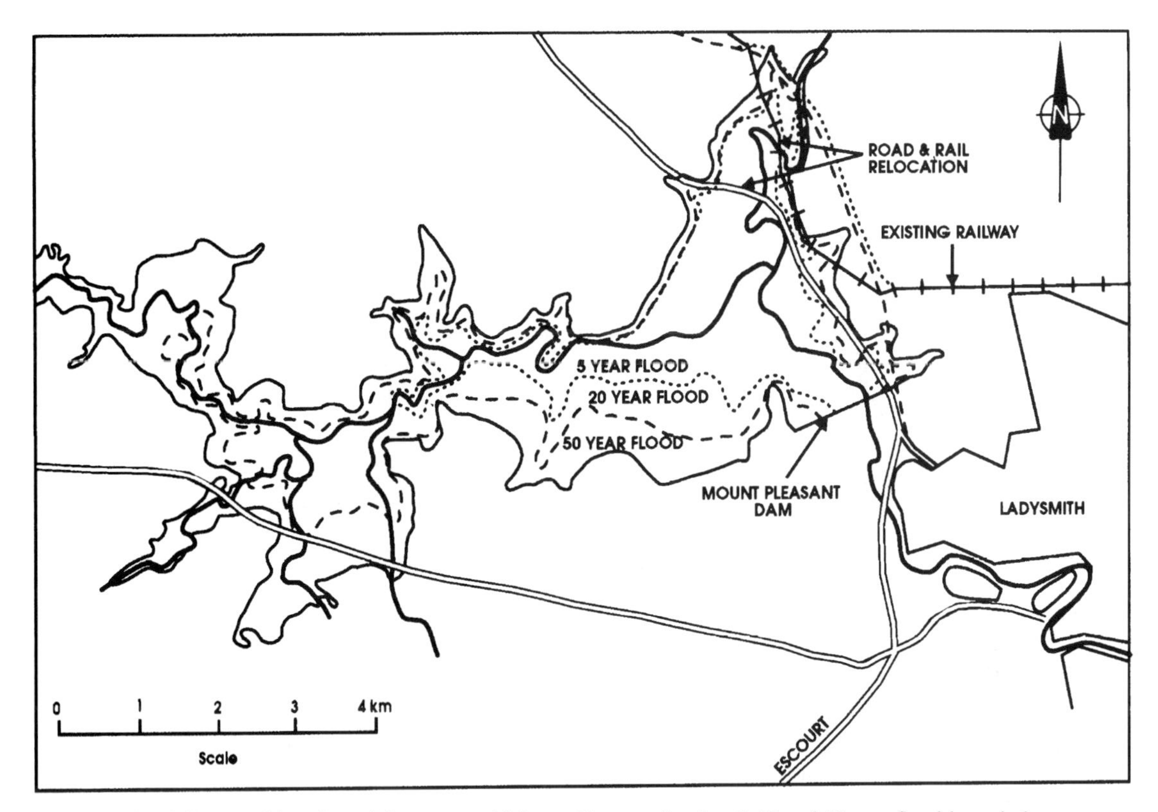

Fig. 6. Proposed location of the proposed Mount Pleasant showing 5, 20 and 50 year flood boundaries.

after the flood. Built-in fuse-plugs are required. Rapid failure of a levee during a flood means that rapidly flowing water discharging from the river can cause damage greater than if no levees were present. Disposal of storm water on the town side of the levees during heavy rainfall would be very difficult.

Canalization increases the bankfull discharge by increasing the velocity of flow and straightening the path of a river. However, there could be the possibility of added erosion and flooding occurring downstream of the canal due to water flowing from it at a higher velocity than the flow in a natural channel. In addition, a canal would have to be continuously maintained and cleaned of vegetation that would grow during periods of low water. In 1994 the cost of canalisation (plus levees) was estimated as around R70 million for protection against a 20 year flood. To accommodate a 50 year flood the cost would rise to over R85 million.

It has been suggested that suitable areas for the relocation of those properties which would be affected by a 20 year flood are available. This would also involve curtailing any further development in designated high-risk areas. The present cost of such relocation is approximately R197 million, which includes purchase of the land and the establishment of the necessary infrastructure as well as the reciting of properties. Figure 7 indicates that the area between the 20 and 50 year flood lines is approximately 10% of the area below the 20 year flood line. Extending the relocation project to the 50 year flood line would cost an estimated extra R20 million, but could be carried out at a later stage. Relocation does not interfere with the natural river system. It could provide flood protection indefinitely for floods with a magnitude of the 50 year flood. The scheme would not require the maintenance cost that the reservoir would in order to keep if functioning. The land vacated could be used as low maintenance cost community facilities such as public parks, sports fields, or an environmental reserve. Such land use is cheaply restored after flooding.

The land vacated would have to be cleared and the rubble waste disposed of in landfills. Because of the rapid rates of urbanization it would be highly likely that informal settlements would spring up on the vacated land. Their removal would present an acute political problem even though the squatters would be in serious danger if the Klip River flooded. Similar squatter settlements in Durban were badly damaged by relatively minor flash floods caused by localized thunderstorms in April 1995. Small dry valleys were instantly transformed into raging torrents. Moreover it would be very difficult

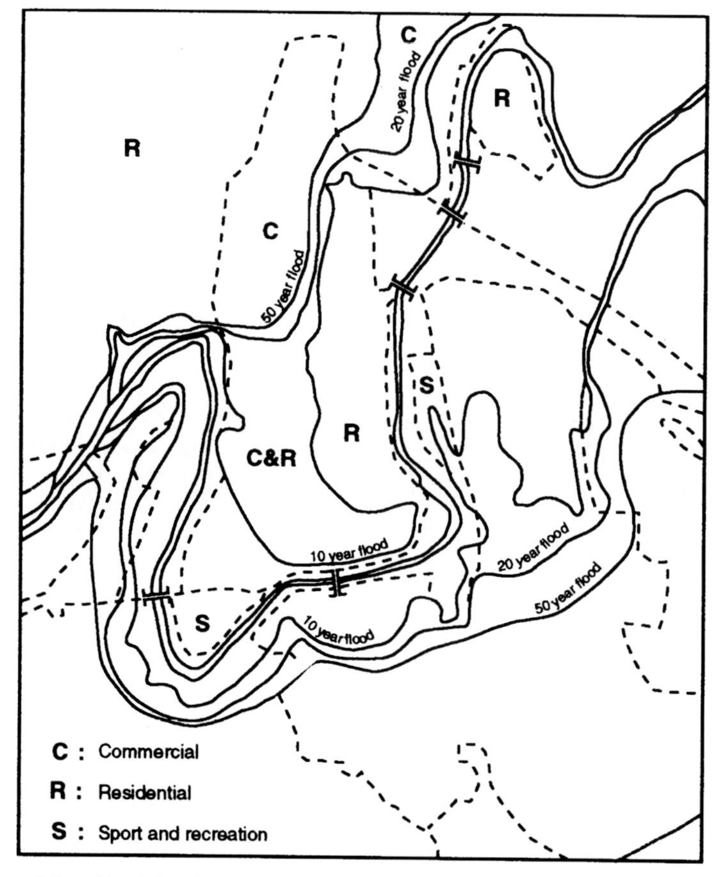

Fig. 7. Commercial, residential and recreational areas of Ladysmith in relation to the 10, 20 and 50 year flood.

to undertake an orderly and prompt evacuation of the area where little formal organisation existed. A potential solution requires a major act of political will allied to enlightened legislation to prevent uncontrolled squatting on recreational land.

Public opinion has not been reliably tested in the past and people affected by flooding presumably would be reluctant to move, preferring to have a flood protection reservoir constructed. In addition, the flat land of a flood-plain is easier to develop and a ready source of water is available to industry from the river. On the other hand, if the public do support the relocation scheme they will need to be involved in the development of the new areas. New development can take place only on the outskirts of Ladysmith.

Conclusion

Ladysmith in the Natal Midlands was located within a meander loop of the Klip River for defensive purposes. Unfortunately, however, this has meant that the town has been subjected to repeated flooding since its establishment in the mid-19th century. After emerging from the Drakensburg Mountains the Klip River flows over a flood-plain which experiences an average of 800–1000 mm of rainfall annually. This falls primarily between October and March, and characteristically as thunderstorms.

The southern part of the town is most seriously affected by flooding, which has occurred more or less once every 4 years. The costs associated with flooding dictate that an adequate system of flood control must be implemented. Several proposed schemes have been advanced, the most popular but most expensive being a flood control reservoir. This scheme would protect the town from a 100 year flood, indeed with the associated levees from a 200 year flood. The other schemes which include the enhancement of the levee system, canalization and relocation of properties are less expensive but would accommodate a 50 year flood. Relocation, however, does not interfere with the natural river system and should involve little future maintenance cost.

A combined geotechnical/geophysical method for the prediction of liquefaction, with particular reference to the Fraser River Delta, British Columbia

James Pyrah, Angela Davis & Dei Huws

School of Ocean Sciences, University of Wales Bangor (UWB), Menai Bridge, Gwynedd, UK

Abstract. Over the past decade or so, much work has been undertaken into the development of various techniques for the measurement of seismic shear wave velocity, offshore, onshore and in the laboratory. Previous research has suggested that the shear wave velocity of an uncemented sand may be a useful index to its liquefaction potential. The research described in this paper has combined the concepts of critical/steady-state soil mechanics and the measurement of shear wave velocity in order to provide a new index of sand consolidation state, enabling the prediction of flow liquefaction in laboratory-prepared samples. Results obtained from the laboratory are then applied to the field situation, in this case, the Fraser River Delta, British Columbia, which is an example of a site currently undergoing a major earthquake stability evaluation. A preliminary analysis of the data indicates that the sediments around Roberts Bank, on the southwestern portion of the delta, are potentially liquefiable. This conclusion is confirmed by other, more conventional, investigative techniques.

Introduction

The liquefaction of non-cohesive sediments due to both static and dynamic loading represents a significant problem for the geotechnical engineer. The 1964 earthquakes in Niigata (Japan) and Anchorage (Alaska) caused spectacular foundation failures, with many buildings suffering large scale tilting, damage to gas, water and electricity utilities and to other infrastructure. More recently, in the Hyogo-ken Nambu (Kobe) earthquake of 17 January 1995, liquefaction was widespread, with most damage due to this sort of failure concentrated around the harbour, where quay walls collapsed and cranes were toppled (O'Rourke 1995). The traditional field investigation method for liquefaction assessment is the standard penetration test (SPT), which is still widely used in many parts of the world despite many well documented deficiencies (Schmertmann 1978). In more recent years, the cone penetration test (CPT) has begun to be used for this type of field evaluation (Robertson & Campanella 1985). In addition, shear wave velocity has been quoted increasingly often as a possible index to liquefaction potential over the past ten or so years (e.g. Seed *et al.* 1983; Stokoe *et al.* 1988; Robertson *et al.* 1992*a, b*) alongside more conventional laboratory or field testing techniques. The incorporation of a geophone (or geophones) into the CPT allows the direct measurement of the seismic shear wave velocity profile during testing (e.g. Robertson *et al.* 1986; Hepton 1989), giving a useful geophysical addition to conventional geotechnical cone penetration techniques.

This paper documents a detailed laboratory and field investigation into the possibility of using shear wave velocity as an index to liquefaction potential, with application to the modern Fraser River Delta, a potentially unstable delta located in the extreme southwestern corner of mainland British Columbia, Canada.

Definitions

As a result of investigations into the phenomenon of liquefaction over the past 30 or so years, two different types may be identified. *Flow liquefaction* occurs due to soil structure collapse in contractive soils, resulting in significant loss of strength and large deformations. *Cyclic liquefaction* occurs in dilative soils where the collapse of soil structure does not occur. Cyclic loading of the soil results in a gradual build-up of pore pressure, which can momentarily reach the *in situ* confining stress. The resulting deformations can be large enough to constitute failure (Robertson *et al.* 1992*b*). Unless otherwise stated, the general term of liquefaction as used in this paper refers to the phenomenon of flow liquefaction.

Steady-state approach to flow liquefaction prediction

The 'critical void ratio' concept of liquefaction prediction was initially described by Casagrande in 1936.

PYRAH, J., DAVIS, A. & HUWS, D. G. 1998. A combined geotechnical/geophysical method for the prediction of liquefaction with particular reference to the Fraser River Delta, British Columbia. *In*: MAUND, J. G. & EDDLESTON, M. (eds) *Geohazards in Engineering Geology*. Geological Society, London, Engineering Geology Special Publications, **15**, 11–23.

A more comprehensive critical state soil model, incorporating many of Casagrande's findings, was developed by Roscoe *et al.* in 1958. Further work on sands, with particular reference to liquefaction, was pioneered by Casagrande (1975) and his co-workers, leading to the development of what is generally regarded as the 'steady-state approach'. Work on this methodology has since been continued by various workers, including Poulos (1981), Poulos *et al.* (1985), Sladen *et al.* (1985), Been & Jefferies (1985), Been *et al.* (1991), and others, partly in response to conventional earthquake loading problems, but also in response to the design and development of artificial sand islands used as a platform for oil drilling in the Canadian Beaufort Sea.

In essence the above methodology is based upon some very simple observations of sand behaviour under shear. Under undrained monotonic loading, a dense sand, or a sand under a sufficiently low confining stress, will tend to dilate, pore-pressures will drop, and the sand will stiffen. In contrast, a loose sand, or one under sufficiently high confining pressure, will tend to contract and raised pore-pressures will develop. This causes a dramatic reduction in strength and subsequent soil structure collapse, producing a distinctive 'flow structure'. Ultimately, a constant residual strength is attained, known as the 'steady state of deformation', which is dependent solely on the initial sample density. The steady state of deformation is defined by Poulos (1981) as:

> that state in which the mass is continuously deforming at constant volume, constant normal effective stress, constant shear stress, and constant velocity. The steady state of deformation is achieved only after all particle orientation has reached a statistically steady state condition and after all particle breakage, if any, is complete, so that the shear stress needed to continue deformation and the velocity of deformation remain constant.

A boundary may be defined between these two different types of behaviour, generally known as the steady-state line, separating sands susceptible to flow liquefaction and those that are not. This boundary is typically defined using consolidated undrained monotonic triaxial tests on loose sand samples to determine the steady state of deformation. If the results of a series of these tests carried out on samples of differing initial confining stresses and densities are plotted in void ratio (e)–log effective confining stress (p') space, the steady-state line can be defined by simply drawing a (usually) straight line through the data points (Fig. 1). This fundamental line divides contractive sand states (above and to the right), from dilative sand states (below and to the left). Been & Jefferies (1985) proposed a normalizing state parameter, ψ, which is simply the difference between the initial void ratio and that at steady state at the same confining stress. Positive states indicate a contractive behaviour and risk of flow liquefaction;

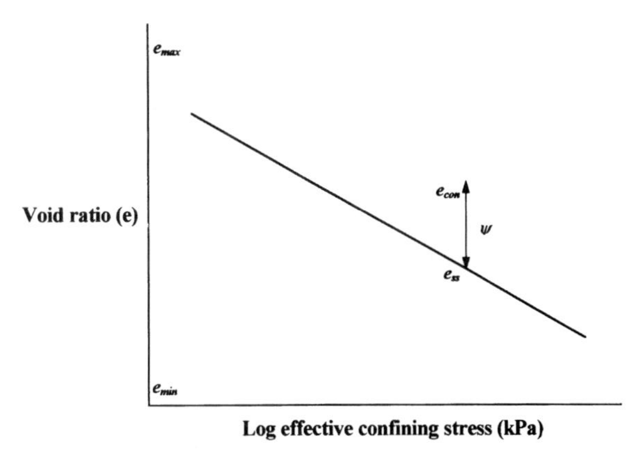

State parameter, $\psi = e_{con} - e_{ss}$

where, ψ = state parameter.

e_{con} = void ratio after consolidation.

e_{ss} = void ratio at steady state.

Fig. 1. Diagrammatic steady-state line, and definition of the state parameter, ψ.

negative states indicate a dilative behaviour, and a strain hardening response.

The above discussion indicates that simply with a knowledge of the *in situ* void ratio, confining stress and the steady-state line for a given material, its *in situ* state can be determined relatively easily. The main practical problem with this approach is the *in situ* determination of the void ratio. This quantity needs to be defined as accurately as possible; steady-state lines tend to be near horizontal, so even a small error in void ratio determination could lead to potentially disastrous consequences for any proposed structure. One current technique for the determination of the *in situ* void ratio is ground freezing (e.g. Yoshimi *et al.* 1979, 1994) which yields high quality undisturbed samples, but is very expensive and is presumably extremely difficult to implement offshore. Another method of determining the *in situ* void ratio is the use of a calibrated cone penetrometer (e.g. Been *et al.* 1986, 1987); this method has been used with some success, although Sladen (1989) suggests that this approach can result in interpretations of sand state that 'if carried through to design could in some cases be catastrophic'. Clearly then, there is a need for some other index for sand state which may be used in addition to conventional techniques.

Shear wave velocity

Seismic shear waves (other names include transverse, or secondary waves) are body waves which propagate by a pure strain in a direction perpendicular to the direction of wave travel. Individual particle motions involve oscillation about a fixed point in a plane at right angles to the direction of wave propagation (Keary & Brooks 1991). Experimentally, the shear wave velocity of a sediment can be shown to be primarily controlled by void ratio and effective stress (Hardin & Richart 1963), although secondary factors such as stress history, ageing effects, shearing strain amplitude, degree of saturation, temperature and overconsolidation ratio (Woods 1991) have some lesser effects, depending on the sediment type involved. Shear wave velocity is relatively simply measured both in the laboratory (e.g. Bennell & Taylor-Smith 1991), and in the field (e.g. Hunter *et al.* 1991). This ease of measurement combined with the fact that the velocity of the shear wave is controlled by principally the same factors that control sand behaviour, make it a promising candidate for the prediction of flow liquefaction in sands and silty sands.

Some of the first attempts to use shear wave velocity as an index of liquefaction are reported by Dobry *et al.* (1981) and Seed *et al.* (1983). These investigations were based on *in situ* shear wave velocity measurements and laboratory cyclic strain-controlled tests. Further empirical correlations have been made between cyclic liquefaction resistance and shear wave velocity by Bierschwale & Stokoe (1984), De Alba *et al.* (1984), Tokimatsu *et al.* (1986, 1988), Tokimatsu & Uchida (1990), Stokoe *et al.* (1988) and Robertson *et al.* (1992*a*, *b*). Until very recently no attempt had been made to combine steady-state concepts with shear wave velocity evaluation (Robertson *et al.* 1995).

Laboratory technique

A brief description of the testing methods and results are presented here, with further details described in Pyrah (1996). All testing was carried out on a Wykham–Farrance strain-controlled triaxial machine; stress (made using a calibrated proving ring), strain, volume and pore-pressure measurements were all logged by hand, and were later entered into a PC based spreadsheet. The triaxial cell allowed the testing of samples (initially) 200 mm in length by 100 mm diameter. The end platens of the cell were modified to incorporate cantilever mounted shear wave 'bender' elements, which allowed the measurement of shear wave velocity down the length of the sample (Schultheiss 1983). The relatively large sample size is advantageous when measuring wave velocities as it allows the P-wave and shear wave generated by the bender transducer to separate by virtue of their different velocities, and also prevents any 'near-field' effects (Viggrani & Atkinson 1995).

Four different sands were tested: a clean beach sand, the same beach sand with 5% kaolinite added, with 10% kaolinite added, and a fine sand with around 8% fines obtained using a vibro-corer on the Fraser Delta, British Columbia (courtesey of the Geological Survey of Canada). All samples were prepared using the moist tamping method, being saturated initially with carbon dioxide and then distilled water. Isotropic consolidation was then performed using high back-pressure techniques (e.g. Sladen *et al.* 1985), ensuring high degrees of saturation and Skempton's 'B' values of greater than 0.95.

The shear wave velocity of the sample was measured over a range of effective confining stresses (p'). The velocity was simply obtained by dividing the travel time by the distance between the tips of the bender transducers. The cell was then drained and the sample kept under a relatively high negative pore-pressure while its dimensions were measured. The cell was subsequently re-assembled, the sample re-consolidated, and an undrained, strain-controlled triaxial test performed, ideally shearing the sample to steady state. Error in determination of the velocities was of the order of $\pm 5\,\mathrm{m\,s^{-1}}$, while those involved in void ratio calculation were around ± 0.006. The results of some 60 such tests were subsequently analysed and are discussed below.

Results

The laboratory data clearly show that, in agreement with the literature, the shear wave velocity of a specific sample is strongly affected by changes in effective confining stress (Fig. 2). A power law regression line can be fitted through the data, the exponent generally having the value of between 0.26 and 0.35. This regression can then be used to estimate the shear wave velocity at any effective confining stress for each sample. If enough of these tests are carried out on samples of differing void ratio, the regression data can be used to illustrate the relationship between shear wave velocity and void ratio, at a specific effective stress (Fig. 3).

The triaxial tests performed during the testing period clearly illustrated the two sorts of sand behaviour described in the introduction above. Loose samples generally attained what may be regarded as steady-state conditions (Fig. 4(a) and (b)). Denser, dilative sands clearly displayed a strain hardening behaviour and tended to fail in shear. Subsequent plotting of the state points obtained for each test in e–$\log p'$ space defined the steady-state line (Fig. 5). The steady-state lines obtained were linear in nature, but generally displayed some scatter around the mean line. The effect of adding fines to the clean sand was to generally lower the steady-state line in e–$\log p'$ space, dramatically increasing the risk of liquefaction, i.e. a far greater density was needed to prevent liquefaction in the sands containing fines than in the clean sand.

Combination of the two data sets allows the definition of a 'critical shear wave velocity line'. Assuming changes

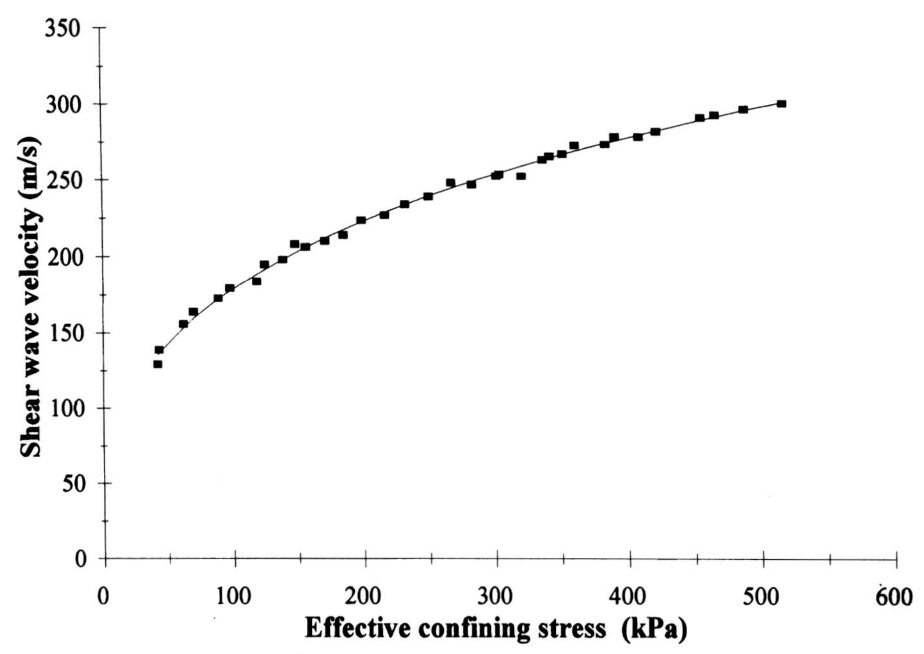

Fig. 2. Relationship between shear wave velocity and effective confining stress.

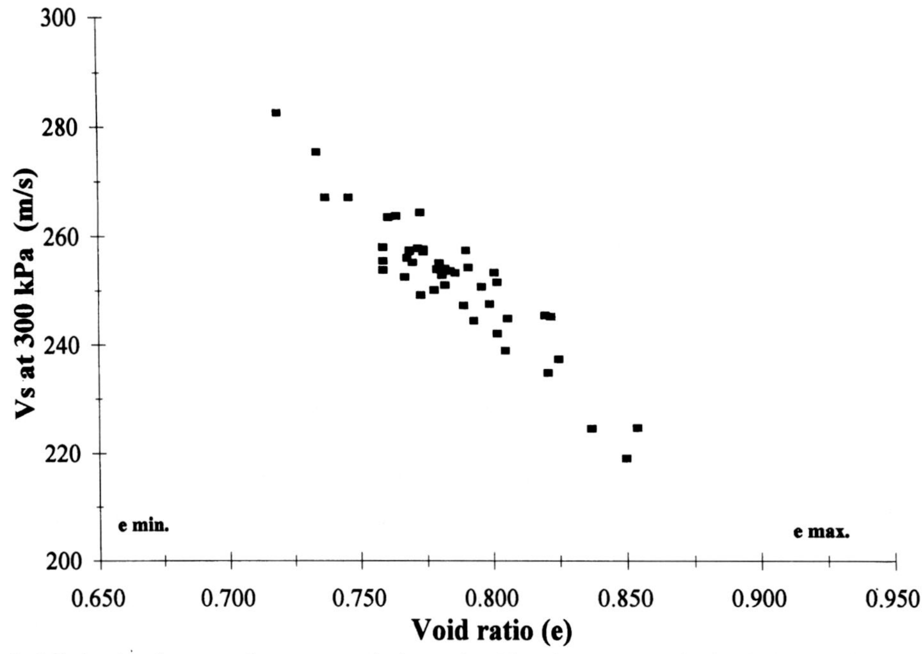

Fig. 3. Relationship between shear wave velocity and void ratio, at a standard effective confining stress of 300 kPa.

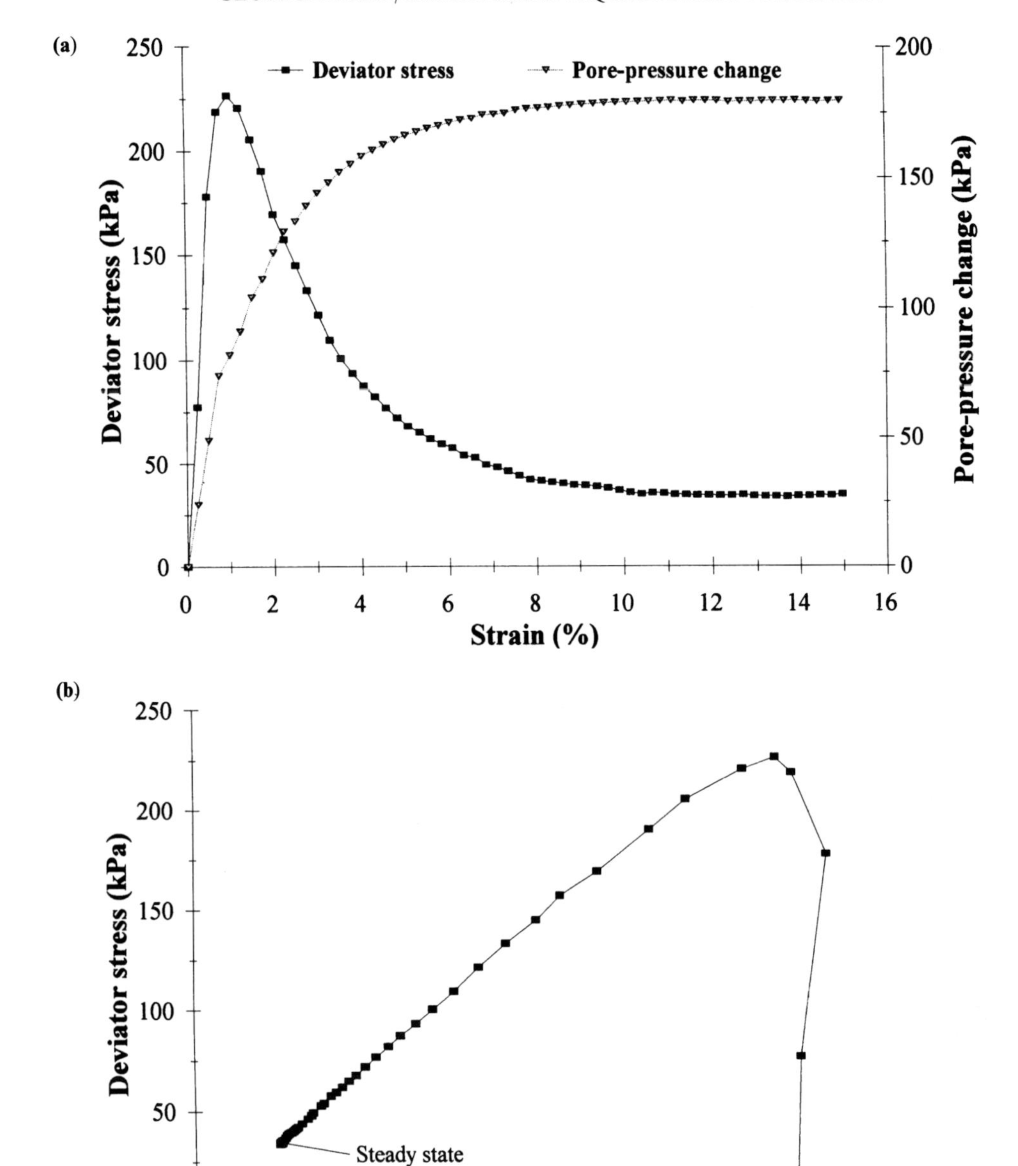

Fig. 4. Monotonic, undrained response of a loose sand, attaining steady-state conditions.

in stress ratio (σ_1'/σ_3') have no significant effect on the shear wave velocity of the sample, it is a simple matter to calculate the velocity at the effective confining stress reached at steady state using the regression data obtained in the first part of the test. These data can then be plotted on a new form of state diagram, with shear wave velocity replacing void ratio on the y-axis (Fig. 6). Any sample whose velocity falls below the critical velocity line at a particular confining stress will tend to strain-soften under shear, and hence liquefy. Samples with a velocity above the critical velocity line will tend to dilate and strain harden under shear.

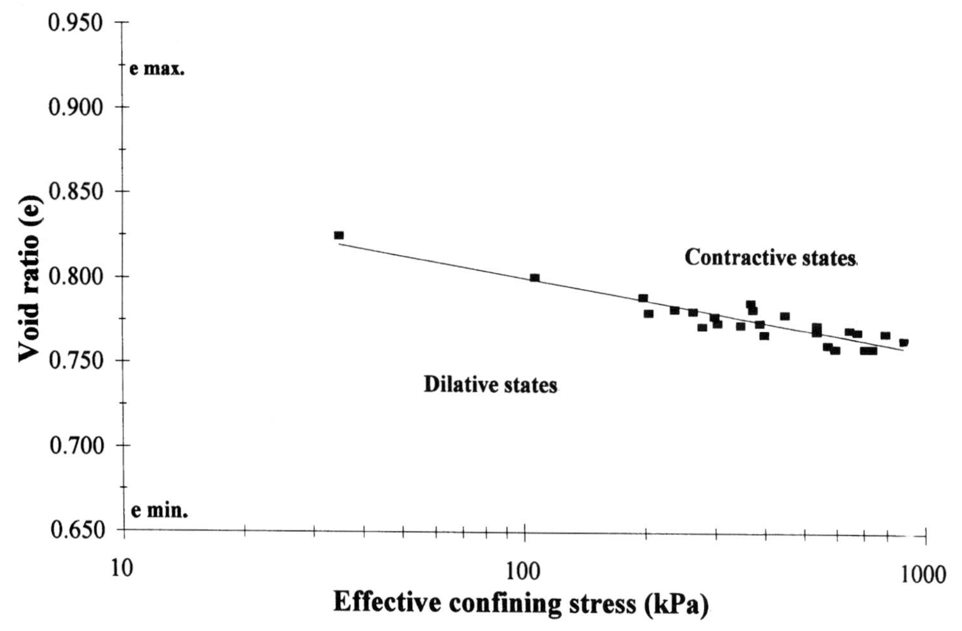

Fig. 5. State diagram defined for a clean beach sand.

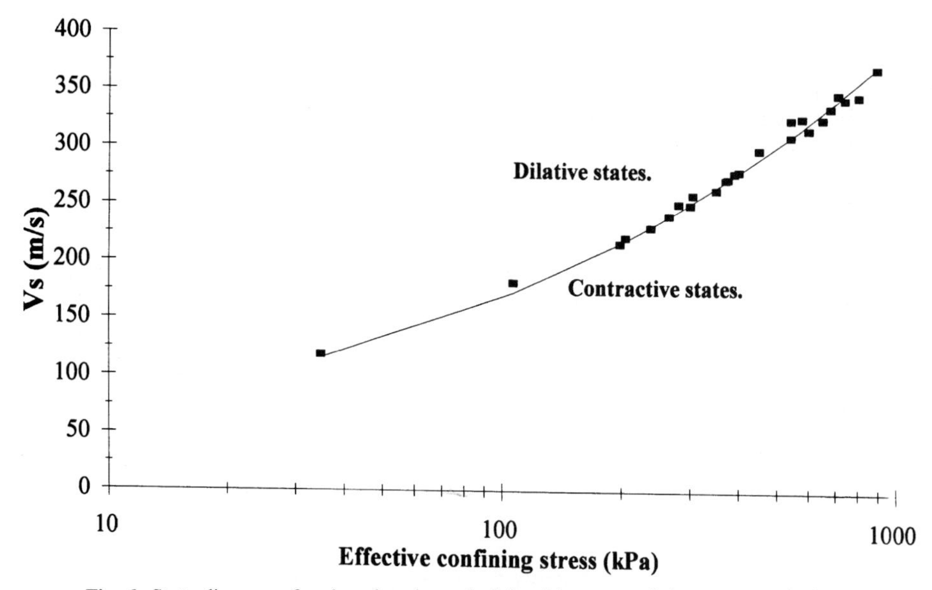

Fig. 6. State diagram of a clean beach sand, defined in terms of shear wave velocity.

Discussion of laboratory results

Definition of a critical velocity separating the two contrasting sand behaviours allows the definition of a new state index. This index is calculated from a knowledge of the *in situ* shear wave velocity and the critical velocity line. It is termed ψ_s for simplicity, and is simply the difference between the shear wave velocity after consolidation and the critical velocity at the same effective confining stress. Figure 7 shows a significant straight line

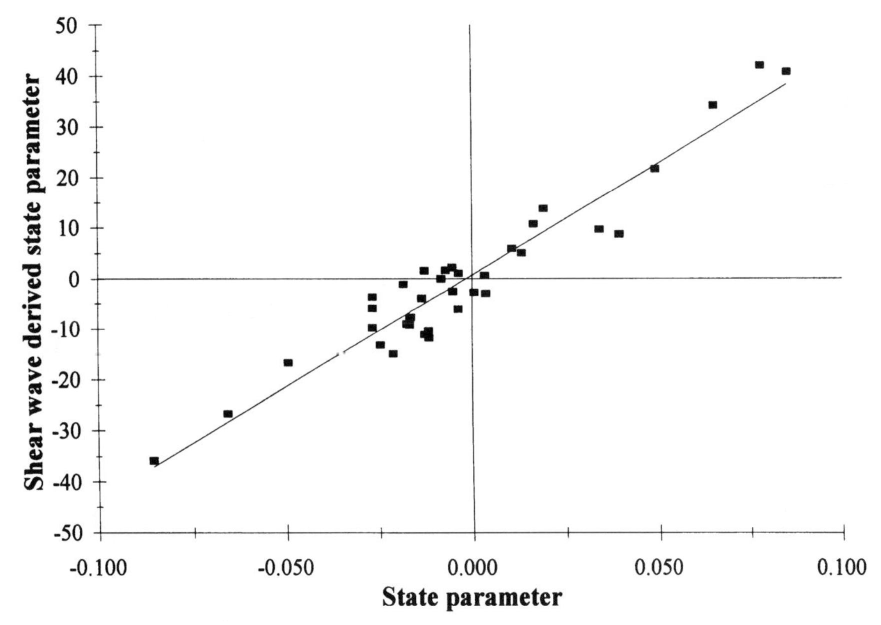

Fig. 7. Comparison between the state parameter, ψ, and the proposed state index ψ_s.

relationship between conventional state as defined by Been & Jefferies (1985) and ψ_s. However, as will also be noticed, the difference between velocities separating contractive and dilative behaviour is not all that great. This is a result of the fact that shear wave velocity is limited in its sensitivity to variations in void ratio, and this remains the main limitation of the proposed technique. Despite this, it can be concluded that shear wave velocity may be used to predict liquefaction potential on laboratory prepared samples where good quality, low error data are relatively simple to collect.

In the field, where errors are more difficult to quantify, the method may still be used, but a degree of caution must be taken when attempting to predict sand state simply from a velocity profile. This would involve assuming homogeneity of the soil profile, which if incorrect, could have potentially disastrous consequences. An improved method would be to combine the measurement of shear wave velocity with other, more conventional CPT techniques (in the form of a seismic cone penetration test). This combination of cone penetration data, field shear wave velocities and laboratory testing on representative soil samples, potentially provides a very powerful method for the prediction of liquefaction potential in the field.

To illustrate the above conclusion in more detail, the following section provides a brief description of the continuing investigation into the stability of the Fraser River Delta, British Columbia. The work forms part of a collaborative research programme between UWB and the Geological Survey of Canada (GSC).

Fraser River Delta: Setting

The Fraser River Delta is located in the southwestern corner of mainland British Columbia, just south of the rapidly growing Vancouver metropolitan area (Fig. 8). The Fraser River, which supplies sediment to the delta, drains an area of over 234 000 km^2, and travels approximately 1400 km from its head-waters; this represents the largest fluvial system on Canada's Pacific coast (Stewart & Tassone, 1989). The mean annual sediment load is 17.3 million t a^{-1}, of which approximately 35% is sand (>0.063 mm), 50% is silt (0.063–0.004 mm) and 15% is clay (<0.004 mm) (Anon 1986).

Vancouver, the Fraser Lowlands and the Fraser Delta itself are all located within a belt of high seismicity that encompasses much of the western coast of British Columbia and the adjoining Washington State, which lies to the south (Milne *et al.* 1978). This seismicity is associated with the convergence of the oceanic America, Juan de Fuca, Explorer and Pacific plates and the continental North American plate. The biggest, recent earthquake occurred in 1946 on central Vancouver Island and had a Richter magnitude of 7.2 (Rogers 1988). Recent research suggests that within the Cascadia subduction zone there may be the possibility of a 'mega-thrust earthquake', with a Richter magnitude of up to 9.2 (Rogers 1988) although this remains a matter of some debate (Campbell & Rotzein 1992).

The delta is of enormous economic importance to the region. Near the edge of the sub-aerial delta is Vancouver

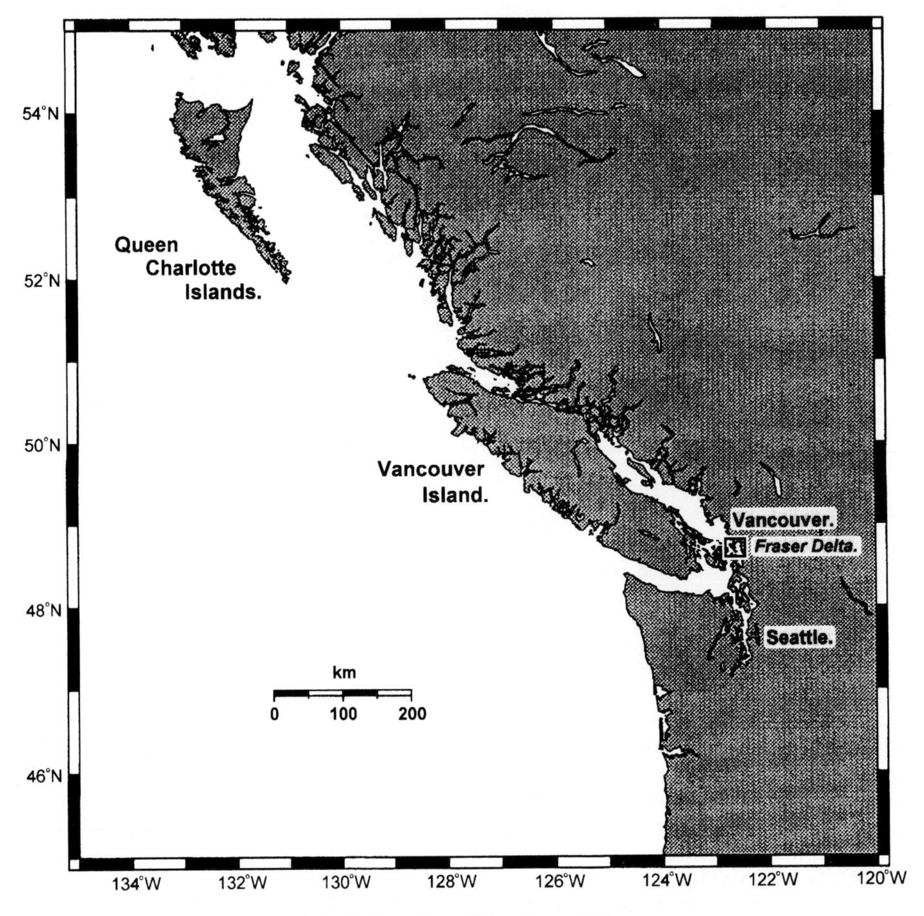

Fig. 8. Location of the Fraser Delta.

International Airport, currently undergoing a $350 million expansion involving large ground stabilization works (Mosher & Barrie 1995). Constructed out over the tidal flats of Robert's Bank, in the southern part of the delta, is the busiest passenger ferry terminal in the world, carrying an estimated 51 million passengers a year (Mosher & Barrie 1995). Adjacent to the ferry terminal lies Tswawwassen Coal Port, Canada's largest coal exporting terminal, which itself is undergoing a major expansion to enable handling of grain and container traffic, involving the biggest ground stabilization effort ever undertaken in Canada (Mosher & Barrie 1995). Crossing the Strait of Georgia, between the delta and Vancouver Island, are numerous power and communications cables.

Fraser Delta: stability issues

The Fraser Delta is composed of thick sequences of loose sands and silty sands underlain by non-lithified glacial/non-glacial material. These Quaternary deposits lie on a Tertiary bedrock characterized by high relief; depths to bedrock range from 100 m up to 1000 m (Harris *et al.* 1995). These loose, unconsolidated sediments pose two main threats to the geotechnical engineer: amplification and liquefaction; only the latter subject will be discussed here. Liquefaction of the sub-aerial delta deposits has received much attention in the last few decades, from a variety of conventional site investigation techniques. The resulting data are summarized by Bryne & Anderson (1987) who suggest that based upon a design acceleration of 0.20 g design velocity, the likely effects of an earthquake of this magnitude on the town of Richmond of would include damage to most buildings due to liquefaction of foundations, disruption of the highway system due to lateral spreading, spreading of bridge abutment fills, cracking of dykes, light to moderate damage to water, sewer, gas, electrical and communication services, and the risk of a major conflagration, due to ruptured gas pipes.

Offshore, efforts to understand the hazard potential have only really begun in the last four years (Mosher &

Barrie 1995). These investigations have involved the use of a variety of marine geophysical techniques (including side-scan sonar, deep towed boomers, multi-channel airgun surveys, electromagnetic surveys, etc.) as well as the collection of a large number of cores. These data have allowed the identification and mapping of a variety of potentially hazardous features on the delta, including a sea-valley extending down the delta slope from the mouth of the Fraser River, with a documented history of mass wasting events, and a possibly dormant or relict

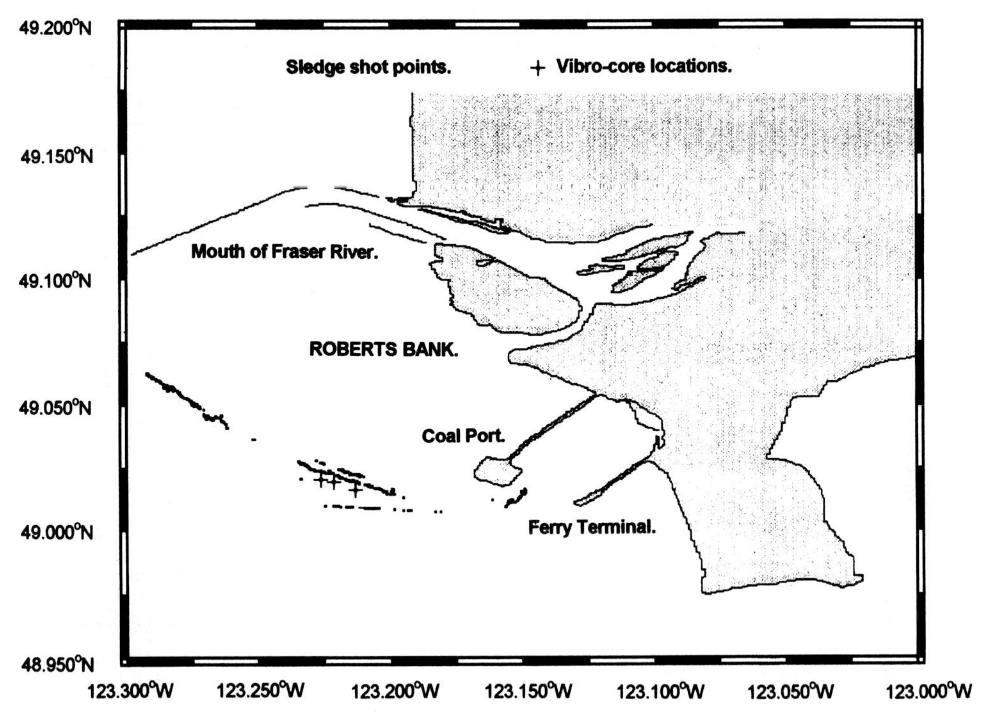

Fig. 9. Detail of delta morphology and data coverage, around Roberts Bank.

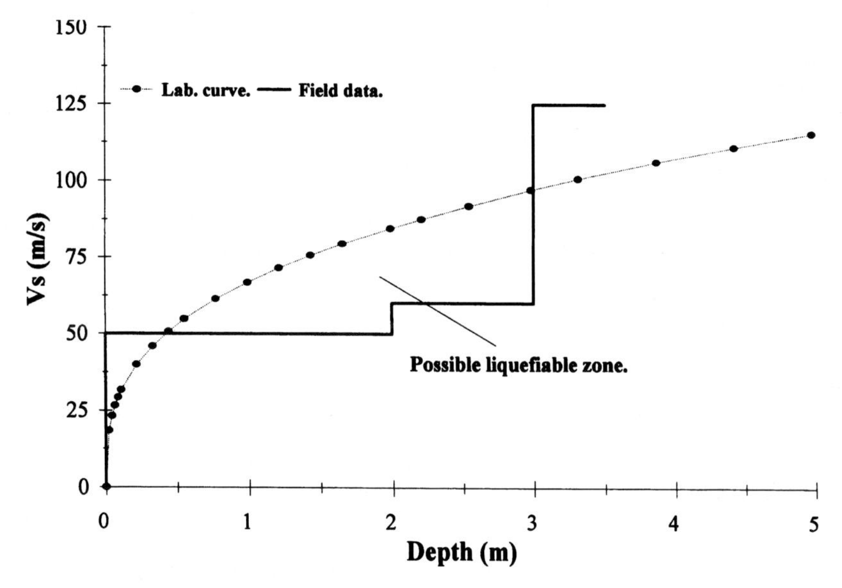

Fig. 10. Combination of shear wave refraction sledge data and the laboratory derived 'critical curve'.

rotational slump feature, located directly offshore the Coal Port (see Luternauer & Finn 1983; Hart *et al.* 1992*a*, *b*; McKenna *et al.* 1992; Hart 1993).

Further specialized geophysical studies include *in situ* shear wave velocity measurements on the sub-marine delta front, using a bottom towed shear wave refraction sled. This device is essentially composed of a towed sled, onto which are mounted two horizontally opposed electromagnetic shear wave sources and an electronics package. These sources generate a shear wave which is detected by six equally spaced gimbal-mounted geophones towed behind the sled in a semi-stiff mat, with a total geophone offset of *c.* 12 m. The system allows calculation of sediment shear wave velocities in the depth range of up to 3–4 m, based on a simple two-layer horizontal refraction model. In a different operational mode, longer geophone offsets are used and depths of investigation can reach 25–30 m (see Davis *et al.* 1989, 1991; Huws *et al.* 1991; Huws 1993). The sled was used during two multidisciplinary cruises aboard the Canadian research vessel, the *John P. Tully* in November 1992 and November 1993. Data coverage is illustrated on Fig. 9. In addition, sediment recovered from a core during an earlier cruise, consisting of fine to silty sand, was subsequently tested using the laboratory techniques described above. This sediment can be regarded as fairly representative of the surficial sediments found around Roberts Bank (P. K. Robertson & H. A. Christian 1995, pers. comm.). Finally, the results of two seismic cone penetration tests (SCPT) collected for the GSC on the tidal flats close to the Coal Port were obtained to aid the analysis.

Discussion

The combination of data collected from field and the laboratory environments provides some interesting

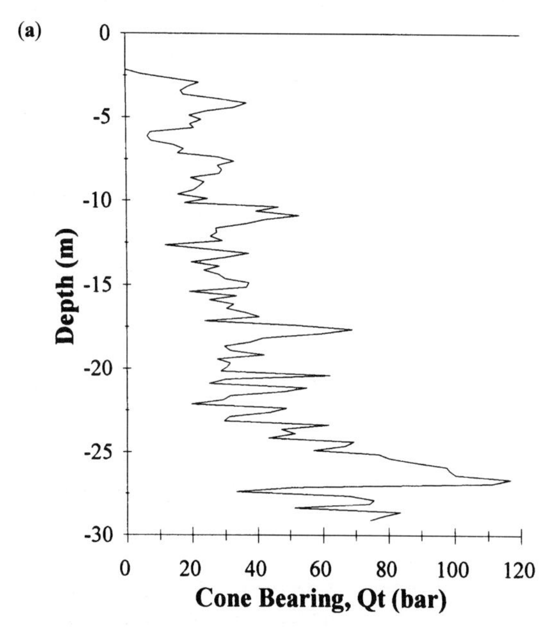

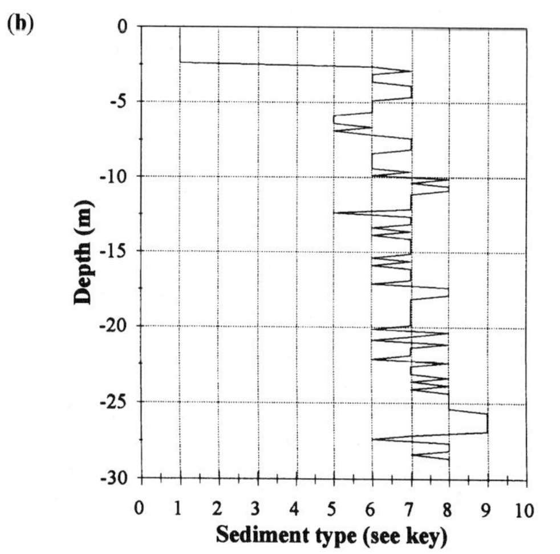

Key:

1. Sensitive fine grained.
2. Organic material.
3. Clay.
4. Silty clay
5. Clayey silt.
6. Sandy silt.
7. Silty sand.
8. Fine sand.
9. Sand.
10. Gravelly sand.

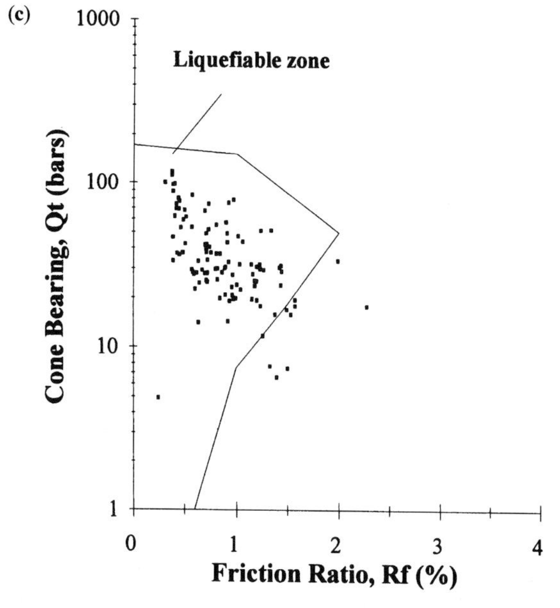

Fig. 11. SCPT data: (**a**) cone bearing profile; (**b**) inferred sediment type; (**c**) liquefaction classification.

(a)

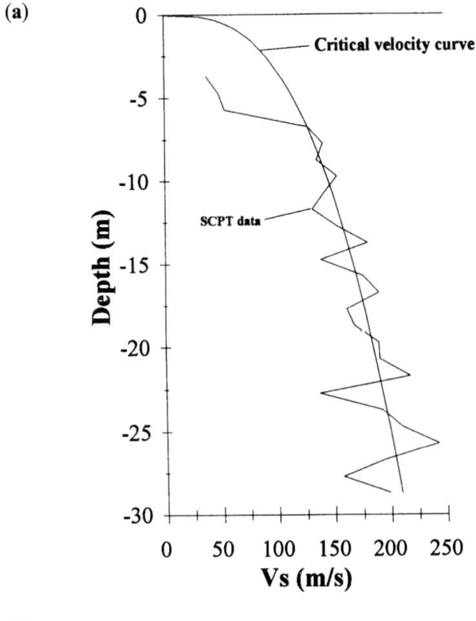

(b)

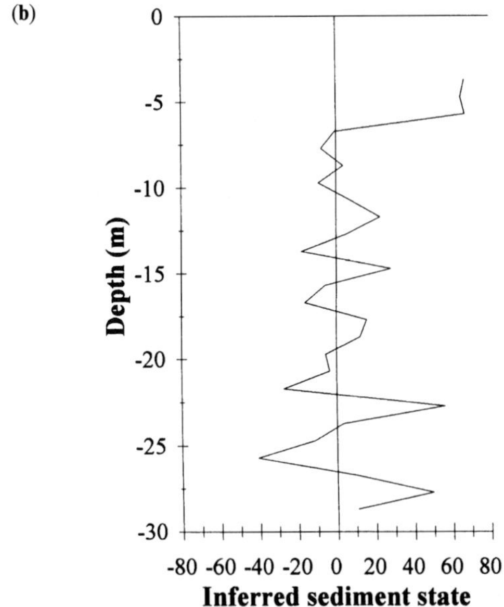

Fig. 12. (**a**) Comparison between SCPT data and laboratory derived critical curve. (**b**) Inferred sediment consolidation state, derived from the field and laboratory derived shear wave results.

Data collected using the shear wave sled amounted to an initial reconnaissance survey, providing a fairly extensive velocity data set which could be used as a platform for further, more detailed investigation. The shear wave velocity data collected during the two cruises on Robert's Bank was relatively consistent, suggesting the surface sediment in the area is fairly homogeneous, in terms of both sediment type and density. An example interpreted shear wave velocity–depth section can be seen in Fig. 10. Assuming that the sea-floor at the measurement site is composed of a sediment similar to that tested in the laboratory, the data appear to indicate a zone liable to liquefaction between approximately 0.5 m and 3 m. In order to extrapolate and estimate sediment states at deeper levels within the sediment column, SCPT data were then referenced (Fig. 11(a)–(c)). Figure 11(a) illustrates the raw penetration resistance measured as the cone was pushed into the sediment. Figure 11(b) illustrates an interpretation of the sediment type for these data, based upon empirical relationships of the type described by Robertson & Campanella (1985). As can be seen from this interpretation, the sediment type appears to consist dominantly of silty sand, generally fining upwards. Figure 11(c) illustrates a liquefaction classification for the same data, once again based upon empirical relationships presented Robertson & Campanella (1985). This classification clearly defines almost all the data as being potentially liquefiable.

Assuming the effects of cementation or strain history have a negligible effect upon the recorded field velocities, a comparison of the laboratory derived 'critical velocity curve' and the SCPT derived field velocities provides an interesting analysis. As can be seen from Fig. 12(a), there appears to be a close correlation between both data sets, suggesting that, in agreement with the cone penetration data, the entire section may be regarded as being close to critical. Figure 12(b) illustrates the inferred sediment state, based upon the velocity data described above. This may be used to further identify zones which may looser, and therefore potentially more liquefiable, than other layers in the same profile.

The proposed approach appears to offer a number of distinct advantages over conventional techniques. Firstly, with a knowledge of the 'critical velocity' derived from laboratory testing, two independent techniques of liquefaction assessment can be performed simultaneously using the SCPT, allowing detailed comparisons and further investigations if the data prove inconclusive. Secondly, shear wave refraction data obtained using the 'seismic sledge' can be used as an aid to correlation between, and extrapolation from cone sites, vastly increasing data coverage and allowing better definition the liquefaction potential of a particular site. The reader is referred to Pyrah (1996) for a more rigorous analysis of these data, which further confirms the relationships described above.

Conclusions

The above laboratory data provide evidence that, in agreement with the literature, the shear wave velocity of uncemented sands is strongly influenced by effective confining stress and void ratio. Integration of geophysical and geotechnical data derived from slightly modified consolidated, undrained triaxial testing shows clearly that shear wave velocity based estimates of liquefaction potential (or consolidation state) correlate well with a more conventional geotechnical measure of liquefaction potential, the state parameter, ψ. The only immediately obvious limitation to the proposed technique is the relative insensitivity of shear wave velocity to void ratio. However, in the laboratory, where boundary conditions can be closely controlled, this limitation is relatively insignificant.

When applied to the field situation of the Fraser River Delta, there appears to be a good correlation between laboratory derived and measured field velocities. In addition, an analysis of the data suggests that, based upon shear wave velocity data alone, there is a significant possibility of liquefaction occurring at certain zones under dynamic or monotonic loading. This evidence is further supported by other geotechnical (primarily from the cone penetration test) data collected at the site and described above. In conclusion, it is the authors' firm belief that predictive methods incorporating shear wave velocity could become an important additional tool for evaluating the liquefaction potential of uncemented sandy deposits, especially when used with other more conventional site investigation methods.

Acknowledgements. The work reported in this paper forms a part of a PhD funded by the Engineering and Physical Sciences Research Council. The authors are also indebted to the Geological Survey of Canada for their financial and logistical support, and their encouragement throughout.

References

ANON 1986. *Sediment Data, British Columbia.* Environment Canada, Inland Waters Directorate, Water Resources Branch, Ottawa, Ontario.

BEEN, K. & JEFFERIES, M. G. 1985. A state parameter for sands. *Geotechnique*, **35**(2), 99–112.

——, CROOKS, J. H. A, BECKER, D. E. & JEFFERIES, M.G. 1986. The cone penetration tests in sands: part 1 state parameter interpretation. *Geotechnique*, **36**(2), 239–249.

——, JEFFERIES, M. G., CROOKS, J. H. A. & ROTHENBURG, L. 1987. The cone penetration test in sands: part 2 general inference of state. *Geotechnique*, **37**(3), 285–299.

——, —— & HACHEY, J. 1991. The critical state of sands. *Geotechnique*, **41**(3), 363–381.

BENNELL, J. D. & TAYLOR-SMITH, D. 1991. A review of laboratory shear wave techniques and attenuation measurements with particular reference to the resonant column. *In*: HOVEM, J. M., RICHARDSON, M. D & STOLL, R. D (eds) *Shear Waves in Marine Sediments.* Kluwer, La Spezia, Italy, 83–93.

BIERSCHWALE, J. G., & STOKOE, K. H. 1984. *Analytical evaluation of liquefaction potential of sands subjected to the 1981 Westmorland earthquake.* Geotechnical Engineering Report GR-84-15, Civil Engineering Department, University of Texas, Austin, Texas.

BRYNE, P. M. & ANDERSON, L. D. 1987. *Earthquake Design in Richmond, British Columbia, Version II.* Soil Mechanics Series No. 109. Department of Civil Engineering, University of British Columbia, .

CAMPBELL, D. D. & ROTZEIN, J. L. 1992. Deterministic basis for seismic design in British Columbia. *In*: *Proceedings of Geohazards '92, Geotechnique and Natural Hazards.* BiTec, Vancouver, 71–78

CASAGRANDE, A. 1936. Characteristics of cohesionless soils affecting the stability of slopes and earth fills. *Journal of the Boston Society of Engineers*, January 1936.

——1975. Liquefaction and cyclic deformation of sands: a critical review. *In*: *Proceedings of the 5th American Conference on Soil Mechanics and Earthquake Engineering.* Buenos Aires, 80–133.

DAVIS, A. M. BENNELL, J. D., HUWS, D. G. & THOMAS, D. 1989. Development of a seafloor geophysical sledge. *Marine Geotechnology*, **8**, 99–109.

——, HUWS, D. G. & BENNELL, J. D. 1991. Seafloor shear wave velocity data acquisition: procedures and pitfalls. *In*: HOVEM, J. M., RICHARDSON, M. D & STOLL, R. D (eds) *Shear Waves in Marine Sediments.* Kluwer, La Spezia, Italy, 329–336.

DE ALBA, P., BALDWIN, K., JANOO, V., ROE, G. & CELIKKOL, B. 1984. Elastic wave velocities and liquefaction potential. *Geotechnical Testing Journal*, **7**(2), 77–87.

DOBRY, R., STOKOE, K. H., LADD, R. S. & YOUD, T. L. 1981. Liquefaction from shear wave velocity. *American Society of Civil Engineers National Convention.* St Louis, Missouri.

HARDIN, B. O. & RICHART, F. E. J. 1963. Elastic wave velocities in granular soils. *Journal of the Soil Mechanics and Foundations Division; Proceedings of the ASCE*, **89**(SM1), 33–69.

HARRIS, J. B., HUNTER, J A., LUTERNAUER, J. L. & FINN, W. D. L. 1995. Earthquake hazards of the Fraser Delta, British Columbia: sediment thickness, shear wave velocity, and site response. Victoria '95, Geological Association of Canada/Mineralogical Association of Canada (GAC/MAC) Annual Meeting, Abstracts **20**, 42.

HART, B. S 1993. Large scale in-situ rotational failure on a low angle delta slope: the Foreslope Hills, Fraser Delta, British Columbia, Canada. *Geo-Marine Letters*, **13**, 219–226.

——, PRIOR, D. B., BARRIE, J. V., CURRIE, R. G. & LUTERNAUER, J. L. 1992*a*. A river mouth submarine channel and failure complex, Fraser Delta, Canada. *Sedimentary Geology*, **81**, 72–87.

——, ——, HAMILTON, T. S., BARRIE, J. V. & CURRIE, R. G. 1992*b*. Patterns and styles of sedimentation, erosion and failure, Fraser Delta slope, British Columbia'. *In*: *Proceedings of Geohazards '92, Geotechnique and Natural Hazards.* BiTec, Vancouver, 365–372..

HEPTON, P. 1989. *Shear Wave Velocity Measurements during Penetration Testing.* PhD, University of Wales, Bangor.

HUNTER, J. A., WOELLER, D. J. & LUTERNAUER, J. L. 1991. Comparison of surface, borehole and seismic cone penetrometer methods of determining the shallow shear wave velocity structure in the Fraser River Delta, British Columbia. *Current Research, Part A Geological Survey of Canada*, Paper 91-1A, 23–36.

HUWS, D. G. 1993. *Measuring and Modelling the in situ Physical Properties of Marine Sediments*. PhD thesis, University of Bangor.

——, DAVIS, A. M. & BENNELL, J. D. 1991. 'Mapping of the seabed via in-situ shear wave (SH) velocities. *In*: HOVEM, J. M., RICHARDSON, M. D & STOLL, R. D (eds) *Shear Waves in Marine Sediments*. Kluwer, La Spezia, Italy, 337–344.

KEARY, P., & BROOKS, M. 1991. *An Introduction to Geophysical Exploration*. Blackwell, Oxford.

LUTERNAUER, J. L. & FINN, W. D. L. 1983. Stability of the Fraser Delta Front. *Canadian Geotechnical Journal*, **29**, 151–156

MCKENNA, G. T., LUTERNAUER, J. L. & KOSTASCHUK, R. A. 1992. Large scale mass wasting events on the Fraser River Delta front near Sands Heads, British Columbia. *Canadian Geotechnical Journal*, **29**, 151–156.

MILNE, W. G., ROGERS, G. C., RIDDIHOUGH, R. P., MCMECHAN, G. A. & HYNDMAN, R. D. 1978. Seismicity of Western Canada. *Canadian Journal of Earth Sciences*, **15**, 1170–1193.

MOSHER, D. C. & BARRIE, J. V. 1995. Constraints to development in the offshore Fraser River Delta. Victoria '95, Geological Association of Canada/Mineralogical Association of Canada (GAC/MAC) Annual Meeting, Abstracts, **20**, 72.

O'ROURKE, 1995. Geotechnical effects. National Centre for Earthquake Engineering Research (NCEER) Response. Special supplement to *NCEER Bulletin*, **9**(1), 6.

POULOS, S. J. 1981. The steady state of deformation. *Journal of Geotechnical Engineering*, **107**(GT5), 553–562.

——, CASTRO, G. & FRANCE, J. 1985. Liquefaction evaluation procedure. *Journal of Geotechnical Engineering*, **111**(6), 777–792.

PYRAH, J. R. 1996. *An integrated geotechnical – geophysical procedure for the prediction of liquefaction in uncemented sands*. PhD Thesis, University of Wales, Bangor.

ROBERTSON, P. K. & CAMPANELLA, R. G. 1985. Liquefaction potential of sands using the CPT. *Journal of Geotechnical Engineering*, **111**(3), 384–403.

——, ——, GILLESPIE, D. & RICE, A. 1986. Seismic CPT to measure in-situ shear wave velocity. *Journal of Geotechnical Engineering*, **112**(8), 791–803.

——, WOELLER, D. J. & FINN, W. D. L. 1992*a*. Seismic cone penetration for evaluating liquefaction potential under cyclic loading. *Canadian Geotechnical Journal*, **29**, 686–695.

——, ——, KOKAN, D. J., HUNTER, J. & LUTERNAUER, J. 1992*b*. Seismic cone techniques to evaluate liquefaction potential. *45th Canadian Geotechnical Conference: Innovation, Conservation and Renovation*. Toronto, Ontario, 5.1–5.9.

——, SASITHARAN, S., CUNNING, J. C., & SEGO, D. C. 1995. Shear-wave velocity to evaluate in-situ state of Ottawa Sand. *Journal of Geotechnical Engineering*, **121**(3), 262–273.

ROGERS, G. C. 1988. An assessment of the megathrust earthquake potential of the Cascadia subduction zone. *Canadian Journal of Earth Science*, **24**, 844–852.

ROSCOE, K. H., SCHOLFIELD, A. N. & WROTH, C. P. 1958. On the yielding of soils. *Geotechnique*, **8**(1), 22–53.

SCHMERTMANN, J. H. 1978. Use the SPT to measure dynamic soil properties? – Yes, but...! *Dynamic Geotechnical Testing, ASTM STP*, **654**, 341–355.

SCHULTHEISS, P. J. 1983. *The influence of packing structure on seismic wave velocities in sediments*. Marine Geological Report No. 83/1, University College of North Wales.

SEED, H. B., IDRISS, I. M. & ARANGO, I. 1983. Evaluation of liquefaction potential using field performance data. *Journal of Geotechnical Engineering*, **109**(3), 458–482.

SLADEN, J. A. 1989. 'Problems with interpretation of sand state from cone penetration test'. *Geotechnique*, **39**(2), 323–332.

——, D'HOLLANDER, D. & KRAHN, J. 1985. The liquefaction of sands, a collapse surface approach. *Canadian Geotechnical Journal*, **22**, 564–578.

STEWART, I., & TASSONE, B. 1989. *The Fraser River Delta: a review of historical sounding charts*. Environment Canada – Inland Waters, Pacific and Yukon Region, Vancouver, British Columbia.

STOKOE, K. H., ROSSET, J. M., BIERSCHWALE, J. G. & AOVAD, M. 1988. Liquefaction potential of sands from shear wave velocity. *In*: *Proceedings of the 9th World Conference on Geotechnical Engineering*. A. A. Balkema, Tokyo–Kyoto, Japan, 213–218.

TOKIMATSU, K. & UCHIDA, A. 1990. Correlation between liquefaction resistance and shear wave velocity. *Soils and Foundations*, **26**(2) 25–35.

——, YAMAZAKI, T. & YOSHIMI, Y. 1986. Soil liquefaction evaluations by elastic shear moduli. *Soils & Foundations*, **26(1)**, 25–35.

——, YOSHIMI, Y. & UCHIDA, A. 1988. Evaluation of undrained cyclic shear strength of soils with shear wave velocity. *In*: *Proceedings of the 9th World Conference on Earthquake Engineering*. A. A. Balkema, Tokyo–Kyoto, Japan, 207–213.

WOODS, R. D. 1991. Soil properties for shear wave propagation. *In*: HOVEM, J. M, RICHARDSON, M. D & STOLL, R. D (eds) *Shear Waves in Marine Sediments*. Kluwer, La Spezia, Italy, 29–40.

VIGGRANI, G. & ATKINSON, J. H. 1995. Interpretation of bender element tests. *Geotechnique*, **45**(1), 149–154.

YOSHIMI, Y., HATANAKA, M. & OH-OKA, H. 1979. Undisturbed sampling of saturated sands by freezing. *Soils and Foundations*, **18**(3), 59–73.

——, TOKIMATSU, K. & OHARA, J. 1994. In-situ liquefaction resistance of clean sands over a wide density range. *Geotechnique*, **44**(3), 479–494.

High-altitude glacial lake hazard assessment and mitigation: a Himalayan perspective

John M. Reynolds

Reynolds Geo-Sciences Ltd, The Stables, Waen Farm, Nercwys, Mold, Flintshire CH7 4EW, UK

Abstract. Glaciers throughout the Himalayas have been receding rapidly over the last few years. In the areas vacated by the ice snouts lakes have formed behind large moraine dams. As with their Andean counterparts, these lakes can pose significant threats to downstream towns, roads and power schemes. For a developing country like Nepal, for example, the prospective loss of economically vital infrastructure and food-producing land can be devastating, in addition to the human misery.

The Tsho Rolpa glacier lake has formed by the retreat of Trakarding Glacier at the head of the Rolwaling Valley in northern Nepal. The lake is now over 3 km long and contains an estimated volume of 80×10^6 m^3. The snout of Trakarding Glacier terminates in the lake which, at its western end, is dammed by an ice-cored moraine. It is feared that the lake could be the source of a major glacier lake outburst flood (known locally as a 'GLOF') which could destroy the nearby villages of Na and Beding, as well as trekking routes, vital bridges, *etc.*, further downstream. At Khimti, 80 km downstream from Tsho Rolpa, a new hydroelectric power scheme is being built and could potentially be at risk.

It is known that the local glaciers in the Rolwaling Valley have been the source of many GLOFs in historical as well as contemporary times. The last GLOF destroyed several houses and valuable farm land at Beding in July 1991.

Following an initial GLOF-risk assessment in September 1994, a trial siphon was installed at Tsho Rolpa in May 1995. This is the first time that such measures have been used in the Nepalese Himalayas. Since then further remediation measures have been proposed and are awaiting funding. The nature of the glacial hazards at Tsho Rolpa and the engineering mitigation measures being proposed are described. Furthermore, it is concluded that an integrated national policy on glacial hazards should be developed urgently.

Introduction

Glaciers in the Himalayas have been receding throughout this century. As the ice fronts retreat up valley, proglacial lakes develop behind terminal moraine dams, some of which are cored with stagnant glacier ice. As the volume of stored water increases, so too does the pressure on the moraine walls. In many cases, the moraine fails catastrophically releasing large amounts of water very quickly. Discharge rates of several thousand cubic metres per second are not uncommon. The released water mixes with morainic material to form a highly mobile and devastating debris flow/flood. The name commonly given to this phenomenon is a *glacier lake outburst flood* (GLOF). In Europe such discharges are known as *débâcles*; in South America they are referred to as *aluviones* (Reynolds 1992). They have also been called, erroneously, *jökulhlaups* (Ives 1986; Benn & Evans 1998); these refer to sub-glacial discharge floods initiated as a result of sub-glacial volcanic activity as occurs in Iceland where the name originated. Although the term GLOF is in essence generic, i.e. it assumes that all such floods originate from glacial lakes, it is perhaps a generalization which does not hold true in all cases.

Glacier lake outburst floods in the Himalayas threaten villages, trekking routes, infrastructure (roads, bridges, etc.), hydroelectric power schemes, and valuable food-producing land. Thankfully, most GLOFs that have occurred in Nepal have been relatively small and the scale of devastation has been limited to areas within the immediate vicinity of the lakes involved. However, one huge catastrophic flood was thought to have occurred about 450 years ago when a lake of some 10 km^2 located behind Machapuckhare broke through its ice-cored moraine dam. The resulting flood inundated the Pokhara basin covering it with 50–60 m of debris (Yamada 1993).

Nepal provides a topographic buffer zone between the low plains of India and the high plateau of Xizang (Tibet) in China. For this reason Nepal has played a vital and very valuable role politically, spiritually and economically between India and Xizang (Tibet). Physiologically, river catchments in Xizang provide water in some of the rivers which flow through Nepal and into northern India

Reynolds, J. M. 1998. High-altitude glacial lake hazard assessment and mitigation: a Himalayan perspective. *In*: Maund, J. G. & Eddleston, M. (eds) *Geohazards in Engineering Geology*. Geological Society, London, Engineering Geology Special Publications, **15**, 25–34.

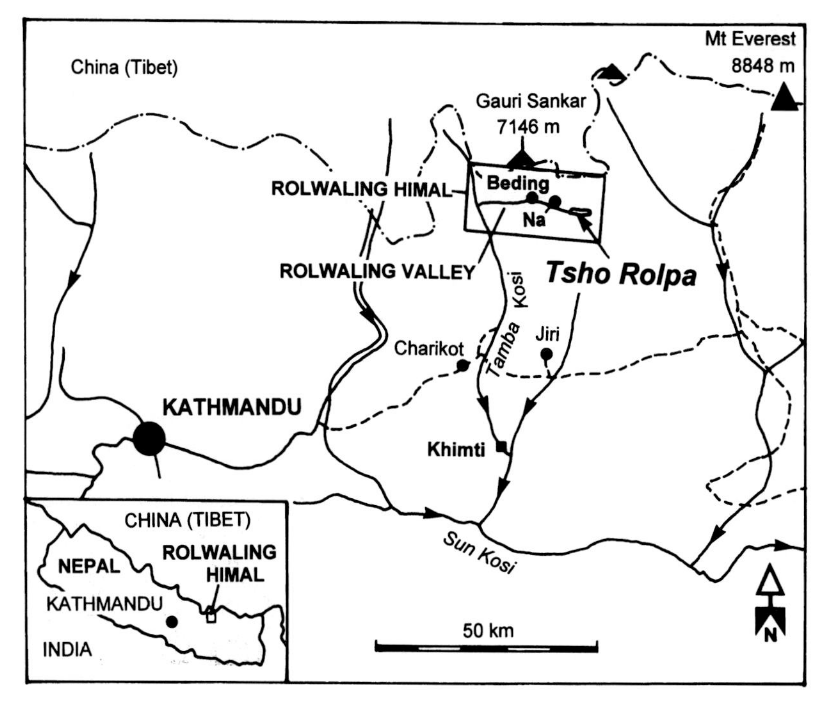

Fig. 1. Location map of Tsho Rolpa in the Rolwaling Valley.

(Chaohai & Sharma 1988). Consequently, catastrophic events in a catchment area in Xizang can have far reaching consequences if the scale of the disaster is large enough. In such cases, run-outs from the source lake may exceed 200 km. Not only does this have economic implications, but also serious political ones too. Furthermore, for an impoverished developing country such as Nepal, which is reliant on food imports to sustain its own population, inward economic investment is crucial. Major infrastructural developments, such as road schemes linking remote areas, and hydroelectric power projects, to name but two, are of critical importance. Yet despite this, there has been little attention paid to the consequences of glacial hazards on such development projects. One reason for this is that the importance of glacial hazards in both the Himalayas and the Andes, for example, is only now starting to be recognized at government level. This recognition is coming about as a result of the increasing amount of literature describing the problems (e.g. Lliboutry *et al.* 1977; Vuichard & Zimmermann 1987; Reynolds 1990, 1992, 1993, 1995, 1998; Grabs & Hanisch 1993) and also because there have been a number of recent glacier lake outbursts which have focused international attention onto the problem. One such example is the potential of a major GLOF from Tsho Rolpa in the Rolwaling Valley (Fig. 1) which is the main subject of this paper. There are others in Nepal which, for reasons of political and commercial sensitivity, will not be addressed here.

Glacier lake outburst floods (GLOFs) in the Himalayas

As mentioned above, there have been many GLOFs in Nepal (Table 1; Yamada 1993). However, experience in the Nepal Himal has shown that the geological importance of GLOFs has not been recognized. Not only have there been many more than those recorded (Reynolds Geo-Sciences Ltd 1997), but the scale of some of them in pre-historical times is significantly larger than any of those recorded from historical or contemporary times. The full extent of GLOFs in the geological record has yet to be realized or fully appreciated. They have probably occurred also in Europe and Scandinavia at the edges of the ice sheet as it receded at the end of the last glaciation. In Nepal, a brief trek along one river valley (the Tamba Kosi; Fig. 1) provided sufficient evidence to suggest that there had been many GLOFs there in the past (Reynolds Geo-sciences Ltd 1994).

One of the GLOFs pertinent to this paper occurred in mid-July 1991. Following three very warm days, an ice avalanche fell into Chubung Lake which burst through

Table 1. *Known occurrences of Glacier Lake Outburst Floods in Nepal*

Year	*Location*	*River system*	*Principal damage*
1980	Phuchan Glacier lake	Sapta Kosi	Damage to forest and river bed, etc.
1964	Gelhaipco lake	Arun/Pumqu	End moraine collapsed due to ice avalanche into the lake; road damaged and 12 trucks lost.
1968	Aycio Lake	Arun/Pumqu	Damage to roads and bridges.
1969	Aycio Lake	Arun/Pumqu	Damage to roads and bridges.
1970	Aycio Lake	Arun/Pumqu	Damage to roads and bridges.
1982	Jinco Lake	Arun/Pumqu	Moraine collapsed due to glacier tongue sliding into lake. Damage to eight villages, livestock killed, fields, roads and bridges damaged.
1964	???	Arun/Pumqu	Damage to forest, bridges and trucks.
1977	Nare Glacier lake	Dudh Kosi	Ice-cored moraine collapsed. Damage to mini-HEP station, road, bridges, fields, etc.
1985	Dig Tsho	Dudk Kosi	Moraine collapse after rock avalanche. Destroyed Namche HEP plant, damaged roads, bridges, fields, houses and caused casualties.
1991	Ripimo Shar Glacier	Tama Kosi	Moraine failure caused damage to trekking routes, killed livestock, damaged fields and houses at Beding village. (See Fig. 2)
1964	Zhangzangbo Glacier	Sun Kosi/Poiqu	Moraine collapsed due to seepage.
1981	Zhangzangbo Glacier	Sun Kosi/Poiqu	Moraine collapse due to glacier front calving. Damage to Arniko Highway, bridges, Sun Koshi HEP station, fields, killed livestock and caused casualties.
1964	Longda Glacier	Trisuli River	No data.

its terminal moraine dam at the southern end of Ripimo Shar Glacier. The breach, which was estimated as being 15 m deep and 20 m wide, is shown in Fig. 2. As is typical of GLOF breaches, a significant debris fan formed immediately downstream of the breach in which the coarser debris load was deposited. The water and finer material travelled downstream past the village of Na, where it killed some livestock and destroyed one bridge, and on through the village of Beding. There it destroyed a number of houses and several valuable potato fields, and caused great alarm amongst the inhabitants.

The event at Ripimo Shar was small in scale yet had sufficient energy to have a run-out of over 10 km. At least four GLOFs have occurred within the Rolwaling Valley in living memory, in addition to that of 1991, although they have not been formally recorded. Evidence of former GLOFs along the Tamba Kosi river valley at distances in excess of 60 km from the nearest glaciers suggests that the area has been affected by much more significant events in the past. There is also possible evidence that GLOF-type flows could have originated within Xizang (Tibet). For example, Lamabaga, a village 5 km N of the confluence of the Rolwaling River and Tamba (Bhote) Kosi, appears to have been built on an alluvial fan/dammed debris flow. If this is the case, then the source of the material must be further north (upstream) from one of the drainage basins inside the Chinese border (Reynolds Geo-Sciences Ltd 1994).

Glacial hazard assessment and mitigation: Rolwaling Valley

Background to the problem at Tsho Rolpa

Following the July 1991 GLOF from Ripimo Shar, the villages of Beding and Na within the Rolwaling Valley sought outside help as they perceived a serious risk of a future GLOF from Tsho Rolpa, a lake many times the size of Chubung, the source of the 1991 flow. As the Rolwaling Valley is on one of the more serious trekking routes to Mt Everest, Western mountaineers soon learned of the plight of the local Sherpa. Furthermore, letters were written to the principal embassies in Kathmandu requesting help. Dr Michiel Damen wrote

Fig. 2. The breach at Ripimo Shar, Rolwaling Valley, after the July 1991 GLOF.

the first formal scientific assessment of the problem in 1992 (Damen 1992). The present author received an invitation to help with the problem at Tsho Rolpa from Dr Damen, and independently from the Water and Energy Commission Secretariat (WECS), Ministry of Water Resources, Kathmandu. WECS had undertaken some preliminary observations in the Rolwaling Valley and around Tsho Rolpa in 1993 (Mool *et al.* 1993), as had various Japanese researchers funded under a Japanese international aid programme (e.g. Yamada 1993). Given the immediacy of the problem, the author was financed by the Emergency Aid Department of the Overseas Development Administration (now the Department for International Development) to visit Tsho Rolpa and to assess the situation, based on previous experience of hazard assessment and mitigation in Peru (Reynolds 1990, 1992, 1993). As a result of this work, it was realized that the moraine complex that dammed Tsho Rolpa was probably cored with ice. As the maximum amount of freeboard of the moraine dam available during the dry season was about 1 m it was felt that the entire moraine complex, coupled with the volume of stored water within the lake (*c.* 80×10^6 m^3; Yamada 1996), formed a highly unstable and potentially lethal situation. It was felt that, unless appropriate remediation works were undertaken, the moraine would fail and the Rolwaling Valley would be inundated, causing widespread loss of life and serious damage to local infrastructure. At Khimti, 80 km downstream from Tsho Rolpa, a new hydroelectric power scheme is under active consideration. Subject to financing provisions, it is anticipated that, at the height of construction, some 1500–2000 workers will be housed in a camp adjacent to the river and could potentially be at risk. It has been estimated by the plant managers that serious GLOF damage could cost in excess of $22 million and put the construction project back by two or more years.

Figure 3 shows the view from Na towards the eastern end of Rolwaling Valley. The breach through which Chubung Lake discharged from Ripimo Shar is obvious. The existing spillway from Tsho Rolpa is indicated by an arrow. The height of the moraines above the valley

Fig. 3. The view towards the eastern end of the Rolwaling Valley from Na, showing the breach in the Ripimo Shar moraine (centre left) and the existing spillway over the Tsho Rolpa moraine (arrowed).

floor is of the order of 200–250 m. The local geography of Ripimo Shar and the western terminal moraine complex of Tsho Rolpa is shown in Fig. 4. More details of the background to this project have been given by Reynolds (1995).

Hazard assessment

The immediate requirement of the field visit in October 1994 was to assess each of three areas; namely, the lake with its moraine complexes, the Trakarding Glacier ice front (which terminates at the eastern end of Tsho Rolpa), and the immediate area surrounding the site. It was important to determine if there was any evidence to suggest the presence of an ice core and its state, to gauge the stability of the moraine complexes (any evidence of slope failure, cracking, anomalous vegetation patterns, differences in material types, etc.) and to inspect the surrounding hanging glaciers as to their common mode of ablation (melt run-off, evaporation, ice calving, etc.). It was also vital to assess the stability of the Trakarding Glacier ice front. Was the glacier grounded or was the tongue afloat in the lake? This question needed to be answered if safe mitigation measures were to be implemented.

In essence, it was concluded from the field visit that the terminal moraine is ice cored in two locations. Evidence for this consisted of hummocky topography, cracked and slumped ground, debris patterns, the presence of sink-holes (drainage through melting ice), etc. Subsequent geophysical resistivity experiments confirmed the presence of massive ice beneath the western end of the lake and into the moraine (OYO Corporation 1995). A simple map of the principal features is shown as Fig. 5, including the interpreted lateral extent of the buried ice. The lateral moraines on the northern and southern sides of the lakes are generally ice-free but are inherently unstable on the lakeside as evidenced by the total lack of vegetation and the constant stream of spalling debris. The outer flanks are much more stable as indicated by almost total vegetation cover.

The Trakarding Glacier ice front was found to be made up of two glaciers that had merged several

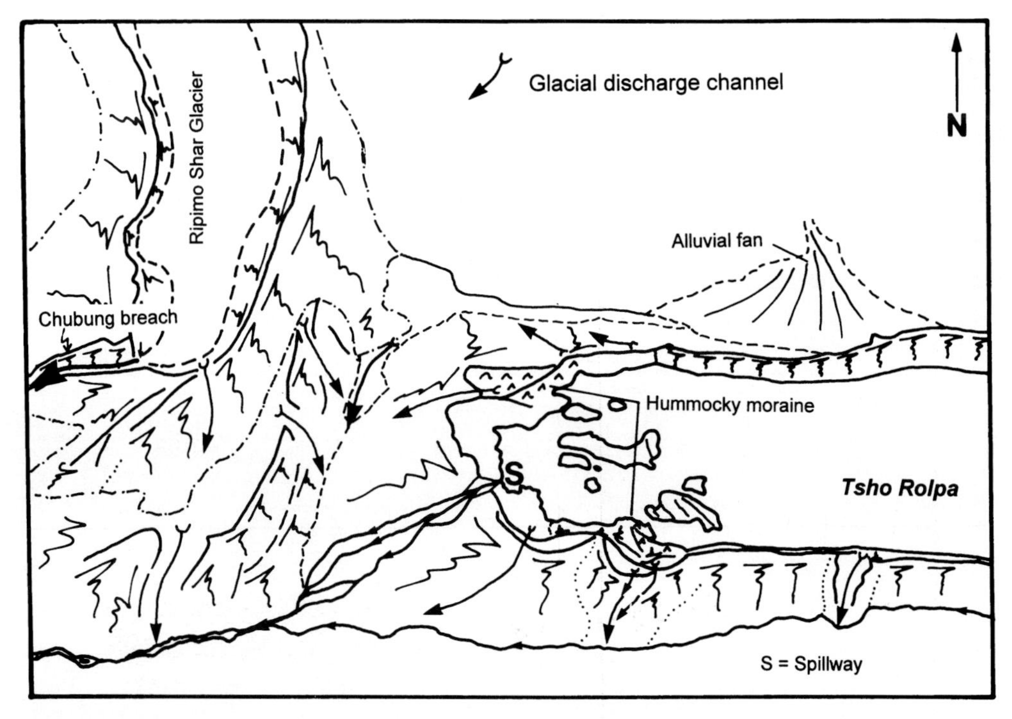

Fig. 4. Interpreted transcription from a colour oblique aerial photograph published by Yamada (1993) (after Reynolds Geo-Sciences Ltd 1994).

kilometres upstream to the east. The northern part of the ice front appeared to be grounded and had a sub-glacial discharge portal evident (Fig. 6; arrowed). The southern part of the ice front was observed to be almost vertical (Fig. 6) and showed signs of cracks parallel to the ice front. Sub-glacial discharge was seen to emanate from below the lake surface at the southern edge of the ice. The form of the ice front indicated its general stability but that relatively small ice avalanches of the order of 1500–5000 m^3 were likely to occur. While these in themselves are not particularly troublesome, the lack of freeboard at the terminal moraine to contain the displacement waves from overtopping the moraine is more serious.

The hanging glaciers on the northern side of the lake were found to be relatively well contained with only sporadic ice avalanches, most of which appeared to be contained locally without direct access to the lake below. The local glaciers on the southern side were found to be set further back from the lake so avalanching was not perceived to be a significant problem. As with any high-altitude mountain range such as the Himalayas or Andes, the extremely steep slopes and the sheer vertical scale provide spectacular snow and ice avalanches that occur frequently most days (Fig. 7).

Hazard mitigation

From the assessment of Tsho Rolpa and its immediate environs it was concluded that mitigation measures should be implemented as soon as was practical. The most cost-effective and practical way to achieve immediate relief from the threat of the moraine being overtopped was by installing a siphon (Grabs & Hanisch 1993; Reynolds Geo-Sciences Ltd 1994). This was the first stage used in a similar project in Peru (Reynolds *et al.* 1998). Whatever was decided, it had to be practically feasible and within the constraints of local transport, i.e. carried on the backs of local porters. With support from WAVIN Overseas BV, a Dutch pipe manufacturer, it was agreed that a trial siphon would be installed before the 1995 monsoon. This was achieved using a 16 cm diameter pipe some 140 m long passing from the southwestern corner of the lake over the lowest freeboard of the moraine and discharging into an existing water course. The discharge flow speed was reported to be about 9 $m\,s^{-1}$, giving an approximate discharge of 140 $l\,s^{-1}$ (van Nes 1996). The successful installation of this siphon is the first time in the Himlayas that such measures have been used. Having demonstrated that the technology works under the prevalent field conditions, additional

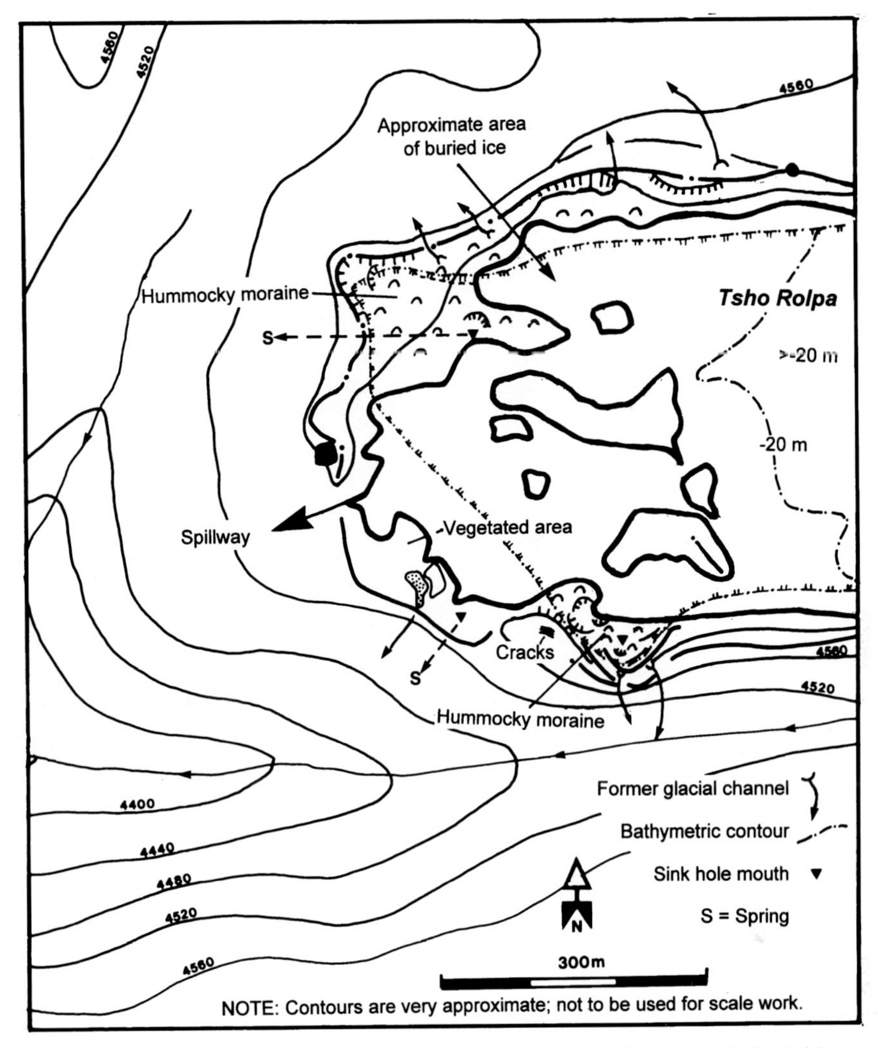

Fig. 5. A simple map showing the locations of principle features. The lateral extent of the buried ice core is also indicated (after Reynolds Geo-Sciences Ltd 1994).

measures are now required to achieve the lowering of the lake water level ultimately by between 15 and 20 m. This is the subject of ongoing political discussions. Given that the ice cores within the moraine are actively melting, it is estimated that, unless the water level can be lowered by 3–5 m by mid-1999, the moraine is still likely to fail catastrophically. For each metre of vertical drawdown achieved, $1–1.5 \times 10^6$ m^3 of water are removed from the volume availabe to form a GLOF. If a 4 m drawdown can be achieved by the 1999 monsoon, for example, the total volume of water that could form a GLOF could be reduced by 20%. Furthermore, such a drawdown will reduce the hydraulic gradient and the hydrostatic pressure head at the moraine dam. Consequently, the onset of a breach through the moraine dam may be delayed sufficiently to permit the installation of more permanent remediation measures, with a final lowering of the lake water level to 15–20 m below its present elevation, assuming sufficient funding is granted in time. However, it is clear that even if the total cost of remediation reaches as much as $8 million over 10 years, this is still significantly less than the potential value of the damage that would result from a catastrophic flood.

Fig. 6. The northern discharge portal of the Trakarding Glacier is indicated by an arrow. The southern part of the ice front is almost vertical (from Reynolds Geo-Sciences Ltd 1994).

The general implications of GLOFs in the Himalayas

To date, the exact number of deaths and the value of damage caused by glacier lake outburst floods in Nepal are not known. However, the frequency with which GLOFs are occurring is increasing (Reynolds Geo-Sciences Ltd 1997), as is the pressure on inward investment in major infrastructure projects, and increasing density of populations within the river valleys in the high Himal. In a nutshell, there are now more people and higher value structures at risk from an increasing number of dangerous glacial lakes. This applies not just to Nepal but across the Himalayas. Indeed, it is also true of the Andes.

It is not sufficient to take a lake-by-lake approach as the implications of a potential major disaster to a country like Nepal are enormous – environmentally, socially, politically and economically. There is a desperate need for an integrated national strategy to deal with GLOFs and related matters such as social and economic development, water resources, infrastructural development, and so on. While external aid can be used to alleviate short-term emergency needs, it is far better to discourage a dependency culture and to develop an indigenous capacity to formulate appropriate responses to such problems. It is pertinent to note that while political obfuscation continues, ice cores within moraines are continuing to melt bringing the likelihood of a major disaster nearer unless appropriate action is taken. For this both the political will and the financial means need to be committed, not just for a short time, but over several decades.

Conclusions

Emergency hazard assessment work undertaken at Tsho Rolpa in 1994 has led to the successful installation of a trial siphon. It is thought that this is the first time that such measures have been taken in the Himalayas. Now that the method has been demonstrated to work under

Fig. 7. A snow avalanche cascades down onto Trakarding Glacier, the lowermost part of which is debris covered (concealed behind the island). The vertical extent shown is over 2 km.

the prevalent conditions, additional initial measures need to be installed by mid-1999 if the lake is to be remediated successfully.

In both the Himalayas and the Andes, the number and sizes of high-altitude mountain lakes are increasing as glaciers recede. With growing populations and increasing numbers of high-cost infrastructure projects, the risk of a major catastrophic flood will also continue to increase unless appropriate action is taken in time.

In addition to dealing with individual cases on an emergency basis, there is a desperate need for a long-term national strategy to deal with glacier lake outburst flood risks and related problems.

Acknowledgements. I am very grateful to the Water and Energy Commission Secretariat (WECS), Ministry of Water Resources, Kathmandu, Nepal, for the invitation to become involved in this work in 1994. I am also very grateful to them and to the Department of Hydrology and Meteorology, Ministry of Science and Technology, for their ongoing support and generous collaboration. My input has been financed by the Emergency Aid Department of the Department For International Development (formerly the Overseas Development Administration), London. The British Embassy in Kathmandu has also been extremely supportive of this work, and to the Ambassador and his staff, I express my gratitude.

References

BENN, D. I. & EVANS, D. J. A. 1998. *Glaciers and Glaciation.* Arnold, London.

CHAOHAI, L. & SHARMA, C. K. (eds) 1988. *Report on first expedition to glaciers and glacier lakes in the Pumqu (Arun) and Poiqu (Bhote-Sun Kosi) river basins, Xizang (Tibet), China.* Lanzhou Institute of Glaciology and Geocryology, Water and Energy Commission Secretariat and the Nepal Electricity Authority, Kathmandu.

DAMEN, M. 1992. *Study on the potential outburst flooding of Tsho Rolpa Glacier Lake, Rolwaling Valley, east Nepal.* International Institute for Aerospace Survey and Earth Sciences, ITC, Enschede, The Netherlands.

GRABS, W. E. & HANISCH, J. 1993. Objectives and prevention methods for glacier Lake Outburst Floods (GLOFs). *In: Snow and Glacier Hydrology (Proceedings of the Kathmandu Symposium, Nov. 1992)*, IAHS Publication No. 218, 341–352.

IVES, J. D. 1986. *Glacial lake outburst floods and risk engineering in the Himalaya.* ICIMOD Occasional Paper No. 5 International Centre for Integrated Mountain Development, Kathmandu, Nepal.

LLIBOUTRY, L., ARNAO, B., MORALES, A., PAUTRE, A. & SCHNEIDER, B. 1977. Glaciological problems set in the control of dangerous lakes in Cordillera Blanca, Peru. Part I: Historical failures of moraine dams, their causes and prevention. *Journal of Glaciology*, **18**(79), 239–254.

MOOL, P. K., KADOTA, T., MASKEY, P. R., POKHAREL, S. & JOSHI, S. 1993. *Interim report on the field investigation on the Tsho Rolpa glacier lake, Rolwaling Valley.* Water and Energy Commission Secretariat Report No. 3/4/021193/1/1, Seq. no. 436, Kathmandu, Nepal.

OYO CORPORATION 1995. *Electrical resistivity exploration at Tsho Rolpa end moraine – final report.* OYO Corporation, Japan.

REYNOLDS, J. M. 1990. Geological hazards in the Cordillera Blanca, Peru. *AGID News*, 61/62, 31–33.

——1992. The identification and mitigation of glacier-related hazards: examples from the Cordillera Blanca, Peru. *In*: MCCALL, G. J. H., LAMING, D. J. C. & SCOTT, S. C. (eds) *Geohazards.* Chapman and Hall, London, 143–157.

——1993. The development of a combined regional strategy for power generation and natural hazard assessment in a high-altitude glacial environment: an example from the Cordillera Blanca, Peru. *In*: MERRIMAN, P. A. & BROWITT, C. W. A. (eds) *Natural disasters: protecting vulnerable communities.* Thomas Telford, London, 38–50. [Also published in 1994 in *Disaster Management*, **6**(1), 29–35.]

——1995. Glacier-lake outburst floods (GLOFs) in the Himalayas: an example of hazard mitigation from Nepal. *Geoscience and Development*, **2**, 6–8.

——1998. Managing the risks of glacial flooding at hydro plants. *Hydro Review Worldwide*, **6**(2), 18–22.

——, DOLECKI, A. & PORTOCARRERO, C. 1998. The construction of a drainage tunnel as part of glacial lake hazard mitigation at Hualcán, Cordillera Blanca, Peru. *In*: MAUND, J. G. & EDDELSTON, M. (eds) *Geohazards in Engineering Geology.* Geological Society, London, Engineering Geology Special Publications, **15**, 41–48.

REYNOLDS GEO-SCIENCES LTD 1994. *Hazard assessment at Tsho Rolpa, Rolwaling Himal, northern Nepal: technical report.* Reynolds Geo-Sciences Ltd, Mold, UK.

——1997. *Ongoing efforts to detect, monitor, and mitigate the effects of GLOFs in Nepal.* Reynolds Geo-Sciences Ltd, Mold, UK.

VAN NES, J. A. 1996. *Trial syphon at lake Tso Rolpa (sic).* WAVIN Overseas, B. V., Dedemsvaart, The Netherlands.

VUICHARD, D. & ZIMMERMANN, M. 1987. The 1985 catastrophic drainage of a moraine-dammed lake, Khumbu Himal, Nepal, cause and consequence. *Mountain Research and Development*, **7**(2), 91–110.

YAMADA, T. 1993. *Glacier lakes and their outburst floods in the Nepal Himalaya.* Water and Energy Commission Secretariat, Kathmandu, Nepal.

——1996. *Report on the investigations of Tsho Rolpa Glacier Lake, Rolwaling Valley.* Water and Energy Commission Secretariat (WECS) and Japan International Cooperation Agency (JICA), August 1996, Kathmandu, Nepal.

On the controls of fluvial hazards in the north Bihar plains, eastern India

Rajiv Sinha

Engineering Geology Group, Department of Civil Engineering, Indian Institute of Technology, Kanpur – 208016 (UP), India (E-mail: rsinha@iitk.ernet.in)

Abstract. The rivers of the north Bihar plains, eastern India, pose three major fluvial hazards: rapid lateral migration, frequent flooding and extensive bank erosion. Lateral shifting of the Kosi, Gandak and several other rivers in the area has been attributed mainly to neotectonic tilting and subsidence of the area, and to some extent, local topography and sedimentological readjustments in the basin. Overbank flooding is a perennial problem, with most of the rivers of the north Bihar plains causing enormous damage to life and property. The construction of embankments along major portions of the rivers is only a short-term solution to mitigate floods not only because of frequent breaches in the embankment due to extremely high discharges during high flows but also because of the fact that these rivers carry a high sediment load causing rapid siltation and thereby raising the water level in a few years' time. Severe bank erosion takes place during the lateral shifting of rivers as well as during high flows. Efforts to prevent these fluvial hazards in the area have largely failed as the geological and geomorphological considerations have not been taken into account.

Introduction

The Indo-Gangetic plains are one of the most extensive tracts of Quaternary alluvial sedimentation in the world. The focus of this paper is on the plains of north Bihar, eastern India (Fig. 1), characterized by two megafans, formed by the Gandak and Kosi river systems, and an interfan area between them drained by the channels of the Burhi Gandak, Baghmati and Kamla-Balan river systems. The main feature of the river systems of north Bihar is the recognition of three different classes based on their source areas. The *mountain-fed* river systems (e.g. Kosi and Gandak) probably first developed where antecedent rivers in the early evolution of the mountain front eroded back preferentially, at points of structural weakness, and captured the evolving drainage of the high mountain terrain to the north. In contrast, the headwaters of the *foothills-fed* river system (e.g. Baghmati) have developed by localized valley erosion on the uplifting foothills, and are probably much younger. The *plains-fed* rivers (e.g. Burhi Gandak) have clearly developed because without them the other river systems are not able to receive the runoff from monsoonal rains. In an earlier work, the different classes of river systems have been shown to exhibit distinctive morphological, hydrological, and sediment transport characteristics (Sinha & Friend 1994). The drainage pattern is typically dendritic with high angles of tributary convergence in the high mountains and foothills (north of the mountain front; see Fig. 1). On the alluvial plains, south of the mountain front, the Gandak and Kosi rivers are characterized by their divergent patterns. The other river systems show a gently converging pattern in the upstream region and are more sinuous in the downstream region.

The plains of north Bihar experience monsoonal rainfall, preceded by pre-monsoon showers. The average annual rainfall in the plains is 100–160 cm whereas the foothills above the plains experience a higher annual rainfall (>200 cm). The main monsoon usually arrives in mid-June in the plains and ceases in September, contributing over 85% of the annual rainfall. In the foothills, the main rain starts a little earlier, usually in late May, and may continue longer, into late October.

Lateral migration, bank erosion and overbank flooding are among the major fluvial processes operating in the area. However, the frequency of these processes is very rapid and the extent is severe; so much so that they are regarded as major 'fluvial hazards' in the area. This papers aims to describe the nature of these hazards, their controlling factors, and a critical review of the measures taken so far to mitigate them.

Lateral migration

Lateral migration is one of the most serious problems posed by the rivers draining the north Bihar plains. The Kosi has changed its course by about 112 km towards west in 250 years (Gole & Chitale 1966; Gohain & Parkash 1990) and has created one of the largest humid megafans in the world. Such channel movements are normally slow and therefore do not attract the immediate attention of the concerned authorities, thereby causing large-scale damage to life and property. The Gandak has also shifted

SINHA, R. 1998. On the controls of fluvial hazards in the north Bihar plains, eastern India. *In*: MAUND, J. G. & EDDLESTON, M. (eds) *Geohazards in Engineering Geology*. Geological Society, London, Engineering Geology Special Publications, **15**, 35–40.

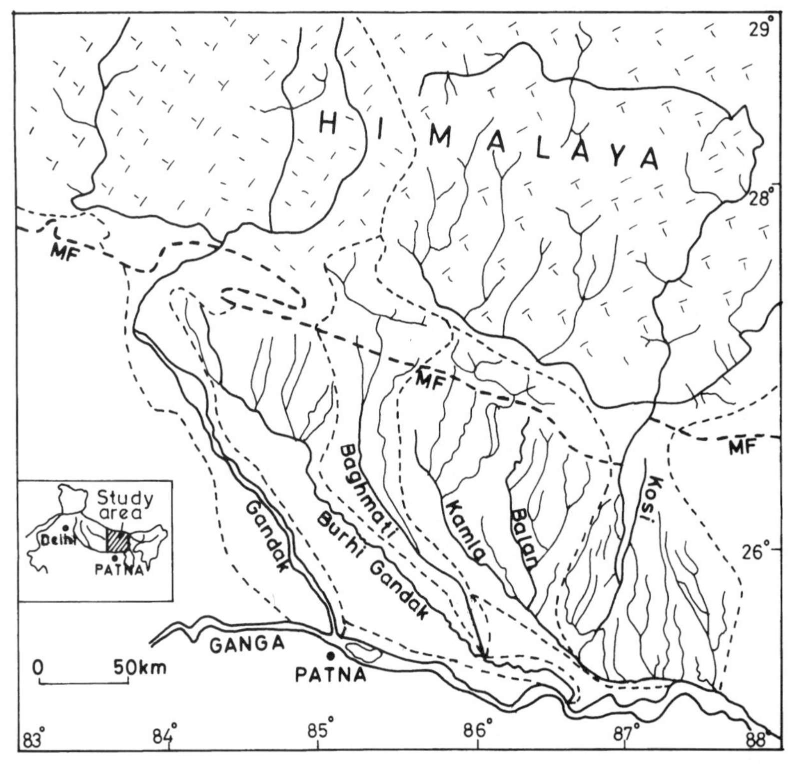

Fig. 1. Study area and major drainage in the north Bihar plains. The Gandak and Kosi are mountain-fed systems with large source areas in high mountains; Baghmati is a foothills-fed system originating in the low hills; Burhi Gandak is an example of a plains-fed system; and the Kamla-Balan is mixed-fed in the sense that it has similar source areas in foothills and plains. MF, mountain front.

by about 105 km in over 80 years (Mohindra *et al.* 1992), the shifting being either side of the river. Other small-scale migrations of the interfan rivers have been recorded by Sinha (1996), which are at times related to the migration history of the fan rivers.

Nature and mechanisms

The nature and scale of migration is extremely variable. For the major trunk rivers originating in high mountains, e.g. the Kosi and the Gandak, the predominant mechanism of lateral shifting is through avulsion, whereas in cases of the smaller rivers, movements through meandering and cut-offs are also frequent.

The creation of megafans such as the one formed by the Kosi River in north Bihar exemplifies long-term migration events where the river emerging from a gorge swings from one side to the other within more or less specified limits, at times between two nodal points. Apart from creating undulations (megafans) in otherwise monotonously flat plains, the migration of major rivers has also resulted in remanant drainage lines accompanied by morphological changes (metamorphosis). The Gandak–Burhi Gandak migration history is one such example where the present course of the Burhi Gandak (meandering type) marks the old course of the Gandak River (braided type) flowing about 105 km to the west at present.

The smaller rivers draining the interfan area also exhibit evidence of relatively rapid migration events mainly through avulsion. The small-scale avulsion events are also accompanied by distinct changes in channel morphology, e.g. increases in wavelength and decreases in sinuosity. The channels are also observed to be in 'underfit' condition at many locations. Earlier work in the area has shown that these changes are mainly in response to changes in channel–floodplain relationships and sedimentological readjustments (Sinha 1996), contrary to the classical belief that these are triggered by regional climatic changes (Dury 1964; Schumm 1969).

Further, the frequent occurrence of meander cut-offs, sinuous palaeochammels and numerous ox-bow lakes on the floodplain suggests local movements of the channels. Meander cut-offs are usually caused by the operation of the same processes of bank erosion which produce lateral channel migration and are in fact the end product of continuous local channel migration (Mosley 1975). Again, differences in morphology and geometry between the new channel and the cut-off are evident, the important differences being a reduction of sinuosities and a decrease in channel length (up to 25%) in about 60 years, as observed in a reach of the Burhi Gandak River.

Controlling factors

The factors governing the avulsive behaviour of the river systems in this area have been studied to some extent (Richards *et al.* 1993; Sinha 1996). It appears that neotectonic movements, regional subsidence and local sedimentological adjustments are among the major factors which trigger avulsion. The Indo-Gangetic basin has witnessed two major earthquakes in the recent past (in 1934 and 1986) which shows that the Himalayan seismicity has a pronounced effect on the fluvial processes operating in the basin. Moreover, the basin has been subsiding since at least mid-Miocene times without any recognizable pause. The total subsidence under the present frontal thrurst is estimated to be up to 6 km over a period of 20–30 Ma, a net rate of 0.2–0.3 m ka^{-1}. The vertical build-up of sediments due to exceptionally high rates of sedimentation (of the order of 9.3 cm a^{-1}) in selected patches is another reason for the sudden and rapid shifting of channels locally. The role of palaeotopography (again a function of rapid aggradation) and the actual position of the channels may also be major factors contributing to channel instability in the region.

Overbank flooding

The rivers of north Bihar cause havoc due to the severity of flooding almost every year. A comparison of mean annual flood and bankfull discharge of rivers at downstream stations (averaged over 15 years, 1975–89) shows that at all stations mean annual flood is distinctly greater than bankfull discharge (Table 1). This implies that overbank flooding (spilling) is a frequent occurrence in the area, at least every second year. Further, mean annual flood at downstream stations is higher than at upstream stations in the case of the Gandak and Burhi Gandak but the reverse trend is shown by the Kosi, Baghmati and Kamla-Balan systems. The downstream increase in mean annual flood presumably reflects the tributary influence. A possible explanation for the down-stream decrease of mean annual flood may be related to the exact position of the stations in relation to the active channel.

Factors controlling overbank flooding

The reasons of freqent overbank spilling are essentially the interplay of meteorological conditions in the region and the hydrological and morphological characteristics of the rivers. The plains of north Bihar are characterized by monsoonal rainfall, and the distribution of rainfall, both in space and time, is extremely uneven which makes the individual floods unpredictable. The heavy precipitation results in an enormous increase in monsoon discharge (almost 40–50 times) with respect to the lean discharge of most of the north Bihar rivers. The shallow alluvial channels between the narrow banks and embankments cannot effectively carry this sudden increase in discharge, resulting in breaches and spilling of banks. Further, the deposition of large quantities of sediments in the plains brought down from their upper catchments reduces the channel cross-sections of the rivers thereby spilling the excess water and inundating vast areas. The high sediment load, mainly as wash load, causes rapid aggradation of the river bed within the embankments. It was observed that at many locations, water level in the channel within the embankments is significantly higher than the general ground level in the surrounding areas.

Figure 2 shows the plot of the values of peak discharge for the period 1975–89 for both upstream and downstream stations for each river. Not only is the actual peak discharge extremely variable from year to year for all rivers, but there seems to be no distinct relationship in pattern or range of variation between the upstream and downstream stations of a particular river. Further, the peak discharge mostly occurs in the month of August but the precise date varies by weeks between

Table 1. *Catchment and flow data for north Bihar rivers (downstream stations)*

	Gandak	*Burhi Gandak*	*Baghmati*	*Kamla-Balan*	*Kosi*
Catchment area above site (km^2)	42987	9580	12973	2945	88480
Mean annual discharge ($m^3 s^{-1}$)	1555	273	189	68	2036
Mean low flow ($m^3 s^{-1}$)	74	23	7	0.1	116
Mean peak flow ($m^3 s^{-1}$)	8271	1417	1194	1058	7246
Highest peak flow ($m^3 s^{-1}$)	13745	2235	2618	1502	10682
Mean annual flood ($m^3 s^{-1}$)	8450	1400	1100	1050	7200
Mean bankfull discharge	5250	950	900	300	5750

All parameters were derived by averaging over 15 years, 1975–89, except for mean low flow for which data are scarce.

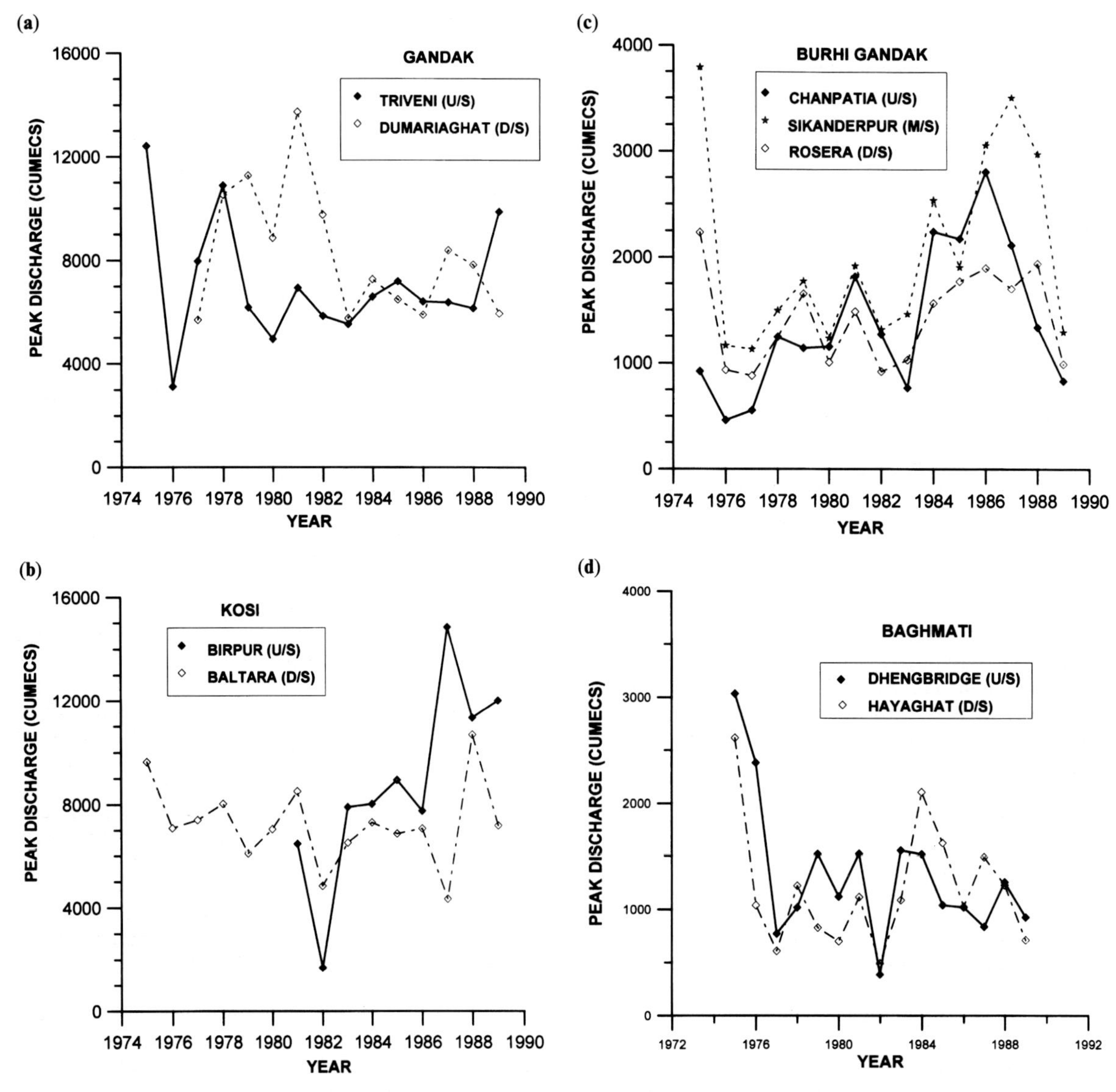

Fig. 2. Variations in peak discharge (maximum discharge observed on a particular day during the whole year) for upstream and downstream stations for the period 1975–89: (**a**) Gandak River, (**b**) Kosi River, (**c**) Burhi Gandak River, (**d**) Baghmati River, (**e**) Kamla-Balan River.

upstream and downstream stations in some years. All this seems to suggest that peak discharge at a particular station is a function of local flood waves created by tributaries apart from the monsoonal rains in the plains.

Flood control measures and their effects

The problem of flooding of north Bihar rivers has long been debated and a variety of flood control measures have been taken. The construction of flood embankments along the left as well as the right bank of the Kosi River has been a major effort. This has changed the morphology of the river from braided to a single deep channel in a significantly long reach because of which translatory tendency of the river does not exist but inundation by spilling still continues in most parts. Along the Gandak River, a series of embankments has been constructed on either side of the river downstream of

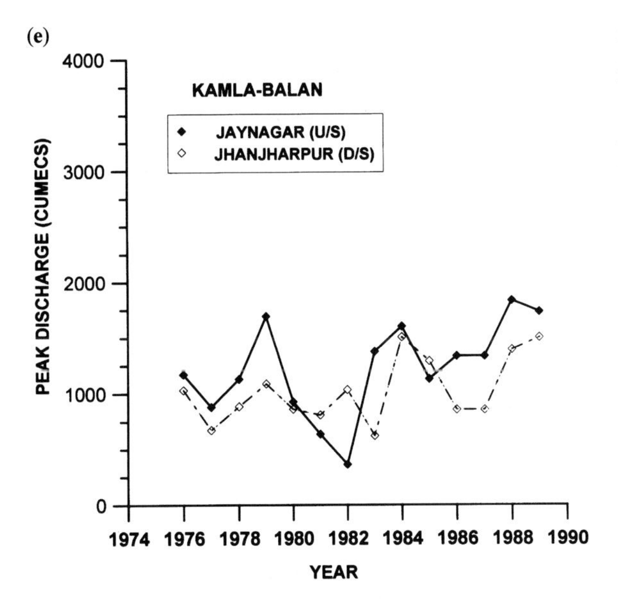

Fig. 2. (*continued*)

Valmikinagar. However, spilling still continues through gaps in the embankments and also due to synchronizing of floods in the Ganga. Breaching of embankments has also occurred at several locations in the past. The situation in no better in other rivers, namely the Burhi Gandak, Baghmati and Kamla-Balan.

The flood control efforts have largely failed as the geological and geomorphological considerations have not been taken into account. Artificial embankments have been constructed in the flood-prone reaches of the rivers but they merely transfer the trouble from one place to another and give a feeling of false security to the people living in the area. Moreover, these embankments interfere with the natural fluvial processes of the rivers. In areas where land is protected from spilling, water-logging and salinity problems have developed. The drainage congestion is a major problem in the lower catchment of the Kosi, particularly in low-lying areas, locally called '*chaurs*'. The downstream reaches of the Gandak in the Vaishali district also have severe water-logging problems and the topsoil is salt-encrusted.

Bank erosion

Significant bank erosion takes place in the area because of the migratory nature of the rivers. My studies of the north Bihar rivers support the idea that there is no relationship between flood dates and bank erodibility, indicating that bank erosion is induced more by seasonal high water than by occasional severe flooding (Handy 1972). Most rivers in the study area are characterized by rapid and frequent changes in stage and discharge which reduce the resistance to erosion of the bank material. The main erosional process operating along the river bank is large-scale slumping, which is known to be more closely related to moisture conditions than to flow conditions, although oscillations in river stage can influence the degree of bank wetting. It appears that following saturation, the bank material becomes unstable and slides along curved shear planes in large blocks. Remains of these blocks are ocassionally seen in the channel intact (only for a short duration); more usually they are broken up and washed downstream, thus contributing significantly to the wash load of the rivers. Undercutting of banks is not necessary for this type of erosion to occur. It does seem, however, that a rise in water level in the channel is conducive to saturation of the bank material and to consequent instability.

Relationship with channel characteristics

Bank erodibility is one of the important parameters controlling the sediment characteristics of the channels, which in turn control the channel pattern. Schumm's (1960) relationship between the width–depth ratio and the silt–clay content indicates that the channel is likely to be narrower and deeper as the silt–clay content in the bed and bank of the channel increases. Brice (1964) concluded that the ability of a bank to confine a stream depends partly on the discharge of the stream. As the discharge of a stream increases, the depth and mean velocity increase, and the banks are not only higher but are also subjected to greater erosive action. For bank material of a given cohesiveness, a small stream would be expected to have a channel of lower width–depth ratio than a large stream.

In the plains of north Bihar, the channels of the fan rivers with very high discharges are characterized by sandy banks and a high width–depth ratio, and their sediment concentration consists of a low percentage of silt–clay. On the other hand, the smaller interfan river channels are characterized by muddy banks, a high concentration of fine (silt–clay) sediments, and a generally meandering channel pattern with a low width–depth ratio. A significant increase in sediment concentration, mainly in terms of wash load, is observed from upstream to downstream reaches, particularly in the interfan river channels which may reflect the consequence of erosion of muddy banks.

Human-induced effects

The construction of the Kosi barrage and flood embankments have certainly checked the westward shifting of the river but has aggravated the problem of erosion on both sides of the embankment. The data suggest a series of breaches between 1961 and 1971 during flooding on the eastern as well as on the western side of the embankment, with or without overtopping depending upon the river stage. Large-scale anti-erosion works have been carried out (e.g. the construction of permeable spurs at vulnerable points) but the problem still remains unabated.

Concluding remarks

A better understanding of the avulsive river systems such as those draining the north Bihar plains is needed before attempting to control the hazards posed by them. For the major fan rivers, the control of the geological factors such as neotectonics and subsidence causing large-scale avulsions is not ruled out. However, studies show that simple localization of sedimentation resulting in subtle elevation differentiation can also drive channel avulsion (Allen 1965). Similar processes may control the avulsion of the individual interfan river channels in which linear sedimentation may result in perching of river channels and then complete switching, driven by overspilling. Such avulsion processes are poorly understood and require a detailed inventory of minor avulsions and the study of three-dimensional geometry of the underlying deposits.

The option of embanking the rivers has been debated as an effective flood control and anti-erosion measure. In Bihar, a strong engineering lobby, with high-level support from concerned organizations in the country, has so far succeeded in convincing the authorities that embankments are an effective means of flood control in the area. There has been an unpopular drive to build embankments as close to the river as possible with a view to protect as much of the population as possible and also to protect banks against erosion. The drive also includes acquiring large areas of land periodically for building retired embankment sections. There have also been suggestions to dredge the major rivers to increase their flood conveyence capacity or to construct more barrages. Most of these suggestions are not only unsound technically but would also involve phenomenal costs. There is now an increasing argument against the embankment strategy on an international level, citing the failure in Mississippi and three major Chinese rivers (Rogers *et al.* 1989). Three major alternatives have been advocated: upstream storage, detention basins within floodplains, and artificial drawdown of groundwater to absorb excess monsoonal rainfall and floodwater. The small-scale irrigation strategy has long been favoured by the World Bank and is now a primary component of Bangladesh's National Water Plan (Brammer 1990). Further, it must be emphasized that the flood control measures in the area should be a well-planned and well-manned effort because most of these rivers continuously readjust their channel in response to flood and sediment deposition. Apart fom considering the new suggestions, hydrometeorological, geomorphological and silt studies in the catchment areas should continue and data should be analysed for future plan of action. The solution to the flooding problem may lie in a better management of catchment conditions, e.g. afforestation and soil conservation practices, rather than in controlling the river flow using embankments.

References

Allen, J. R. L. 1965. A review of the origin and characteristics of recent alluvial sediments. *Sedimentology*, **5**, 89–191.

Brammer, H. 1990. Floods in Bangladesh II. Flood migration and environmental aspects. *The Geographical Journal*, **156**, 158–165.

Brice, J. C. 1964. Channel patterns and terraces of the Loup rivers in Nebraska. *USGS Professional Paper*, **422-D**, D1–D41.

Dury, G. H. 1964. Principles of underfit streams. *USGS Professional Paper*, **452-A**, A1–A67.

Gohain, K. & Parkash, B. 1990. Morphology of Kosi Mega fan. *In*: Rachocki, A. H. & Church M. (eds) *Alluvial Fans: A Field Approach.* Wiley, Chichester, 151–178.

Gole, C. V. & Chitale, S. V. 1966. Inland delta building activity of Kosi river. *Journal of the Hydraulics Division, Proceedings American Society of Civil Engineers*, **92**, 111–126.

Handy, R. L. 1972. Alluvial cutoff dating from subsequent growth of a meander. *Geological Society America Bulletin*, **83**, 475–480.

Mohindra, R., Parkash, B. & Prasad, J. 1992. Historical geomorphology and pedology of the Gandak megafan, middle Gangetic plains, India. *Earth Surface Processes & Landforms*, **17**, 643–662.

Mosley, M. P. 1975. Meander cut-offs on the river Bollin, Cheshire in July 1973. *Revue de Geomorphologique Dynamique*, **24**, 21–31.

Richards, K., Chandra, S. & Friend, P. F. 1993. Avulsive channel systems: characteristics and examples. *In*: Best, J. L. & Bristow, C. S. (eds) *Braided Rivers.* Geological Society, London, Special Publication, **75**, 195–203.

Rogers, P., Lyndon, P. & Seckler, D. 1989. Eastern waters study: strategies to manage flood and drought in the Ganges–Brahmaputra basin. ISPAN, USAID, Washington.

Schumm, S. A. 1960. The shape of alluvial channels in relation to sediment type. *In*: Schumm, S. A. (ed.) *Erosion and Sedimentation in a Semiarid Environment.* Geological Society Professional Paper, Washington, 17–30.

——1969. River metamorphosis. *Journal of the Hydraulics Division, Proceedings American Society of Civil Engineers*, **95**, 255–273.

Sinha, R. 1996 Channel avulsion and floodplain structure in the Gandak–Kosi interfan, north Bihar plains, India. *Zeitschrift für Geomorphologie*, N.F., Suppl.-Bd **103**, 249–268.

—— & Friend, P. F. 1994. River systems and their sediment flux, Indo-Gangetic plains, northern Bihar, India. *Sedimentology*, **41**, 825–845.

The construction of a drainage tunnel as part of glacial lake hazard mitigation at Hualcán, Cordillera Blanca, Peru

J. M. Reynolds,[1] A. Dolecki[2] & C. Portocarrero[3]

[1] Reynolds Geo-Sciences Ltd, The Stables, Waen Farm, Nercwys, Mold, Flintshire CH7 4EW, UK
[2] Rust Environmental, 29 Cathedral Road, Cardiff, South Glamorgan CF1 9HA, UK
[3] Unidad de Glaciología y Recursos Hídricos, Huaraz, Ancash, Peru

Abstract. At Hualcán in the Cordillera Blanca is a high-altitude glacier lake dammed by a moraine. Local glaciers regularly produce ice avalanches. In 1988 it was confirmed that the moraine was ice-cored. The rate of melting of the ice was sufficiently fast that, unless mitigation measures had been undertaken rapidly, the moraine would have collapsed. This would have resulted in the inundation downstream of Carhuaz, a town with a population of 25 000 people. Following the successful installation of siphons in 1988–89 to reduce the water level by 8 m, it was decided to undertake more permanent engineering works to ensure that the lake could never again pose a threat. It was proposed to construct a 2-m diameter tunnel, 155 m long beneath a rock bar below the moraine dam, to lower the lake level by a further 20 m. This would create sufficient freeboard to contain possible displacement waves.

Work on the tunnel was started in May 1993 using compressed air drilling and blasting with hand excavation. The initial tunnel design consisted of a single tunnel drive 135 m long plus a 20-m inclined drive under the lake. A second, near-vertical shaft was constructed for ventilation and access. Had the proposed method of breakthrough from the tunnel to the lake been carried out, it would have resulted in a rockburst leading to the catastrophic discharge of the lake. The tunnel design was changed on site to include three additional inclined drives from the main shaft, reaching the lake at vertical intervals of 5 m. Using this method, the lake level was lowered 20 m safely.

The objective of this paper is to describe in detail the geotechnical engineering aspects of the tunnel construction, the reasons for the change in its design, and the results of the mitigation work.

Introduction

On 13 December 1941, the town of Huaraz in the Callejón de Huaylas, Cordillera Blanca, was severely damaged by a catastrophic flood/debris flow caused by the rupture of lakes Palcacocha and Acoshcocha to the east. Over five thousand people were killed and considerable damage was caused to the town. Although this was far from being the first known disaster of its kind in the Cordillera Blanca (the earliest known in historical times occurred in 1702), it hit at the very heart of the Ancash Department, the town of Huaraz. Consequently, Corporación Peruana del Santa (CPS) was established in order to evaluate the safety of mountain lakes and to utilize the naturally stored water in the generation of hydroelectric power. CPS later became known as ElectroPerú in which the Unidad de Glaciología y Recursos Hídricos was established in 1966. One of the products of this work was the first glacier inventory of the Cordillera Blanca (Ames 1988), and some 40 hazard assessment and mitigation projects have been undertaken by ElectroPerú throughout the country. The problems of these catastrophic floods have been reviewed in detail by Lliboutry *et al.* (1977) and by Reynolds (1990, 1992, 1993).

The Cordillera Blanca is the largest glacier-covered area in the tropics, with some 722 glaciers covering a total area of 723 km^2 (Kaser *et al.* 1990). It is now well established that glaciers have been receding throughout the last 50 years (Hastenrath & Ames 1995), although reduced rates of recession and minor re-advances have occurred between 1974–79 and 1985–86. As the ice fronts retreat up-valley, lakes tend to form behind the moraines that marked the furthest advance positions of the ice tongues. Some of these moraine dams are ice-cored and are thus vulnerable to catastrophic collapse, releasing many thousands of cubic metres of water in very short time periods, typically several hours only. These catastrophic floods are known in South America as 'aluviones' (singular 'aluvión').

There are four main factors which, when combined, may lead to the eventual collapse of a natural moraine dam with significant impact on the local community.

- the volume of water within the lake;
- the presence of hanging glaciers (sources of ice avalanches);

Reynolds, J. M., Dolecki, A. & Portocarrero, C. 1998. Construction of a drainage tunnel as part of glacial lake hazard mitigation at Hualćan, Cordillera Blanca, Peru. *In*: Maund, J. G. & Eddleston, M. (eds) *Geohazards in Engineering Geology*. Geological Society, London, Engineering Geology Special Publications, **15**, 41–48.

- the structure and characteristics of the moraine dam; and
- the morphology of the land downstream, coupled with the presence of vulnerable communities and infrastructure.

For a lake to be considered to form a significant risk, there has to be sufficient volume of water within it which, if released catastrophically, would have the capacity to cause major damage downstream.

Hanging glaciers either act as a source of ice avalanches or may themselves actually slide down the mountainside into the lake. The displacement waves arising from the impact of the falling ice create seiche or standing waves within the lake. Depending upon the amplitude of the seiche wave, the local moraine dam may be over-topped and the overflow may cause erosion on the distal side so compromising the integrity of the dam. If the first seiche wave damages the dam, successive waves may aggravate the weakness until the moraine fails.

Moraines are complex masses of heterogeneous materials formed by many processes: deposition, thrusting, slumping, ablation of material, downwasting by waterflow, fluvio-glacial processes, slope instabilities, etc. Some moraines are affected by piping whereby water is able to permeate through the moraine through conduits from which fine material is removed. Continued attrition of material from within the moraine weakens it until failure occurs. Furthermore, inherent weaknessses within a moraine can be exacerbated by earthquakes. If an ice core is present, formed by the burial of stagnant glacier ice, its melting results in the gradual lowering of the moraine dam until the freeboard is reduced to nothing and the lake overtops the moraine.

If the above phenomena occur in a remote area, potential breaches through the moraine may have only a limited effect on the immediate locality and there may be little risk to communities. However, with increasing populations causing growing communities particularly at the confluences of rivers which offer good development sites in otherwise very steep mountainous regions, a growing number of people are at risk from inundation from aluviones. Also, vital infrastructure (roads, railways, bridges, etc.) may be at risk, and in some cases, hydroelectric power installations may be vulnerable. Should such structures be severely damaged or totally destroyed the consequential economic loss may be very

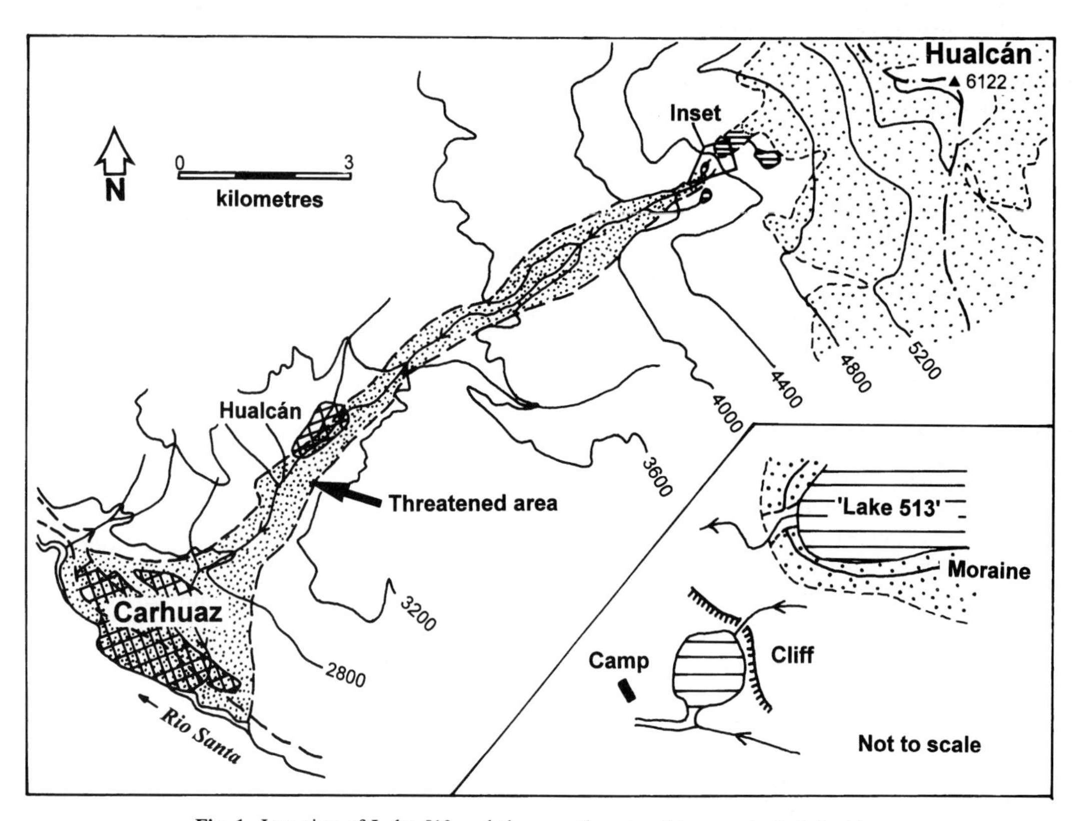

Fig. 1. Location of Lake 513 and the area threatened by a potential aluvión.

substantial, not only in the immediate aftermath of a disaster, but for periods in excess of decades (Reynolds 1992). The total economic losses can easily mount to billions of dollars. For developing countries such as Peru, such huge financial losses could have a very significant impact on the entire financial state of the country over many years.

As a means of reducing the potential risk to communities vulnerable to possible alluviones, hazard assessment has been and is being undertaken. Once a glacier lake has been identified as being potentially dangerous, and for what reasons, a mitigation strategy is devised. In Peru, four different approaches have been taken to alleviate the dangers of catastrophic floods.

- excavation of open cuts in the moraine dam to lower the lake's water level;
- construction of siphons;
- tunnelling through bedrock into the lake to drain it; and,
- construction or restitution of the natural dam.

Aspects of moraine structures and methods of remediation have been discussed in detail by Lliboutry *et al.* (1977) and by Reynolds (1992). Open cuts in a moraine can be excavated during the dry season when a lake's water level is lower than that during the wet season. Such a method is risky as any displacement wave arising from an ice avalanche can rip through the cut and breach the moraine. This should only be attempted where there is no risk of avalanches into the lake.

Siphons are attractive in that they are readily transportable, relatively easy to install, and can be very effective, as will be described later. They are used as a first-stage method of reducing water level prior to undertaking more substantive measures.

Tunnelling can only be carried out through competent rock beneath or beside a moraine dam. The costs of such a method are such that tunnelling is only undertaken where the expense can be justified. Unfortunately, not all moraine dams are suitable for tunnelling as the local materials may be inappropriate for such a technique.

Reconstitution of a moraine dam can be achieved by making a cut into the freeboard, placing a substantial culvert and recovering it to restore the freeboard. This exposes the moraine to the risks from avalanches, but if the work can be carried out quickly, the method can be very effective.

One aspect of hazard mitigation which is often overlooked but which is absolutely vital to the safe remediation of any glacial lake is that the whole glacier/lake system should be monitored carefully throughout the mitigation work. This permits the possible detection of any potential problems arising in the local catchment.

While this paper concentrates on describing a specific example of glacial hazard mitigation in Peru, catastrophic floods are known to occur elsewhere, such as in the Alps (Tufnell 1984; Dutto *et al.* 1991), the Himalayas (Ives 1986; Reynolds 1995, 1998), and in North America (Mt Rainier, Washington State (O'Connor & Costa 1993); Canadian Cordillera (Clague & Evans 1994); Yukon (Clarke 1982)), amongst many others.

Background to Hualcán

Physical and geological setting

In August 1988 a field visit was arranged to an unnamed lake, referred to as Lake 513, at the head of a river valley approximately 12 km NE of Carhuaz (Fig. 1). This town has a population of *c.* 25 000, and lies within the Callejón de Huaylas, some 30 km NW of Huaraz. ElectroPerú had previously investigated this lake in 1985 and again in 1988. Over this time interval the ice-cored moraine damming the lake had lowered by about 4 m at an average rate of 11 cm per month. At the time of the authors' (J.M.R. and C.P.) visit in August 1988, water from the lake was already filtering over the ice core within the moraine (Fig. 2) and draining via two springs on the down-valley side. There was less than 1 m of freeboard. It was thought that if nothing had been done to mitigate the situation, the moraine would have failed within two months, probably destroying Carhuaz with the loss of many thousands of lives.

History of hazard assessment and initial mitigation

In August 1988, it was concluded that the immediate course of action was to install a siphon to lower the lake level by at least 8 m and preferably 12 m to below the level of a natural bedrock rim beneath the moraine. By the end of October 1988, one siphon had been constructed and was discharging at a rate of 190 l s^{-1}. However, by Christmas, the water level had not changed even though about 1 million m^3 had been discharged. Fortunately, funding for a further siphon was arranged and installation was completed by the end of January 1989. The combined rate of discharge was 500 l s^{-1}. By 31 March, the water level had fallen 2 m and fell a further 2 m by June 1989. By June 1990, the water level had been lowered by a total of 5 m and it was considered safe to excavate a channel through the moraine to ensure that the lake level could not rise above this level. More details of the hazard mitigation phases have been given by Reynolds (1990, 1992).

In 1991, a significant ice avalanche occurred causing displacement waves up to 2 m high which overtopped the moraine through the excavated channel causing regressive erosion of the moraine and an aluvión ensued. This resulted in a channel some 20–25 m deep being eroded through the outer moraine. Fortunately, deposition of flood debris was contained locally within a flat-lying infilled lake downstream of Lake 513. Flood water and

Fig. 2. The moraine dam at Lake 513 prior to any remediation works (August 1988). The 1991 aluvión broke through in the area indicated between the arrows.

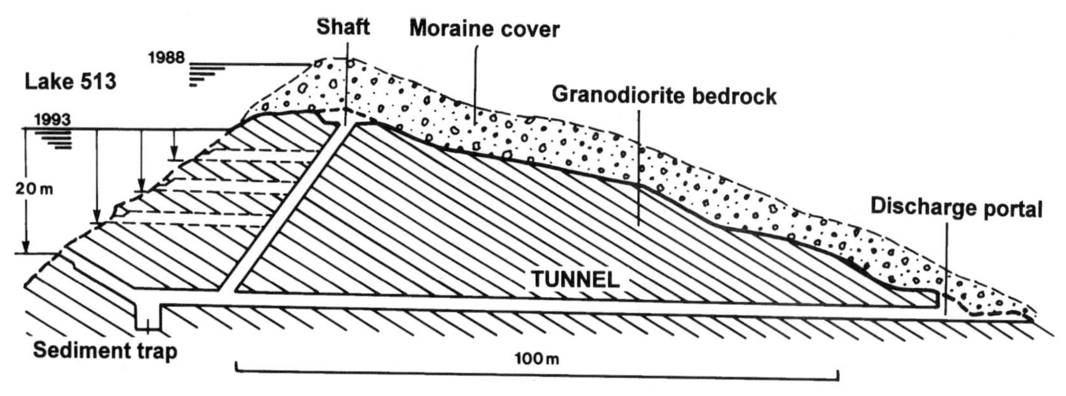

Fig. 3. The basic design of the 2-m diameter tunnel. The moraine cover observed in 1988 is indicated approximately. The revised tunnel design (with three additional sub-horizontal drives) is indicated by dashed lines.

some debris caused damage to bridges several kilometres downstream but no lives were lost. The occurrence of this aluvión was a stark reminder of the potential power of such floods. It demonstrated that there was still a real risk of a more substantial aluvión occurring should a larger ice avalanche take place. Consequently, it was decided to construct a 2-m diameter tunnel, 155 m long, through the rock bar to 20 m below the then water level in order to drain the lake still further. The basic design of the tunnel is shown in Fig. 3.

The tunnel

Engineering geology

Surface geological mapping by geologists from Electro-Perú provided initial information on rock characteristics and the nature of discontinuities. No intrusive investigation or laboratory analysis was undertaken. Electro-Perú had decided to drive the 2-m diameter drainage tunnel using drill-and-blast techniques with hand excavation. The majority of the tunnel was likely to be constructed through the granodiorite bedrock, with the final rock 'plug' masked by morainic debris on the floor and sides of the lake. Engineering geological mapping of the initial discharge portal area of the tunnel confirmed the strength of the rock mass and anticipated support requirements for the tunnel.

The conditions exposed light grey, coarse-grained massive and fresh, very strong granodiorite rock with generally widely spaced joints. The rock mass was intersected by a number of low-angle normal faults and numerous randomly orientated, generally tight, low persistence, generally rough but planar joints. At several locations within the tunnel, significant water inflows were encountered, generally emanating through open, vertical and mineralized joints (probably leakage through interconnection with the lake floor). These flows did not affect the overall stability within the tunnel.

Rock mass classification systems were employed to confirm tunnelling condition and to predict support. Joint orientation can affect blasting, span widths, roof support and groundwater flows, and it was vital that this information was collected and assessed as tunnelling progressed. The Peruvian geologists were introduced for the first time to the Norwegian Geotechnical Institute (NGI) system of Barton *et al.* (1974, 1992). On the basis of an evaluation of a large number of case histories of underground excavation stability, Barton *et al.* (1992) proposed an index for the determination of a tunnelling quality of a rock mass. The Tunnelling Quality Index, Q, is based on an assessment of six parameters relevant to the consideration of the rock discontinuity fabric, groundwater conditions and stress regime. The numerical value of this index Q is given by:

$$Q = (\mathrm{RQD})/J_n \times (J_r/J_a) \times (J_w)/(\mathrm{SRF}) \quad (1)$$

where RQD = rock quality designation; J_n = joint set number; J_r = joint roughness number; J_a = joint alteration number; J_w = joint water reduction factor; and SRF = stress reduction factor.

The Q system can be used to predict behaviour and support requirements by the use of an additional parameter, D_e, equivalent dimension, where:

$$D_e = \frac{\text{span, diameter or height of excavation (m)}}{\text{excavation support ratio (ESR)}} \quad (2)$$

where the excavation support ratio is related to the use for which the excavation is intended and the extent to which some degree of instability is acceptable. Q values in excess of 10, representing good quality rock, were generally encountered along the tunnel route. For the majority of the Lake 513 tunnel, the equivalent dimension was taken as $2\,\mathrm{m}/1.6 = 1.25\,\mathrm{m}$.

Hence, given the tunnel width/excavation span and function of the tunnel, the assessment confirmed that the tunnel was unlikely to require support for the construction duration time and full service life. Particular areas required special consideration, but no remedial measures

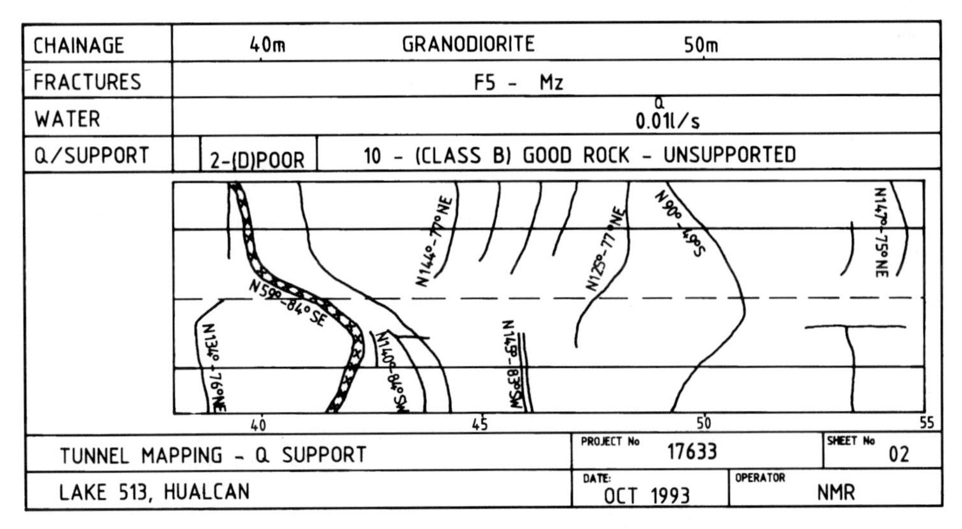

Fig. 4. Example of a tunnel log at Lake 513.

were needed in practice. An example log of the engineering geology of the tunnel is presented in Fig. 4.

Construction

Excavation of the tunnel using hand-held compressed-air drilling equipment, blasting of explosives and 'mucking out' by hand was a slow process, but was the only available technique given the extreme site conditions (altitude, remoteness and equipment availability in relation to site access).

The typical blasting pattern and sequence, shown in Fig. 5 provided the optimum rock breakage and debris pile formation for the particular rock type, strength and cross-sectional area of the tunnel.

A charge of approximately 5 kg m^{-3} per 1.3 m depth of tunnel round using burn cut with a total of 38 holes per round and half a second delay period was used for most of the length of the tunnel. Ventilation was by compressed air blown through tubes to the tunnel face area, with clearance subsequently travelling throughout the entire tunnel length.

On return to the face, visual inspection and scaling of loose rock was carried out prior to 'mucking out' with shovel and wheelbarrow. The rate of advance with one to two rounds per day was typically between 10 m and 13 m per week.

Breakdowns were very common due to prevailing conditions, particularly the thin air at altitudes of 4500 m. No support or permanent lining was utilized. The downstream tunnel portal is shown in Fig. 6. The view is towards Lake 513. The shaft gantry above the ventilation shaft is arrowed. The smooth bedrock was exposed by the 1991 aluvión.

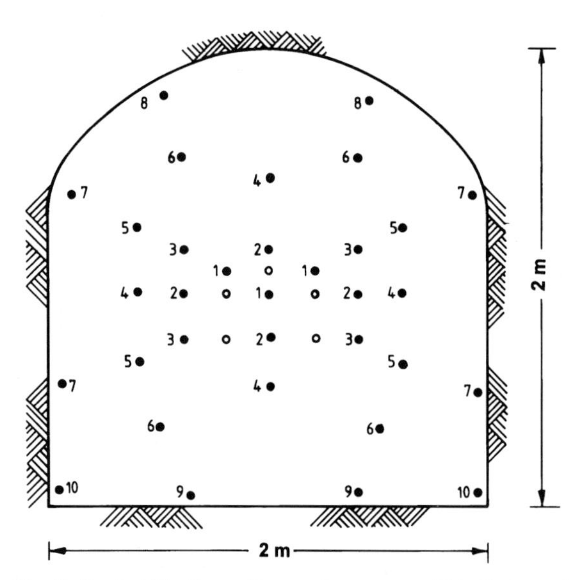

Fig. 5. Typical blasting pattern and sequence within the tunnel at Lake 513.

Breakthrough into the lake

It was originally intended by ElectroPerú that the breakthrough from the tunnel into the lake should be achieved by a single explosive blast at a level 20 m below the then water level. Had this procedure been followed, it was estimated that the *static* water pressure alone within the tunnel would have exceeded the local overburden pressure; the dynamic pressure effect of the water driving through the tunnel immediately after the explosion would have added to the local in-tunnel pressure. The combined effect would have undoubtedly resulted in a rock burst at the discharge portal. This would have led to an uncontrolled discharge of the lake, with potentially catastrophic consequences downstream.

During the visit to the site by all three authors in October 1993, an alternative design was devised which required only additional explosives. All other potential options were either far too costly or were considered impractical under the prevalent field conditions. It was decided that a series of three additional sub-horizontal drives (Fig. 3) should be constructed using the same excavation techniques as for the rest of the tunnel. The work was undertaken between October 1993 and May 1994. In October 1993, ElectroPerú engineering geologists and engineers were given brief but necessary training in new techniques of tunnel construction and safety (e.g. that of Barton *et al.* 1992).

Each additional drive terminated within bedrock several metres from the estimated base of the lake and at approximately 5 m vertical increments. The uppermost tunnel was connected with the lake by a small explosive charge producing a 0.9 m diameter portal into the lake. The lake discharged through this until the water level reached the lower lip of the portal. The second tunnel was then connected to the lake by the same process and the lake drained a further 5 m. Problems occurred with the charges within the end of the third additional tunnel and this was abandoned. The lake-side portal of the main tunnel was exploded through under 10 m head of water. Using this staged approach, the structural integrity of the bedrock bar was retained, and the dynamic pressures within the tunnel during drainage were reduced relative to those anticipated had the original plan been followed. Furthermore, the rates of water flow and the peak volumes were all within the tolerances within the existing river system downstream, thereby minimizing damage to the river bed during the controlled drainage. By May 1994, the lake level had been lowered safely by 20 m.

An additional benefit of this design is that, should the lowermost tunnel become blocked, and the lake level

Fig. 6. The downstream portal of the tunnel during construction in 1993. See the text for details.

rise again, the discharge portals at higher elevations would still be able to provide safe discharge of the lake. Maintenance of the higher portals was possible once the lake achieved its reduced level. This also increased the freeboard of the rock bar against any possible displacement waves arising from large-scale ice or rock avalanches.

Final discussion and conclusions

The present case history has demonstrated the successful implementation of both short-term and long-term mitigation measures following appropriate glacial hazard assessment. Installation of two siphons provided almost immediate relief, allowing time to implement a larger-scale engineered solution, namely the tunnel. The two greatest risks throughout the project were a potential lack of finance for the mitigation measures within an appropriate time scale, and the possible occurrence of a significant ice or rock avalanche into the lake during the construction of mitigation measures.

The original tunnel design was demonstrated to be potentially hazardous but alternative solutions were adopted that resulted in the lowering of the lake level by 20 m safely. Revised training of local geological and engineering staff has raised their awareness of new tunnel construction and safety techniques.

From initial awareness of there being a problem at Lake 513 through to the completion of safety works took 9 years. Had the initial siphons not been installed when they were, there would have undoubtedly been a major aluvión with the loss of thousands of lives.

For glacier hazard mitigation to be successful, there has to be a monitoring programme in place that facilitates the recognition of potentially dangerous lakes. There then has to be the political and financial support on a short enough time scale for the implementation of immediate mitigation measures. At the same time, plans can be developed for the long-term mitigation work leading to the final completion of lake remediation.

There are thought to be in excess of 600 potentially dangerous lakes within the Peruvian Andes alone. To cope with what is a growing problem, given the continuation of glacier recession and lake development through global warming, there is an urgent need for a national government strategy for glacier hazard assessment and mitigation. This project has demonstrated that, given the timely intervention and appropriate mitigation measures, dangerous lakes can be made safe relatively inexpensively. The costs of mitigation are a fraction of those of disaster relief. Donor countries, who typically spend vast sums of money on relief programmes following disasters, could contribute money more effectively by supporting mitigation programmes in order to reduce the risk of potential disasters. This approach is better for both the donor country and the recipient nations. It is perhaps appropriate that this fact is emphasized at the mid-way point in the International Decade for Natural Disaster Reduction.

Acknowledgements. J.M.R. and A.D. are both grateful to the Emergency Aid Department of the Overseas Development Administration for funding the visit in November 1993. ElectroPerú has provided excellent facilities, hospitality and logistical support throughout this work.

References

AMES, A. 1988. *Glacier inventory of Peru. Part 1. Huaraz.* HIDRANDINA, SA, Unit of Glaciology and Hydrology.

BARTON, N., LIEN, R. & LUNDE, J. 1974. Engineering classification of rock masses for the design of tunnel support. *Rock Mechanics*, **6**, 189–236.

——, GRIMSTAD, E., AAS, G., OPSAHL, O. A., BAKKEN, A., PENDERSEN, L. & JOHANSEN, E. D. 1992. Norwegian method of tunnelling. *World Tunnelling*, June, 231–238; August, 324–330.

CLAGUE, J. J. & EVANS, S. G. 1994. Formation and failure of natural dams in the Canadian Cordillera. *Geological Survey of Canada Bulletin*, **464**.

CLARKE, G. K. C. 1982. Glacier outburst floods from 'Hazard Lake', Yukon Territory, and the problem of flood magnitude prediction. *Journal of Glaciology*, **28**(98), 3–21.

DUTTO, F., GODONE, F. & MORTARA, G. 1991. L'écroulement du glacier supérieur de Coolidge (Paroi nord du Mont Viso. Alspes occidentales). *Revue de Géographie Alpine*, (2), 7–18.

HASTENRATH, S. & AMES, A. 1995. Recession of Yanamerey Glacier in the Cordillera Blanca, Peru, during the 20th century. *Journal of Glaciology*, **41**(137), 191–196.

IVES, J. D. 1986. *Glacial lake outburst floods and risk engineering in the Himalaya.* ICIMOD Occasional Paper No. 5 International Centre for Integrated Mountain Development, Kathmandu, Nepal.

KASER, G., AMES, A. & ZAMORA, M. 1990. Glacier fluctuations and climate in the Cordillera Blanca, Peru. *Annals of Glaciology*, (4),136–140.

LLIBOUTRY, L., ARNAO, B., MORALES, A., PAUTRE, A. & SCHNEIDER, B. 1977. Glaciological problems set in the control of dangerous lakes in Cordillera Blanca, Peru. Part I: Historical failures of moraine dams, their causes and prevention. *Journal of Glaciology*, **18**(79), 239–254.

O'CONNOR, J. E. & COSTA, J. E. 1993. Geologic and hydrologic hazards in glacierized basins in North America resulting from 19th and 20th century global warming. *Natural Hazards*, **8**, 121–140.

REYNOLDS, J. M. 1990. Geological hazards in the Cordillera Blanca, Peru. *AGID News*, 61/62, 31–33.

——1992. The identification and mitigation of glacier-related hazards: examples from the Cordillera Blanca, Peru. *In:* MCCALL, G. J. H., LAMING, D. J. C. & SCOTT, S. C. (eds) *Geohazards.* Chapman & Hall, London, 143–157.

——1993. The development of a combined regional strategy for power generation and natural hazard assessment in a high-altitude glacial environment: an example from the Cordillera Blanca, Peru. *In:* MERRIMAN, P. A. & BROWITT, C. W. A. (eds) *Natural Disasters: Protecting Vulnerable Communities.* Thomas Telford, London, 38–50.

——1995. Glacier-lake outburst floods (GLOFs) in the Himalayas: an example of hazard mitigation from Nepal. *Geoscience and Development*, (2), 6–8.

——1998. High-altitude glacial lake hazard assessment and mitigation: a Himalayan perspective. *In*: MAUND, J. G. & EDDLESTON, M. (eds) *Geohazards in Engineering Geology.* Geological Society, London, Engineering Geology Special Publication, **15**, 25–34.

TUFNELL, L. 1984. *Glacier Hazards.* Longman, London.

Coastal cliff behaviour and management: Blackgang, Isle of Wight

R. Moore, A. R. Clark & E. M. Lee

Rendel Geotechnics, 61 Southwark Street, London SE1 1SA, UK

Abstract. In January 1994 coastal erosion and landslide activity at Blackgang, Isle of Wight, resulted in a dramatic retreat of the coastal cliffs and landward extension of ground movement. The initial impact of the event involved the destruction of two cottages, an access road, several cars and caravans, and around 12 homes had to be evacuated. The event raised considerable local and national media attention.

The initial landslide response was co-ordinated by South Wight Borough Council, with advice from Rendel Geotechnics, and was aimed at ensuring public safety and security in the area. A detailed investigation was subsequently carried out to identify the extent and causes of coastal instability and cliff recession in the context of 'cliff behaviour units'. An understanding of the characteristic cliff behaviour units has proved the key to assessing future landslide and cliff recession potential as well as identifying options for coastal cliff management at Blackgang.

The coastal cliffs have been significantly oversteepened by the latest events and appear to be very sensitive to rainfall, with only relatively moderate to high winter rainfall totals expected to cause further cliff-top retreat and landslide reactivation. Consequently, in the absence of a financially and environmentally acceptable slope stabilization and coast protection scheme, continued managed retreat of cliff-top development and land use is seen as the only viable option.

Introduction

Cliff recession presents a significant hazard to planners and managers along many stretches of the British coast (Rendel Geotechnics 1995*a*). The nature of the problems that are encountered varies according to the geological setting, and range from the effects of rapid cliff retreat, as on the Holderness coast (Humberside), to relatively slow ground movement on unstable coastal slopes, as at Ventnor, Isle of Wight, and threats to public safety from cliff falls on hard rock coasts. However, the recession process is important for maintaining conservation sites such as geological exposures, and in supplying sediment to beaches, shingle ridges, sand dunes, saltmarshes and mudflats. These landforms absorb wave energy arriving at the coast and can form important components of flood defence solutions, either alone or where they front flood embankments or sea walls. It is one of the challenges of shoreline management to ensure that soft and semi-soft engineering solutions for coastal defence are sustainable by maintaining a continued supply of sediment of particular grain sizes from eroding cliffs on neighbouring coastlines (Lee 1995).

Cliff recession is a four-stage process involving *detachment* of particles or blocks of material, its *transport* through the cliff system, its *deposition* on the foreshore and its *removal* by marine action. It is a complex and uncertain process, controlled by conditions and processes operating on both the foreshore and the cliff. For this reason the concept of a 'cliff behaviour unit' has been developed for the Ministry of Agriculture, Fisheries and Food to provide a framework for the prediction of cliff recession (Rendel Geotechnics 1995*b*). These units (CBUs) span the foreshore to the cliff top and are coupled to adjacent CBUs within the framework provided by littoral cells or coastal process units.

This paper illustrates how the identification and characterization of cliff behaviour units at Blackgang, Isle of Wight, has provided the context for the measurement and prediction of cliff recession and the selection of appropriate cliff management strategies. The coastal cliffs at Blackgang clearly demonstrate that the recession process is often complex, with episodic phases of activity followed by periods of relative calm, which present unique problems to coastal managers.

The site

Blackgang is situated on the southern coast of the Isle of Wight (Fig. 1(a)). The area has a long history of coastal instability. The community at Blackgang has considerably diminished in size since the 1850s when the area was developed as one of the Isle of Wight's main tourist attractions. The progressive retreat of the coastal cliffs and periodic land movements have destroyed houses and infrastructure and caused extensive loss of land in the past. The Blackgang Chine Theme Park has had to adapt to the impact of land instability by relocating attractions,

MOORE, R., CLARK, A. R. & LEE, E. M. 1998. Coastal Cliff behaviour and management: Blackgang, Isle of Wight. *In*: MAUND, J. G. & EDDLESTON, M. (eds) *Geohazards in Engineering Geology*. Geological Society, London, Engineering Geology Special Publications, **15**, 49–59.

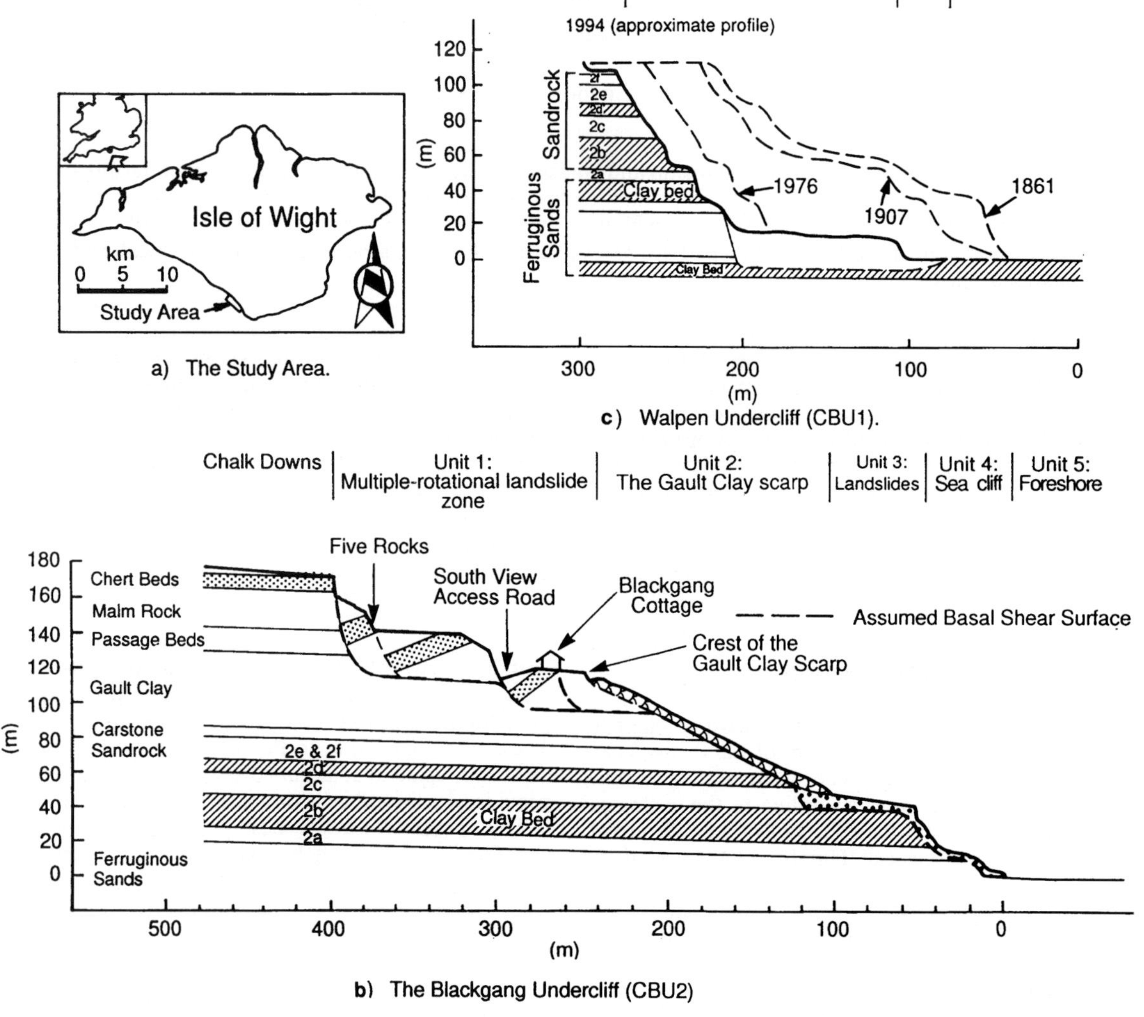

Fig. 1. Location and physical character of coastal cliffs: (**a**) the study area; (**b**) the Blackgang Undercliff (CBU 2); (**c**) Walpen Undercliff (CBU 1).

reinstating footpaths and undertaking regular safety inspections in vulnerable areas. The most recent landslide event occurred on the evening of 12 January 1994, and which resulted in the evacuation of around 12 homes, the destruction of two cottages, an access road, several cars and caravans and a large section of coastal footpath. The landsliding was triggered by exceptional winter rainfall during December 1993 and the early weeks of January 1994, and was the most recent expression of the ongoing retreat of the sea cliffs due to marine erosion (Hutchinson *et al.* 1981; Bray 1994).

The site lies on the southwestern limb of the outlier of Cretaceous rocks which form the Southern Downs of the Isle of Wight. The geological succession is shown in Fig. 1(b). The Lower Greensand units comprising the Carstone and Sandrock are well exposed in the 110-m high sea cliffs to the southwest of Blackgang. The Upper Greensand and Gault are best exposed to the southeast, at the 170-m high Gore Cliff, which forms part of the rear-scarp of The Undercliff. The Undercliff is a local name for the complex of ancient coastal landslides, over 12 km in length, which extend from Blackgang in the west to Luccombe in the east. Superficial deposits include hillwash derived from the Chalk and Upper Greensand which mantle the subdued slopes of a relic coombe valley at Blackgang. The geological structure of the Lower Cretaceous beds of the Southern Downs dip gently 1°–2° to the SSE (White 1921).

Identification of cliff behaviour units

A geomorphology map (Fig. 2) of the coastal slopes and foreshore conditions was prepared during a field reconnaissance survey, which made use of aerial photographs taken after the landslide event in January 1994 to obtain detail of the inaccessible areas and coastal cliffs.

The map identifies a variety of coastal landslide features comprising a complex pattern of mass movement types arranged within a series of landslide systems. These landslide systems are the dominant features on the coastal cliffs and have, hence, been used to define the nature and extent of individual cliff behaviour units. Two CBUs have been recognized: the Walpen Undercliff and The Undercliff.

The Walpen Undercliff (CBU 1)

The 110-m high cliffs of the Walpen Undercliff are developed in weak Lower Greensand rocks, comprising a series of lithologically controlled benches and undercliffs which correspond with clay and sand horizons respectively. Cliff recession occurs through a combination of seepage erosion, small-scale slides and rockfalls. The cliff-top zone is gently sloping and is part of the relic coombe valley feature which still drains the natural surface and sub-surface groundwater away from the Southern Downs.

The Undercliff (CBU 2)

The 12-km length of The Undercliff comprises a series of ancient deep-seated landslides (Hutchinson *et al.* 1991; Lee & Moore 1991; Rendel Geotechnics 1995*b*). These landslides involve different failure mechanisms at three levels within the geological column involving multiple-rotational failure within the Gault clay (forming an upper tier), mudslides developed on a Gault clay scarp, and compound block failures developed on the clay-rich horizons within the Sandrock (forming a lower-tier), each separated by steep cliffs. Three interrelated landslide systems (2a–c; Fig. 2) have been identified within The Undercliff at Blackgang.

Past landslide events and erosion rates

The history of past landslide events and coastal erosion at Blackgang has been established from a variety of sources including national and local newspaper reports, past editions of Ordnance Survey maps, local knowledge and archive aerial photography. Maps and photographs are the main sources of spatial information that have been used to identify former cliff-top and cliff-base positions. Press cuttings provide the temporal information on the occurrence of events which have had a major impact on development at Blackgang.

A systematic review of newspaper articles dating back to the early part of this century indicates that coastal instability has frequently affected the Blackgang area throughout this period. The historical database has been used to identify episodes of activity in each CBU and their component landslide systems.

Erosion of the Walpen Undercliff (CBU 1) has been widely reported, especially since 1960. Retreat of the cliff top has resulted in the loss of at least six houses, with recent key events in 1960, 1961, 1963, 1966, 1974 and 1975. In 1982 rapid retreat of the cliff top at Blackgang Chine was reported (Anon 1982). Hutchinson *et al.* (1981) provide an account of the nature and causes of the retreat of this stretch of coast.

Landslide reactivation in The Undercliff at Blackgang (CBU 2) has affected each of the landslide systems (2a–c) described above. The South View landslide system (2a) was affected by large cliff falls from the rear-scarp in 1952 and 1959. In 1978 a major landslide on the coastal slopes destroyed at least three houses, around 13 chalets and a number of caravans. Bromhead *et al.* (1991) provide a detailed geotechnical account of this event. The Cliff Cottage landslide system (2b) has been affected by repeated land movements which have caused damage to the 'Old Blackgang Road' and led to the destruction of at least three houses. Major events were experienced in 1928, 1936, 1960, 1970 and 1978. The Blackgang Cottage landslide system (2c) was affected by movements in the winter of 1960 and in May 1989. The gardens of several cottages subsided in 1960 and January 1994.

A comparison of past editions of the Ordnance Survey maps of the area provides additional evidence of the extent of cliff retreat and land instability since 1862. When early maps are used in conjunction with aerial photographic evidence, considerable changes in prominent features such as the cliff-top and cliff-base positions can be determined.

The retreat of the *cliff base*, which is developed in the lowermost unit of the weak Sandrock and underlying Ferruginous Sands, totalled some 96 m between 1862 and 1994. Map evidence indicates that the retreat of the cliff base has accelerated from an average rate of 0.34 m a^{-1} between 1861 and 1907, to 0.63 m a^{-1} between 1907 and 1980 and 1.57 m a^{-1} between 1980 and 1994 (see also Hutchinson *et al.* 1981). A major consequence of the retreat of the cliff base has been to narrow the lower undercliff benches from an average width of over 100 m in the mid-nineteenth century to its

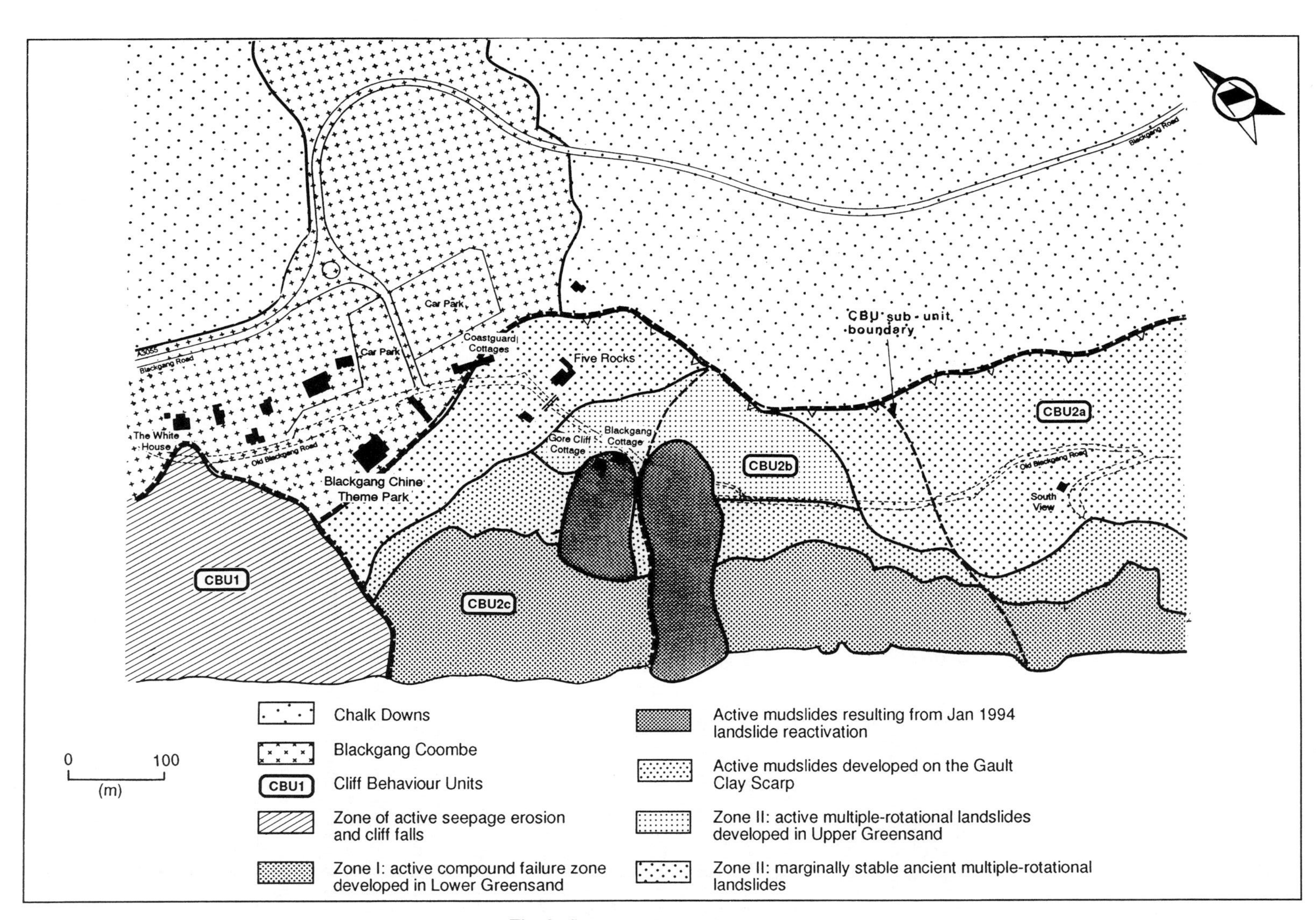

Fig. 2. Summary geomorphology map.

present width of around 20 m. In places, the lower undercliff has been almost entirely removed.

Cliff-top recession of the Walpen Undercliff (CBU 1) occurs largely in response to the retreat of the cliff base through marine erosion, seepage erosion and relatively shallow mass movements (Hutchinson *et al.* 1981) and totalled some 140 m between 1938 and 1994. As noted with the cliff base, the rate of retreat appears to have increased in recent years, with 70 m cliff-top recession recorded between 1980 and 1994 at an average rate of 5.0 m a^{-1}. In January 1994, a shallow slip at the crest of the high cliffs at Blackgang Chine, resulted in the landward extension of movement by around 40 m, demonstrating the importance of episodic events in cliff-top recession.

The retreat of the *Gault clay scarp* in The Undercliff (CBU 2) has been around 35 m (or 2.5 m a^{-1}) since 1980. Higher rates of recession were apparent during the late 1970s when a major landslide event occurred (Bromhead *et al.* 1991). During this phase of activity the Gault clay scarp receded by around 60 m emphasizing again that major changes are brought about by episodic events rather than being uniform in space or time. The *rear scarp* of The Undercliff is not believed to have been affected by significant recession for many thousands of years.

The 1994 landslide event

The landslide activity of January 1994 affected both CBUs and led to considerable concern for the safety of a small community living on the coastal slopes (including 'New Age' travellers) and tourists visiting Blackgang Chine Theme Park. The emergency landslide response, causal factors and initial investigations are described by Rendel Geotechnics (1994) and Clark *et al.* (1995). Both CBUs were affected by first-time or renewed instability, some to a greater extent. One consequence of the event was that landslide system 2c was almost entirely reactivated by ground movement, involving new and reactivated compound block failures on the lower-tier benches and the initiation of mudslides upon the Gault clay scarp (Fig. 1(b)). The movements resulted in considerable loss of land and amenities within the Blackgang Chine Theme Park. The mudslides were responsible for the eventual undermining of Blackgang and Gore Cliff Cottages, which were situated on a bluff between landslide systems 2b and 2c.

The most notable land movements occurred in landslide system 2b, where a large 150 m by 40 m wide section of ancient rotational landslides, forming the upper tier of The Undercliff, suddenly dropped 15 m. This sudden movement caused the dramatic tilt and damage to Blackgang and Gore Cliff Cottages reported on the evening of 12 January, when the alarm was first raised. An occupant had to be rescued from one of the cottages as the windows and doors were jammed by the distortion to the structure. Another effect of the deep-seated rotational movement was to push thousands of tonnes of predominantly Upper Greensand debris, along with the Old Blackgang Road and several cars and caravans, over the Gault clay scarp, where charged with high groundwater levels, mudslides rapidly transported the debris to the lower tier and beach.

Cliff behaviour

Development of an evolutionary model from the geomorphological and historical evidence provides the basis for understanding the past and recent behaviour of the various cliff units and gives an indication of their possible future development. This can be demonstrated with reference to CBU 2 The Undercliff landslide systems, although it is important to appreciate that there is also a strong integration of process and form in CBU 1 the Walpen Undercliff, albeit achieved by a different range of recession mechanisms.

The Undercliff is part of a relic landslide complex which was formed by a range of mass movement activity, stimulated by past fluctuations in sea level and climate (Rendel Geotechnics 1995*c*). The landslide section taken through The Undercliff is dominated by three sub-parallel cliffs separated by two broad platforms mantled by varying depths of landslide debris. Detailed geomorphological mapping of these features has revealed five interrelated units which largely reflect geological controls (Fig. 1(b)).

- *Unit 1*: an upper tier of multiple rotational landslide blocks and debris in front of an almost vertical cliff (rear-scarp) which marks the inland limit of The Undercliff landslide complex. New failures involving detachment of blocks from the rear-scarp are extremely rare, probably occurring at 1000–10 000 year intervals (Hutchinson *et al.* 1991). In the intervening periods relatively small-scale failures and slow degradation of the existing landslide mass occur, mainly during periods of high groundwater levels. If the unit is undermined by activity on the Gault clay scarp, large-scale reactivation of the upper tier may occur.
- *Unit 2*: the seaward limit of unit 1 is defined by a steep slope developed largely in the lower silty beds of the Gault clay (the Gault clay scarp). *In situ* Carstone may be visible at the base of the scarp where recent landslide activity has exposed the bedrock. The slope is affected by mudslides which transport debris from unit 1 to unit 3 below. The Gault clay scarp is generally mantled by landslide debris and is largely a subdued feature throughout much of The Undercliff. However, where the scarp is affected by active

mudslides, debris is rapidly removed from the scarp exposing *in situ* material beneath. Mudslide activity causes the recession of the Gault clay scarp.

- *Unit 3*: a lower tier or platform comprising old compound landslide blocks and debris. Current landslide activity involves the reactivation of pre-existing landslide blocks and debris seaward across the lower tier as a consequence of unloading caused by the retreat of the sea-cliffs. At Blackgang the width of the lower tier benches has decreased as a result of sea-cliff recession.
- *Unit 4*: near-vertical sea-cliffs developed in *in situ* Lower Greensand and capped by landslide debris. These cliffs fail through a combination of rock falls, wedge failures, localized seepage erosion, and particle detachment induced largely by marine undercutting. Cliff retreat is generally at a uniform rate accomplished by small-scale processes, involving up to 0.5–1 m of cliff loss in each event.
- *Unit 5*: a sandy and boulder covered foreshore cut in Sandrock, sloping at 2°. The sea-cliffs and landslide slopes regularly supply material to the back of the shore, much of which is rapidly removed by storm waves. However, more resistant boulders of Carstone and Upper Greensand tend to remain on the foreshore.

There appears to be a notable degree of integration between the form of these units and the processes operating throughout the landslide complex. The integration can be described in terms of the following evolutionary model (Fig. 3):

1. Retreat of the sea-cliff (unit 4) leads to loss of material in the seaward section of the lower tier (unit 3) stimulating landslide activity upslope through large-scale movements of the pre-existing compound landslides.
2. Continued seaward movement of the compound block failures causes unloading of the Gault clay scarp, which consequently fails through mudslide activity.
3. The headward extension of the mudslides leads to unloading of the seaward section of the upper-tier pre-existing multiple rotational landslide zone.
4. The rate of supply of landslide debris to the sea-cliff and foreshore increases over time as the compound landslides, mudslides and extension of activity into the multiple rotational failure zone become more active and well established. Ultimately, the failure of the rear-scarp may take place. This behaviour results in the build-up of debris in the seaward portion of the lower tier and on the foreshore which may inhibit further recession through the protection of the cliff base from direct wave attack.
5. A reduction in sea-cliff recession would lead to reduced activity upslope, with a consequent reduction in debris supply to the foreshore. This in turn would result in the sea-cliffs becoming increasingly vulnerable to wave attack through the loss of cliff-base protection, thereby restimulating landslide activity upslope.

It is readily apparent that slope failure and landslide activity throughout the system are promoted by marine erosion (removal of debris from the foreshore and sea-cliff recession) and the consequent unloading of the slopes above. The timing of events, however, is usually related to factors such as heavy rainfall and high groundwater levels. Field evidence suggests that the system has been undergoing stages 1–4, outlined above, characterized by the recession of the sea-cliff, lower tier and Gault clay scarp and the deep-seated failure of the multiple rotational failure zone above. Thus, the effects of sea-cliff recession through relatively small-scale events have been transmitted inland to the Gault clay scarp and upper tier leading to seasonally active movements and intermittent large-scale events.

By analogy with other major landslide complexes in soft rocks, such as Black Ven on the Dorset coast, it may be expected that the erosion of the sea-cliff and landslide response could be expected to show a broad process balance (dynamic equilibrium) in the long term, whereby the average rate of recession throughout the CBU should broadly be comparable with the supply of material from units 1, 2, 3 and 4 and its removal from the foreshore (unit 5).

The situation in CBU2 at Blackgang suggests that almost continuous retreat of the sea-cliff through relatively small-scale events can be expected to give rise to intermittent large-scale recession in the landward parts of the system. During the 132 year period for which records are available there may have been as many as 100 small (0.5–1 m size) failures of the sea-cliff at any one point. The effect of these failures has been the gradual unloading of the Gault clay scarp above, which has led to the large-scale failure of the upper-tier multiple rotational failure zone. Subsequent recession of inland slope features takes place at a much faster rate than the sea-cliff, but these rates may be expected to diminish over time to the very slow rates or inactivity of the preceding periods. In general the historical recession rates match this pattern, although it appears that the retreat of the crest of the Gault clay scarp has not kept pace with sea-cliff recession, leading to a reduction in width of the lower-tier compound failure zone.

The relationship between the recession of the Gault clay scarp and the extension of the reactivated failures upslope appears to be very complex. Between 1980 and 1984 the Gault clay scarp in landslide system 2b retreated by around 35 m, or 2.5 m a^{-1}. Even greater recession was apparent during the late 1970s when major deep-seated landslide activity was reported. During this period of landsliding the Gault clay scarp receded by

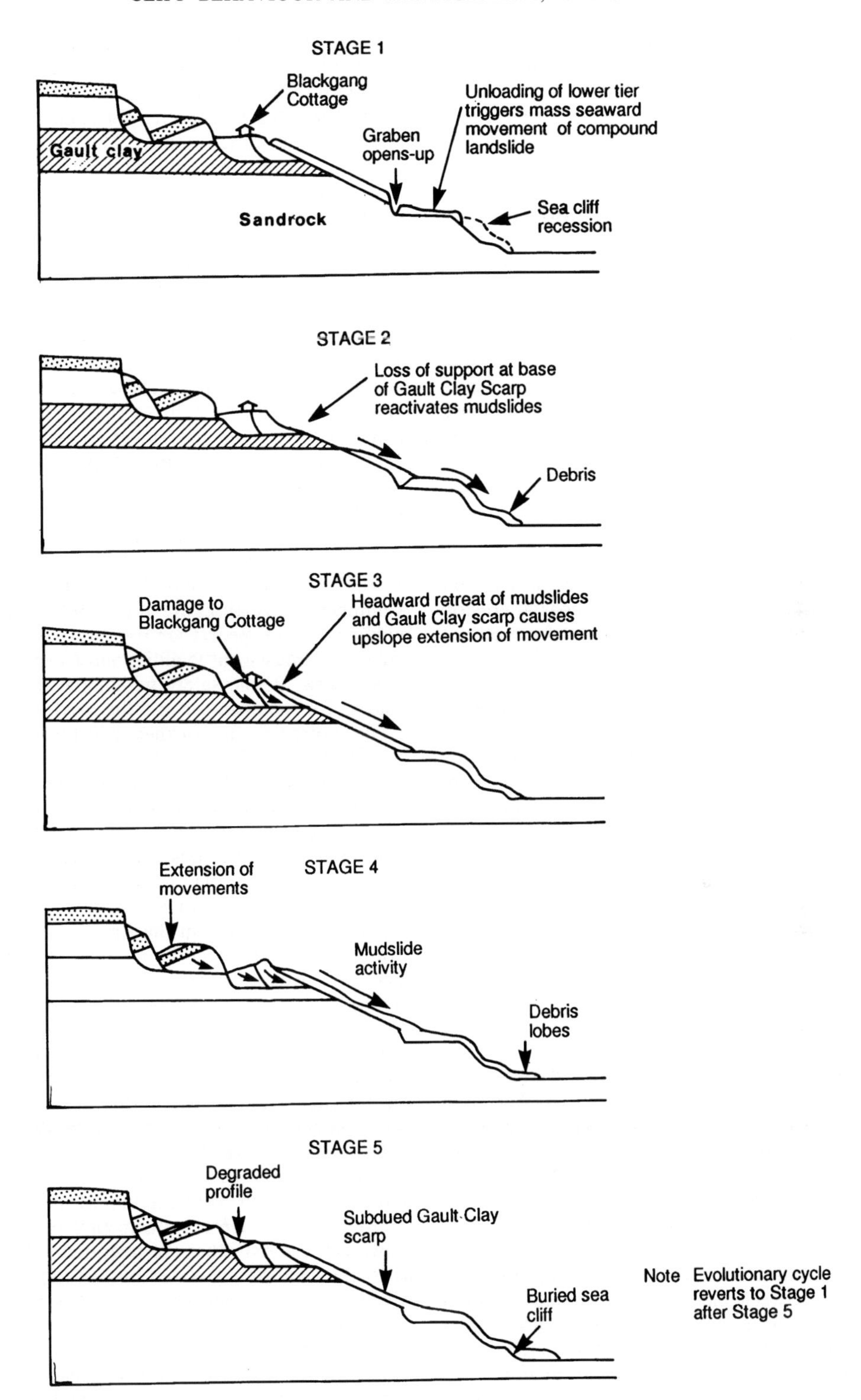

Fig. 3. Evolutionary model of cliff recession in CBU 2.

around 60 m emphasizing that major changes are brought about by episodic events rather than being uniform in time.

Landslide hazard potential and risk

The events in January 1994 were the latest stage in the intermittent and progressive reactivation of the CBUs. It seems likely that future landslide behaviour will continue to follow this pattern, with activity extending landwards into areas currently not affected by ground movement. The key factor in determining the future evolution of the Undercliff is the relative long-term recession rates of the sea-cliff and Gault clay scarp. Three possible scenarios have been identified which define the range of possible future trends over the next 50 years or so:

- retreat of the Gault clay scarp is equal to sea-cliff retreat in the long term (dynamic equilibrium) so that there is a balance between the rate of coastal erosion and mass movement and preservation of form or cliff morphology;
- retreat of the Gault clay scarp is less than sea-cliff retreat in the long term so that there is a tendency towards the development of a narrower steeper Undercliff (slope steepening) eventually leading to the merger of the sea-cliff and Gault clay scarp to form high coastal cliffs such as at Blackgang and The Landslip;
- retreat of the Gault clay scarp is greater than sea-cliff recession in the long term so that there is a tendency towards the development of a wider, shallower Undercliff profile (slope reduction).

Using the geomorphological map boundaries it is possible to highlight those areas in landslide systems 2a–c where the potential extension of landslide activity is most likely (Fig. 4).

In *landslide system 2a*, activity is currently confined to the lower Undercliff and Gault clay scarp. Although land movements have been more extensive here in the past (Bromhead *et al.* 1991) the latest activity will have led to the unloading of the marginally stable rotational landslide zone above, making the landslides more susceptible to rainfall than in the past. Potential ground movements could involve the settlement of the multiple rotational blocks and the opening of large tension cracks between blocks. There is also potential for small rockfalls from the rear-scarp where open joints can be observed in the cliff-face, although such events are generally rare. The area remains occupied and future activity may lead to damage to the remaining buildings and lightweight chalets and caravans.

In *landslide system 2b*, activity now occupies much of The Undercliff up to the talus slope beneath the rear-scarp. Seasonal degradational movements of the affected area can be expected for some years, especially during the wet winter months. The recent subsidence of the multiple rotational landslides beneath the rear-scarp may lead to the development of rockfalls and the collapse of the scarp-face and cliff-top footpath, giving rise to concerns for public safety in the area.

Landslide system 2c was extensively affected by land movements in January 1994. The rates of movement were generally slow which allowed for the evacuation of several homes that were subsequently destroyed by the event. The considerable volumes of debris which have been mobilized will ultimately lead to the unloading of the remaining marginally stable upper tier and a tendency for ground movement to extend back to the rear-scarp (Fig. 4). Potential slope movements within the Blackgang Chine Theme Park are likely to be restricted to the Gault clay scarp in the short term. However, the retreat of the scarp into Blackgang coombe can be expected in the medium to long term if the margin of stability of the coastal slopes continues to decline. A number of structures are at risk in the area, including buildings, a steel pedestrian overbridge, an access road, and many amusement facilities and footpaths within the theme park. Public safety within the theme park is a concern for the operator and the local authority.

The potential for further cliff-top retreat in the Walpen Undercliff (CBU 1) can be assessed from an understanding of the mechanisms of failure and the spatial and temporal projection of historical rates of cliff-top recession into the future. The rapidly retreating high cliffs are very unstable. Cliff-top recession of $5\,\mathrm{m\,a^{-1}}$ has not only led to the dramatic retreat of the cliffs, but also the recent failure of superficial deposits above the cliff top within Blackgang coombe, extending 40 m inland and undermining buildings and the Old Blackgang Road. There remains considerable potential for the further retreat of the cliff top and adjacent land which will cause future damage to existing development. A further consequence of the rapid recession of these cliffs would be the potential undermining of the main coastal road (A3055) within the next 30 years or so, based on current cliff-top retreat rates (Fig. 4).

The relative risk between the different landslide systems can be readily assessed by considering the nature, value and vulnerability of land use in each landslide system. Here, a distinction needs to be made between those risks that are tolerated voluntarily (such as by the local residents and businesses) and involuntarily (such as by visitors); it is widely recognized that individuals exposed to an involuntary risk may be considerably more wary of the consequences than if the risk was voluntarily accepted. Hence, the risk to the public within the theme park (landslide system 2c) is a major concern. The concentration of property, services

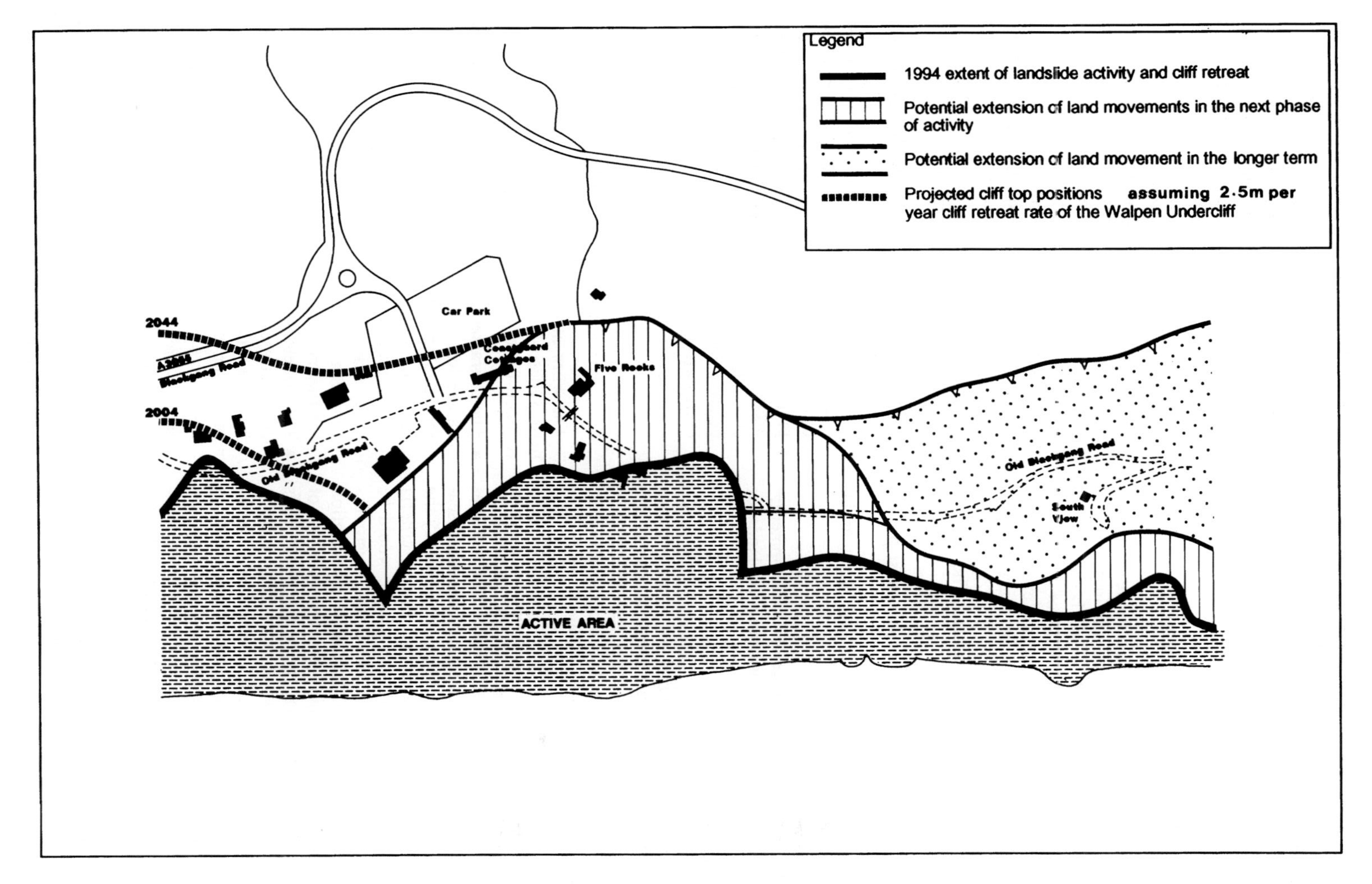

Fig. 4. Landslide potential at Blackgang

and infrastructure in and around landslide systems 2c and 2d make these the most high-risk settings, with potential landslide activity imposing a significant constraint on future land use.

Management options

Although cliff recession and ground movement have always been a problem at Blackgang, the impact of the latest landslide event in January 1994 has been seen as a significant deterioration in the stability of the coastal slopes. The potential for further damaging events is very high. The threat to the community at Blackgang has prompted a careful consideration of the options for managing the problems in the future. The options include:

- the control of cliff-base retreat through coast protection measures;
- the control of groundwater and surface water drainage into unstable areas;
- slope stabilization measures;
- managed retreat of infrastructure and development;
- monitoring to provide early warning of landslide events;
- doing nothing.

An assessment of the viability of these options requires an appreciation of the nature of coastal cliff behaviour, the elements at risk, and the benefits and costs of undertaking such works. When considering the elements at risk, the safety of the public is by far the greatest concern at Blackgang, particularly where they are allowed access on or adjacent to coastal slopes which are subject to active ground movement or liable to sudden collapse. Although coastal protection schemes could be developed to protect the Lower Greensand sea-cliffs, it is considered unlikely that they would satisfy the Ministry of Agriculture, Fisheries and Food's criteria for allocating grant aid, i.e. that the scheme would be technically and environmentally sound and economically viable (MAFF/WO 1993). Landslide stabilization measures to improve the stability of slopes may be of benefit in some areas. However, the height of the cliffs and the magnitude and extent of instability make a large-scale engineered solution impractical. At Blackgang it is recognized that any scheme may have a significant impact on landscape and geological conservation interests. Managed retreat and monitoring were therefore seen as the most appropriate and cost-effective management options which allow for the optimum use of land based on a knowledge of coastal cliff behaviour and yet ensure the safety of the public through the installation of continuous monitoring and early warning equipment.

The landslide hazard potential and associated risks are now taken into account by the theme park operator and the local authority, through their enforcement responsibilities under the Health and Safety at Work Act and with respect to public entertainment licensing of specific events. The aim of this is to ensure that unsafe areas are avoided or that adequate precautions are taken to minimize the risks to the public. With this in mind the operator commissioned a more detailed (1:1000 scale) survey of slope hazard within the theme park, to delineate unsafe areas from those considered at less risk, along with the identification of land that may be suitable for theme park development in the short term. It was intended that the slope hazard map would be regularly updated and therefore provide the framework for implementing the following actions:

- avoidance of areas affected by the active movements in January 1994;
- relocation of theme park attractions away from unsafe areas;
- provision of an early warning of instability through regular inspections of the slopes and the installation of automatic tiltmeters in high-risk areas;
- preparation of emergency action plans for use in the event of a landslide incident to enable the fast and safe evacuation of the theme park, to co-ordinate emergency repairs and to establish procedures for post-event safety inspections.

Conclusions

The work at Blackgang has demonstrated that three main types of scientific and technical information are required for effective and sustainable cliff management:

- The characteristics of the cliff behaviour unit and the nature of contemporary processes. At Blackgang the historical perspective indicates that recent events are not abnormal, but part of a sequence of events that have been shaping the coastal cliffs for many centuries. An assessment of the geotechnical character of the different CBUs is vital for determining the nature of risk.
- The development of an evolutionary model which demonstrates the integration of process and form between the various elements of the CBU. Of particular importance is the need to understand how marine erosion at the cliff base is transmitted inland. At Blackgang, for example, almost continuous sea-cliff retreat through relatively small-scale events gives rise to occasional large-scale recession of the Gault clay scarp and upper-tier slopes. A key issue that needs to be addressed is whether the CBU has been in a state of dynamic equilibrium over the period of the historical record or whether there is a trend towards

the development of a new characteristic form. This will shed light on the possible scenarios for future cliff behaviour and assessment of risk, and may form the basis of predictions of coastal recession.

- Consideration of the full range of cliff management strategies, including 'do nothing', and an assessment of the environmental impacts of these. At Blackgang, for example, options such as coast protection were ruled out at an early stage on the grounds of economic viability and environmental impact. This enabled efforts to be directed towards identifying and developing pragmatic strategies which sought to reduce the level of risk whilst maintaining the environmental and amenity interest of the coastal cliffs.

Acknowledgements. The landslide response and investigation work was commissioned by SWBC. The assistance of R. G. McInnes, South Wight Borough Council, and co-operation of Messrs Debell, Blackgang Chine Theme Park, are gratefully acknowledged.

References

ANON 1982. *Isle of Wight County Press*, 20 October 1982.

BRAY, M. 1994. On the edge. *Geographical Magazine*, April 1994.

BROMHEAD, E. N., HUTCHINSON, J. N. & CHANDLER, M. P. 1991 The recent history and geotechnics of landslides at Gore Cliff, Isle of Wight. *In*: CHANDLER, R. J. (ed.) *Slope Stability Engineering, Developments and Applications*. Proceedings of the International Conference on Slope Stability organised by the ICE, Shanklin, Isle of Wight, 189–196.

CLARK, A. R., MOORE, R. & MCINNES, R. G. 1995 Landslide response and management: Blackgang, Isle of Wight. *In*: *Proc. 30th MAFF Conference of River and Coastal Engineers*, Keele, 6.3.1–6.3.23.

HUTCHINSON, J. N., CHANDLER, M. P. & BROMHEAD, E. N. 1981. Cliff recession on the Isle of Wight SW coast. *In*: *Proc. of the 10th Int. Conf. on Soil Mechanics and Foundation Engineering*, Stockholm, 429–434.

——, BRUNSDEN, D. & LEE, E. M. 1991. The geomorphology of the landslide complex at Ventnor, Isle of Wight. *In*: CHANDLER, R. J. (ed.) *Slope Stability Engineering, Developments and Applications*. Proceedings of the International Conference on Slope Stability organised by the ICE, Shanklin, Isle of Wight, 213–218.

LEE, E. M. 1995. Coastal cliff recession in Great Britain: the significance for sustainable coastal management. *In*: HEALY, M. G. & DOODY, P. J. (eds) *Directions in European Coastal Management*. Samara, Cardigan, 185–193.

—— & MOORE, R. 1991 *Coastal Landslip Potential Assessment: The Ventnor Undercliff, Isle of Wight*. Department of the Environment, London.

MINISTRY OF AGRICULTURE FISHERIES AND FOOD/WELSH OFFICE 1993. *A Strategy for Flood and Coastal Defence in England and Wales*. MAFF, London.

RENDEL GEOTECHNICS 1994 *Assessment of landslide management options, Blackgang, Isle of Wight*. Report to South Wight Borough Council, March 1994.

——1995*a*. *The Occurrence and Significance of Erosion, Deposition and Flooding in Great Britain*. HMSO, London.

——1995*b*. *Soft cliffs: prediction of recession rates and erosion control techniques*. Literature review and project definition study, unpublished report to MAFF, London.

——1995*c*. *The Undercliff of the Isle of Wight: a review of ground behaviour*. South Wight Borough Council.

WHITE, H. J. O. 1921. *A Short Account of the Geology of the Isle of Wight*. Memoir of the Geological Survey of England and Wales, HMSO, London.

The performance of some coastal engineering structures for shoreline stabilization and coastal defence in Trinidad, West Indies

Russell J. Maharaj

Landslide Section, Disaster Prevention Research Institute, Kyoto University, Uji, Gokasho, Kyoto 611, Japan and Institute of Marine Affairs, Hilltop Lanes, Chaguaramas, Trinidad and Tobago, West Indies

Abstract. Coastal erosion in Trinidad has caused a loss of land and subsequent damage to many coastal engineering structures. Damage includes roadway failures, loss of agricultural land, coastal flooding and damage to property. In many areas, structural engineering measures have been implemented to prevent these from recurring. An example of erosion and structural measures used in a segment of rapidly eroding coastline along the east coast of the island is presented. The site is part of a coastal section of the largest wetland on the island, the Nariva Swamp, through which drains the largest river system, the Nariva River. Structural measures used include steel sheet piled revetment, steel-reinforced concrete retaining walls, gabion basket retaining structures, boulder splash aprons, a steel-reinforced concrete pile cluster, concrete columns and blocks and boulder rip-rap. All these an show signs of structural failure, and some have already collapsed. Failures include spelling and cracking of concrete; rebar corrosion; steel pile corrosion; foundation settlement under retaining walls and structures, with subsequent surfical cracking; basal foundation sediment scour and undermining; wave and fluvial induced removal of rip-rap and splash apron boulders, and seasonal burial and sedimentation of the splash apron and concrete pile cluster. Based on this study, it appears that construction-induced high energy conditions and oversight in design and construction were partly responsible for failures.

Introduction

Trinidad is a generally flat, low-lying tropical island located at the southeast corner of the Caribbean Sea and the Caribbean Plate. It lies between 5 and 7 km E of Venezuela and the South American mainland, and south of the Lesser Antilles Island Arc. The island is geologically young, consisting largely of Tertiary clastic sediments, some carbonates and low-grade regional metamorphics, which became emergent in the late Tertiary, and in some areas in the Quaternary (Barr & Saunders 1965). Consequently, slopes in many inland and coastal areas have not attained a state of equilibrium and are severely eroded.

The island is surrounded by three water masses: the open ocean conditions of the mid-west Atlantic to the east and south, the Caribbean Sea to the north and northwest, and the semi-enclosed, estuarine and shallow Gulf of Paria to the west (Fig. 1). The island is therefore subject to a variety of coastal and nearshore circulation patterns and dynamic processes. The coastal areas are also affected by high, July to December discharge from the Orinoco River draining the interior of Venezuela (Cooper & Bacon 1981).

The problem

Due to the diverse and varying geological, geomorphological, hydrological, oceanographic and climatic conditions, the coastline is subject to considerable stress. This causes severe erosion of low lying and cliff areas (Fig. 2(a)). Sometimes, erosion is exacerbated by human activities, causing considerable damage to infrastructure lifeline structures and facilities. In some areas, coastal flooding is also common. This occurs as a result of wave set-up during high water spring tides in combination with high stream discharge. These cause frequent isolation of settlements and towns and reduce economic activity in nearby areas. Along the east coast of the island, coastal roadway flooding is a serious problem because there is only one major roadway linking the north and south of the island in this area. Further, this roadway services important oil and gas recovery facilities located in offshore sections of the Atlantic continental shelf, and therefore is of considerable economic importance.

To reduce coastal erosion, many shore-stabilization structures were constructed. These include concrete sea walls; steel, sheet piled revetment; gabion basket retaining walls; reinforced concrete pile clusters; rip-rap;

MAHARAJ, R. J. 1998. The performance of some coastal engineering structures for shoreline stabilization and coastal defence in Trinidad, West Indies. *In*: MAUND, J. G. & EDDLESTON, M. (eds) *Geohazards in Engineering Geology*. Geological Society, London, Engineering Geology Special Publications, **15**, 61–69.

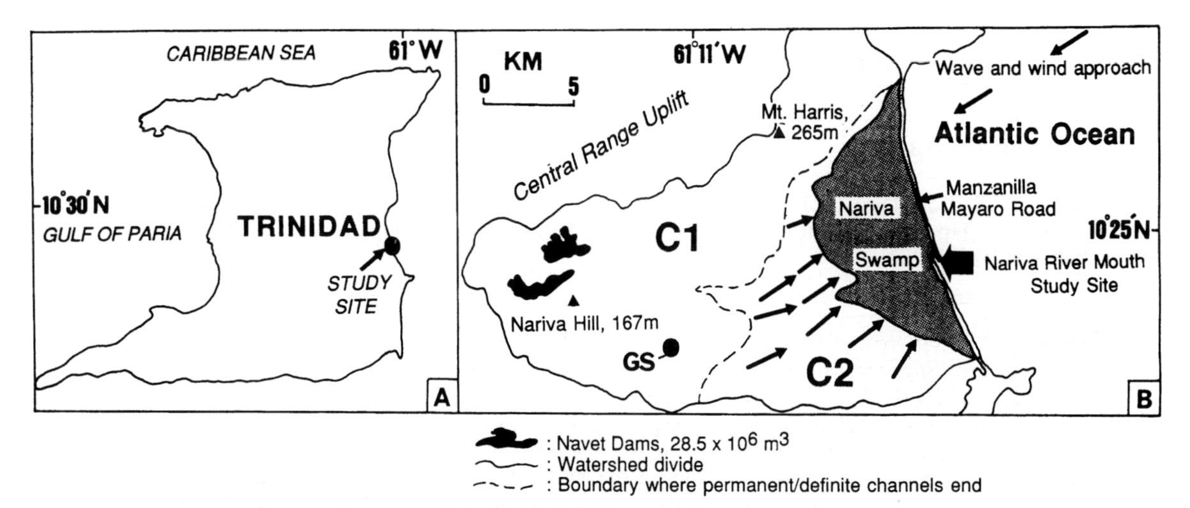

Fig. 1. (**a**) Location map of Trinidad and the study site. (**b**) Some physical characteristics of the Nariva Swamp and adjacent watershed. GS, river gauging station; C1, main catchment area; C2, marshy interface with indefinite channels. Arrows show the direction of river flow into the swamp. Hydrological information is based on Water Resources Agency (1989) data. See text for further details.

concrete retaining walls; dimension stone or masonry walls; gabion basket groynes and boulder splash aprons (e.g. Fig. 2). Many areas were filled and reclaimed to facilitate the construction of these structures. However, many of these structures have failed, while others show initial signs of failure, which over engineering time, may eventually collapse. Consequently, this presents serious erosion, engineering and land use management problems for these coastal and adjacent inland areas.

This paper presents some examples of such erosion, flooding and structural failures from a segment of the Atlantic coastline and the mouth of a major river (Fig. 1). This study is the first part of investigations into the dynamics of coastal areas and an assessment of the effectiveness of shoreline stabilization and erosion control structures in Trinidad.

The study area

The study site is located at latitude 10°23′40″N and longitude 61°1′30″W (Fig. 1), at the mouth of the Nariva River. The Nariva River is the main river draining the adjacent coastal wetland, the Nariva Swamp. At the site, there is a wooden, single-lane bailey bridge, with hollow, cylindrical (*c.* 0.3 m diameter) steel piled foundation support. This links the northern and southern sections of the Manzanilla Mayaro Road, a double commuter roadway and the only one in this part of the island (Fig. 1). Sections of this road are constructed on reclaimed swamp deposits, while the road sub-base consists of yellow, Tertiary marine limestones. The road is a flexible bituminous pavement, with an average thickness of 2.5 cm.

Near the Nariva River mouth, the seaward aspect of the road has been 'protected' by the construction of several retaining structures and backfilled with coarse gravel and some yellow limestone boulders. The details of these structures will be described in a subsequent section. On the landward aspect of is the road is the open swamp and primarily red mangrove forest.

Methodology

Data were obtained from field observations and a literature review. Climatic data are from Schwerdtfeger (1976) and Cooper & Bacon (1981). Geological data are

Fig 2. Some physical characteristics of the site and coastal stabilization structures. (**a**) Erosion of the beach adjacent to the site, showing an undercut and a 1.75 m scarp cut in weathered Holocene beach deposits. (**b**) Erosion of the seaward aspect of the river ball (E), tilting and failure of sheet piles (T), settlement and warping of masonry walls (S) and undercutting by river outflow (U). The arrows R and W are river outflow and wave advance respectively in this figure. (**c**) Dry season sedimentation of the concrete pile cluster. (**d**) A failed segment of sheet piles (S), marbled rip-rap (RR), settlement and warping of gabion baskets (SW), seaward tilting of sheet piles (T) and eddy circulation (E); e: collapsed sections of sheet piles (F) alongside the road (Rr). (**f**) Dry season erosion (E) at the concrete pile cluster, tension cracks in pile members (TC) and recent mortar paste applied to the cracked and spelled concrete surface (M). (**g**) Settlement of concrete walls and blocks (S and CC), cracking in concrete (C), seaward tilting of sheet piles (T) and settlement and warping (SW) of gabion baskets alongside the bridge (B).

a

b
Er
W
T
U
S
R
E

c

RR
T
SW
W
R
E
d

F
Rr
F
W
R
e

I
Er
f

T
SW
B
C
S
S
CC
g

from Kugler (1959) and Barr & Saunders (1965). Hydrological parameters were measured in the field using methods described by US Army Corps of Engineers (1984), from the Land and Water Development Authority, Ministry of Agriculture, Government of Trinidad and Tobago, Water Resources Agency (1989) and 1:12 500, black and white, vertical, panchromatic stereopairs. Sediment data are from Chenery (1952), Government of Trinidad and Tobago (1966, 1971*a*, *b*, 1972), Ahmad & Davis (1971), Ahmad & Wilson (1992), and field engineering geological descriptions using ASTM (1988) standards. Wave climate data are field observations taken from Bertrand *et al.* (1991), and measurements based on methods described by the US Army Corps of Engineers (1984) and Herbich (1991). Topographic data were collected from the Lands and Survey Division, Government of Trinidad and Tobago. The petrography of rip-rap materials are based on standard geological methods.

Climate

Trinidad has two seasons, a January–June dry season and a July–December rainy season. The island is affected by the seasonally strong, Northeast Trade Winds. These bring considerable moisture and cause high intensity showers in coastal areas in the east and northeast of the island (Schwerdtfeger 1976; Cooper & Bacon 1981). Mean annual rainfall in this area and the headwaters of the Nariva Swamp vary between 250 and 300 cm, with 150–200 cm at the study site. Rainfall may also be associated with convectional downpours following diurnal heating, or associated with storms, tropical depressions and hurricanes. These are common from July to October in this part of the Caribbean (Schwerdtfeger 1976). Although the island is located to the south of many storm and hurricane paths, higher rainfall, stronger winds and abnormally severe coastal oceanographic conditions still develop here. The Northeast Trade Winds are also the primary winds responsible for generating offshore waves and their subsequent propagation in the eastern coastal areas.

Due to the high rainfall received, rivers in the eastern part of the island show extremely high discharge during the rainy season. These rivers also flow to the east and therefore drain into the Atlantic. In addition, the Orinoco River also demonstrates high discharge during this same period. This discharge contributes coastal sediments to the Atlantic coastal areas, possibly also affecting nearshore circulation and dynamics.

Site engineering geology and dynamics

The site is located in the central part of the seaward aspect of the Nariva Swamp. The Nariva Swamp is the largest wetland on the island, with an area of approximately 71 km^2, i.e. 20% of the total watershed. Water is supplied from several river systems in the catchment area, but mainly from the Navet River Basin. This river is dammed in two places, the Navet Dams, which have a capacity of 28.5×10^6 m s^3. A gauging station (GS) along this river recorded a mean annual runoff of 1.2 m s^{-1}, for a drainage area of 46.6 km^2, between 1967 and 1987 (Fig. 1(b)). The main catchment area of the swamp, C1, is approximately 199 km^2, i.e. 55% of the watershed. This is separated from the swamp by a marshy area, C2, characterized by indefinite channels and watercourses, with an area of 94 km^2, i.e. 25% of the watershed (Fig. 1(b)).

The Nariva Swamp is of Holocene age (Kugler 1959) and consists of interlayered and interfingered, sometimes lens-shaped deposits of sand, mud and plant detritus in various stages of decomposition. This wetland is characterized by poor drainage and tidal-induced flushing. This is due to the partly confined conditions created by seaward sand bar deposits to the north of this site, beach development and sand sedimentation. Several, shoreline-parallel, emergent and regressive, Holocene shelf sand bars in the interior of the swamp also cause poor flushing. This is exacerbated by poor vertical drainage due to the presence of low-permeability mud and organic deposits and a high water table due to its near-sea elevation. The wetland flora in the hinterland also impedes high stream discharge from the main rivers flowing into the swamp. This is due to high drag coefficients of water plants, which cause high water and energy dissipation. Consequently, the discharge of this river flows over a larger area within the swamp, which due to its poor drainage, causes almost permanent, waterlogged conditions to develop.

As a result of its geographical location, the inflow of sea water from the adjacent nearshore areas and associated wave climate and tidal conditions affect the site. The mouth of the Nariva River here is concave seawards, with an angular change varying between 55° and 70°. This angle changes with season and stream discharge, wave climate and tidal effects. In response to these changes, the site also has variable water depths and seaward channel gradients. Consequently, wave conditions reaching the seaward boundary of this site also vary. During the high rainy season stream discharge, the maximum channel depth can vary from 4.8 m at the initial point of inflection, to 5.8 m at the end of this inflection area, with an approximate seaward channel slope of 2.29°. The minimum depth measured here was 1.5 m. The driving force associated with peak river discharge is quite high, and several fibreglass surveying staffs were broken here during efforts to measure water depths. Efforts to sample sediments were also difficult here. During peak river discharge, the water level is higher than the road elevation.

The maximum free channel surface varies between 20 and 45 m. It is wider during the rainy season due to river

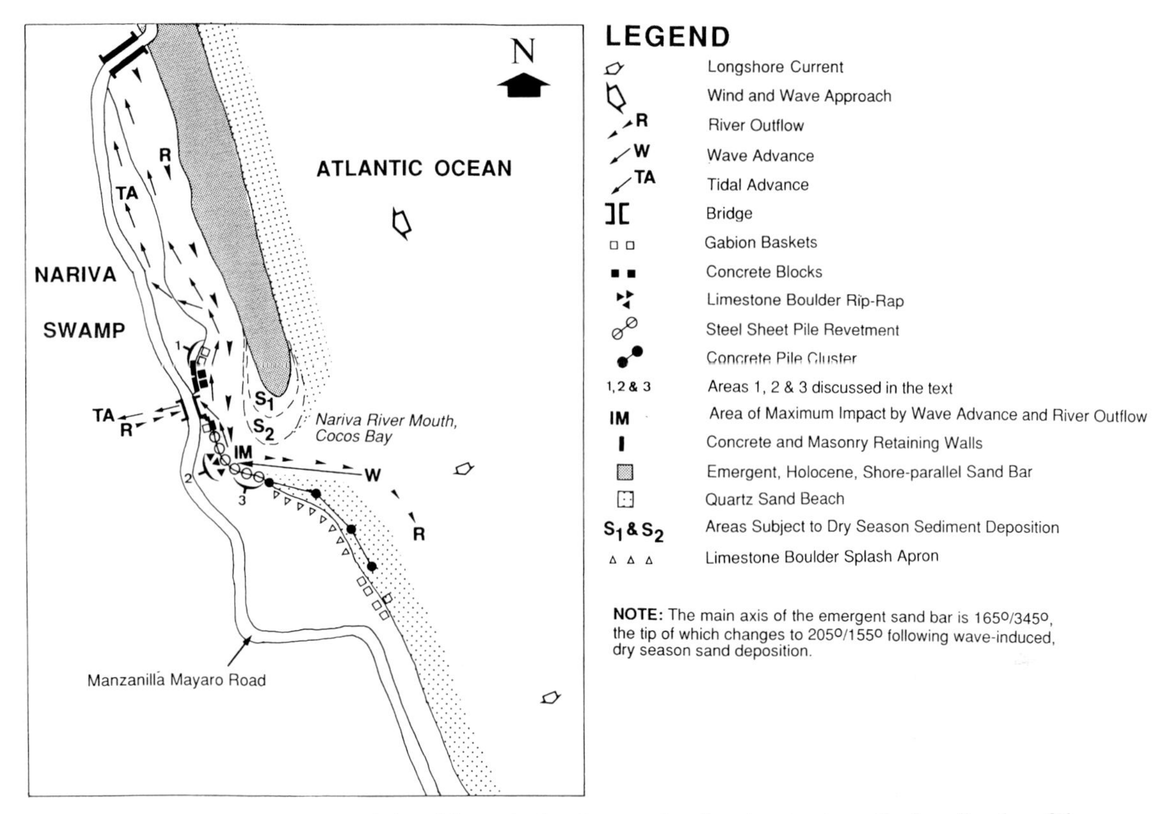

Fig. 3. Some physical characteristics of the study site, the coastal engineering structures, the flow direction of the Nariva River and wave approach into the river.

bank erosion. Peak surface velocity during this season varies between 0.5 and $1\,m\,s^{-1}$ in the final straight channel reach, to $1–3\,m\,s^{-1}$ in the curved section and where the river enters the surf zone. Using floating buoys, it was found that the critical area of concentration of stream discharge is on the roadway bank of the channel inflection (IM, Fig. 3). Advancing wave energy is also concentrated here. Minimum velocities of nil were recorded in the dry season, with dry season maximums of $0.1–0.2\,m\,s^{-1}$ during falling tide. Clockwise eddys (due to the more dominant wave effect) develop in response to river outflow and wave advance. These cause a partly wake area to develop on the opposite side of t he channel. The water surface in this curved channel section is also very uneven, with distinct jumps and an asymmetrical ripple-like surface.

Sediments are fine, subangular quartz sands, with some shell fragments, and with interlayered and interfingered plant debris and mud. They are soft and compressible. They are also saturated with water and have a high hydrogen sulphide content, with a distinct foul scent being found from as shallow as 0.75 m (Area 1, Fig. 3). This suggests a high sulphide content. These soils may also be acidic sulphate soils (Ahmad & Wilson 1992). Deep horizons in such environments are also characterized by framboidal pyrites crystallization. These soils oxidize and dry immediately when exposed to air and produce a very bright orange-red stain. Due to the high plant content and swamp conditions here, peat layers are also common. The Nariva Swamp is also known to contain natural oil seepage sites. These are associated with late Tertiary mud volcanoes, e.g. Bois Neuf (Kugler 1959) These oil seeps may contribute additional sulphides to the soil and water systems. In swamps such as this one, organic acids are also released from decaying plant residue. Ahmad & Wilson (1992) also note that some smectite may be associated with these types of soils.

To the south of the site, the soils on land are highly weathered Holocene beach sand with clay layers. These belong to the Cocal soil series. Some bivalve shell layers are present in these soils. In the beach area, these soils underlie the beach sediments. In the upper beach, they are less than 1 m below the beach sediment surface. Slopes here are less than 5°. To the west (in the swamp), deep hydromorphic peaty clays, with impeded internal drainage, are present. These belong to the Macaw soil series. Slopes here are also less than 5°.

The wave climate in this channel mouth varies with season. Wave approach is from 067°, due to the orientation of the channel mouth. However, from offshore areas, initial propagation is from 045°, which refracts to the east closer to land, then a bit south as it enters the channel. During the rainy season and high water spring tides, advancing waves have a height of 1.50 m. This height may be exceeded during storms. These waves are plunging breakers, with significant battering forces exerted on the coastal structures here. These also cause overtopping, splash and spray, which flood the roadway (up to 0.5 m deep) and then flow into the adjacent swamp. The water depth during these conditions is sometimes up to 2.0 m in Area 3 (Fig. 3). During the dry season, waves are usually 0.3–0.5 m high, and only in the central part of the channel. At this time, Area 3 is covered with sand and shelly gravels. Short wave periods in the rainy season vary between 4 and 6 s, with longer 12–15 s sky periods in the dry season.

The southern end of the site is affected by annual, rainy season erosion of sediment (Fig. 2(f)), and annual, dry season sedimentation of quartz-rich sand and shelly (bivalve) gravels (Fig. 2(c)). The origin of this sand is not known, but an offshore shelf source is possible, as it represents the only source of this type of material on the east coast. Due to variations in sedimentation, water depth also varies in response to sediment removal and erosion. The maximum thickness of sediments deposited here is 1.55 m, over an area 125 m long and 25 m wide. This represents a wedge of sediment greater than 4800 m^3.

During the dry season, tidal advances into the river mouth and further inland are more significant than during the rainy season (due to low stream discharge). Since the Nariva River is dammed, this further reduces stream discharge in the dry season. Consequently, for about half of the year, tidal-dominated conditions are present here.

Wind direction varies, but is generally from the east and east-northeast and is usually less than 1 $m\,s^{-1}$. Near to the bridge, an east-southeasterly wind is common during the rainy season.

The coastal water in this area is light brown during the rain season, due to high tannin and organic acid concentrations (derived from decaying red mangrove bark). During the dry season, the water is a light green colour due to high phytoplankton concentrations.

Characteristics and performance of coastal structures

The coastal structures present at the site include steel sheet piled revetment; reinforced steel concrete piles; concrete columns and blocks; gabion basket retaining structures; boulder splash aprons and boulder rip rap (Figs 2 and 3).

Steel sheet piling is found along the roadway in the central part of the site and in the outer concave segment of the channel (Figs 2(e) and 3). Sheet piles are approximately 12 m long, 0.6 m wide and 1 cm thick. They are fluted/grooved along their edges to facilitate interlocking between adjacent pile members. These are not electrochemically protected, painted or coated to protect them from corrosion. Consequently, the subaerial sections are extremely corroded and completely disintegrated. The are sections in the intertidal zone are fouled with a few barnacles and bivalves and are severely corroded. They have a distinct, bright red-orange staining. Many sections in the centre of this section of revetment are severely eroded and some have collapsed completely (Figs 2(d) and (e)). The top of each pile is extremely jagged due to corrosion. Many other adjacent sections show tilting into the river as channel of up to 30° from the vertical (Figs 2(b), (c) and (g)). The intertidal sections are also extremely abraded by sand and shingle particles entrained in high energy waves, with most of the corroded steel being eroded. They are also highly pitted. This causes a further reduction in metal thickness, as the passive rust layers are removed, while new and more reactive surface area become exposed. The removal of old rust removes some temporary protection afforded against further and new corrosion. The rate of removal of this material is not known at this time. Below the ground, the buried section may be in good condition, due to the reducing, oxygen-free environment present. It was not possible or practical to corroborate this assumption.

The area of maximum tilting of these piles into the river channel is slightly deeper than the zone of maximum corrosion. The channel aspect of this revetment is extremely scoured, with variations in water depth due to sand removal and deposition. These cause the effective exposed length within the water column to vary. The most stable section of this revetment is Area 3 (Fig. 3).

Reinforced steel concrete piles are found at the southern end of the site. They are pre-cast and were driven on-site (Fig. 2(f)). These are isometric in cross-section, with approximately 25 cm sides. Rebars are embedded more than 6 cm within the structural units from the sides, but less than 6 cm from the pile head in some units. The aggregate in the concrete is exposed in many spalled sections. It consists of medium to fine sand (with some quartz), and angular, medium gravel fragments (including some limestone and quartz). These piles show many tension cracks, with apertures up to 2.0 cm, and nearly 1 m long. These cracks are generally linear, sometimes wavy (Fig. 2(f)). On the head of the piles, they sometimes radiate from the central axis or are concentric. These have caused penetration of water and air and extensive rebar corrosion and subsequent iron oxide staining (Fig. 2(f)) on the exposed 1.75 m sections. As a result, the concrete shows spalling and cracking in many areas. Extensive abrasion by sand and shingle is also common in these exposed area. Consequently, intense surface pits on the seaward aspects are present. On the seaward aspect and around the base of each pile, scouring is aggressive. This causes considerable removal of beach sediment and offshore transport in the rainy season. To the land area behind the pile cluster, sand is also continuously reworked (Fig. 2(f)).

Concrete blocks and walls are found to the north of the site and the wooden bridge. Concrete blocks are isometric (0.5 m sides) and they are found adjacent to the channel area and held in place by their high density. These show considerable settlement and suggest foundation compressibility. A maximum vertical settlement (from an assumed original horizontal surface) of 0.35 m was measured in Area 1 (Figs 2(g) and 3). The concrete structures here also show extensive cracking and foundation settlement, with maximum settlement in Area 1 (Fig. 2(g)). In this area, some yellow limestone boulder rip-rap has been placed to fill the depression created by settlement. However, some of this has been displaced by rolling into the river channel and by the erosion associated with overtopping during extreme wave conditions. Backfilled material used in the construction of an embankment (during construction of the bridge) has been covered with concrete. This also shows cracking, especially near to the northern end of the bridge.

Gabion baskets are found in both the northern and southern sections of the site (Fig. 3). These are almost completely destroyed due to abrasion and corrosion of the steel wire mesh used to construct the baskets. Although plastic-coated wire was used, this is of no real advantage, as it usually becomes brittle from alternate wetting and drying (in extreme sunlight), and subsequently easily abraded. As a result, the boulders are easily removed. In the curved section of the channel where gabion baskets are present, foundation settlement and extensive surface deformation have occurred (Figs 2(d) and (g)). Here, the gabion baskets are covered with an 8 cm thick layer of concrete. This is completely broken. To the extreme south of the site, gabion basket groynes and retaining structures are present. These have also suffered abrasion and wire corrosion, but, more significantly, sedimentation and complete burial. They are hardly visible, even during the wet season, i.e. the erosive period.

Rip-rap is found mainly in the central section of the sheet pile revetment (Fig. 2(d)).The boulders are low-grade Neogene metamorphic marbles, with a well-developed fissility, mineralogically controlled by deformed, lens-shaped calcite grains, white micas and chlorite. The mica content is generally less than 10%. They are also derived from a highly fractured geological terrain and they contain microcracks and mesoscopic-scale cracks. The boulders are generally ablate and have a maximum length of 1.0 m, with a short axis of 0.3 m. These were placed to prevent the further erosion of the roadway fill and foundation where sections of the sheet piling have collapsed. Rip-rap was placed here on several occasions. However, it is frequently removed during high water spring tides due to rolling and sliding. Consequently, the rip-rap does not prevent the ingress of sea or roadway erosion. In this area, the roadway fill also shows settlement.

A 2–3 m wide, boulder splash apron is found near the southern end of the site, landward of the concrete pile cluster (Fig. 3). This consists of Tertiary, reefal, yellow limestone boulders, with an average maximum diameter of 25 cm. The boulders are placed on soft sand overlying clayey sands and soils. These are partly removed due to scouring by high water tides, or sometimes covered by sand during the dry season. Scouring also removes the underlying *in situ* soils. Sand is sometimes deposited on land by waves or onshore winds.

Failure modes and mechanisms

Several failure modes are associated with the diverse set of coastal defence works at this site. One primary mode is scouring and basal erosion. This occurs along the seaward aspect of the sheet pile revetment, the concrete pile cluster, the gabion baskets and the concrete blocks and walls.

Scouring is associated with peak stream discharge and tidal outflow from the swamp and river. This is concentrated where the channel begins to curve and along the entire western section of the channel south of this point of inflection. Since this site is part of a near sea level swamp and the sediments are largely sandy and organic with shells, they can become waterlogged. Their densities may therefore be low, possibly lower than their optimum value. Sediments deposited here during the dry season are dumped, as is common in these types of environment, and may therefore have low densities. The sediments in the channel also cannot attain high densities due to fluctuating energy conditions. Therefore, they will continuously be eroded and transported in suspension.

The above is also true for Area 3 of sheet piled revetment. At the onset of the rainy season and with increases in wave activity, larger waves are generated. These cause rapid removal of sand from Area 3 due to rapid loading, liquefaction and their re-suspension in the water column. At these times, incoming waves progressively become larger in this area due to increase in sediment removal and progressive increases in water depth. Therefore, the critical wave heights impacting on the revetment here increase. With this increase in wave height and energy is the downward deflection of some wave energy on the sediment surface. This causes rapid vertical loading of the sand, which in the undrained saturated state, liquefies and loses its strength. In the water column, the sediments become suspended and are removed.

To the seaward aspect of the pile cluster, a similar, but less pronounced effect is envisaged. In addition, localized scouring due to eddy generation at the base of each pile occurs. During high tide, significant eddying also occurs behind the pile cluster in the area of the splash apron. Scouring effects are more pronounced here due to the drag effect created by a shallower, rougher bottom surface. This removes a large amount of sand between the boulders, dislodges many boulders and erodes the underlying soils. Due to the near sea elevation of this site, wave advance is not arrested by this splash apron. Further, the elevation of the top of this splash apron is less than the elevation of the land here. Therefore, waves also advance on land here.

Scouring also causes another problem alongside the roadway edge of the channel. Due to sand removal, some passive pressure generated by the foundation on the seaward side, and which also stabilizes the revetment and walls, is partly removed. This can cause a relative increase in active earth pressure on the opposite side of the revetment. In conjunction with basal scour by stream discharge, failure of the foundation and settlement can result. Surface structures will therefore collapse and show extensive cracking. This is best exemplified in the section of sheet pile shown in Fig. 2(d) and in the undercut area (U) in Fig. 2(b). At this latter site, erosion by outflow and inflow via the small tributary here is also partly responsible for this effect. Floating mangrove debris transported on the water surface shows a preferred outflow toward this area, while floating coconuts transported by incoming waves are also directed here. These suggest a concentration of river and wave energy here.

Corrosion is another primary reason for failure of the steel piles. As mentioned earlier, some sections are completely corroded. Since no forms of electrochemical protection or coating are used, corrosion is inevitable.

In the concrete piles, it is the author's belief that microcracks may have developed in response to piling. This is based on interpretation of crack geometry and size distribution (sometimes radiating from the pile head and with a decrease in their width and length away from the head). This may have subsequently led to an ingress of water and air and eventually to corrosion of the rebars. With oxidation of the steel and a subsequent increase in the effective volume of the rebars, the concrete will crack (due to low tensile strength). Sulphides and organic acids derived from the swamp soils and chloride ions from seawater can further contribute to concrete deterioration. The concrete also consists of variable aggregate grain sizes and shapes. This may have influenced the permeability of the final casted members, but at this time, not enough data are available to form a conclusion on this aspect.

Effect of road construction on site dynamics

Before construction, it is possible that some erosion of the channel may have occurred at the southern end. However, the dissipation of stream and tidal flow and energy by swamp vegetation and mangroves along the channel would have minimized such erosion. During this time, stream discharge may have also been over a larger area. Further, prolific growth of mangroves along the channel sides prior to site clearance may have been significant enough to reduce the forceful impact of advancing waves. This is supported by the fact that channel sections immediately adjacent to this site are lined with natural luxuriant growth of mangroves and are stable. Therefore, the site may also have been a relatively stable one.

To facilitate the construction of the roadway, fill and reclamation was necessary, in combination with bridge construction and coastal/river defence works. This removed wetland flora and reduced the effective channel area. The road and related structural works also created an artificial barrier to free channel and tidal flowed. As a result, outflow stream energy was redirected and concentrated in a smaller area, on the roadway (barrier) and related structural stabilization works. This increased the effective erosion at the site and the energy exerted on the artificial barrier. Road construction also restricted tidal advance into this area, with a net concentration of tidal and wave energy on the barrier. Consequently, the site developed post-construction, localized, higher energy conditions. As a result, significant failures have resulted.

Conclusions and recommendations

Based on the foregoing observations and analysis, the study site can be considered an extremely dynamic one. Such dynamism is seasonally controlled by rainfall and discharge from the Nariva River, tidal conditions, wave climate and coastal sedimentation. The site was also modified by road construction and coastal engineering works. In response to these modifications, channel and wave energy appears to have been concentrated in a smaller area. Consequently, negative and damaging effects are exerted on the coastal defence works here.

Based on the failure types observed and possible failure mechanisms, the structures described here have performed poorly due to oversights in planning, design and/or during construction. This suggests that a more systematic, integrated approach should be taken in the analysis of such coastal engineering problems. Further, the hill range of existing, possible or anticipated conditions should be considered in design and analysis, especially with regard to possible construction-induced problems. Planning and use of coastal and river bank areas require a comprehensive analysis of the dynamics and characteristics (engineering, physical, chemical and biological) of these environments, especially if both coastal and riverine conditions co-exist. Therefore, the need for an integrated approach is important.

It is significant to note that minimal erosion and high river bank stability exist in adjacent mangrove-lined channel sections of this river, adjacent rivers and other coastal areas. This raises the question of the possible use of biotechnical and big-engineering stabilization measures. These can employ the use of on-site wetland flora, in conjunction with structural engineering works. In this way, an optimum environment-friendly solution to such complex problems can be reached.

In addition, comparison of the stable conditions which exist in these natural mangrove systems and the unstable conditions created by construction show that human-made structures can be subjected to construction-induced, higher energy conditions. In the case of the study site, this was manifested by a reduction of stream channel section area, with a corresponding increase in discharge per unit area of channel. This later caused intense basal scouring, bank erosion and concentration of wave and tidal energy on the river and roadway stabilization works. These human-made conditions can be deleterious to the performance of structures and may lead to their partial or complete failure through engineering time. Consequently, planning, design and construction activities should consider the additional effects of creating higher energy conditions at the construction site.

It must be noted that the near-sea elevation of this flat coastal and alluvial area is not a favourable characteristic for infrastructure development. Therefore, if any construction is planned for this area, then some important consideration should be given to this characteristic.

Acknowledgements. This work was done while the author was based at the Institute of Marine Affairs, Trinidad, West Indies. The financial and logistic support of a Japanese Monbusho Scholarship to prepare this paper is gratefully acknowledged. Thanks are due to P. Maharaj for carefully reading and editing the manuscript.

References

AHMAD, N. & DAVIS, C. E. 1971. The effect of drying on release of native and added potassium of six West Indian soils with contrasting mineralogy. *Soil Science*, **112**, 100–106.

—— & WILSON, H. W. 1992. Acid sulphate soils of the Caribbean region – their occurrence, reclamation and use. *Soil Science*, **153**, 154–164.

ASTM 1988. *Annual Book of ASTM Standards. Soil and Rock, Building Stones; Geotextiles*, Vol. 04.08. ASTM, Philadelphia.

BARR, K. W. & SAUNDERS, J. B. 1965. An outline of the geology of Trinidad. *In*: *Trans. 4th Carib. Geol. Conf., 1965*, Trinidad.

BERTRAND, D., O'BRIEN-DELPESH, C., GERALD, L. & ROMANO, H. 1991. Coastlines of Trinidad and Tobago – coastal stability perspective. *In*: CHAMBERS, G. (ed.) *Coastlines of the Caribbean*. American Society of Civil Engineers, New York, 1–16.

CHENERY, E. M. 1952. *Soils of Central Trinidad*. Government Printing Office, Trinidad.

COOPER, G. E. & BACON, P. R. 1981. *The Natural Resources of Trinidad and Tobago*. Edward Arnold, London.

GOVERNMENT OF TRINIDAD AND TOBAGO 1966. *Soil and Land Capability Study of Trinidad*. Government of Trinidad.

——1971*a*. *Land Capability Survey, Trinidad*. Government Printing Office, Trinidad.

——1971*b*. *Soil Map of Trinidad, 1:150 000*. Government Printing Office, Trinidad.

——1972. *Land Capability Map of Trinidad, 1:150 000*. Government Printing Office, Trinidad.

HERBICH, J. B. (ed.) 1991. *Handbook of Coastal and Ocean Engineering*, Vols 1 and 2. Gulf, New York.

KUGLER, H. B. 1959. *Geological Map of Trinidad and Geological Sections, 1:100 000*.

SCHWERDTFEGER, W. (ed.) 1976. *Climates of Central and South America. World Survey of Climatology*, Vol. 12. Elsevier, Amsterdam.

US ARMY CORPS OF ENGINEERS 1984. *Shore Protection Manual*, Vols 1 and 2, 4th edition. Coastal Engineering Research Center, US Government Printing Office.

WATER RESOURCES AGENCY 1989. *Legend and Hydrogeological Map of Trinidad, 1:200 000*. Government of Trinidad and Tobago.

Assessing erosion of sandy beaches due to sea-level rise

Robert J. Nicholls

Flood Hazard Research Centre, Middlesex University, Queensway, Enfield EN3 4SF, UK

Abstract. The prospect of accelerated sea-level rise in the coming century makes beach erosion more likely. Given the concentration of human habitation and infrastructure adjacent to sandy beaches, it is prudent to consider the implications of such changes. To select the most appropriate approach to evaluate this problem, three levels of increasingly complex assessment can be distinguished: (1) screening assessment (SA); (2) vulnerability assessment (VA); and (3) planning assessment (PA). SA and VA are amenable to analysis with a Bruun rule-based approach. SA can be quickly and effectively accomplished using a simple proportionality approach that shoreline recession is in the range of 100 to 200 times the sea-level rise scenario. For VA, a range of shoreline response scenarios can be estimated by selecting appropriate low and high offshore boundary conditions. PA ideally requires a comprehensive sediment budget approach. SA may trigger VA, which may in turn, trigger PA, depending on the results. The limitations of this structured approach are discussed.

Introduction

This paper presents a framework for the assessment of possible beach erosion due to sea-level rise based on experience in a number of countries (Nicholls & Leatherman 1995). Given the concentration of human activities in many beach locations, a long-term erosional trend has serious implications. As the beach width progressively declines, so upland areas behind the beach will be increasingly exposed to wave action during storms, and in developed coastal areas, significant losses might occur during coastal storms.

Globally, it is estimated that 70% of sandy shores were eroded over the last few decades (Bird 1985), so long-term beach recession is already a major geohazard. At the same time, global sea-level rise has been estimated to be in the range of 0.1–0.2 m over the last century (Warrick & Oerlemans 1990). While the causes of the erosional trend of beaches are much debated, it is argued by some authors that this global sea-level rise is an important factor (e.g. Vellinga & Leatherman 1989).

It is predicted that global sea-level rise will accelerate significantly in the coming decades due to anthropogenic global warming. The most likely rise is about 0.5 m by 2100, with a rise of up to 1.0 m by 2100 being possible (Wigley & Raper 1992). This prospect has given rise to a number of local, national and even global scale assessments of vulnerability to sea-level rise (e.g. McLean & Mimura 1993; O'Callahan 1994; Nicholls & Leatherman 1995). The likelihood of more widespread and more rapid beach erosion is a major concern of these studies.

Sandy beaches and sea-level rise

Sea-level rise promotes the erosion of beaches (National Research Council 1987). However, sea level is only one factor determining shoreline position and erosion is an integrated response to the entire coastal sediment budget of any area (SCOR Working Group 89 1991). Thus, beach erosion should not be expected to be a universal response to sea-level rise. However, as more than 70% of the world's beaches are already eroding, 20% to 30% are stable, and less than 10% are accreting (Bird 1985), any acceleration in sea-level rise will increase this existing tendency towards a sediment deficit. Therefore, an erosional response can be expected, with some exceptions.

There are two distinct effects of sea-level rise on beach evolution, depending on geomorphic setting (Stive *et al.* 1990). On uninterrupted sandy coasts with no inlets, rising water levels induce profile disequilibrium and hence, erosion due to profile adjustment. This direct effect of sea-level rise is often known as 'the Bruun rule' (Bruun 1962, 1988) and this approach has received most attention in assessments of the effects of sea-level rise. On interrupted sandy coasts with inlets, sea-level rise causes additional erosion of the shoreline as the inlet/bay system will provide an additional sink for sand. This was termed the indirect effect of sea-level rise by Stive *et al.* (1990).

Levels of assessment for sea-level rise impacts

Given our imperfect understanding of beach response to sea-level rise, a structured framework is essential for any assessment. *Vulnerability* to sea-level rise and other coastal implications of climate change embraces both (1) the physical and human *susceptibility* of the coastal zone to sea-level rise and (2) the ability of society to cope with those changes (e.g. IPCC 1992). After the completion of vulnerability assessments, the *evaluation of policies* to respond to sea-level rise (and other coastal

NICHOLLS, R. J. 1998. Assessing erosion of sandy beaches due to sea-level rise. *In*: MAUND, J. G. & EDDLESTON, M. (eds) *Geohazards in Engineering Geology*. Geological Society, London, Engineering Geology Special Publications, **15**, 71–76.

Table 1. *Levels of assessment for sea-level rise impact and response studies (adapted from Hoozemans & Pennekamp 1993; WCC 1994)*

Level of assessment	*Time-scale*	*Precision*
Screening assessment (SA)	2–3 months	Lowest
Vulnerability assessment (VA)	1–2 years	
Planning assessment (PA)	ongoing	Highest

implications of climate change) will require further, more comprehensive, analysis.

Therefore, three levels and time-scales of assessment can be usefully distinguished (Table 1). Screening assessment (SA) is a reconnaissance-level approach which only aims to identify susceptibility with limited consideration of possible responses, while vulnerability assessment (VA) is a more comprehensive analysis, including some preliminary assessment of the ability of society to respond to the projected changes. The aim of both SA and VA is to focus attention on critical issues concerning the coastal zone, given sea-level rise, rather than supplying precise predictions. In general, previous SA/VA studies have made the simplifying assumption that sea-level rise is the only cause of shoreline change to facilitate the analysis. Planning analysis (PA) of different responses to sea-level rise within integrated coastal zone management (ICZM) is part of a continuous management process which ideally aims to integrate responses to all the existing and potential problems of the coastal zone, including minimizing vulnerability to the long-term effects of all aspects of climate change (WCC 1994). Therefore, the modelling precision required increases as analysis progresses from SA to VA, and on to PA (Table 1). Further, the completion of each level of assessment may trigger a higher level of assessment. The focus of this paper is reconnaissance approaches suitable for SA and VA.

Uncertainty is an unavoidable element of assessing the implications of sea-level rise. As already discussed, projections of future global sea-level rise encompass a large range of possible rises. Given the present consensus that global sea-level rise will be ≤ 1 m by the year 2100, it is best to assess a range of sea-level rise scenarios including 0.3, 0.5 and 1.0 m. Lastly, local land uplift or subsidence must also be considered, so that the global sea-level rise scenarios can be transformed into relative (or local) sea-level rise scenarios – a beach responds to local rather than global changes.

Erosion modelling on uninterrupted coasts

The Bruun rule

To estimate erosion due to sea-level rise on uninterrupted coast, the Bruun rule is generally utilized (e.g. Leatherman 1991; Nicholls *et al.* 1995). The fundamental assumption is that the profile conserves its 'average' or 'equilibrium' shape relative to sea level. Therefore, using appropriate boundary conditions, the profile shape defines all the information necessary to predict the response to sea-level rise. This key assumption has been utilized by later authors as part of more comprehensive long-term shoreline evolution models (e.g. Everts 1985; Stive *et al.* 1990). The original formulation applies a two-dimensional cross-shore balance of sediment: to maintain the profile shape, the upper part of the profile erodes and the lower part of the profile accretes, translating the shoreline landward. This gives a simple equation (Fig. 1):

$$R = G(L/H)S \quad \text{(1a)}$$

where

$$H = B + h_* \quad \text{(1b)}$$

and R is the shoreline recession due to a sea-level rise S; h_* is the depth at the offshore boundary (henceforth, the depth of closure); B is the appropriate land elevation, defined by the landward boundary; L is the active profile width between the boundaries; and G is the inverse of an overfill ratio (US Army Corps of Engineers 1984). G represents the grain size of the eroded material – material that is too fine for the beach is lost, increasing the predicted recession. It has often been mistakenly omitted in previous studies.

The inverse of the beach slope (L/H) multiplies any sea-level rise, giving a large recession for a small rise in sea level. Therefore, flatter beaches are predicted to show greater sensitivity to sea-level rise than steeper beaches. A number of laboratory and field studies confirm an erosional beach response to rising water levels (e.g. Hands 1983). However, on careful inspection, the application of the Bruun rule raises some fundamental questions and any predictions of shoreline recession need careful interpretation (SCOR Working Group 89 1991).

One major limitation is response time. As Bruun (1962) noted in his original paper, sea-level rise does not change the shape of the profile; it only creates a potential for erosion which the availability of wave energy realizes. Therefore, profile adjustment to higher sea levels is expected to take time (SCOR Working Group 89 1991). The major implication of this observation is that the Bruun rule is best applied over long time-scales (decades to centuries). Another major and related problem is that an offshore boundary is difficult to specify. Based on wave climate, the likelihood of larger waves increases with time-scale, introducing a time dependency for this offshore boundary (Stive *et al.* 1992).

Application of the Bruun rule

The study coast needs to grouped into longshore sections which have similar wave and profile characteristics. This may be accomplished by using geomorphic principles, or

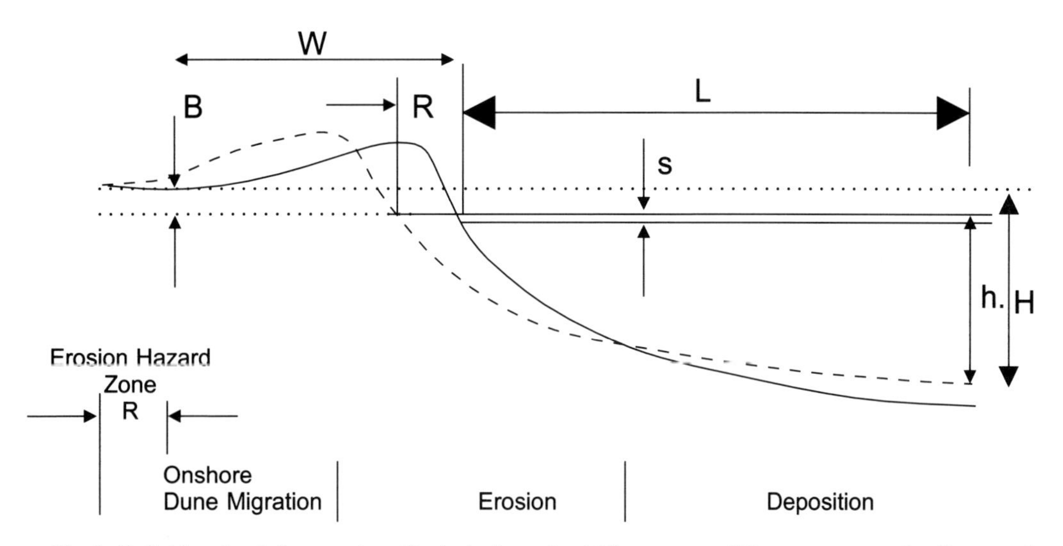

Fig. 1. Definition sketch for equation (1). A rise in sea level (S) causes an offshore movement of sediment and a shoreline recession (R).

regular sampling along the coast. The aim is to define longshore segments which can be represented by a single cross-section.

Rather than determine a single estimate of possible beach recession, a range of shoreline change scenarios which aim to encompass the likely shoreline response can be determined (Nicholls *et al.* 1995). It is important to assess the composition of the eroded material (Equation 1). If this is uncertain, composition scenarios should be estimated, although this will increase the range of the projections. The boundary conditions as discussed below are recommended.

Offshore boundary

Two depths of closure are utilized. They are defined by consideration of the range of possible time-scales and hence encompass the likely shoreline response to a given sea-level rise scenario. Hallermeier's (1981) estimate of the annual depth of closure d_l was generalized to a time-dependent form $d_{L,y}$, where y is the return period in years. The low estimate is the annual depth of closure as originally defined by Hallermeier (1981):

$$d_{L,1} = 2.28H_s - 68.5(H_s^2/gT_s^2), \quad (2)$$

where H_s is the annual exceeded wave height in a 12 h period, and T_s is the associated wave period. Landward of this boundary, adjustment to any rise in sea level is expected to take less than a year.

The high estimate is a depth of closure typical of a century ($d_{L,100}$). Based on the limited data that are available, Nicholls *et al.* (1995) suggested that

$$d_{L,100} = 1.75d_{L,1}. \quad (3)$$

An appropriate reference depth for $d_{L,1}$ and $d_{L,100}$ appears to be 1 m above low water (Nicholls *et al.* 1995). Given the typical exponential form of the beach/nearshore profile (Fig. 1), a larger depth of closure estimates a lower beach slope, so $d_{L,100}$ will usually predict greater beach recession than $d_{L,1}$.

Onshore boundary

The onshore boundary in Bruun rule calculations is often specified as the dune crest. However, as the dune is progressively removed so its height will decline. Therefore, to ensure continuity, it is generally better to assume that the dune is conserved as the beach retreats and, therefore, the average elevation of the land behind the dune defines B. Addition of the dune width (W) to the profile width and a reduced height B, as shown in Fig. 1, will allow the dune to maintain its height relative to sea level and migrate onshore:

$$R = G((L + W)/H)S \quad (4)$$

The predicted recession rates will be a little higher than Equation (1), but given that $W \ll L$, and H will be little reduced, the increase in recession using Equation (4) will be small, except in areas with relatively high dunes. Allowing the dune to be maintained will also preserve existing environmental values and the existing standard of protection against storms and flooding. It should be noted that this approach defines the origin of the erosion hazard zone as the landward side of the dune – the hazard zone extends a distance R inland (Fig. 1). Other assumptions about the onshore boundary are possible, but they should be clearly stated. If the dune is assumed to be lost or degraded, the implications of declining storm and flood protection should be assessed.

In areas without dunes, such as sites where seawalls have been constructed, other realistic boundary conditions are required (e.g. Dean & Maurmeyer 1983).

Wave and profile data

Wave climate is often a large uncertainty in these calculations. Given the limited wave data that are often available, longshore interpolation and extrapolation of the available closure depths was essential. In future studies, considerable improvements might be possible by using wave hindcasts and/or refraction modelling to construct more realistic wave climates. Consistent wave data collection using new technology such as satellite altimetry would be especially useful.

Once boundary locations are defined, the corresponding profile width needs to be determined. In the absence of cross-shore profiles, this can be scaled from the best available bathymetric charts.

Sea-level rise effects near inlets

The shorelines around inlets are often the most dynamic areas of the coast. The Bruun rule was designed for coastlines some distance from inlets (Bruun 1962). Given sea-level rise, the inlet/bay system acts as an additional sink for sand. This indirect effect of sea-level rise can make the processes described by the Bruun rule negligible (Stive *et al.* 1990).

The possible role of these sinks in the regional sediment budget and the relative sensitivity of the coast should be considered at all levels of assessment. The high susceptibility of shorelines near inlets should be noted and communicated to policy-makers as part of any assessment. Historical experience provides some indication of the minimum activity that might be expected given accelerated sea-level rise. For VA, an equilibrium response can be considered as a worst-case scenario. The sink term is the area of the inlet and associated lagoons multipled by the sea-level rise scenario. The source of this sand is additional erosion along the open coast. The longshore extent of influence is more difficult to determine, but values of around 10–20 km are sensible estimates.

Future analyses

The procedure described gives reconnaissance estimates of land susceptible to erosion given a rise in sea level. Despite significant research activity (e.g. WCC 1994), knowledge concerning much of the world's coast remains limited, while in more studied areas, more detailed assessments may be required. Therefore, a continued need for reconnaissance and simple level approaches is anticipated. An approach for future analysis is suggested in Table 2. The results of SA may trigger VA, which in turn may trigger PA.

For SA, studies should be completed quickly, so the Bruun rule calculations described previously are too sophisticated and time-consuming. Based on four national studies, predicted recession for $d_{L,1}$ and $d_{L,100}$ is better approximated by 100 and 200 times the sea-level rise scenario, respectively (Table 3). (This compares with the more often-quoted rule of thumb for the Bruun rule of 50 to 100 times the rise in sea level.) Therefore, for any sea-level scenario (S), the area $100S$ to $200S$ landward of the dune must be considered susceptible, with a wider zone near inlets. This type of approach lends itself to low-cost oblique videography techniques of assessment (Nicholls *et al.* 1993; Leatherman *et al.* 1995).

For VA, the Brunn rule can be applied as already described and coupled with other elements of a vulnerability assessment (societal impacts and possible responses). If information on other terms in the sediment budget is easily available, such as rates of longshore transport, this should be included in the analysis (e.g. Titus *et al.* 1985). As with SA, inlets present a significant problem.

For PA within ICZM, a moe comprehensive sediment budget approach is desirable. An existing model for such a study is Stive *et al.* (1990), which was used to examine long-term coastal development of the Netherlands with and without accelerated sea-level rise. These results have already been integrated into long-term coastal manage-

Table 2. *Modelling approaches versus level of assessment*

Level of assessment	*Type of coast*	*Modelling approach*
SA	no inlets	$100S$–$200S$*
	inlets	?? qualitative estimate
VA	no inlets	Bruun rule
	inlets	?? worst-case estimate
PA		sediment budget

* S is the sea-level rise scenario.

Table 3. *Average recession at a national scale using the Bruun rule with* $d_{L,1}$ *and* $d_{L,100}$*, respectively. (Data from Nicholls & Leatherman 1995)*

Country	*Recession (m)*	
	Low	*High*
Senegal	$110S$*	$170S$
Uruguay	no data	$150S$
Venezuela	$80S$	$122S$
Argentina	$100S$	$210S$

* S is the sea-level rise scenario.

ment decisions and a national policy of dynamic defense of the shoreline has been selected (Louisse & Kuik 1991; Hillen & DeHaan 1993). It should be noted that VA and PA may be considered part of a continuum and certain policies, such as enlarged building setbacks, may be shown to have sufficient merit based on a VA level of assessment. Such judgement needs to be applied at the individual study level.

Discussion and conclusions

Assessment of beach response to sea-level rise is facilitated by considering the level of prediction required. More simple approaches are appropriate to more simple questions such as SA and VA. Further, SA/VA studies may trigger more detailed assessment as appropriate. In terms of response costs, the erosion estimates provide a first-order estimate of the sand volume necessary to counter the erosion and hence the potential costs of beach nourishment (Nicholls *et al.* 1995).

However, the limitations of the SA/VA results should be noted; they are first-order estimates of shoreline change which often ignore significant factors such as longshore transport of sand. In addition, other climate change factors such as changes in wave climate and storminess could be just as significant as sea-level rise, but presently appropriate local and regional scenarios cannot be developed due to the resolution of the available climate models. (Arbitrary scenarios of change can be evaluated for sensitivity purposes.) If important societal values are identified as being threatened by SA or VA studies, further more detailed studies are triggered (*cf.* Volonté & Nicholls 1995). Therefore, their role as triggers for further analysis is emphasized.

Significant improvements at the VA level of assessment are likely in the near future. The Bruun rule assumes an equilibrium response to sea-level rise above the depth of closure, and no response below it. This is an unrealistic assumption as there is a gradient in sediment transport across the shoreface. Onshore feed from the shoreface to the active zone can be measured in the Netherlands, countering to some extent the erosion predicted by the Bruun rule (Stive *et al.* 1990). However, in general, the shoreface cannot be distinguished as a sink or source of sand without extensive observations. Further development of shoreface models will probably reduce this data requirement (Stive & DeVriend 1995).

For the precision required to investigate and formulate policy (PA), the role of all major processes in the coastal sediment budget, including human influence and other impacts of climate/global change such as changing storm frequency or direction, or changing sediment supply, must be considered. If future problems of beach erosion are to be successfully managed, a comprehensive understanding of the sediment budget at any site will be essential.

Acknowledgements. This paper was largely derived from research funded by the US Environmental Protection Agency (Project Officer, J. Titus). The author would like to acknowledge his former colleagues: S. Leatherman, C. Volonte, K. Dennis and G. French; and all the overseas collaborators who made this research possible. He would also like to acknowledge important conversations with R. Hallermeier, W. Birkemeier, M. Stive and H. DeVriend.

References

BIRD, E. C. F., 1985. *Coastline Changes – A Global Review.* Wiley, Chichester.

BRUUN, P. 1962. Sea level rise as a cause of shore erosion. *Journal of Waterways and Harbors Division, ASCE*, **88**, 117–130.

——1988. The Bruun rulse of erosion by sea-level rise: a discussion on large-scale two- and three-dimensional usages. *Journal of Coastal Research*, **4**, 627–648.

DEAN, R. G. & MAURMEYER, E. M. 1983. Models for beach profile response. *In*: KOMAR, P. D. (ed.) *Handbook of Coastal Processes and Erosion.* CRC, Boca Raton, FL, 1551–166.

EVENTS, C. H. 1985. Sea level rise effects on shoreline position. *Journal of Waterway, Port, Coastal and Ocean Engineering, ASCE*, **111**, 985–999.

HALLERMEIER, R. J. 1981. A profile zonation for seasonal sand beaches from wave climate. *Coastal Engineering*, **41**, 253–277.

HANDS, E. B. 1983. The Great Lakes as a test model for profile responses to sea level changes. *In*: KOMAR, P. D. (ed.) *Handbook of Coastal Processes and Erosion*, CRC, Boca Raton, FL, 167–189.

HILLEN, R. & DEHAAN, TJ. 1993. Development and implementation of the coastal defence policy for the Netherlands. *In*: HILLEN, R. & VERHAGEN, H. J. (eds) *Coastlines of the Southern North Sea.* American Society of Civil Engineers, New York, 188–201.

HOOZEMANS, F. M. J. & PENNEKAMP, H. A. 1993. *A general approach to vulnerability assessments.* Delft Hydraulics, Discussion Paper (unpublished).

INTERGOVERNMENTAL PANEL ON CLIMATE CHANGE 1992. *Global Climate Change and the Rising Challenge of the Sea.* Report of the IPCC Coastal Zone Management Subgroup. Rijkswaterstaat, The Netherlands.

LEATHERMAN, S. P. 1991. Modelling shore response to sea-level rise on sedimentary coasts. *Progress in Physical Geography*, **14**, 447–464.

——, NICHOLLS, R. J. & DENNIS, K. C. 1995. Aerial videotape-assisted vulnerability analysis: a cost-effective approach to assess sea-level rise impacts. *Journal of Coastal Research*, Special Issue No. 14, 15–25.

LOUISSE, C. J. & KUIK, T. J. 1991. Future coastal defence in the Netherlands: strategies for protection and sustainable development. *Journal of Coastal Research*, **7**, 1027–1035.

MCLEAN, R. F. & MIMURA, N. (eds) 1993. *Vulnerability Assessment to Sea Level Rise and Coastal Zone Management.* Proceedings of the IPCC Eastern Hemisphere Workshop, Tsukuba, Japan.

NATIONAL RESEARCH COUNCIL 1987. *Responding to Changes in Sea Level: Engineering Implications.* Academic, Washington, DC.

NICHOLLS, R. J. & LEATHERMAN, S. P. (eds) 1995. Potential impacts of accelerated sea-level rise on the developing world. *Journal of Coastal Research*, Special Issue, **14**.

——, DENNIS, K. C., VOLONTÉ, C. R. & LEATHERMAN, S. P. 1993. Methods and problems in assessing the impacts of accelerated sea-level rise. *In*: BRAS, R. (ed.) *The World At Risk: Natural Hazards and Climate Change*. AIP Conference Proceedings, American Institute of Physics, New York, 193–205.

——, LEATHERMAN, S. P., DENNIS, K. C. & VOLONTÉ, C. R. 1995. Impacts and responses to sea-level rise: qualitative and quantitative assessments. *Journal of Coastal Research*, Special Issue, **14**, 26–43.

O'CALLAHAN, J. (ed.) 1994. *Global Climate Change and the Rising Challenge of the Sea*. Proceedings of the IPCC Workshop held at Margarita Island, Venezuela, March 1992. National Oceanic and Atmospheric Administration, Silver Spring, MD.

SCOR WORKING GROUP 89. 1991. The response of beaches to sea-level changes: a review of predictive models. *Journal of Coastal Research*, **7**, 895–921.

STIVE, M. J. F. & DEVRIEND, H. J. 1995. Modelling shoreface profile evolution. *Marine Geology*, **126**, 235–248.

——, ROELVINK, J. A. & DEVRIEND, H. J. 1990. Large scale coastal evolution concept. *In*: *Proceedings 22nd Coastal Engineering Conference*. ASCE, New York, 1962–1974.

——, DEVRIEND, H. J., NICHOLLS, R. J. & CAPOBIANCO, M. 1992. Shore nourishment and the active zone: a time scale dependent view. *In*: *Proceedings 23rd Coastal Engineering Conference*. ASCE, New York, 2464–2473.

TITUS, J. G., LEATHERMAN, S. P., EVERTS, C. H., KREIBEL, D. L. & DEAN, R. G. 1985. *Potential impacts of sea-level rise on the beach at Ocean City, Maryland.* US Environmental Protection Agency, EPA 230-10-85-013, Washington, DC.

US ARMY CORPS OF ENGINEERS 1984. *Shore Protection Manual*, Vol. 1 (4th edition). Coastal Engineering Research Center, Waterways Experiment Station, Vicksburg, MS.

VELLINGA, P. & LEATHERMAN, S. P. 1989. Sea-level rise, consequences and policies. *Climate Change*, **15**, 175–189.

VOLONTÉ, C. R. & NICHOLLS, R. J. 1995. Sea-level rise and Uruguay: potential impacts and responses. *Journal of Coastal Research*, Special Issue, **14**, 262–284.

WARRICK, R. A. & OERLEMANS, H. 1990. Sea-level rise. *In*: HOUGHTON, J. T., JENKINS, G. J. & EPHRAMUS, J. J. (eds) *Climate Change: The IPCC Scientific Assessment*. Cambridge University Press, Cambridge, 257–281.

WIGLEY, T. M. L. & RAPER, S. C. B. 1992. Implications for climate and sea level of revised IPCC emission scenarios. *Nature*, **357**, 293–300.

WCC 1994. *Preparing to meet the coastal challenges of the 21st century*. Report of the World Coast Conference, Noordwijk, 1–5 November 1993. The Hague, Ministry of Transport, Public Works and Water Management.

SECTION 2

VOLCANIC AND SEISMIC GEOHAZARDS

Volcanic hazards and their mitigation

W. J. McGuire

Benfield Greig Hazard Research Centre, Department of Geological Sciences, University College London, Gower Street, London WC1E 6BT, UK

Abstract. Over the past century, a range of volcanic hazards, particularly pyroclastic flows and debris flows, have claimed over 60 000 lives, and between 1980 and 1990 alone, have detrimentally affected the day-to-day lives of over 600 000 people. Although a battery of mitigation measures are now available to reduce the impact of volcanic hazards, rapidly increasing populations near active volcanoes in the developing world will ensure increasing vulnerability to hazardous volcanogenic phenomena into the next millenium, and new initiatives are required to prevent a contemporaneous rise in the numbers of volcanic disasters. Alongside increased monitoring, partly through improved satellite observations, a programme of public education and training is needed to ensure that both civil authorities and local populations are able to respond rapidly and appropriately to a developing volcanic emergency. This in turn requires greater focus on improved communication between scientists and the responsible civil authorities, and the formulation of workable contingency plans based upon reliable and informed scientific opinion.

Introduction

Unlike other geohazards, such as floods or earthquakes, volcanoes are unique in generating a range of destructive phenomena of widely differing characteristics (Table 1 and Fig. 1(a)). These have the potential not only to cause injury and loss of life on a major scale, but also to seriously disrupt the local or regional economy for several years, and to make large tracts of land unusable for centuries. Since AD 1700 volcanic activity has been responsible for the deaths of over a quarter of a million people (IAVCEI IDNDR Task Group 1990), and a series of volcanic disasters since 1980 have alone resulted in nearly 30 000 deaths, although (as discussed later) over two-thirds of this number lost their lives needlessly in a single event.

Between 1980 and 1990, over 600 000 people were also detrimentally affected by volcanic activity, through damage to property, inundation of agricultural land, evacuation and resettlement, and related effects (UNESCO 1993). Table 2 lists some of the major volcanic eruptions involving loss of life since the eighteenth century, draws attention to the range of hazards responsible for volcano-related deaths, and highlights the particularly destructive potential of pyroclastic flows, debris flows, and volcanogenic tsunami. Interestingly, the largest death toll over this period results from famine and disease; a secondary consequence of a major volcanic eruption which is not simply mitigated. It is estimated (Simkin & Siebert 1994) that there are currently around 550 active volcanoes, with over 10% of the world's population living sufficiently close to be at risk from eruptions (Peterson 1986). With many of these volcanoes located in developing countries around the Pacific rim and in Southeast Asia experiencing rapid population growth, the potential exists for more eruptions to have disastrous consequences, unless mitigation measures can be made more effective.

This recognition has led, particularly since the Nevado del Ruiz (Colombia) catastrophe in 1985 (Voight 1988, 1990), to changes in emphasis in volcanic hazard mitigation. In particular, it is now the consensus view that being able to communicate effectively with civil authorities and vulnerable populations, and educate them about the threat posed by their local volcano, is at least as important as understanding how that volcano works and forecasting when it will next erupt. Failure to accomplish this will inevitably – as at Nevado del Ruiz – result in eventual and needless loss of life.

Table 1. *Principal volcanic hazards and associated effects*

Hazards	*Effects*
Lava flows	Famine
Tephra	Water pollution
Pyroclastic flows and surges	Disease
Lahars and floods	Disruption of the social and economic infrastructure
Landslides	
Directed blasts and atmospheric shock waves	
Volcanic gases	
Tsunamis	
Climate modification	

McGuire, W. J. 1998. Volcanic hazards and their mitigation. *In*: Maund, J. G. & Eddleston, M. (eds) *Geohazards in Engineering Geology*. Geological Society, London, Engineering Geology Special Publications, **15**, 79–95.

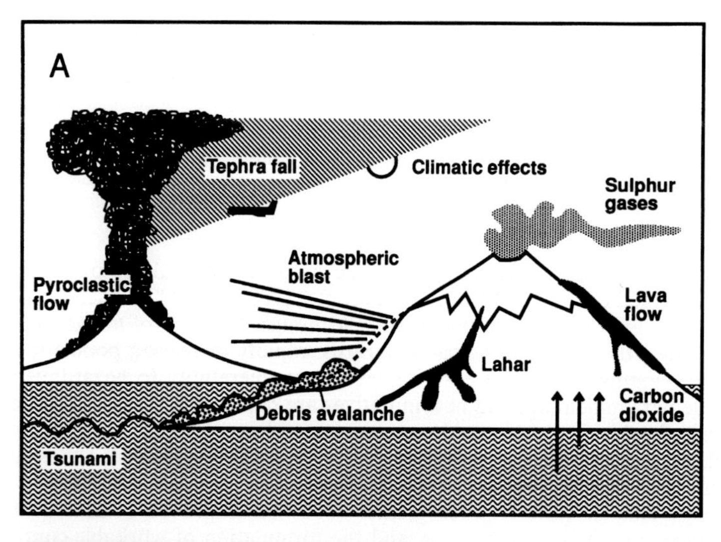

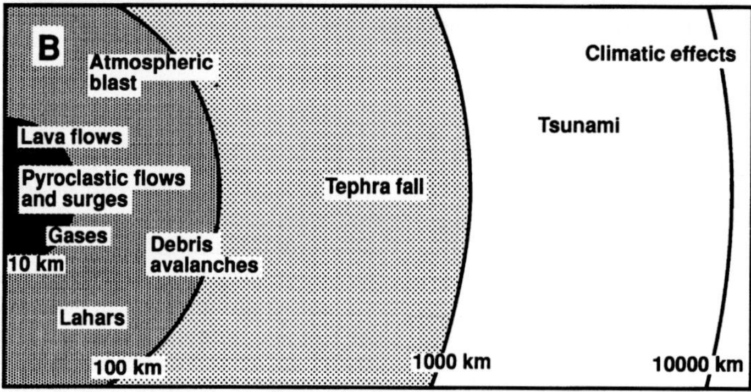

Fig. 1. Volcanoes generate a broad spectrum of destructive phenomena of widely different characteristics (**a**), the impact of which may be confined to the local area or extend across the entire planet (**b**).

Table 2. *Selected volcanic disasters since the eighteenth century*

Volcano (country)	*Year*	*Fatalities*	*Cause*
Laki (Iceland)	1783	10 500	Famine
Unzen (Japan)	1792	15 188	Tsunami
Tambora (Indonesia)	1815	92 000	Mainly famine
Krakatau (Indonesia)	1883	36 417	Tsunami
Mt Pelée (Martinique)	1902	29 000	Pyroclastic flows
Kelut (Indonesia)	1919	5 110	Debris flows
Lamington (PNG)	1951	2 940	Pyroclastic flows
El Chichon (Mexico)	1982	1 700	Pyroclastic flows
Nevado del Ruiz (Columbia)	1985	25 000	Debris flows
Lake Nyos (Cameroon)	1986	1 746	Volcanic gases
Pinatubo (Philippines)	1991	500	Various

Such issues are emphasized in the IAVCEI (International Association of Volcanology and Chemistry of the Earth's Interior) report, 'Reducing Volcanic Disasters in the 1990s' (IAVCEI IDNDR Task Group 1990), and their importance is further acknowledged in the UN Decade Volcano and European Union Laboratory Volcano initiatives (Table 2). As part of the UN-designated International Decade for Natural Disaster

Table 3. *Volcanoes selected for special study during the UN IDNDR*

Decade volcanoes (UN sponsored)
Colima (Mexico)
Galeras (Columbia)
Mauna Loa (USA)
Merapi (Indonesia)
Mount Rainier (USA)
Nyirogongo (Zaire)
Sakurajima (Japan)
Santa Maria (Guatemala)
Ta'al (Philippines)
Ulawan (Papua New Guinea)
Unzen (Japan)
Vesuvius (Italy)
Laboratory volcanoes (EU sponsored)
Etna (Sicily, Italy)
Furnas (Azores, Portugal)
Krafla (Iceland)
Piton de la Fournaise (Réunion Island, Indian Ocean, France)
Teide (Tenerife, Spain)
Santorini (Greece)

Reduction (IDNDR), both initiatives are designed to support the testing of new monitoring methods and mitigation measures at a range of different types of potentially destructive volcano. Lessons learned during the Decade and Laboratory Volcano projects can then be applied to the remainder of the world's active volcanoes, with the goal of ensuring that the detrimental impact of volcanic eruptions in the future is minimized.

A review of volcanic hazards and associated effects

The varying scales of volcanic eruption ensure that the hazardous phenomena generated have the potential to affect the local area immediately surrounding the volcano, the geographical region within which the volcano is located, and, in the largest events, even the entire planet (Fig. 1(b)). In most cases, hazards are associated with explosive rather than effusive eruptions and are characterized by relatively high VEI (Volcano Explosivity Index) numbers. The VEI (Newhall & Self 1982) is a logarithmic scale from 0 to 8 which is based primarily on the amount of mass ejected during an eruption, and the size of the eruption column. Generally speaking, hazardous phenomena generated by VEI 1–3 events are localized, while the effects of VEI 4–5 eruptions have the potential to disrupt everyday life on a regional scale. VEI 6 and above events are capable of affecting the population of the entire planet through their impact on the global climate. The intensively studied Mount St Helens eruption of May 1980 (Lipman & Mullineaux 1981) had a VEI of 5 while the 1991 eruption of Pinatubo in the Philippines (Wolfe 1992), with a VEI of 6 was one of the largest eruptions this century. VEI 7 and 8 events are rare, with return times of thousands or tens of thousands of years. The last VEI 7 was the 1815 eruption of Tambora (Indonesia); an event which led to huge loss of life through post-eruption famine and disease (see Table 2), and caused dramatic, short-term climate changes in North America and Europe (Harington 1992). There is no record of a VEI 8 eruption during the Holocene, and the last appears to have been the cataclysmic 'super-eruption' of Toba (Sumatra, Indonesia) around 74 ka BP (Rampino & Self 1992). Estimates of the temperature falls resulting from the ejection into the stratosphere of 10^{15} g of fine ash and sulphur gases are of the order of 12–15°C for some parts of the globe, and the eruption is proposed by Rampino and Self as the final trigger which accelerated an already cooling planet into full ice-age.

Predominantly effusive eruptions have low (normally 0, 1 or 2) VEI numbers, and their primary associated hazard involves the relatively quiet extrusion of lava flows (e.g. Kilburn & Luongo 1993) (Fig. 2), often accompanied by mild to moderate explosive activity. In contrast, explosive eruptions are capable of generating an entire spectrum of destructive phenomena, sometimes simultaneously, making it particularly difficult to mitigate their effects. During a major explosive event, heavy ash-fall from the eruption column (e.g. Blong 1981) (Fig. 3) may spread over tens of thousands of square kilometres, disrupting travel and communications, causing structural damage to buildings, damaging crops and contaminating water supplies, and affecting the health of both livestock and humans. In the immediate vicinity of the volcano, partial collapse of the eruption column typically results in the generation of pyroclastic flows (e.g. Davies *et al.* 1978; Fisher & Heiken 1982) (Fig. 4); gravity-driven mixtures of high-temperature volcanic gases, pumice, hot ash, and coarser debris, which are capable of travelling at velocities in excess of 100 km h^{-1}, causing major loss of life and the total destruction of settlements in their path. Where pyroclastic flows come into contact with a body of water, they are often transformed rapidly into debris flows which have a similarly devastating destructive capability. As evidenced by the Pinatubo experience (Pierson 1992), the precipitation-related reworking of pyroclastic deposits may also generate debris flows (Fig. 5) capable of disrupting life in the surrounding area for years after the end of the eruption.

Many of the above hazardous phenomena may be generated near-simultaneously when a volcano experiences structural failure and the consequent instantaneous depressurization of a subsurface magma body. The climactic eruption of Mount St Helens on 18 May 1980 provides the best documented account of such an

Fig. 2. Aa lavas formed during the 1983 eruption at Mount Etna (Sicily). Lava flows characterize effusive rather than explosive eruptions, and cause problems through the destruction of property and inundation of cultivated land, rather than due to injury and loss of life.

event (Lipman & Mullineaux 1981). Here a magnitude 5 earthquake triggered the detachment of the northern flank of the volcano which had been outdomed and destabilized due to the intrusion of fresh magma. The removal of the north flank in the form of a gigantic ($2.5\,km^3$) landslide (Voight *et al.* 1981; Glicken 1991) initiated a major explosive eruption characterized by heavy and widespread ash fallout and the generation of extensive pyroclastic and debris flows (Janda *et al.* 1981). An additional hazard was the lateral blast triggered by the unroofing of the magma body, which devastated a zone extending as far as 30 km from the volcano, and completely flattening fully grown forests over 10 km distant (McClelland *et al.* 1989).

Collapsing volcanoes pose a further serious threat if they are located near a body of water, and the generation of tsunami due to masses of volcanic debris entering a lake or ocean are well documented at a number of volcanoes, including Unzen (Japan) where, in 1792, over 14 000 lives were lost in this manner (Tsuji & Hino 1993). Such collapses can reach staggering proportions, with prodigious landslides from the Hawaiian volcanoes estimated at involving up to $5000\,km^3$ (Moore *et al.* 1994) and capable of generating Pacific-wide tsunami (Young & Bryant 1992). Giant waves have also been generated during explosive eruptions at island volcanoes, and such an event at Krakatoa in 1883 took the lives of 36 000 people on the coasts of neighbouring islands (Simkin & Fiske 1983).

Volcanic eruptions, particularly of the explosive type, are preceded and accompanied by seismic activity which may be sufficiently strong to cause structural damage to buildings. They are also the source of noxious gases, particularly SO_2 and CO_2 which, in the latter case, although not eruption related, resulted in over 1700 deaths around the crater lakes of Monoun (1984) and Nyos (1986) in the Cameroon (e.g. Kling *et al.* 1987; Baxter & Kapila 1989). As demonstrated at Poás

Fig. 3. Ash fall associated with predominantly explosive eruptions can cause widespread disruption to travel, communications, and power supplies, in additon to causing health problems and damage to crops and livestock. At Montserrat (Caribbean), ash fall associated with dome collapse at the Soufriére Hills volcano has necesitated evacuation of Plymouth (illustrated), the main town on the island.

Fig. 4. Pyroclastic flows, such as those generated at the Soufriére Hills volcano, constitute one of the most destructive and life-threatening of all eruptive phenomena. The only effective way of mitigating their impact is through operation of a judicious land management policy which avoids threatened areas, combined with timely evacuation of vulnerable people and livestock.

volcano (Costa Rica), SO_2 can be a persistent problem at actively degassing open-vent volcanoes, causing damage to crops, and respiratory problems in the local population. In the stratosphere, it is also the sulphur aerosols that have a cooling effect on global climate by absorbing incoming solar radiation which is thereby prevented from reaching the Earth's surface.

Mitigating the effects of volcanogenic gravity flows

Lava flows

Lava flows represent the principal hazard at basaltic volcanoes, and are also commonly generated at andesitic types (Kilburn & Luongo 1993). Due to their low velocities, ranging normally from a few metres to several kilometres per hour, lava flows are rarely life-threatening. Nevertheless, yield strengths are sufficient to give all but the most fluid lavas the ability to demolish buildings, while volumes may be sufficient to inundate tens to hundreds of kilometres of usable land. Primarily on the basis of surface textures, lava flows are broadly categorized into aa or pahoehoe types. Aa lavas typically form flow-fields with clinkery surfaces which are fed by tube or channel systems, and which tend to move forward on a relatively narrow front(s). Predominantly pahoehoe flows, in contrast, have generally lower viscosities than their aa counterparts, and consequently move more rapidly and on a wider front.

Because aa flows tend to move forward on one or more relatively narrow fronts, and are fed by a discrete tube or channel system, opportunities for diversion and blockage are greater than for pahoehoe flows. A number of artificial means have been tried to intervene in the development of an aa lava flow field (Chester *et al.* 1985), with varying degrees of success (Table 5, Fig. 6),

Fig. 5. Volcanogenic debris flows have been responsible for over 20 000 deaths during this century alone, and can result in serious disruption for several years after an eruption through the remobilization of tephra and pyroclastic flow deposits. Around Pinatubo (illustrated), for example, debris flows continue to clog up the lower reaches of river valleys draining the volcano, causing the flooding of neighbouring settlements and cultivated land.

including (i) aerial bombing (Lockwood & Torgerson 1980); (ii) cooling the flow front with water (Williams & Moore 1973); (iii) disrupting the levées of solidified lava which border the main feeder channel to cut off the supply of fresh lava to the flow front (e.g. Abersten 1984; Barberi *et al.* 1993); (iv) blockage of the main channel or tube, again to cut off the lava supply (Barberi *et al.* 1993); and (v) constructing diversion or holding barriers to protect vulnerable properties (Macdonald 1962; Colombrita 1984; Barberi *et al.* 1993)

Mount Etna (Sicily) has been a particular focus for experimenting with different techniques to modify aa flow-field development, both during the 1983 eruption (Abersten 1984; Colombrita 1984) and again in the major effusive eruption of 1991–93 (Barberi *et al.* 1993), which threatened inhabited parts of the volcano. During the earlier event, explosives were used to blast through the main feeder channel levée, reducing the flow of lava to the front by diverting part of it into an artificially constructed channel. The operation was only partially successful, but the eruption ceased before further intervention could be undertaken.

Between December 1991 and April 1993, one of the largest volume eruptions on Etna for 300 years threatened the town of Zafferana on the east flank of the volcano. A 21 m high, 230 m long, artificially constructed earth barrier was successful in holding back the flow front for several months, but was overtopped in the spring of 1992, allowing lava to reach the outskirts of the town where several buildings were destroyed. Shortly afterwards, Italian scientists and explosive experts successfully blocked the main feeder tube near its source, cutting off the supply to the flow front and causing new magma to travel over the surface of the

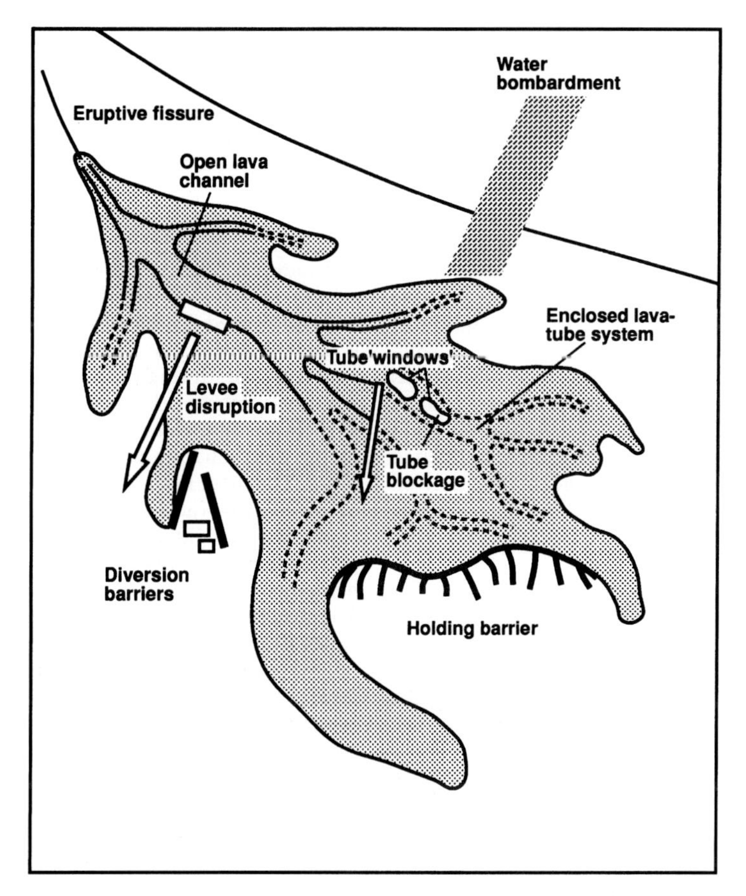

Fig. 6. A number of measures have been tried to dam or divert lava flows, particularly during the 1983 and 1991–93 eruptions of Mount Etna, with varying degrees of success. In the case of tube- or channel-fed aa flows, some success has been achieved through the construction of holding dams, the explosive breaching of marginal levées, and the blockage of feeder tubes.

flow-field, causing thickening rather than lengthening (Barberi *et al.* 1993; GNV 1993). Although the intervention undoubtedly worked, and can be regarded as a milestone in terms of lava hazard mitigation, the flow length was already near the maximum for Etnean aa flows, and doubts have been expressed about whether Zafferana would have been further threatened. Furthermore, the diversion was made easier by the fact that the flow-field was confined within the walls of a cliff-bounded depression (the Valle del Bove), so ensuring that no other settlements were threatened as a result of the intervention.

As lavas flows are gravity driven, their paths tend to be constrained by the local topography, and can therefore be predicted. Consequently, effective hazard zonation maps can be relatively easily constructed, highlighting those areas most at risk from flow-related damage, and those which are topographically shielded. As has been accomplished for Etna (Guest & Murray 1979; Chester *et al.* 1985), this method permits the definition of lava flow 'catchment areas' for individual settlements and provides some estimate of the relative risk to each. During a continuing eruption, the development of the flow-field may also be modelled by applying, to a digital terrain model (DTM) of the volcano, an appropriate algorithm describing the behaviour of the flow. This technique was also utilized at Etna during the 1991–93 eruption to forecast the growth of the flow-field which posed a threat to the town of Zafferana (Barberi *et al.* 1993), with a resulting good correlation between predicted and observed flow boundaries. Despite a continuing lack of concensus with regard to which factors are most important in determining the final length of a lava flow, recent advances have been made in this area. In particular, Kilburn (1996) has demonstrated that because of solidification during advance, lava flows naturally structure themselves to evolve within a restricted number of regimes, and that for the most common regime (aa flows) the controlling factors combine so that maximum flow lengths can be estimated directly from the mean slope in the path of the flow. At some volcanoes, it may also prove possible to predict the final volume of a flow-field soon after the start of the eruption. This has been accomplished for dome-related lavas at Unzen volcano (Japan) by Chen *et al.* (1993), who identified a correlation between flow volume and the neodymium isotopic composition of the lavas; the latter reflecting the degree of assimilation of continental crust into mantle-derived basalt magma during storage. Identification of this relationship has important hazard mitigation implications, as it may constitute a useful predictive tool for eruption duration (volume/effusion rate) at other subduction zone-related volcanoes where mixing of crustal- and mantle-derived magmas is common.

Pyroclastic flows and surges

Pyroclastic flows and surges (lower density versions of pyroclastic flows) are typically generated at andesitic to rhyolitic volcanoes (e.g. Mont Pelée, Martinique; Mount St Helens, USA; Unzen, Japan; Pinatubo, the Philippines; Soufriére Hills, Montserrat) by means of a range of mechanisms. These include lava dome collapse or explosion, and the gravitational collapse of a large eruption column. Pyroclastic flows, such as those generated at Unzen since 1991 (Nakada 1992) and, more recently, by dome collapse at the Soufriére Hills volcano on Montserrat (Fig. 4) (Cole *et al.* 1996), are also known as 'block and ash flows', and tend to be smaller scale than the voluminous, pumice-rich 'ignimbrites' associated with column collapse, which characterized the Pinatubo eruption.

A combination of high velocities (tens to hundreds of metres per second), and high temperatures (300–800°C) ensures that pyroclastic flows and surges are amongst the most life-threatening of all volcanogenic hazards. They are also capable of near-total structural damage (the degree being dependent on the density and temperature of the flow) and the inundation of large areas of usable land. Furthermore, substantial pyroclastic flow deposits may source explosively-generated secondary flows, and may be remobilized to form equally destructive debris flows.

Due to their destructive potential, the only effective way of mitigating the effects of pyroclastic flows (Table 5) is through pre-evacuation of the threatened area based upon a combination of hazard mapping and contemporary monitoring. Except in cases of large-scale column collapse, pyroclastic flows are commonly confined to river valleys and topographic lows, although the lower density surge component may overspill the valley sides. The threat to local inhabitants may thus be significantly reduced by adoption of a judicious building policy. In cases of surges and the lower density margins of pyroclastic flows, survivors' accounts from activity at Unzen during 1991 supports the notion that buildings and motor vehicles may provide some protection from low density pyroclastic surges.

Debris flows and floods

Volcanogenic debris flows (sometimes referred to as *lahars*) and floods are responsible for nearly all the volcano-related deaths over the last 15 years, and may be generated during the course of an eruption (e.g. Janda *et al.* 1981) or, during the post-eruption period (e.g. Pierson 1992), by reworking of pyroclastic (either tephra or pyroclastic flow) debris. Both scenarios may have a major destructive impact on the surrounding communities. The most recent and tragic example of the former occurred at Nevado del Ruiz (Columbia) in 1985 (Herd *et al.* 1986; Voight 1988, 1990), when a relatively minor pyroclastic flow/surge eruption melted part of the permanent snow and ice field that caps the volcano. The resulting flash flood rapidly travelled down a river valley draining the upper slopes, stripping the regolith off the valley sides and evolving quickly into a debris flow. At the valley entrance, over 40 km from the summit, the debris flow spread out to engulf the town of Armero with the loss of over 20 000 lives.

In contrast to the Columbian situation, debris flow formation in the Pinatubo region constitutes a much longer-term problem which continues at the time of writing (mid-1998), five years after the eruption. Here, thick (in excess of 100 m) pyroclastic flow deposits covering the upper flanks of the volcano provide the source for the debris flows which are reworked downslope by heavy rains, causing silting up and over-spilling of the lower stretches of river valleys draining the volcano.

Despite their potentially devastating impact, mitigating the effects of volcanogenic debris flows is made relatively simple by the fact that they are strongly topographically constrained (even more so than lava and pyroclastic flows) and commonly confined to river courses. Mapping the distribution of older deposits can provide invaluable information on the likely paths of debris flows generated during a future eruption, allowing the preparation of hazard zonation maps and the definition of high-risk areas. As illustrated by the Nevado del Ruiz situation, however, the availability of such information is no guarantee that it will be heeded by the civil authorities.

At volcanoes where debris flows are recognized as a problem, a range of measures are available to reduce their impact (Table 5, Fig. 7). In the upper reaches of valleys draining the volcano, seismometers and trip-wires can provide sufficient warning to permit evacuation of settlements downslope, while a combination of baffles and sediment dams can reduce the mass content

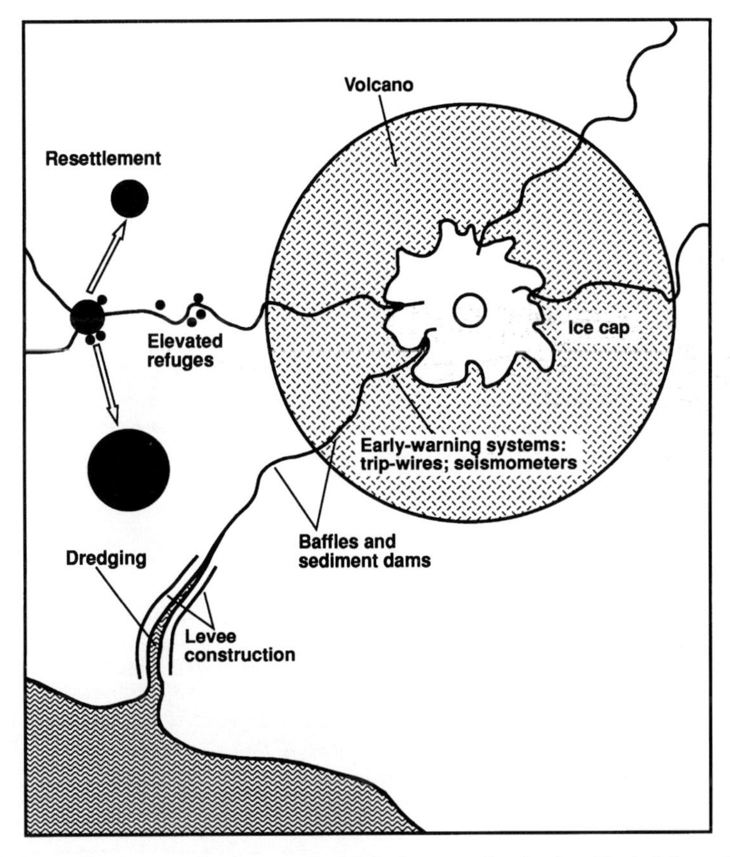

Fig. 7. The life-threatening impact of volcanogenic debris flows can be simply mitigated through a sensible building policy together with development of an early warning system and the construction of elevated refuges. Flooding of adjacent land due to overspill from clogged rivers can be alleviated to some extent by regular dredging and the construction and maintenance of levées.

of the flows thereby reducing their destructive potential. Artificially constructed earth mounds may also provide elevated refuges for the inhabitants of vulnerable settlements. Problems with flooding caused by the silting-up of the lower reaches of river valleys may be resolved by a combination of continuous dredging and the construction of levées (Fig. 8), although the cost of having to maintain such activities for several years, as at Pinatubo, can be prohibitively expensive for developing world countries.

Debris avalanches

Volcanogenic debris avalanches (disaggregated volcanogenic landslides) (Siebert 1984, 1992; McGuire 1995) may be associated with an eruption or may occur during a period of inactivity (McGuire 1996). The collapse of the north flank of Mount St Helens on 18 May 1980 provides the most closely studied example of the former, while the 'cold' collapse of part of the flank of Unzen volcano in 1792 resulted in the highest known death toll due to the lateral collapse of a volcano (Tsuji & Hino 1993). Both the volumes and the runout distances of the Mount St Helens and Unzen collapses were small when compared to those of older volcanogenic debris avalanches (Table 4), highlighting the potentially catastrophic impact of a similar-scale event occurring in the future at a densely populated volcano.

The destabilization of part of a volcanic edifice, which invariably precedes collapse and debris avalanche formation, may occur progressively over thousands of years or rapidly over a few weeks to months (McGuire & Saunders 1993; McGuire 1996). The former situation is more common at basaltic volcanoes, and is often associated with the repeated intrusion of magma along persistent paths (rift zones), which eventually fail when the accumulating internal stresses can no longer be

Fig. 8. Levée construction and continuous dredging of the lower reaches of the rivers draining Pinatubo, during the years following the 1991 eruption, failed to prevent serious flooding of adjacent towns and the inundation of large tracts of agricultural land.

accommodated (e.g. Swanson *et al.* 1976; McGuire *et al.* 1990). Rapid destabilization, in contrast, is typical of more silicious volcanoes, where intrusive plug-shaped masses of viscous magma (cryptodomes) can lead to flank-tumescence rates in excess of several metres a day, causing eventual collapse. Intrusion of a cryptodome generated the 1980 collapse of Mount St Helens, and may also be the reason for collapse of part of the old crater wall surrounding the dome complex on 26 December 1997 at the Soufriére Hills volcano in Montserrat.

Due to the velocities of debris avalanches, which may be in excess of $100\,\mathrm{m\,s^{-1}}$ ($360\,\mathrm{km\,h^{-1}}$), pre-evacuation is the only realistic means of minimizing loss of life (Table 5). This requires, however, (i) prior recognition that a landslide is likely in the short term, (ii) a forecast of the timing of the collapse, and (iii) delineation of the threatened area. Identification of landslide-prone slopes may be accomplished initially by detailed geological and structural mapping, followed by deformation monitoring of any identified mobile zones. As demonstrated at Mount St Helens (McClelland *et al.* 1989), however, forecasting the timing of a volcanogenic landslide has proved problematical. Although the rate of deformation might be expected to accelerate prior to collapse, this does not appear to have happened during destabilization of the north flank during the spring of 1980. Instead, a linear deformation trend was observed, using the electronic distance measurement (EDM) method, until immediately before the north flank was seismically detached from the rest of the volcano. This illustrates that at dynamic structures such as active volcanoes, where seismicity is common, 'external' factors may short-circuit the deformation process and lead to collapse earlier than would otherwise be expected. This argues strongly for the earlier – rather than later – evacuation of an area threatened by a newly mobilized slope at an active volcano.

Table 4. *Volumes and runout distances of selected subaerial volcanic debris avalanches (from McGuire 1996)*

Volcano	*Deposit*	*Volume (km^3)*	*Runout (km)*
Nevado di Colima		22–33	120
Socompa		17	35
Volcán de Colima		6–12	43
Shasta		26	50
Popocatapetl		28	33
Chimborazo	Riobamba	8.1	35
Mawenzi		7.1	60
Akagi	Nashikizawa	4	19
Galunggung		2.9	25
Mount St. Helens (1980)		2.5	24
Fuji	Gotenba	1.8	24
Shiveluch (1964)		1.5	12
Bandai-san (1888)		1.5	11
Egmont	Pungarehu	0.35	31
Unzen (1792)		0.34	6.5
Asakusa	Migisawa	0.04	6.5

Delineating the area likely to be affected by a debris avalanche may be relatively simply accomplished. The runout distance can be broadly determined through application of a relationship which links distance travelled to collapse height and, to lesser extent, volume of the collapsing mass (Ui 1983). As for lava flows, a more detailed forecast of the spatial distribution of the debris avalanche may be obtained by utilizing an appropriate algorithm for the avalanche mechanism superimposed on a DTM of the theatened area. This suffers, to some extent, from a lack of consensus about exactly how debris avalanches travel, particularly those long-runout landslides which travel farther than expected.

Where the collapse is associated with the intrusion of fresh magma, effective hazard mitigation is made considerably more difficult by the fact that the event is likely to trigger an eruption through depressurization of underlying magma. At Mount St Helens, the resulting atmospheric shock wave, pyroclastic flows, debris flows, and extensive ash fall affected a considerably greater area than the debris avalanche, and the impact of such phenomena must therefore be included in any comprehensive debris avalanche hazard mitigation plan.

Mitigating the problem of tephra and ballistic projectiles

Although less dramatic than other volcanic hazards, tephra – the all-encompassing term for volcanic ejecta – can be one of the most disruptive of all (Table 5), particularly during sub-Plinian and Plinian explosive eruptions where ejected volumes may exceed $1 \times 10^8\ m^3$. Typically, ash fall represents the most extensive and disruptive form of tephra, and recent significant ash-producing eruptions have occurred at Mount St Helens (1980); El Chichon, Mexico (1982); Pinatubo (1991); Cerro Hudson, Chile (1991); Rabaul, Papua New Guinea (1994); and, to a lesser extent, at Soufriére Hills, Montserrat (1996).

Table 5. *Volcanic hazards: principal mitigatory measures*

Lava flows	Damming and/or diversion; flow-front water cooling
Pyroclastic flows and surges	Judicious siting of settlements; pre-evacuation
Debris flows and floods	Judicious siting of settlements; construction of elevated refuges; construction of sediment dams and baffles; dredging and levée construction; seismometer and trip-wire warning systems
Tephra	Evacuation of poorly constructed buildings; removal of accumulating tephra from roofs; availability of facc masks/protective headgear; appropriate medical care for respiratory problems; contingency plans for power, communication and; transport disruption; availability of uncontaminated water supplies; measures to minimize crop and livestock damage; warnings to air traffic
Landslides and debris avalanches	Identification of collape-prone areas; slope-stability monitoring; pre-evacuation
Directed blasts and shock waves	Pre-evacuation
Volcanic gases	Gas monitoring; resettlement if a persistent problem; pre-evacuation if episodic and predictable; public safety guidelines construction of elevated refuges where appropriate
Tsunami	Identification of unstable slopes adjacent to water bodies; slope-stability monitoring; pre-evacuation; establishment of tsunami warning network

In the developing world, the immediate problem from tephra arises from ash accumulation on the roofs of poorly constructed buildings. Particularly when wet, less than 10 cm of ash can cause collapse of corrugated iron roofs, and many of the deaths during the eruption of Pinatubo were caused in this way as families sheltered from the ash fall and the heavy rains brought by a contemporaneous typhoon. The problem can be simply alleviated by educating the local population to ensure that their roofs are kept swept clear of accumulating ash. This, however, also caused a problem at Pinatubo, with a number of deaths and many injuries – often to children – resulting from falls from and through roofs during sweeping.

Significant ash fall may also rapidly hinder travel, communications, and power supplies, respectively through making roads impassable, by downing telephone lines, and by damaging power stations and power lines. These problems can be alleviated by having appropriate contingency plans in place, in the form of available earth-moving equipment, repair crews on stand-by, and filters in place on sensitive power-generating equipment. Similarly, measures should be taken to ensure sufficient supplies of uncontaminated drinking water, for both humans and livestock, and to provide face masks to prevent ash inhalation which may cause both short-term (asthma) and long-term (silicosis) respiratory problems.

In the longer term, heavy ash fall has the potential to cause economic disaster through crop destruction and livestock deaths from eating contaminated vegetation. Following future large ash-producing eruptions in small developing countries, external food and agricultural aid may be essential, without which the situation could deteriorate rapidly into famine. After the Tambora eruption of 1815, for example, most of the resulting 90 000 deaths were due to famine caused by crop destruction, while following the 1783 Laki (Iceland) eruption (Thordarson & Self 1993), widespread starvation caused the deaths of a quarter of the population along with over 200 000 livestock.

Fig. 9. During directed explosions from the Soufriére Hills lava dome on the night of 17 September 1996, over 50% of the buildings in the evacuated settlement of Long Ground were damaged by ballistic projectiles. Impact craters produced during these events measured up to 5 m across and 1.5 m deep.

Over the last 15 years, over 80 civil aircraft have suffered serious damage due to flying into ash-laden eruption columns (Casadevall, 1994 and papers therein), including serious incidents over Alaska during 1989–90 and Indonesia in 1982 which almost resulted in crashes. Most recently, an Air Canada jet flew into the 14 km high ash column generated by the 17 September 1996 eruption of the Soufrière Hills volcano, causing engine and windscreen damage. Fears by the airline industry of the costs (which could exceed $1 billion for a full Boeing 747, including litigation costs) of a future crash have prompted a number of initiatives which should reduce the number of ash encounters in the future. These include a satellite-based warning system which detects and tracks eruption columns, standardized response protocols for aircrew who encounter ash, and the development of on-board ash-detecting radar for routes which cross regions with large numbers of active volcanoes (e.g. Kamchatka, Alaska and Southeast Asia).

In contrast to ash fall, ballistic projectiles are only a problem close to the eruption site (normally within a few kilometres). As ejected blocks may have fallen from heights of several kilometres and exceed a metre or more across, no human-made structures are strong enough to prevent penetration, and the only effective mitigation can be pre-evacuation of threatened areas. Defining the extent and distribution of these will depend upon the nature and scale of the likely activity, and will constitute a different problem at each volcano. The impact of ballistic projectiles on human-made structures was recently well demonstrated (Fig. 9) by the 17 September 1996 directed explosions at the Soufriére Hills volcano, which damaged over 50% of the buildings in the evacuated settlement of Long Ground, fortunately with no loss of life.

Minimizing the impact of other hazardous volcanic phenomena

In addition to gravity flows, tephra and ballistics, a number of other hazardous volcanic phenomena have potential for injury, loss of life, and destruction of property, most particularly noxious gases, volcanogenic earthquakes and tsunami. Noxious gases may be either a persistent or episodic problem. In the former case, open vent volcanoes may continuously discharge sulphur-rich gases, which are often carried in the same direction by prevailing winds causing both health problems and damage to crops and livestock downwind. Such a situation is currently causing problems at Poás volcano (Costa Rica), and at Soufriére Hills, where extensive areas of vegetation have been destroyed by constant degassing and by related acid rain. Linked to noxious gas production is the transport and precipitation of poisonous trace elements such as fluorine (e.g. Notcutt & Davies 1989). This element may become concentrated in animal feed leading to illness and death, and in local water supplies, causing damage to the teeth and bones of the local population.

Where sulphur-rich gases, and the elements they carry, are a persistent problem, their effects can only really be mitigated by resettlement. In the case of less reactive, but equally dangerous gases, such as carbon dioxide and carbon monoxide, however, measures can be taken to effectively mitigate their impact. Such gases are typically released through the soils on active and dormant volcanoes, and high concentrations have been detected at Vulcano (Aeolian Islands) (Baubron *et al.* 1990) and Sao Miguel (Azores) (Baxter, P. J., pers. comm.). Due to their densities, these gases concentrate in basements and ground floors of buildings where concentrations may be sufficiently high to cause death to young children and even sleeping adults. The problem can, however, be reduced by regular monitoring of CO_2 and CO levels in buildings, and by a public education policy which highlights the risks and provides appropriate guidelines. As previously discussed, periodic convective turnover may also release large quantities of CO_2 from crater lakes, as happened at Lakes Monoun and Nyos in the Cameroon. Completely removing such a threat to the inhabitants of lakeside villages can only be accomplished by permanent resettlement, although continuous monitoring of CO_2 levels, combined with controlled degassing of the lake waters (Walker *et al.* 1992) and the construction of elevated refuges may also prove effective in significantly reducing the risk of death and injury.

Volcanogenic earthquakes may accompany eruptive activity or may simply reflect the sub-surface movement of magma without eruption. In both cases, magnitudes are rarely as high as those associated with tectonic events, but the activity may persist for much longer (hours to days, or semi-continuously for weeks or more). Consequently, such earthquakes are capable of structural damage, particularly in developing world countries where buildings may be poorly constructed. In certain circumstances, increased levels of seismicity may be accompanied by persistent ground deformation, a combination which is very effective in destabilizing human-made structures. This occurred during the periods of unrest at the Campi Flegrei caldera (Bay of Naples) between 1982 and 1984 (e.g. Barberi *et al.* 1984) causing significant structural damage to buildings in the town of Pozzuoli near the epicentre of activity, and necessitating its evacuation. At some volcanoes, aseismic creep along active faults is also a problem. At Mount Etna (Sicily), for example (Lo Giudice & Rasà 1992; Rasà *et al.* 1996), magmatic stresses within the volcano are relieved by annual centimetric rates of movement along a number of active fault strands which cross the heavily populated eastern flank. These cause damage to roads, which require repeated repair, and also to buildings which, in some cases, have collapsed as a result (Fig. 10).

Mitigating the effects of volcanogenic tsunami represents a major challenge to volcanologists and to society in general, primarily because the largest such events have the potential to devastate a major part of the planet. The biggest volcanogenic tsunami probably represent the greatest and most destructive waves ever generated, barring those formed by major impact events in the oceans, and volcano collapse-related tsunami deposits in the Hawaiian islands (Moore & Moore 1984) and eastern Australia (Young & Bryant 1992) testify, respectively, to local (to the source) run-up heights of 375 m, and destructive effects even at distances of several thousand kilometres.

The local impact of relatively small-scale collapse-generated tsunami may be reduced by a combination

Fig. 10. Aseismic creep and discrete earthquakes at Mount Etna combine to damage structures built on or near active faults which cut the eastern flanks of the volcano, and which may play a role in dissipating magma-related stresses through their movement.

of measures (Table 5) including identification and monitoring of potentially unstable or creeping slopes adjacent to bodies of water, and pre-evacuation of threatened areas during eruptions or episodes of accelerated deformation. The wider threat may be reduced by implementation of a regional tsunami warning scheme. Where the predicted volume of the collapsing mass is of the order of tens to hundreds of cubic kilometres, the potential exists for the generation of giant tsunami with sufficient energy to cross oceans, and effective mitigation becomes considerably more problematical. Based upon run-up heights produced by past collapses of the Hawaiian volcanoes, and on preliminary modelling studies, collapse of the unstable western half of the Cumbre Vieja volcano (La Palma, Canary Islands), for example, may generate a destructive tsunami which would pose a serious threat to the eastern seaboard of the United States. Only an Atlantic tsunami warning system combined with effective evacuation plans for threatened areas could reduce injury and loss of life in such circumstances.

The roles of monitoring and hazard mapping

Monitoring has a critical role to play in reducing the impact of volcanic hazards, particularly through constraining the timing of a forthcoming eruption. While an entire battery of geophysical, geochemical and geodetic, techniques are now available (McGuire *et al.* 1995; Scarpa & Tilling 1996), the 'core' methods of seismic and deformation monitoring continue to provide the best means of tracing the movement and accumulation of magma during reactivation of a volcano, and of forecasting the onset of the succeeding eruption. A comprehensive seismic array, in combination with appropriate post-processing software, will permit the location of earthquake epicentres associated with magma-related rock fracturing with sufficient accuracy to determine the rate of ascent of magma from depth. At the same time, a range of geodetic monitoring methods, including EDM (electronic distance measurement), GPS (global positioning system), precise levelling, and ground-tilt measurements, may constrain the volume of freshly emplaced magma through determination of the degree and extent of pre-eruption tumescence. Seismic and ground deformation data can be supplemented by the results of microgravity, magnetic, electrical and gas-geochemistry surveys, to provide a more detailed picture of the pre-eruption disposition of magma within a reactivated volcano.

Although a comprehensive monitoring network may allow the timing of the onset of an eruption to be better constrained, it is less effective in forecasting the nature and extent of an impending eruption. These parameters may be estimated, however, from studies of the products of past eruptions. Geological mapping of, for example, older pyroclastic flow or debris flow deposits, may help to delineate the paths favoured by such phenomena and permit the construction of hazard-zonation maps which can be used by civil authorities both in the development of contingency plans for future eruptions, and in the formulation of longer-term land management policies. While hazard zonation maps should never be regarded as offering a precise guide to future activity, together with a comprehensive monitoring programme they do offer the best means of reducing injury and loss of life, primarily through ensuring timely evacuation of threatened areas.

The importance of education and communication

The destruction by volcanogenic debris flows of Armero and neighbouring settlements on the slopes of Nevado del Ruiz (Colombia) in 1985 constituted the worst volcanic disaster of the last 95 years. It also demonstrated graphically the importance of communicating the seriousness of a threat to the civil authorities, and via them to the local population. No matter how comprehesive the monitoring programme or how precise forecasts of the timing and nature of an impending eruption, this information is worthless unless, firstly, it can be imparted with sufficient urgency to the appropriate authorities and, secondly, that it is acted upon rapidly and effectively. In contrast to Nevado del Ruiz, the more recent eruptions at Pinatubo (1991) and Rabaul (1994), have demonstrated how good communications between scientists and the civil authorities can lead to a more responsive population during a volcanic crisis, and on both occasions this undoubtedly minimized the loss of life. At Pinatubo, scientists and authorities combined to ensure that tens of thousands of inhabitants living in the vicinity of the volcano were made aware of the destructive phenomena they might face during the impending eruption. This was accomplished through the distribution and viewing of a volcanic hazards video made for the IAVCEI (International Association of Volcanology and Chemistry of the Earth's Interior) by the late volcanologists/film-makers, Maurice and Katja Krafft, killed while filming on Unzen in 1991. There is no question that the film had a much greater impact than either written material or public lectures by scientists, and its message was crucial in persuading over a quarter of a million people to leave the threatened area, probably saving in excess of 10 000 lives. At Rabaul, an extended seismic and ground deformation crisis during the late 1980s was instrumental in encouraging scientists and authorities to develop and practise contingency plans which would ensure a more effective public response during a forthcoming eruption.

Although no activity occurred during the 1980s crisis, a major explosive eruption occurred in September 1994, following only 27 h of premonitory activity. Fortunately, the local population was now well educated and trained in terms of how to respond to such a crisis, and a rapid and well-organised self-evacuation resulted in minimal loss of life.

Even successful communications between scientists and authorities have to date been largely the result of a response to a developing crisis. If the numbers of volcanic disasters are to be reduced in the future, however, increasing attention must be paid to educating public officials and local populations prior to any eruptive crisis, so that both know what to expect during future activity. This is particularly important at volcanoes which have not erupted within living memory, and where, as a consequence, there is little or no understanding of the threat. The IAVCEI have gone some way towards this by issuing a second video which focuses on the mitigation of the different volcanic hazards and promotes the formulation of contingency plans at all active volcanoes. Alongside videos, PC-based eruption simulators are also planned which will utilize a computer-game format to present public officials with a range of eruption scenarios, thereby familiarizing them with both the hazards and the logistical problems they can expect to face during a future eruption.

The future

As we approach the year 2000 it looks as if the death toll for the twentieth century from volcanic hazards will total somewhere between 60 000 and 70 000. If this toll is not to rise dramatically during the coming few decades, alongside the rapidly increasing numbers of people living close to active volcanoes, then a number of measures must be adopted. Some of these are technically or scientifically based initiatives, which are likely to be led by volcanologists, while others rely upon increased co-operation between scientists, local authorities, and national and international organizations with an interest in mitigating the effects of natural hazards.

At present little more than 20% of the 550 or so active volcanoes are monitored to any extent. It is therefore of primary importance that greater numbers of the world's active volcanoes are monitored as we enter the next millenium, if only at the baseline level. This may be accomplished, to some degree, through increased utilization of satellite platforms capable of observing all the world's volcanoes using sensors which can detect pre-eruptive ground deformation (e.g. Massonet *et al.* 1995) and thermal anomalies (Rothery *et al.* 1995). On the ground, cheaper and simpler instrumentation must be made available for the monitoring of volcanoes in the developing world, while at the same time increasing the numbers of training courses aimed at familiarizing local scientists and technicians in their use and repair.

Volcano awareness programmes should be initiated at all active volcanoes, targeted at making the local population appreciate and understand the threat posed by their volcano. These should be accompanied by pre-emptive and responsive contingency plans formulated by the civil authorities, on the advice of informed scientific opinion, and designed to ensure that a future volcanic crisis does not become a disaster.

In conclusion, it should be appreciated that modern society has yet to face a volcanic eruption on the largest possible scale. The Toba 'super-eruption' occurred around 74 Ka BP, and, given a time-averaged VEI 8 return period of around 50 ka (Simkin & Siebert 1994), a similar sized event can be expected at any time. The consequences of another 'super-eruption' could be catastrophic on a global scale, with the potential for a global temperature fall of 3–5°C (Rampino & Self 1992). Furthermore, such an event may be beyond our abilities to effectively mitigate, particularly given the short-sighted and short-termist approaches of most national governments and international political organizations.

References

ABERSTEN, L. 1984. Diversion of a lava flow from its natural bed to an artificial channel with the aid of explosives: Etna 1983. *Bull. Volcanol.*, **47**, 1165–1177.

BARBERI, F., CORRADO, G., INNOCENTI, F. & LUONGO, G. 1984. Phlegrean Fields 1982–84: a brief chronicle of a volcano emergency in a densely populated area. *Bulletin Volcanologique*, **47**, 175–85.

——, CARAPEZZA, M. L., VALENZA, M. & VILLARI, L. 1993. The control of lava flow during 1991–1992 eruption of Mount Etna. *Journal of Volcanology and Geothermal Research*, **56**, 1–34.

BAUBRON, J. C., ALLARD, P., & TOUTAIN, J. P. 1990. Diffuse volcanic emissions of carbon dioxide from Vulcano Island, Italy. *Nature*, **344**, 51–53.

BAXTER, P. J. & KAPILA, M. 1989 Acute health impact of the gas release at Lake Nyos, Cameroon 1986. *Journal of Volcanology and Geothermal Research*, **39**, 265–275.

BLONG, R. J. 1981. Some effects of tephra falls on buildings. *In*: SELF, S. & SPARKS, R. S. J. (eds) *Tephra Studies*. Reidel, Boston, 405–420.

CASADEVALL, T. J. (ed.) 1994. *Volcanic Ash and Aviation Safety: Proceedings of the First International Symposium on Volcanic Ash and Aviation Safety*. US Geological Survey Bulletin 2047. US Government Printing Office, Washington, DC.

CHEN, C.-H., DEPAULO, D. J., NAKADA, S. & SHIEH, Y.-N. 1993. Relationship between eruption volume and neodymium isotopic composition at Unzen volcano. *Nature*, **362**, 831–834.

CHESTER, D. K., DUNCAN, A. M., GUEST, J. E. & KILBURN, C. R. J. 1985. *Mount Etna: The Anatomy of a Volcano*. Chapman & Hall, London.

COLE, P. D., SPARKS, R. S. J., ROBERTSON, R., STEVENS, N. F., YOUNG, S. R., NORTON, G. E. & HARFORD, C. 1996. Eruption mechanisms and physical parameters of the explosive eruption of the 17th September 1996 Soufriére Hills volcano, Montserrat. *In*: WADGE, G. (ed.) *The Soufriére Hills Eruption, Montserrat* (extended abstracts volume). The Geological Society of London, 39–40.

COLOMBRITA, R. 1984. Methodology for the construction of earth barriers to divert lava flows: the Mount Etna 1983 eruption. *Bulletin Volcanologique*, **47**, 1009–1038.

DAVIES, D. K., QUEARRY, M. W. & BONIS, S. B. 1978. Glowing avalanches from the 1974 eruption of volcano Fuego, Guatemala. *Geological Society of America Bulletin*, **89**, 369–384.

FISHER, R. V. & HEIKEN, G. 1982. Mount Pelée, Martinique: May 8 and 20 1902, pyroclastic flows and surges. *Journal of Volcanology Geothermal Research*, **12**, 339–371.

GLICKEN, H. 1991. *Rockslide-Debris Avalanche of May 18th 1980, Mount St Helens Volcano, Washington.* US Geological Survey Professional Paper 1488, Washington, DC.

GNV (Gruppo Nazionale di Vulcanologia) 1993. *Operazione Etna '92.* Video cassette with narration.

GUEST, J. E. & MURRAY, J. B. 1979. An analysis of hazard from Mount Etna volcano. *Journal of the Geological Society, London*, **136**, 347–354.

HARINGTON, C. R. (ed.) 1992. *The Year Without a Summer: World Climate in 1816.* Canadian Museum of Nature, Ottawa.

HERD, D. G. & THE COMITE DE ESTUDIOS VULCANOLOGICOS 1986. The 1985 Ruiz volcano disaster. *Eos*, **67**, 457–460.

IAVCEI IDNDR TASK GROUP 1990. Reducing volcanic disasters in the 1990s. *Bulletin of the Volcanological Society of Japan*, **9**, 215–236.

JANDA, R. J., SCOTT, K. M., NOLAN, R. M. & MARTINSON, H. A. 1981. Lahar movement, effects, and deposits. *In*: LIPMAN, P. W. & MULLINEAUX, D. R. (eds) *The 1980 Eruptions of Mount St Helens, Washington.* US Geological Survey Professional Paper 1250, 461–478.

KILBURN, C. R. J. 1996. Patterns and predictability in the emplacement of subaerial lava flows and flow fields. *In*: SCARPA, R. & TILLING, R. I. (eds) *Monitoring and Mitigation of Volcano Hazards.* Springer, Berlin.

—— & LUONGO, G. 1993. *Active Lavas.* UCL, London.

KLING, G. W., CLARK, M. A., COMPTON, H. R. *et al.* 1987. The 1986 Lake Nyos gas disaster in Cameroon, west Africa. *Science*, **236**, 169–75.

LIPMAN, P. W. & MULLINEAUX, D. (eds). 1981. *The 1980 Eruptions of Mount St Helens.* US Geological Survey Professional Paper 1250. Washington, DC.

LOCKWOOD, J. P. & TORGERSON, F. A. 1980. Diversion of lava flow by aerial bombing – lessons from Mauna Loa volcano, Hawaii. *Bulletin Volcanologique*, **43**, 727–741.

LO GIUDICE, E & RASÀ, R. 1992. Very shallow earthquakes and brittle deformation in active volcanic areas: the Etnean region as an example. *Tectonophysics*, **202**, 257–268.

MCCLELLAND, L., SIMKIN, T., SUMMERS, M., NIELSON, E. & STEIN, T. C. 1989. *Global Volcanism 1975–1985.* Prentice Hall, New Jersey.

MACDONALD, G. A. 1962. The 1959 and 1960 eruption of Kilauea volcano, Hawaii, and the construction of walls to restrict the spread of the lava flow. *Bulletin Volcanologique*, **2**, 249–294.

MCGUIRE, W. J. 1995. Volcanic landslides and related phenomena. *In*: *Landslides Hazards Mitigation With Particular Reference to Developing Countries.* The Royal Academy of Engineering, London, 83–95.

——1996. Volcano instability: a review of contemporary themes. *In*: MCGUIRE, W. J., JONES, A. P. & NEUBERG, J. (eds) *Volcano Instability on the Earth and Other Planets.* Geological Society, London, Special Publications, **110**, 1–23.

——, PULLEN, A. D. & SAUNDERS, S. J. 1990. Dyke-induced block movement at Mount Etna and potential slope failure. *Nature*, **343**, 357–359.

—— & SAUNDERS, S. J. 1993. Recent earth movements at active volcanoes: a review. *Quaternary Proceedings*, **3**, 33–46.

——, KILBURN, C. R. J. & MURRAY, J. B. 1995. *Monitoring Active Volcanoes: Strategies, Procedures, and Techniques.* UCL, London.

MASSONET, D., BRIOLE, P. & ARNAUD, A. 1995. Deflation of Mount Etna monitored by spaceborne radar interferometry. *Nature*, **375**, 567–570.

MOORE, J. G. & MOORE, G. W. 1984. Deposit from a giant wave on the island of Lanai, Hawaii. *Science*, **226**, 1312–1315.

MOORE, J. G., NORMARK, W. R. & HOLCOMB, R. T. 1994. Giant Hawaiian landslides. *Annual Reviews of Earth and Planetary Sciences*, **22**, 119–144.

NAKADA, S. 1992. Volcanic hazard at Unzen: (1) 1990–1992 eruption of Unzen volcano. *Landslide News*, **6**, 2–4.

NEWHALL, C. G. & SELF, S. 1982. The volcanic explosivity index (VEI): an estimate of explosive magnitude for historical volcanism. *Journal of Geophysical Research*, **87**, 1231–1238.

NOTCUTT, G. & DAVIES, F. B. M. 1989. Accumulation of volcanogenic fluoride by vegetation: Mount Etna, Sicily. *Journal of Volcanology and Geothermal Research*, **39**, 329–333.

PETERSON, D. W. 1986. Volcanoes: tectonic setting and impact on society. *In*: *Studies in Geophysics: Active Tectonics.* Panel on Active Tectonics, National Academy Press, Washington DC, 231–246.

PIERSON, T. C. 1992. Rainfall-triggered lahars at Mt Pinatubo, Philippines, following the June 1991 eruption. *Landslide News*, **6**, 6–9.

RAMPINO, M. R. & SELF, S. 1992. Volcanic winter and accelerated glaciation following the Toba super-eruption. *Nature*, **359**, 50–52.

RASÀ, A., AZZARO, R. & LEONARDI, O. 1996. Aseismic creep on faults and flank instability at Mount Etna volcano, Sicily. *In*: MCGUIRE, W. J., JONES, A. P. & NEUBERG, J. (eds) *Volcano Instability on the Earth and Other Planets.* Geological Society, London, Special Publications, **110**, 179–192.

ROTHERY, D A., OPPENHEIMER, C. & BONNEVILLE, A. 1995. Infrared thermal monitoring. *In*: MCGUIRE, W. J., KILBURN, C. R. J. & MURRAY, J. B. (eds) *Monitoring Active Volcanoes: Strategies, Procedures, and Techniques.* UCL, London.

SCARPA, R. & TILLING, R. I. 1996. *Monitoring and Mitigation of Volcano Hazards.* Springer, Berlin.

SIEBERT, L. 1984. Large volcanic debris avalanches: characteristics of source areas, deposits, and associated eruptions. *Journal of Volcanology and Geothermal Research*, **22**, 163–197.

——1992. Threats from debris avalanches. *Nature*, **356**, 658–659.

SIMKIN, T. & FISKE, R. S. 1983. *Krakatau 1883: The Volcanic Eruption and its Effects*. Smithsonian Institution, Washington, DC.

—— & SIEBERT, L. 1994. *Volcanoes of the World*. Smithsonian Institution, Washington, DC.

SWANSON, D. A., DUFFIELD, W. A. & FISKE, R. S. 1976. *Displacement of the South Flank of Kilauea Volcano: The Result of Forceful Intrusion of Magma into the Rift Zones*. US Geol. Surv. Prof. Paper 963.

THORDARSON, T. & SELF, S. 1993. The Laki (Skaftar Fires) and Grimsvotn eruptions in 1783–85. *Bulletin Volcanologique*, **55**, 233–263.

TSUJI, Y. & HINO, T. 1993. Inundation heights of the tsunami accompanied with the landslide of Mayuyama Hill on Shimbara peninsula in 1792 on the east coast of Ariake Bay. *In*: *Proceedings of Tsunami '93*. IUGG/OUC International Tsunami Symposium, Wakayama, Japan, 727–739.

UI, T. 1983. Volcanic dry avalanche deposits – identification and comparison with non-volcanic debris stream deposits. *Journal of Volcanology and Geothermal Research*, **18**, 135–150.

UNESCO 1993. *Disaster Reduction*. Environment and Development Briefs 5. Banson, London.

VOIGHT, B. 1988. Countdown to catastrophe. *Earth and Mineral Sciences* **57**, 17–30.

——1990. The 1985 Nevado del Ruiz volcano catastrophe – anatomy and retrospection. *Journal of Volcanology and Geothermal Research*, **42**, 151–188.

——, GLICKEN, H., JANDA, R. J. & DOUGLASS, P. M. 1981. Catastrophic rockslide avalanche of May 18. *In*: LIPMAN, P. W. & MULLINEAUX, D. R. (eds) *The 1980 Eruption of Mount St Helens, Washington*. US Geological Survey Professional Paper 1250, 347–377.

WALKER, A. B., REDMAYNE, D. W. & BROWITT, C. W. A. 1992. Seismic monitoring of Lake Nyos, Cameroon, following the gas release disaster of August 1986. *In*: MCCALL, G. J. H., LAMING, D. J. C. & SCOTT, S. C. (eds) *Geohazards*. Chapman & Hall, London.

WOLFE, E. W. 1992. The 1991 eruption of Mount Pinatubo, Philippines. *Earthquakes and Volcanoes*, **23**, 5–35.

WILLIAMS, R. S. & MOORE, J. G. 1973. Iceland chills a lava flow. *Geotimes*, **18**, 14–17.

YOUNG, R. W. & BRYANT, E. A. 1992. Catastrophic wave erosion on the south-eastern coast of Australia: impact of the Lanai tsunamis ca. 105 KA? *Geology*, **20** 199–202.

The distal impact of Icelandic volcanic gases and aerosols in Europe: a review of the 1783 Laki Fissure eruption and environmental vulnerability in the late 20th century

John Grattan

The University of Wales, Institute of Geography and Earth Sciences, Aberystwyth SY23 3DB, UK
(E-mail: jpg@aber.ac.uk)

Abstract. This paper presents detailed documentary evidence for volcanic pollution of the atmosphere over many parts of Europe during the eruption of the Laki Fissure in Iceland during AD 1783. The environmental impact of this event appears to have been severe, with reports of an acid dry fog in many parts of Europe, and often of associated damage to vegetation. It is argued that the meteorological conditions that led to this event are precisely the same as those which are associated with modern-day air pollution episodes in European conurbations, and the potential addition of millions of tonnes of volcanic gaseous material to the air over already polluted cities poses a rare, but in health terms potentially major, risk.

Introduction

Assessment of the distance from source at which volcanic gases may have an impact on soils, vegetation, animals and human health frequently emphasizes the proximal nature of these events (Garrec *et al.* 1977; Le Guern *et al.* 1988). However, recent works suggest that distal impacts may also be significant. Tephra fall (Dugmore 1989; Pilcher & Hall 1992), and archaeological and palaeoenvironmental evidence (Baillie & Munro 1988; Burgess 1989; Blackford *et al.* 1992; Charman *et al.* 1995) from Britain and Ireland suggest that Icelandic volcanic eruptions have exerted a considerable influence upon the environment in the past. The precise nature of the forcing mechanism has, however, been open to speculation and consideration has been given to both climatic and non-climatic mechanisms. This paper presents documentary evidence which demonstrates that the gases emitted in Icelandic and Italian volcanic eruptions may be transported great distances and in sufficient concentration to have a severe impact upon the environment. The environmental impact of transported acid gases is dependent on the sensitivity of the environment to external forcing, rather than the magnitude of the forcing mechanism alone. In prehistoric times the impact of volcanic gases and aerosols appears to have been greatest in those areas where the soils and ecology were vulnerable to acid deposition; namely the base depleted podzols and gley soils of upland and northern areas of Britain and Ireland (Grattan & Charman 1994; Grattan & Gilbertson 1994). In the late 20th century anthropogenic activity has created an environment which may be particularly sensitive to the introduction of acid gases and aerosols emitted in volcanic eruptions, namely the great urban conurbations with their often marginal air quality. Air quality in European cities is frequently observed to exceed the thresholds where deterioration of human health may be expected (Holman 1989; Klidonas 1993; O'Riordan 1993; Bower *et al.* 1994). Attempts to mitigate this problem are based on calculations of health thresholds and attempts to reduce car and industrial emissions. This paper proposes that a rare but significant risk is posed to human health in Europe where the already marginal air quality measured in many locations, albeit periodically, may be significantly enhanced by the addition of gases emitted in an Icelandic or Italian volcanic eruption. This hypothesis is illustrated by a consideration of a major volcanic pollution event in AD 1783, the year of the Laki Fissure eruption in Iceland.

The Laki Fissure eruption: volatile emission and dispersal

The Laki Fissure eruption began on 8 June 1783 and continued until early February 1784 (Thórarinsson 1969). During this period approximately 9.9×10^{13} g of acid were emitted, of which 9.19×10^{13} g were sulphuric acid (Thordarson & Self 1993), with hydrochloric and hydrofluoric acids included in the remainder (Pétursson *et al.* 1984). Thordarson & Self (1993) suggested that 60% of the total volume was discharged over the first 48 days in five eruptive episodes, the first three of which occurred during 8–14 June. Estimates of the discharge of SO_2 based on this work indicate a daily acid discharge of 1.38×10^{12} g during June and July. In Europe this

GRATTAN, J. 1998. The distal impact of Icelandic volcanic gases and aerosols in Europe: a review of the 1783 Laki Fissure eruption and environmental vulnerability in the late 20th century. *In*: MAUND, J. G. & EDDLESTON, M. (eds) *Geohazards in Engineering Geology*. Geological Society, London, Engineering Geology Special Publications, **15**, 97–103.

period coincides with descriptions of a dry fog and obscured Sun. Benjamin Franklin's description of this in Paris is well known (Franklin 1784; see also Lamb 1970; Sigurdsson 1982), but the description recorded by Gilbert White, an English country parson and naturalist, was equally dramatic:

> The summer of 1783 was an amazing and portentous one, and full of horrible phenomena; for besides the alarming meteors and thunder-storms that affrighted many counties of this kingdom, the peculiar haze or smokey fog, that prevailed for many weeks in this island and in every part of Europe, and even beyond its limits, was a most extraordinary appearance, unlike anything known within the memory of man. By my journal I find that I had noticed this strange occurrence from June 23 to July 20 inclusive, during which period the wind varied to every quarter without making any alteration in the air. The sun, at noon, looked as blank as a clouded moon, and shed a rust coloured ferruginous light on the ground, and floors of rooms; but was particularly lurid and blood coloured at rising and setting. All the time the heat was so intense that butchers meat could hardly be eaten on the day after it was killed; and the flies swarmed so in the lanes and hedges that they rendered the horses half frantic, and riding irksome. (White 1789)

White's account indicates the presence of volcanic gases and aerosols in the atmosphere, but unlike the more familiar explosive or Plinian eruptions which eject volcanic material into the stratosphere (Newhall & Self 1982), fissure eruptions rarely possess sufficient energy to penetrate the stratosphere (Tripoli & Thompson 1988). Thordarson & Self (1993) concluded that the majority of emitted material was confined to the troposphere at heights of up to 5 km. The dispersal of material ejected to these altitudes depends on the speed and direction of high level rather than surface winds. While low level winds travel towards a low pressure centre, high altitude winds converge on high pressure areas. Anticyclonic weather dominated Europe's weather between 21 June and

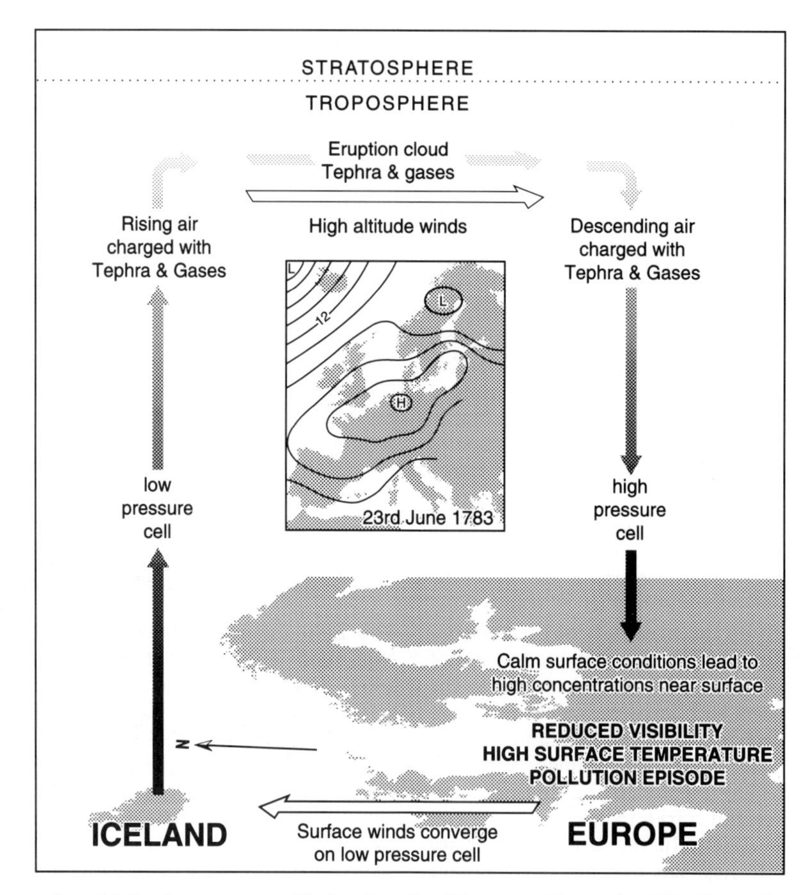

Fig. 1. Proposed model for the transport of Icelandic volcanic gases to Europe in 1783. (Synoptic map adapted from Kington 1988.)

20 July 1783 while a low-pressure area existed in the in the vicinity of Iceland (Kington 1988). Volcanic material may therefore have been transported to Europe by high altitude convergent winds. Descending air within the high pressure cell then appears to have concentrated volcanic gases near the ground surface (Fig. 1), descriptions of a dry fog being found found in many parts of Europe.

The Laki Fissure eruption: distribution and descriptions of a 'hot dry fog'

The following have all been extracted from British documentary sources. The European picture will be enhanced by studies of European material now under way, nevertheless a clear picture is presented of the presence of a 'hot dry fog' across large areas of Europe (Fig. 2).

In Britain:

> ...so long in a country not subject to fogs, we have been cover'd with one of the thickest I remember. We never see the sun but shorn of his beams, the trees are scarce discernible at a mile's distance, he sets with the face of a hot salamander and rises with the same complexion. (Cowper, 29 June 1783, in King & Ryskamp 1981).

> Two thirds of July was the same thick blue air as June ended with the month mainly hot, sometimes very hot. (Barker 1783)

> I am sorry your Ladyship has suffered so much by the heat... I am tired of this weather... it parches the leaves, makes the turf crisp... and keeps one in a constant mist that gives no dew but might as well be smoke. (Walpole, in Cunningham 1938)

> The state of the atmosphere for this week past has been more remarkably close and thick than was ever observed at this season. Such a haziness has prevailed, that the hills, at two or three miles distance, have not been discernible, and the appearance of the sun has been like that of a faint ball of fire, without a ray darting from it. (*Bristol Journal* 19th July 1783)

In Europe:

> Advices from the continent: France, Spain, Italy and other parts, showed it had equally overspread all the

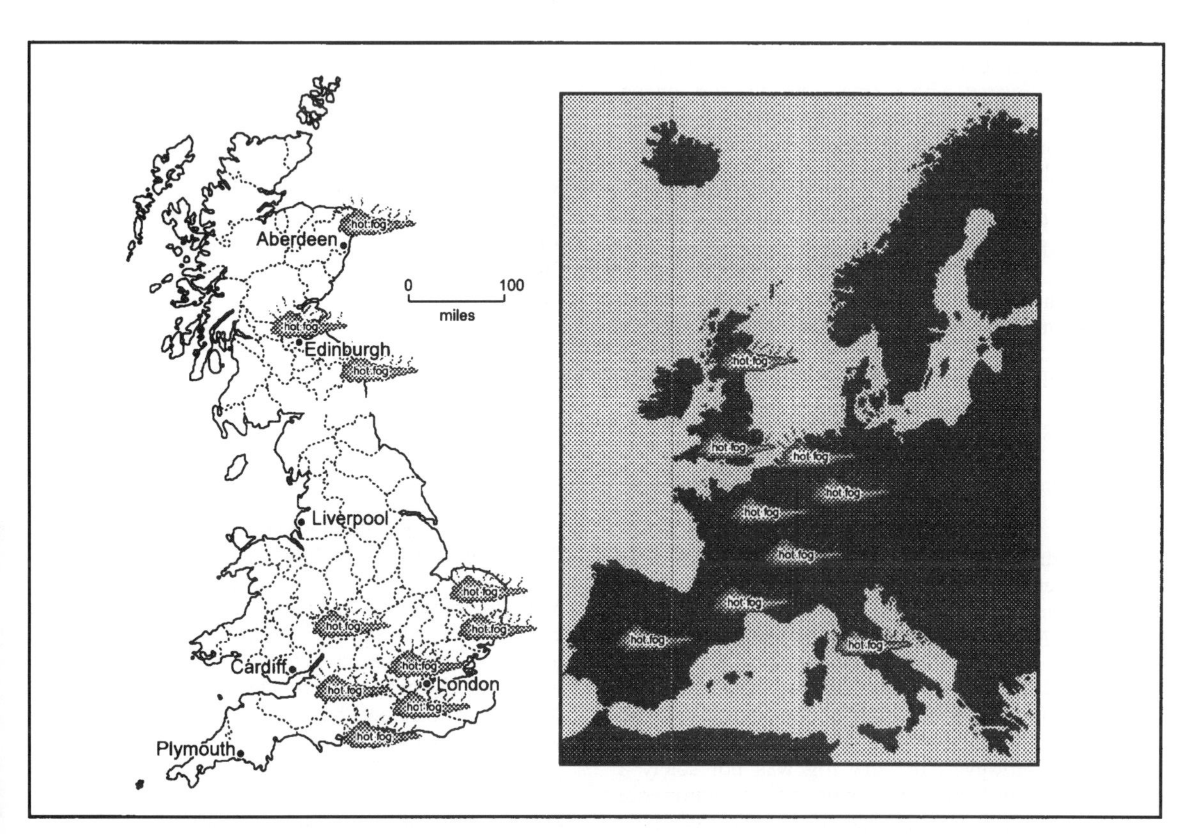

Fig. 2. The distribution of the 'hot dry fog' over Europe in 1783.

countries of Europe and by degrees it was found to be universal over the face of the globe. (Gilpin 1783)

...the moon appeared ruddy and...the sun could be looked at without being blinded. The fog was hot, dry and dense, and this phenomenon was observed not only by us, but also elsewhere in Italy, Germany and France. (Gennari 1783, in Camuffo and Enzi 1995).

...during several of the summer months of the year 1783, when the effect of the sun's rays to heat the earth in these northern regions should have been greatest, there existed a constant fog over all Europe... this fog was of a permanent nature; it was dry, and the rays of the sun seemed to have little effect towards dissipating it. (Franklin 1784)

Paris, July 4th: For a considerable time past the weather has been very remarkable here; a kind of hot fog obscures the atmosphere, and gives the sun much of a dull red appearance which the wintry fogs sometimes produce...those who are come lately from Rome say, that it is as thick and hot in Italy, and that even the top of the Alps is covered with it, and travellers and letters from Spain affirm the same. (*Bristol Journal* 19 July 1783)

Paris, July 8th: Our atmosphere has, for many days past, been covered with a thick and dry fog, owing to the succession of great heat after long and heavy rains. (*The Morning Herald and Daily Advertiser* 15 July 1783)

A report from Salon de Provence in the south of France, dated 11 July, spoke of 20 days of continuous fog:

The sun, although hot, does not dissipate it, neither day nor night...the countryside appears whitish grey...and the fog sometimes emits a strong odour and is so dry, it does not tarnish a looking glass and instead of liquifying salts, it dries them. (*Aberdeen Journal*, 18 August 1783)

Reports from Emden, northern Germany, acknowledged that

the thick dry mist which has continued so long, seems spread over the whole of Europe...during the day it veils the sun and in the evening there is a tainted odour. (*Morning Herald and Daily Advertiser* 5 August 1783)

In southern Italy, the persistent dry fog may have been further reinforced by emissions from Stromboli, Vulcano and Vesuvius which were also active in 1783 (Simkin *et al.* 1981). However, the dry fog was not solely a meteorological curiosity for in many places its presence is associated with damage to vegetation and occasionally to livestock.

The Laki Fissure eruption: environmental impacts

The impact upon the environment of the toxic material emitted during the Laki Fissure eruption in Iceland was severe, and is well documented in Thórarinsson (1979) and Gunnlaugsson *et al.* (1984). While less severe elsewhere in Europe, the impacts were none the less serious. Sir John Cullum has left a particularly detailed account of the damage witnessed in his garden near Bury St Edmunds, Suffolk:

The aristæ of the barley, which was coming into ear, became brown and weathered at their extremities, as did the leaves of the oats; the rye had the appearance of being mildewed; so that the farmers were alarmed for those crops.... The Larch, Weymouth Pine, and hardy Scotch fir, had the tips of their leaves withered; the first was particularly damaged and made a shabby appearance the rest of the summer. The leaves of some ashes very much sheltered in my garden suffered greatly. A walnut-tree received a second shock... which completed the ruin of its crop. Cherry-trees, a standard peach tree, filbert and hasel-nut-trees, shed their leaves plentifully, and littered the walks as in autumn. The barberry-bush was extremely pinched, as well as the *Hypericum perforatum* and the *Hirsutum*: as the last two are solstitial and rather delicate plants, I wondered the less at their sensibility; but was much surprised to find that the vernal black-thorn and sweet violet, the leaves of which one would have thought must have acquired a perfect firmness and strength, were injured full as much. All these vegetables appeared exactly as if a fire had been lighted near them, that had shrivelled and discoloured their leaves. (Cullum 1783)

Similar accounts can be found elsewhere in Britain and Europe:

Throughout most of the eastern counties there was a most severe frost between the 23rd and 24th June. It turned most of the barley and oats yellow, to their very great damage; the walnut trees lost their leaves and the larch and firs in plantations suffered severely. (*Sherborne Mercury* 14 July 1783)

Monday night last [23 June], a very sudden and extraordinary alteration in the appearance of the grass and corn growing in this neighbourhood...in so much that the grazing land, which only the day before was full of juice and had upon it the most delightful verdure, did, immediately after this uncommon event, look as if it had dried up by the sun, and was to walk on like hay. The beans were turned to a whitish colour, the leaf and blade appearing as if dead. (*Cambridge Chronicle and Journal* 5 July 1783)

> On Wednesday 25 June it was first observed here, and in this neighbourhood, that all the different species of grain, viz, wheat, barley, and oats, were very yellow, and in general to have had all their leaves but their upper ones in particular, withered, within two or three inches at their ends; the forward barley and the oats most so... their awns appeared... withered also. Many of the oats'... chaff husks were withered in like manner;... About this time, and for 3 days both before and after, there was an uncommon gloom in the air, with a dead calm. The dews were very profuse. The sun was scarce visible, even at mid-day, and then entirely shorn of its beams so as to be viewed by the naked eye without pain. (*The Ipswich Journal* 12 July 1783)

In Hampshire, Gilbert White recorded similar phenomena:

Monday, 23 June	The blades of wheat in several fields are turned yellow and look as if scorched by the frost.
Saturday, 5 July	Leaves fall much from many trees and hedges.

In Scotland, dead fish were observed floating in waters outside Edinburgh:

> On Wednesday night we had a great storm of thunder and lightning, accompanied by a very heavy fall of rain... next morning... there were found in the dam above the sawmills on the water of Leith, a number of different kinds of fish floating on the surface of the water. (*The Caledonian Mercury* 5 July 1783)

Accounts from northern Germany and the Netherlands report not only the 'infectious smell' of the fog, but also that 'all the trees on the borders of the Ems have been stripped of their leaves in one night' (*Ipswich Journal* 9 August 1783), and Symons (1888) reported that in Holland a large number of plants were damaged by the blighting influence of the fog. In central Germany, trees also lost their leaves:

> Mount Gleichberg... affords at present a singular and terrible phenomenon; the vapours which continually surround it are increased much, and form a thick mist which extends 8 leagues. This mist, which has destroyed the verdure of our woods, and has been substituted with a whitish tint, is, without doubt, by the scent, formed of sulphureous exhalations. (*The Morning Herald and Daily Advertiser* 12 August 1783)

All the symptoms described are consistent with damage by acid deposition (Lang *et al.* 1980; Wisniewski 1982). In modern studies, cereal crops are frequently noted as being adversely affected, with damage to barley, oats and rye, but wheat is relatively unscathed. Studies of the direct effects of acid precipitation on these plants show that wheat is less susceptible than barley to reactions with sulphuric acid (Craker & Bernstein 1984). The damage to the leaves of the trees and in particular the 'Scotch fir' (*Pinus sylvestris*) is typical of the damage caused by the absorption of sulphur dioxide (Caput *et al.* 1978). Leaf lesions may be observed at a $pH < 3.5$ and serious leaf damage will occur if $pH \leq 2.8$ (Watt Committee on Energy 1984). The shedding of leaves is a classic response to concentrations of fluorine and hydrofluoric acid, and charring is typical of damage caused by a sulphuric acid aerosol. These effects can be enhanced within a forest microclimate (Unsworth 1984). All the reported symptoms suggest that acid volatiles were present in sufficient concentration to cause serious plant damage. The Laki Fissure eruption is the most likely source for these.

Volcanic pollution in Europe: the role of regional climate in 1783

Kington (1988) has produced daily synoptic charts for Europe in the 1780s. These identify the presence of a relatively stable high-pressure air cell situated over Europe between late June and July 1783, and it was during this period that the dry fogs and crop damage were observed. This association demonstrates the intimate relationship which exists between regional weather patterns and the transport and deposition of toxic volatile material in Europe in June 1783. This evidence emphasizes the role of stable high-pressure cells which had the effect of concentrating volcanic emissions in a column of descending air for several days. In 1783, many contemporary English accounts describe calm air, a hazy atmosphere and heavy dews before and during the pollution episodes (Cullum 1783; White 1789). The atmospheric conditions which concentrated the emissions of the Laki Fissure to such dramatic effect in 1783 are similar to those which concentrate anthropogenic emissions in cities today. The prevailing synoptic and local climatic situation is therefore also likely to (variously) ameliorate or concentrate the impact downwind of the toxic volatile gases and aerosols discharging from an Icelandic eruption.

Discussion: volcanic pollution in Europe and vulnerability in the late 20th century

The documentary material presented above demonstrates that erupted volcanic gases may be concentrated within the troposphere by regional air circulation and concentrated to a sufficient degree to cause severe environmental pollution at great distances from the eruptive source. Were the events of 1783 to be repeated today, the volcanic material transported to Europe would necessarily enhance the already polluted air found in European conurbations. While it is clear that eruptions on the scale of the Laki Fissure are rare, there have been five in Iceland during the entire Holocene (Hjartanson 1994), and relatively minor eruptions have also been associated with similar environmental damage. Camuffo & Enzi (1995) have documented frequent and severe environmental impacts following the eruption of relatively minor Italian volcanoes and 19 episodes of crop damage and dry fogs were documented between 1374 and 1819. The concentration of gases emitted in non-eruptive degassing ought also to be considered as this may enhance the sulphur content of air in European cities. In non-eruptive degassing, Mt Etna annually emits SO_2 equal to the entire sulphur production of France (Garrec *et al.* 1977; Andres *et al.* 1993). Much of this material is transported to the east by prevailing winds and research is currently under way to investigate the relationship, if any, between sulphur emission, regional air circulation and severe air pollution events in Athens.

The exceedence of safety guidelines for atmospheric pollution in European cities is common, particularly under stable atmospheric conditions (UNEP/WHO 1992; Klidonas 1993; Bower *et al.* 1994; Shahgedanova & Burt 1994), precisely the conditions which favour the concentration of volcanic gases in toxic concentrations. Ecological literature emphasizes the concept of 'Critical thresholds' when estimating the likely impact of a pollutant. The concept assumes that there is a damage threshold for the response of a system to externally supplied stress. This may vary between different ecosystem or species (pollution 'receptors'). The critical load for a particular receptor–pollutant combination is defined as the highest deposition load that the receptor can withstand without long-term damage occurring (Bull 1991: 30). In the urban air environment, human activity has created a system which periodically approaches and sometimes exceeds its critical threshold in terms of human health. The research outlined above has identified a major episodic point source for air pollution in the European environment. Should a sulphur-producing volcanic eruption occur in Iceland or Italy, at a time when human activity and atmospheric circulation have already created a stressed or marginal urban air environment then a major health hazard may occur, and in addition panic may become widespread in the general population (Grattan & Brayshay 1995). The addition of volcanogenic sulphur gases and/or aerosols to an already marginal environment may have severe consequences for that environment.

References

ANDRES, R. J., KYLE, P. R. & CHUAN, R. L. 1993. Sulphur dioxide, particle and elemental emissions from Mt Etna, Italy during July 1987. *Geologisches Rundschau*, **82**, 687–695.

BAILLIE, M. G. L. & MUNRO, M. A. R. 1988. Irish tree rings, Santorini and volcanic dust veils. *Nature*, **322**, 344–346.

BARKER, T. 1783. *Meteorological Register, Made at Lyndon Hall, Rutland.* Unpublished manuscript, available in the National Meteorological Library and Archive, Bracknell, UK.

BLACKFORD, J. J., EDWARDS, K. J., DUGMORE, A. J., COOK, G. T. & BUCKLAND, P. C. 1992. Hekla-4: Icelandic volcanic ash and the mid-Holocene Scots Pine decline in northern Scotland. *The Holocene*, **2**(3), 260–265.

BOWER, J. S., STEVENSON, K. J., BROUGHTON, G. F. J. *et al.* 1994. *Air Pollution in the UK 1992/93.* Warren Spring Laboratory, Stevenage.

BULL, K. 1991. Critical load maps for the U.K. *NERC News*, July, 31–32.

BURGESS, C. 1989. Volcanoes, catastrophe and the global crisis of the late second millenium BC. *Current Archaeology*, **117**, 325–329.

CAMUFFO, D. & ENZI, S. 1995. Impacts of the clouds of volcanic aerosls in italy during the last seven centuries. *Natural Hazards*, **11**, 135–161.

CAPUT, C., BELOT, Y., AUCLAIR, D. & DECOURT, N. 1978. Absorption of sulphur dioxide by pine needles leading to acute injury. *Environmental Pollution*, **16**, 3–15.

CHARMAN, D. J., GRATTAN, J. P., KELLY, A. & WEST, S. 1995. Environmental response to tephra deposition in the Strath of Kildonan, northern Scotland. *The Journal of Archaeological Science*, **22**, 799–809.

CRAKER, L. E. & BERNSTEIN, D. 1984. Buffering of acid rain by leaf tissue of selected crop plants. *Environmental Pollution*, A **36**, 375–382.

CULLUM, J. 1783. Of a remarkable frost on the 23rd of June, 1783. *Philosophical Transactions of the Royal Society of London.* Abridged Volume **15**, 1781–1785, 604.

CUNNINGHAM, P. 1938. *The letters of HORACE WALPOLE. Volume VIII.* Richard Bentley, London.

DUGMORE, A. J. 1989. Icelandic volcanic ash in Scotland. *Scottish Geographical Magazine*, **105**(3), 168–172.

FRANKLIN, B. 1784. Meteorological imaginations and conjectures. *Memoirs of the Literary and Philosophical Society of Manchester*, **2**, 373–377

GARREC, J. P., LOUNOWSKI, A. & PLEBIN, R. 1977. The influence of volcanic fluoride emissions on the surrounding vegetation. *Fluoride*, **10**, 153–156.

GILPIN, W. 1783. *An historical account of the weather during twenty years from 1763–1785.* Bodleian MS, Eng. Misc. d. 564.

GRATTAN, J. P. & BRAYSHAY, M. B. 1995. An amazing and portentous summer: environmental and social responses in Britain to the 1783 eruption of an Iceland Volcano. *The Geographical Journal*, **161**(2), 125–134.

—— & CHARMAN, D. J. 1994 Non-climatic factors and the environmental impact of volcanic volatiles: implications of the Laki fissure eruption of AD 1783. *The Holocene*, **4**(1), 101–106

—— & GILBERTSON, D. D. 1994. Acid-loading from Icelandic tephra falling on acidified ecosystems as a key to understanding archaeological and environmental stress in northern and western Britain. *The Journal of Archaeological Science*, **21**(6), 851–859.

GUNNLAUGSSON, G. A., GUDBERGSSON, G. M., THÓRARINSSON, S., RAFFNSON, S. & EINARSSON, T. 1984. *Skßftareldar 1783–1784.* Mal Og Menning, Rekyavik (in Icelandic with English summaries).

HOLMAN, C. 1989. *Air Quality and Health.* Friends of the Earth, London.

HJARTANSON, Á. 1994. Environmental changes in Iceland following the great Þjórsá Lava eruption 780014C years BP. *Münchener Geographische Abhandlungen*, **B12**, 147–156.

KING, J. & RYSKAMP, C. 1981. *The letters and prose writings of WILLIAM COWPER*, vol. II. Clarendon Press, Oxford.

KINGTON, J. A. 1988. *The Weather for the 1780s over Europe*. Cambridge University Press, Cambridge.

KLIDONAS, Y. 1993. The quality of the atmosphere in Athens. *The Science of the Total Environment*, **129**, 83–94.

LAMB, H. H. (1970). Volcanic dust in the atmosphere; with a chronology and assessment of its meteorological significance. *Philosophical Transactions of the Royal Society of London, Series A*, **266**(1170), 425–533.

LANG, D. S., HERZFELD, D. & KRUPA, S. V. 1980. Responses of plants to submicron acid aerosols. *In*: TORIBARA, T. Y., MILLER, M. W. & MORROW, P. E. (eds) *Polluted Rain*. Plenum, New York and London, 273–290.

LE GUERN, F., FAIVRE-PIERRET, H. & GARREC, J. P. 1988. Atmospheric contribution of volcanic sulphur vapour and its influence on the surrounding vegetation. *Journal of Volcanology and Geothermal Research*, **35**, 173–178.

NEWHALL, G. C. & SELF, S. 1982. The volcanic explosivity index (VEI): an estimate of explosive magnitude for all volcanism. *Journal of Geophysical Research*, **87**(C12), 1231–1238.

O'RIORDAN, T. 1993. Industrial pollution control in the UK. *The Science of the Total Environment*, **129**, 39–53.

PÉTURSSON, G., PALSSON, G. A. & GEORGSSON, G. 1984. Um Eiturahrif. *In*: GUNNLAUGSSON, G. A., GUDBERGSSON, G. M., THORARINSSON, S., RAFFNSON, S. & EINARSSON, T. (eds) *SKAFTAR ELDAR 1783–1784*. Mal Og Menning, Rekyavik (in Icelandic with English summaries).

PILCHER, J. R. & HALL, V. A. 1992. Towards a tephrochronology for the north of Ireland. *The Holocene*, **2**(3), 255–259.

SHAHGEDANOVA, M. & BURT, T. P. 1994. New data on air pollution in the former Soviet Union. *Global Environental Change*, **4**, 201–227.

SIGURDSSON, H. 1982. Volcanic pollution and climate: the 1783 Laki eruption. *EOS. Transactions, American Geophysical Union*, **63**(32), 601–603.

SIMKIN, T., SIEBERT, L., MCCLELLAND, L., BRIDGE, D., NEWHALL, C. & LATTER, J. H. 1981. *Volcanoes of the World*. The Smithsonian Institution, Washington.

SYMONS, G. J. 1888. *The eruption of Krakatau and subsequent phenomena: report of the Krakatau Committee of the Royal Society of London*. Trubner, London.

THÓRARINSSON, S. 1969. The Lakagigar eruption of 1783. *Bulletin Volcanologique*, **33**(3), 910–929

——1979. On the damage caused by volcanic eruptions with special reference to tephra and gases. *In*: SHEETS, P. D. & GRAYSON, D. K. (eds) *Volcanic Activity and Human Ecology*. Academic Press, New York, 125–159.

THORDARSON, TH. & SELF, S. 1993. The Laki [Skaftßr Fires] and Grømsvötn eruptions in 1783–85. *Bulletin Volcanologique*, **55**, 233–263.

TRIPOLI, G. J. & THOMPSON, S. L. 1988. A three-dimensional numerical simulation of the atmospheric injection of aerosols by a hypothetical basaltic fissure eruption. *In*: *Global Catastrophes and Earth History*, Abstract volume LPI and NAS Conference, Snowbird, Utah, 20–23 October 1988, 200–201.

UNEP/WHO 1993. *Urban Air Pollution in the Megacities of the World*. Blackwell, Oxford.

UNSWORTH, M. H. 1984. Evaporation from forests in cloud enhances the effects of cloud deposition. *Nature*, **312**, 262–264.

WATT COMMITTEE ON ENERGY 1984. *Report number 14. Acid rain*. The Watt Committee on Energy Limited.

WHITE, G. 1789. *The Natural History of Selbourne*, reprinted 1977. Penguin, London.

WISNIEWSKI, J. 1982. The potential acidity associated with dews, frosts and fogs. *Water, Air, and Soil Pollution*, **17**(4), 361–377.

Lessons from the Kobe earthquake

Philip Esper[1] & Eizaburo Tachibana[2]

[1]Sir Alexander Gibb Ltd., Gibb House, London House, Reading RG6 1BL, UK
[2]Department of Structural Mechanics, Osaka University, Osaka, Japan

Abstract. The Kobe earthquake is considered to be one of the most devastating and costly natural disasters in recent history considering the number of buildings destroyed, the number of people killed and injured, the size of the affected area, and the extent and severity of damage to a wide range of structural types. As a result, important questions have been raised about earthquake preparedness, disaster response, seismic design and codes of practice, and upgrading of earthquake-resistant structures. This paper presents an overview of this earthquake, investigates the extent and types of damage caused, studies the factors behind each type of damage, and highlights important lessons learnt from this earthquake. Protective measures and future research that should be undertaken in Japan, and all countries that are at seismic risk, are recommended in order to reduce damage and casualties in future seismic events.

Introduction

Prior to the Hyogoken-Nanbu earthquake, Kobe was the second largest port of Japan, with a population of around 1.4 million that live on a narrow 4 km wide strip of land between Osaka Harbour and the Rokko Mountains (see Fig. 1). The Kobe earthquake occurred at 05:46 hours (local time) on Tuesday 17 January 1995 and is considered to be the greatest natural disaster to have struck Japan since the 1923 great Kanto earthquake that devastated large areas of Tokyo and Yokohama and killed approximately 143 000 people (fire was the main cause).

The earthquake was assigned a magnitude of 7.2 by the Japan Metrological Agency (JMA) and its epicentre was located approximately 20 km SW of Kobe city centre, just north of the Awaji Island (identified by a star in Figs 1 and 2). According to the *Ashai Newspaper* of 25 January 1995 (Anon 1995*a*), 89% of the casualties from this earthquake were caused by building collapses, the majority of which were wooden houses, and most people were at home when the earthquake struck. Nearly 55 000 houses collapsed and approximately 32 000 were partially destroyed. These figures exclude damage caused by fire, which exceeded 7500 houses. It wrecked elevated roadways and railways, initiated landslides and destroyed ports and harbour facilities. As a result of this severe damage and fire, over 5400 people were killed, as many as 34 000 were injured, and around 350 000 were made homeless. The official estimates, which were released one week after the earthquake, of direct losses in the Hyogoken prefecture alone exceeded $100 billion; current estimates have reached US$200 billion, and indirect losses will undoubtedly raise the total amount even more (see Fig. 3).

Architectural and planning background

Kobe has been a port since the 13th century and developed in the 19th century to become the second largest and busiest port in Japan, as it is today. It lies on a narrow coastal plain facing the Inland Sea and backs up against Mount Rokko. While Kobe developed before World War II into a thriving city, the Allied bombing destroyed the region. Recovery started in 1945 and for a 15-year period beginning in the late 1950s, Japan enjoyed phenomenal economic growth. Concurrent with this development, a number of important architectural works were appearing in Japan, and an important movement, known as the 'Metabolists', caught the attention of architects everywhere around the world. This movement has probably helped Japan to be one of the rare cases in which the metropolis is still alive and doing well.

Most Westerners who have visited Japan would agree that the Japanese city is a livelier and somehow more human place than most Western cities. Although it is the innate patience and politeness of the Japanese that is believed to make this work, another factor could be that residential areas in Japan are very densely built up. A 1970 comparison by Morris (1970) of the most dense areas/cities showed the following:

- London (Paddington): 131 persons acre^{-1}
- New York (Manhattan): 120 persons acre^{-1}
- Tokyo: 116 persons acre^{-1}

ESPER, P. & TACHIBANA, E. 1998. Lessons from the Kobe earthquake *In*: MAUND, J. G. & EDDLESTON, M. (eds) *Geohazards in Engineering Geology*. Geological Society, London, Engineering Geology Special Publications, **15**, 105–116.

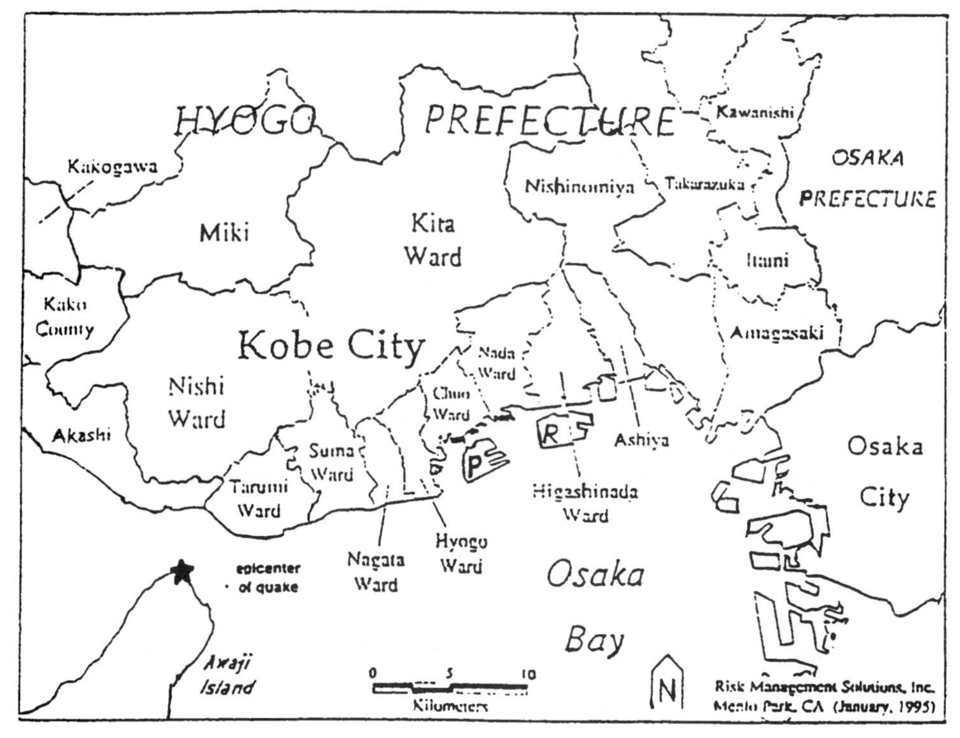

Fig. 1. Map of the affected area – Kobe and vicinity. (P: Port Island, R: Rokko Island).

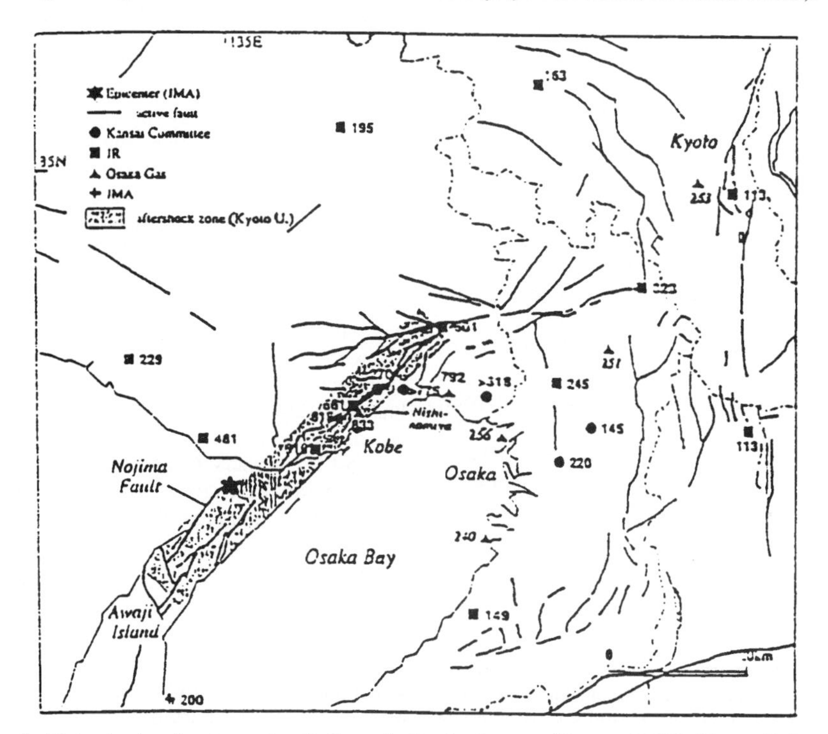

Fig. 2. Main shock epicenter, active faults and aftershock zone. (Source: DPRI, Kyoto University).

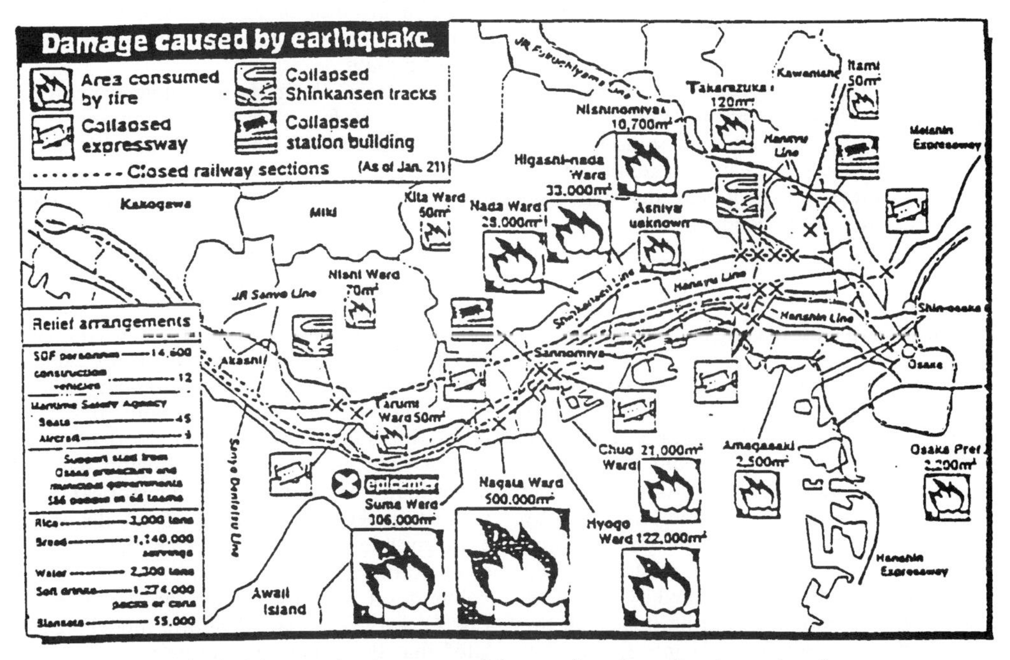

Fig. 3. Diagram showing extent of damage four days after the earthquake.

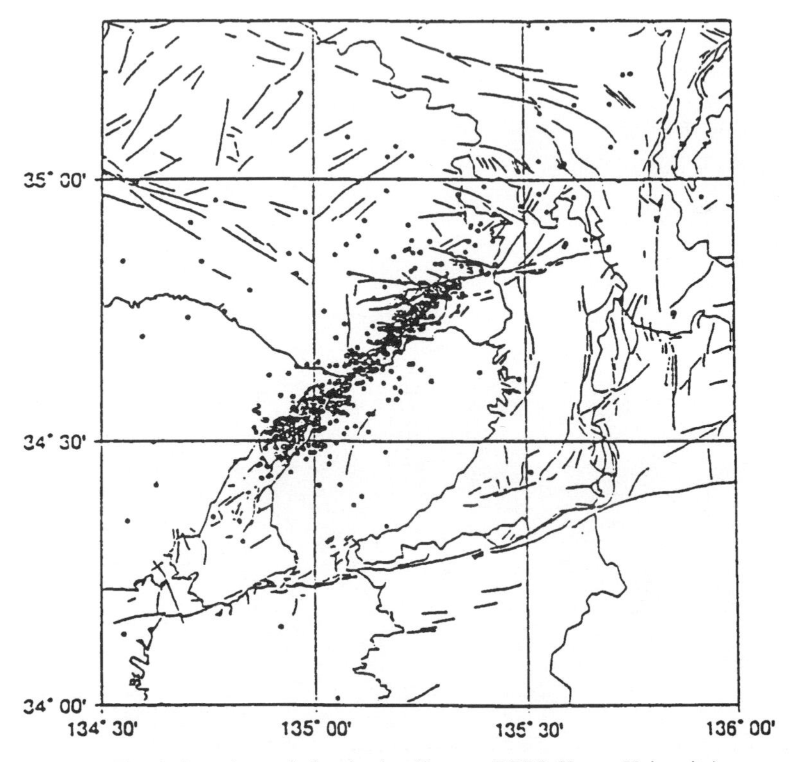

Fig. 4. Locations of aftershocks. (Source: DPRI, Kyoto University).

The importance of the above figures is reflected in the fact that while high densities are achieved in London and New York by medium- and high-rise buildings, in Tokyo (and Japan in general) people live in one- and two-storey dwellings. This, apparently, has been a more likeable way of living in Japan. It is these houses that were affected most by the Kobe earthquake, the simple reason being that these houses lacked the structural versatility to withstand seismic loading.

Geological and seismological aspects

The cities of Kobe, Ashiya and Nishinomiya are located between the Rokko Mountains (north) and Osaka Bay (south) as shown in Fig. 1. Mount Rokko is composed primarily of granite and is crossed by many faults. Near the ground surface the granite has weathered to become decomposed granite soil, which is prone to erosion by rain. Moderate slopes exist at the foot of the mountain, consisting of a series of alluvial fans of decomposed granite soils from Mount Rokko deposited by landslides and mud flows of sand and stone during floods. In the urban plains, more than ten rivers flow into Osaka Bay within a distance of 22 km. Alluvial fans, composed mainly of gravel and sand, are found all over the Hanshin plain.

Some areas along the coastline were reclaimed by placement of fill, typically decomposed granite soils. Examples are Port Island and Rokko Island (see Fig. 1). At Port Island, the thickness of soft clay is around 7 m in the northern area and 16 m in the southern area. These will be discussed in more detail in the next section.

Japan sits on the intersection of four tectonic plates that are under constant seismic risk. Kobe, which is located relatively far from the main tectonic collision zones, was considered (before this earthquake) as less seismically active than other areas in Japan, such as Tokyo. The earthquake on 17 January, however, was not along the major inter-plate faults but rather on the intra-plate fault zone (see Fig. 2) that spread within one of these four plates. This region has experienced magnitude 7 or greater events in historical times; in 1596 an earthquake of magnitude 7.5 hit the area; in 1916, a magnitude 6.1 event occurred at almost the same epicentral location.

Based on the distribution of the aftershocks (see Fig. 4) and teleseismic waveform modelling, it is postulated that the strike-slip rupture was bilateral from the

Fig. 5. Scarp of the Nojima fault on Awaji Island showing both vertical and horizontal offset.

hypocentre, with a total length of between 30 and 50 km. The epicentre of the magnitude 7.2 event was located approximately 20 km southwest of the city centre of Kobe in the strait between Awaji Island and the south central part of Honshu, the main island in Japan. The rupture had the following characteristics, as reported by Yokohama City University:

Focal mechanism:	strike = 233°NE
	dip = 85°
	slip = 165
Focal depth:	$D_F = 14.5$ km
Epicentre location:	34°36.4′N latitude
	135°2.6′E longitude
Seismic moment:	$M_o = 2.5 \times 10^{26}$ dyne.cm
Fault area:	$S = 40 \times 10$ km^2
Relative displacement:	$U = 2.1$ m
Stress drop:	$D_S = 100$–200 bar
Duration of main rupture:	$T = 11$

The earthquake occurred in a region where a complex system of active faults had been previously mapped (see Fig. 2). Large ground movements were observed and associated with fault ruptures in Awaji Island; a maximum movement of 1.70 m of right lateral strike slip and 1.30 m of vertical slip were observed at a scarp of the Nojima Fault on Awaji Island. Figure 5 shows both the vertical and the horizontal offset of this scarp. The rupture propagated through the city centre of Kobe and directly adjacent to the cities of Nishinomiya and Ashiya. The damage was mainly concentrated along the line of fault fracturing. Local site conditions may have influenced the level of ground shaking, and investigations showed that many structures situated on alluvial soils were destroyed.

The ground motion recorded at the Kobe Maritime Metrological Observatory (KMMO), which is the closest to the epicentre, is shown in Table 1. This record represents one of the strongest ground motions observed. The duration of the earthquake was measured at the Observatory at just less than 15 s. Velocity–time histories of rock sites elsewhere in the Kobe–Osaka region are consistent with this observation. Some soft soil sites, however, show durations of strong shaking as long as 100 s.

Figure 6 shows ground motions record at the Matsumuragumi Technical Research Center (MTRC) which is located approximately 35 km from the epicentre. At foundation level, peak ground accelerations were 272 gal (0.2 g) in the N–S direction, 265 gal (0.27 g) in the E–W direction, and 232 gal (0.24 g) in the U–D (vertical) direction. This shows that the peak vertical acceleration was approximately 0.9 times the peak horizontal acceleration, which exceeds typical values observed in past earthquakes. Both recording stations – KMMO and MTRC – are located on hard ground near the epicentre.

Table 1. *Ground motion recorded at the Kobe Maritime Meteorological Observatory (1 gal = 1 cm s^{-1})*

Direction	*Acceleration (gal)*	*Displacement (mm)*
North–south	818	180
East–west	617	180
Up–down	332	100

Geotechnical aspects

The major geotechnical aspects associated with this earthquake are described below.

Large ground movements were associated with fault ruptures in Awaji island; the maximum movement observed was 1.70 m of right lateral strike slip and 1.30 m of vertical slip. This ground movement was limited elsewhere on the island since the geological profile on the island is granite.

Massive soil liquefaction was observed in many reclaimed lands in Osaka Bay including two human-made islands: Port Island (8 km^2) and Rokko Island (6 km^2). Both islands have been constructed with fill derived from decomposed granite. SPT on the fill gave a value for $N = 6$. The grain size of the fill varies from gravel and cobble-sized particles to fine sand, with a mean grain size of approximately 2 mm. In Port Island soil, liquefaction of this fill material (the thickness of submerged fill in this island is about 12 m) resulted in ground settlement between 0.50 m and 0.75 m. Many of the quay walls are gravity wall-type structures which have displaced laterally by approximately 1–7 m (see Fig. 7). Consequently, derricks and cranes were either severely damaged or made unusable and 179 out of 186 of the heavy shipping berths were made inoperative.

Soil liquefaction also occurred along the port of Kobe and caused extensive damage to many industrial and port facilities such as tanks, wharves, quay walls, cranes, and the collapse of the Nishinomiya Harbour Bridge girder.

Several underground subway stations constructed by cut-and-cover suffered major damage to their reinforced-concrete pillars. One particular example, which is of a great importance (see below), is the collapse of Daikai Station on the Kobe Rapid Transit Railway (KRTR), which resulted in large surface settlements with a maximum value of 2.5 m. The collapsed portion of this station is approximately 100 m long. It is constructed as a reinforced-concrete box structure of a 5.50 m high and 15.30 m wide section. The distance from the street surface to the top of the structure is 4.60 m. Preliminary information (Anon 1995*b*) indicates that the station was constructed on sands underlain by stiff clay. The central reinforced-concrete columns of the station failed, apparently by a combination of shear and axial loading. More evident is the fact that there was only nominal confining

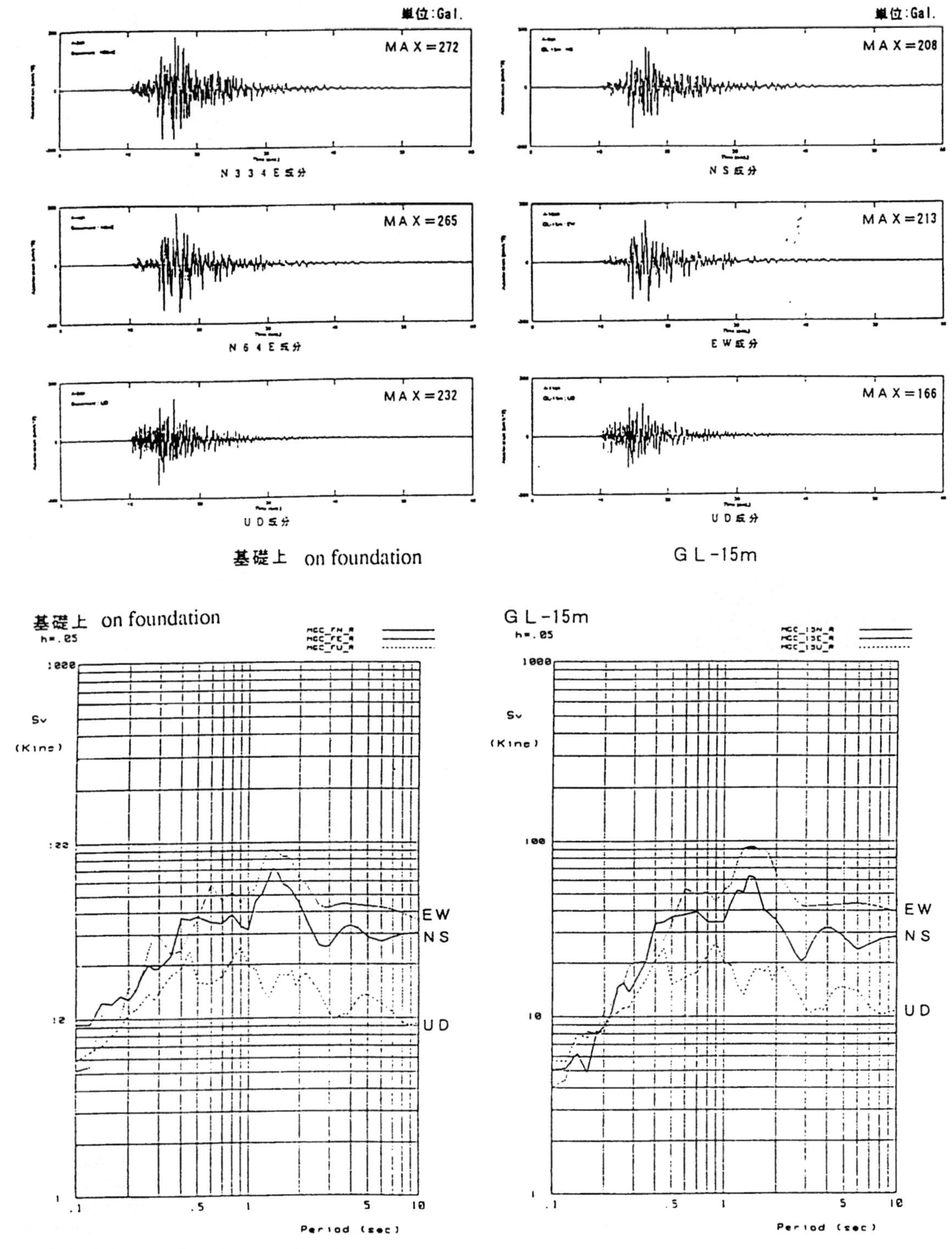

Fig. 6. Ground motion records and velocity response spectra near the epicenter. (Source: Matsumuragumi Technical Research Center, MTRC).

Fig. 7. Liquefaction and lateral spreading at Port Island.

steel: 9 mm diameter stirrups on 180 mm spacing, and 20 mm diameter longitudinal rebar on 140 mm centres in the concrete.

Numerous landslides and rockfalls resulted along the hillside of Nishinomiya area. In Nikawa, a landslide buried 34 people.

Damage to building structures

Table 2 shows statistics of the damage caused by this earthquake in terms of both residents and houses. These quantities are based on statistical data of an inspection made before 20 January 1995 and collected by the Architectural Institute of Japan (AIJ), Kyoto University, Osaka Institute of Technology (OIT), Fukui University and Hiroshima University (Nakashima & Bruneau 1995). It is clear from the above table that the damage seems to be concentrated in the Hyogo prefecture. Most of the severe damage in and around Kobe is concentrated along a narrow band 1–2 km wide and approximately 30 km long, beginning at the northern tip of Awaji Island and extending across the Akashi channel in the ENE direction along the coastline (see Fig. 3). The aftershock zone overlaps the heavily damaged region which was assigned a magnitude of 7 on the Shindo scale. Table 3 shows the seven levels of intensity of the Japanese Shindo scale and their corresponding levels on the Modified Mercali Intensity (MMI) scale.

The majority of the collapsed buildings were unbraced, or lightly braced, one- and two-storey post and beam traditional wooden houses with heavy clay tiled roofs. These are known as Shinkabe and Okabe. Little nailing is used in either type and they rely on wood joinery for connections. This type of heavy roofing has been widely used in Japan as it provides excellent resistance to the strong gusty winds of typhoons which frequently sweep across the country. Most recently, stud-wall construction has been entering the Japanese market, but it is not yet a major influence.

Structural damage occurred also in reinforced, steel and composite structures. Generally, no material type was immune from damage, but most of the severe damage occurred in non-ductile steel and concrete structures that were built prior to the implementation of requirements for ductile detailing. As per data collected in the year 1989, timber, steel, concrete and

Table 2. *Damage statistics in Hyogo, Osaka and Kyoto prefectures*

	Impact on residents		*Damage to houses*				
	Deaths	*Missing*	*Due to ground shaking*			*Due to fire*	
			Collapse	*Severe*	*Minor*	*Destroyed*	*Damaged*
Hyogo Prefecture	5394	2	81 206	62 826		7119	337
Kobe city			54 949	31 783		7046	331
Nishinomiya			17 716	13 474		48	4
Ashiya			2 543	1 519		13	2
Takarazuka			1 339	3 718		2	
Awaji			73	583			
Osaka Prefecture	18		881	5 190	32 617		
Osaka city	12		190	1 785	8 759		
Toyonaka	4		654	2 842			
Kyotu Prefecture	1		3	3	1 109		
Kyoto city			1		500		

steel-encased reinforced concrete (referred to as SRC) account for approximately 30%, 40%, 20% and 10% respectively of all square metres of construction in Japan (Nakashima & Bruneau 1995). With respect to modern buildings, it should be noted that those designed according to the 1981 Building Standard Law – which mandated ductile detailing for the first time – generally performed well. A recent study by Ohbayashi Corporation on the performance of residential and office buildings constructed by this Corporation in the Kobe earthquake is summarized in Table 4. Such a global evaluation of seismic performance of buildings must be used to evaluate the effectiveness of design codes.

As far as seismic resistance is concerned, an important revision to the Japanese Building Code (JBC) took place in 1971, when requirements necessary to enhance the shear behaviour of reinforced-concrete members were first considered. A major overhaul of the JBC then occurred in 1981, when a two-level design procedure was implemented, along with requirements to explicitly consider ductile behaviour. These two periods provide important milestones in the Japanese history of earthquake-resistant design.

Table 3. *The Japanese Shindo Intensity Scale and its equivalent levels on the MMI scale*

The Japanese Shindo scale	*Equivalent MMI scale levels*
1	I–II
2	II–IV
3	IV–V
4	V–VII
5	VII–VIII
6	VIII–IX
7	IX–XII

Over 100 reinforced mid-rise buildings that were constructed during the 1960s and 1970s have failed, in many cases catastrophically. Most failures appear to have been shear failures of columns that had very light transverse reinforcement (the spacing of stirrups exceeded 200 mm in many cases) (see Fig. 8). Another reason is that stirrups in columns, quite often were not welded at the ends or anchored into the concrete, as is usually the case in Japanese design and construction practice. This indicated bad supervisor of works at the time of construction, which was supported by other examples; for instance, some fillet-welds that were encountered during investigation works on some buildings were found to be specified as butt-welds on design plans.

A number of mid-height single-storey pancake collapses were observed,with the 1960's vintage eight-storey Kobe City Hall being the best example. While this building sustained a complete collapse of the sixth floor (see Fig. 9), the neighbouring 1980s vintage 30-storey New City Hall was undamaged. The soft storey collapse of the sixth floor in the former building was because this floor formed a transition storey from an SRC system to a pure RC system. This type of construction was common in Japan until some 30 years ago, purely for economy. Hence, this is effectively the first earthquake where this type of collapse has been observed.

Damage to infrastructure

Many transportation facilities failed and were out of operation for a long period of time. Railway tracks snapped in many locations, and trains flipped on their sides. Many decks and girders of elevated roadways and railways shifted from their original location or even dropped to the ground. The Kobe city main highway

Table 4. *Performance of buildings constructed by Ohbayashi Corporation*

	Green tags (negligible damage)	*Yellow Tags (moderate damage)*	*Red Tags (severe damage)*
Pre-1971 old Seismic Design Code	42%	22%	36%
1972–80 Transitional Period	72%	17%	11%
Post-1981 New Seismic Design Code	84%	10%	6%

Fig. 8. Collapse of a reinforced concrete column due to inadequate transverse reinforcement.

disintegrated in many places. Many modern bridges, such as the Akashi Suspension Bridge, however, sustained no or little damage.

In the case of the elevated Hanshin Expressway – which links Kobe, Osaka and beyond – huge single hammerhead reinforced-concrete piers, which were holding up the girders, broke off at their bases failing in shear and/or bending, resulting in a 500 m segment to be tipped onto the ground. This elevated segment was constructed in 1968–69 under older seismic provisions and was scheduled for retrofitting in the future by the Hanshin Expressway Public Corporation. Also, at this location of the elevated expressway the superstructure changed from steel to concrete, increasing the mass of the structure, and hence the earthquake forces acting on it, sufficiently to push this section over and cause it to collapse transversely. Lack of enough transverse reinforcement was also observed in some of the collapsed columns.

Rail facilities were particularly hard hit in this earthquake. Three main lines – JR West, Hankyu, and Hanshin – which run in general on elevated embankments, all sustained embankment failures, overpass collapses, distorted rails and other severe damage. In Kobe, the Shinkansen (Bullet Train) is generally in a tunnel through Rokko Mountain. At the east portal of the tunnel, the line is carried on an elevated viaduct, built in the 1960's. For a length of 3 km, this viaduct was severely damaged with a number of the longer spans collapsing due to shear failure of the supporting columns.

Fig. 9. Pancake (or soft-storey) collapse of the sixth floor of the city hall.

Electric power and telecommunications performed relatively well in the earthquake, with little damage and reduction in service. Gas and water supplies, however, were not restored fully until around the middle of March. In spite of the sincere efforts that both Kobe Water Department and Osaka Gas Company made to implement anti-seismic measures to prevent such service interruptions from occurring, 725 000 customers were without gas supply and 29 000 were without water until around the middle of February. The water and gas systems sustained numerous breaks in their underground pipelines and distribution systems, with a general lack of service in the cities of Kobe, Ashiya and Nishinomiya.

Fire started by the earthquake

Large numbers of pipe breaks, including natural gas and water supplies, occurred in soil-liquefied areas; this led to fires in many locations in Kobe and Nagata district. Approximately 100 fires broke out within minutes after the earthquake, primarily in densely built-up low-rise areas of the central city: Nada, Higashinada, Hyogo and Suma wards, which are comprised of mixed residential–commercial occupancies, predominantly of wood construction (see Fig. 3). Water for fire-fighting purposes was available for only two to three hours, including use of underground cisterns. Subsequently, water was available only from tanker trucks. Wind, however, was calm and fire advance was subsequently slow. The final burnt area in Kobe is estimated at 1 million m^2, with 50% of this being in the Nagata ward.

Lessons learned

Tremendous damage was observed in this earthquake, some of which has not been encountered in past earthquakes. The following important lessons should be recognized. Local site conditions had a major impact on the level of shaking. It is therefore vital to identify the influence of subsurface soil conditions on the amplification of ground motion.

The area of Shindo 7 intensity (very severe shaking) extends from Kobe through Nishinomiya, approximately 20 km long and 1 km wide. The percentage of buildings that collapsed in this area is high. Hence, the concentration of damage to this area may be a consequence of the amplification of vibration due to topographic irregularities and/or the presence of soft soil underlying this area.

Ground improvement by the vibro-rod method was originally used in many locations on both Port and Rokko Islands; preliminary investigations showed that this improvement was very effective in reducing ground settlement.

Earthquake-resistant berths at Maya Futo wharf performed very well in this earthquake with almost no lateral spreading.

Reinforced geotextile walls along the JR railway line in Higashi-Nada district performed relatively well compared to the level of damage to the surrounding site.

The Kobe area port facilities were constructed under the Ministry of Transportation's Construction Standard B (as opposed to more stringent A and super-A standards). Importance factors were not considered in the design as these simply do not exist in the Japanese design codes. Considering the importance of Kobe's facilities in that they handle 30% of Japan's container freight traffic, this is now considered substandard.

The failure of the Daikai station of the Kobe Rapid Transit Railway is highly significant because it is the first instance of severe earthquake damage to a modern tunnel for reasons other than fault displacement and instability near the portal.

The peak vertical ground acceleration was approximately equal to the peak horizontal acceleration, as recorded in some stations. This is much higher than what was observed in past earthquakes.

Matsumuragumi Technical Research Centre and Computer Centre of Ministry of Post and Communications (CCMPC) are the only two buildings in the affected area (approximately 35 km from the epicentre) that are founded on base-isolators. The former was visited by the two authors and no signs of damage exist at all throughout the whole building. The same was reported by others on the CCMPC building.

In subsequent studies and investigations of structural damage, characteristics of the fault rupture, proximity to the rupture surface, dynamic properties of the subsurface soil conditions and local and regional geographic features must be taken into account.

While hospitals are intended to provide relief to disaster victims, the NHK television news reports (Anon 1995*c*) showed victims being carried away from a damaged hospital after the earthquake. This type of structure must be designed with greater resistance than ordinary structures.

Most failures in reinforced mid-rise buildings that were constructed during the 1960s and 1970s appear to have been shear failures of columns that had very light transverse reinforcement and/or unwelded stirrups. This indicates bad supervision of works at the time of construction, which was supported by other examples such as some fillet-welds being encountered during investigation works on some buildings that were found to be specified as butt-welds on design plans.

A report in 1990 presented to the Hanshin Expressway Corporation (HEC) showed that, chemical deterioration inside the concrete pillars had been taking place for some time due to what is known as alkali-aggregate reaction causing the pillars to crack. Although this was acted on immediately with regular maintenance schemes involving the injection of a special resin into the cracks to stop deterioration, it was reported after the earthquake that the strength of the pillars immediately before the earthquake was only around 42% of their original strength.

Conclusions and recommendations

It has to be made clear that Kobe's destruction was not a failure of Japanese technology. Evidence indicates that the majority of the buildings designed according to the present Japanese codes sustained little or no damage. Each earthquake is unique. What really matters is that engineers, decision makers and politicians should learn from this recent disaster so that future hazards can be made less damaging. Some recommendations are made below.

There is an urgent need to establish reliable and cost-effective techniques to evaluate, repair and strengthen/retrofit existing structures and buildings in Japan that suffered minor or localized damage. This particularly applies to structures, buildings and old wooden houses, elsewhere in Japan, that were constructed before current seismic design regulations were implemented.

A global evaluation of seismic performance of buildings must be used to evaluate the effectiveness of design codes in all countries that are at seismic risk. Current Japanese seismic codes do not consider importance factors, as the design and construction of Kobe port facilities indicated. A review of these codes is now considered necessary.

High values of peak vertical acceleration were experienced in this earthquake. This was also suggested by certain types of damage observed in various structures. Hence, a review of all design codes of practice is needed in order to take this factor into account.

Soil–foundation–structure interaction studies could not be restricted to heavy or special structures. Evidence provided by this earthquake showed that this interaction was significant in the seismic behaviour of different types and sizes of structures.

Structures that are required to provide essential services after disasters, such as hospitals and fire stations, must be designed with greater resistance to seismic forces than ordinary structures. In this respect, methods such as base-isolation should be taken more seriously as a powerful means for enhancing the seismic performance of these structures.

Based on the Californian Northridge earthquake damage (Anon 1994) hidden (undetected damage) is likely to exist following this earthquake, particularly in steel and SRC buildings. A need to establish consistent methods for estimating the extent of damage and the residual strength of these structures is urgent now, in order to determine if these structures have the capacity to withstand future earthquakes.

Construction on fill material needs to be re-examined along with the phenomena of soil liquefaction. Reliable methods and techniques should be used to control settlements of structures founded on fill materials.

Pancake collapse, or soft storey collapse, in building structures was mainly due to using SRC for the lower storeys and RC for the upper storeys. This formed a discontinuity in construction material and column sizes at the level of transition, forming a brittle weak region in the building. Research is needed to quantify the damage exhibited by these structures using proper dynamic response analyses with models representative of these damaged structures. The results of this study? then, should be fed back to the seismic design codes of practice.

Good supervision and maintenance are two important factors that should be maintained during the construction stage and throughout the intended life of the structure. This will ensure that the seismic resistance of the structure is not impaired from, at least, what it was designed to be.

Acknowledgements. The authors express appreciation for the support of Osaka University, Disaster Prevention Research Institute at Kyoto University, Matsumuragumi Technical Research Center, Taylor Woodrow Construction Holdings Ltd, University of Westminster, and Griffiths Cleator and Associates.

References

NAKASHIMA, M. & BRUNEAU, M. 1995. *Preliminary Reconnaissance Report of the 1995 Earthquake*, Architectural Institute of Japan (AIJ), DPRI, Kyoto University, Kyoto.

ANON 1995*a*. *Ashai Newspaper*, Japan. 25 January.

ANON 1995*b*. National Centre for Earthquake Engineering Research, *NCEER Response*, Special Issue, January. State University of New York at Buffalo, New York, 1995.

MORRIS, A. E. G. 1970. *Urban Japan – Image and Reality*. London.

ANON 1995*c*. *NHK Television News Reports*, 28 January.

ANON 1994. *New Civil Engineer*. 29 September.

Earthquake-mail (E-mail?) for low-seismic zone earthquake hazard assessment

A. den Outer & P. M. Maurenbrecher

Department of Engineering Geology, Sub-Faculty of Applied Earth Sciences, Delft University of Technology, Mijnbouwstraaat 120, 2628 RX, Delft, The Netherlands

Abstract. The Roermond earthquake of 13 April 1992 in the Netherlands, of magnitude 5.8 on the Richter scale, although claiming only one fatality through a heart-attack, caused significant damage estimated at £30–40 million. The low recurrence determined at 135 years for an earthquake of this magnitude, does not encourage the funding of research despite the large quantity of data assembled, especially in connection with surveys to assess compensation grants from the earthquake hazard fund set up by the Netherlands government.

Electronic mail (E-mail) is a tool made used both to transfer information and solicit information. An earthquake mail facility could be integrated in a Seismological Engineering Information System (SEIS). Such a facility is being set up to provide links between users and suppliers of the data and analysis methods in the SEIS. An earthquake discussion box, on E-mail, concerned with problems in earthquake engineering in low-seismicity countries in northwest Europe is a cautious preliminary initiative to provide a communication platform between researchers in seismology and earthquake engineering.

Introduction

Earthquake engineering research in the Netherlands received a significant impetus after the Roermond earthquake in April 1992. The magnitude of the earthquake was 5.8 on the Richter scale. Initial studies on the earthquake were reported by van Eck & Davenport (1995). The earthquake initiated new research in the Netherlands and re-affirmed the relevance of former research.

Earthquake research in a low-seismicity country consequently means low budgets, no national organization focusing on earthquakes, few specialists working full-time on the problem, and sporadic communication between the individual researchers working in the earthquake field.

The main benefit of good communication is that it prevents reinventing the wheel; encouraging technological progress rather than duplication. Many possible contact channels exist between specialists in earthquakes. The diversity of the disciplines concerned with earthquakes can cause an imbalance of emphasis. The *Geologie en Mijnbouw* publication is weighted more towards seismological papers than to earthquake engineering subjects. Communication between the different partners in earthquake sciences involves the translation of specialist knowledge up to (or down to) the level and relevancy in other disciplines. In the UK, the Geological Society of London has an Engineering Group, a Tectonics Group and an Information Group, all three of which could contribute towards earthquake sciences. One way to encourage a cross-flow of information is to hold joint meetings. These need careful and formal planning. How can one encourage more intensive exchange of knowledge and information without having to resort to elaborate planning arrangements?

Networking

To ensure the transfer of information and knowledge between different locations, times, time zones, researchers or disciplines successfully, it is necessary to be complete and clear in reporting the source data and methods used. Interactive communication can allow for expanding on aspects poorly understood or, if such expansion cannot take place, a message can be left in a discussion box asking for clarification.

The newly developed ‘Internet’, ‘electronic highway’ or ‘cyberspace’ with, amongst others (see Table 1), E-mail facilities, which have become almost as common as the telephone, offer a solution to the communication links. The links are short and the computer will park messages until the other party is available.

The discussion boxes especially form a link between people with common interests. This link allows groups of researchers to enjoy, follow and/or join in discussions on specific topics.

The solution for fast communication between widely spread researchers with the same interests might very

DEN OUTER, A. & MAURENBRECHER, P. M. 1998. Earthquake-mail (E-mail?) for low-seismic zone earthquake hazard assessment. *In*: MAUND, J. G. & EDDLESTON, M. (eds) *Geohazards in Engineering Geology*. Geological Society, London, Engineering Geology Special Publications, **15**, 117–122.

Table 1. *Internet facilities for communication*

Type of facility	*Description*	*Comments*
E-mail	Electronic mail is used to directly send a personal message via the Internet. The sender has to know the E-mail number of the addressee.	The tool is useful for private communication with a short time-delay (about 10 min, if both parties are available). The tool is less useful for sending files, software and graphs.
Mail boxes or discussion boxes	The internet boxes are meeting places of people with the same interest. Messages sent to the boxes will be relayed to every member of the box, so contact with yet unknown interested persons can be made.	The boxes are useful for relaying research work and ideas to colleagues. However, since the boxes are public, privacy of ideas and work is not possible, unless the boxes are screened for new entries. Again this tool is not suitable for transferring files, etc.
Bulletin Boards	Bulletin boards (BBs) are often public, sites where any type of information in files, sometimes even sorted on topic, has been or can be dumped. The files are left behind for public use. The BBs sometimes have menu systems to find your way through themes or latest information.	BBs are useful for obtaining software (only shareware), and comments on software by users and data. However, a regular update of articles or software is the responsibility of the contributors and therefore is sometimes not available.
World Wide Web	The World Wide Web (WWW or 3W) is an improved (especially with respect to standard access and lay-out) version of the BB. A Windows-supported environment makes searching very easy. The trend on the Internet is developing towards 3W, which forces a lot of companies to ensure up-to-date information on their home-page on 3W.	3W allows the user to quickly find his/her way through the electronic highway. The transfer of all kind of data in various formats is more or less bug-proof. However, using 3W one needs the same experience and tricks as using a library system, as not all institutes, topics or other information can be found easily.
CD	As if to simulate networks a CD (compact disc) can contain a tremendous amount of information. These can be extensive bibliographic databases such as Geodex, or more entertaining encyclopaedias such as MicroSoft's ENCARTA.	On the subject of 'earthquake', over 4000 words of text are given, with cross-referencing on keywords such as 'geology', 'seismology' and 'plate tectonics'. Several illustrations are provided such as photographs and the action pictures on the motion of faults.

well be this type of discussion box. The Tectonics Group of the Geological Society, for example, have set up such a box.

Internet communication is increasingly through WWW (World Wide Web) and Bulletin Boards (BBs). Written text, graphs, maps, photographs and computer programs are accessible all over the world. The access of these 'electronic sites' is relatively easy when using user-interfaces like Mosaic, Netscape, Gopher or ftp. Although it is possible to send all sorts of files with E-mail to another person, user-interfaces will reduce the likelihood of accumulating irrelevant or superfluous information.

Networks also allow access to database systems that are usually bibliographic. Many libraries offer access without passwords, whereas other systems such as PASCAL or GEODEX require passwords, obtained through paid membership. Use of this facility is similar to use of CD systems. The large storage allows for a complete encyclopædia with photographic and animated cinematic diagrams (e.g. fault kinematics; Encarta 1995).

Lessons in communication from historical earthquakes

History

'Around AD 130 the Chinese scholar Chang Heng, reasoning that waves must ripple through the Earth from the source of an earthquake, constructed an elaborate bronze vessel to record the passage of such waves. Eight balls were delicately balanced in the mouths of eight dragons placed around the circumference of the vessel; a passing earthquake wave would cause one or more of the balls to drop' (Encarta 1995). Presumably a number of these devices were installed at the main centres in China so that Chang Heng could plot the intensity attenuation of an earthquake by the number of balls dropped by the vessels. Communication was required to collect information about a particular event not much different to present seismological stations exchanging seismograph data to determine hypocentres, properties of the Earth's interior and attenuation of the various waves. Further communication is required to determine the MKS or similar intensity scale (MMS, JMA or USGS) by sending out questionnaire mailings to municipalities believed to have registered a particular earthquake event so as to determine the spatial distribution of intensities.

Connecting causes and consequences

Both historical and present research have focused more on predicting and determining the causes of earthquakes than on mitigating the damage caused by them by building better structures. Again the distribution of papers in *Geologie en Mijnbouw* (van Eck & Davenport 1995) bears this out: scientific articles on earthquakes not only out number articles on earthquake damage but also appear to have a higher level of scientific endeavour.

Scientific earthquake research did little to prevent the recent deaths in the Sakhalin Island M7.5 earthquake, where most buildings collapsed in the town of Neftegorsk, and which was considered to be the worst earthquake ever in Russia. Two other recent earthquakes were in Greece. The most recent, a M6.1 earthquake, affected Corinthia (15 June 1995, Greece) and caused considerable damage to buildings and loss of life. A month earlier (13 May 1995), a M6.6 earthquake in central-north Greece caused severe damage to housing amounting to 700 buildings that collapsed totally. As prevention would appear less expensive than remedial work it is surprising that this element is given relatively little attention.

A combination of factors result in the mistaken belief that our faith in science will ensure that earthquakes, like some diseases, will be prevented or, at least, that ample warning such as for storms will be given to allow potential victims to move to a safer place. The most recent earthquakes, such as the M6.5 event on 25 June 1995 in Taiwan, prove that this notion has not developed much further since Chang Heng began monitoring earthquakes.

Pertinent earthquake information

Despite these misgivings, the science of earthquake engineering has progressed sufficiently to allow for proper design of structures against earthquakes. The difficulty appears to be in channelling the information to those who must be informed of the dangers and how a building should be made safe. Such information was not lost on the Intel Corporation (who design and manufacture the semiconductors which form the central hardware to most computers). The Loma Pietra earthquake of 1989 caused damage amounting to $800 000 to their buildings and loss of revenue. Had no precautions been taken, the damage was estimated at $25 million (Anon 1995). Based on the figures given in *The Economist*, it appears that for every dollar invested in preventing earthquake damage, $10 is saved in losses should an earthquake strike. The saving is not only in remedial work but also in lost business as a result of plant being out of action.

In 1995, the Kobe earthquake was considered to have occurred in a relatively quiet area for Japan. The magnitude of 6.9 was regarded as relatively 'middling' (Anon 1995) compared to other Japanese earthquakes. It was a combination of the design of traditional housing (though modern structures were also affected) and the soft foundation soils that caused the high casualties.

Research emphasis

The Economist article (Anon 1995) significantly drives the message home: learn more about how structures and soil conditions are influenced by earthquakes and pass that knowledge on to the user. When earthquakes occur, a laboratory opportunity arises as structures are subjected to low-frequency high amplitude compression, shear and surface waves, often up to failure. The forces act on a matrix consisting of a multitude of designs together with variable soil conditions.

It is in low-seismicity countries that such relationships could be studied rigorously: earthquake codes for structures usually do not exist and, as most buildings remain intact, the initiating weaknesses in buildings can be recorded. Low-seismicity countries, though, lack the infrastructure to bring the appropriate disciplines together to set up projects and obtain finance.

Significant effort was put into collecting and analysing data shortly after the Roermond earthquake event of 13 April 1992. The work culminated into a symposium held at Veldhoven nine months after the event and most presentations are given in (Van Eck & Davenport 1995). Despite this initial activity, a surprising amount of data still requires dissemination for lack of interest (and hence finance) on the part of the government, possibly because two significant floods have diverted hazard attention away from earthquakes. New strategies have to be sought to salvage what Anon (1995) emphasizes: the engineering consequences of earthquakes irrespective of predictions.

Engineering earthquake infrastructure

The strategy sought makes appropriate use of Intel's electronics: to set up an infrastructure through network communications. The latest technology may aid a national or regional organization to perform research in earthquake engineering. Secondly, we suggest that such a communications facility is given the name: SEIS (Den Outer *et al.* 1994): Seismic Engineering Information System. Pin the name, for the time being, to all the information contained, thus far, on databases and files on the computer and to a discussion box in the network.

The Seismology Section of the Royal Netherlands Meteorological Institute (KNMI) are keen to have such a system operational. Hopefully other institutes such as the Dutch State Geological Survey (RGD) will also participate in the system.

Ideally, the SEIS should consist of, in order of priority, a database system, a geographic information system (GIS), and an expert support system, to supply the basis of earthquake engineering research and data storage. Furthermore, the SEIS should contain a communication module that is able to use E-mail and/or fax to make short communication links between the users of the SEIS possible.

The development of the SEIS is a long process and requires a lot of effort from each of the disciplines in earthquake engineering. There are 'many rivers to cross' regarding the ownership of data and information, as well as problem-solving routines, and the supply of computer time and space.

SEIS activity is expected in 1996/1997 and then it very likely only be completed for the part of engineering geology. It is not clear when (or if) the SEIS will also become available for the rest of northwest Europe.

Initially this was tried in 1993 using the network E-mail facilities to transfer information towards the University of East Anglia for a joint EEC proposal together with the KNMI, the University of Liege, The Royal Observatory at Uccel in Belgium and the University of Potsdam, Germany to carry out research on the Roermond earthquake. The facility was also used to approach researchers in Greece for possible participation, but lack of time and complexity of such EEC proposals prevented this possibility.

Earthquake electronic communication: results

SEIS is intended to become a complete and useful tool for earthquake engineering. Such a system may take a number of years to reach its full potential of integrated communication between earthquake engineers in the Netherlands and the rest of northwest Europe becomes available.

Recent developments in E-mail and discussion boxes for E-mail make it possible for any person with a specific interest to request his/her own discussion box. Advertising to other interested persons will result in a group of people discussing the same topic via this discussion box.

It may be praiseworthy to start such a discussion box, but such initiatives may come to little if there is a lack of interest. To determine if there was any interest for an earthquake engineering discussion box and to test the effectiveness of discussion boxes, a message was left on a GEOLOGY discussion box on the MAILBASE system in the UK.

It became clear that, after asking this question on the GEOLOGY discussion box on the MAILBASE system in the UK, there was an interest, indicated by the positive response of about 25 people to the request. This response is somewhat incomplete as discussion boxes on seismology or other related topics (i.e. dynamic loading) to earthquake engineering have not been included. This would only increase the number of interested people.

Even though it was clearly stated in the request that the discussion box was only concerned with northwest Europe, people from other parts and outside of Europe also responded. These people originated from both low- and high-seismicity countries and will be a welcome addition to the discussion box.

People from other low-seismicity countries will probably recognize the problems in the Netherlands and can contribute with their experience in solving them. Researchers from high-seismicity countries have very likely progressed further through a national organization in establishing an earthquake engineering research platform in their country, but could contribute with their experience and expertise.

Some respondents informed us that there are already discussion boxes on the MAILBASE and other E-mail systems. However, these discussion boxes have a rather general character.

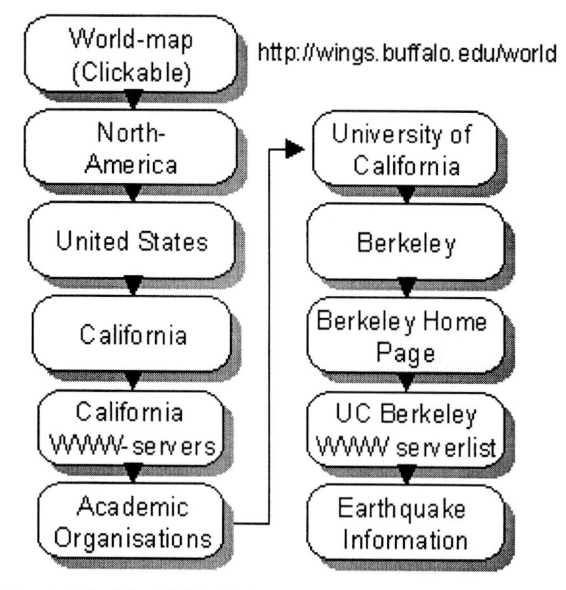

Fig. 1. The WorldWideWeb-path to earthquake engineering information at the University of California (Berkeley).

For example, the Earthquake Engineering information centre of Berkeley, which can be found on the WWW (see Fig. 1), gives information, from all over the world, on recent earthquake positions and strengths. This information is important but insufficient to work on a local level like northwest Europe, especially in earthquake engineering where causes and prediction of structural damage are of more concern.

Previous experiences with discussion boxes

Since discussion boxes have already existed for quite some time, it is easy to apply for them and become a member. The principal drawback remains that it is often very difficult to find out what the topic is that is being discussed in a specific discussion box. Often only the name of the discussion box gives an indication of the supposed contents of it. However, very often the name of the box contrasts with the messages it contains the messages are often unrelated and diverse in subject. One way to ensure that the discussion stays within the purpose of the discussion box, would be to screen membership applications.

The discussion boxes can be and often are fully automatically operated. This means that the owner of the discussion box (the person that first requested the discussion box) does not have a lot of work in monitoring the actions that take place. For example, members that do not submit messages for half a year are excluded automatically and have to apply again.

A problem with this is that there is no way that the owner has control of who discusses what on the box.

If the discussion box is semi-automatic then the owner will have to monitor the applications to the discussion box. In the case of the earthquake engineering box, this would mean that everybody is welcome initially, but those caught discussing late-night movies would be removed. Often, new members do not receive adequate explanation of what the rules and contents of the discussion box really are.

The earthquake engineering box will supply every new member with information about the object of the box, previous discussion topics and names of persons that played an important role in those discussions. In this way the mailbox will not be a collection of repetitions, and new members interested in earlier discussions can directly address the people involved in the discussion.

Results from use of the Internet and E-mail

After the Kobe earthquake on 17 January 1995, the amount of scientific information available shortly after the tremor in the newspapers was neither large nor of good quality.

Two days after the earthquake it was possible via the WorldWideWeb (WWW) to retrieve the recorded seismic signal as well as stereographic aerial colour photos from the harbour site. This information was obtained from Kobe University in Japan, which had already made a separate entry for specific information about the earthquake.

It is clear that the speed of data transfer is highly advantageous with respect to obtaining preliminary results, allowing others to make comparisons and validate such information. Contact persons were clearly mentioned in the WWW pages so any queries about the information could be transmitted and replied to within a few hours.

This way of using the Internet may be typical for countries with an earthquake tradition, as the following examples will show.

Neither the twin-quake in Greece in spring 1995 nor the M4.8 earthquake of June 1995 that Belgium experienced near the border with France could be located on the WWW. Information could only be obtained through the newspapers, radio and television. This information included little more than the location, time, strength, number of casualties and amount of damage.

One item of information worth considering, as an earthquake engineer, was a remark on the radio that earthquakes appear every 10 to 20 years in this part of Europe and that the last one appeared in Roermond, The Netherlands, in 1992. Although the reporter did not realize the flaw of this information, it is a major discussion point between researchers, responsible institutes and government. How valuable are the recurrence statistics in our discipline? A problem that can hardly be solved in low-seismicity countries, due to the lack of information, and which can often cause deception as was the situation for the relatively 'low-seismicity zone' of the Kobe area.

Apart from some newspaper clippings, details on the Sakhalin earthquake in Russia (the biggest so far in that area) could not be found anywhere on the WWW.

However, interested earthquake engineers can easily obtain information on the most recent earthquakes anywhere around the world. By sending an E-mail message to *qedpost-request@neis.cr.usgs.gov*, containing the message: *subscribe qedpost firstname lastname*, one will receive a daily update of all earthquake events.

Conclusion

A need exists to increase communication between workers on earthquake engineering, especially in low-seismicity countries. The initial step is to start an earthquake engineering discussion box focusing on problems in northwest Europe.

At the time of writing the earthquake discussion box for northwest Europe was not available. Some management steps still have to be taken to make the box available to everyone that is interested.

The proposed discussion box will be: EQUAKE-ENG-NWEUR. To subscribe to this discussion an e-mail message should be sent to: *mailbase@mailbase.ac.uk*, containing the text: *subscribe EQUAKE-ENG-NWEUR @mailbase.ac.uk*. All necessary information will be sent to applicants once the subscriber's request has been received.

References

ANON, 1995. *The Economist*, 22 April, 3–11.

DEN OUTER, A., MAURENBRECHER, P. M. & VAN ECK, T. 1994. Development of a seismological engineering information system (SEIS) for the April 13th, 1992 Roermond earthquake, the Netherlands. *In*: *Proceedings of the International GeoInfo V Conference*, June 1994, Prague, Czech Republic.

VAN ECK, T. & DAVENPORT, C. A. (eds) 1995. Seismotectonics and seismic hazard in the Roer valley graben: with emphasis on the Roermond earthquake of April 13th, 1992. *Geologie en Mijnbouw*, **73**, 2–4.

ENCARTA 1995. An Earthquake. CD Encyclopaedia extract.

SECTION 3

SLOPE STABILITY HAZARDS

Engineering hazards in the Taroko Gorge, eastern Taiwan

D. N. Petley

Department of Geology, University of Portsmouth, Burnaby Building, Burnaby Road, Portsmouth PO1 3QL, UK

Abstract. The population of Taiwan is concentrated on the low-lying coastal strips, separated by the Central Mountains. Whilst communication across the island is economically essential there are only three usable roads. This paper investigates natural hazards on the easternmost section of one of these roads, the Central Cross Island Highway. In the study area the road follows the base of the 600 m deep Taroko Gorge in a series of tunnels and rock ledges. The road is heavily used for transport whilst the gorge is one of the nation's premier tourist attractions.

Maintenance of the road in Taroko Gorge represents a profound challenge to the engineering geologist. The walls of the gorge are composed of deformed marble, gneiss and schist and are prone to failure under the intense seismic activity. Additionally the area is prone to an average of three to four tropical cyclones per annum, with up to 1200 mm of rainfall in each.

A detailed geomorphological and engineering survey of the gorge has been undertaken. The data are analysed along with information obtained from the highway maintenance board. It is demonstrated that during typhoons the area suffers an unusually severe set of geological hazards. The heavy rainfall initiates rockfalls and landslides along the gorge. Flooding occurs widely both as a result of the increased discharge of the river and as a result of water cascading onto the road from the adjacent cliffs. The destruction of important bridges may also occur. An examination is made of the magnitude of these hazards, and techniques used by the highway authority are described and analysed in relation to the use of the road for both transport and tourism.

Introduction

This paper reviews and discusses the engineering hazards associated with a part of the Central Cross Island Highway in Taiwan. The road is one of only three routes across the centre of Taiwan and is therefore of great strategic importance. Additionally it is subject to intense tourist pressures due to the beauty of the landscape through which it passes. However, parts of the road are subject to severe natural hazards including earthquakes and typhoons, presenting the engineering geologist with profound challenges.

Background

Taiwan is an independent island state of China located approximately 300 km off the eastern coast of the mainland. Whilst it is world-renowned for its economic power and the strength of its manufacturing industry, the island is predominantly mountainous, with the majority of the population and the industrial base being located in narrow coastal plains (Fig. 1). The Central Mountains extend the length of the island, rising to a maximum altitude of 3997 m. Lines of communication linking the heavily populated coastal plains have to cross extremely difficult terrain. The severity of the topography is such that only three major roads have been built across the range (Fig. 1). The Central Cross Island Highway, on which this study is concentrated, links the major city of Taichung with the central sector of eastern Taiwan, including the major cities of Hualien and Suao (Fig. 1). Alternative routes across the island involve long detours and are of poor quality.

The Central Cross Island Highway, which splits into two on the western side of the Central Mountains (Fig. 1), was completed with American assistance in 1960 at a cost of 450 lives. The road is about 200 km in length, reaching a maximum altitude of 2570 m. The easternmost section runs through the Taroko National Park, initially traversing relatively gentle highland slopes of the Central Mountains before descending along the walls of valleys carved by the upper tributaries of the Li-Wu River. From Tien-Hsiang the road follows the Taroko Gorge for 20 km before emerging at the coast at the mouth of Taroko Gorge. It is the section of road through Taroko Gorge on which the current study is centred.

Physical setting

Taiwan is located in a highly complex tectonic setting at the junction of the Eurasian and Pacific plates (Fig. 2).

PETLEY, D. N. 1998. Engineering hazards in the Taroko Gorge, eastern Taiwan. *In*: MAUND, J. G. & EDDLESTON, M. (eds) *Geohazards in Engineering Geology*. Geological Society, London, Engineering Geology Special Publications, **15**, 125–132.

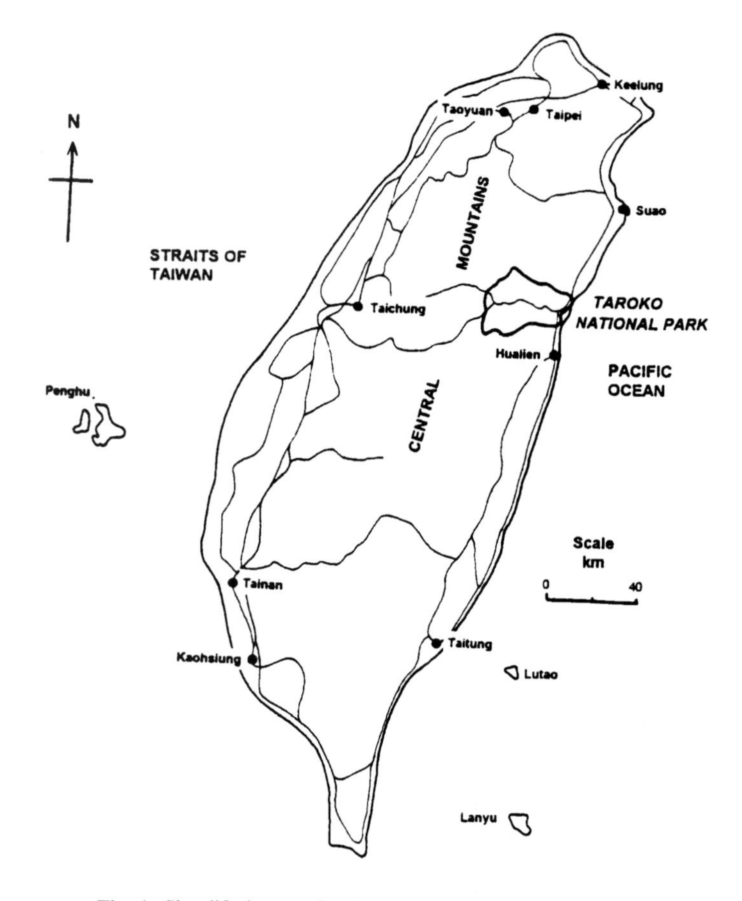

Fig. 1. Simplified map of Taiwan (after Longley *et al.* 1992).

The central mountains of Taiwan are predominantly composed of metamorphic rocks, with the grade of metamorphism increasing to the east (Ho 1987). The metamorphic rocks are believed to have been emplaced during a western underthrust of Pacific (oceanic) crust beneath the continental Eurasian plate during the Mesozoic. However, about 4 Ma the polarity of the subduction in the Luzon Arc, including southern Taiwan, appears to have reversed with easterly subduction of apparently oceanic material attached to the Eurasian plate beneath the northwesterly moving Pacific plate (Fig. 2). This subduction appears now to have ceased as a result of contact between the Luzon Arc and the continental plate. Although active subduction has now apparently ceased, the collision zone is still experiencing compression and strike-slip movement along the boundary between the two plates, inducing contemporary uplift rates of 1.5–5.5 mm^{-1} in the Central Mountains (Lin 1991). The continued plate movement also accounts for the frequent shallow earthquakes felt in Taiwan. Taroko Gorge is one of the most seismically active parts of Taiwan, with an average of 116 earthquakes per year being recorded. Most of the shocks are relatively minor, although historical records suggest that Taiwan was affected by 95 major earthquakes in the period 1624–1895 for example (Tsai *et al.* 1987).

The high rates of uplift, the steep mountains, the highly faulted geology and regular seismic activity promote difficult conditions for the engineering geologist. However, the picture is further clouded by the extremely high denudation rates to which Taiwan is subject. These are primarily a consequence of the high uplift rates, the highly faulted geology, and the impact of tropical cyclones.

Taiwan is prone to an average of three to four tropical cyclones (typhoons) per annum (Hung 1987). Typhoons normally develop in an oceanic area around Guam in which sea surface temperatures exceed 27°C, thereafter tending to travel in a northwesterly direction at about 20 km hr^{-1} for 4–11 days before dispersing over the continental land-mass. They typically have a diameter of

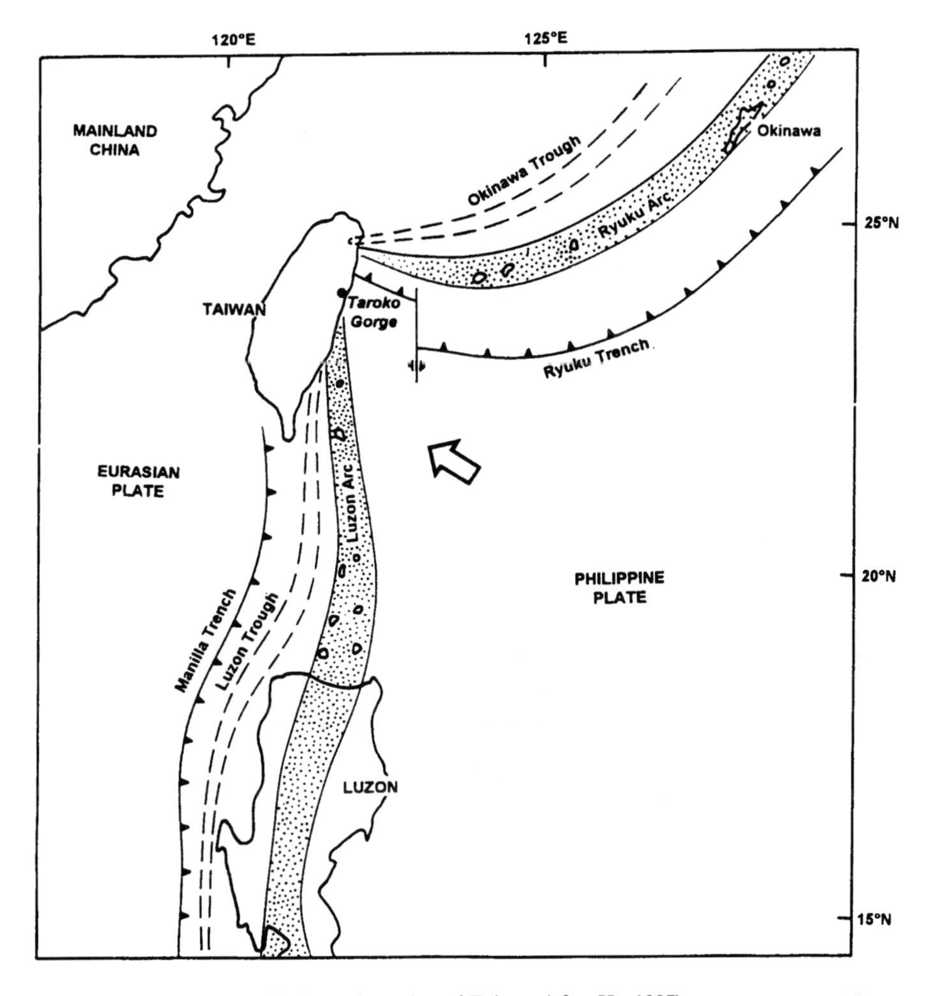

Fig. 2. Tectonic setting of Taiwan (after Ho 1987).

150–300 km, with intense winds rotating around a low pressure 'eye'. Locally, gusts of wind of up to 400 km hr^{-1} (111 m s^{-1}) and precipitation rates of over 1200 mm in a 24 h period may occur. Typhoon Lynn, which struck Taiwan on 13–27 October 1991, deposited 1900 mm of rain on parts of Taiwan, including a maximum of 1100 mm rain in a 24 h period.

Taroko Gorge

Taroko Gorge is a site of great natural beauty (Fig. 3). It has formed as a result of the high uplift rates and large magnitude climatic events, allowing the Li-Wu River to carve a ravine up to 600 m in depth through marble, gneiss and schist. The Central Cross Island Highway runs between 10 and 50 m above the river bed at the base of the gorge, cut on a series of marble ledges connected with bridges and tunnels (Fig. 4). The highway provides ready access to the spectacular scenery and is consequently an extremely popular tourist attraction. Responsibility for the road lies jointly between the Highway Maintenance Bureau and the Taroko National Park Authority. These two agencies have to contend with serious difficulties associated with the use of the road both as a major communication link and as a recreational resource. Visitors to the National Park tend to arrive either by car or by coach, and usually remain within their vehicles for most of their visit, with the exception of certain readily-identifiable locations such as the exhibition at the National Park headquarters and a number of viewpoints. However, guided walking expeditions across Taiwan are popular with large groups of school children in the summer vacation. Such trips

Fig. 3. A view of a section of the Taroko Gorge, showing the steep marble cliffs. Arrowed is a section of the CCIH, sited on a ledge before entering a tunnel. For scale, the tunnel is sufficiently large to (just) accept two double-decked coaches side-by-side.

include the walk along the road running at the base of the gorge. Whilst the nature of the conflict over the use of the road for both tourism and transport is beyond the scope of this paper, the duality of use is of relevance to the methods that the highway agencies use to maintain the road and surrounding area. Clearly it is in the national interest to keep the road open for as much of the year as possible; clearing the road of debris is a major priority once a typhoon has passed. However, there is also a need to ensure that the visitors are offered the maximum protection to ensure that they are not endangered, whilst still providing access to the scenery.

Three major hazards exist on the road. Firstly, rockfalls, either in the form of isolated blocks or in larger masses, are common. Secondly, there is some landslide activity in parts of the highway. Finally, there is a hazard from flooding during high intensity precipitation events. The remainder of this paper is concerned with the nature of these hazards and the techniques used by the highway agencies to engineer against them in light of the demands made by the road users.

Methodology

A detailed analysis of the Taroko Gorge section of road was made during a prolonged field visit in the summer of 1991. Three main surveys were undertaken, in conjunction with other members of a team of Earth scientists. These were a geological survey, a geomorphological survey and a survey of engineering structures within the gorge. These were backed up with a desk survey and a review of the effects on the highway of Typhoon Amy on 18–19 July 1991. The geomorphological survey identified important features such as cliffs, landslides, debris chutes and channels, and allowed an understanding to be gained of the processes that act to form the landscape. A section of the geomorphological map is presented in Fig. 5 whilst the geological survey allowed correlation between the types of hazard and the underlying geology. The engineering survey allowed an analysis of the efficacy of the technological solutions used by the highway authorities.

Results

The geological survey

The Taroko Gorge predominantly consists of marble, schist and gneiss (Fig. 4). These rocks, of Palaeozoic and Mesozoic age, underwent considerable deformation during the Cretaceous and Tertiary as a result of the complex tectonic history.

The majority of the area (approximately 65%) consists of massive Permian marble. The hardness of this rock is readily apparent as it forms most of the extremely high cliffs found along the gorge. However, there is also considerable faulting and jointing, resulting in a high incidence of small rockfalls. These predominantly occur during typhoon events when large volumes of water are able to enter the discontinuities. Approximately 20% of the area is composed of a coarse gneiss. Additionally, two types of schist, black and green, were identified, comprising about 15% of the area. The black schist readily weathers into angular fragments allowing the development of colluvium. The green schist is located in relatively isolated bands within the some of the marble units. It preferentially weathers, controlling the local drainage pattern and allowing the development of caves and notches.

There are also some isolated areas of Quaternary deposits overlying the solid geology. Such deposits

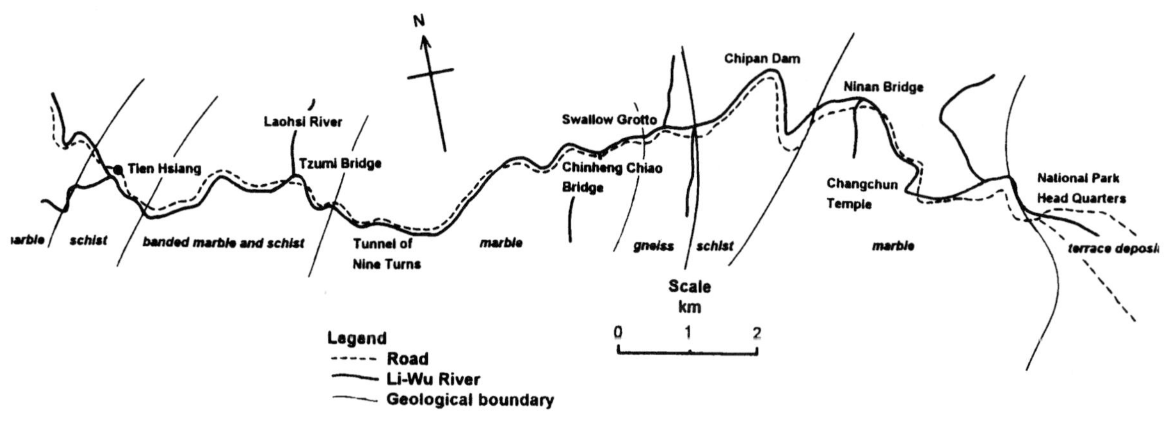

Fig. 4. Map of Taroko Gorge including geological boundaries.

include colluvium, which results from the accumulation of debris from higher slopes, and fluvial deposits. Some of the fluvial deposits are found at considerable altitude; for example, thick, fluvial, Quaternary deposits are found at about 200 m above the valley floor at Pulowan.

Geomorphology and engineering geology

Flood hazards. The high precipitation rates that may accompany typhoons can induce a massive increase in flow in the Li-Wu River. Typhoon Ofelia, which occurred in 1990, caused over $100\,mm\,h^{-1}$ of rainfall for two consecutive hours at Taroko. As a result, peak river flows in the Li-Wu measured $1600\,m^3\,s^{-1}$, with a lag time of 5 h. For reference, base flow is approximately $40\,m^3\,s^{-1}$.

It is clear that such major typhoons may induce major engineering hazards in the Taroko Gorge. During such an event the highway is closed to all traffic as the roadway becomes inundated by the extremely large volumes of water that both fall directly onto the road surface and also cascade from the adjacent slopes. During smaller typhoons the road may not become flooded although considerable volumes of water still flow. The Li-Wu River itself does not tend to affect the highway, which is a minimum of 10 m above the river bed. However, the four bridges across the river are more vulnerable to high magnitude flood events. The majority of the crossings are made on girder or suspension bridges with no support from the river bed, but the main crossing of the coastal highway, near to the river mouth, is sufficiently wide to require support from the river bed. During Typhoon Ofelia this bridge was destroyed. A replacement was in service within a year of the destruction of the original bridge, but the loss of such a major crossing caused considerable disruption to communication links. During reconstruction of the bridge, all traffic was required to cross the river via a temporary ford located 2 km upstream. However, the now-disused ford was destroyed in the first typhoon of 1991, although the new bridge was undamaged.

In addition to the main river crossings, the road also has to cross two major tributaries of the Li-Wu River, at Tzumu Bridge and Chinheng Chiao Bridge. At the time of the survey, Chinheng Chiao Bridge did not show any signs of major damage as a result of flood events, although inundation of the deck of the bridge as a result of poor drainage did occur during periods of heavy rain. Both the supports and the deck of Tzumu Bridge, which is approximately 5 m above the floor of the tributary, showed signs of typhoon damage. This was probably caused by large boulders transported during high flow events. At the time of the visit, damage to the bridge had been repaired with temporary supports.

Some attempts have been made to protect against the effects of flood events. Approaches include the use of high level and girder bridges such that the supports and superstructure are clear of the channel. This has prevented the loss of the majority of the bridges. The bridge under construction east of Changchun Temple uses a single span to avoid flood hazards.

Keeping the road passable during typhoons is probably not a viable option given the nature of the topography both within the gorge and further to the east. Therefore there is little to be gained in preventing flooding of the road surface, although this could be reduced by using the geomorphological map (Fig. 5) to deduce the areas that are most prone to inundation from the surrounding cliffs.

In conclusion, flood events represent a considerable problem in major typhoons. Temporary flooding may render the road impassable. Greater difficulties result from structural failures of bridges along the highway. In consequence, flooding is a major problem for road users after a typhoon but does not represent a major hazard for tourists as the gorge is inaccessible during major typhoons.

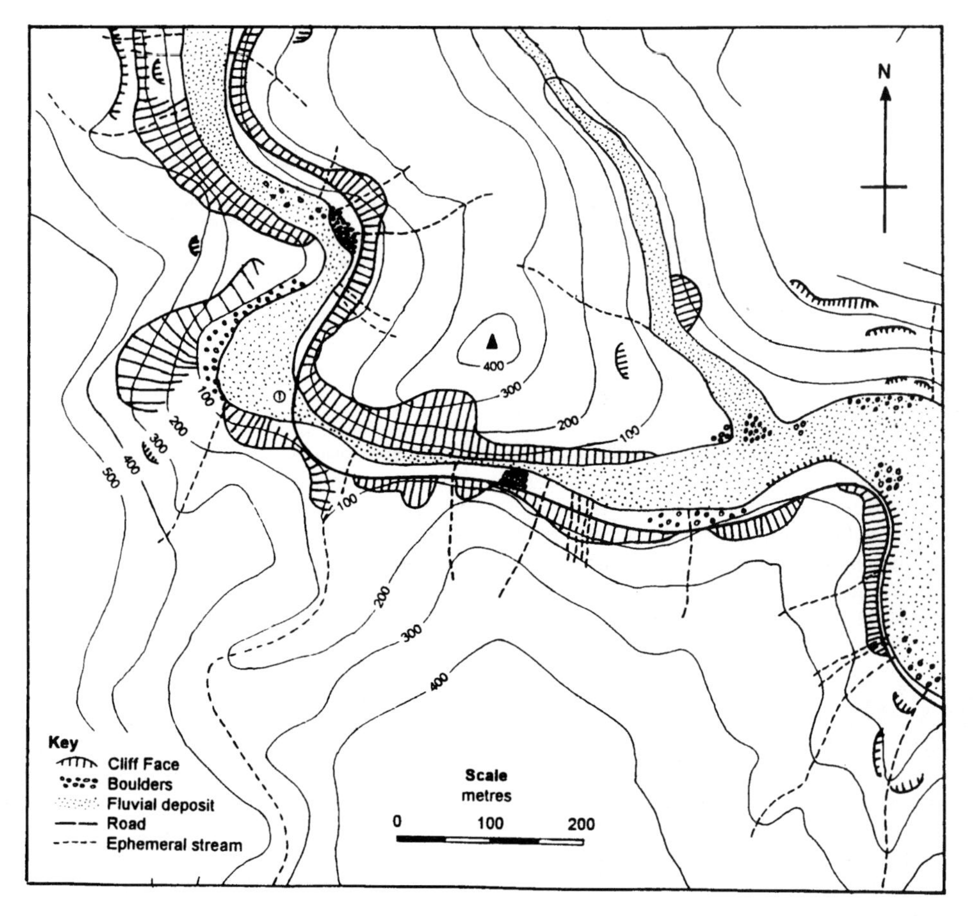

Fig. 5. Geomorphological map of the area around Changchun Temple.

Mass movements. The major hazard to the highway and its users results from the instability of slopes (Fig. 6). Primarily the hazard is from slopes above the road, although in a few places the road bed may be removed by landslide activity. It is difficult to classify the types of hazard into distinct groups as activity represents a continuum from dry rockfalls through debris flows to landslides. However, for simplicity, the problem of mass movements will be considered in two sections: rockfalls, including debris flows, and landslides.

Rockfalls. Taroko Gorge consists of high, steep rock walls in heavily disturbed rock masses. As a result the incidence of rockfall activity is very high. Two types of rockfalls commonly occur in the gorge. Firstly, there are major movements involving large volumes of material. Secondly, there are minor rockfalls, in which single lumps of material become detached and fall. Much of this activity occurs during typhoons when water may enter discontinuities in the rock mass or blocks are transported by ephemeral channels that terminate at the edge of the cliff. A further possible trigger for rockfalls is the almost constant seismic activity. Fatal rockfalls occasionally occur; for example, three tourists were killed in a major fall close to Changchun Temple in 1990. After a typhoon the highway may be closed as a result of rockfalls; the highway was closed for a month after Typhoon Ofelia in 1990.

The incidence and impact of rockfalls has been analysed on the basis of two sets of data. Firstly, the geomorphological survey, coupled with monitoring carried out during the summer of 1991, allows the mechanisms of rockfall events to be elucidated. Secondly, Longley *et al.* (1992) analysed data on the frequency of road clearances collected by the Highway Maintenance Bureau.

The data from the Highway Maintenance Bureau suggest that during the period 1986–91, 67% of rockfalls were the result of typhoons (Table 1 after Longley *et al.* 1992). A further 23% of rockfalls occurred during

Fig. 6. A rockall scar in a vegetated slope. Note the house located directly below the scar; the road is located between the house and the cliff-face.

periods of heavy rainfall. The remainder of the rockfalls (*c.* 5%) were the result of minor seismic activity. The area is also prone to periodic major seismic shocks. At present it is not possible to estimate the effects of a very major earthquake, which would probably induce extensive damage. The geomorphological survey revealed that considerable amounts of loose material, including boulders with a diameter in excess of 15 m, are located in debris chutes on the slopes above the gorge. A major earthquake would probably induce at least some of these to fall, inducing massive geomorphological change and causing a severe hazard to road users. However, the occurrence of a major earthquake is so unpredictable that it is almost impossible to legislate for. Possible measures to protect users could include the lining of some of the numerous tunnels (at present, most tunnels are unlined) to provide shelter against falling rocks.

The data have also been analysed in terms of the geology of the rock mass (Table 2). Rockfalls mainly occur in areas with marble or gneiss bedrock. These materials are able to sustain steep slopes (50°–90°) and therefore form steep cliffs. Weathering of these materials is relatively slow; most denudation appears to occur as a result of rockfalls.

These data suggest that marble is more likely to be the cause of major rockfalls and that such falls will include larger volumes of material than those in gneiss. One engineering approach to reducing hazards on the highway would be to protect areas of road below marble cliffs by emplacing a layer of shotcrete or a steel mesh over the slope. However, in many places this is precluded by the height of the cliffs and the requirement to maintain the natural beauty of the area. Alternative measures include the construction of avalanche-type shelters, as has been constructed approximately 1 km E of Chipan Dam (Fig. 4), where 75 m of road have been protected. The efficacy of this shelter is evident as a large amount of debris has accumulated on its roof although observation during Typhoon Amy suggested that the unprotected area immediately east of the shelter is also badly affected by rockfalls. A second solution is the avoidance of hazardous stretches with the use of tunnels. An active programme of re-routing more hazardous stretches of the road through tunnels is being undertaken. The primary justification for re-routing is to increase pedestrian safety on congested stretches of road, but a secondary advantage will be to reduce damage to the roads from rockfalls. Perhaps the simplest option is to continue to allow the use of such stretches of the road but to prevent pedestrian access. However, at present such an option is not deemed acceptable due to the demand for walking tours through the gorge.

A large proportion of the slope failures occur in the schist. Rockfalls in this material tend to have a greater volume than those of other materials (Table 2). The schist undergoes rapid weathering in the semi-tropical climate as a result of its soft, layered structure. Slopes in the schist tend to be to have developed a layer of colluvium and to have become vegetated. Consequently failures in the schist tend to be in the form of relatively low strain rate translational landslides within the colluvium, activated during typhoons. As the road has been built on ledges cut into the bedrock, most of the landslides occur on the slopes above the road. The low rates of movement do not represent a serious hazard to road users but do induce frequent blockage of the highway. The large volumes of material involved (averaging 2494 m^3) require a considerable effort to clear.

One landslide, just west of Tien Hsiang (Fig. 4), causes very significant disruption to the road. This landslide has an arcuate scar approximately 100 m in width and 150 m high and has a shear surface at about 3 m below the surface (Longley *et al.* 1992). Movement during typhoons occurs along steeply dipping planes of foliation as a result of two effects. Firstly, the construction of the road has over-steepened the slope and induced failure; secondly, the channel of the Li-Wu is eroding the toe of the landslide. Movement occurs below the level of the road, inducing subsidence, and above the road, leading to the deposition of material onto the road surface. The landslide threatens the road during every typhoon; for example, the road was blocked for three days after Typhoon Amy in 1991.

The highway authorities have taken very direct action to reduce the hazard posed by the Tien Hsiang landslide. During the visit in 1991 a new tunnel was being driven to by-pass this area. The new road will have the added advantage of separating tourists, who will probably still use the old road, and other road users, who will prefer to utilize the new tunnel. The continued use by tourists of the road across the Tien Hsiang landslide will require that the road is maintained; however, there will not be the urgency to repair it in the event of a major typhoon.

Many other small landslides were noted adjacent to the road. Some of the larger examples include translational slips at Tzumu Bridge and at Chipan Dam (Longley *et al.* 1992). All of the landslides represent a hazard in the form of blockage of the road during typhoons but are generally not hazardous to the users.

Discussion

As outlined above, the Taroko Gorge section of the Central Cross Island Highway represents an environment with a unique set of challenges for the engineering geologist. The location of the road near to the base of the gorge induces hazards from flooding, rockfalls and landslides. It is fortunate that the narrowness of the gorge eliminates any hazard from the strong winds that are also associated with typhoons. The use of the road for both communications and tourism leads to a conflict in requirements and places a high level of obligation on the highway authority. Since the creation of the National Park in 1986 there has been immensely high investment in improvements to the road, including the construction of four new sections of tunnel, one of which has required a new river crossing. Each of these tunnels will have the effect of protecting the road against blockage and of reducing construction. Avalanche shelters have been built in areas where there is a particularly notable hazard from rockfalls; generally these have been highly successful, although in places they need to be extended. The use of retaining walls to prevent shallow landslides in the schist have been widely used. Generally this approach is efficient in that it prevents large masses of material from slipping onto the road. However, during heavy rainfall the ground behind them becomes saturated, inducing saturated overland flow, which transports large volumes of debris onto the road surface. The amount of debris deposited during such an occurrence is small and is relatively easy to clear.

Finally, the highway authorities have used shotcrete and rock-bolting in a number of places; most notably where the road has to pass beneath overhangs or very jointed rock masses. This approach has been used sparingly but has been highly successful; the level of stability of treated faces appeared to be very high. However, the use of this technique is extremely limited in this area (it is widely used on other stretches of the highway) due to the poor aesthetic quality of treated faces.

Overall the highway authorities conduct a remarkable operation in maintaining the road. The level of hazard to road users appears to be relatively low considering the hazardous nature of the terrain. The authority use three main approaches to maintain the road:

- the construction of new stretches of road within tunnels to bypass areas with either a high incidence of problems or with severe construction;
- the use of engineering techniques to protect the existing stretches of road, including avalanche shelters, retaining walls and rock-bolting;
- continual clearance of the road after rockfalls and landslides, particularly as a result of typhoons.

Allied to this is an acceptance that no amount of engineering will allow the use of the road during typhoons themselves; instead the authority has concentrated on measures that will allow the re-opening of the road at the earliest opportunity. There is clearly a conflict between the need to undertake the construction programme to protect the road and the need to avoid environmental damage. Generally the authorities have balanced this problem admirably.

Acknowledgements. The author would first like to thank the Taroko National Park for their continued assistance to this research project. The fieldwork was conducted in collaboration with E. Longley, M. Ibsen, M. Pendry, D. North-Lewis and Chen-Hui Fan whose assistance was greatly appreciated. Financial support for this project was provided by The Royal Geographical Society, Sediment Deformation Research and NERC as a part of studentship GT/4/90/GS/86. Finally the assistance of the University of Portsmouth Department of Geology is gratefully acknowledged.

References

Ho, C. D. 1987. A synthesis of the geologic evolution of Taiwan. *Memoir of the Geological Society of China*, **9**, 1–18.

Hung, J.-J. 1987. Landslides and related researches in Taiwan. *Memoir of the Geological Society of China*, **9**, 23–44.

Lin, J. C. 1991. *Neotectonic landforms of the coastal range, eastern Taiwan*. PhD Thesis, University of London.

Longley, E., Fan, C-H., Ibsen, M-L. North-Lewis, D., Pendry, M. & Petley, D. N. 1992. *The environmental impact of tropical cyclones*. Final report of the University of London Expedition to Taiwan

Tsai, C-C., Loh, C-H. & Yeh, Y. T. 1987. Analysis of earthquake risk in Taiwan based on seismotectonic zones. *Memoir of the Geological Society of China*, **9**, 447–464.

Integrated use of Landsat TM and SPOT panchromatic imagery for landslide mapping: case histories from southeast Spain

Richard Eyers,[1] John McM. Moore,[2] Javier Hervás[3] & J. G. Liu[2]

[1] De Beers Consolidated Mines, GeoScience Centre, PO Box 82232, Southdale 2135, South Africa
[2] Geological Remote Sensing Group, T.H.Huxley School of the Environment, Earth Science and Engineering, Imperial College of Science Technology & Medicine, London SW7 2BP, UK
[3] Institute for Remote Sensing Applications, Joint Research Centre, European Commission, 21020 Ispra, Varese, Italy

Abstract. This paper demonstrates some of the ways in which digital space imagery can be used for landslide mapping. Landslides ranging in size from a few metres to several hundred metres across have been studied at a test site in semi-arid terrain in Almeria Province, SE Spain. The study illustrates the capabilities and limitations imposed by spatial and spectral resolution in Landsat Thematic Mapper (TM) and SPOT panchromatic space imagery. The results demonstrate the scales at which, landsliding which can be studied, using SPOT P and Landsat TM imagery.

Spectral and textural features of landslides were enhanced digitally. Textural discrimination was improved by linear contrast stretching combined with edge enhancement by adding back the results of Laplacian filtering. Spectral features of landslides, including soil moisture dependent vegetation and rock debris types, are well-displayed in supervised RGB colour composites and band ratio images.

A data integration technique, the Brovey Transformation, was used to produce colour composites combining textural and spectral features from SPOT Pan and Landsat TM imagery. Landslides with a minimum width of 250 metres can be identified in space images. Interpretation of results revealed the presence of several previously unrecorded landslides.

Daedalus 1268, Airborne Thematic Mapper (ATM) digital imagery has been used to demonstrate the value of improved spatial and spectral resolution imagery for landslide mapping on the scale of the Spanish test sites. The results illustrate the future system requirements for more detailed landslide mapping and demonstrate the improvements of scale which will occur when space imagery with pixel sizes of five metres becomes more readily available.

Introduction: Rio Aguas Valley landslides

The flanks of the Río de Aguas Valley, east of Sorbas village, in Almeria Province, contain numerous landslides. Most of these structures originated during a period of rapid valley incision related to Pleistocene and Holocene drainage base level change. The current semi-arid climate and sparse vegetation cover, particularly on south facing slopes, have caused some slides to reactivate during periods of excessive rainfall and flash flooding and when slopes are artificially over-steepened by engineering work, as has happened near the study area recently, during major road building projects.

The Rio de Aguas Valley contains incised meanders but runs approximately parallel to the strike of a series of northward dipping Miocene (Tortonian and Messinian) molasse strata and gypsum beds. The strata can be divided into three packets of marls and sands, separated by limestone units and gently dipping, massive gypsum beds, containing an extensive karst. The beds dip northward, across the river valley, at angles between 5° and 25°. Landsliding instability has in several cases been caused through loading of unconsolidated marls (low shear strength) by massive gypsum and limestone beds in areas of water seepage. Instability on the south side of the valley is dominantly combined rotational and dip slippage, accompanied by flow in the lower marls and sandstones, caused in part by loading from the overlying limestone unit. In the upper statigraphical unit, local dip slope movement of marls has been induced by overlying massive sandstone beds. On the northern flank of the valley, toppling failure along the escarpment of a massive gypsum unit several tens of metres in thickness is linked to flowage in the middle unit marls. The result has been to create several marl debris flows, carrying large gypsum blocks. A recent (1991) dual carriageway motorway has been routed across at least one of these flows.

Two test site localities for the dip slope, and two for the scarp slope conditions were chosen for study (Fig. 1). A dataset of panchromatic SPOT, Landsat Thematic Mapper (TM) digital space imagery and Daedalus Airborne Thematic Mapper (ATM) imagery was used. For

Eyers, R., Moore, J. McM., Hervás, J. & Liu, J. G. 1998. Integrated use of Landsat TM and SPOT panchromatic imagery for landslide mapping: case histories from southeast Spain. *In*: Maund, J. G. & Eddleston, M. (eds) *Geohazards in Engineering Geology*. Geological Society, London, Engineering Geology Special Publications, **15**, 133–140.

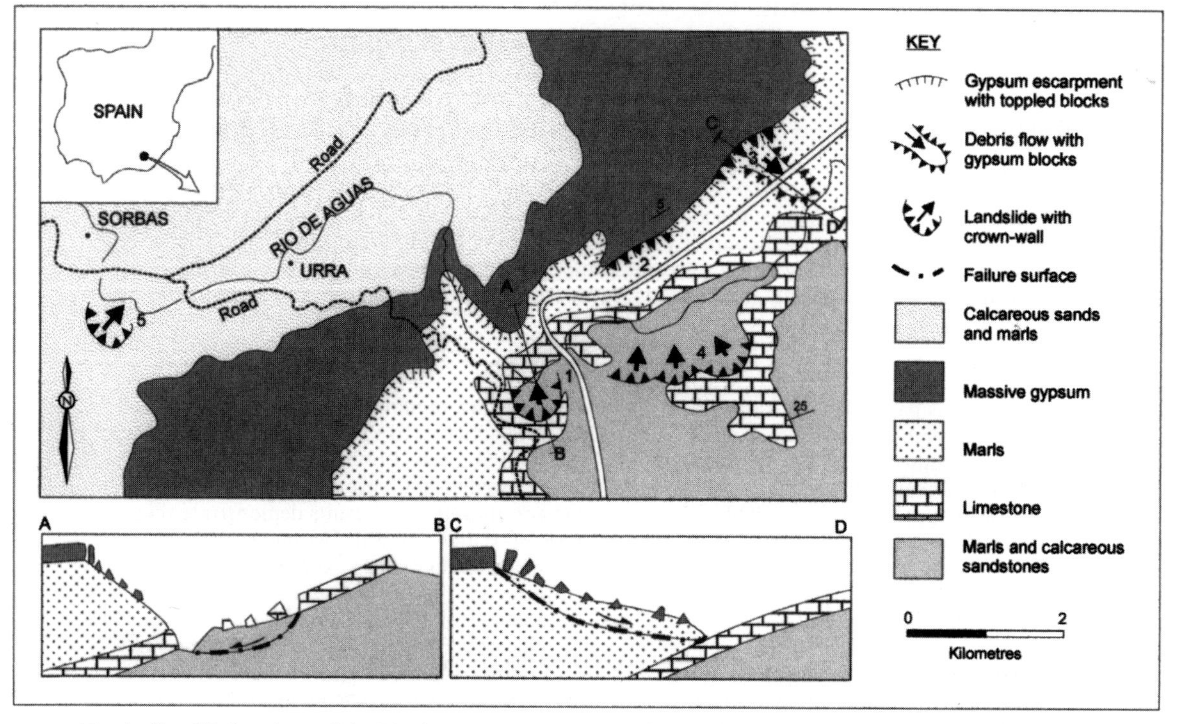

Fig. 1. Simplified geology of the Rio de Aguas area showing the landslide localities: (1) Carrasco Landslide, (2) El Tesoro Escarpment, (3) Marchalico Viñicas Debris Flow, (4) Los Perales Landslides and (5) Maleguica Landslides. Sketch sections A–B and C–D show the generalized form of landslides 1, 2 and 3.

each of these localities, enhanced images have been studied to assess the capabilities of space and aircraft systems for display of textural and spectral information useful for landslide recognition and mapping.

One objective was to demonstrate the optimum ways in which image texture information on landslide form, crown-wall escarpment shape and slip material surface topography can be combined with the spectral properties of rocks and vegetation. The resulting images show the limits of scale to which space imagery can be used in landslide mapping and the improvements in representation of detail which will become possible by improving spatial and spectral resolution in future space platform imaging systems.

Digital imagery

The visible and short wave (vis/swir) infrared and thermal (mir) infrared space imagery used for this study came from the Landsat 5 and SPOT satellites. Currently, these are the space image types most widely used for regional geotechnical mapping. In addition, Daedalus 1268 (ATM) imagery was used for more detailed site investigation studies. The characteristics of the aircraft and space platform systems are summarized in Table 1.

ERS-1 synthetic aperture radar (SAR) imagery is also available for the study area but after preliminary appraisal of other research results (Mason 1993), it was decided that ERS-1 SAR was not suitable for landslide study in the Río de Aguas valley terrain because of layover distortion and shadowing effects.

Spectral enhancement techniques

Certain rock types in the banks of the Río de Aguas valley area can be identified using combinations of digital number (DN) values from three or more of the seven TM bands or eleven ATM channels. Individual lithologies are not easily identifiable in the monochrome SPOT panchromatic imagery.

A simple and successful approach to lithology discrimination is to use combinations of bands whose response to rock and vegetation is known from theory or experience (McKean *et al.* 1991). Identification of a landslide using spectral characteristics depends on the displaced material having a different spectral signature

Table 1. *The optical imaging system specifications*

	Pixel dimension	*No. of bands*	*Scene size*	*Spectral range (μm)*
SPOT pan	10 m	1	60 km × 60 km	0.51–0.73 (vis)
Landsat TM*	30 m	7	185 km × 185 km	0.45–2.35 (vis/swir) 10.4–12.5 (mir)
Daedalus ATM†	7.5 m	11	3.5 km × 5 km‡	0.42–2.35 (vis/swir) 8.5–13.0 (mir)

* The thermal channel (TM6) pixel dimension is 120 m.
† The spatial resolution and scene size are altitude dependent.
‡ Extracted from a 20 km swathe.
Information: Barrett & Curtis (1992) and NERC.

to the surrounding undisturbed terrain. In the Rio Aguas area this occurs only in a few slides which transport anomalously large quantities of gypsum or limestone debris away from their source outcrops.

Colour composite images

One of the best and simplest images for lithological mapping of the Sorbas basin is a false-colour composite image showing TM bands 5 3 & 1 in displayed in red, green and blue (Crósta & Moore 1989). A TM4, 5 and 7 (RGB) colour composite image gives acceptable lithological discrimination and highlights vegetation in red. The 30 metre pixel size of TM means that there are many pixels with mixed spectral signatures in the Landsat images of the Rio Aguas test sites and the TM images were too coarse to allow the identification of specific landslides (Fig. 2 upper part). This reduces the usefulness of more sophisticated enhancement techniques for terrain discrimination.

Colour composite images produced from three band combinations of ATM bands 9 5 and 2 ATM8, 9 and 10 and ATM8, 10 and 11 (RGB) are excellent products for basic interpretation (Wright 1995).

Directed spectral enhancement techniques

A characteristic type of mass movement in the study area is transport of gypsum blocks mixed with marls in a debris flow. Band difference techniques can enhance, selectively, terrain types with diagnostic spectral properties e.g. vegetation, gypsum, clays and ferruginous soils. Moore *et al.* (1993) advocate image band differencing for mapping gypsum using TM7-5.

The 7.5 metre pixel size ATM (aircraft) imagery reduces problems of mixed spectral signature pixels. The eleven band ATM sensor also permits improvement in discrimination of terrain types using more complete spectral signatures than are provided by spaceborne TM imagery. Wright (1995) used ATM spectral signatures to design a compound difference image to discriminate gypsum and vegetation, in which each pixel was analysed according to the following conditional statement:

$$\text{IF ATM2} < 90 \text{ or ATM 10} > 130 \text{ THEN NULL}$$
$$\text{ELSE (ATM 2} + \text{ATM 8)} - \text{(ATM 10} + \text{ATM 5)}$$

The resulting image was used to examine the distribution of gypsum debris in the Marchalico Viñicas debris flow (site 3).

Enhancements incorporating thermal data

The TM band 6 and ATM band 11 sensors (8–13 μm spectral range) differ from the visible and short wave infra red detectors of Landsat TM and the Daedalus ATM instrument. The thermal (mid-infra-red) sensor responds to emitted thermal radiation and can therefore, in principle, be used to identify terrain with anomalous thermal properties, including areas with unusual soil moisture content. Shikada *et al.* (1993) used Landsat TM thermal data in a landslide study. Unfortunately, the 120 metre pixel size of the TM thermal (Band 6) imagery effectively precludes its application to landslide mapping on the scale of features in the Rio Aguas area.

The ATM thermal band 11 data is recorded as pixels of the same size as the visible and *swir* bands. Consequently, the ATM can be used to assess the value of high-resolution thermal imagery for landslide related feature mapping. Gypsum outcrops in the Sorbas area have high DN values in ATM band 11. Chan (1994) used this to separate gypsum from marl in ATM colour composite images, to display the form of the gypsum escarpment and the Marchalico Viñicas debris flow which contains numerous gypsum blocks. Both gypsum and marl have similar spectral signatures in the visible and near infra red (reflected radiation) ATM bands. ATM 11 thermal information greatly improved the spectral discrimination of terrain types in the Rio de Aguas area (Fig. 3).

Fig. 2. Parts of Landsat TM (scene 099/34) and SPOT pan (scene 039/276). Images of the Rio de Aguas area. The top image is TM band 4; the lower part is a Brovey image from a Laplacian edge enhanced SPOT/TM merge using TM band 4.

Textural enhancements

Many features of the Sorbas landslides are textural and are most easily identified in higher spatial resolution imagery. Using the test-sites, it was possible to assess the relative value of different contrast enhancement techniques and texture filters as aids to structure mapping.

The three most evident textural features of the target landslides are arcuate crown-wall scarps, lobate toes and hummocky, broken or fissured surfaces on translated material. Movement inevitably alters the textural character of the slide material. Although individual debris blocks are not identifiable in the SPOT or ATM imagery, their textural effects are commonly expressed as topographic shadow patterns.

Contrast enhancement

Contrast enhancement alters the tonal range of an image and is the most useful basic enhancement technique for monochrome images. Many diagnostic features of landslides are revealed by irregular topographic shadows. Inter-active piecewise linear stretching (PLS) is one of the most effective ways to improve shadow display for qualitative interpretation. When preparing imagery for Brovey data integration it is essential that all input bands are stretched to occupy the full digital number (DN) range available (Niblack 1986).

Filtering

The irregular boundaries and surface texture of landslides are suited to enhancement using non-directional convolution filters. Much of the detailed texture of an image is contained in high frequency information and 'edges' between DN regimes, e.g. shadow boundaries or contact lines between spectrally different terrain types. Laplacian convolution filtering enhances high-frequency edge information. Adding the data from a Laplacian filter image to a properly weighted original image, produces an image with a similar DN range and contrast to the original but with sharpened edges.

Filtering is carried out by passage of a 'kernel' or 'box' filter across the image A typical addback

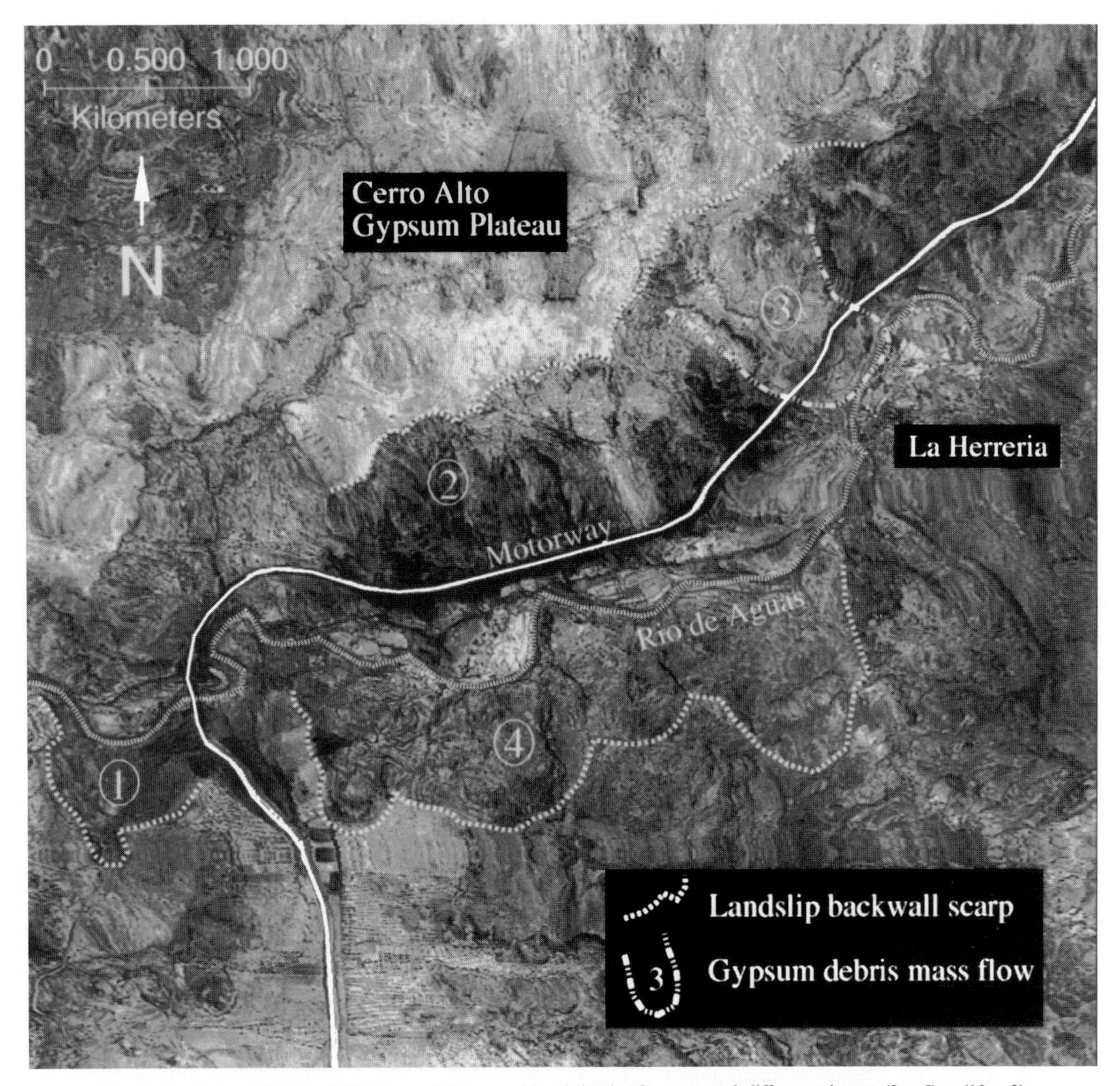

Fig. 3. Daedalus ATM band 11 (thermal) image with addback of compound difference image (9 + 5) – (10 + 3) and Laplacian edge enhancement. Lighter tones in the northern part of the image correspond to the massive gypsum plateau. The compound difference image was used to highlight gypsum debris in the Marchalico Viñicas Landslide (site 3).

Laplacian filter is defined thus:

$$\begin{matrix} 0 & -1 & -0 \\ -1 & 4 & -1 \\ 0 & -1 & 0 \end{matrix} + \begin{matrix} 0 & 0 & 0 \\ 0 & x & 0 \\ 0 & 0 & 0 \end{matrix} = \begin{matrix} 0 & -1 & 0 \\ -1 & 4+x & -1 \\ 0 & -1 & 0 \end{matrix}$$

The user specifies the weighting factor 'x' most appropriate for any particular case. The normal case is '$x = 1$'. Reduction or increase in the value of 'x' will increase or reduce the edge sharpening effect. An edge-sharpened image can be manipulated in exactly the same way as an unfiltered image.

Data integration

Merging images can be used to produce a composite product which has the combined spatial and spectral resolution of its constituent images. A merged TM and

SPOT image was produced for the Rio de Aguas valley area, using the Brovey Transformation algorithm within the ER Mapper software package (Fig. 2, lower part). The merged image combines the spectral resolution of TM with the spatial resolution of SPOT Panchromatic imagery.

The Brovey process is mathematically equivalent to a hue, saturation and intensity (HSI) data integration transformation (Equations 1–3) in which SPOT Panchromatic image is substituted for the intensity component in a transformed TM image. The result is a combined image with the hue and saturation (spectral) of the TM image and intensity (textural information) of a 10 metre pixel SPOT Pan image. It is possible to prepare the TM RGB images, using band combinations in input channels, specifically to enhance particular spectral features, before HRGB transformation.

Two merged images were produced using the Brovey transformation to combine the TM colour composites with SPOT Panchromatic data. In these images the scalloped form of El Tesoro escarpment (site 2) was enhanced by improving the spectral contrast between the gypsum and underlying marl.

$$\text{red channel} = \left(\frac{\text{TMi}}{\text{TMi} + \text{TMii} + \text{TMiii}}\right) * \text{SPOT pan} \quad (1)$$

$$\text{green channel} = \left(\frac{\text{TMii}}{\text{TMi} + \text{TMii} + \text{TMiii}}\right) * \text{SPOT pan} \quad (2)$$

$$\text{blue channel} = \left(\frac{\text{TMiii}}{\text{TMi} + \text{TMii} + \text{TMiii}}\right) * \text{SPOT pan}. \quad (3)$$

Brovey transformation combinations of SPOT Pan and simple TM three-band colour composite images are the best product for combined textural and spectral interpretation of Landsat and SPOT imagery. Substituting a Laplacian filtered SPOT image for the original SPOT Pan image in the Brovey Transformation further improves the textural information display in the Brovey SPOT/TM combined image.

Landslide test-site targets

The appearance of the landslide targets in the field and in enhanced Landsat TM, SPOT Pan and ATM imagery is discussed below.

Carrasco landslide (1)

The Carrasco landslide (Fig. 1, Section A–B) is a dip-slope slide in which the Miocene limestone unit has been dislocated by rotational slip and flowage in the lower marls and sandstones. The result is a limestone debris strewn slide whose toe is undercut by a Rio de Aguas river meander.

Texture enhanced, Laplacian addback SPOT Panchromatic imagery reveals the form of the arcuate, crown wall scarp in limestone and the hummock texture of the slip debris. The Rio de Aguas river channel is also clearly shown in the SPOT imagery, giving a clear indication of the relationship between the destabilising river meander course, and the slip toe (Fig. 3).

El Tesoro escarpment (2)

The escarpment at the edge of an extensive plateau made up of gently northward-dipping, massive gypsum beds, marks the northern boundary of the Rio de Aguas Valley. There has been widespread gypsum block toppling failure along the escarpment, as a result of the rapid erosion and gravity flow in underlying marls. Despite widespread field evidence of movement, there are no obvious textural or spectral features in images which can be related to individual slips at the El Tesoro locality.

The arcuate, scalloped form of the gypsum escarpment is well shown in the ATM and TM images and is a clear indication of the instability of the area. The scalloped escarpment edge is particularly well shown in ATM monochrome and colour composite images incorporating the ATM 11 (thermal) band (Fig. 3, Loc. 2)

Marchalico Viñicas debris flow (3)

The Marchalico Viicas landslide is a marl debris flow carrying numerous large, gypsum blocks. The flow, three kilometres in length, is traversed by the motorway and reaches Rio Aguas, where the toe is eroded by the river. It originates from an area where groundwater springs and resurgences emerge from the base of the gypsum escarpment. These resurgences and springs, from the gypsum karst, were the water source for the now abandoned hamlet of Marchalico Viñicas. Seepage from the resurgences may have contributed to viscosity reduction in the marls and lubrication of failure surfaces during the main period of mass debris flow.

The Marchalico Viñicas flow is the only landslide in the Rio Aguas District with a distinctive spectral character. The size of the flow and the large number of incorporated gypsum blocks give a spectral signature in the thermal bands of both Daedalus ATM and Landsat TM imagery. Spectral enhancement of both TM and ATM imagery by ratio and difference images, effectively highlights the gypsum debris material and vegetation around the sites of springs and resurgences.

Los Perales landslides (4)

The several large landslide features on the southern flank of the Rio Aguas Valley above the hamlet of Los Perales are dip-slope slides affecting the lower marls and sandstones together with some areas of limestone outcrop. Undercutting and erosion of the slide toes by the Rio Aguas appears to have been the trigger mechanism for movement. An unusual feature of the slide material southwest of Los Perales, is the presence of numerous, extensional 'gull' type fissures in destabilized, northward dipping, limestone beds.

The slides are compound and delineation of their exact form is made more difficult by their lack of distinctive spectral signature, higher density of vegetation and shadows on north facing slopes. Nevertheless, textural evidence from both SPOT Pan and ATM imagery allows the morphology of both crown wall and the texture of fractured and slipped ground to be identified (Loc. 4) (Figs 2 & 3).

Maleguica landslides (5)

There are landslides which are not obvious from the road, in the upper sands and marls on the Rio de Aguas valley sides near Sorbas town. Interpretation of the texturally enhanced SPOT images and edge-enhanced Laplacian addback Brovey SPOT pan/TM (4,3,2 and 5,4,1) images produced for study of the lower valley test sites, revealed the previously unnoticed Maleguica slides.

Field checks showed the area has been widely affected by compound rotational slip and flow of marls and intercalated, massive sandstones. The Maleguica slides serve to illustrate the value of the approach described in this paper for the discovery of unknown landslides, even in a well-investigated area.

Gor Valley, Granada Province

Using the SPOT Pan/Landsat TM Brovey image merge methodology, developed for the training area, a series of landslides with similar form to those of the Rio de Aguas Valley were subsequently identified in the Gor Valley, Granada Province (Eyers 1994). The landslides occur in a valley incised through a sequence of Quaternary silt, clay and conglomerate layers, lying conformably above Tertiary sediments. Clay horizons in the series of sediments may be lubricated by precipitation. These offer potential sliding surfaces. A slight northeast dip in the sequence indicates that failure is more likely on the southern side of the valley.

Conclusions and discussion

The studies reported in this paper demonstrate some of the ways in which enhanced SPOT and Landsat TM imagery can be used, together, for spectral and textural feature recognition in unstable terrain. SPOT Panchromatic imagery, merged with Landsat TM, in an edge enhanced Brovey transformation image is one of the most effective image products for landslide mapping at scales to 1:30 000. Combining textural and spectral features in Brovey images permits identification and mapping of landslides where the width of the displaced mass, as defined by the IAEG (1990), is greater than 250 m. This is a cautious specification and it is based on the knowledge that identifiable features in imagery are not continuous. Individual features with lengths less than 250 m are identifiable in many many, if not all, cases. Texture in Laplacian edge enhanced, SPOT Pan images shows crown-wall shape, the form of translated masses and the texture of disturbed ground in all the Rio Aguas slides.

Spectral enhancements, including TM and ATM difference and ratio images are useful for rock type discrimination, for example, massive gypsum and limestone, both *in situ* and as part of a landslide debris trains. Spectrally enhanced TM difference and ratio images can also be used to locate water sources, e.g. springs and resurgences, indirectly, using vegetation indicators. Information from the thermal bands of both Daedalus ATM and Landsat TM can be help in distinguishing certain rock types and highlighting the distribution of anomalous soil moisture.

The Daedalus 1268 Airborne Thematic Mapper (ATM) instrument, although an interesting research tool with extended spectral capabilities in both the near infra-red and thermal ranges, suffers from serious problems of geometric distortion. Resolution depends on aircraft altitude during acquisition but can be as detailed as 1 metre pixels. ATM products illustrate how future improvement of spatial and spectral resolution in space imagery will allow geological and engineering landslide remote sensing to become more detailed. In the future, landslides with dimensions comparable with those identifiable in 1:10 000 air photography will be mappable, regionally, using space imagery.

Developments in hyperspectral and high spatial resolution sensors following the NASA Lewis and Clark sensor experiments will undoubtedly provide improved imagery for geotechnical mapping. Stereoscopic SPOT Panchromatic and Russian KFA1000 digitized space photography (nominal 5 metre resoluion) is the already available for parts of the world and allows Digital Elevation Models (DEMs) with 10 metre contour spacing to be created for terrain slope analysis, complementing the spectral and textural information discussed in this paper.

Data from Landsat and SPOT are available for large areas of the world and older scenes can be purchased relatively cheaply. Landslide identification using the techniques described in this paper, can be incorporated in a regional scale geohazard mapping project at relatively low cost.

Acknowledgements. The authors acknowledge with thanks, the provision of Daedalus 1268 Airborne Thematic Mapper image data by the UK Natural Environment Research Council (NERC), Landsat TM imagery from the European Union Joint Research Centre (JRC), Italy, and the Imperial College remote sensing data archive. Digital processing was carried out using Earth Resources (ER) Mapper software which forms part of the University of London Inter-Collegiate Research Services (ULIRS) facilities at Imperial College. The authors wish to express their appreciation of the contributions made to this study by MSc students Vitus Chan and Robert Wright in discussion and through their unpublished project reports. Mr Alex Davis is acknowledged for assistance in production of illustrations for this paper.

References

BARRETT, D. & CURTIS, L. F. 1992, *Introduction to Environmental Remote Sensing*. 3rd edition, Chapman & Hall, London.

CHAN, V. 1994. *Landslide Mapping Using ATM Imagery in Rio Aguas, Almeria Province, Spain*. MSc Project report, University of London.

CRÓSTA, P. A. & MOORE, J. MCM. 1989. Geological mapping using Landsat Thematic Mapper imagery in Sorbas Province, south-east Spain. *International Journal of Remote Sensing*. **10**, 3, 505–514.

EYERS, R. 1994. *The Use of Satellite Imagery for the Detection of Landslides in Southeast Spain*. MSc Thesis, University of London.

IAEG 1990. Suggested nomenclature for landslides, IAEG Commission on Landslides, *Bulletin of the International Association of Engineering Geology*, No. 41, Paris.

MASON, P. J. 1993. *An investigation of ERS-1 Synthetic Aperture Radar as a tool for the interpretation of geological structure, in conjunction with thematic mapper multispectral imagery*. MSc Thesis, University of London.

MCKEAN, J., BUECHEL, S. & GAYDOS, L. 1991. Remote sensing and landslide hazard assessment. *Proceedings of the Eighth Thematic Conference on Geologic Remote Sensing*, April 29th–May 2nd, Denver, Colorado, USA, 792–742.

MOORE, J. MCM., LIU, J. G., COPP, D. L. & YOUNGS, K. 1993. Mineral discrimination by directed band difference techniques using TM and ATM data. *Proceedings of the Ninth Thematic Conference on Geologic Remote Sensing*, 8th–11th February, Pasadena, California, USA, 821–831.

NIBLACK, W. 1986. *An Introduction to Digital Image Processing*. Prentice-Hall, London.

SHIKADA, M., KUSAKA, T., KAWATA, Y. & MIYAKITA, K. 1993. Extraction of characteristic properties in landslide areas using thematic map data and surface temperature. *Proceedings of IGARSS: Better understanding of earth environment*, August 18th–21st, Tokyo, Japan, 103–105.

WRIGHT, R. 1995. *An analysis of slope movements and associated lithologies using ATM data*. MSc Project report, University of London.

Landslides and their control in the Chinese Loess Plateau: models and case studies from Gansu Province, China

Xing-min Meng & Edward Derbyshire

Department of Geography, Royal Holloway, University of London, Egham, Surrey TW20 0EX, UK and Geological Hazards Research Institute, Gansu Academy of Sciences, Lanzhou, China

Abstract. Over much of the western part of the Loess Plateau of north China, aeolian silts averaging 150 m in thickness drape the mountainous pre-Pleistocene bedrock. The semi-arid climate with its violent monsoonal rainstorms, and the recurrent earthquakes associated with the continuing uplift of the adjacent Tibetan Plateau, contribute to frequent and rapid failure of the loess-mantled slopes. Losses of livelihood and life are amongst the most serious in all China. Landslides in eastern Gansu total more than 40 000 and affect over 27% of the total area of the Province. Accelerating population growth and associated expansion in both industry and agriculture, especially since the 1950s, has added urgency to the need to improve control and management of the landslide hazard. Several approaches to control and prevention have evolved, tailored to failure type and stage of development. Some of these are discussed, particular attention being paid to certain successful procedures employed in the control of two large landslides in the major city of Tianshui. It is suggested that these slides provide a useful model for future landslide control and prevention strategies in similar urban agglomerations in the Chinese Loess Plateau.

Introduction

Eastern Gansu Province, in northwestern China, makes up the western part of the Loess Plateau (Fig. 1). Here loess deposits (predominantly wind-lain silts) occur as a drape on the pre-Pleistocene bedrock. This is the greatest pile of wind-blown dust on Earth, with an average thickness of 150 m and a maximum thickness (*c.* 330 m) near the city of Lanzhou (Derbyshire 1983). The great thickness of the loess in this region is attributed to strong deflation of dusts from the extensive deserts to the west and north of the loess region during glacial periods of the Pleistocene, when an enhanced Mongolian high pressure system maintained northwesterly winds across a cold and dry continental interior. During the many warmer and moister climatic phases of the Pleistocene (including interglacials and interstadials), dust accretion rates declined, with a complementary increase in weathering and pedogenesis that produced up to 37 ancient soils (palaeosols) interbedded with the loess units. Vigorous uplift of the adjacent Tibetan Plateau has given rise to a landscape in which mountains make up about 75% of Gansu's total area of 400 000 km^2 (Derbyshire *et al.* 1991). This loess-mantled region thus has a relative relief of between 200 and 700 m with steep slopes (27° to 40°; *cf.* Dijkstra *et al.* 1993). The climate is semi-arid, with annual precipitation in the range 200 to 600 mm, up to 70% of which is usually concentrated in the months of July, August and September, and up to 40% has been known to fall in a single storm (Derbyshire *et al.* 1991; Derbyshire & Meng 1995). In this environment the loess slopes are very susceptible to extensive slope failure and a tendency to progressive development of natural subsurface piping systems.

Gansu Province has a long history of disastrous landsliding which has taken several hundred thousand lives in this century alone. The sites of more than 40 000 large-scale landslides are now known. Their extent covers about 27% of the total area of Gansu Province (Derbyshire *et al.* 1991; Derbyshire & Meng 1995). With the dramatic rise in population and the growth of both industry and agriculture over recent decades, awareness of the threat posed to individual communities and the national economy by the landslide hazard has become widespread. Chinese scientists have responded to this trend by devoting much more attention to landslide research and to the collection of critical data, including case studies. Much attention is currently focused on the spatial and temporal distribution of slope instability, and upon the development of programmes for hazard assessment, leading to mitigation strategies.

Meng, X. & Derbyshire, E. 1998. Landslides and their control in the Chinese Loess Plateau: models and case studies from Gansu Province, China. *In*: Maund, J. G. & Eddleston, M. (eds) *Geohazards in Engineering Geology*. Geological Society, London, Engineering Geology Special Publications, **15**, 141–153.

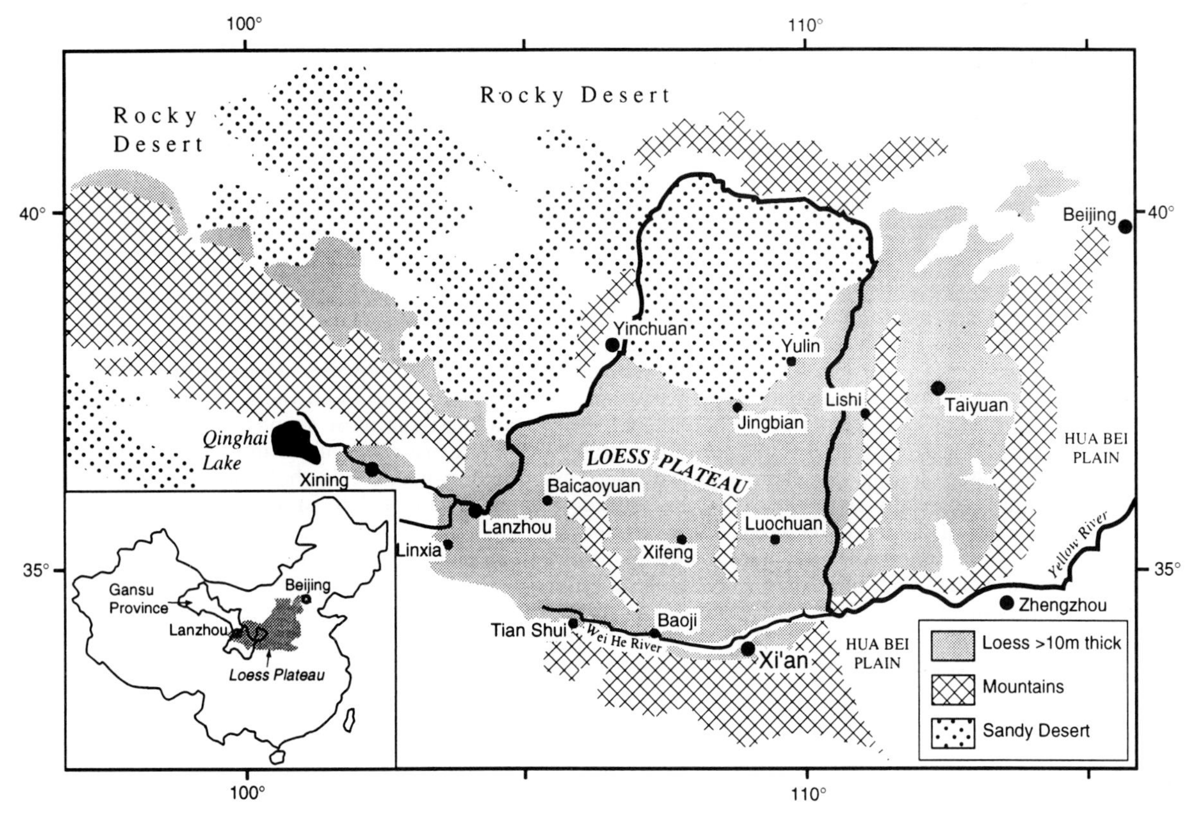

Fig. 1. Distribution of loess (>10 m) in north-central China.

Models of landslides in loess

On the basis of our experience over the past ten years in Gansu Province, landslides involving loess may be classified according to the following five models (Fig. 2).

Sliding of loess along the bedrock surface ('bedrock contact landslide')

This is the commonest type of loess landslide throughout the Chinese Loess Plateau. Bedrock frequently consists of impermeable argillite or mudstones, so that groundwater movement is impeded and a zone of saturation builds up at the loess/bedrock interface. The surface of the bedrock is often weathered, and the argillites in particular may contain the clay mineral smectite (Derbyshire *et al.* 1995*a*). Repeated wetting of this zone results in a progressive decrease in strength which may ultimately lead to slope failure, especially when the gradient of the loess/bedrock interface is greater than 15°. Undercutting of slopes associated with gullying often increases the incidence of slope failure. In the right geological and geomorphological conditions, landslides in this group may be very large. In Tian Shui City, for example, there is a slope (made up of argillites overlain by 30–100 m loess) 12 km wide and 800 m high, no part of which has escaped landsliding at some time in the recent past.

Loess slides along surfaces of intercalated palaeosols ('palaeosol contact landslide')

Numerous palaeosols, developed during former periods of warm and humid climate (interglacials and interstadials), are common in Chinese loess successions. They are characterized by finer mean grain sizes (clay contents 30–50%), much lower permeabilities and much greater compaction values than are found in the loess. Most palaeosols lie broadly parallel to the present plateau-like ('*yuan*') surfaces, i.e. they are sub-horizontal. Underground pipe systems, which develop by a combination of local liquefaction, collapse and abstraction during rainfall events, are strongly controlled by the abundant fissures or joints found within the loess. Many of these pipes terminate or, at least, are deflected where they meet the buried soil horizons. Thus vadose waters are impeded by the palaeosol horizons resulting in localized enhancement of soil moisture content and consequent

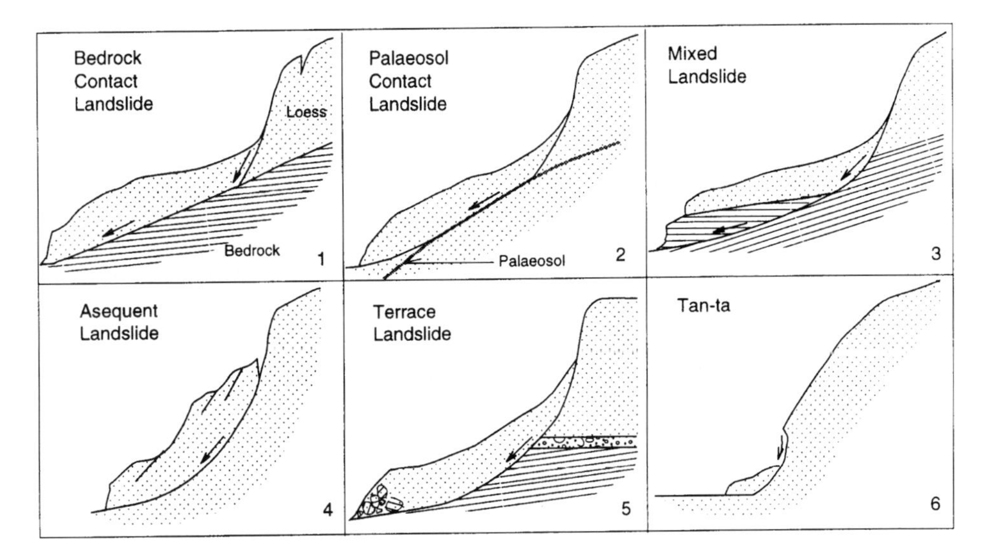

Fig. 2. Principal types of landslides involving loess.

reduction in soil strength. The likelihood of failure is even greater where undercutting of slopes arises from river incision or artificial excavation. The 1982 loess landslide at Taibei Cao in Qingyan county provides a good example of a failure triggered by excavations associated with roadworks.

Combined loess and bedrock landslide along bedding planes or joints within the bedrock ('mixed landslide')

When loess overlies permeable bedrock, especially when jointing is well developed, groundwater moving down through the overlying loess is not impeded at the bedrock contact. Normally, such loess-covered slopes are stable. However, intercalated mudstones or the presence of shallow faults or clay-lined shear zones may cause the slope to fail. This was the case at Argan, about 22 km S of Lanzhou, where intermittent sliding occurs on slopes consisting of loess overlying Cretaceous sandstone intercalated with thin mudstone. The contact between loess and bedrock is visible about 1.5 m above the valley floor, but there is no evidence of movement of the loess along the bedrock contact. In fact, the slip surface occurs within a thin mudstone member dipping at 10°–25° in the same sense as the slope. The narrowness of the valley restricts the amount of sliding movement to less than 20 m. Accumulated slide masses are removed by water bodies, built up following torrential rainfall events, that transform the slide masses to debris or mud flows. This is a cyclic process. The last recorded movement occurred at 2020 hours on 13 September 1984. Sliding speeds were slow, the toe extending for a distance of 16 m in a total sliding period of just 3 h. The slow rate of movement allowed both residents and animals to be evacuated before the houses finally collapsed.

Landslides with slip surfaces developed along bedrock joints are uncommon in the Loess Plateau, occurring only in conditions of very steep slopes and high relative relief. The reason for this is that the shear strength along an insequent joint system is normally higher than that along the bedding. Out of a total of 220 landslides investigated in the Lanzhou area, only two sites of this type have been recorded. Both are on Gaolan Mountain, overlooking downtown Lanzhou. The mountain consists of 215 m of loess overlying Neogene argillite and sandstone. Relative relief is 586 m and slope gradients are very steep (35°–41°). The major slip surfaces of the two landslides follow a well-developed joint system dipping down slope at 34°. Slope modelling and stability analysis suggest that the bedrock slope is stable under normal conditions but on the basis of a number of instability analyses for north China incorporating the effects of earthquakes, it is considered that failure would occur given an earthquake shock greater than 8 on the Richter scale. Both of these landslides have been attributed to earthquakes in the historical period (Meng & Zhang 1989).

Landslides entirely within loess (asequent landslides)

Landslides developed exclusively in loess are rare. Even during the frequently severe rainstorms, failure within the mass of the loess (in contradistinction to zones of jointing) is unusual due to low rates of infiltration within intact loess (see Muxart *et al.* 1994). Nevertheless, landslides within loess do occur. One such slide, at Baitashan near Lanzhou, was a direct result of over-steepening by manual excavation. It killed nine people and destroyed seven homes. The low moisture content of less than 12% coincides with the highest residual shear strength (between 10% and 4%; Dijkstra *et al.* 1994) resulting in relatively stable soil. Consequently, loess is not generally prone to sliding.

Terrace landslides

Landsliding along the margins of river terraces is a common occurrence in the Yellow River drainage system. Terraces generally consist of a planed rock surface overlain by fluvial or colluvial sediments and then, in turn by aeolian or reworked loessic silts sometimes 10 or more metres thick. The older the terrace, the thicker the loess cover. Terrace risers are steep (38°–45°). If the terrace basement is composed of impermeable argillite or mudstone, the fluvial or colluvial gravels act as an aquifer below the loess. Springs are common along such zones and are frequently marked by the presence of re-precipitated salts washed out of the overlying loess. Such eluviation of salts and fine particles in the presence of relatively high moisture contents results in softening and strength loss in the basal loess, which is then subject to deformation from the loess overburden stresses. Deformation may result in fissuring within both loess and bedrock which, in turn, provides new avenues for water movement. When the strength of the argillite bedrock declines to a critical point and the cracks develop throughout the loess, failure is inevitable. The thicker loess landslides are usually found on terraces. The initial displacement occurs in the bedrock foundation of the terrace so that, with propagation of the failure plane within the loess, shear strength falls suddenly from peak to residual values. As a result, this landslide type usually exhibits very high sliding speeds and is often catastrophic. Since almost all development and infrastructure occurs on terraces, even small terrace landslides may prove disastrous. In the case of large slide masses, air-lubrication may build up during sliding, enhancing the sliding distance. A classic example of such a situation is provided by the Sale Shan landslide, located some 78 km S of Lanzhou. The loess here is about 150 m thick and overlies bedrock of Neogene argillite intercalated with some thin conglomerates and clayey limestone layers. The argillites have been subjected to long-term softening due to concentration of high groundwater levels in the contact zone between loess and argillites. This reached a point at which they were no longer able to support the overlying loess. At 1746 hours on 7 May 1983, 31 million m^3 of loess in a mass 1000 m wide and 170 m high moved over a distance of 1600 m in less than one minute, killing 273 people and overwhelming three villages.

Tan-ta

In addition to the landslide types discussed above, small adjustment failures on loess slopes occur in response to changing local conditions. These are known in China as tan-ta (in both singular and plural) and are generally excluded from landslide classifications because they contain no clear failure plane. Tan-ta are generally less than 10 m in diameter, and occur predominantly on steep slopes in loess. Initiation of the slides is influenced by a variety of factors including infiltration of snow-melt or rain water, undercutting of the slope and Earth

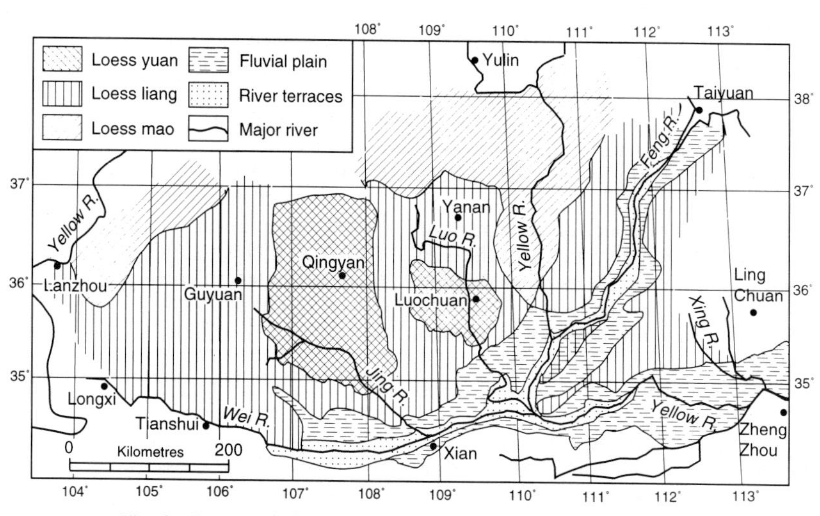

Fig. 3. Geomorphological regions of the Chinese Loess Plateau.

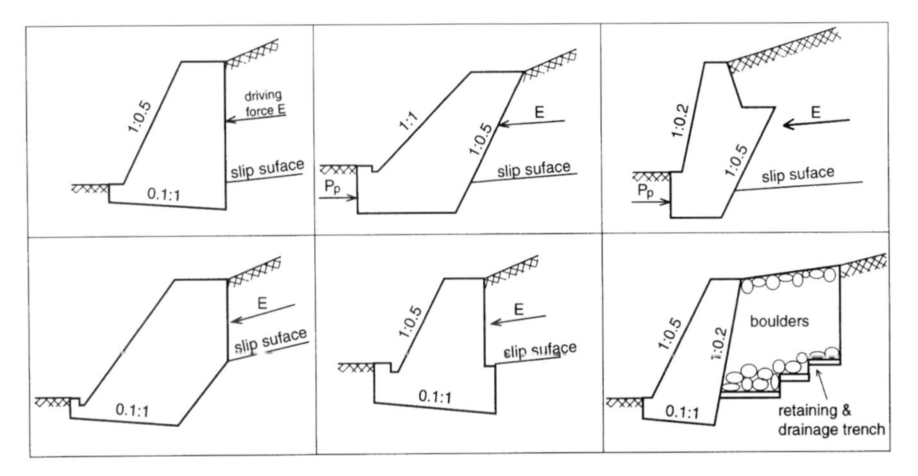

Fig. 4. Some basic design forms for gravity retaining walls used in landslide control.

tremors, and human activities, such as irrigation and excavations for roads and railways. Once started, these slides disintegrate rapidly, often having high velocities (Derbyshire *et al.* 1993). Tan-ta may trigger adjacent landslides, particularly when they occur in the toe zone of a relict landslide or on a slope in critical condition.

Distribution of loess landslides

It is becoming evident that the different geomorphological units of the loess tend to contain particular landslide types. In the Chinese Loess Plateau, four major geomorphological units of the loess, now widely known by

Fig. 5. Temporary retaining wall consisting of boulders and cobbles emplaced on the toe of the Jiaoshuwan landslide, Tianshui, in November 1990.

their Chinese names *yuan* (plateau), *liang* (ridge), and *mao* (rounded hill), to which is added the *terrace* category, are classified as described below (Fig. 3).

Loess yuan landforms are widely developed, especially in the 'classic' loess terrrain of the south-central part of the Loess Plateau. The underlying bedrock is of simple structure over large areas, outcrops being confined to deeply incised valleys. Sinkholes and underground pipe systems (known, appropriately, as *loess karst*) are well developed on yuan slopes, and are an important element of the loess hydrological system. In cases where bedrock lithology on yuan margins is of argillite or mudstone, both the terrace- and mixed-type of landslide tend to be concentrated.

Loess liangs are the dominant geomorphological feature over extensive parts of the Loess Plateau, where relative relief often exceeds 600 m. Loess thicknesses range from about 150 m on liang crests down a few metres on the slopes. When the dip of the argillite bedrock conforms to the surface, landslides frequently cover all parts of a liang slope (Meng *et al.* 1988). Bedrock contact type landslides are the commonest in liang terrain, with occasional occurrence of the mixed type.

Loess mao, distinctive round or eliptical hills, occur mainly in the northern part of the Loess Plateau (Fig. 3), where mean annual precipitation is generally less than 300 mm. Mao slopes are predominantly convex in profile. The loess here is thinner than in the yuan and liang areas, the dense valley system being broad and shallow. Rock outcrops are rare. Much of the loess has been reworked because of the effect of the rare but high intensity rainstorms which enhance erosion of the poorly vegetated loess surface characteristic of much of the Mao area. Landslides are rarely seen, except where active uplift has stimulated valley incision or where the surface has been artificially excavated, in which case some asequent, contact or mixed landslide types may be found (see Meng *et al.* 1988).

The *loess terraces* of the Yellow River and its tributaries are impressive, some being many kilometres in width. Their distinctive stratigraphy (loess overlying fluvial gravels resting on planed bedrock) is an important determinant of hydrology and slope stability. Loess karst is general, and terrace-type landslides common, especially where irrigation is practised (Derbyshire *et al.* 1995*a*, *b*).

Landslide control

The loess country of northern China contains many examples of the application of engineering measures at varying levels of sophistication to maximize the area of

Fig. 6. Installed concrete and steel piles (4 × 4 m) being inspected by Wang Jingtai (second from left) at the Jiaoshuwan landslide.

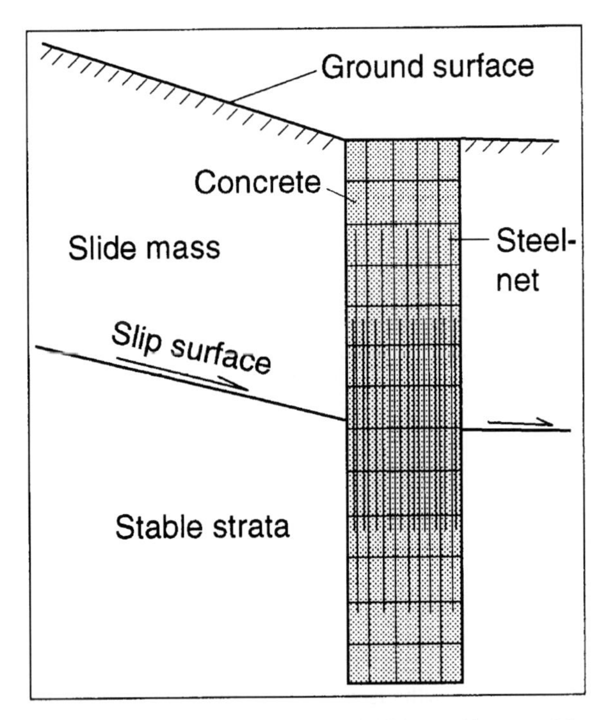

Fig. 7. Schematic diagram showing the form and layout of the 4 × 4 m concrete and steel piles.

usable land in the face of population and economic pressures. Gansu Province has a long and costly history of struggling with loess landslides. In the early 1950s, largely because of a general lack of experience of loess landslide control, many retaining structures failed to prevent slope failures and some engineering works even made matters worse. For instance, in 1953 it was believed that a section of railway and its associated buildings were threatened by a potentially catastrophic terrace landslide at Tianshui in Gansu Province. The early signs consisted of several tan-ta on the toe slope. Instead of installing preventive structures based on a detailed site assessment, only a small retaining wall was installed. Very soon, this blocked the groundwater drainage lines and failure of the whole slope followed soon afterwards, destroying the railway line and many buildings. Such hard-won experience has led, in recent decades, to much higher success rates using controlling structures and to a number of practical methods that are generally closely attuned to both the natural conditions and the local economy.

Landscaping

Artificially cut slopes are an inevitable response to development work in mountainous regions. On slopes consisting entirely of loess, appropriate re-shaping and the installation of a drainage system has proved to be an effective approach to slope failure minimization, especially in the case of tan-ta and asequent landslides. Four models of landscaped loess slopes are in use, each having its own particular advantages and disadvantages.

Rectilinear slopes are the simplest to design and operate, involving less excavation, although the upper part of the slope may collapse rather readily, and the whole of the slope surface remains vulnerable to water erosion. Nevertheless, this form is suitable for loess slopes with a total height of less than 18 m.

Concave slopes are similar in form and function to the scar section of an already-failed slope, with the upper part exposing a smaller area to erosion. However, construction of a slope with this form usually involves greater excavation than in the case of a straight slope. The concave form has been found quite suitable for high slopes, especially when there is some groundwater seepage from the toe.

The *convex slope* form, matching many natural slopes in loess terrain, nevertheless requires more extensive excavation than the landscaping of concave slopes. Any slope inflexion on loess slopes is soon smoothed by the action of surface water. The convex form works quite well in the case of high slopes (15–25 m) consisting of relatively highly consolidated older loess in the lower part of the made slope and looser, younger loess above.

Stepped slope landscaping of the lower parts of slopes serves to reduce overburden pressures more effectively than any other slope form, giving an impression of greater stability. Installation of a drainage system in the middle of the slope may considerably reduce slope erosion, although the steps themselves remain vulnerable to collapse. An obvious disadvantage of creating a stepped slope is the substantial excavation it involves. Stepped landscaping has been used with some success on quite high slopes, as the senior author's design for a slope 57 m high in the Lanzhou area shows, over 6 years after its construction. Stepped slopes provide the further advantage of providing access to personnel for maintenance work in mid-slope.

Retaining structures

Retaining walls. Aggregates and cement are abundant and hence cheap in northwestern China, so it is not surprising that retaining walls and similar structures figure highly as local solutions to slope instability situations (Figs 4 and 5). The size, shape and location of retaining walls used to control landslides are determined on the basis of the slide-triggering process (the driving force), the thickness of the slide mass, the foundation conditions at the site of the proposed wall and so on.

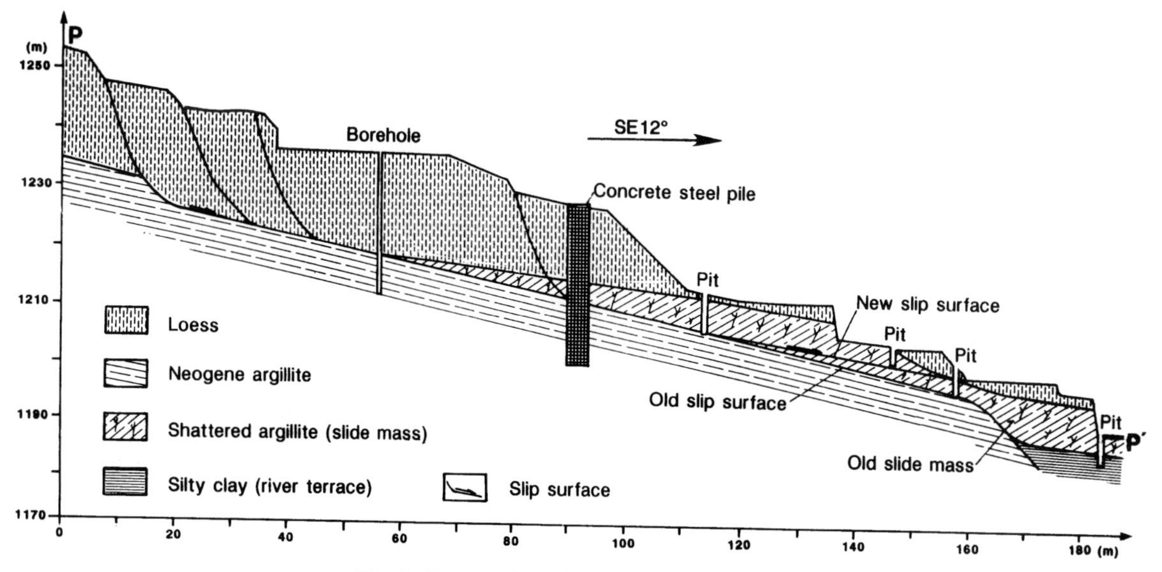

Fig. 8. Cross-section of the Jiaoshuwan landslide.

Underground drainage trenches. Underground drainage trenches are quite effective in reducing movement in many groundwater-rich but shallow landslides (less than 10 m in thickness). They can be used to divert groundwater supply and improve the flow of springs. A particularly successful strategy is to engineer such trenches in the lower part of a slide mass and align them parallel to the slide movement direction so that they satisfy the dual function of drainage and slope retention. Such trenches are commonly used together with a retaining wall.

Retaining piles. Retaining piles have been used widely in China in recent years, particularly because they are simple (although frequently dangerous) to construct, partly because they offer a wider choice of site locations than retaining walls and, not least, because they have proved effective in controlling landslides. The pile may be made of wood, steel piping, structural steel, and steel-reinforced concrete. Reinforced concrete piles are much used to control loess landslides in China where they can claim some success. For example, the authors successfully controlled a large landslide in Tian Shui city using a design involving reinforced concrete piles 4×4 m thick and 27 m long (Figs 6 and 7). Other schemes have used reinforced concrete piles up to $3.5 \times 7.0 \times 50$ m to control landslides in loess. The design of these very large concrete and steel piles derives from bridge-pier designs (*cf.* Chinese Railway Ministry 1983). Unfortunately, in the present economic state of much of China, any such technical success is dependent upon installation of these piles within hand-dug pits.

Using the methods mentioned above, most shallow and moderately thick landslides can be effectively controlled. Most bedrock contact landslides occur because of accumulation of groundwater above a relatively impermeable stratum, causing a reduction in shear strength of the material around the contact zone. A combined structure consisting of a retaining wall or piles and drainage trenches has proved to be the best means of controlling them. In the case of mixed landslides, a series of short retaining piles pinning together the relatively displaced layers is probably the best approach. Palaeosol contact landslides are best dealt with by sinking a series boreholes through the buried soils to facilitate through drainage. The tendency of slopes to fail when being undercut or excavated is best compensated for by building retaining structures. Terrace landslides can sometimes prove extremely difficult to handle, especially where slopes are steep and the loess is thick. The best amelioration strategy is probably to reduce the slope angle and, where there is abundant groundwater, it may be necessary to install some horizontal borehole drainage. A serious terrace landslide occurred in 1989 on the fourth terrace of Yellow River about 30 km S of Lanzhou City. Here the loess is over 50 m thick and overlies Cretaceous argillite and sandstone. The surface had been subjected to intensive irrigation for a number of years, and the terrace slopes were beginning to show signs of impending failure. Our first act was to reduce the slope angle. However, because of the abundance of groundwater and zones of loess liquefaction induced by slight slope displacement, the whole slope threatened to develop

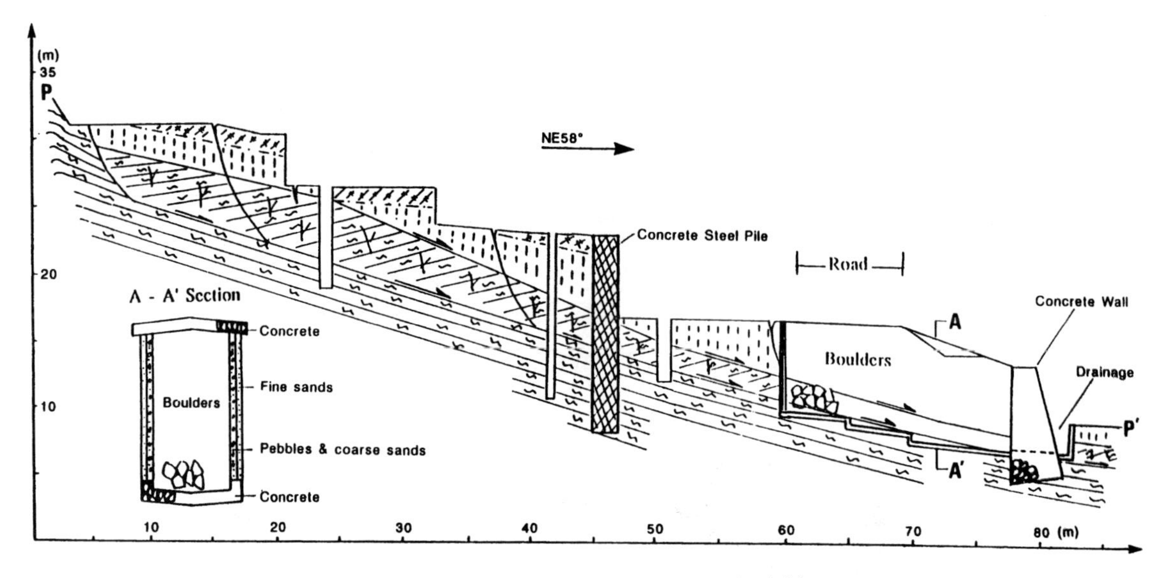

Fig. 9. Cross-section of the Taishanmiao landslide.

into a huge retrogressive slide. An underground drainage trench system was then designed, but the local construction team found it very difficult to install and operate. These slopes were still on the move at the time of writing.

Case studies

On 11 August 1990, a heavy rainstorm occurred at Tianshui. A total of 107 mm of rain fell in two and a half hours, giving rise to floods, debris flows and landslides. The two case studies considered here, known as Jiaoshuwan and Taishanmiao, are particularly hazardous situations induced by that rainstorm. Their initial movements were very slow but by October of 1990 they had become quite rapid, the maximum rate of movement recorded being 77 mm day^{-1}. It was evident that were the two landslides to accelerate suddenly, over 2000 local residents would be threatened (Derbyshire & Meng *et al.* 1995). As a short-term measure during November 1990, two temporary retaining walls, consisting of boulders, were built on the toes of both landslides, and several drainage wells were dug across the slide mass. These measures greatly reduced the rate of movement (to about 1 mm day^{-1}), and thus provided valuable time for engineering geological investigations and for the planning of permanent remedial works. Both sites lie within relict landslides and both have the properties of bedrock contact slides (along a bedrock surface of Neogene argillite) and mixed landslides with both loess and bedrock sliding along bedding planes (Figs 8 and 9). The driving forces used for the design of retaining

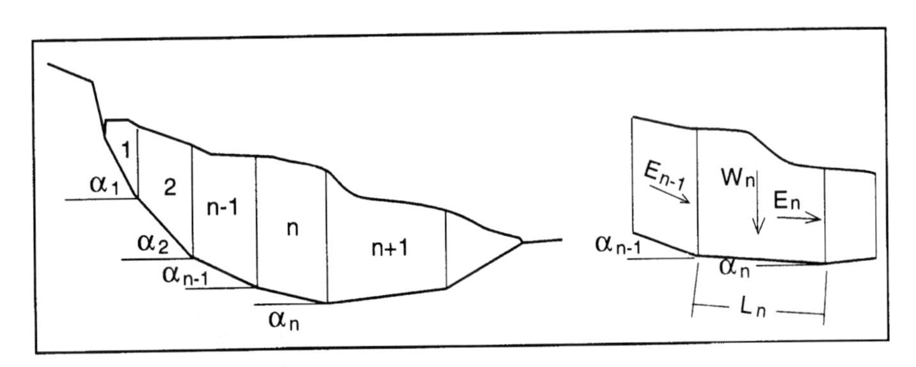

Fig. 10. Diagram to illustrate driving force analysis using the method of slices.

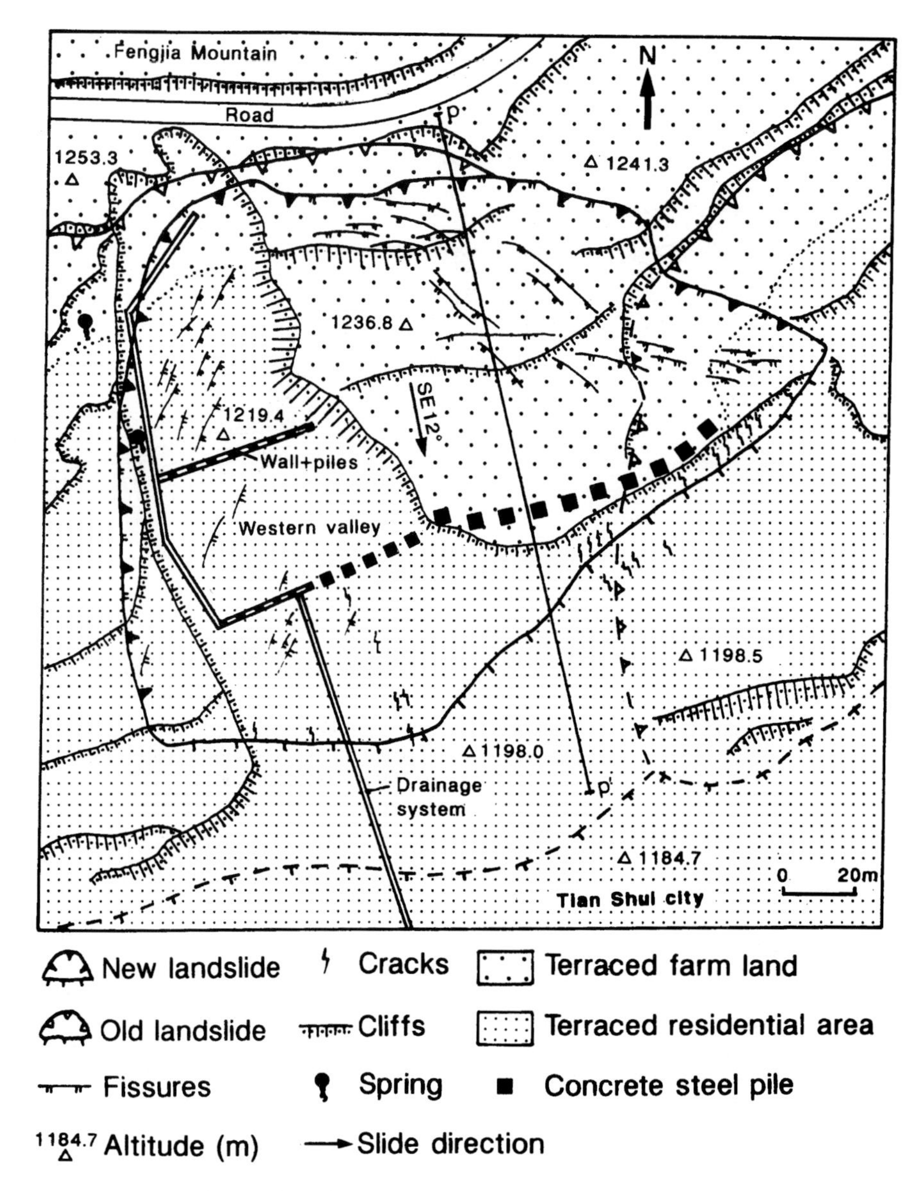

Fig. 11. Distribution of control structures installed in the Jiaoshuwan landslide.

structures are calculated using the slice analysis method (Anon 1984), the equation being (Fig. 10):

$$E_n = F \times W_n \times \sin\alpha_n - W_n \times \mathrm{con}\,\alpha_n \times \mathrm{tg}\,\phi_n$$

$$- C_n \times L_n + E_{n-1} \times [\mathrm{con}(\alpha_{n-1} - \alpha_n)$$

$$- \sin(\alpha_{n-1} - \alpha_n) \times \mathrm{tg}\,\phi_n]$$

where by E_n is the driving force ($\mathrm{t\,m^{-2}}$) at n block, F is the safety factor, W_n is the weight of n block (t), α_n is the inclination angle of the slip surface of n block, ϕ_n is the friction angle on the slip surface of n block, C_n is the cohesion on the slip surface of n block ($\mathrm{t\,m^{-2}}$), L_n is the length of the n block slip surface (m), E_{n-1} is the residual driving force above the n block ($\mathrm{t\,m^{-2}}$), and α_{n-1} is the inclination of the slip surface of the block above the n block.

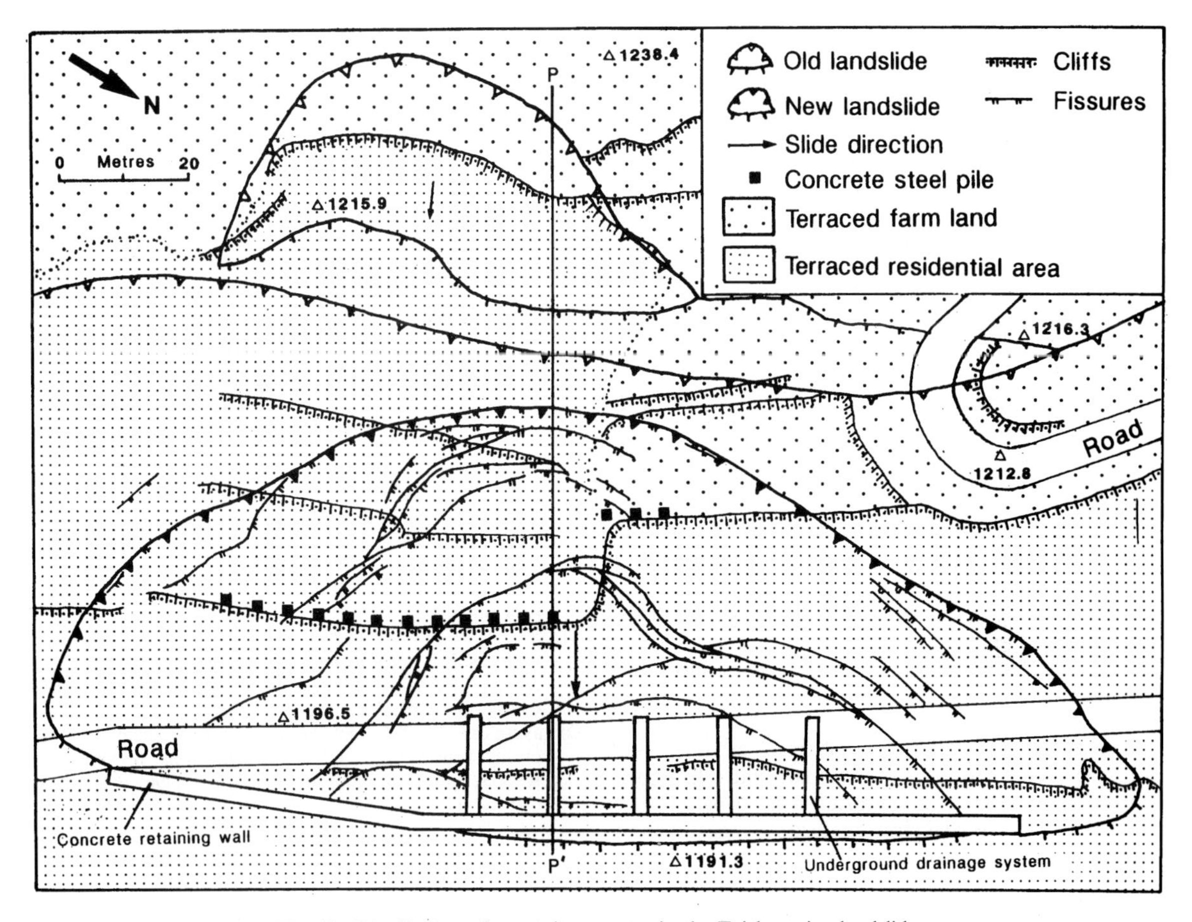

Fig. 12. Distribution of control structures in the Taishanmiao landslide.

The driving force values for Jiaoshuwan and Taishanmiao of $120\,\mathrm{t\,m^{-2}}$ and $36\,\mathrm{t\,m^{-2}}$, respectively, were derived and used in the calculations for the remedial measures.

Landslide control at Jiaoshuwan

Given that the principal reason for the reactivation of the landslides at Jiaoshuwan was human excavation and impedance of groundwater outflows, the control design measures included the following:

1. A row of 10 concrete and steel piles was built at the toe of the landslide to resist the sliding forces, the size of individual piles ranging between $4 \times 4 \times 27$ m and $4 \times 4 \times 11$ m (Fig. 11).
2. An underground drainage system, made up of trenches with a cross-section of 2×5 m and a length of 129 m was set up in the middle of the western valley of the landslide where groundwater was concentrated. The drainage trenches were filled with boulders, and filter layers were placed on their flanks, consisting of pebbles on the outside, coarse sands in the middle and fine sands in the core. The bottom of the drainage trenches were lined with concrete and cut below the slip surface. This drainage serves not only to reduce the water content in the slide mass but also to create a resistance against the sliding forces.
3. In order to avoid secondary slides in the western valley, two relatively small rows of concrete steel piles combined with retaining concrete walls were installed. At the back of the retaining walls, filter layers similar to those in the underground drainage system were installed.
4. An extensive drainage system consisting of channels was engineered on the slope surface to direct rainfall water.

5. After these measures had been completed all cracks on the slope surface were filled and tamped to minimize infiltration.

Landslide control at Taishanmiao

Because of the retrogressive nature of the Taishanmiao slide and the concentration of groundwater in the lower part of the slide mass, control measures were mainly concentrated in the lower half of the feature. They were as follows:

1. A concrete retaining wall was built on the toe, with an underground drainage system consisting of five branches, each being $2 \times 7 \times 15$ m (Fig. 12). It is thus a similar structure to that used in the western valley of the Jiaoshuwan landslide.
2. In order to avoid the failure of the upper (artificially cut) part of the slide, a row of 22 concrete and steel piles, each $2 \times 2 \times 12$ m was built.
3. An extensive drainage system was constructed on the surface of the slope.
4. All cracks on the slope surface were filled and tamped.

Since the completion of these measures in June 1992, no further deformation of the two landslides has been recorded by a series of monitoring nets using SINCO digital inclinometers and electron distance measurement (EDM) equipment.

Conclusions

Study of several thousand landslides made up of varying proportions of loess in Gansu Province, China, have been used to derive a simple classification based on the location of the principal shear plane and the presence of water-conducting structures. While there is some evidence that different geomorphological forms favour particular landslide types, it is clear that the majority of landslides threatening life and property in the western Loess Plateau are caused, or at least enhanced, by human interference, notably in oversteepening slopes by excavation and by interfering with the groundwater system, especially by introducing large-scale irrigation. Landslide control measures in this region include landscaping and a variety of retaining structures including artificial drainage systems. These approaches have met with some success in the past decade, although some large slides, located on the thick loess found on many river terraces, caused by long and intensive irrigation, remain a major problem. On the basis of studies of several thousand landslides in north China, it is fair to say that the majority of landslides threatening life and property are caused, or at least enhanced, by human interference, notably in oversteepening slopes by excavation.

Acknowledgements. We record our deep gratitude to the late Wang Jingtai of the Geological Hazards Research Institute, Gansu Academy of Sciences, China, for his direction of the vital stages of the work reported here and for his constant encouragement. X.M.M. would like to thank J. Rose of the Department of Geography, Royal Holloway, University of London for his ready support and encouragement. The research programme from which these results are drawn was funded by the Commission of the European Communities (Contracts CI.1.0109.UK(H) and CI1*-CT92-0004) and the Government of Gansu Province, China.

References

ANON 1984. *Technical Handbook, Volume 4: Railway Engineering Design*. Chinese Railway Publishing House, Beijing (in Chinese).

CHINESE RAILWAY MINISTRY 1983. *Design and Calculation of Retaining Piles against Landsliding. No. 2 Survey and Design Academy*. Chinese Railway Ministry, Beijing, China (in Chinese).

DERBYSHIRE, E. 1983. On the morphology, sediments, and origin of the Loess Plateau of central China. *In*: GARDNER, R. & SCOGING, H. (eds) *Mega-geomorphology*. Clarendon, Oxford, 172–194.

—— & MENG, X. M. 1995. The landslide hazard in north China: characteristics and remedial measures at the Jiaoshuwan and Taishanmiao slides in Tian Shui city, Gansu Province. *In*: MCGREGOR, D. F. M. & THOMSON, D. A. (eds) *Geomorphology and Land Management in a Changing Environment*. Wiley, Chichester, 89–104.

——, WANG, J. T., JIN, Z., BILLARD, A., EGELS, Y., JONES, D. K. C., KASSER, M., MUXART, T. & OWEN, L. 1991. Landslides in the Gansu loess of China. *In*: OKUDA, S., RAPP, A. & ZHANG, L. Y. (eds) *Loess: Geomorphological Hazards and Processes. Catena Suppl.*, **20**, 119–145.

——, DIJKSTRA, T. A., BILLARD, A., MUXART, T., SMALLEY, I. J. & LI, Y. J. 1993. Threshholds in a sensitive landscape: the loess region of central China. *In*: THOMAS, D. S. G. & ALLISON, R. J. (eds) *Landscape Sensitivity*. Wiley, Chichester, 97–127.

——, MENG, X. M., WANG, J. T., ZHOU, Z. Q. & LI, B. X. 1995*a*. Collapsible loess on the Loess Plateau of China. *In*: DERBYSHIRE, E., DIJKSTRA, T. & SMALLEY, I. J. (eds) *Genesis and Properties of Collapsible Soils*. Kluwer, Dordrecht, 267–293.

——, VAN ASCH, T., BILLARD, A. & MENG, X. M. 1995*b*. Modelling the erosional susceptibility of landslide catchments in thick loess: Chinese variations on a theme by Jan de Ploey. *Catena*, **106**, 1–17.

DIJKSTRA, T. A., DERBYSHIRE, E. & MENG, X. M. 1993. Neotectonics and mass movements in the loess of north-central China. *In*: OWEN, L. A., STEWART, I. & VITA-FINZI, C. (eds) *Neotectonics: Recent Advances. Quaternary Proceedings*, **3**, 93–110.

——, Rogers, C. D. F., Smalley, I. J., Derbyshire, E., Li, Y. J. & Meng, X. M. 1994. The loess of north-central China: geotechnical propeties and their relation to slope stability. *Engineering Geology*, **36**, 153–171.

Meng, X. M. & Zhang, S. W. 1989. Characteristics and mechanism of Gaolanshan landslides. *Journal of Gansu Sciences*, **2**(1), 48–53.

——, Wang, D. K. & Hong, J. X. 1988. *Landslides and debris flows in Lanzhou Region, Gansu Province, China*. Internal report (in Chinese).

Muxart, T., Billard., A., Derbyshire, E. & Wang, J. 1994. Variation in runoff on steep, unstable loess slopes near Lanzhou, China: initial results using rainfall simulation. *In*: Kirkby, M. J. (ed.) *Process Models and Theoretical Geomorphology*, Wiley, Chichester, 337–355.

Inspection and risk assessment of slopes associated with the UK canal network

G. J. Holland & M. E. Andrews

British Waterways, Wellington Park House, Thirsk Row, Leeds LS1 4DD, UK

Abstract. Canals in the UK typically comprise artificial channels, with a wide range of embankments and cuttings. These structures often suffer from a range of problems unique to the canal environment including age, wide variation in construction methods and, in the case of embankments, a requirement to retain water. It is the responsibility of British Waterways to regularly inspect and evaluate these slopes with regard to long-term safety and identification of maintenance requirements.

A system of slope inspection procedures has been developed which is tailored to aspects of slope stability related to canal structures. This is used in conjunction with a risk evaluation procedure to identify areas of concern on a proactive basis. The results are used to co-ordinate maintenance effort, and identify requirements for more detailed ground investigation.

Introduction

British Waterways (BW) manages and maintains over 3000 km of canals and river navigations, and many thousands of associated structures including bridges, tunnels and embankment dams. The canal system stretches from northwest Scotland to southern England and traverses a wide diversity of topography and geological terrains. A simplified map of the system is shown in Fig. 1.

One feature which distinguishes a canal from other water courses is the requirement to retain water, often at the top of a slope, within an artificial channel. The majority of the UK's canals were constructed between 1761 and 1830 by private venture and using methods which were at the forefront of the construction techniques of the day. The materials used for construction were nearly always locally obtained and of extremely variable quality, the methods and quality of construction often depended on the contractor and on the diligence of the engineers (some of whom, such as Brindley, Rennie and Telford, have become household names). Construction records have only rarely survived to the present day.

Following the advent of the railways in the mid-19th century, the use of the canal system for freight transport gradually declined and so did the amount of maintenance carried out. The last 20 years have seen a reversal of the decline with the appreciation that the canals are an integral part of the UK's transport and land drainage system, and with a significant rise in the volume of leisure traffic. The change in boat types from the traditional horse-drawn narrowboat to more modern, faster, powered craft, coupled with the relative lack of attention for many years, has left British Waterways with a legacy of maintenance problems which it is striving to overcome.

In times of limited financial and human resources, methods are being developed to direct resources to areas of greatest need commensurate with both public safety and the operational requirements of the waterways

Philosophy of inspection

In common with other geographically widespread organizations, British Waterways inspection and recording processes developed over many years, frequently locally and often in response to particular local problems. This led to a diversity of types and styles of inspection for many different structures including slopes. In an attempt to rationalize and co-ordinate certain types of structure inspection, instructions were given in 1973 and in 1988 to allow the systematic evaluation and monitoring of changes to the physical condition of structures, and to identify work required. The system utilized a simple checklist of 11 key points for slopes whereby each point could be categorized as a condition ranging through five categories from very good to bad.

The frequency of embankment inspections was determined as half yearly, to be carried out by local engineering personnel (often with little or no geotechnical background). 1992 saw the production of the *British Waterways Engineering Inspection Procedures Manual*, within which inspection procedures were further formalized in terms of frequency, competency of the inspector, and reporting procedures. For embankments it was determined that a *Principal Inspection* should be carried out at a maximum interval of five years by a chartered engineer or equivalent, an *Intermediate Inspection* at yearly intervals by an engineering supervisor and a *Length Inspection* by a local foreman at no more than monthly intervals.

Holland, G. J. & Andrews, M. E. 1998. Inspection and risk assessment of slopes associated with the UK canal network. *In*: Maund, J. G. & Eddleston, M. (eds) *Geohazards in Engineering Geology*. Geological Society, London, Engineering Geology Special Publications, **15**, 155–165.

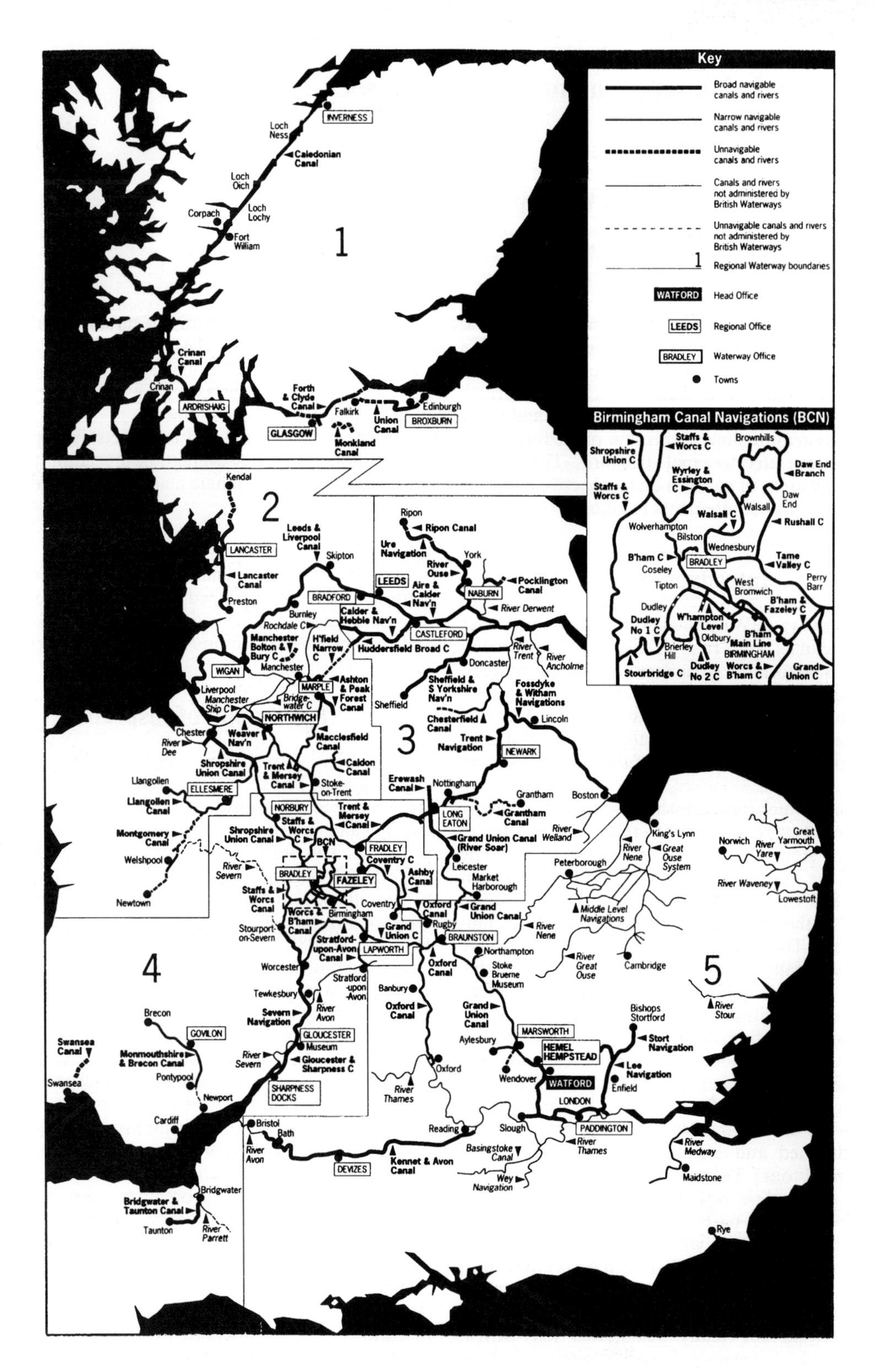

Fig. 1. Plan of the UK waterway system.

The purpose of inspections has changed over the years. In the past, inspections were frequently carried out when problems had already occurred, and were designed to identify the need for technical assistance, ground investigation and remedial works (the fire fighting approach). British Waterways have in recent years adopted a far more proactive attitude, allowing the use of preventative maintenance techniques to repair defects before the situation becomes serious enough to result in the closure of a canal or to prejudice public safety.

Slope inspections are now more concerned with providing a comprehensive database of information (both embankments and cuttings) which allows the monitoring of change, and more importantly the rate of change of the slope condition, and therefore allows the targeting of resources and financial planning into the future.

The slope inspection process is constantly evolving. Although there is much fine tuning to be done, the process is already proving valuable in developing *asset management plans* which are defining standards, policies and practices, providing information on the performance and condition of each asset, and identifying shortfalls. Schemes to remedy identified shortfalls are being prioritized and included in long-term (10 year) investment plans.

The scale of the problem

It is estimated that of the 3000 km of navigable waterways administered by British Waterways, at least 1000 km are either on embankment or in cutting, or partly cut into sidelong ground with a human-made embankment on the downslope side. Many slopes are in rural locations, which results in difficulties of access, and are often heavily vegetated making the inspection process arduous and time-consuming. Others are in inhabited areas where the consequences of failure of a slope, and particularly of a breach of an embankment, can potentially be great in terms of the area and number of persons affected. In such locations the inspection process must be extremely rigorous and take into account the condition of associated structures such as retaining walls, culverts and aqueducts and the location of adjacent developments (which can and do have a fundamental effect on the stability of canal slopes).

Against this background it must be appreciated that inspection resources are scarce. A small geotechnical team, as well as other conventional duties, is charged with carrying out the initial Principal Inspection on each slope (and sometimes follow-up Principal Inspections) and providing the basic data, a brief report and recommendations. For reporting purposes, a standard structure report form is used to provide a text summary of the inspection, details of defects and recommendations.

A photographic record is also attached to allow a visual assessment to be made of changes over time. Intermediate inspections are carried out as part of the general duties of a local engineering supervisor (of which there are 25 nationally) who must also inspect all other structures on their waterway at least annually.

Principal Inspections

Looking back over the history of slope problems associated with the canals, it is clear that in nearly every case tell-tale signs of impending failure were present well before the event. However, the significance of these signs may not always have been understood, particularly in the absence of previous comprehensive inspection records.

It has long been realized that the key to success in the inspection process is speed and simplicity, but to accurately record features and their relevance in a form that can be easily used, often by the geotechnically untrained. Previous attempts to record detail in text form have proved lacking and inappropriate to linear structures often extending for many hundreds of metres.

It has been found that the most appropriate method is to record slope and structure details in a standard format on 1:1250 scale Ordnance Survey base plans. An example is given in Fig. 2. The scale has been chosen as a compromise between the need to record fine detail, the availability of standard 1:2500 scale OS sheets, enlargement facilities and quality of reproduction.

At present, data are recorded manually onto the base plans, using the standard symbol set shown in Fig. 3, although with the advent of Geographical Information Systems, digitally based plans and scanning technology, this will shortly become more efficient and of higher quality. The standard symbol set is based where possible on the Engineering Group Working Party (Anon 1972), although because of conflict with other symbols in use it has been found necessary to create many new ones.

The inspection process takes into account the following:

- Local knowledge of the length involved, including the history of maintenance and problems. In this respect continuity of records is extremely important.
- Inspection of the full canal structure (banks, bank protection, weirs) and of the slopes associated with it including the crest, slope face and toe, and adjacent land which may well be outside BW ownership. This obviously has to take in both sides of the canal and not simply an inspection from the more easily accessible towpath.
- It must be capable of being carried out visually, safely and in a reasonable time but must be comprehensive.
- The consequences of a failure must be assessed in terms of its potential impact on local communities, infrastructure and canal users.

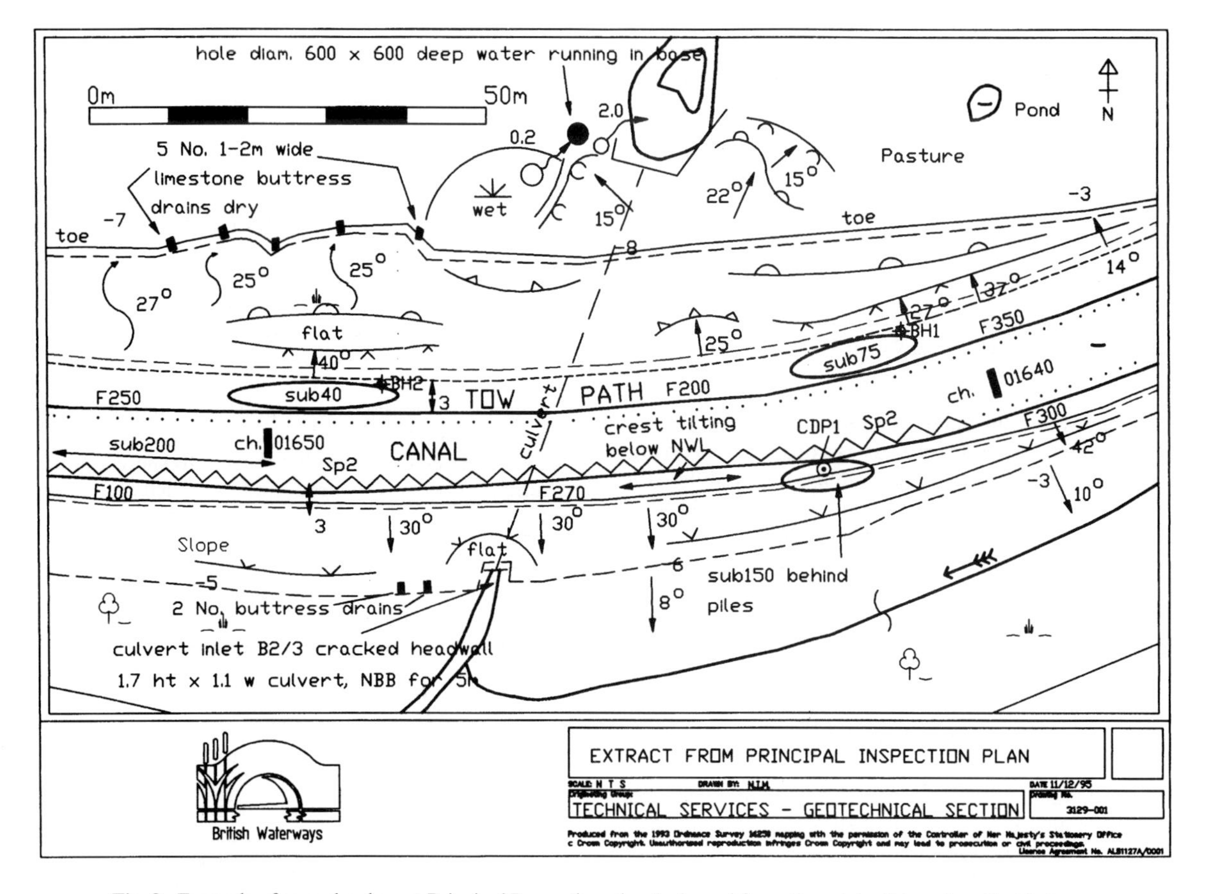

Fig. 2. Example of an embankment Principal Inspection plan (enlarged from the original) based on the Ordnance Survey 1:1250 map and reproduced with the permission of the controller of Her Majesty's Stationary Office, Crown Copyright reserved.

- The results of previous ground investigations, monitoring, survey information and reports.
- Other structures associated with the canal, including feeders, drainage, culverts and retaining walls.

The Principal Inspection process requires a trained geotechnical eye and a geomorphological mapping ability as well as experience of structure inspections.

Techniques for geomorphological mapping have been described by Waters (1958) and Savigear (1965), and there are numerous case studies involving their application. However, although some features are common to all slopes, many are peculiar to the waterways. Features recorded on the 1:1250 base plans and their relevance are as follows:

- Standard slope features such as breaks in slope, angles, heights, scarp faces and bulges. In the absence of detailed topographical survey information, it has proved useful to relate levels of features to a datum of canal water level as shown in Fig. 4 such that features below the canal are given a negative value, and above a positive value. Relative levels are generally estimated but, given practice, have proved to be surprisingly accurate and more than adequate.
- Features indicative of canal leakage (or sometimes naturally occurring water flow), including issue points and an estimate of flow, wet or ponded areas, flow in drainage courses, discoloration of water, deposition of suspended solids and hydrophilic vegetation. Canal leakage can cause wholesale saturation of an embankment resulting in loss of strength, subsidence or mass gravitational movement. Because of the variability of construction materials, discrete leakage paths can result in local piping and collapse, as can fractured service pipes (frequently situated in the canal towpath). As much of the canal structure is beneath water level, routine visual inspection of the canal bed is not always possible, and indeed physical examination of

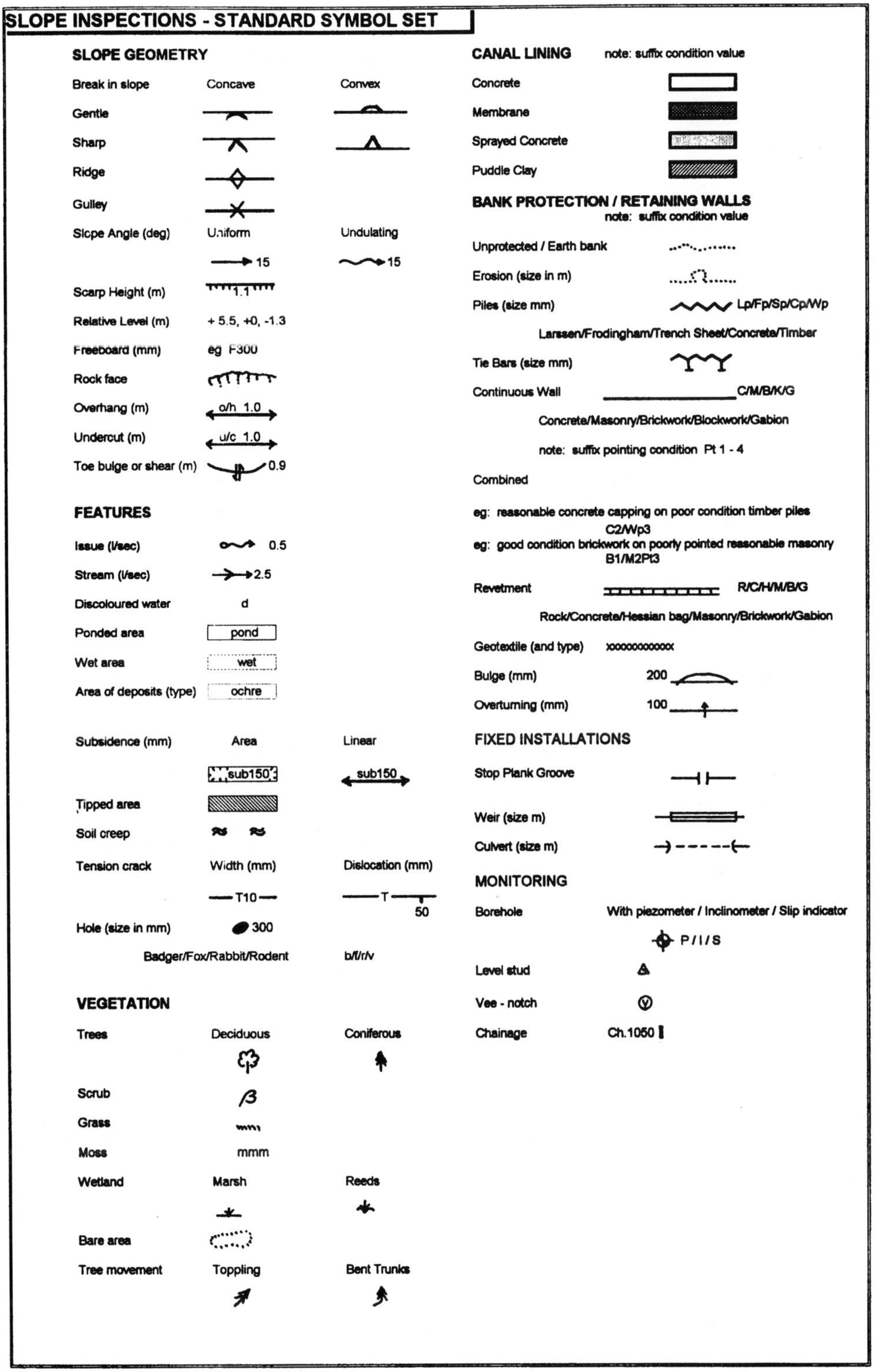

SLOPE INSPECTIONS - STANDARD SYMBOL SET

SLOPE GEOMETRY

Break in slope	Concave	Convex
Gentle		
Sharp		
Ridge		
Gulley		
Slope Angle (deg)	Uniform 15	Undulating 15
Scarp Height (m)	1.1	
Relative Level (m)	+ 5.5, +0, -1.3	
Freeboard (mm)	eg F300	
Rock face		
Overhang (m)	o/h 1.0	
Undercut (m)	u/c 1.0	
Toe bulge or shear (m)	0.9	

FEATURES

Issue (l/sec)	0.5	
Stream (l/sec)	2.5	
Discoloured water	d	
Ponded area	pond	
Wet area	wet	
Area of deposits (type)	ochre	
Subsidence (mm)	Area	Linear
	sub150	sub150
Tipped area		
Soil creep		
Tension crack	Width (mm)	Dislocation (mm)
	T10	T 50
Hole (size in mm)	300	
	Badger/Fox/Rabbit/Rodent	b/f/r/v

VEGETATION

Trees	Deciduous	Coniferous
Scrub		
Grass		
Moss	mmm	
Wetland	Marsh	Reeds
Bare area		
Tree movement	Toppling	Bent Trunks

CANAL LINING

note: suffix condition value

- Concrete
- Membrane
- Sprayed Concrete
- Puddle Clay

BANK PROTECTION / RETAINING WALLS

note: suffix condition value

Unprotected / Earth bank	
Erosion (size in m)	
Piles (size mm)	Lp/Fp/Sp/Cp/Wp
	Larssen/Frodingham/Trench Sheet/Concrete/Timber
Tie Bars (size mm)	
Continuous Wall	C/M/B/K/G
	Concrete/Masonry/Brickwork/Blockwork/Gabion
	note: suffix pointing condition Pt 1 - 4

Combined

eg: reasonable concrete capping on poor condition timber piles C2/Wp3

eg: good condition brickwork on poorly pointed reasonable masonry B1/M2Pt3

Revetment	R/C/H/M/B/G
	Rock/Concrete/Hessian bag/Masonry/Brickwork/Gabion
Geotextile (and type)	xxxxxxxxxxxx
Bulge (mm)	200
Overturning (mm)	100

FIXED INSTALLATIONS

- Stop Plank Groove
- Weir (size m)
- Culvert (size m)

MONITORING

Borehole	With piezometer / Inclinometer / Slip indicator
	P / I / S
Level stud	
Vee - notch	V
Chainage	Ch.1050

Fig. 3. Standard symbol set for Principal Inspections.

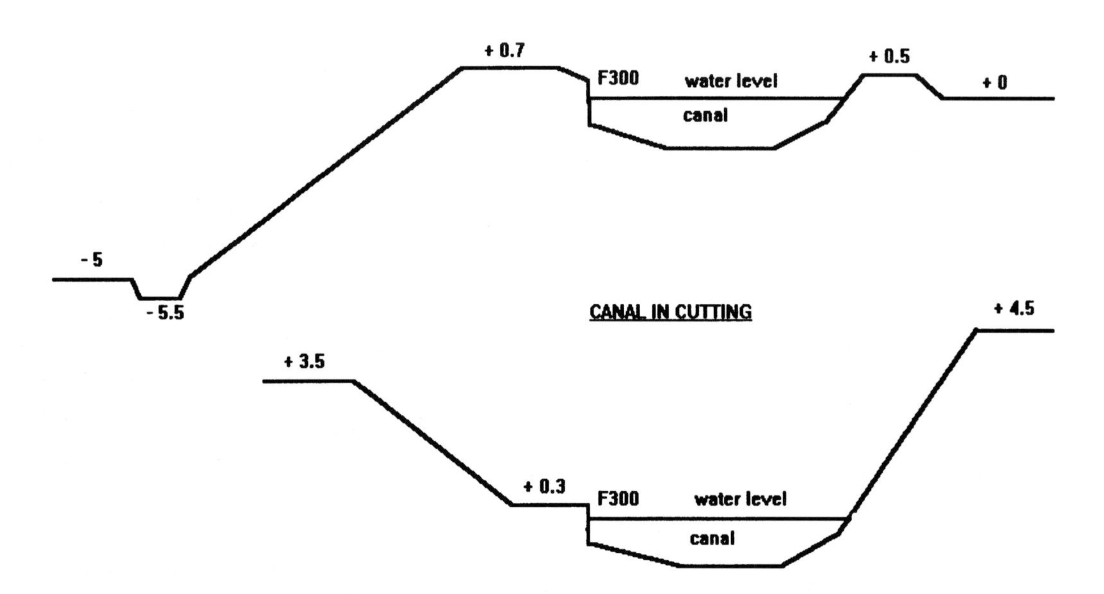

Fig. 4. Diagrams showing indicative levels relative to canal water level datum.

the bed can cause damage to its water-retention properties. However, in still conditions sometimes enough can be seen to identify tell-tale bed depressions, or leaf debris drawn into leakage areas.

- The risk of embankment failure is increased if water levels in the canal rise above normal. Canal water levels are usually artificially controlled by means of weirs or sluices, but feeders (except from reservoirs) are generally uncontrolled. Blockages of weirs or trash screens or storm discharges can result in unacceptable raising of water levels which at best results in 'top water' leakage into the slope, and at worst leads to overtopping and rapid erosion at some low point in an embankment. Inspection therefore records the amount of 'freeboard' along the canal and also the nature of the catchment.
- Deterioration of canal bank protection is frequently a precursor to slope stability problems, due to slope saturation or fines removal. The inspection process records the type and position of the various types of bank protection and canal lining (if any) and ascribes a condition value (see Table 1) of between 1 and 4 to each type depending on the perceived degree of deterioration. This is later used to guide decisions on maintenance work.
- Slope toe deterioration, such as undercutting within the canal or river erosion. This is a common cause of slope oversteepening and eventual stability problems.
- Movement of slopes can manifest itself in many ways. Careful note is made of any indications of settlement at the crest (such as tension cracks, subsidence of canal bank walling or pile lines) and on the slope itself. Features such as disturbed fence lines, bent and leaning trees, backscars, toe bulging and heaving (particularly into toe drain ditches) are all recorded.
- The effect of animal activity on slopes is often underestimated. The burrowing of small animals into canal banks is a common occurrence. Although this is generally at a level slightly above normal water level in the canal, a rise in water level can promote significant leakage and erosion along these paths. Major burrowing by badgers, particularly into the reworked Lias Clay embankments of the Midlands, has caused substantial problems in the past, including the collapse of setts. Access to the slope crest for grazing animals in search of water can lead to trampling, erosion and reduction of freeboard to such an extent that overtopping can occur.

The speed of inspection varies considerably depending on the type and dimensions of the slope, its accessibility, vegetation density, weather conditions and the number of features to be recorded. However, in practice an average of two slopes per day per person is achievable. Given the corridor of inspection, this compares well with the geomorphological mapping rates of 150 000–300 000 m^2

Table 1. *Structure condition values*

Values	*Condition*	*Description*
1	Good	As new or with only slight deterioration; no work required
2	Reasonable	Some deterioration evident but not requiring attention
3	Poor	Deterioration to such an extent that repair is required
4	Bad	Deterioration to such as extent that urgent repair is required

Note: In attaching a condition value a judgement must be made regarding the relative risk and consequences of the condition.

per day per person quoted by Brunsden *et al.* (1975) for highway studies in South Wales.

In order to prioritize slopes for future Principal Inspections, a scaled down process known as a *Quick Inspection* can be used. This allows in the order of 6 linear km of canal per day per person to be inspected and the risk of failure to be rapidly assessed. It is not rigorous, it is subjective and is no substitute for a Principal Inspection. However, in the absence of other records it is an essential preliminary and allows the preparation of simple waterway hazard maps. It also familiarizes the inspector with overall conditions along a particular length and gives the inspector some background against which to determine priorities.

It has been found that there is no 'best time of year' to carry out inspections. Summer inspections have the advantage of good weather, long daylight hours and generally dry slopes on which the tell-tale signs of leakage can easily be seen. However, dense vegetation is a problem, as is boat traffic which tends to raise canal sediment into suspension and reduces visibility for bed and bank inspections.

The situation is reversed in the winter, and although higher water levels in the canal amplify top water leakage problems, their location on a seasonally wet slope, often coupled with full toe drainage ditches, can be difficult.

The use of aerial photographs is not possible in most cases due to the propensity for tree growth along the canal line. British Waterways are in a difficult position regarding vegetation clearance, and the requirements of inspection are often incompatible with those of conservation. Wholesale vegetation clearance for inspection is clearly not an option in most cases, nor is it desirable geotechnically due to the consequent destruction of root systems, the increase in slope pore pressures and runoff erosion. In particularly dense vegetation, corridors in the order of 1m wide have been cut to aid inspection, although it is recognized that large areas of the slope remain unseen. In these situations a water flow in a toe drain ditch can often be the only visual indication of impending problems.

Slope risk evaluation

In 1991 a document was prepared entitled *risk management: embankment breach rapid assessment*, which was intended for use throughout British Waterways to identify lengths of embankments vulnerable to breach failures (Schlegel *et al.* 1991).

Although the system was used successfully in Scotland to 'score' embankments on the Union Canal, there were some problems during usage, particularly in its application to other types of slopes where a breach situation was not relevant (for example, a cutting failure affecting third-party property at the top of the slope). A revised system has now been developed to allow the evaluation of all types of slopes using a Principal Inspection record to provide basic data, together with some knowledge of the countryside through which the canal passes, and the water control facilities.

The evaluation is carried out in three parts:

1. slope failure hazard
2. slope failure magnitude
3. consequences

The risk factors are calculated separately and then combined on a proforma (Fig. 5) to give an overall risk value. Various weightings are used in the scoring system, based mainly on experience, and engineering judgement.

Slope failure hazard

It is acknowledged that there is a considerable difference between the risk attached to a slope failure below the canal which could result in a breach, and that of a failure above the canal where individual lives or third-party property may be affected. This is reflected in the scoring system.

For slopes and retaining walls a risk factor of 1 is applied if Fig. 6 is examined and it is considered that there is no necessity for an evaluation.

Slopes have been subdivided on whether they comprise essentially rock or soil/made ground. Values chosen as cut-offs (such as 11° and 19°) are based on experience and on a wide range of studies of various materials. Where slopes contain retaining walls then both are assessed and the highest risk factor used.

Due to the variations in length of slopes and the possible changes of circumstances within their lengths, some may have two or more risk factors, allowing the slope to be divided into risk factor zones.

Magnitude of failure

The magnitude of failure depends on three main elements.

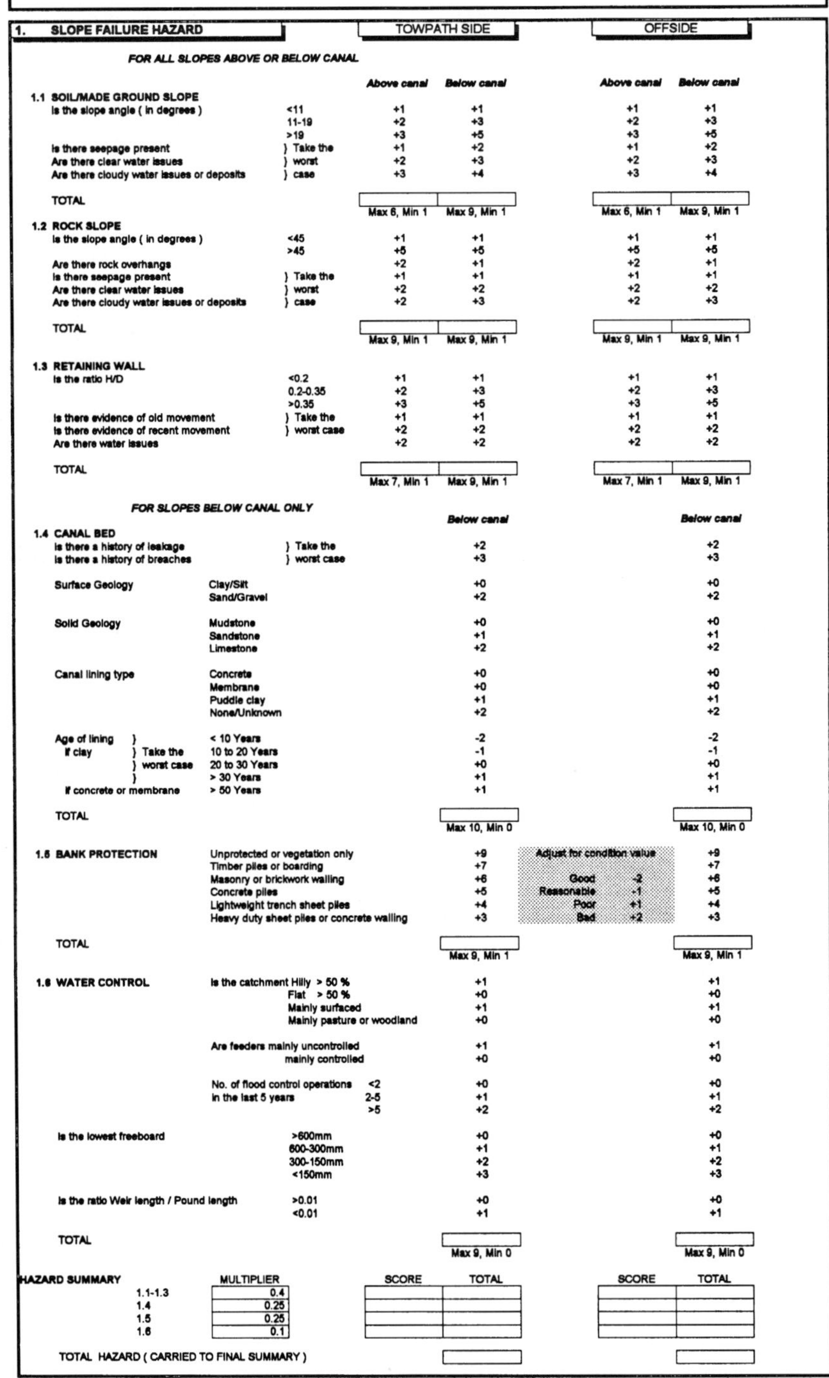

SLOPE RISK EVALUATION

SITE NAME: **NGR / CHAINAGES:**

INSPECTOR: **INSPECTION DATE:**

1. SLOPE FAILURE HAZARD — TOWPATH SIDE / OFFSIDE

FOR ALL SLOPES ABOVE OR BELOW CANAL

			TOWPATH SIDE *Above canal*	TOWPATH SIDE *Below canal*	OFFSIDE *Above canal*	OFFSIDE *Below canal*
1.1 SOIL/MADE GROUND SLOPE						
Is the slope angle (in degrees)	<11		+1	+1	+1	+1
	11-19		+2	+3	+2	+3
	>19		+3	+5	+3	+5
Is there seepage present	} Take the		+1	+2	+1	+2
Are there clear water issues	} worst		+2	+3	+2	+3
Are there cloudy water issues or deposits	} case		+3	+4	+3	+4
TOTAL			Max 6, Min 1	Max 9, Min 1	Max 6, Min 1	Max 9, Min 1
1.2 ROCK SLOPE						
Is the slope angle (in degrees)	<45		+1	+1	+1	+1
	>45		+5	+5	+5	+5
Are there rock overhangs			+2	+1	+2	+1
Is there seepage present	} Take the		+1	+1	+1	+1
Are there clear water issues	} worst		+2	+2	+2	+2
Are there cloudy water issues or deposits	} case		+2	+3	+2	+3
TOTAL			Max 9, Min 1	Max 9, Min 1	Max 9, Min 1	Max 9, Min 1
1.3 RETAINING WALL						
Is the ratio H/D	<0.2		+1	+1	+1	+1
	0.2-0.35		+2	+3	+2	+3
	>0.35		+3	+5	+3	+5
Is there evidence of old movement	} Take the		+1	+1	+1	+1
Is there evidence of recent movement	} worst case		+2	+2	+2	+2
Are there water issues			+2	+2	+2	+2
TOTAL			Max 7, Min 1	Max 9, Min 1	Max 7, Min 1	Max 9, Min 1

FOR SLOPES BELOW CANAL ONLY

			TOWPATH SIDE *Below canal*		OFFSIDE *Below canal*
1.4 CANAL BED					
Is there a history of leakage		} Take the	+2		+2
Is there a history of breaches		} worst case	+3		+3
Surface Geology	Clay/Silt		+0		+0
	Sand/Gravel		+2		+2
Solid Geology	Mudstone		+0		+0
	Sandstone		+1		+1
	Limestone		+2		+2
Canal lining type	Concrete		+0		+0
	Membrane		+0		+0
	Puddle clay		+1		+1
	None/Unknown		+2		+2
Age of lining }	< 10 Years		-2		-2
If clay } Take the	10 to 20 Years		-1		-1
} worst case	20 to 30 Years		+0		+0
}	> 30 Years		+1		+1
If concrete or membrane	> 50 Years		+1		+1
TOTAL			Max 10, Min 0		Max 10, Min 0
1.5 BANK PROTECTION	Unprotected or vegetation only		+9	Adjust for condition value	+9
	Timber piles or boarding		+7		+7
	Masonry or brickwork walling		+6	Good -2	+6
	Concrete piles		+5	Reasonable -1	+5
	Lightweight trench sheet piles		+4	Poor +1	+4
	Heavy duty sheet piles or concrete walling		+3	Bad +2	+3
TOTAL			Max 9, Min 1		Max 9, Min 1
1.6 WATER CONTROL	Is the catchment Hilly > 50 %		+1		+1
	Flat > 50 %		+0		+0
	Mainly surfaced		+1		+1
	Mainly pasture or woodland		+0		+0
	Are feeders mainly uncontrolled		+1		+1
	mainly controlled		+0		+0
	No. of flood control operations	<2	+0		+0
	in the last 5 years	2-5	+1		+1
		>5	+2		+2
Is the lowest freeboard		>600mm	+0		+0
		600-300mm	+1		+1
		300-150mm	+2		+2
		<150mm	+3		+3
Is the ratio Weir length / Pound length		>0.01	+0		+0
		<0.01	+1		+1
TOTAL			Max 9, Min 0		Max 9, Min 0

HAZARD SUMMARY

	MULTIPLIER	TOWPATH SIDE SCORE	TOWPATH SIDE TOTAL	OFFSIDE SCORE	OFFSIDE TOTAL
1.1-1.3	0.4				
1.4	0.25				
1.5	0.25				
1.6	0.1				
TOTAL HAZARD (CARRIED TO FINAL SUMMARY)					

Fig. 5. Slope risk evaluation pro-forma.

2. SLOPE FAILURE MAGNITUDE		TOWPATH SIDE		OFFSIDE	
FOR ALL SLOPES ABOVE OR BELOW CANAL					
		Above canal	*Below canal*	*Above canal*	*Below canal*
2.1 PROXIMITY OF VULNERABLE STRUCTURES OR PERSONS	<10m	+7	+9	+7	+9
	10-20m	+6	+8	+6	+8
	20-30m	+4	+7	+4	+7
	30-50m	+1	+5	+1	+5
	50-70m	+0	+4	+0	+4
	70-100m	+0	+2	+0	+2
	100-150m	+0	+1	+0	+1
	>150m	+0	+0	+0	+0
TOTAL		Max 7, Min 0	Max 9, Min 0	Max 7, Min 0	Max 9, Min 0
FOR SLOPES BELOW CANAL ONLY					
			Below canal		*Below canal*
2.2 HYDRAULIC FACTOR					
Broad canal / navigation	} Take the		+6		+6
Narrow canal	} worst case		+4		+4
Length of pound between locks	< 1 km		+0		+0
	1-3 km		+1		+1
	3-7 km		+2		+2
	> 7 km		+3		+3
Nearest stop plank grooves	<0.5 km		-2		-2
	0.5-1 km		-1		-1
	> 1 km		+0		+0
Staff response time	< 2 hrs		-2		-2
	2-4 hrs		-1		-1
	> 4 hrs		+0		+0
TOTAL TOTAL			Max 9, Min 0		Max 9, Min 0
2.3 TOPOGRAPHY BELOW CANAL	Narrow valley		+9		+9
Note: Assessed values allowed	Flat plain		+3		+3
	Effective barrier		+2		+2

MAGNITUDE SUMMARY	MULTIPLIER	SCORE	TOTAL	SCORE	TOTAL
Proximity	0.33				
Hydraulic	0.33				
Topography	0.33				
TOTAL MAGNITUDE (CARRIED TO FINAL SUMMARY)					

3. CONSEQUENCES	Towpath side	Offside
FOR ALL SLOPES ABOVE OR BELOW CANAL		
Hazard to community, many people, communication link	+9	+9
Hazard to small groups of houses, people, property	+8	+8
Hazard to isolated houses, people, communications	+7	+7
Hazard to individual life	+6	+6
Hazard to isolated non - residential property	+5	+5
Hazard to high value land, SSSI	+4	+4
Hazard to low value land, minor access route	+3	+3
Hazard only to low value land	+2	+2
No hazard except disruption to navigation	+1	+1

FINAL EVALUATION SUMMARY	Towpath side	Offside
HAZARD		
MAGNITUDE		
CONSEQUENCES		
TOTAL RISK FACTOR		

COMMENTS/NOTES

Signed:

Date:

Fig. 5. *(continued)*

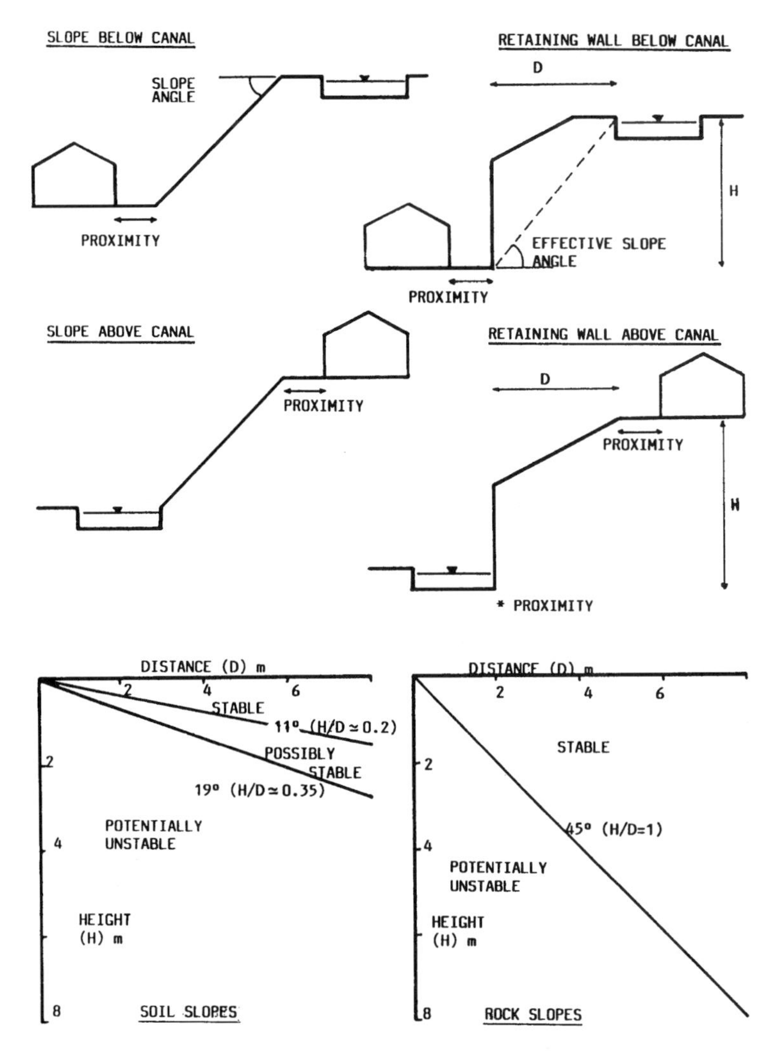

Fig. 6. Slope risk evaluation – stability appraisal chart.

Proximity to failure. In the event of a failure of a slope above the canal, the proximity of properties or persons to the crest of the slope is the critical factor. In the event of a failure of a slope below a canal resulting in a breach situation, water will flow from the canal down the remains of an embankment towards people and property at risk. The magnitude of the threat to them will be dependent on how close they are to the breach site, those immediately adjacent being at greatest risk and those a distance away being less.

Hydraulic factor. This factor considers those circumstances which may affect the flow from a breach once that breach has occurred, and relates to the consequences of a breach rather than the risk of an embankment breaching. It takes into account the maximum available volume of water (depending of the length of the canal pound and the width and depth of the channel) and the ability to prevent or restrict the outflow (position of bridge narrows, stop planks and staff response time).

Topography. The nature of the land downstream of an embankment will affect the way in which water discharging from a breach will be concentrated on people and property. For example, an embankment across a narrow valley once breached will have the total discharge carried down that valley and any community within it will be at risk. On the other hand, a long embankment across a flat area when breached will discharge over a considerable area to a shallow depth;

Table 2. *British Waterways slope Principal Inspections and evaluations 1994–95*

BW region	*No. of embankment Principal Inspections*	*No. of cutting Principal Inspections*	*No. of days per person*	*Recommended works/investigations (£k)*
Scotland	19	7	12	1340
Northeast	30	9	19	2183
Northwest	2	0	2	200
South	13	1	7	842
Midlands/Southwest	8	1	3	380

the numbers affected may be more, although the level of risk to which they are exposed will be less.

In addition, other features such as raised roads, railways, walls, river channels, etc., may direct any discharge away from communities. In determining a score, values of +9, +3, etc., are for guidance and scores between these limits are allowed depending on the assessed topography.

Consequences of a failure

In the event of a failure (with or without a breach), lives, and properties of varying values, may be put at risk above, within and below the canal system and in any area of flooding. The following system is used to assess the varying levels of consequence:

9 Hazard to a community or main communication link. More than 20 people will be at risk. Examples such as eight or more houses, hotel, hospital, motorway/trunk road or main line railway.
8 Hazard to non-residential buildings, factories, offices, small groups of houses and people.
7 Hazard to isolated houses and people or communication link.
6 Hazard to individual lives (towpath users, boaters).
5 Hazard to isolated non-residential buildings.
4 Hazard to areas of high quality farmland, livestock, arable, etc.
3 Hazard to low quality land and minor access routes.
2 Hazard only to low quality land.
1 No hazard, except to operation of the waterway.

Conclusions

The use of a combination of geotechnical mapping techniques, structure condition assessment and slope risk evaluation can be used successfully to identify potential slope problems, to monitor slope performance and to direct human and financial resources to those slopes with the greatest potential risk. Table 2 shows that over a period of less than one year a total of 90 embankments and cuttings were inspected and evaluated by a two-person geotechnical team. The risk evaluations were used to prioritize works totalling nearly £5 million for a 10 year period, and set out future inspection regimes for local engineering staff.

In order to minimize the use of staff time on inspections, it has been essential to train geotechnical engineers in structure inspection techniques, as well as civil engineers in slope inspection techniques. This has invariably resulted in a greater understanding of the processes involved, a more effective inspection process, greater staff flexibility, and focused design of ground investigation and remedial works. It has also gone some way towards allaying the fears expressed by Fookes (1986) that risk evaluation depends to a significant extent on the discipline or profession of those quantifying the risk.

It is believed that future developments including electronic data acquisition, and linked GIS and database software, will further enhance the system, and the recording and reporting quality. The techniques described are particularly appropriate for linear water-retaining structures, including embankment dams, and are highly cost-effective.

References

ANON 1972. The preparation of maps and plans in terms of engineering geology. *Quarterly Journal of Engineering Geology*, **5**, 293–381.

BRUNSDEN, D., DOORNKAMP, J. C., FOOKES, P. G., JONES, D. K. C. & KELLY, J. M. 1975. Large scale geomorphological mapping and highway engineering design. *Quarterly Journal of Engineering Geology*, **8**, 227–254.

FOOKES, P. G. 1986. Land evaluation and site assessment (hazard and risk). *In*: CULSHAW, M. G., BELL, F. G., CRIPPS, J. C. & O'HARA, M. (eds) *Planning and Engineering Geology*. Geological Society, London, Engineering Geology Special Publications, **4**, 273–282.

SAVIGEAR, R. A. G. 1965. A technique of morphological mapping. *Annals of the Association of American Geographers*, **55**, 514–538.

SCHLEGEL, W. A., FIRTH, A. & MARTIN, W. S. 1991. *Risk management: embankment breach rapid assessment*. Internal British Waterways document.

WATERS, R. S. 1958. Morphological mapping. *Geography*, **43**, 10–17.

Small is beautiful: investigations and remedial works for minor slope failures

J. M. Reid

Transport Research Laboratory, Crowthorne, Berkshire RG45 6AU, UK

Abstract. Minor slope failures receive much less publicity than major failures such as the Scarborough landslide, but can cause considerable damage and inconvenience. The engineering geologist is often required to carry out investigations into the failure and decide on appropriate remedial works in a very short timescale and with a limited budget. Difficulties of access to the slipped area limit the use of conventional ground investigation methods, but specialized lightweight methods such as hand-operated shell and auger boring can be very useful. It is important to define the geological situation as soon as possible, in order to understand the nature of the failure and the processes driving it. Remedial works which are appropriate can then be designed. Two case studies are presented to illustrate the process of investigation, design and construction; the collapse of a sewer on the bank of the River Tay in Perth, and the subsidence of a minor road on a steep hillside near Lanark. A gabion wall and mattress was adopted in the first case; a bored pile retaining wall in the second.

Introduction

Engineering geologists go about their work away from the glare of publicity most of the time. Occasionally, however, a major event occurs which brings the profession to the centre of public attention. These events are often associated with major and spectacular ground failures such as the Scarborough landslide or the Heathrow tunnel collapse. Yet for every case that makes the headlines there are a far greater number of minor ground failures which go largely unreported but which can cause great inconvenience locally. For example, Perry (1989) describes the occurrence of slope failures on motorway earthworks in England and Wales. The investigation and remediation of these minor failures, particularly slope failures, is the theme of this paper.

Major failures are generally followed by detailed ground investigations and back analysis, which may lead to changes in design methods. The Heathrow tunnel collapse is leading to a reappraisal of the use of the New Austrian Tunnelling Method (NATM), for example (Anon 1994), and the failure of Carsington dam led to a greater understanding of the role of progressive failure (Skempton & Vaughan 1993). This process is time-consuming and expensive, but is regarded as worthwhile because of the risk to public safety if the failure is repeated. These resources are seldom available to the engineering geologist dealing with a minor failure; he or she is often required to carry out investigations and design remedial works in a very short timescale and with a limited budget. Clear thinking and a good grasp of engineering geological principles are therefore essential.

In dealing with any slope failure, there are three basic questions to be answered:

- What has happened?
- Why has it happened?
- What can be done about it?

It is important to avoid answering the third question first, or the resulting remedial works may be inappropriate and may even exacerbate the situation.

Answering the first two questions involves carrying out an investigation into the failure. A desk study should be carried out, focusing not only on the geology but also on the previous history of the site. Areas of landslipping may be indicated on old maps, as may old river courses, marshes and other features. Aerial photographs can be very helpful in this respect. Local knowledge on the history of slope movement can be very valuable but should be regarded with caution; people's memories are seldom exact. A detailed site walkover should be carried out at an early stage. The slope failure should be examined in detail, noting the geometry of the failure, position and nature of the back scar, any seepages or tension cracks and other significant features. The surrounding area should be examined for any signs of general instability such as leaning trees, uneven ground, springs and seepages and tension cracks. A study of the surrounding area can often provide clearer information on the nature of the preconditions to failure. In most cases a detailed topographic survey will be required, not only for investigation and analysis of the failure but also for design and construction of the remedial works.

REID, J. M., 998. Small is beautiful: investigations and remedial works for minor slope failures. *In*: MAUND, J. G. & EDDLESTON, M. (eds) *Geohazards in Engineering Geology*. Geological Society, London, Engineering Geology Special Publications, **15**, 167–176.

Fig. 1. Hand-operated boring equipment showing bracehead.

Fig. 2. Hand-operated boring on Perth sewer collapse site.

Investigation of the failure may involve the use of unconventional techniques. Normal light cable percussive boreholes may be carried out outwith the slipped area to establish the stratigraphy. It is seldom possible to undertake these boreholes in the failure zone, yet it is important to establish the ground conditions in order to define the nature of the failure; is it shallow or deep, circular or planar, and where is the water table? Without direct investigation it is difficult to answer these questions.

Lightweight equipment can be used to advantage in these situations. Dynamic probing is valuable in that it gives a continuous profile through the soil, which may enable identification of the base of the slip zone. However, the lack of samples is a major drawback. The recent development of a lightweight automated boring rig, which allows recovery of disturbed and small undisturbed samples and the execution of SPTs, promises to be of great value in small failure investigations. However, there is a technique, widely used in Scotland but almost unknown in England, which is ideal in situations of difficult access, i.e. hand-operated shell and auger boring.

Hand-operated boring consists of scaled-down light cable percussive equipment on a string of shortened SPT rods with a wooden cross frame or bracehead at the top. It is operated by three or four people. The bracehead is thrust down, pushing the tools into the soil, and turned several times. The rods are withdrawn and unscrewed, the shell emptied, and the process repeated. The borehole is cased with either 76 mm or 125 mm casing, which is pushed down by the bracehead. Disturbed and undisturbed samples can be taken and SPTs carried out. The apparatus is illustrated in Figs 1 and 2. The rate of progress is slower than for a cable percussion rig because of the methods employed, allowing clearer definition of the strata. There is a better 'feel' (literally) for the material encountered. Hand rig teams are often drillers who specialize in this type of work and acquire considerable expertise. With an engineering geologist supervising and logging samples, a detailed picture can be built up of the stratigraphy of the slip zone. Hand-operated boring was used successfully in the two case studies described below.

A further advantage of hand-operated boring is that instrumentation can be located at selected depths in the slip zone. Inclinometers can provide information on the depth of the slip zone and the direction and rate of any ongoing movement. Piezometers can indicate water pressures in the various strata. Since many slope failures are triggered by water, this information is important for analysis of the failure and design of remedial measures.

Slope failures are by nature unstable, and even with lightweight equipment it may not be possible to carry out investigation work safely. An assessment of the stability of the failure should be made during the walkover survey. This should include what, if any, type of investigation can be carried out and what measures are required to monitor the slope for ongoing movement. The health and safety of workers should not be put at risk because of the desire to obtain information. A risk assessment should be carried out before any work is undertaken. If the failure zone is still unstable, it may be necessary to obtain information solely from boreholes outwith the slipped area.

Whatever the methods of investigation employed, the aim is to establish the geometry of the failure and the porewater pressure regime. When this has been done, stability analysis should be carried out to determine conditions on the slope at the time of failure. Geotechnical parameters for the strata may be obtained from laboratory tests if time and budget allow. If not, they may be estimated from correlations with index properties such as percentage clay or the liquid limit, or by comparison with published values. The values should be refined during the stability analysis. Any proposed remedial measures can then be analysed to see how the factor of safety would be improved. The effect of the work on adjacent slopes and properties should also be considered.

Having answered the first two questions, 'What has happened?' and 'Why has it happened?', the third can now be addressed: 'What can be done about it?' The answer will depend on a number of factors, depending on the situation. These factors will include the nature of the failure, economic factors, health and safety considerations, the space available and the importance of any structure or process threatened by the failure. If a major road or the water supply to an industrial process is threatened by the failure, more resources are likely to be made available than if the consequences of failure are negligible. The measures adopted should address the nature of the failure and aim to provide an appropriate solution.

A wide range of remedial measures are possible. These may range from 'do nothing' if the hazard posed by the failure is minimal, to fairly elaborate structural designs if reinstatement is considered essential. A list of possible measures might include:

- do nothing;
- dig out failed material, install drainage and reinstate slope;
- regrade slope to safe angle with suitable material;
- regrade and reinforce slope with soil nails or geotextiles;
- install gabion wall at toe and regrade slope;
- install sheet pile wall at toe and regrade slope;
- install retaining wall and regrade slope;
- install bored pile supported retaining wall and regrade slope;
- install contiguous bored pile wall and regrade slope;
- abandon site.

For shallow slips, simple measures such as excavation of the failure zone and replacement with granular fill, or regrading and installation of drainage may be adequate. Drainage will be an important aspect of all remedial works. A range of remedial measures for slope failures on highway embankments is described by Johnson (1985). For larger scale failures, structural measures such as sheet piling or gabions may be necessary to ensure stability. In the case of major, deep-seated failures it may not be economically possible to design a safe solution, and the site may have to be abandoned. An example of this last case is the A625 road at Mam Tor, Derbyshire, which has had to be closed because of repeated landslipping.

Remedial measures are thus likely to be site-specific, each site having to be considered individually. The process of investigation and remediation is illustrated by the following case histories.

Sewer collapse on riverbank

In the mid-1980s, Perth was Scotland's fastest growing port. The main access road to the harbour ran along the top of a steep bank above the River Tay. A 1100 mm diameter sewer had been installed along the slope in 1968. In March 1986, a failure occurred on the slope over a length of about 40 m breaking the sewer pipe and threatening the stability of the road. The road was temporarily closed and an assessment of the failure carried out. As luck would have it, a series of road repairs were under way in the centre of Perth at the time, so there was pressure to re-open the harbour road as soon as possible.

Inspection of the site revealed the back scar of the slip only 1–2 m below the verge of the road. A topographic survey was carried out and a ground investigation organized. Two light cable percussion boreholes were carried out on flat ground at the top of the slope outwith the slipped area, three hand-operated shell and auger boreholes were carried out in the slip zone, and four hand-operated boreholes were carried out at the foot of the slope on the shore of the River Tay. Inclinometers and piezometers were installed in a number of the boreholes. One of the hand-operated boreholes is shown in Fig. 2.

The sequence of strata present on the minor bank is summarized below and shown on Fig. 3.

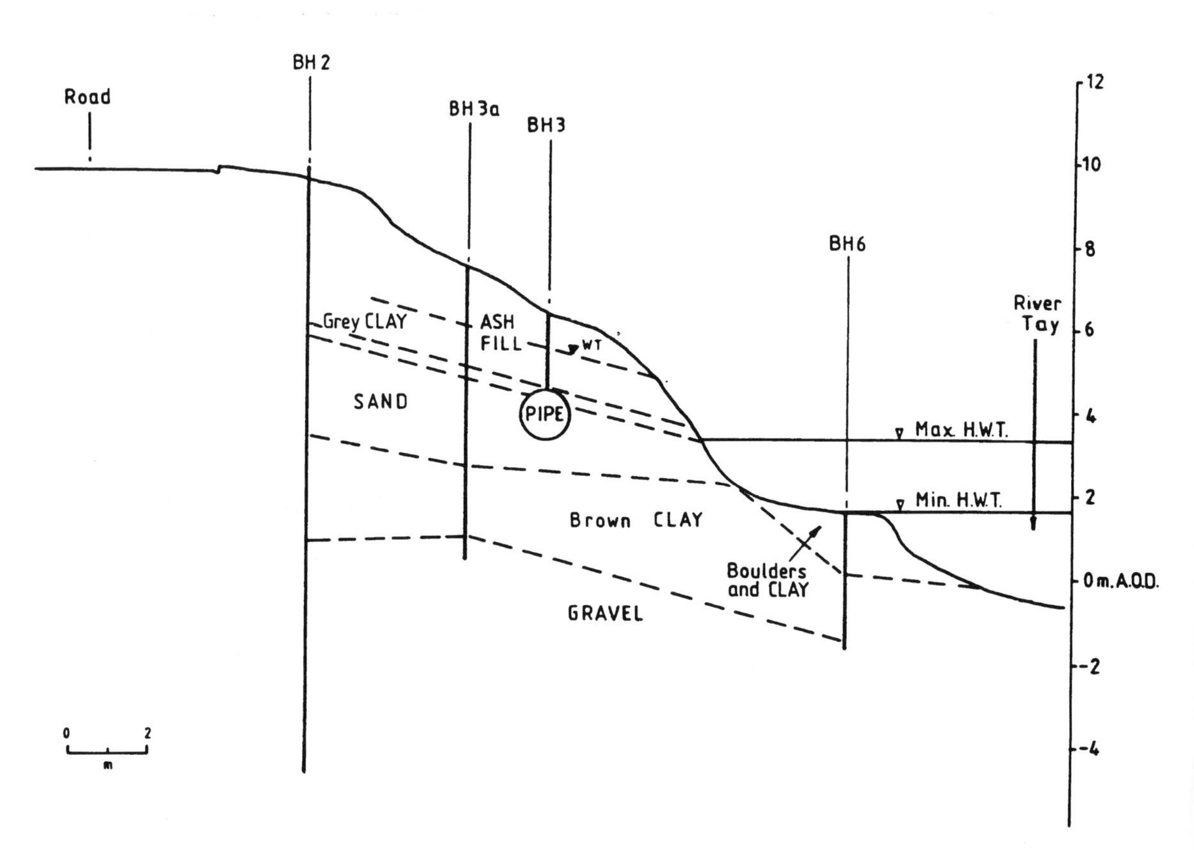

Fig. 3. Cross-section through the slip zone, Perth.

- *Made ground*: loose ash, gravel, sand and clay
- *Clay*: soft grey silty clay, 0.5 m thick (alluvium)
- *Sand*: loose gravelly sand (alluvium)
- *Clay*: soft, brown, laminated silty clay (late glacial lacustrine deposit)
- *Gravel*: dense sandy gravel (glacial meltwater deposit)

The thin layer of soft clay below the made ground was identified in the hand-operated boreholes on the slope, but not in the light cable percussive boreholes at the top of the slope. The clay may genuinely not have been present in these boreholes, or may not have been identified as a separate stratum because of mixing into the granular materials above and below. This illustrates the better definition of strata possible in a well-controlled hand-operated borehole.

Identification of the clay layer provided the key to understanding the nature of the failure. It provided a potential failure plane and accounted for the observed dampness of the lower layers of the made ground in the boreholes. A non-circular analysis using Janbu's method showed that ponding of water in the made ground above the clay could reduce the factor of safety of the slope to less than 1.0. By contrast, the factor of safety for a deep-seated failure through the underlying laminated brown clay was 1.37. The analyses were carried out using shear strength parameters estimated from a limited number of SPT, hand vane, grading, and liquid and plastic limit tests and represent typical peak shear strengths for the various materials. Readings from the inclinometers showed that minor movements were still ongoing. These were confined to the made ground and the grey clay, confirming that the position of the failure plane was in the grey clay. It was concluded that the failure was due to a rise in the water table in the made ground to a critical level during a storm event, accompanied by erosion at the toe by the River Tay.

Having identified the mechanism of failure, appropriate remedial measures had to be devised. The options were constrained by the road at the top of the slope and the river at the toe. Discussions with staff who had installed the original sewer indicated that the river bank had eroded by several metres over the years. Some form of protection against erosion was therefore required. Because of the risk of erosion to the areas on either side of the failure, the remedial works had to be extended

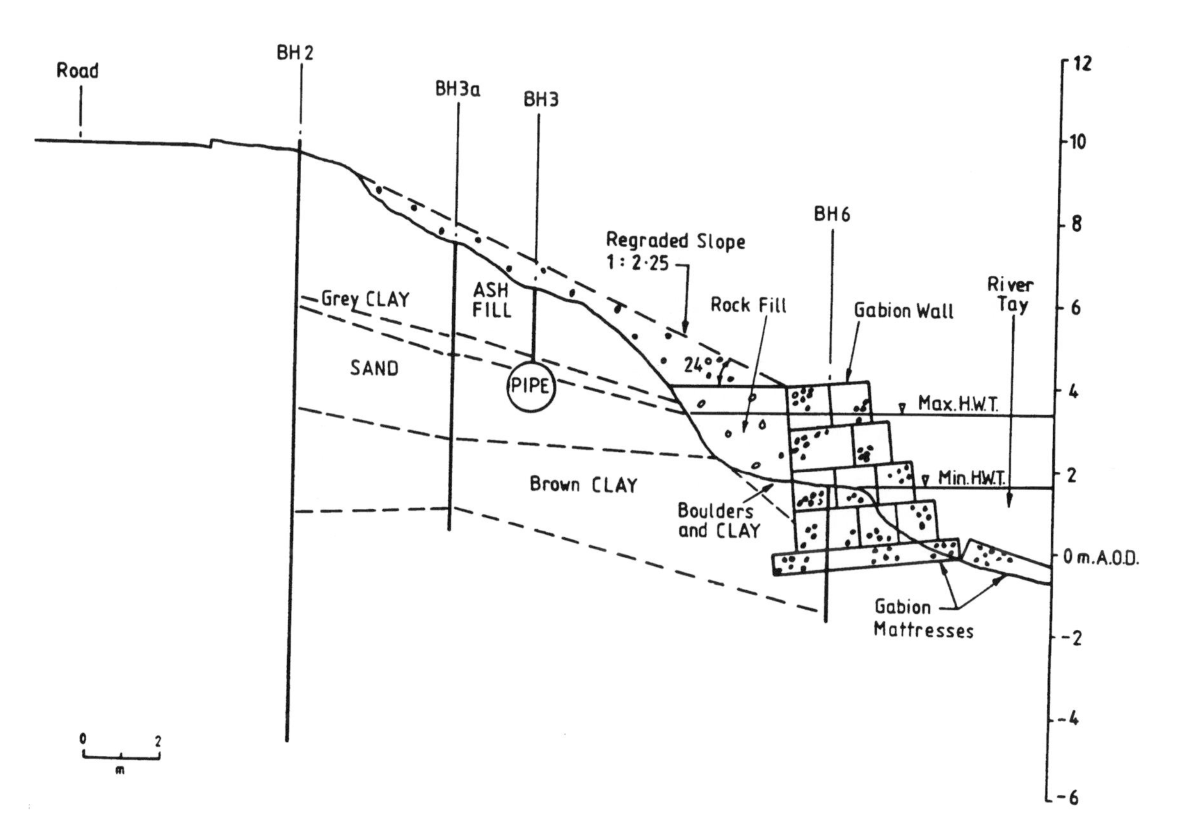

Fig. 4. Gabion wall remedial works, Perth.

over a length of 100 m to ensure adequate protection of the slope.

A sheet pile wall driven to the gravel below the brown clay was considered. This would have presented difficulties of access for the necessary plant, and the vibration could have triggered further movement of the slope. There was also the possibility that further erosion on the river side could remove support for the piles in the long term. This option was therefore rejected.

The chosen option was to construct a gabion wall at the toe of the slope, with gabion mattresses in front to provide protection against erosion, and to regrade the slope with rockfill. The proposals are shown on Fig. 4. This solution increased the factor of safety of the slope to 1.44, which was considered satisfactory. The work was carried out in the late summer and autumn of 1986 and completed successfully. The completed works are shown in Fig. 5.

Monitoring of the inclinometers showed that very little further movement was occurring on the slope. It was therefore decided to re-open the lane of the harbour access road furthest from the slope in April 1986, subject to no further movement occurring on the slope. This greatly relieved the pressure of traffic on the centre of Perth. Monitoring of the inclinometers was continued until they were destroyed during the remedial works, but no further significant movements were recorded.

Minor road on steep hillside

The valleys of the River Clyde and its tributaries north of Lanark are steep-sided, 'U'-shaped troughs oversteepened by glacial action. There is a history of slope instability in the area, and a number of roads are affected by minor movements. It would not be economically viable to stabilize all of the affected areas. Minor repairs are carried out where necessary under routine maintenance. Only where failure of the road has occurred are remedial works carried out.

The B7056 road runs east from Crossford in the Clyde Valley up the side of one such steeply sloping valley occupied by the Mashock Burn. Ground conditions consist of fluvioglacial deposits of fine sand and silt. Minor subsidence had occurred at a section near the top of the slope during the 1980s. In 1988, a 25 m long reinforced concrete slab had been installed along the edge of the carriageway nearest the slope. This method is frequently used to strengthen the verge of the road, as a major cause of damage is heavy vehicles running off the pavement. However, further minor subsidence occurred on part of the slope about one year later. In February 1990, remedial works were undertaken, comprising excavating the slope below the road and placing rockfill. A slope failure then occurred during a spell of very wet weather. Rainfall measurement showed this to be the

Fig. 5. Completed gabion wall, Perth.

Fig. 6. Failure on minor road, Crossford.

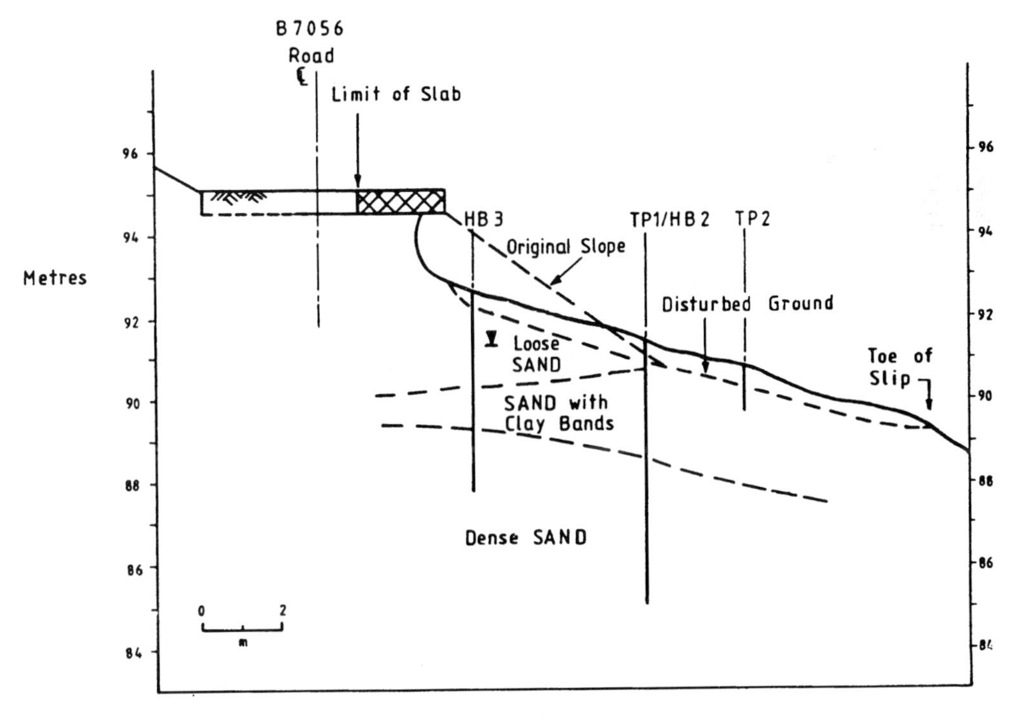

Fig. 7. Cross-section through the slip zone, Crossford.

wettest January and February on record. A mass flow occurred over a larger area than the previous subsidence, undermining the concrete slab. The road was closed, leading to a lengthy diversion for local traffic. The site shortly after the failure is shown in Fig. 6.

A detailed assessment of the slope was carried out following the failure. This included a topographic survey of the site, a desk study, a walkover, and a ground investigation. The investigation comprised five hand-operated shell and auger boreholes and four hand-excavated trial pits. SPTs were carried out in the boreholes and piezometers were installed. This method of investigation allowed quick and easy access to the failure zone in a way that would only have been possible with expensive staging for a light cable percussive rig. The boreholes were taken to a depth of over 6 m.

The topographic survey revealed that the natural slope of the hillside is very steep, ranging from 1:2 to 1:2.5. The road is bordered by a hedgerow on the outer side. This hedge appears curved in section suggesting slow ongoing movement of the slope. At the slip site, the soils consisted of uniform silty medium to fine sand with bands of soft laminated silty clay and sandy silt. The strata were loose to a depth of 4–5 m becoming medium dense to dense below this depth. A cross-section of the slip is shown in Fig. 7.

Groundwater was encountered at shallow depth in two of the boreholes and two trial pits but the remainder were dry. Piezometers installed in the base of the boreholes remained dry. However, one piezometer installed above a silt and clay horizon gave water levels about 1.3 m below ground level. This suggests a perched water table is present near the surface due to the presence of bands of soft silt and clay in the sand, but that the strata are free draining below this level.

The slip was caused by the combination of soil and groundwater conditions at the site. Slope stability analysis was carried out using typical peak shear strength parameters estimated from SPT results and grading tests and a perched water table derived from the piezometer readings. The results indicated a factor of safety close to unity. The entire hillside probably has a fairly low factor of safety in view of the fine-grained nature of the soils and the steep slope, but the added factors of oversteepening due to road construction and a localized perched water table rendered the failure zone particularly susceptible. The actual failure was triggered by a combination of a prolonged wet spell, and a heavy storm causing surface run-off onto the slope from the carriageway.

In view of the marginal stability of the slope as a whole, the failure of the earlier rockfill replacement works, the perched water table, the loose nature of the near-surface strata and the disturbance caused by the earlier slip, an earthworks solution was now no longer appropriate. A structural solution was required, with the aim of ensuring the stability of the carriageway in the area of the existing slip. The solution had to include the following features:

- provision of drainage to the slope;
- transfer of load to the dense sand at depth;
- preservation, if possible, of the existing concrete slab, to avoid further disruption to the carriageway;
- provision of adequate support to the carriageway under all weather conditions;
- provision of drainage to the carriageway to prevent surface water running onto the slope during storm events.

The chosen solution was a concrete retaining wall founded on 0.30 m diameter bored piles. The piles were taken to a depth of 8 m to ensure transfer of load to the more competent strata at depth. Rockfill was used as a free-draining backfill behind and below the retaining wall. French drains were installed on the face of the slope, which was regraded with existing material. A pipe drain was provided from the site to the foot of the slope, then leading across several fields to an outfall on the Mashock Burn. The carriageway drainage was improved, using kerbing to direct surface water into the gullies and keep it off the slope. The remedial works were carried out between January and May 1991. In the event it proved impossible to retain the concrete slab, as further erosion in the interval between investigation and construction had left it seriously undermined. It was removed and replaced with rockfill. The site is shown with the slab removed and the piles installed in Fig. 8. In all other respects, however, the works were constructed in accordance with the design. The completed retaining wall is shown in Fig. 9. The road was reopened to traffic in May 1991.

Conclusions

The investigation and remediation of minor slope failures is an important part of the work of the engineering geologist. Keen observation and an ability to assess the geological situation rapidly are required, as is a willingness to use unconventional methods to obtain information on the nature of the failure. Remedial measures should be designed to address the causes of the failure, and should not be undertaken until this has been established. The only exceptions are measures required to ensure the safety of the site or adjacent structures.

Minor failures offer a stimulating challenge to the engineering geologist. They combine elements of detective work and problem solving and can be interesting and enjoyable to deal with. There is no need for any feelings of inferiority when compared with some of the major disasters which hit the headlines from time to time.

Fig. 8. Remedial works: bored pile foundation.

Fig. 9. Remedial works: completed retaining wall.

Acknowledgements. The ground investigations for both case studies were carried out by Ritchies, Glasgow Road, Kilsyth. The gabion wall at Perth was constructed by Tawse Construction Ltd and the retaining wall at Crossford by MacKenzie Construction Ltd. The author would like to thank the Director of Water Services, Tayside Regional Council and the Director of Roads, Strathclyde Regional Council for permission to publish this paper. The two case studies were carried out while the author worked for the Babtie Group, Glasgow. The assistance of colleagues, particularly W. A. Wallace, is gratefully acknowledged.

References

ANON 1994. *New Civil Engineer*, 3 November.

JOHNSON, P. E. 1985. *Maintenance and repair of highway embankments: studies of seven methods of treatment.* Research Report 30, Transport and Road Research Laboratory, Crowthorne.

PERRY, J. 1989. *A survey of slope condition on motorway earthworks in England and Wales.* Research Report 199, Transport and Road Research Laboratory, Crowthorne.

SKEMPTON, A. W. & VAUGHAN, P. R. 1993. The failure of Carsington Dam. *Geotechnique*, **43**, 151–173.

Rock slope hazard assessment: a new approach

P. McMillan & G. D. Matheson

Transport Research Laboratory Scotland, Craigshill West, Livingston EH54 5DU, UK

Abstract. Uncontrolled bulk blasting and the application of 'standard' designs have left a legacy of many unstable highway rock slopes, some of which are potentially hazardous. Effective management of these slopes requires knowledge of their location and the risk posed to the road user. Existing rock slope stability assessment and risk evaluation systems use various approaches to data collection, and different scales and nomenclature for results. In addition, these assessments are usually undertaken on a reactive basis, prompted by rockfalls. As a consequence road users may be exposed to risk before problems are addressed, comparison of results is very difficult, budgetary problems arise as incidents are largely unforeseen, and prioritization of funds is impossible.

As a result of a research programme undertaken for the Scottish Office Industry Department (SOID) National Roads Directorate, a new, proactive approach to identifying and classifying rock slope hazards has been developed. Application of this new approach is likely to improve risk management of hazardous rock slopes.

The first stage of this approach derives a rock slope hazard index from rapid, standardized field data collection. This is used to classify rock slopes into four categories depending on the requirement for future action. The second stage derives a rock slope hazard rating from detailed field surveys. The rating is a measure of the risk of a vehicle incident being caused by rockfall.

The results of a field trial of the rock slope hazard index, carried out on a section of trunk road in the Scottish Highlands, are described.

Introduction

The uncontrolled use of bulk blasting techniques and application of 'standard' designs have left a legacy of many unstable highway rock slopes. Some unstable slopes pose are a risk to the road and road user and remedial action is therefore required. Effective management of these requires a knowledge of the location of all unstable rock slopes and an indication of the level of risk posed to the road user. It is then possible to prioritize future action.

At present, rock slope stability assessment and risk hazard evaluation are usually undertaken on a reactive basis, often prompted by rockfalls. Such a reactive approach does not address problems until after road users have been exposed to the risk, and presents considerable budgetary problems as incidents are inevitably unforeseen. A proactive approach to minimize the risk to road users prior to incidents occurring and allow effective priority based budgeting of maintenance funds. A standard, repeatable and rapid method of rock slope risk assessment is required for such a proactive approach. At present there are several subjective assessment schemes in use by different specialist consultants. All of these schemes have their limitations and none meet the criteria for an effective proactive system.

In 1993 a research project was commissioned by the Scottish Office Industry Department (SOID) to investigate rock slope risk assessment and if necessary develop new methods of assessing rock slope risk. This paper outlines the philosophy behind these new methods.

Background to risk assessment

Risk assessment has been the subject of considerable recent investigation (Chowdhury 1992; Skipp 1993; Hambly & Hambly 1994). There are many definitions of risk assessment, some simple and others complex. Chowdhury (1992) provided an overview of risk assessment as applied to geomechanics and Table 1 is taken from that paper.

In general, it is difficult to apply quantitative, probability-based risk assessment to highway rock slopes as many elements of rock slope instability are difficult to quantify. In particular, predictions of likely failure timing and frequency are difficult and unreliable. This is in part due to poor historical data on rock slope failures. Therefore, any assessment of risk associated with rock slopes is, to some extent, dependent on the subjective judgements.

McMillan, P. & Matheson, G. D. 1998. Rock slope hazard assessment: a new approach. *In*: Maund, J. G. & Eddleston, M. (eds) *Geohazards in Engineering Geology*. Geological Society, London, Engineering Geology Special Publications, **15**, 177–183.

Table 1. *Definitions of risk (after Chowdhury 1992)*

Simple definitions
1. Risk = probability of failure
2. Risk = probability of failure multiplied by the loss from failure
   = expected loss, expected damage or expected cost
3. Risk = $PC + f(P, C, x)$
   where P is probability, C is consequence and f is some function of P, C and other relevant variables x (x could be a measure of fairness of risk, its nearness today, how catastrophic it is)[a]

Symbolic definitions
1. Risk = uncertainty + damage
2. Risk = (hazard/safeguards)
3. Risk consequence/unit time = frequency events/unit time × magnitude consequence/event
4. Risk = [(event/unit time) (consequences/event)]
5. Risk = [(events/unit time) (consequences/events)^k]
   where $k > 1$ is used to amplify the importance of events with large damages

A person or society which acts on the basis of: $f = 0$ is called risk neutral, $f > 0$ is called risk averse, $f < 0$ is called risk prone.

The current research has attempted to distil existing knowledge, inject new thinking based on expertise at TRL Scotland, and develop a new approach to risk assessment for rock slope instability. In this approach unstable rock and the level of risk posed by such instability are identified and evaluated. The process has been termed *hazard assessment*, the level of risk being identified and quantified in terms of a *hazard index* and a *hazard rating*. In this context, a hazard is defined as unstable rock which could be a danger to the road or to road users. The hazard index is a simple measure of the potential for failure and the consequences of that failure, and the hazard rating is a more detailed and comprehensive measure of the probability of failure and the consequences. The hazard index and hazard rating, as used in this paper, are therefore defined as:

$$\text{hazard index} = \text{potential for failure} \times \text{consequences of failure}$$

$$\text{hazard rating} = \text{probability of failure} \times \text{consequences of failure}$$

Factors influencing rock slope failure and risk

The likelihood of failure from a particular rock slope and the risks to a road and road users are dependent on a number of factors which fall into four broad groups:

- *Geotechnical factors.* These are factors associated with the rock mass and the host material properties and include discontinuity properties (dip, azimuth, trace length, principle spacing, dilation, infill, weathering, roughness and planarity), rock material properties (strength, weathering, mineralogy) and groundwater. These factors contribute to the type, size and severity of potential failures.
- *Geometric factors.* Geometric factors are those associated with the geometry of rock slopes, verges and adjacent roads, and include slope height, angle and profile, position and size of berms, angle of natural slope above cutting, rock trap size and shape, carriageway width, sight lines and cutting type, the proximity of open water, buildings, steep slopes and services. These factors influence the consequences of failure.
- *Remedial work factors.* Many unstable rock slopes have been subject to some form of remedial action in an attempt to reduce the hazard presented by rockfall. The influence of these on hazard is dependent on the scope and effectiveness of the remedial works.
- *Traffic factors.* The volume and behaviour of traffic on a road influence the consequences of a rockfall in terms of the likelihood of a vehicle incident. Traffic volumes and speed flow relationships are important factors in evaluating this influence.

It is clear that determining the risk posed by rock slopes is complex, given the large number of factors influencing rock slope failure and consequent risk, and the likely complex relationships that exist between many of these factors. It is not possible within the scope of this paper to describe how such relationships have been evaluated. However, the logic underlying the new methods of hazard assessment is summarized in the following sections.

Objectives

The aim of the research was to develop a system of assessing risk on rock slopes which would allow effective management of rock slopes on the Scottish trunk road network. Such a system had to be proactive, allow priority based budgeting and be applicable to most, if not all, rock slopes on the trunk road network. It therefore had to involve a rapid assessment stage so that this would not be prohibitively expensive and had to make it possible to classify slopes on the basis of future action. To achieve these objectives a two-stage approach has been developed.

Rock slope hazard index and rating

Overview

The first stage derives a rock slope hazard index from a rapid field assessment. Rock slopes are classified into one of four action categories based on the value of this index. This acts as a coarse sift, eliminating slopes with a low

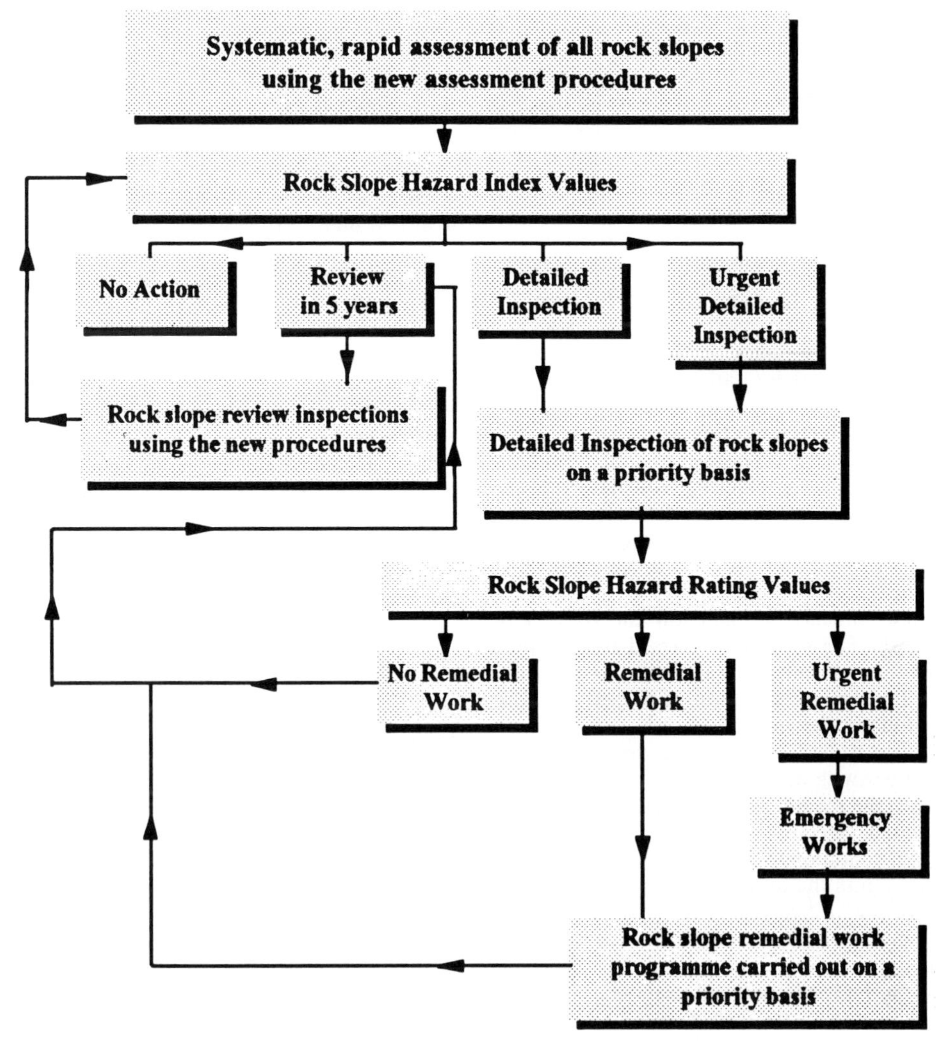

Fig. 1. Rock slope hazard index and rating systems.

risk potential from a later, detailed assessment stage. Detailed assessment involves deriving a rock slope hazard rating and can be carried out on a priority basis as budgets allow. Rock slopes are placed into one of three action categories based on the value of this rating. The relationship between the hazard index and rating is illustrated in Fig. 1.

Data collection

Standardized data collection is an essential feature of any hazard assessment system. Achieving repeatability in the collection of geotechnical data from rock slopes by different engineering geologists is notoriously difficult. In an attempt to overcome this, the data collection procedure for the hazard index and rating surveys both use standard forms that define the type of data required. In the case of the index there are three forms, one each for geotechnical, geometric and remedial work data. Almost all data are visually estimated and recorded by choosing from a set of options for each parameter. This is a rapid process and even in geotechnically complex rock slopes can be carried out in less than 30 man minutes. It is therefore possible to collect data from more than 14 rock slopes in a single man day.

An important part of the hazard index survey involves numbering and referencing the location of each rock slope to the trunk road link and section reference system. This facilitates subsequent identification. All rock slopes are also photographed.

In the case of the hazard rating survey there are five forms, one for each type of potential failure and one for discontinuity mapping. Some data are recorded as actual measurements and others are estimated and recorded by choosing from a set of options. The hazard rating requires detailed measurements and, in most cases, involves accessing the rock slope. Rating surveys therefore require more site time and effort than index surveys. Rock slope reference numbers from the index surveys are retained for the rating surveys.

The rock slope hazard index system

The rock slope hazard index is a relatively simple estimate of the potential for failure on highway rock slopes hazard and the risks posed, and is derived from rapid, standard field data collection procedures.

Parameter values have been derived that reflect the influence of geotechnical, geometric, remedial work and traffic factors on rock instability and the level of risk posed. The index is derived by following a standard calculation procedure using these parameter values as input. The calculation process follows a logical route dictated by the influence of these on the rock slope hazard risk level. The rock slope hazard index values derived from these calculations are used to prioritize future action through classification of slopes into four categories as follows shown in Table 2.

The rock slope hazard index is intended to act as a course sift. Slopes with an index of <1 are unlikely to present a risk to the road or road user and therefore fall into the 'no action' category. An index of between 1 and 10 indicates that conditions on a rock slope are such that a risk may develop in the future. These slopes therefore fall into the 'review in five years' category and require only minimal maintenance commitment.

Rock slopes with an index of greater than 10 could present a risk to the road and road users and therefore require action to investigate their nature and severity. Prioritization of action is achieved by subdividing these slopes into the two categories of 'detailed inspection' and 'urgent detailed inspection'. Slopes in these categories are considered to require a detailed inspection, which may reveal the need for remedial action.

Derivation of the rock slope hazard index. Derivation of the hazard index involves attributing values to the many individual parameters that influence the risk posed by rock slope instability and using these values in a standard calculation procedure. A list of the parameters used in the index is as follows:

- evaluated failure potential (plane, wedge and toppling, based on Matheson 1983)
- factor of safety for failure
- discontinuity principle spacing, trace length and dilation
- observations of potential potential failure (plane, wedge, toppling, ravelling)
- potential failure size and position on rock slope
- rock material strength and weathering
- groundwater
- rock trap size and shape
- slope profile and berms
- carriageway width
- road sight lines
- cutting type and associated hazards (steep slopes, buildings, services, open water)
- remedial works
- traffic volume

Derivation of the hazard index from these parameters is illustrated in Fig. 2.

Derivation of the rock slope hazard index involved establishing numerical values for each parameter. Although a full discussion is outside the scope of the present paper, the underlying logic is described below.

A range of values were determined for each parameter to reflect the range of influence of the parameter on the level of rock slope risk. This was done with reference to published relationships, where available, and from experience and specialist knowledge. In general, each data input category was allocated a corresponding parameter value. However, in some situations (e.g. for evaluated failure potentialvalues and rock trap size and shape) parameter values were calculated from more complex relationships. Parameter values also reflected their relative importance in influencing the level of risk. The complete set of parameter values is referred to as the *parameter library*.

There are two types of parameter used to derive the rock slope hazard index: primary parameters and secondary parameters. *Primary parameters* establish the potential for failure and *secondary parameters* influence the likelihood, severity and consequences of failure. There are three sets of primary parameter values in the hazard index: those related to the discontinuity – rock slope geometry relationships; those related to the potential failure observations; and those related to the potential for failure on the natural slope above the cutting. Primary parameters are additive in that they influence the derivation of the index by addition.

Secondary parameters are multiplicative and influence derivation of the index by multiplication. A parameter value of unity indicates a neutral effect on the index,

Table 2. *Categories of slope*

Action category	*Rock slope hazard index value*
1. No action	<1
2. Review in 5 years	1–10
3. Detailed inspection	10–100
4. Urgent detailed inspection	>100

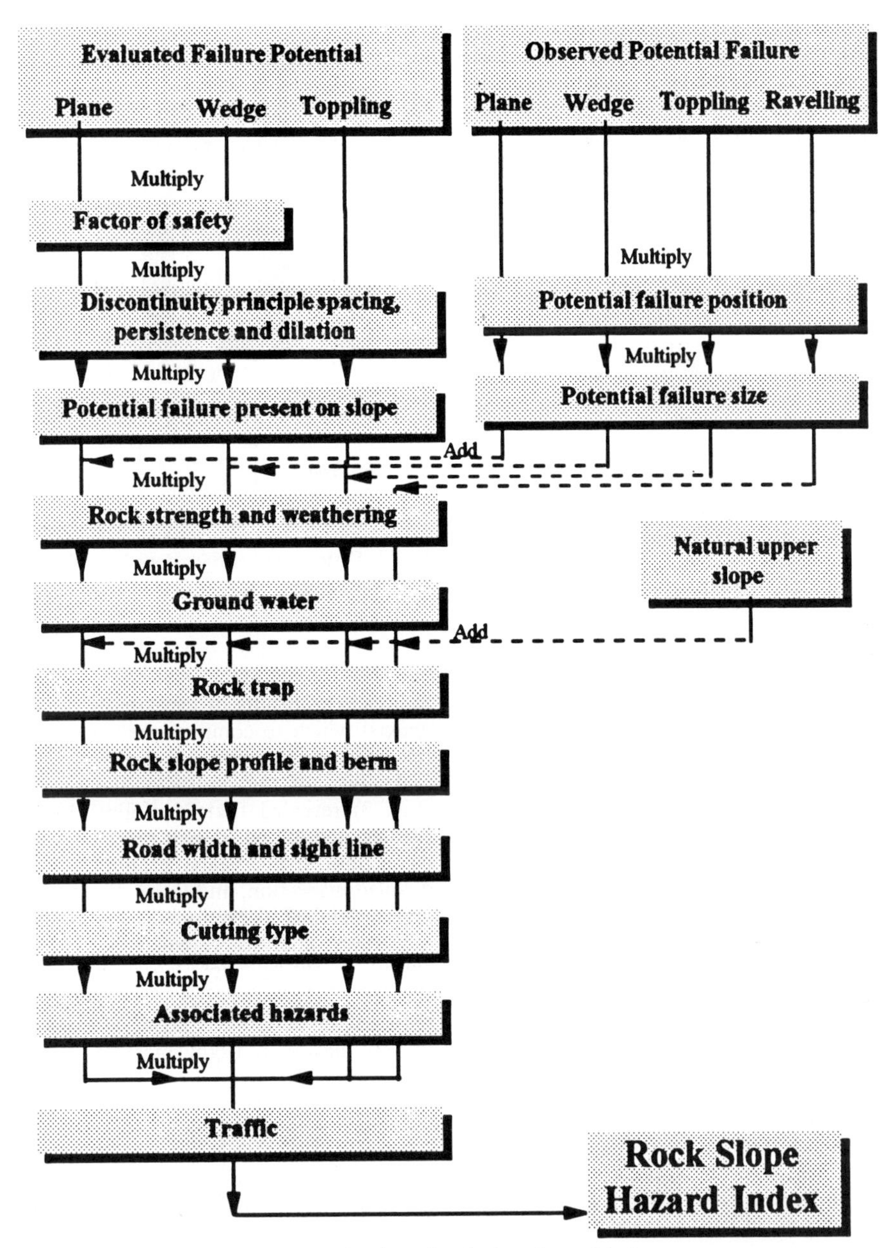

Fig. 2. Derivation of rock slope hazard index.

whilst a value of greater than unity increases the index, and a value of less than unity decreases the index.

The logic of the relationships illustrated in Fig. 2 is that primary parameter values are derived according related to the potential for each type of failure on a slope. Each type of failure is assumed to act independently. They therefore follow separate calculation paths in deriving the index. Calculation of the index involves multiplying the primary parameters by successive secondary parameters and adding other relevant primary parameters.

The rock slope hazard rating system

The rock slope hazard rating is a more detailed and comprehensive measure of the risk posed by a unstable rock slope and is derived from the sum of the individual risks posed by all of the potential failures on that slope. It is a semi-probabilistic method based upon the principles of quantitative risk analysis but without complete historical, statistical data to back up all of the assumptions. The rating is still under development and is therefore only briefly discussed below.

The rating calculations employ conventional deterministic models for calculating *factor of safety* (FoS) for plane, wedge and toppling failure in probabilistic analyses. Such analyses derive a probability of failure for each potential failure on a rock slope. These are then factored to account for slope geometry, trap geometry, road geometry and traffic to derive a value which is indicative of the probability of a vehicle incident occurring as a result of rock failure. Combining all of these values for a rock slope and expressing the combined value in terms of the number of vehicle incidents per 10^8 hours of exposure produces the rock slope hazard rating.

The rock slope hazard rating acts as a fine sift to separate potential instability requiring remedial work from that requiring no remedial work. Those slopes requiring action are split into two action categories of *remedial work* and *urgent remedial work*. Slopes requiring no remedial work are included in the next five year review using the index system (see Fig. 1).

Field trials of the rating system are still required in order to establish the threshold values which separate the three categories of 'no remedial action', 'remedial action' and 'urgent remedial action'.

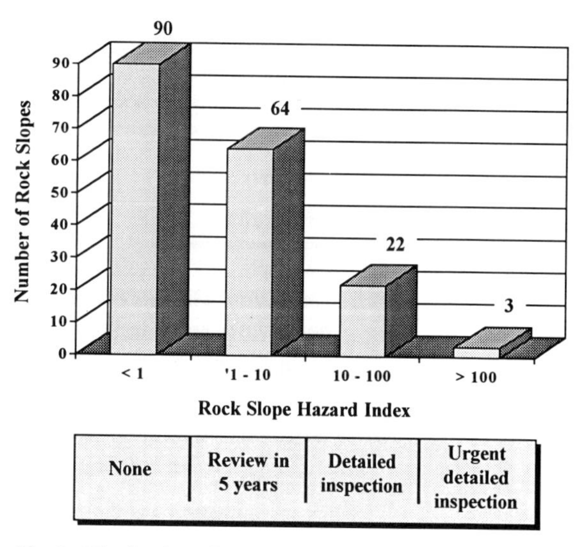

Fig. 3. Distribution of rock slope hazard index values obtained from the trial.

Field trial of the rock slope hazard index

A field trial was undertaken during July 1994 to evaluate the performance of the rock slope hazard index. A 50 km section of trunk road in the Scottish Western Highlands was chosen for the trial.

Fieldwork was carried out in two stages. Firstly, all rock slopes (179) were surveyed using the new standard system developed for the hazard index. Secondly, an independent subjective stability assessment of a randomly selected number (34) of the slopes was carried out.

The first stage of fieldwork using the new procedures and standard forms proved very successful and it was possible for a two-person team to survey between 30 and 40 rock slopes per day. Traffic data for the study section were obtained from the Highland Region, Regional Traffic Flow Plan, 1993 provided by the SOID National Roads Directorate. The maximum seasonal flow values were used in calculating the rock slope hazard indices.

During the second stage of the fieldwork an independent expert carried out subjective hazard assessments for a randomly selected number (34) of rock slopes within the study area. This expert was not involved in either the first phase of the fieldwork or deriving the rock slope hazard indices. The results of this survey provided a reference against which to compare the rock slope hazard index results were compared.

Analysis of the results of the hazard index survey (Fig. 3) revealed that 90 slopes (50.3%) required no further action, 64 slopes (35.8%) required a review inspection, in five years, 22 slopes (12.3%) required a detailed inspection, and three slopes (1.7%) required an urgent detailed inspection.

Further analysis revealed a relationship between cutting height and action required (Fig. 4). The data showed that, as slopes get higher, the percentage requiring no action decreased. The percentage requiring review showed a slight increase and the percentage

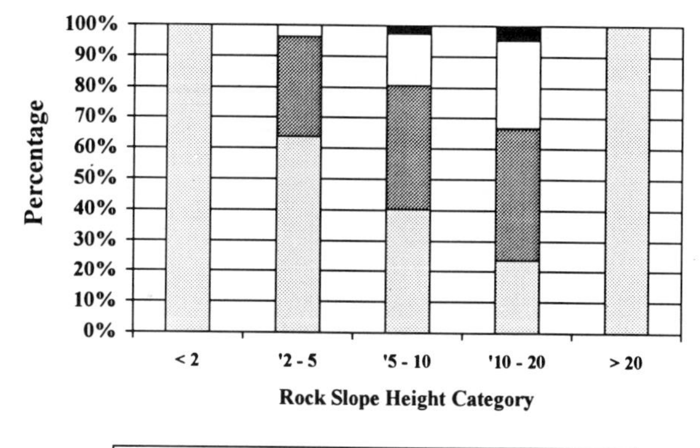

Fig. 4. Variation in required action with slope height category.

Table 3. *Comparison of hazard index and subjective action categories*

Slope no.	*Rock slope hazard index category*	*Subjective category*
6	1	1
15	1	1
25	2	2
28	1	1
32	2	1
41	1	1
42	2	1
43	3	3
46	3	3
48	3	3
53	3	3
66	2	1
69	1	1
70	1	1
71	4	4
72	1	1
73	4	4
74	3	3
75	4	4
80	3	3
98	1	1
101	2	1
106	2	3
123	3	3
125	2	2
127	1	1
137	2	2
140	1	1
142	2	2
144	2	2
147	2	1
159	1	2
169	1	1
172	2	3

requiring detailed inspection showed a marked increase. These trends could not be reliably applied to slopes in the categories less than 2 m and greater than 20 m because of the small number slopes in each (one and five respectively).

Comparison of the action categories derived from the rock slope hazard index values with those derived from the subjective field assessment shows 79% agreement (Table 3). The discrepancies mainly occur between the no action and review categories (15%) with only two discrepancies (6%) between the review and detailed inspection categories in which the subjective assessment gave the higher category.

Summary and conclusions

Improved methods of identifying and classifying the risk from unstable rock slopes are rock slope are required to allow more effective management of highway rock slopes. Existing methods employ variable approaches to data collection and use different scales and nomenclature for presenting results. These make comparison of the results almost impossible.

A new two-stage approach to rock slope risk assessment, involving the calculation of a hazard index and a hazard rating, has been developed.

A trial of the first stage of this new approach, the rock slope hazard index, illustrated that the fieldwork and analyses could be carried out rapidly and that the results were are compatible with those arrived at by an independent expert using subjective judgement. The rock slope hazard index could therefore be applied as a management tool to allow priority based budgeting of detailed rock slope assessments.

Acknowledgements. The authors would like to thank A. Blair for his assistance in fieldwork and with data analysis, and the SOID Roads Directorate Network Management, in particular I. Ross and D. Bannerman, and Highland Regional Council, Lochaber District for their co-operation with the trial of the hazard index.

The research reported in this paper is part of a project for the Scottish Office Industry Department. The paper is published by permission of the Director of Roads, Scottish Office and the Chief Executive, TRL.

References

CHOWDHURY, R. N. 1992. Probabilistic risk analysis in geomechanics and water engineering. *In*: *Geomechanics and Water Engineering in Environmental Management*. AA Balkema, Rotterdam.

HAMBLY, E. C. & HAMBLY, E. A. 1994. Risk evaluation and realism. *Proceedings of the Institute of Civil Engineers, Civil Engineering*, **102**, 64–71.

MATHESON, G. D. 1983. *Rock stability assessment in preliminary site investigations*. TRRL Laboratory Report 1039, Department of the Environment, Department of Transport, Transport and Road Research Laboratory, Crowthorne.

SKIPP, B. O. (ed.) 1993. *Risk and Reliability in Ground Engineering*. Thomas Telford, London.

Silica gels: a possible explanation for slope failures in certain rocks

Christine Butenuth,[1] Marie-Luise Frey,[2] Michael Henry de Freitas,[1] Nikolaos Passas[1] & Carlos Forero-Duenas[1]

[1] Department of Civil and Environmental Engineering, Imperial College of Science, Technology and Medicine, London SW7 2BU, UK
[2] Verbandsgemeinde Gerolstein – GEO-PARK, Kyllweg 1, 54568 Gerolstein, Germany

Abstract. Studies of a weak sandstone which disaggregates in water reveal that particles of colloidal dimensions are released as the sand-sized particles separate. Colloids are the building blocks of gels and their presence suggests that gels (in this case silica gels) could be a component of this material. Samples of the disaggregated sandstone were collected soon after a landslide within it had occurred and, when left to stand in the ground water collected with them, naturally recemented themselves. Such behaviour implies that gels are not only present in this sandstone but also contribute to its strength. The role of gels may therefore be significant to those geohazards arising from a loss of strength.

Introduction

Rocks are mechanical mixtures, frequently of different minerals, pore gases and/or pore fluids. Sedimentary rocks are deposits of detritus and/or chemical, and/or biogenic materials and so-called 'cement'. The investigative techniques commonly used to study the solid phases of such mixtures are petrological thin sections and X-ray diffraction. However, many cements are either not crystalline in nature or have embedded crystalline 'phases' that are too small to be detected as individuals possessing a lattice order: they are thus beyond the range of analyses by these methods. Normal element chemical analyses do not help here either as only the element composition is determined and not the nature of the materials that they form. Quantitative analyses of the nature of cement in sedimentary rocks, or of sediments exhibiting a cohesion that can be attributed to the presence of a bonding agent, or indeed of any rock where a cement is considered to exist, is often found to be a problem. This being the case, evidence for the nature of cement needs to be considered from many sources, not least from a study of the behaviour of a cemented material. This paper describes such behaviour for a sandstone and proposes that the behaviour witnessed points towards the role of silica gels as a cementing agent.

Case history

In 1993, after a period of long and heavy rainfall, failure occurred in a slope of Bunter Sandstone in the Eifel region (Butenuth *et al.* 1994, 1995*a*, *b*). The sandstone disintegrated into a suspension of fine, sandy material in which small rounded pieces of sandstone were carried: this slurry 'flowed' downhill, blocking a railway line as well as a road, and damming a small stream to form a lake.

A sample of the totally disintegrated material was taken, poured into a glass test-tube and left standing undisturbed under its own aqueous solution for a few days. During this period the granular material sedimented itself into the lower part of the test-tube and developed a distinguishable interface between itself and the water column above it. The test-tube was then tilted to see if the previously uncemented sand would flow in the tube under the force of gravity: it did not. The previously horizontal interface between the sediment and the water column above it was rotated by the same angle as that of the test-tube. The previously uncemented, totally disintegrated, material had recemented itself to a certain degree.

Some pieces of the original Bunter Sandstone taken from the site were put into a glass vessel which was then topped up with 'tap water' at an ambient temperature of about 20°C. The same phenmena as those observed in the field were seen. The piece of sandstone disintegrated into uncemented quartz grains, with a few rounded pieces of rock remaining whose disintegration proceeded further but only very slowly.

From all these observations it is concluded, that

- the sandstone which failed was 'weak', in as much as it disintegrated when immersed in water (Dobereiner & de Freitas 1986);

Butenuth, C., Frey, M. L., De Freitas, M. H., Passas, N. & Forero-Duenas, C. 1998. Silica gels: a possible explanation for slope failures in certain rocks. *In*: Maund, J. G. & Eddleston, M. (eds) *Geohazards in Engineering Geology*. Geological Society, London, Engineering Geology Special Publications, **15**, 185–191.

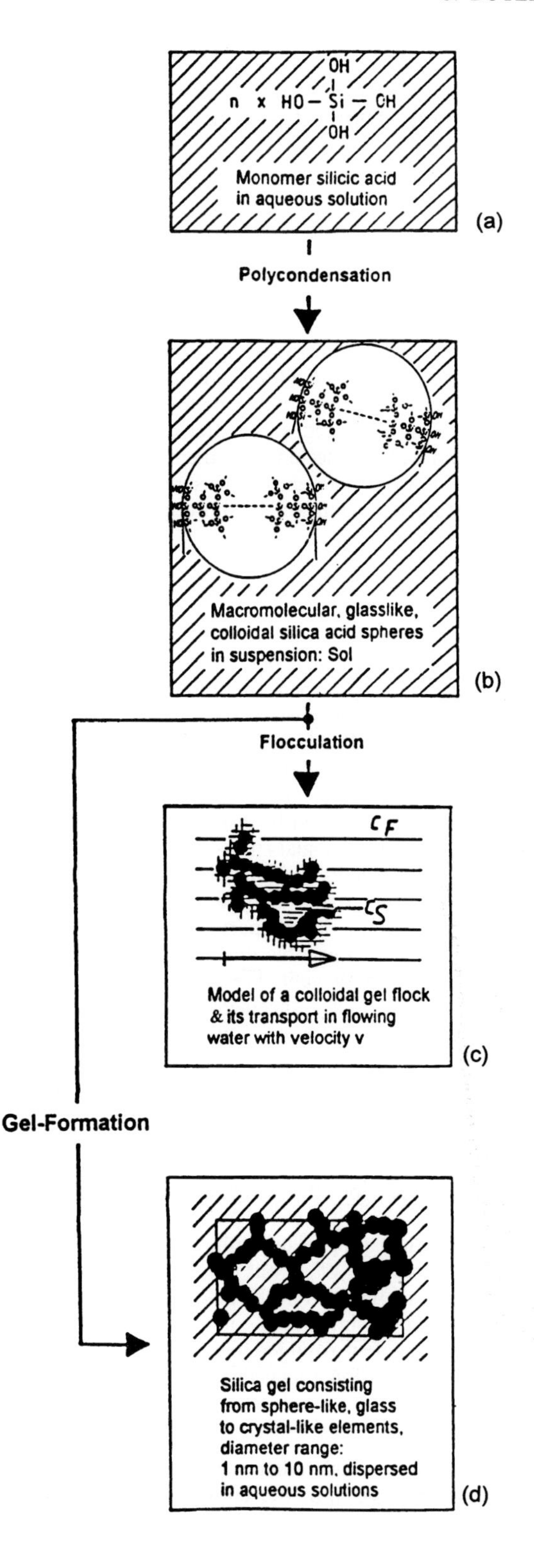

- as its aggregates of grains disintegrated, colloidal material was released (its presence being betrayed by the visibility of light as rays passing through the supension: the Tyndal effect) implying that such material was present in this sandstone, in addition to the solid crystalline grains; and
- these disintegrated grain aggregates, when left in the presence of the aqueous solution (analyses of which confirms the presence of silicon, aluminium and iron) with particles of colloidal dimensions, regain sufficient strength to resist gravitationally induced shear stresses.

These phenomena are characteristics also exhibited by gels, which suggests that gels may have played a role as 'cement' in this sandstone. It is therefore relevant to consider how cementing agents such as gel may behave.

Sols and gels

When gelatin dessert powders are dispersed in hot, but not boiling water, the solution so produced, when cooled, sets to a semi-rigid state described as a gel. Figure 1 illustrates the established model for the occurrence of an inorganic silica gel.

From a solution of monomolecular silicic acid, H_4SiO_4 (Fig. 1(a)), sphere-like particles of colloidal dimensions are formed first (Fig. 1(b); Iler 1979). In this case the polymerization of H_4SiO_4 is accompanied by the split off of water. Such processes are due to the polycondensation reactions. Such a system, i.e. containing free movable particles of colloidal dimensions, is called an incoherent system; a sol.

Sols are unstable systems, even though a certain level of stability can be achieved by adsorption at the surface of its particles of either electrical charges or electrically charged substances. Removal of these charges may lead to either flocculates or to gels of the sol. To produce a silica gel, chains from these glass-like spheroidal particles are formed (Figs 1(c) and (d)). These continue to coalesce with time, forming a coherent system, i.e. one in which a network of spheres is created throughout the whole system. The gel can continue to loose water even under water (see e.g. Stuart 1953). With time and continued dewatering, very small crystallites of silica 'phases' form.

The properties of gels depend on their history. In particular their properties are time and environment dependent. Figure 2 shows an example of the elastic

Fig. 1. The formation of an inorganic sol and gel (see text). Note that (c) illustrates the flocculation of these colloidal gel particles as in river water. c_S = silica concentration inside the flocks; by c_F = silica concentration in the river water: $c_F < c_S$.

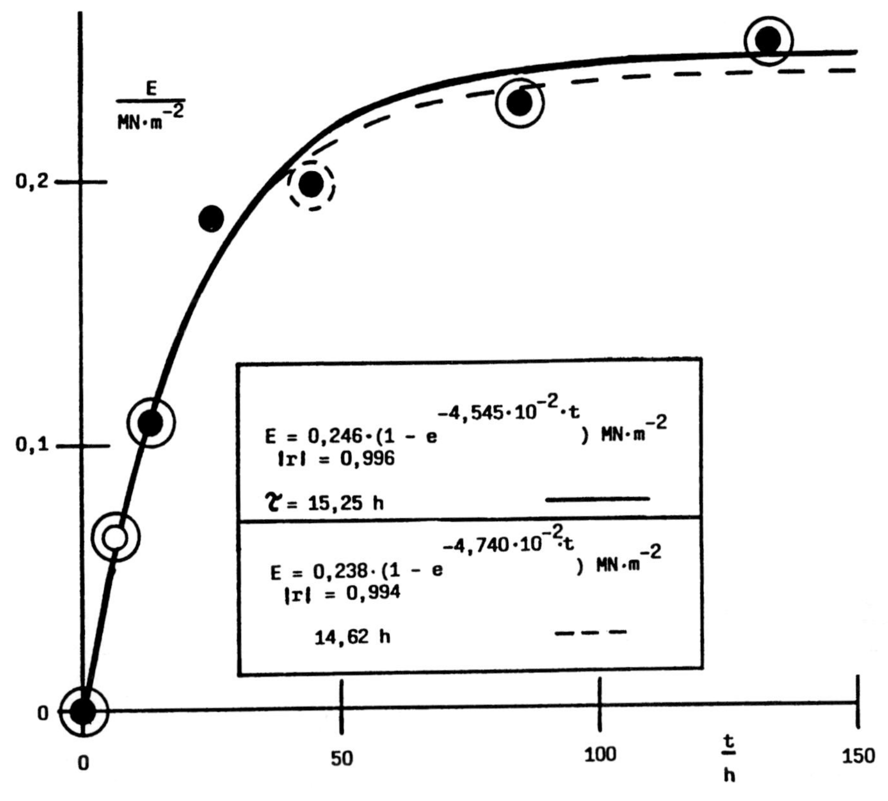

Fig. 2. The gain in value of the elastic modulus with time exhibited by a (after Hatschek 1932).

modulus of a gel increasing with time (Hatschek 1932). Delicate experiments reported by Denisov & Reltov (1961) demonstrated how individual quartz grains increased their bond-strength to a quartz plate with time; they attributed this to the action of silica gels. Mitchell & Solymar (1984) used this evidence to explain the increase in strength observed with time in alluvial sands.

With continued ageing of the gel, very rigid substances may eventually be produced; indeed, silica gels are used in a wide range of conventional applications, e.g. in the production of high quality silica for the electronics industry.

Sols and gels are thus substances made up of particles of colloidal dimensions, whose properties can change with time (Stauff 1960). They are known to occur ubiquously in natural systems and to occur, for example, as opals, flints, volcanic ashes, silicidified woods and in soils; but this is not their only form of occurrence, because they also occur in springs, groundwater and rivers (e.g. Ostwald & Fischer 1917; Jones & Biddle 1966; Butenuth 1982, 1986*a*, *b*, 1989, 1990).

Gels and the loss of strength

The inverse of gel formation is peptization. This process forms the particles of a sol. The dissolution of sol particles to H_4SiO_4 can only be achieved by the inverse reaction of polycondensation, i.e. by hydrolysis. Denisov & Reltov (1961) considered that the disruption of the surface of the silica particles in their experiments was the result of hydrolysis, accompanied by a reversion to silica gels (see also Butenuth *et al.* 1994). A juvenile silica gel with a high water content is able to swell and with this process high pressures have to be expected. These pressures are usually much higher than the tensile strength of sandstone.

Figure 3 illustrates the time-dependent loss of strength measured in sandstones in the presence of water vapour, in a fixed environment (Wiid 1970). Table 1 lists the wet and dry strength of a number of different sandstones after certain fixed times. Note that not only does strength decrease with time (time of observance was *c.* 40 days), but the amount by which it decreases changes from sandstone to sandstone. This is an interesting piece of

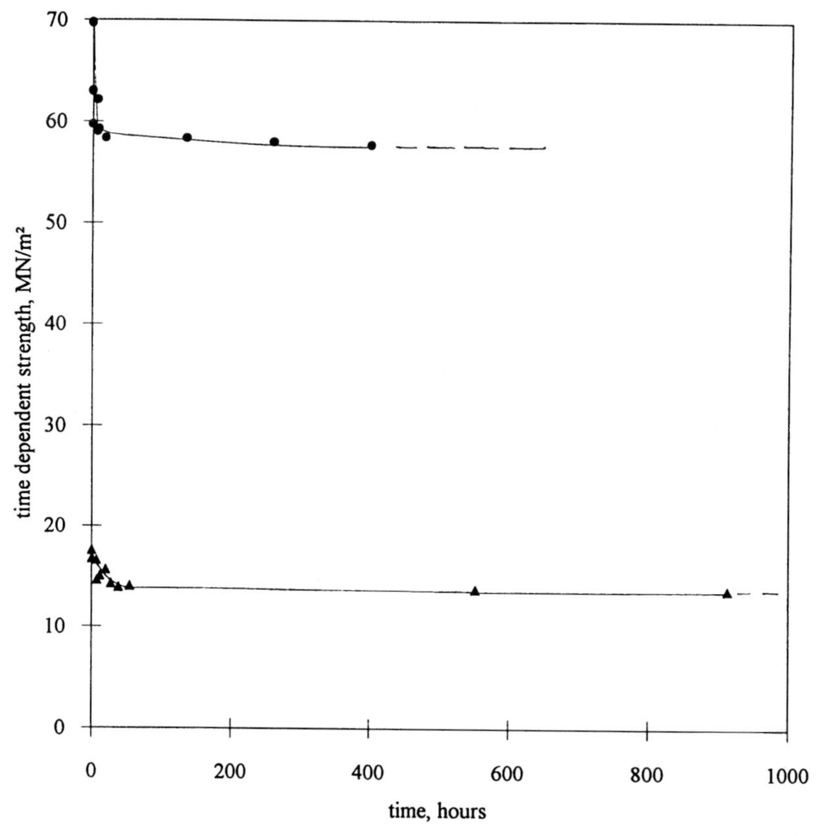

Fig. 3. The loss in unconfined compressive strength with time, under a constant environment (after Wiid 1970). ▲, wet sandstone (dry bulk density, 2140 kg m^{-3}; true porosity, 23%; modulus of elasticity, 12 GN m^{-2}). ●, wet sandstone (dry bulk density, 2110 kg m^{-3}; true porosity, 26%; modulus of elasticity, 26 GN m^{-2}).

Table 1. *Strength loss of sandstones from dry to wet conditions (after Dyke & Dobereiner 1991)*

Reference	*Rock*	*Saturated strength/dry strength*
Colback & Wiid (1965)	sandstones and shales	50%
Wiid (1970)	sandstones	60%
Kitaowa *et al.* (1977)	sandstones	64%
Bell (1978)	Fell Sandstones	68%
Hassani *et al.* (1979)	sandstones	70%
Ferreira *et al.* (1981)	Bauru Sandstone	50%
Priest & Selvakumar (1982)	Bunter Sandstone	65%
Koshima *et al.* (1983)	Bauru Sandstone	50%
Pells & Ferry (1983)	sandstone	41%
Dobereiner (1984)	Waterstones	83%
Dyke (1984)	Waterstones	49%
	Bunter Sandstone	71%
	Penrith Sandstone	66%
Gunsallus & Kulhawy (1984)	coarse crystalline sandstone	98%
	coarse clastic sandstone	81%
	fine grained sandstone	49%
Denis *et al.* (1986)	limestones	48%
Howarth (1987)	sandstones	67%

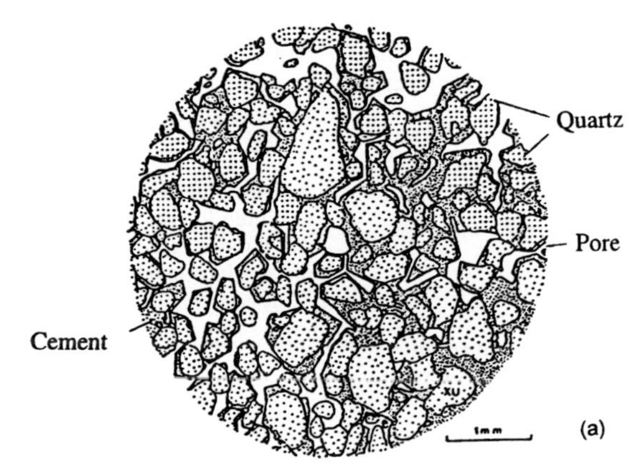

Xu S., de Freitas M.H. and Clarke B, 1988

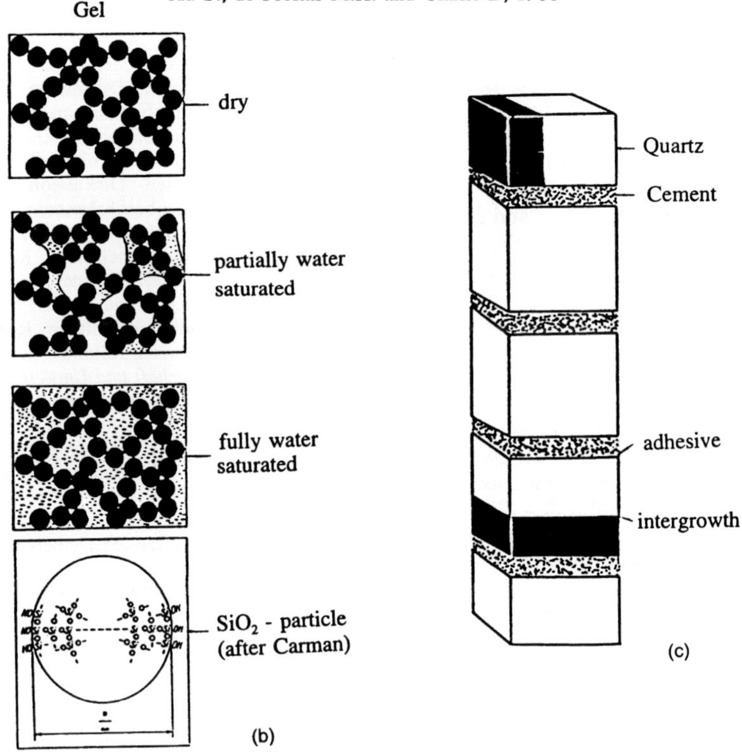

Fig. 4. A conceptual model for sandstone. (**a**) Petrological character of Penrith sandstone: the 'cement' is considered to be at the boundaries of the quartz overgrowth on the original quartz grains. (**b**) The Carman model of a colloidal particle, forming the basis of a sol, and the gel formed from of such particles (this gel can exist under fully saturated, partially saturated and dry conditions). (**c**) Conceptual model of a sandstone where the cement is silica gel: black and white both represent quartz grains in different crystallographic orientations, which have intergrown.

evidence because it raises the question of why this difference should occur: conventionally such differences would be attributed to differences in the area of grain to grain contact, variable porosity, and the spatial distribution of the cement (Dobereiner & de Freitas 1986), all of which might occur at different points, but there is also evidence to suggest that qualitative differences of the cement may also have to be considered. Figure 4 illustrates a sandstone known to many in the UK and a conceptual model of its particle relationships. With such a model the behaviour of the 'cement', if it were a substance with colloidal dimensions, would readily explain the differences in loss of strength and varying conditions observed in different sandstones.

Gels as a geohazard?

Landslides are commonly associated with periods of rainfall yet an inspection of the detailed evidence, plots of rainfall and slope movement on a common time axis, commonly shows no clear evidence of how the two are related.

It is interesting to note that in the *Proceedings of the 6th International Symposium on Landslides* (1992) conflicting opinions are recorded on the action of rainfall. Some consider antecedent rainfall to be critical (e.g. Gillon *et al.* 1992; Lacerda & Schilling 1992), whilst others, most notably Brand (1992), believe that far too much emphasis has been placed on antecedent rainfall and that short-term rainfall intensity is the critical parameter. Could it be that the confusion arises from the difference in the time-dependent behaviour of gels that may be present? The situation is not clear and at present the differences are generally attributed to the hydraulic conductivity of the ground: in the highly conductive material rainfall intensity is considered the key whereas in material of low hydraulic conductivity antecedent rainfall may be the key.

All authors in the 1992 *Landslide Proceedings* attribute slope instability associated with rainfall events to increases in pore water pressure and a loss of effective stress, rather than to a reduction in the effective strength of the materials involved. However, one case history is of note: Janbu *et al.* (1992). Janbu describes a spate of slides in northern Norway that followed an unusually warm summer and evolved during later periods of intense winter rain (quick clays were not involved). A slide at Skei killed two people and Janbu observes: 'The most frightening experience in this case is that a compacted fill of fairly well graded material can become so liquid, by external excess pore pressure build up, that it can flow on almost horizontal ground with such a speed that it damages well built houses 20 m to 30 m away so severely that lives are lost'. It is clear that Janbu *et al.* are surprised by the response of the ground, yet if gels are involved such a response is entirely possible.

Conclusion

Evidence is presented to conclude that silica gels and, generally speaking, other colloidal fine materials are natural constituents of sandstones, and probably of other rocks rich in silicate minerals, and that their time-dependent behaviour has the potential to be a serious hazard. Further work is currently under way to study the loss of strength that occurs between quartz particles in the presence of sols and gels.

Acknowledgements. One of the authors, C.B. expresses her gratitude to the Ministerium für Wissenschaft und Forschung des Landes Nordrhein Westfalens for a Lise Meitner Stipendium, which made this work possible. We dedicate this work to Professor G. Butenuth.

References

BUTENUTH, G. 1982. Löslichkeit von Modifikationen der Kieselsäure und von Silikaten, Untersuchungen an Thermalwässern von Neu-Seeland und Island. *Wissenschaft und Umwelt (ISU)*, 1, 14–20.

——1986*a*. Physikalisch-chemische Bedingungen in hydrogeologischen Räumen: Diskussion von Beobachtungen an tieftemperierten oberflächennahen Grundwässern, Teil I. *Wissenschaft und Umwelt (ISU)*, 1, 1–10.

——1986*a*. Physikalisch-chemische Bedingungen in hydrogeologischen Räumen: Diskussion von Beobachtungen an tieftemperierten oberflächennahen Grundwässern, Teil II. *Wissenschaft und Umwelt (ISU)*, 2, 66–78.

——1989. Physikalisch-chemische Reaktionsbedingungen in hydrogeochemischen Räumen, Beobachtungen an Flüssen, Teil I. *Wissenschaft und Umwelt (ISU)*, 4, 143–155.

——1990. Physikalisch-chemische Reaktionsbedingungen in hydrogeochemischen Räumen, Beobachtungen an Flüssen, Teil II. *Wissenschaft und Umwelt (ISU)*, 3, 141–156.

BUTENUTH, C., FREY, M.-L. & DE FREITAS, M. H. 1994. Kieselsauregele als Bindemittel in Buntsandsteinproben der Südeifel. Vorläufige Beobachtungen. *Poster Gesellschaft für Umwelt Geologie*, Heidelberg, Germany.

——, SCHETELIG, K. & DE FREITAS, M. H. 1995*a*. Bildung, Struktur und Festigkeit von Sandsteinen sowie qualitative Versuche zu ihrer Modellierung. *Geotechnik*, special edition for the 10th Nationale Tagung für Ingenieurgeologie 17–18 May 1995, Freiberg, Germany, 272–279.

——, DE FREITAS, M. H. & SCHETELIG, K. 1995*b*. Newer observations and experiments on interactions between mineral surfaces and aqueous solutions in geological systems. *Wissenschaft und Umwelt (ISU)*, **1**, 43–52.

BRAND, E. W. 1992. Slope instability in tropical areas. *In*: BELL, D. H. (ed.) *Proceedings of the 6th International Symposium*, Christchurch, New Zealand, **3**, 2031– 2051.

DE FREITAS, M. H. 1993. Weak arenaceous materials. *In*: CRIPPS *et al.* (ed.) *The Engineering Geology of Weak Rocks*. Engineering Geology Special Publication 8, Balkema, Rotterdam, 115–123.

DENISOV, N. V. & RELTOV, B. F. 1961. The influence of certain processes on the strength of soils. *In*: *Proceedings of the 5th International Conference Soil Mechanics and Foundation Engineering*, **1**, 75–78.

DOBEREINER, L. & DE FREITAS, M. H. 1986. Geotechnical properties of weak sandstones. *Geotechnique*, **36**, 79–94.

DYKE, C. G. & DOBEREINER, L. 1991. Evaluating the strength and deformability of sandstones. *Quarterly Journal of Engineering Geology*, **24**, 123–134.

GILLON, M. D., RILEY, P. B., HALLIDAY, G. S. & LILLEY, P. B. 1992. Movement history and infiltration, Cairnmuir Landslide, New Zealand. *In*: BELL, D. H. (ed.) *Land-slides, Proceedings of the 6th International Symposium*, Christchurch, New Zealand, **1**, 103–109.

HATSCHEK, E. 1932. The study of gels by physical methods. *Journal of Physical Chemistry*, **36**, 2994–3009.

ILER, R. K. 1979. *The Chemistry of Silica*. Wiley, New York.

JANBU, N., NESTVOLD, N. & GRANDE, L. 1992. Winterslides a new trend in Norway. *In*: BELL, D. H. (ed.) *Landslides, Proceedings of the 6th International Symposium*, Christchurch, New Zealand, **3**, 1581–1586.

JONES, J. B. & BIDDLE, J. 1966. Opal genesis. *Nature*, **210**, 1353–1354.

LACERDA, W. A. & SCHILLING, G. H. 1992. Rain induced creep-rupture of Soberbo Road. *In*: BELL, D. H. (ed.) *Landslides, Proceedings of the 6th International Symposium*, Christchurch, New Zealand, **1**, 142–152.

MITCHELL, J. K. & SOLYMAR, Z. V. 1984. Time-dependent strength gain in freshly deposited or densified sand. *Journal of Geotechnical Engineering*, **110**, 1559–1576.

OSTWALD, W. & FISCHER, M. H. 1917. *Theoretical and Applied Colloidal Chemistry*. Wiley, New York.

STAUFF, J. 1960. *Kolloidchemie*. Springer, Berlin, Göttingen and Heidelberg.

STUART, H. A. 1953. *Die Physik der Hochpolymeren*. Springer, Berlin, Göttingen and Heidelberg.

WIID, B. L. 1970. The influence of moisture on the pre-rupture fracturing of two rock types. *In*: *Proceedings of the 2nd International Congress of the International Society of Rock Mechanics*, Beograd, **2**, 3–35.

XU, S., DE FREITAS, M. H. & CLARKE, B. 1988. The measurement of tensile strength of rock. *In*: *Proceedings of the International Society of Rock Mechanics and Power Plants*, Madrid, 125–132.

An extension of probabilistic slope stability analysis of china clay deposits using geostatistics

D. M. Pascoe,[1] R. J. Pine[2] & J. H. Howe[3]

[1] Camborne School of Mines, University of Exeter, Redruth, Cornwall TR15 3SE, UK
[2] Golder Associates (UK) Ltd, Landmere Lane, Edwalton, Nottinghamshire NG12 4DG, UK
[3] EEC International Ltd, John Keay House, St Austell, Cornwall, PL25 4DJ, UK

Abstract. An understanding of the reasons for the distribution of highly kaolinized zones in the St Austell granite is useful to the china clay industry because a better understanding of shear strength distribution spatially should lead to improved models and give better insight into the progressive development of failure surfaces.

A field cone penetrometer survey of a bench in a working china clay pit has been correlated with laboratory shear strengths of samples from the same slope. A 3-D block model produced using a commercial integrated mining software package was then incorporated into a slope stability analysis.

The results were compared with analyses in which the spatial variability of the same shear strength data were incorporated using geostatistics. The factors of safety from the probabilistic analysis were log-normally distributed but those from the two geostatistically generated analyses were outside the extreme ranges of the probabilistic data. This emphasizes the importance of any concentrated zones of weak or strong material within a section with the same overall average frictional shear strength values. This is not correctly captured by a purely random strength assignment. The use of other geostatistical analysis techniques such as indicator kriging and conditional simulation are also under consideration to accommodate extreme variations in shear strength as occur over short distances in the china clay host material, kaolinized granite.

Introduction

In open pit mining there is pressure to steepen slopes to maximize the economic returns. This must be balanced with safety considerations, especially in the long term for final slope profiles. China clay in the St Austell granite is extracted from open pits typically 50–100 m deep. Some of the more critical slopes are within nearly fully kaolinized granite and are eventually planned to be 150 m deep. Typical slopes have overall angles of 25° to 30°. Slopes are divided into benches about 15 m high with bench face angles of 40° (see Fig. 1). In highly kaolinized areas slope angles have to be reduced with a resulting loss in production. An understanding of the distribution of highly kaolinized zones is desirable for the following reasons:

- to reduce uncertainty regarding shear strength distributions;
- to develop an understanding of the effect of specific shear strength distributions on slope failure development;
- to enable a more realistic estimation of china clay reserves (a mineralogical rather than a stability consideration.

Slopes have been designed to date using a 'conventional' approach based on limit state factor of safety analysis incorporating laboratory test data on shear strength and unit weight and *in situ* groundwater level measurements. All the data are variable but field inspection shows that the greatest variability is in the distribution of the degree of alteration, hence shear strength of the kaolinized granite. The approach shown in this paper is based on a geostatistical analysis of shear strength distribution derived from closely spaced penetrometer tests. Patterns of shear strength and resulting factors of safety for particular failure surfaces are determined from the geostatistical model and compared with random strength models based on Monte Carlo simulation.

Site investigation

The St Austell granite is the most extensively kaolinized of several plutonic exposures associated with the Cornubian batholith, which extends for over 250 km from the Isles of Scilly in the southwest to Dartmoor in the northeast. Low temperature (70–150°C), low salinity fluids of the latest stage of fluid activity in the St Austell granite are thought to have led to kaolinization (Alderton & Rankin 1983), with meteoric water being incorporated by convective fluid movement. A literature

Pascoe, D. M., Pine, R. J. & Howe, J. H. 1998. An extension of probabilistic slope stability analysis of china clay deposits using geostatistics. *In*: Maund, J. G. & Eddleston, M. (eds) *Geohazards in Engineering Geology*. Geological Society, London, Engineering Geology Special Publications, **15**, 193–197.

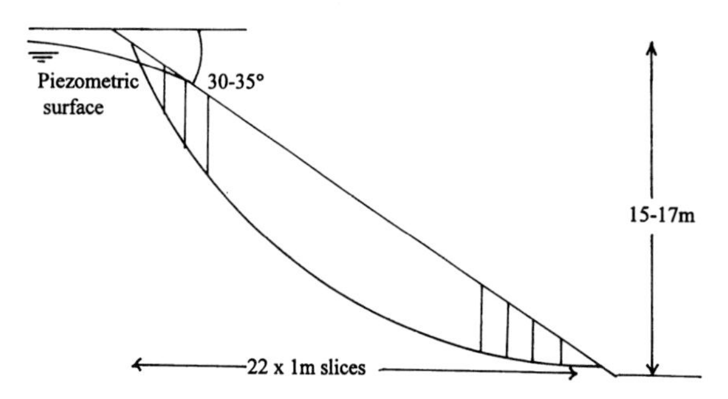

Fig. 1. Typical bench section showing slices for Bishop analysis.

survey of the structural geology of the region and, on a smaller scale, mapping of the orientation and spacing of discontinuities within the research area has been carried out and this demonstrates a relationship between the network of interconnected, fluid conducting joints and the pattern of kaolinization. Three major discontinuity sets have been mapped in the pit, striking approximately at 050°, 100° and 180°.

Data for the slope stability analyses were obtained from a slope in one of the china clay pits in the St Austell granite which has been monitored in detail for possible slope movement (Pine *et al.* 1994). Fifteen boreholes were drilled for soil sampling and piezometer installation in the slope to depths between 20 and 50 m. Triaxial shear tests on 29 samples were conducted over a range of effective confining stresses from 30 to 150 kPa, similar to stresses expected at the locations of critical failure surfaces. The results gave an overall friction angle of 37° with a standard deviation of 4°. Similar results were obtained from shear box tests.

The permeability at each of the piezometer positions was determined using falling head tests giving a range of values between 8.9×10^{-9} and 9×10^{-7} m s^{-1}, emphasizing the variability in the degree of kaolinization. It is planned to use this data, along with measured pore pressures and rainfall data for the last ten years, to model

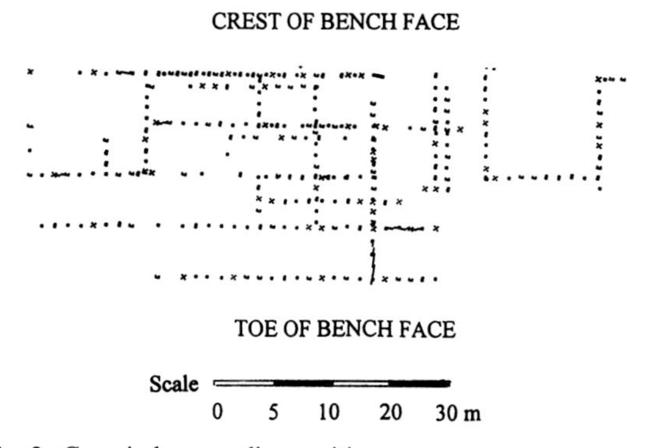

Fig. 2. Cone index sampling positions.

the response of the pore pressures in the slope to periods of prolonged rainfall and drought. The extreme values will be used for further stability assessments.

In the current work, cone index testing was undertaken using a hand-held penetrometer. The penetrometer was pushed into the ground at a steady rate with cone index (CI) readings taken at 75 mm intervals to a depth of 600 mm. A grid of 18 cone index sampling points in a zone of approximately 55×24 m was accurately surveyed, with intermediate sampling points being measured by tape (see Fig. 2). The cone index survey enabled geostatistical modelling of the spatial variability of kaolinization and incorporation into subsequent stability analyses.

To date, the cone index values have been used to assign frictional shear strengths based on the laboratory triaxial tests. It is planned to calibrate the cone index values using the results of shear testing of samples obtained from the same locations using the Leeds shear box (Hencher 1994, pers. comm.). With this box, samples are obtained in the field using the laboratory shear box fitted with a temporary cutting edge to minimize sample disturbance.

Geostatistical analysis of data

To demonstate the effect of incorporating the spatial variability of the shear strength parameters into the slope stability analysis, geostatistical models of the shear strength were made using DATAMINE (1993) integrated mining software. Each position at which a cone index test was carried out was digitized into a surface points file and readings at successively deeper levels were processed to give the true XYZ coordinates. Results from the 225 mm level were chosen for the analyses in this paper. The geostatistical analyses produced estimates of parameter values for blocks of ground in three dimensions taking into account the spatial variability of the deposit.

Anisotropy of shear strength/cone index values was modelled by representing the range of influence in each direction by the major and minor axes of an ellipsoid which coincide with axes of geometric anisotropy. The major axis of the ellipse, in an 80° direction, has a range of 14 m and the minor axis a range of 11 m. These dimensions are an indication of the upper measure of the scale of the kaolinization pattern. Sampling points are restricted to the plane of the slope, thus only two axes of the full 3-D ellipsoid can be defined. The third axis was assumed to be equal to the minimum in order to allow a demonstation of the geostatistical techniques, whereas in reality this may not be the case. This problem might be overcome by sampling on an adjacent perpendicular bench.

Fitting of an optimum model variogram to the experimental variogram resulted in a two-structure spherical

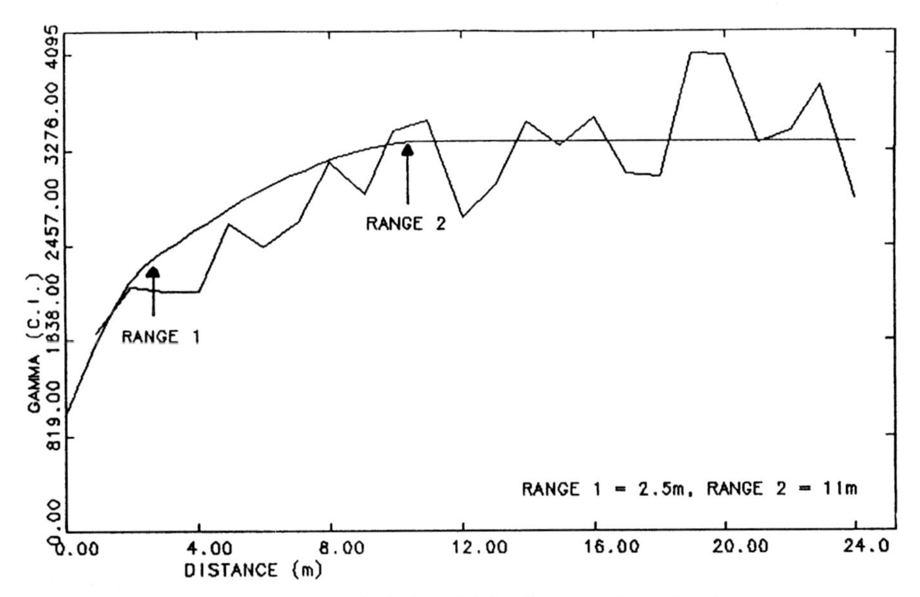

Fig. 3. Two-structure spherical model fitted to experimental variogram.

model being chosen as shown in Fig. 3. Cross-validation was used to assess the validity of the variogram model by point kriging each sampled location. Kriging is a process by which weighting factors are assigned to surrounding *in situ* parameter values to give the best linear unbiased estimate of value at an intermediate point. The program removes each observed value in turn (jackniffing) and kriges all adjacent values in the search neighbourhood whilst temporarily excluding the measured value at that location from the calculation. The resulting estimate is then compared with the actual sample value. By minimizing the kriging variance, the distribution of the kriged data may be made to approximate the sample data distribution more closely. Interpretation of these results is quite subjective however. A comparison of the distribution of kriged values with that of actual sample values shows that the tails of the distribution are poorly represented in the kriged data (Ravenscroft & Armstrong 1990).

A kriged block model comprising $22 \times 10 \times 55$ 1-m^3 blocks was made of the bench where the cone index tests were carried out. Two vertical sections were taken through the model mapping the kriged shear strength values onto the corresponding slip surface for stability analysis.

Limit state slope stability analysis

Probabilistic risk analysis

Limit state slope stability analyses were carried out using the program JACOB (developed by the second author) which incorporates the Bishop method. A slice through a single bench of the pit was simulated in the form of 22 1-m wide slices to which 100 different sets of randomly selected friction angle values were assigned. The friction angles were taken from a normal distribution matching the triaxial data (mean 37°, standard deviation 4°). The slip surface involved a whole bench and a high piezometric surface was included. An average unit weight of 22 kN m^{-3} was used throughout for the slope material.

Results of the probabilistic analysis are summarized in Fig. 4. The mean factor of safety was 1.27 and the standard deviation was 0.035. The distribution was slightly log-normal (close to normal). The probability that F was less than 1.25 was about 36%. Of the 100 values of F, the absolute minimum was 1.20 and the absolute maximum was 1.38. These values would be slightly different for other simulations using the same input data.

Stability analyses based on kriged strength values

Using the kriged data model, two sections were selected representing the extremes of high/low shear strength distributions. In case A the shear strengths in the lower part of the slope were generally low, and in case B they were generally high, but the averages for the whole slope were similar to the overall average friction angle of 37°, as in the probabilistic analysis. The resulting factors of safety were 1.18 and 1.39, slightly beyond the extremes of the probabilistic analysis.

Current modelling using strength values of the blocks intersected by a fixed postulated slip circle may be

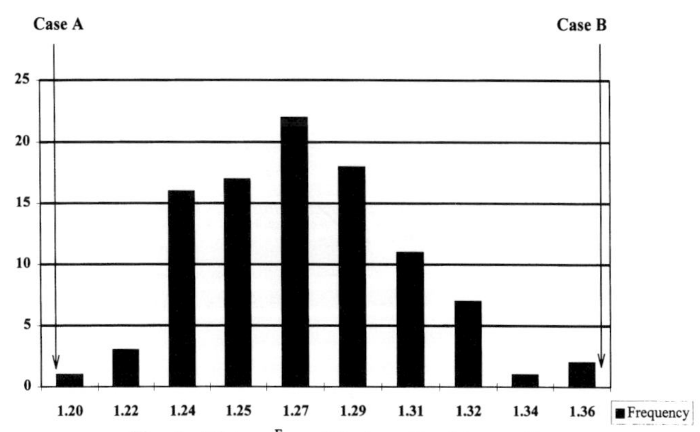

Fig. 4. Histogram of factor of safety results.

improved upon by allowing the path of slip surface to vary by passing through the weakest adjacent blocks, thus modelling more closely the actual development of the surface during failure. It is planned to use both limit state analysis as above and the computer program FLAC (1991) for this. FLAC is a finite difference continuum model which allows the development of a slip surface due to material failure (in pseudo-time) which closely models the observed development of progressive failure.

Discussion and conclusions

Deterministic slope stability analysis using single average values for each geotechnical parameter is improved upon by carrying out a probabilistic analysis, based on a Monte Carlo simulation for example. Using this method, the shear strength parameters may be treated as stochastic variables. The Monte Carlo approach, however, assumes that the simulated values of each block are independent, thus ignoring the spatial correlation, which is important in such deposits as kaolinized granite. By incorporating the spatial variability of shear strength parameters into the model using geostatistical methods, more realistic results are likely.

In the limit state example presented it is clear that although the average value for friction angles assigned to individual slices may be similar from case to case, the distribution is important. In particular, where there is a concentration of low friction values near the toe of the slope, where confining stresses are low, lower overall shear strength will apply. This will result in lower values for the factor of safety than for a completely random set of friction values. It is anticipated that this will be shown more clearly with the proposed FLAC modelling.

Kriging techniques, which are seen to smooth the data, are not entirely suited to geotechnical modelling because the tails of the distributions, which may be the most critical, are poorly represented. In the present case an important feature of the high strength tail is the presence of small inliers of intact unkaolinized granite which cannot be adequately characterized by cone index testing. However, this may be overcome by the use of indicator kriging to define different ranges of anisotropy at different cut-offs. In a preliminary test, a histogram of the cone index data was used to choose cut-offs at cone indices of 100, 200 and 300. An arbitrary value of 500 was assigned to solid granite. At each cut-off, indicator files were made in which values below the cut-off were assigned the value 0, and values above the cut-off, the value 1. It was found that the ranges of influence at the 100 CI cut-off, i.e. the highly kaolinized material, were smaller than the ranges at the 300 CI cut-off. This was not altogether unexpected as the continuity of the highly kaolinized zones seen in the slope is much less than for the less kaolinized areas. This may reflect the mode of formation of the kaolin deposits where discrete fractures control the kaolinization.

In order to address the problem of inadequately represented tails of the distribution in kriged block modelling, conditional simulation is currently being assessed using the program SGSIM developed by Dowd (1992). Using this technique, the simulated values have the same distribution as the real ones, and also have the same spatial correlation between values as that estimated for the real values. The results are conditional upon the simulation having the same values at sampled points as the sample values themselves. This method is also ideally suited to probabilistic analysis as each simulation is a unique realization and by making many such simulations and incorporating the results into stability analyses the probability of failure may be determined and compared with results of random probabilistic slope stability analyses.

Acknowledgements. The authors wish to acknowledge the co-operation of English China Clays International, Europe, and their permission to publish the material in this paper. D. Tucker and M. Newton of Datamine International have provided valuable assistance with the geostatistical applications of Datamine.

References

ALDERTON, D. H. M. & RANKIN, A. H. 1983.The character and evolution of hydrothermal fluids associated with the kaolinized St Austell granite, SW England. *Journal of the Geological Society, London*, **140**, 297–309

DATAMINE 1993. *DATAMINE*, Version 3.5.3. Datamine International, Old Deanery Stables, Cathedral Green, Wells, Somerset BA5 2UE.

DOWD, P. A. 1992. A review of recent developments in geostatistics. *Computers & Geosciences*, **17**, 1481–1500.

FLAC 1991. *FLAC Version 3.01 User Guide*. Itasca Consulting Group Inc., Minneapolis, MN.

PINE R. J., PASCOE, D. M. & HOWE, J. H. 1994. Assessing the risk of failure of pit slopes in the china clay deposits of SW England. *5th International Mine Water Congress, Nottingham, England, 19–23 September, 1994.* IMWA (International Mine Water Association), 467–477.

RAVENSCROFT, P. J. & ARMSTRONG, M. 1990. Kriging of block models – the dangers re-emphasised. *In*: *APCOM 90, Proceedings of the 22nd International Symposium, Berlin, West Germany, 17–21 September 1990.* Technische Universitat Berlin (Kongresse und Tagungen, Heft 51), **11**, 577–587.

Use of landslide inventory data to define the spatial location of landslide sites, South Wales, UK

P. J. Jennings[1] & H. J. Siddle[2]

[1] Halcrow Asia Partnership Ltd, Central Plaza, Wanchai, Hong Kong
[2] Sir William Halcrow and Partners Ltd, 31–33 Newport Road, Cardiff CF2 1AB, UK

Abstract. This paper presents and analyses the findings of a landslide inventory survey undertaken in deeply dissected upland valleys in layered Carboniferous lithology, Rhondda Valleys, South Wales. This area has one of the highest recorded concentrations of landslides in the UK.

The paper analyses the terrain factors identified at landslide sites (lithology, slope angle, geographical setting) with factors defining the landslide (such as type/depth of movement, age, activity, material, areal extent, geometry). Using these data, a composite profile of probable landslide sites is developed. This provides a valuable tool in further quantifying likely landslide hazard at sites, giving an indication of the potential of, for example, first-time failures or whether degraded relict landslides may be present.

Discussion is provided on the findings, particularly landslide locations within the valley and their age and current state of activity.

Introduction

In the mid-1980s the Department of the Environment (DOE), UK began sponsoring research into the development of a methodology for preparing landslide potential maps (Halcrow 1986, 1990). This research was primarily centred on the Rhondda Valleys, South Wales (see Fig. 1). As a necessary part of this work, a large database of landslide sites was compiled. Interpretation and extension of this database to include additional terrain factors provides a valuable insight into the nature of the landsliding mechanisms and can provide details of the sites most likely to be affected by first-time failures or where degraded relict failures might exist.

Physiology and geological setting

The Rhondda, located in the central portion of the South Wales Coalfield, consists of two valleys: the Rhondda Fach and the Rhondda Fawr. Like nearly all the valleys

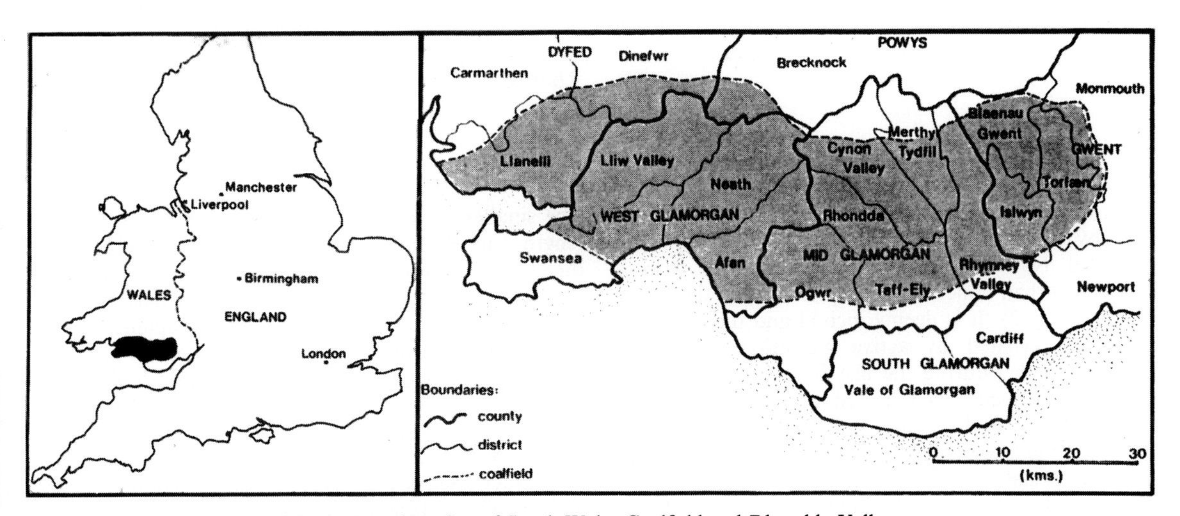

Fig. 1. Location plan of South Wales Coalfield and Rhondda Valleys.

JENNINGS, P. J. & SIDDLE, H. J. 1998. Use of landslide inventory data to define the spatial location of landslide sites, South Wales, UK. *In*: MAUND, J. G. & EDDLESTON, M. (eds) *Geohazards in Engineering Geology*. Geological Society, London, Engineering Geology Special Publications, **15**, 199–211.

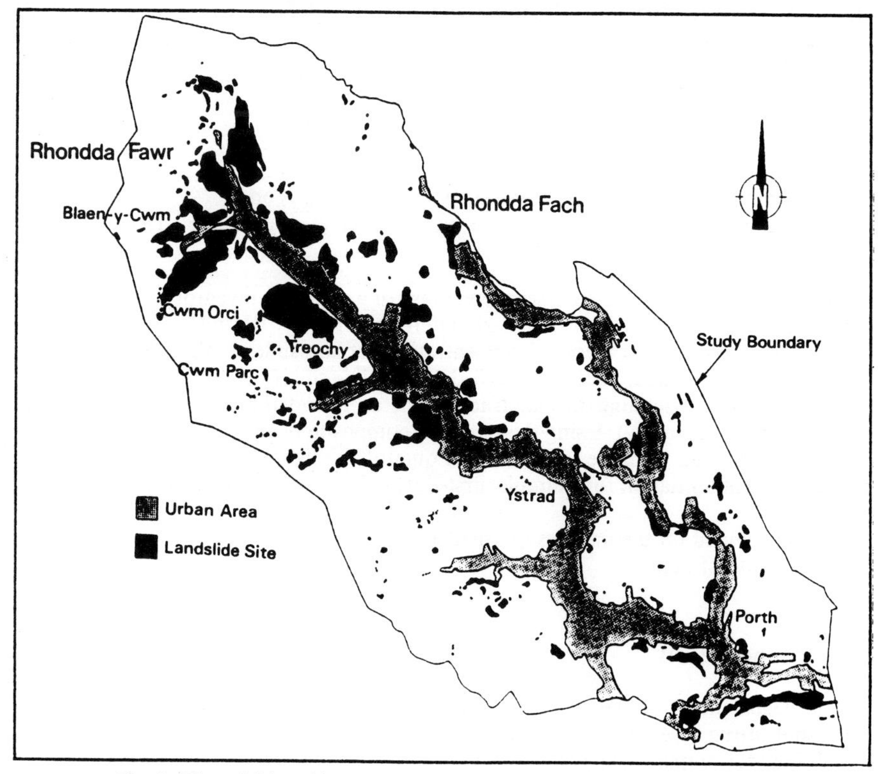

Fig. 2. Plan of Rhondda Valleys showing landslide sites and urban areas.

in the coalfield they are deeply incised with, in places, the valley floor being 350 m below a high upland plateau, resulting in a very constricted steep-sided valley. The Rhondda first came to prominence in the latter half of the 19th century with coal mining. Today, large urban areas are still located throughout the valleys.

Rhondda Fawr

The topographical watershed of the Fawr covers some 67.6 km^2, which includes upland plateaux but is dominated by the Rhondda Fawr, a narrow steep-sided valley orientated NW–SE, some 20 km in length, with a significant urban concentration spread along the valley floor (see Fig. 2). It is deeply incised and this has been further exacerbated by active glaciation originating within the valley during the last Ice Age. The layered lithological sequence is reflected in the characteristic bench features in the valley sides, known as 'slacks', indicating the presence of relatively thin argillaceous horizons (usually including a coal seam). Several tributary valleys feed into the Fawr, the most significant being three northeast-facing, high steep-walled corries to the south. The valley floor is generally narrow (less than 1 km wide), except at Treorchy where softer mudstones are exposed in the valley floor and lateral erosion has widened the valley significantly. Further downstream at Ystrad the valley becomes constricted, only starting to widen again at Porth, at the confluence with the Rhondda Fach.

Rhondda Fach

The smaller of the two valleys, for the majority of its length the Rhondda Fach is aligned in a NW–SE direction. The valley's entire length is within the more resistant Pennant Measures, resulting in a narrow valley with high steep stepped-sides with no significant valley bottom widening. There are no corries although several incipient ones or possibly niviation hollows are present along the northeast-facing slopes. The present valley cross-section is markedly asymmetrical, the river tending to flow down the eastern side. The western side is characterized by deeper deposits of boulder clay, whereas the eastern side is underlain mostly by glacial sands and gravels deposited by the past concentration of glacial meltwaters. The overall drainage pattern is similar to that of the Rhondda Fawr, with tributary steams flowing from the plateaux having a steep descent to the main river.

Geology

The South Wales Coalfield is an elongated oval-shaped asymmetric basin about 90 km from east to west and 30 km from north to south, covering an area of about 2700 km^2. Preserved within the basin is the stratigraphy of some 2500 m of Carboniferous sedimentary rocks, known traditionally as the Coal Measures (for a detailed description see Woodland & Evans 1964). The stratigraphy of the Coal Measures is divided into three dominant divisions: Upper, Middle and Lower Coal Measures, where the latter two are more similar in nature having predominantly argillaceous sequences as opposed to the arenaceous stratigraphy of the Upper Measures (Fig. 3).

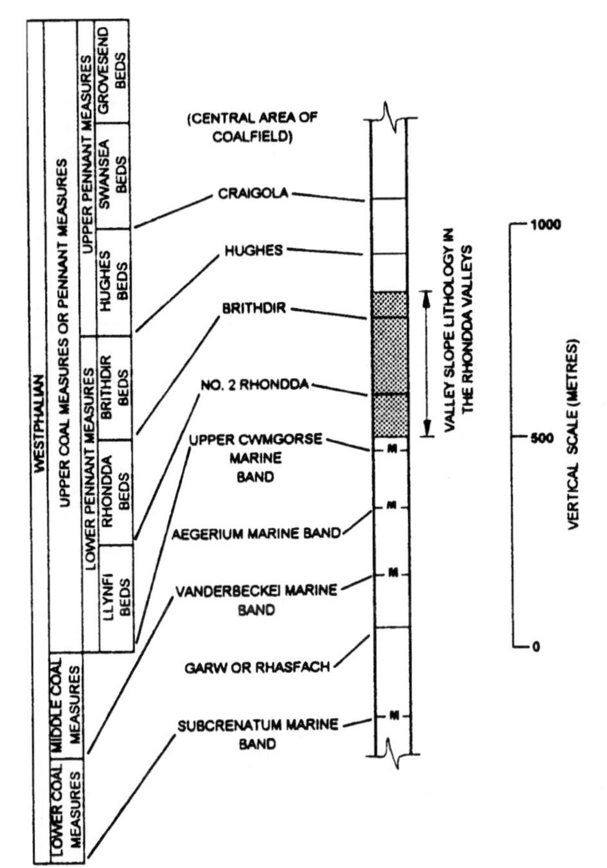

Fig. 3. Typical valley side lithology.

The Lower and Middle Coal Measures comprise mainly of mudstone and siltstone sequences with some sandstone beds common in places. Within these measures there is generally an orderly sequence of lithologies which represents a repetitive cyclical depositional nature. The Upper Coal Measures are dominated by poorly sorted sub-greywacke sandstone, locally called Pennant Sandstone, containing thin argillaceous beds. The depositional cycle, which can range in thickness from 30 m to 150 m, consists of about 80% sandstone. Towards the top of the cycle there is a gradational change into siltstone, mudstone, seatearth and a thin coal seam. These measures commonly form the valley sides within the coalfield and it is the inclusion of the generally thin argillaceous bed that gives the valleys their characteristic benched profile.

The most common structural faults in the coalfield are the cross-faults which trend generally NW–SE. These in turn have exerted an influence on the main joint system of the Pennant Sandstone which trends generally in the same direction.

Geology of the Rhondda Valleys with respect to slope instability

The Upper Coal Measures are the most significant valley-side lithological formation and within the Rhondda they can be subdivided into Brithdir, Rhondda and Llynfi Beds (Fig. 3).

The Pennant Sandstone units comprise a significant proportion of these beds and are the dominant lithology in the upper valley side. They are composed of a uniform medium to coarse-grained, generally feldspathic, sandstone which is extremely hard and resistant to weathering. *In situ*, individual beds can range from 50 m to 150 m and are generally well-bedded and 'massive' in nature. The lithology is markedly well jointed with the trend of the main joint sets following the major fault swarms. Within the Rhondda this is parallel to the valley sides. Therefore, any instability affecting the Pennant Sandstone tends to cause blocks to fail successively along pre-existing discontinuities, such as joints. A precursor to deep-seated failures can be indicated by a dilation of vertical discontinuities in the sandstone beds at the surface, resulting in large cracks opening parallel to the valley sides.

The discontinuities within the Pennant Sandstone provide a high secondary permeability. This allows the downward transmission and storage of water along well-developed planes. Argillaceous beds underlying the sandstones have a very low relative permeability and tend not to have as pronounced joint sets and therefore act as aquicludes preventing the vertical move-ment of water. In such instances, a reservoir of groundwater can build up above the argillaceous bed, with a subsequent increase in pore water pressures. Where these beds crop on the valley side they tend to be marked by a spring line. Denness (1972) and Conway (1974) have described this situation and its relationship with slope instability. Within the coalfield there is a recognizable spatial occurrence of landslides at or below sandstone units overlying relatively thick argillaceous beds.

Superficial deposits along the valley sides are dominated by head deposits; these are undifferentiated deposits originating from mainly solifluction and slope weathering processes. Reactivation of relict solifluction shear surfaces is a common cause of shallow landslipping. Boulder clay deposits formed of unstratified material deposited during deglaciation generally mantle the valley bottom and lower slopes. These deposits can have a much lower permeability than the underlying rock and can prevent groundwater emerging, causing a build up of pore water pressures (Bishop *et al.* 1969).

Mining activity within the area over the last century has contributed to hillslope instability. Such visible effects are the dilation of joints and tilting of beds, which have in turn altered significantly the hydrological regime (Daughton *et al.* 1977). The siting of spoil on the slopes has also induced catastrophic instability (Bishop *et al.* 1969).

The landslide inventory survey

The principal objectives of the inventory were to identify, map and classify areas of natural landslides and to assess their state of activity in order to establish an accurate and complete database of past and present slope movements in the study area. This consisted of a desk study, where areas of known instability were delimited together with areas requiring more investigation, and a field visit, which verified and refined the information obtained from the desk study. For details, see Halcrow (1986).

A fundamental necessity in any landslide inventory is the adoption of a suitable classification system. A previous survey of landslides within the entire South Wales Coalfield (Conway *et al.* 1980) used a simple classification system based on the type and depth of movement derived from the classifications of Varnes (1978) and Skempton & Hutchinson (1969). This classification, with some modification, has been adopted for this latest survey. Two major movement types are used to classify the landslides, namely rotational and translational slides. These have been further subdivided by depth of failure into 'shallow' and 'deep-seated'. An additional all-encompassing heading of 'complex' is also considered necessary; this includes all deep-seated failures that cannot satisfactorily be placed into another group, such as those that comprise two or more significant major movements. This results in six movement categories as shown in Fig. 4.

Other factors used to characterize the landslide sites were obtained from several sources principally during the desk study stage and verified on the site visit. Table 1

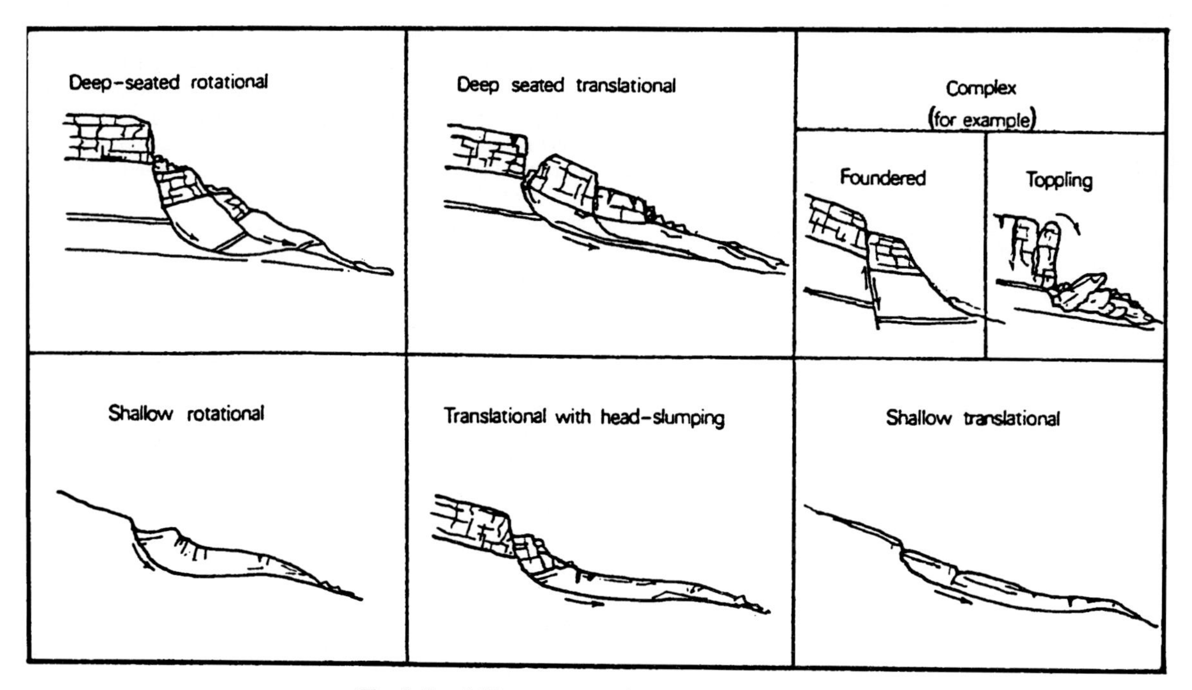

Fig. 4. Landslide movement classification (six types).

Table 1. *Factors used to characterize landslides sites*

Factors used to classify landslide site		*Objectivity rating*
Classifier	*Sub-classifier*	
Depth and type of movement	Deep-seated rotational Deep-seated translational Deep-seated complex Shallow rotational Shallow translational Shallow translational with head-slumping	Medium to good
Activity	Active Recently active Partly active Dormant	Medium
Age	Historic (now to 100 BP) Prehistoric (100–10 000 BP) Immediate post-glacial	Poor to medium
Geographical location	Simple description given of site with respect to valley.	Good
Material	Superficial deposit Bedrock	Good
Areal extent	None	Good
Slope angle	None	Good
Downslope length	None	Good
Local lithological association	Names of local argillaceaous beds, particularly coal seams	Good

shows the factors collated. This is not an exhaustive list but it is considered sufficient to reasonably assess the potential hazard arising from a landslide event. Each factor also has an objectivity rating indicating the degree of interpretation required to evaluate the factor, ranging from poor, medium to good. In general it proved more difficult to assess activity, and in particular, age of occurrence. This has an obvious limiting effect when attempting to construe likely return periods for landslide events.

The range and definitions of activity used are based on visual examination and relate to the 'freshness' of the morphology (details are provided in Halcrow 1986). Altogether four groups are considered; these being active, recently active, dormant and a group which recognizes that parts of a landslide complex may be currently in a different state of activity termed 'partly active'. This last group essentially applies to dormant landslides that show evidence of some recent activity.

The only reliable source of dating landslide events for the inventory has been through published information obtained in the desk study. This can account for the 'recently active' failures, placing them as occurring in the last 50 to 100 years conterminous with the opening up of the area with the discovery of coal; whereas 'dormant' failures cannot be dated as easily, occupying a time span ranging from perhaps 100 BP to the end of the last ice age.

The terms 'deep-seated' and 'shallow' are assessed from visual inspection of the landslide. The distinction is arbitrary and generally if the failure has penetrated more than 5 m into the bedrock it is termed deep-seated. Shallow failures would typically involve superficial deposits with no significant amount of bedrock. Although this distinction is arbitrary the terms have been implied (Terzaghi 1950) or used (Ward 1945; Skempton 1953) in many classification systems.

The two broad categories used to define the material involved in the slope movement are 'superficial deposits' and 'bedrock', equivalent to the 'engineering soils' and 'bedrock' as used by Varnes (1978). Superficial deposits, especially on the hillslopes, are not easily defined and therefore only two broad divisions are recognized: 'undifferentiated', containing an ill-sorted mix of slope deposits; and 'glacially derived' deposits. Bedrock has been subdivided into the various coalfield lithological units. Only the dominant material involved in the movement is recorded.

The geometry of a landslide site is defined using areal extent, slope angle and downslope length. These have been selected from the host of morphological terms that are available to describe the geometry of a landslide (e.g. Blong 1973; Brunsden 1973; Crozier 1973; Varnes 1978).

From previous work it is apparent that there are associations between landslides and certain coal seams. In particular, coal seams that crop at the head or upper portion of the landslide are recorded.

Results of the landslide inventory

A summary of the inventory's findings shows that within the area, 347 landslides were recorded, ranging in size from 65.9 ha to 0.03 ha. Altogether some 9.16 km^2 of landslides were identified, representing 9.4% of the total survey area. This concentration of landslides is one of the highest recorded for any area in the UK.

The distribution of landslide area along the Rhondda Fawr's length is shown in Fig. 5. This distribution shows multiple 'jumps' where a particularly large landslide concentration exists. Within the valley, about 75% of the total landslide area is located in the upper valley, which coincides with the exposure of the predominantly argillaceous Middle Coal Measures. Both shallow and deep-seated landslides show a bias towards the upper valley and particular concentrations of shallow landslides are found in the Cwm Parc and Cwm Orci tributaries. About 60% of the deep-seated landslide area

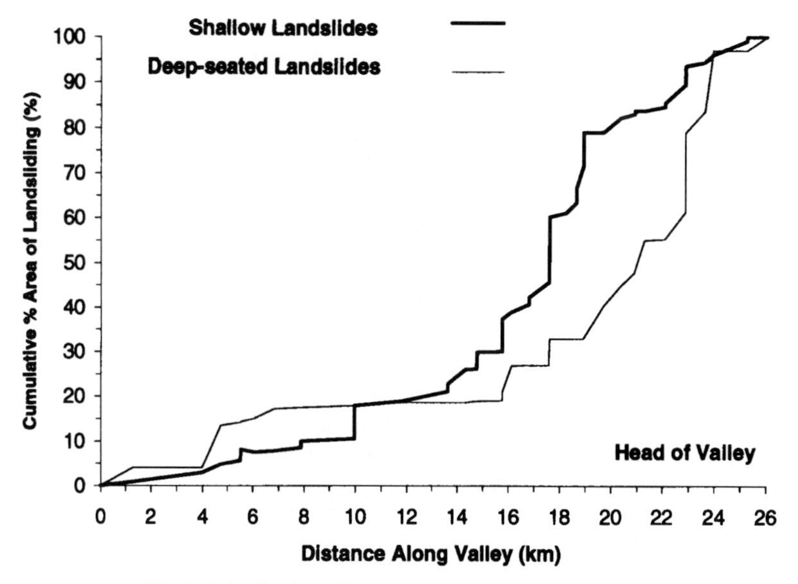

Fig. 5. Distribution of landslide area along the Rhondda Fawr.

is located in the upper valley. There is also a noticeable concentration of areas of deep-seated failures within the Blaen-y-Cwm tributary.

The distribution of landslide area along the Rhondda Fach is shown in Fig. 6. The valley has fewer landslides than the Fawr but the distribution is similar, with an increase towards the upper portion of the valley corresponding to an exposure of argillaceous beds in the lower hillside. One particular landslide accounts for 20% of the failed area; this consists of a length of hillside that has failed in a succession of translation slides.

Figure 7 shows the frequency of occurrence and area of the six landslide types. Shallow landslides are especially significant, accounting for over 85% of the total number, particularly the shallow translational slides (58.4%) and the shallow translational slides with slumping at the head (20.8%). The deep-seated landslides only account for 15% of the total number but

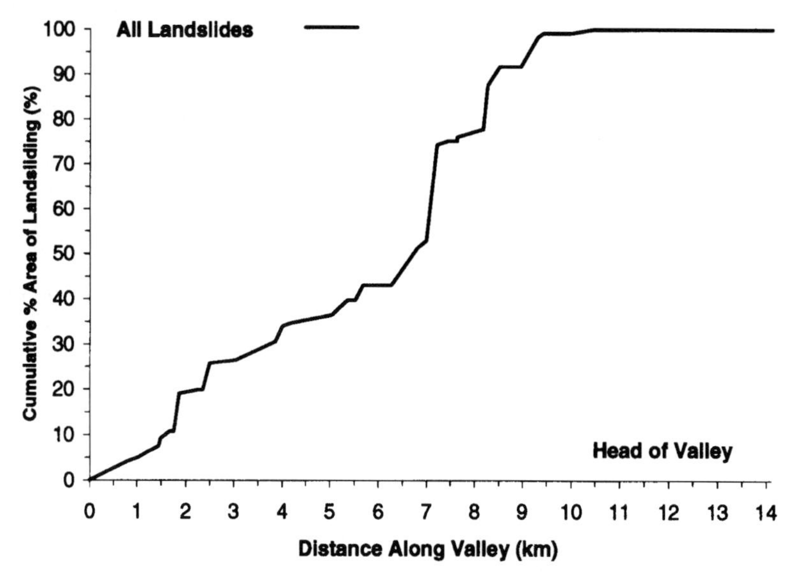

Fig. 6. Distribution of landslide area along the Rhondda Fach.

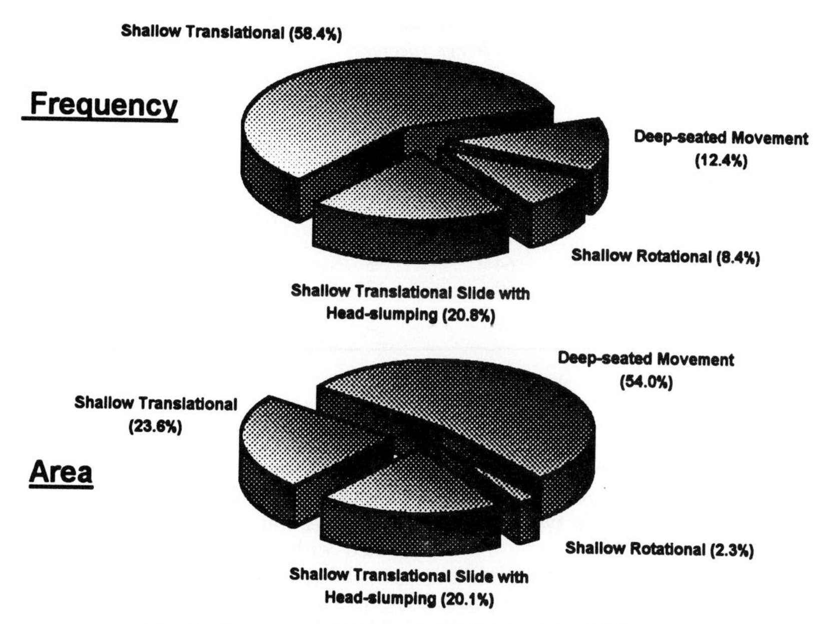

Fig. 7. Frequency of occurrence and area of landslide types.

occupy 54% of the total landslide area, with an average size of 11.28 ha. The largest is a deep-seated complex failure known as Gelli Goch, covering 65.85 ha and located towards the head of the Rhondda Fawr where the majority of deep-seated failures are situated.

Landslide activity classification data (Table 2) show that the majority of the failures (61%) are dormant, with 30% showing signs of having been recently active and 5% partly active. The active landslides only account for 4% of the total.

Although the immediate hazard to the local community is represented by the active landslides, the recently active landslides account for 30% of the total number and are in a potentially unstable state.

In terms of area, the active landslides represent 9% of the total area (84.3 ha) and the recently/partly active landslides 33% (or 300.5 ha).

The slope angles from each landslide site were calculated from elevations at the head and toe obtained from topographical maps. The results (Table 3) show a mean value ranging from 23.7° to 26° for the various landslide movement types. The slope angle distribution (Fig. 8) shows a slight positive skew. The most prevalent slope angle is 19°. This is attributable to the fact that over 85% of the landslides are shallow and have occurred in superficial deposits on the lower and middle slopes where the slope angle is less.

Table 2. *Activity state of landslide types*

Landslide type	*Active*	*Recently active*	*Dormant*	*Partly active*	*Total*
Shallow					
rotational	2	18	9	–	29
translational	5	54	138	5	202
translational with head-slumping	6	29	32	5	72
all shallow	13	101	179	10	303
Deep-seated	2	3	33	6	44
All landslide types	15	104	212	17	347
Percentage	4.3	30.0	61.1	4.9	–

Table 3. *Estimated original slope angles at landslide sites*

Landslide type	*Number*	*Range (degrees)*		*Standard deviation*	*Mean (degrees)*	*Mode (degrees)*
		max	*min*			
Shallow						
rotational	29	64	8	14.15	23.7	30
translational	202	56	5	10.52	23.8	19
translational with head-slumping	72	53	9	9.32	26.0	21
all shallow	303	64	5	10.66	24.4	19
Deep-seated	44	69	13	10.54	25.6	25
All landslide types	347	69	5	10.64	24.6	19

Table 4 shows the relationship between landslide movement type and downslope length. There is a wide range of distances (5–800 m) and a large standard deviation for all movement types. The distribution of downslope length for the shallow landslides is typically less than 90 m.

The majority of landslides occur in undifferentiated superficial material (Table 5). The bedrock has been subdivided into the individual lithological units: Brithdir Beds are only present on the upper slopes/plateaux and are involved in very few failures; Rhondda Beds are involved in the majority of failures comprising bedrock but this would be expected as it is the prevalent slope-forming lithology; Llynfi Beds and the Middle Coal Measures are represented in only a few failures, being essentially located in the valley floor. Some of the larger failures occupy a significant proportion of the valley side and therefore incorporate several lithological sequences. This is especially true of the deep-seated failures.

Table 6 shows the occurrence of landslides with coal seam crops at the head or within the upper part of the failure. A minority of landslides (31%) are not associated with any seams but this does not discount the possibility that they are located on other argillaceous horizons. This, however, was not recorded. The majority of landslides appear to be located on a significant coal seam. The Rhondda Nos. 1 and 2 seams are associated with over half of all failures. These seams are laterally persistent and generally underlie a considerable depth of Pennant Sandstone, which can be a substantial aquifer.

If landslide activity is shown against associated coal seam (Table 7), the seams which generally crop on the upper/middle hillside are seen to be located with the active landslides.

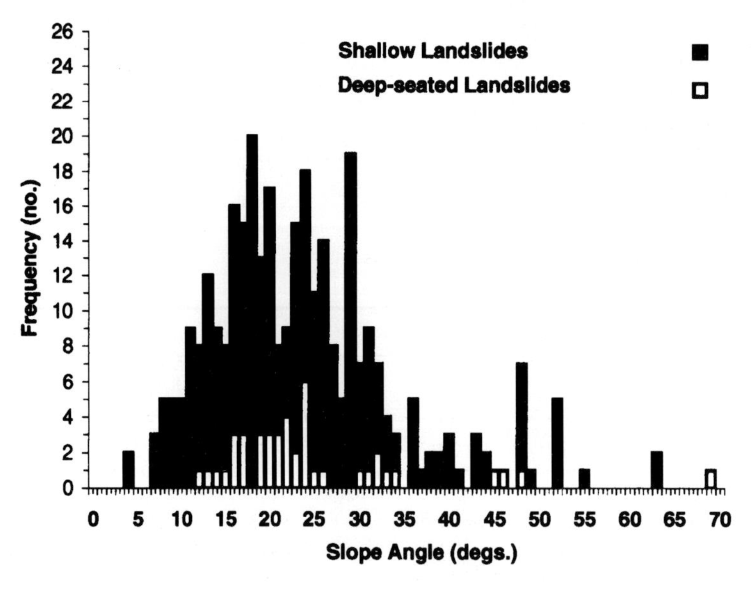

Fig. 8. Distribution of slope angles at landslide sites.

Table 4. *Showing downslope length for the various landslide types*

Landslide type	*Number*	*Range (degrees)*		*Standard deviation*	*Mean (metres)*	*Mode (metres)*
		max	*min*			
Shallow						
rotational	29	210	10	62.61	56.2	10
translational	202	320	5	63.49	78.9	60
translational with head-slumping	72	510	5	101.19	142.1	50
all shallow	303	510	5	80.39	91.7	50
Deep-seated	44	800	22	188.20	262.7	200
All landslide types	347	800	5	115.28	113.34	50

Discussion of landslide inventory results

A principal consideration in analysing the landslide inventory data is to acquire a better appreciation of the landslide system within the valleys. This endeavour should result in a more effectively defined landslide management policy. The findings of the inventory show that the most common type of landslide are shallow movements, accounting for 87% of the total number. Of the individual landslide types, the shallow translational slide represents 58.4%. Another significant but similar group, comprising 20.8%, are the translational slides with slumping at the head of the failure.

Deep-seated movements, whilst only comprising 13% of the total number of occurrences, accounted for 54% of the landslide area. Shallow movements were generally small, with an average size of 1.39 ha.

The majority of the large deep-seated landslides are dormant and considered to be of great antiquity, assumed to be associated with the end of the glacial or post-glacial epoch occurring at about 10 000 BP (Woodland & Evans 1964; Woodland 1985). Within the survey area a dormant shallow translational slide with head slumping has been dated at about AD 1250 (Conway *et al.* 1980), coinciding with the end of the 'Little Optimum' era when a milder climate existed. Conversely, the 'Little Ice Age' between the 15th and 19th centuries was a time of substantial climatic deterioration and numerous of the now dormant landslides within the survey were undoubtedly active or initiated during this period.

Palmquist & Bible (1980) have put forward several model scenarios on the distribution of landslides within valley watersheds. For example, the 'incision model' assumes that with a fall in sea level, a 'knickpoint' migrates upstream. This wave of erosion initiates hillslope failures as it passes, resulting in older failures situated downstream with progressively younger failures upstream. The Rhondda Fawr has the older deep-seated landslides at the head of the valley and this can perhaps be related to a glacial retreat model. The valley is believed to have developed its own glacier during the last ice age which during the climatic amelioration would have retreated into the head of the valley. Here a glacial environment would have existed for longer and would have had its maximum erosive effect on the hillslopes. The result of this, in combination with a hillslope geology prone to instability, would be large late glacial/immediate post-glacial landslides concentrated at the heads of the valley. This model does not operate singly; various models operate on the valley at any time and this can eventually obliterate the landslide distribution created by any one model.

The majority of landslides have been classed as dormant, with only 4% being in an active state. This latter group represents the immediate hazard to the local community, although the immediate hazard from the recently active landslides cannot be discounted. All categories of slope movements except shallow rotational slides have the significant proportion of their number as dormant. In terms of area, 88.7% of deep-seated movements are dormant, with 8.1% active. In comparison, 43.8% of shallow movements are dormant, and 9.5% are active, but a larger proportion (46.7%) are classed as recently active, i.e. in a marginally unstable condition.

Active landslides are in an unstable condition, i.e. they are undergoing continuous or intermittent movements, having not yet reached an equilibrium state. Recently active landslides can be considered to be in a marginally stable state where the forces causing the initial failure may still attain a level where reactiviation cannot be discounted. Dormant landslides are those considered to be stable, where there is no discrete movement of the displaced material and the margin of safety of the slope is sufficiently high to accommodate all existing forces. Dormant landslides would have been initiated under conditions which no longer prevail today, e.g. a wetter climate.

As the majority of landslides are shallow, superficial deposits comprise the majority of the failed material. Undifferentiated superficial deposits were the main material involved in 63% of landslides. The dominant cropping bedrock is the Rhondda Beds, which account for 14.5% of the overall landslides.

Table 5. *Showing the dominant material involved in the various landslide types*

Landslide type	*Bedrock lithology →*						*Superficial deposits*		*Total*
	Upper slopes					*Lower slopes*			
	Brithdir beds	*Rhondda beds*	*Llynfi beds*	*Middle Coal measures*	*Rhondda/ Llynfi beds*	*Llynfi/Middle coal measures*	*Undifferentiated*	*Glacially derived*	
Shallow									
rotational	–	4	–	1	–	–	16	8	29
translational	2	11	10	–	–	–	159	20	202
translational with head-slumping	–	12	9	1	2	–	43	5	72
all shallow	2	27	19	2	2	–	218	33	303
Deep-seated	–	23	6	3	11	1	–	–	44
All landslide types	2	50	25	5	13	1	218	33	347
Percentage	0.6	14.4	7.2	1.4	3.7	0.3	62.8	9.5	–

Table 6. *Showing number of landslide types occurring on various coal seams*

Landslide type	*Coal seams*[a] →												
	Upper slopes									*Lower slopes*			
	Brithdir Rider	*Brithdir*	*No. 1 RhR*	*No. 1 Rh*	*Daren Rhestyn*	*No. 2 RhR*	*No. 2 Rh*	*No. 3 Rh*	*Tmd*	*Hafod*	*Abergorky*	*Thin coal*	*No coal seam*
Shallow													
rotational	–	–	2	6	1	–	2	–	4	1	1	–	13
translational	5	12	16	45	7	1	27	–	14	3	2	2	78
translational with head-slumping	-	5	6	18	3	1	15	1	10	1	1	1	16
all shallow	5	17	24	69	11	2	54	1	28	5	4	3	303
Deep-seated	–	–	5	10	3	–	11	–	4	5	1	1	4
All landslide types	5	17	29	79	14	2	65	1	32	10	5	4	111
Percentage (%)	1.3	4.5	7.8	21.1	3.7	0.15	17.3	0.3	8.6	2.7	1.3	1.1	29.7

[a] RRh: Rhondda Rider, Rh: Rhondda, Tmd: Tormyndd.

Table 7. *Summary of terrain factors associated with landslide sites*

Landslide movement type	*Average area (ha)*	*Average downslope length (m)*	*Average slope angle (degrees)*	*Dominant material*	*Significant coal seams*	*Geographical location within valley*	*Activity*	*Age*	*Frequency (no.)*
Shallow									
Rotational	0.7 (s.d. = 1.28)	56.2 (s.d. = 62.61)	23.7 (s.d. = 14.2)	Bedrock (17%); superficial (73%)	Rh1 (21%) Tmd (14%) none (65%)	Localized groupings are concentrated within both valleys, but generally not widely distributed		Historic	29
Translational	1.1 (s.d. = 1.7)	78.9 (s.d. = 65.5)	23.8 (s.d. = 10.5)	Bedrock (11%); superficial (89%)	Rh1 (21%) Rh2 (13%) none (40%)	Clustering in upper–middle Rhondda Fawr; and well distributed in the Rhondda Fach	For all shallow landslides about 4% are currently active with 33% showing signs of recent activity. The majority though are dormant	Prehistoric/ historic	202
Translational with head-slumping	0.6 (s.d. = 3.5)	142.1 (s.d. = 101.2)	26.0 (s.d. = 9.3)	Bedrock (33%); superficial (76%)	Rh1 (24%) Rh2 (20%) Tmd (13%) none (23%)	Widely scattered groups in the Rhondda Fawr; few occurrences in the Rhondda Fach		Prehistoric/ historic	72
Deep-seated									
Rotational	10.4 (s.d. = 17.6)			Bedrock	Rh1 (21%) Rh2 (58%) none (21%)			Prehistoric	24
Translational	10.7 (s.d. = 15.3)	For all types of deep-seated landslides: 262.7 (s.d. = 188.2)	25.6 (s.d. = 10.5)	Bedrock	RhR1(40%) Rh1 (20%) Rh2 (20%) none (20%)	For all types of deep-seated landslides there is a significant concentration towards the head of the Rhondda Fach	For deep-seated landslides the majority (75%) are dormant. However, for large failures there may be some active elements, particularly in the downslope margins	Prehistoric	10
Complex	14.9 (s.d. = 20.3)			Bedrock	Rh1 (30%) Rh2 (30%) none (40%)			Prehistoric	10

[a] Rh1: No. 1 Rhondda, Rh2: No. 2 Rhondda, RhR1: No. 1 Rhondda Rider, Tmd: Tormyndd. Percentages in brackets refer to the number of landslides associated with a coal seam. s.d.: standard deviation.

It has long since been recognized in the South Wales Coalfield that there is an association between landslides and the springlines formed at the crop of argillaceous horizons underlying sandstone aquifers (Knox 1927). In particular, the Rhondda Nos. 1 and 2 seams were highlighted then as they are now as being associated with a large number of failures. Most landslides (69%) occur on the crop of the significant argillaceous beds within the (predominantly sandstone) Upper Coal Measures. These argillaceous beds usually incorporate a coal seam. The most significant of these are the Rhondda Nos. 1 and 2 seams along whose crop 38% of all landslides occurred. The postulated failure mechanism associated with the outcrop of these beds on the valley side is: water intercepted from the sandstones above soaks into the weathered argillaceous rock, softening and weakening the rock sufficiently to induce shear failure due to the weight of the overlying sandstone blocks (Woodland 1985). Bentley *et al.* (1980) point out that the presence of closely spaced, randomly orientated discontinuities in the argillaceous beds will facilitate the progressive migration of a curved failure surface.

The slope angle distributions for the shallow landslides recorded by Conway *et al.* (1980) in the 'South Wales Coalfield Landslip Survey' indicated an hierarchical progression of different movement types with slope angle. This trend showed rotational failures occupying the highest mean slope angle of 21.8°, followed by translational slide with head slumping (20.6°), translational slides (18.5°), and flows (17.7°). The results from this survey do not show the same correlation with landslide type (see Table 3).

The survey adopted landslide movement as the basis for describing the failures. Altogether six different classes of movement were used and these were identified in the field by a team of engineering geologists each surveying a different area. Obviously, not all landslides are immediately identified and as easily classified; the observer has to use a degree of subjective judgement which will result in a conditioned response. This problem of a conditioned response concerned Jones *et al.* (1961) who employed a uniformity test to examine the degree of subjectivity. In the case of correctly identifying landslide sites, Wieczorek (1984) for example uses a certainty of identification criteria ranging from 'definite' to 'questionable'.

Landslide hazard profile

A breakdown summary of the terrain factors identified at all the landslide sites is given in Table 7. Considering the shallow and deep-seated failures separately using the information presented, a profile of the sites that may be affected by landsliding and what type of landslide may occur, or may already be present, can be established.

The sites most likely affected by shallow landslides are located on undifferentiated superficial deposits where the slope angle is about 24° or above, with a tendency to be situated downslope of the crop of the Rhondda Nos. 1 or 2 coal seams. A typical failure at this site would occupy about 1 ha, with a downslope length of about 90 m. A small proportion of the failures are currently active, though 33% show signs of recent activity.

The deep-seated failure sites are not as easily characterized. The sites they are most likely to affect are located within the head of the Rhondda Fawr, particularly where the Middle Coal Measures crop on the lower slope. A typical failure is likely to extend over 10 ha and involve the entire vertical height of the hillside, having a downslope length in excess of 250 m. These failures are considered to have originated under a different climatic regime to that which exists today and therefore the majority of the displaced mass is now dormant. First-time failures would be considered rare in the present day

The main landslide hazard is from reactivation of pre-existing failures, the results indicating that relatively few landslides are currently active. Whilst dormant landslides are in a relatively stable state under today's conditions, the disturbance of these failures by local human activity (e.g. cutting into the toe) could result in its reactivation. The likelihood of this happening would be increased where the landslide was particularly degraded and not readily recognizable from surface morphology alone. Therefore geotechnical investigations at sites should not only determine the existing stability of the hillside but should be directed at identifying whether a site has been affected by previous instability.

Acknowledgements. The interpretation of the results presented here is based on work sponsored by the Department of the Environment and the Welsh Office and undertaken by Sir William Halcrow and Partners from 1985 to 1990. The contribution of the above is gratefully acknowledged, in particular the survey teams ably led by A. Thomas.

References

BENTLEY, S. P., COOPER, L. M. & GEDDES, J. D. 1980. Slope instability in the Upper Coal Measures: South Wales. *In: Cliff and Slope Stability, South Wales.* University College Cardiff, 111–134.

BISHOP, A. W., HUTCHINSON, J. N., PENMAN, A. D. M. & EVANS, H. E. 1969. *Geotechnical investigation into the cause and circumstances of the disaster of 21 October 1966.* In a selection of technical reports submitted to the Aberfan Tribunal, HMSO, London.

BLONG, R. J. 1973. A numerical classification of selected landslides of the debris-avalanche flow type. *Engineering Geology*, **7**, 99–114.

BRUNSDEN, D. 1973. The application of system theory to the study of mass-movement. *Geol. Applicata e Idrogeol.*, **3**, 185–207.

CONWAY, B. W. 1974. *The Black Van Landslip, Charmouth, Dorset*. Report of the Institute of Geological Science 74/3.

——, FORSTER, A., NORTHMORE, K. J. & BARCLAY, W. J. 1980. *South Wales Coalfield Landslip Survey*. Institute of Geological Science Special Survey Division, Engineering Geology Unit, Report EG 8041.

CROZIER, M. J. 1973. Techniques for the morphometric analysis of landslips. *Zeitschrift für Geomorphologie*, **17**(1), 78–101.

DAUGHTON, G., NOAKES, J. S. & SIDDLE, H. J. 1977. Some hydrologeological aspects of hillsides in South Wales. *In*: *Proceedings of the Conference on Rock Engineering*. University of Newcastle-upon-Tyne, British Geotechnical Society, 423–439.

DENNESS, B. 1972. *The reservoir principle of mass movement*. Report of the Institute of Geological Science 72/7.

HALCROW, SIR WILLIAM & PARTNERS 1986. *Rhondda landslip potential assessment: interim report*. Report for the Department of the Environment and Welsh Office.

——1990. *Assessment of landslip potential: South Wales*. Final report for the Department of the Environment and Welsh Office.

JONES, F. O., EMBODY, D. R. & PETERSON, W. L. 1961. *Landslides Along the Columbia River Valley North-eastern Washington*. US Geological Survey Professional Paper 367.

KNOX, G. 1927. Landslides in the South Wales Valleys. *Proc. Trans. S. Wales, Inst. Engs.*, **43**, 161–247, 257–290.

PALMQUIST, R. C. & BIBLE, G. 1980. Conceptual modelling of landslide distribution in time and space. *Bulletin of the International Association of Engineering Geologists*, **21**, Krefeld, 178–186.

SKEMPTON, A. W. 1953. Soils mechanics in relation to geology. *Proceedings of the Yorkshire Geological Society*, **29**(1),

—— & HUTCHISON, J. N. 1969. Stability of natural slopes and embankment foundations. State-of-the-Art Report, 7th International Conference on Soil Mechanics and Foundation Engineering, Mexico, State-of-the-Art, Vol. 2, 291–340. 33–62.

TERZAGHI, K. 1950. Mechanism of landslides in application of geology to engineering practice. *Geological Society of America*, Berkey Volume, 83–123.

VARNES, D. J. 1978. Slope movement types and processes. *In*: SCHUSTER, R. L. & KRIZEK, R. J. (eds) *Landslides: Analysis and Control*. Special Report 176. Transporta-tion Research Board, National Academic of Sciences, Washington, DC, 11–33.

WARD, W. H. 1945. The stability of natural slopes. *Geographical Journal*, **105**, 170–191.

WIECZOREK, G. F. 1984. Preparing a detailed landslide inventory map for hazard evaluation and reduction. *Bulletin of the Association of Engineering Geologists*, **XXI**(3), 337–342.

WOODLAND, A. W. 1985. The South Wales Coalfield – an introduction to its geology and landslips. *In*: *Proceedings of a Symposium on Landslides in the South Wales Coalfield*. Polytechnic of Wales, Pontypridd, 9–18.

—— & EVANS, W. B. 1964. *The Geology of the South Wales Coalfield, Part IV, The Country around Pontypridd and Maesteg*. Memoir of the Geological Society, 3rd edition. HMSO, London.

SECTION 4

HAZARD MAPPING

Graphical methods for hazard mapping and evaluation

G. J. Smith[1] & M. S. Rosenbaum[2]

[1] Wardell Armstrong, Lancaster Building, High Street, Newcastle, Staffordshire ST5 1PQ, UK
[2] Department of Civil & Structural Engineering, The Nottingham Trent University, Burton Street, Nottingham NG1 4BU, UK

Abstract. Graphical techniques are presented as a practical approach to hazard assessment. These were initially developed for the evaluation of potential collapse associated with old chalk mine workings. However, the underlying methodology is free-standing and now capable of being tailored to other ground engineering situations. This places a strong emphasis on the inclusion of abstract factors, otherwise excluded from mechanistic forms of assessment by traditional methods. The ways in which such varied information can be analysed and then synthesized in an assessment are discussed. The graphical approach lends itself to both overview and audit, and is therefore relevant for project evaluation as well as geohazard assessment.

Introduction

The theme of underground chalk mine instability is used throughout this paper as a vehicle for the development of a new approach to hazard assessment incorporating graphical presentation. Existing methods of appraisal tend to be based on factor weightings, algebraic expressions and logic-based constructs (Smith & Rosenbaum 1994). However, there is currently no method which can synthesize all the various factors while retaining and conveying the technical context.

Factors and fuzziness

The disparate nature of factors contributing to small-scale instability of old chalk mines is illustrated in Fig. 1.

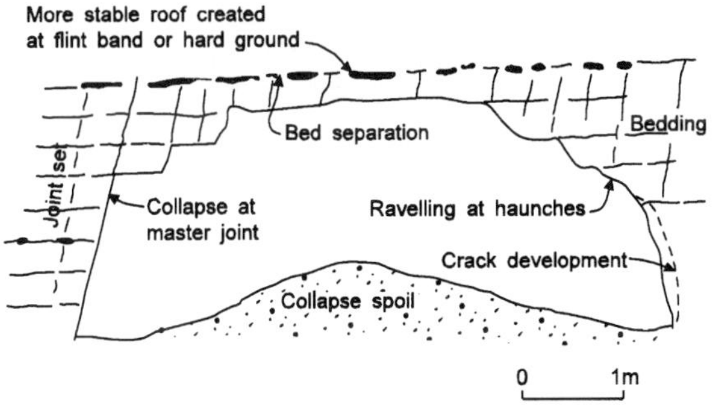

Fig. 1. Earlham Road chalk mine: contributory factors leading to instability of larger tunnel sections.

This shows how graphical display can effectively convey a variety of spatial information which would tend to be obscured by, for example, a numerical rating value derived as the main product from rock mass classification schemes, such as the Q and RMR systems.

For situations controlled by multiple parameters, the comparative representation of factors forms an important basis for assessment. Regardless of the eventual complexity of the hazard assessment, the elements of such a basic descriptive scheme need to be represented in such a way as to preserve their spatial context. This should emphasize and clarify the most important factors in the evaluation. The conventional style of graphical display using a geological section to describe the ground conditions around a mine can be taken as the basis for such an approach. This is extended in Fig. 2 to incorporate both

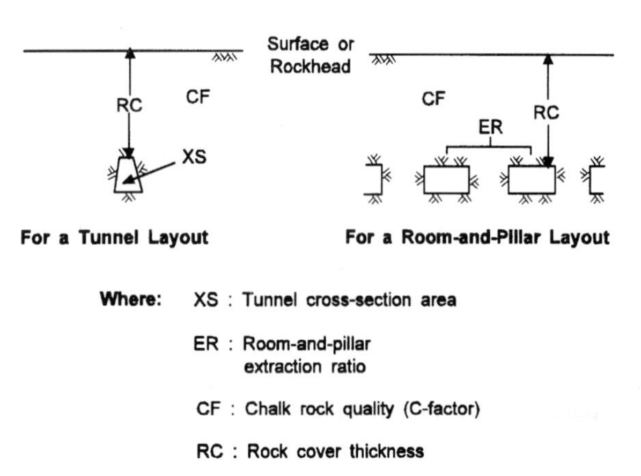

Fig. 2. Elements of a basic descriptive scheme.

SMITH, G. J. & ROSENBAUM, M. S. 1998. Graphical methods for hazard mapping and evaluation. *In*: MAUND, J. G. & EDDLESTON, M. (eds) *Geohazards in Engineering Geology*. Geological Society, London, Engineering Geology Special Publications, **15**, 215–220.

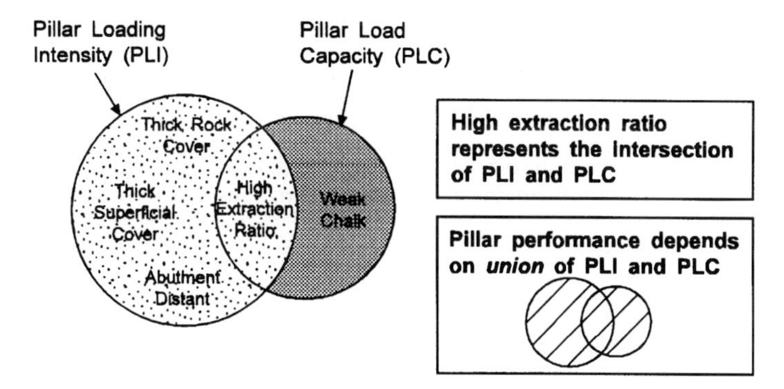

Fig. 3. Example of a Venn diagram used for displaying performance factors relevant to a room-and-pillar mine layout.

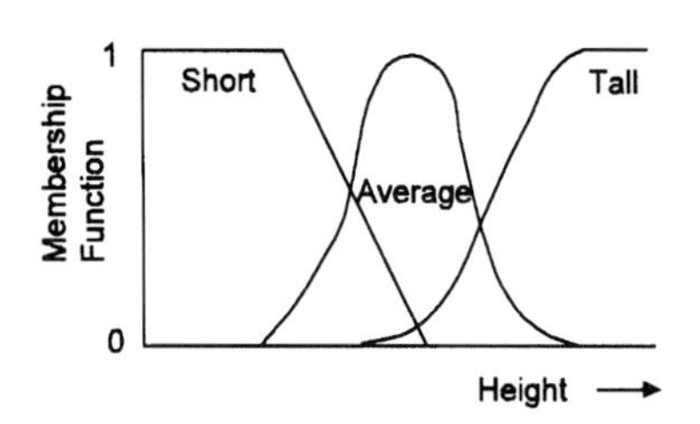

Fig. 4. Examples of types of membership functions for fuzzy sets for the parameter 'height'.

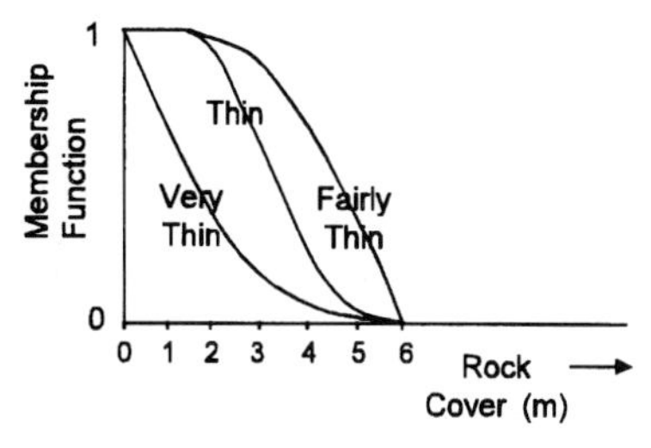

Fig. 5. Example of a fuzzy mine parameter modified to take account of finer semantic distinctions.

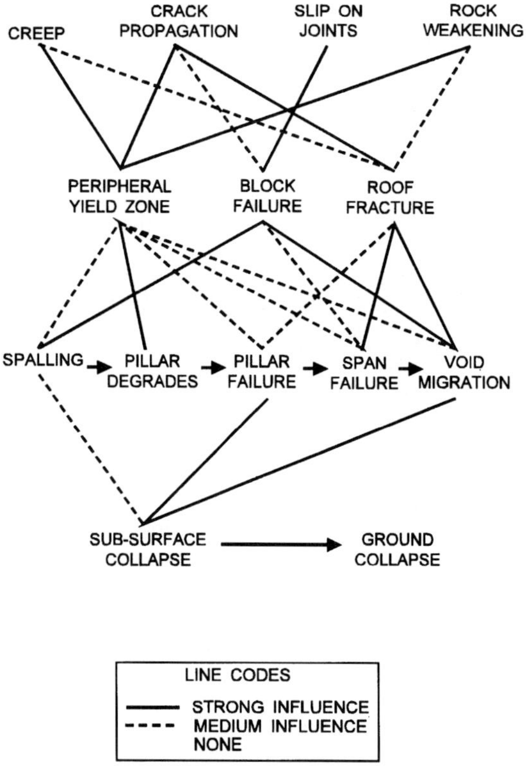

Fig. 6. Example of a mechanism influence network.

geometrical and qualitative factors which can be quantified as information becomes available.

A particular set of factors might well be expected to play a key role in influencing the principal mechanisms of ground performance, and so it is important to identify which these might be (Hudson *et al.* 1991). Such factors can be represented using Venn diagrams, as illustrated in Fig. 3 for a room-and-pillar mine layout. Here the factor 'high extraction ratio' has been highlighted as being the most important within the overall domain for the pillar performance. This domain encompasses both pillar loading intensity and pillar load capacity. In this case each factor has been qualified but the scale for such descriptions is not rigid, and indeed the thresholds may not be sharply defined. It is therefore attractive to now introduce the idea of fuzzy sets, which is able to incorporate an element of possibility with the graphical approach. An associated problem will be the definition and calibration of the semantic descriptions employed, such as small/large and good/bad. This has led to the development by others (Dubois & Prade 1980) of a concept concerning the degree of membership of one or

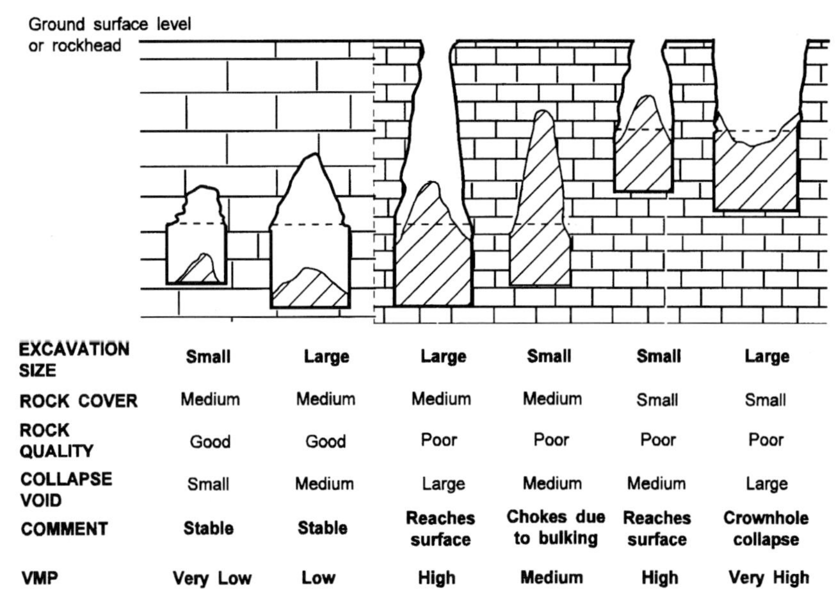

Fig. 7. Illustration of aspects of Void Migration Potential (VMP).

more particular fuzzy sets. This is expressed by values which lie between 1 (full membership) and 0 (not a member) which apply to each membership function. Figure 4 shows this condition for three such functions, concerning height, so conveying the qualities of 'short', 'average' and 'tall'.

Quantification in terms of fuzzy set membership can then be modified to take account of finer semantic distinctions such as 'very' and 'fairly', as shown in Fig. 5. The effect of such modifiers is to decrease ('very thin') or increase ('fairly thin') the value of the membership function concerning the unqualified descriptor 'thin'.

Networks and assessment flowcharts

For situations characterized by physical processes, involving multiple criteria as factors, it is also necessary to consider the potential interrelationships between the criteria. Figure 6 illustrates how a network of mechanism influences can be used to represent these, within which each link represents connections based on cause and effect.

Many types of ground behaviour are feasible and each needs to be considered. Void migration above chalk mines can be taken as one such example: a process which is controlled by a number of different factors (Smith & Rosenbaum 1993*a*). The way in which the factors control void migration can be represented schematically, as shown in Fig. 7. Such an approach ideally requires the use of quantitative terms but the semantic scheme is nevertheless capable of clarifying the relative importance of each factor in the process being considered.

In order to establish the data necessary for undertaking the assessment, the basic graphical representation can be supplemented with qualitative information and by the results of direct measurement. An example of how this approach can be portrayed for a chalk mine is shown in Fig. 8 where the extraction ration reached 59% and the likelihood of void migration breaching the cover has to be assessed. Whether for the case of a specific mine or for a more general ground engineering situation, it is similarly possible to produce rating values for each of the feasible deleterious processes by employing a systematic approach represented as a graphical flowchart, described in detail elsewhere (Smith & Rosenbaum 1994). This embodies the nature of the dominant interactions which could potentially lead to surface collapse, an example of which is shown in Fig. 9. Within each box of the flowchart, the right-hand side rating

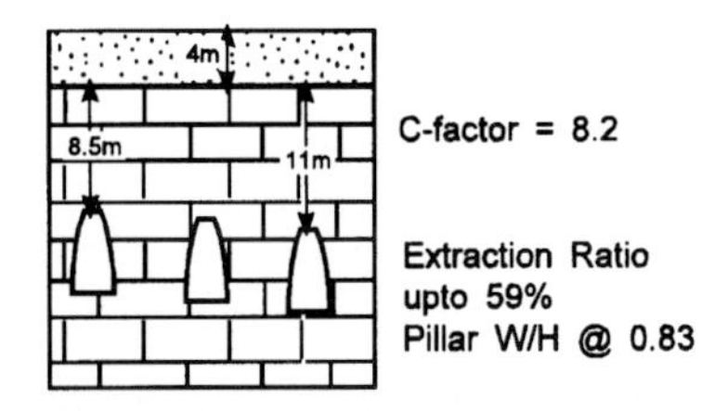

Fig. 8. Example of a graphical chalk mine summary.

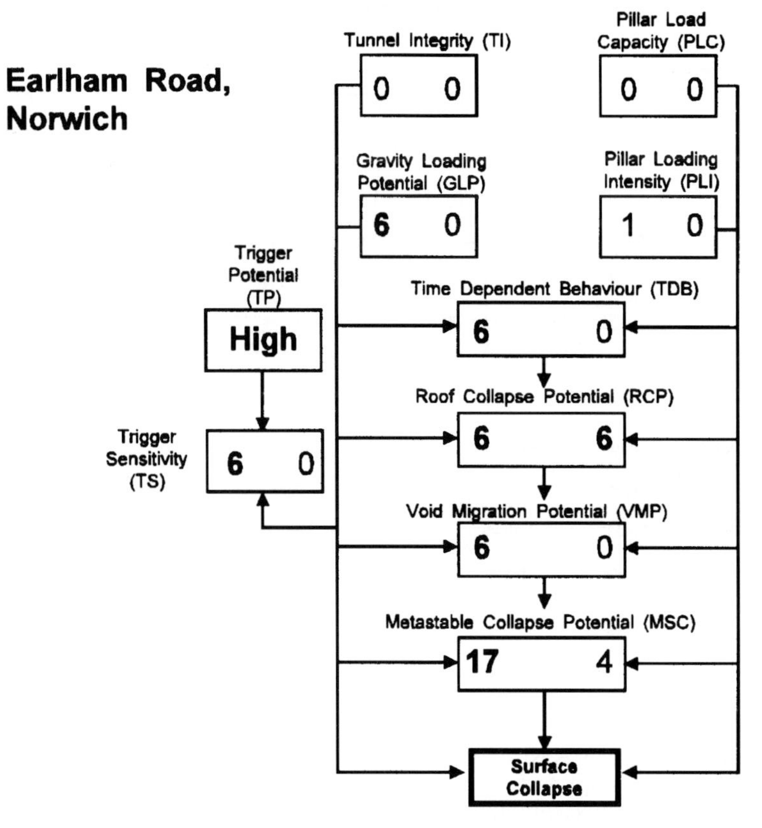

Fig. 9. Flowchart representing Performance Factors and their rating values for Earlham Road chalk mine, Norwich; applicable for stability assessment.

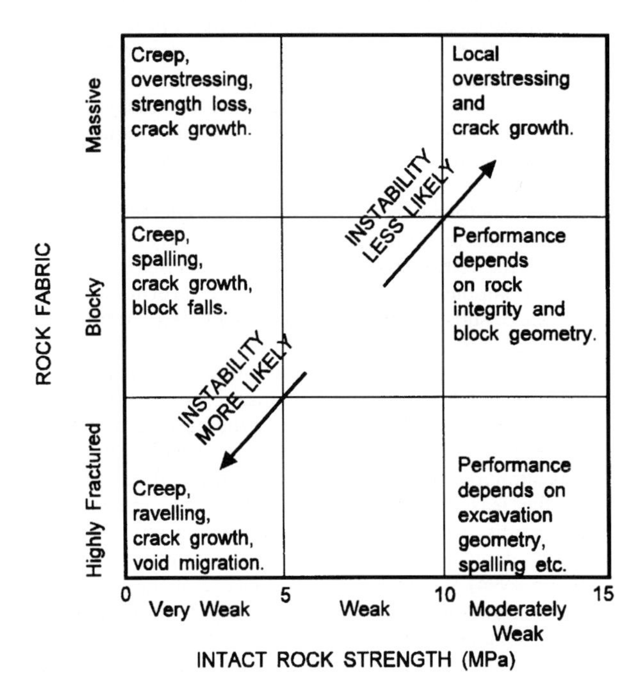

Fig. 10. Chart portraying notional extremes of performance related to rock fabric and strength.

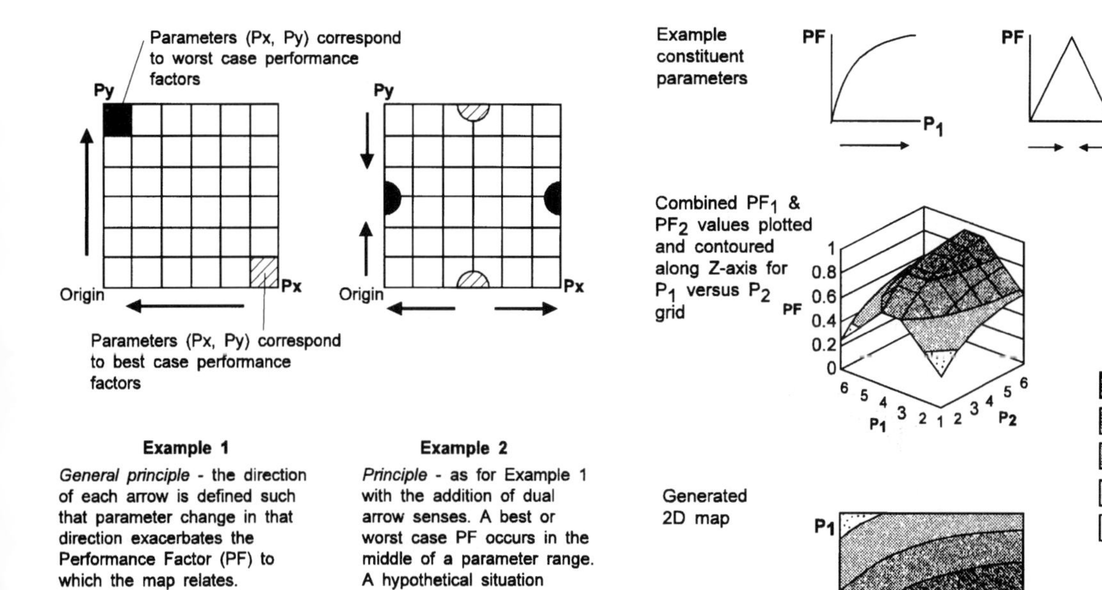

Fig. 11. Principles portrayed as 2D Performance Factors (PFs) and as a parameter map.

Fig. 13. Example of 2D parameter maps generated by specified Performance Factor (PF) functions.

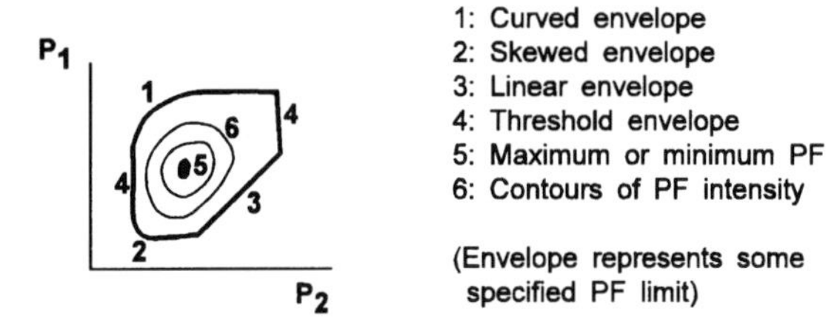

Fig. 12. Specific considerations for 2D parameter maps – in this case a Performance Factor (PF) is influenced by parameters P_1 and P_2.

Performance Factor (PF) = Roof collapse potential

Worst case Best case

Parameters -
TX = Tunnel cross-section
CQ = Chalk Quality
RC = Rock Cover

PF level -
H = High
M = Medium
L = Low

Fig. 14. Component 2D parameter maps to describe a 3D example.

values represent beneficial indicators whereas the left-hand side values represent prejudicial indicators.

Mapping of parameter influence

Just two parameters can combine to produce a marked influence on rock mass performance. The degree to which they can achieve this is shown in Fig. 10 as a chart portraying the controls on specific classes of behaviour, together with their interrelationships. The stability implications of any particular situation can then be readily inferred. Such a chart-based graphical approach may be extended into multiple dimensions and may be readily produced for a study area, portrayable in conventional map form or as a raster image within a GIS to describe the hazard (Smith & Rosenbaum 1993*b*).

The chart portrayed in Fig. 10 is just one specific example of its type. However, the principle may be made more generally applicable for any pair of performance factors by generating a parameter map, as shown in Fig. 11. This gives the potential of combining the (two-dimensional) parameter map with the concept of fuzzy membership introduced earlier and can be portrayed as a suite of contours representing performance factor intensity. Figure 12 shows how such contours can be used to define the parameters of boundary limits and thresholds.

A performance factor rating can then be obtained by reading of the relative influence of each parameter from the map. The combined effect, here produced simply by addition with equal weighting, can be portrayed as a surface using a three-dimensional perspective diagram, as shown in Fig. 13.

In practice, roof collapse is found to be a necessary precursor to the process of void migration. An assessment of the hazard arising from roof collapse is therefore but one example of the many mechanisms which must be addressed when assessing the general problem of void migration above old mines. Results of the analysis for each such mechanism can then be collectively portrayed in chart graphical form, as illustrated in Fig. 14. In this way, the worst-case scenarios can be readily identified and those factors of greatest importance regarding mine instability evaluated.

References

DUBOIS, D. & PRADE, H. 1980. *Fuzzy Sets and Systems – Theory and Applications*. Academic, New York.

HUDSON, J. A., ARNOLD, P. N. & TAMAI, A. 1991. Rock engineering mechanisms information technology REMIT: Part 1 – The basic method; Part 2 – Illustrative case examples. *In*: *Proceedings of the 7th International Congress on Rock Mechanics, Aachen*, Vol. 2. Balkema, Rotterdam, 1113–1119.

SMITH, G. J. & ROSENBAUM, M. S. 1993*a*. Abandoned shallow mineworkings in Chalk: a review of the geological aspects leading to their destabilisation. *Bulletin of the International Association of Engineering Geology*, **48**, 101–108.

—— & ——1993*b*. Abandoned mineworkings in chalk: approaches for appraisal and evaluation. *Quarterly Journal of Engineering Geology*, **26**, 281–291.

—— & ——1994. Arithmetic and logic – at the boundary between geological data and engineering judgement. *In*: *Proceedings of the 7th International Congress of the International Association of Engineering Geology*, Lisbon, Portugal, 5–9 September 1994. Balkema, Rotterdam, **6**, 4517–4526.

The role of engineering geology in the hazard zonation of a Malaysian highway

J. R. Cook,[1] A. McGown,[1] G. Hurley[2] & Lee Eng Choy[3]

[1] Dept. of Civil Engineering, University of Strathclyde, Glasgow, G4 0NG, UK
[2] Soil andRock Engineering, Wisma Sin Heap Lee, SO 400, Kuala Lumpar, Malaysia
[3] DPI Konsult, JLM 557/2 P.J. Selangor, Malaysia

Abstract. The University of Strathclyde and the Institut Kerja Raya Malaysia (Public Works Institute) have undertaken a joint research project into the investigation, description, classification and geotechnical analysis of tropically weathered *in situ* materials (TWIMs) in Malaysia in general and on a major highway in particular. The research project has developed a general approach to TWIM profile characterization and classification which allows specific systems to be developed to deal with particular project requirements.

A key part of the research is to provide practical engineering geological and geotechnical advice to a hazard identification and risk assessment project on the 113 km long East–West Highway in Kelantan, northern Peninsular Malaysia. This highway has an ongoing history of persistant slope failure and contains a large number of significant earthworks in generlly mountainous terrain underlain by a ange of tropically weathered bedrocks including sedimentary, meta-sedimentary and igneous materials. The 3-year long-term project produced hazard maps for the earthworks along the highway. These maps provided a more scientific basis for cost-effective planning and design of maintenance and preventative remedial works along the highway.

This paper outlines the main elements of the project and indicates the key influence that the engineering geological character of the tropically weahered soil-rock profiles has had on the assessment of slope hazard.

Introduction

General background

The Institut Kerja Raya Malaysia (Public Works Research Institute) and the University of Strathclyde have undertaken a joint research study into the investigation, description, classification and geotechnical analysis of tropically weathered *in situ* materials (TWIMs) in Malaysia. A key part of the study was to provide practical engineering geological and geotechnical advice to a concurrent hazard identification and risk assessment project on a highway in northern Peninsular Malaysia.

The East–West Highway (EWH) is approximately 113 km long and runs through mountainous terrain between Gerik, in Perak, to Jeli in Kelantan (Fig. 1). The highway is the only link between the eastern and western states in the northern section of the country and is, therefore, an important social and economic transportation link across Peninsular Malaysia. The highway was completed in 1982 and was constructed without the benefit of a detailed geotechnical investigation and, for strategic reasons, largely without structures. The alignment runs for considerable lengths along sidelong ground and involves a large number of earthworks, a large proportion of which are of a cut–fill nature. Between 1982 and 1994 approximately £125 million has been spent on slope rehabilitation and maintenance works to keep the highway open. A further £24 million has recently been allocated for 25 slopes which are in need of repair.

A 3 year long-term project was implemented by the Institut Kerja Raya Malaysia (IKRAM) to produce hazard and risk maps for the highway. These maps will provide a more scientific basis for cost-effective planning and design of maintenance and preventative remedial works along the highway. This study began in October 1993 and continued until the end of September 1996.

Project aims

The principal aims of the EWH hazard and risk project were as follows:

- to develop hazard and risk maps which will enable improved preventive remedial works planning;
- to provide a technical database for short-term remedial design works;
- to develop a single long-term management database containing site investigation information, slope data,

COOK, J. R., MCGOWN, A., HURLEY, G. & LEE ENG CHOY 1998. The role of engineering geology in the hazard zonation of a Malaysian highway. *In*: MAUND, J. G. & EDDLESTON, M. (eds) *Geohazards in Engineering Geology*. Geological Society, London, Engineering Geology Special Publications, **15**, 221–229.

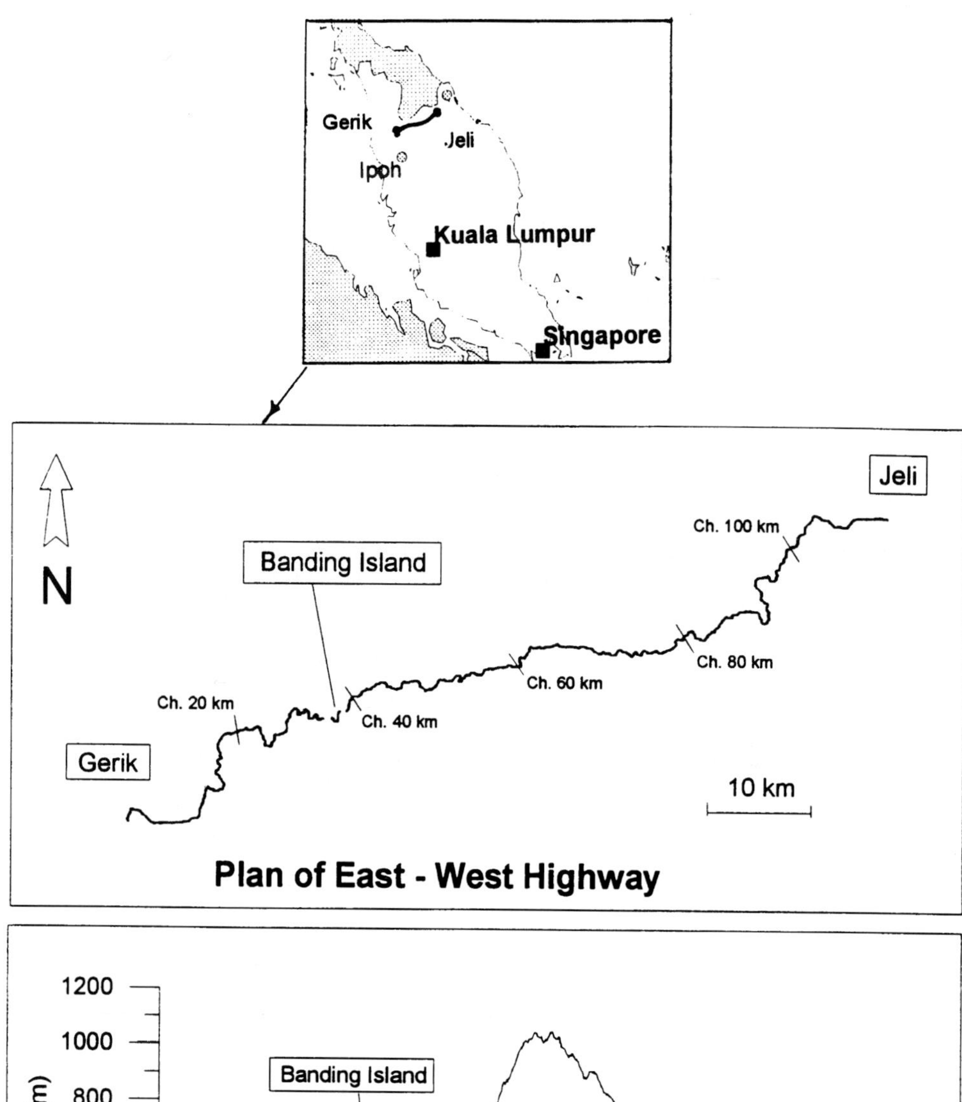

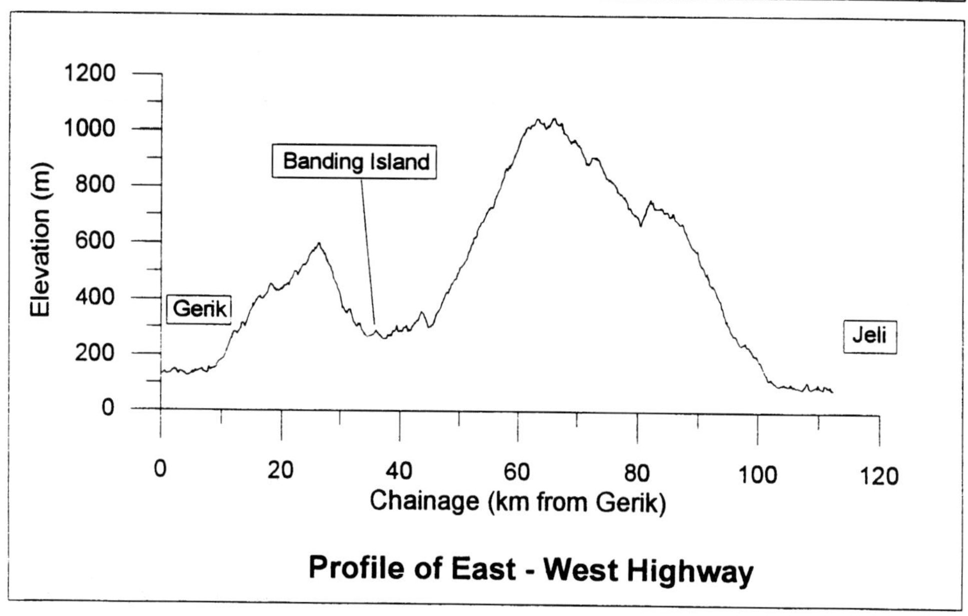

Fig. 1. East–West highway location.

construction reports, geological data, and records of maintenance works;
- to develop a systematic approach which can be applied to other regional planning projects in Malaysia.

The overall aim of the engineering geological research study was to set up a systematic approach to the investigation of TWIMs based on a clearly established knowledge of their engineering geological nature and their geotechnical behaviour both *in situ* and in the laboratory. This work is set out in a series of 'Geoguides' published by IKRAM, covering site investigation, laboratory investigation, and geotechnical characterization as they relate to the occurrence and general nature of TWIMs.

Hazard and risk assessment of the East–West Highway

General

In broad terms the general methodology for the production of the hazard and risk maps has been to statistically analyse large sets of field slope data in association with the engineering performance history of the road. From this it has been possible to identify those factors most relevant to slope failure and hence rate each slope with regard to hazard and risk.

For reasons of practical engineering application, the hazard analysis programme carried out along the EWH has been in terms of the resolution of individual slopes rather then the definition of hazard areas. For the purposes of this particular project the following definitions were adopted:

- *Hazard maps*: delineate where there is a finite *probability* of slope instability during the lifetime of the highway.
- *Risk maps*: quantify the *vulnerability* of the hazard in terms of potential damage to the road and the affect on traffic access.

The majority of the main project work has been undertaken by a team composed of local consultants (Jurutera Perunding ZAABA) in conjunction with an overseas specialist geotechnical organization (Soil and Rock Engineering). This team was supported by the University of Bristol in terms of the development of the Geographical Information System (GIS) and computer modelling, and the University of Strathclyde with respect to engineering geology and the geotechnical characterization of tropically weathered materials.

The statistical approach

Data from each of the 1125 slopes along the East–West Highway have been gathered by a multidisciplinary field team comprising geotechnical engineers, engineering geologists and geomorphologists using specially developed proformas which enabled information to collected in a computer-friendly coded format. The principal data sets are listed in Table 1. These site recordings form part of a database that has been linked to the SPANS GIS.

Table 1. *Principal slope inventory data sets*

Type	General geology	Culvert condition
Location	Rock profile type	Natural drainage type
	Colluvium	Natural drainage size
Slope height	rock exposure area	Natural flows
Slope shape	Rock type	Erosion protection
Slope angle	Discontinuity pattern	
Slope strike	Rock condition	Types of erosion
No. of benches	Soil exposure area	Erosion severity
Bench geometry	Soil type	Erosion gulley geometry
Crest length	Soil origin	
	Corestone %	Instrumentation
Distance to ridge/gulley	Surface crusting	
Topographic setting		Condition of earthwork
Catchment area	Drainage types present	Failure modes
	Drain condition	% Failure
Vegetation cover	Blockage %	Distance road-scarp
Artificial cover	Extent of drainage	Backscarp length
Logging activity	Weepholes	Failure area
Distance to tree line	Shute drain condition	
Pavement condition	Bench drain condition	
Extent of cracking	Batter drain condition	
Causes of cracking		

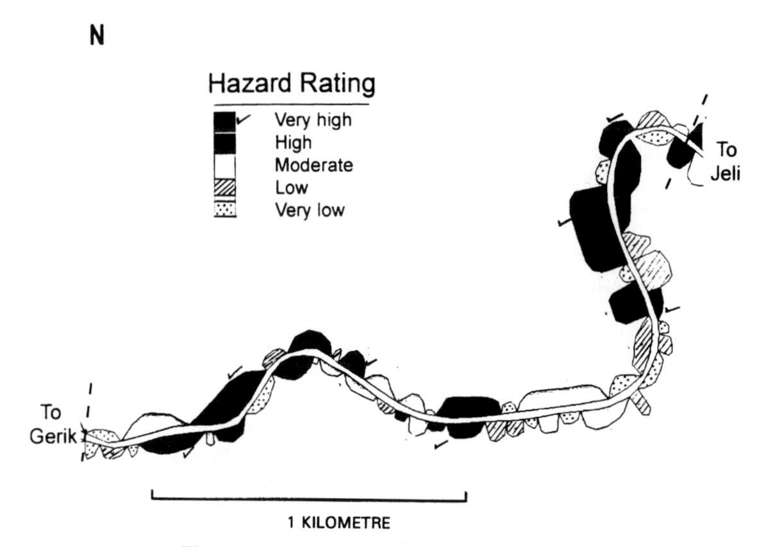

Fig. 2. Typical section of preliminary hazard map.

Table 2. *Engineering geological input to main project*

Main EWH project activity	*Engineering geological activity*
Desk study review	Review of existing engineering geological knowledge of EWH and tropically weathered soil-rock profiles
EWH database formation	File structure for compatible geotechnical database
Development of fieldwork procedures	Engineering geological proforma: location; exposure; material and correlation with slope forms
Main slope inventory fieldwork	Geological mapping; geology and terrain units
Discriminant analysis (DA) of collected slope inventory data	Assembly of geological units on the basis of perceived character and performance into statistical groups
Factor overlay analysis (FOA) using results of DA	Identification of behaviour patterns for typical profiles and the relative effects of weathering
Hazard map production and verification	Engineering geology as part of an overall engineering judgement filter to the statistical outputs
Development of risk ratings	Identification of the geological and terrain relationships having an influence on the failure risk
Analysis of high-risk areas	Input of geotechnical characteristics from detailed soil-rock TWIM profile studies

Statistical analyses have been carried out using the SSPX software on 25 slope variables collected by the field team. These analyses, together with applied engineering judgement, have been used to identify the most significant parameters that contribute to instability. The significant factors were then given numerical values and combined through a factor overlay analysis to give *hazard values*. Finally, these values are converted to *hazard ratings*. The ratings vary from 'very low hazard' to 'very high hazard'. Figure 2 presents a typical hazard map for a section of the highway.

The conversion of hazard to risk involved the identification of significant parameters, the assignment of a weighting to each of them, and the assessment of risk rating for each slope. Identified parameters were as follows:

- likely maximum size of a failure
- how much the potential will impinge on the road
- potential ability of a failure to close the road
- ease of construction of a failure by-pass
- cost of reconstruction

The engineering geological input

Table 2 outlines the principal geological and engineering geological inputs into the analysis of hazard and risk. In general terms the engineering geological fieldwork inputs may be summarized in terms of scale, i.e. at a geological/terrain unit level, at a profile/mass level and at a material level.

Initial inputs were in the form of background information and in the planning and development of field procedures. Of particular importance to the latter was an appreciation of the most significant engineering geological parameters that could be rapidly collected by the slope inventory field teams without operator bias, given the general bedrock types and the tropical weathering environment.

Table 3. *Summary of geological units along the EWH*

Unit	*Geological Formation*	*Description*
1	Bd	B: Baling Group
2	Bb	a: predominantly arenaceous
3	Bd	b: pyroclastic (predominantly fine)
4	Ba	d: predominantly argillaceous
5	Bd	
6	Ba	ST: Sungei Tiang Schists
7	Bd	a: amphibolite schist
8	Ba	b: quartz mica schist
9	Bd	
10	Bb	SM: Sungei Mangga Beds
11	Bd	a: argillaceous
12	MRGf	b: arenaceous
13	MRGp	MRG: Main Range Granite
14	STb	f: fine/foliated
15	SMb	p: medium coarse/porphyritic
16	SMa	
17	PG	PG: Porphyritic Pergau Granite
18	STb	
19	JG	JG: Jeli Granite/Granodiorite

The recognition of geological and terrain units likely to behave as statistically valid groupings was identified as being of fundamental importance. Surface geological mapping allowed the division of the route into a series of 19 units based on lithological character. Table 3 lists these geological units as they relate to the regional geological formations. Analysis of the initial geological units revealed that, because of their variable size and the amounts of earthwork within them, amalgamation of the lithologically based units was required for reasons of statistical validity. Table 4 lists the revised statistical geology groups.

Although arguable on pure geological grounds, this amalgamation was seen as fully justifiable in terms of the stability project, given that the engineering geological research project would be able to input more detailed information when required.

Preliminary statistical analysis indicated that variables significant to slope instability are different for the different geological groups. The results also indicated that the most discriminating variable for rocks cuts generally related to slope geometry and material and mass geotechnical character. Figure 3 illustrates the use of the statistical geological groups in analysing failure occurrences.

The results of the research into the geotechnical characterization of representative soil-rock profiles and the classification of their weathering patterns was seen as having an important role in the practical evaluation of the statistical data.

The engineering geological characterization of the East--West Highway

General approach

The aim of the engineering geological programme, for both research and practical purposes, was to firstly,

Table 4. *Statistical geological groups*

Statistical group	*Description*
BaF1	Predominantly interbedded shales and siltstones with some pyroclastic argillaceous material and occasional fine sandstones.
BaM	Mixed group of interbedded argillaceous to arenaceous sedimentary rocks. Locally lightly metamorphosed.
BaF2	Group of argillaceous metasediments from phyllite to schist in character. Sub-group of pyroclastic chlorite schists.
MetS1	Predominantly schistose rocks with gradations to phyllites and locally meta-sediments
MG	Fine to coarse granite. Subdivision fine/coarse.
MetS2	Predominantly schistose rocks; locally phyllites and meta-sediments.
PG	Predominantly coarse porphyritic granite.
MetS3	Predominantly schistose rocks; locally phyllites.
JG	Jeli complex of schists and f.m. granite/granodiorite intrusion.

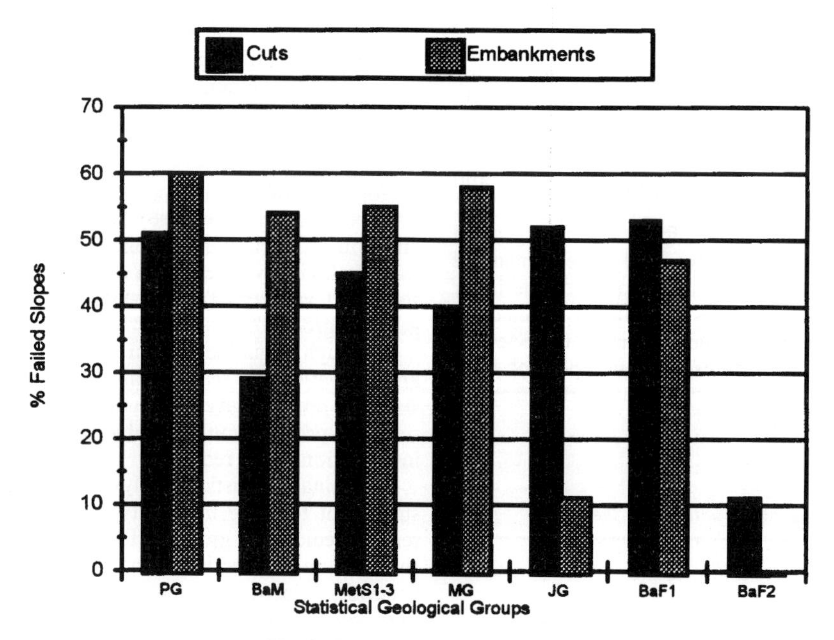

Fig. 3. Failure versus geology group.

define geology and terrain units of similar nature along the EWH and then to characterize typical soil-rock profiles from these units bearing mind their tropically weathered character. The steps in the engineering geological programme are summarized as follows:

1. identification of principal geology divisions
2. examine and define typical soil-rock masses
3. division of road into terrain and geology units
4. sampling and testing trials
5. selection of suitable representative profiles
6. detailed description, sampling and testing of profiles
7. suitable laboratory testing programme

The engineering geological fieldwork programme on the EWH differed from the main slope inventory in that rather than obtaining readily available data from each slope it sought to gather more detailed information from a strictly limited number of representative soil-rock profiles. It was felt that using this approach in conjunction with the EWH database would provide a powerful semi-empirical tool for hazard definition and the consequent risk assessment.

Geological setting

The EWH runs through a variable terrain underlain by bedrock ranging from granite to metamorphic and sedimentary materials, all of which have been subjected to the tropical weathering process. The landform varies from low undulating hills to steep-sided high hills or mountains. The most severe terrain occurring within the central belt where the road alignment runs across natural slopes of 30°–35° with a topographic variation of up to 500 m.

Four principal granites occur along the route: three are part of the Main Range Granite and the fourth, the Jeli Granitoid, which has a granodioritic composition, may be associated with the Stong Metamorphic complex (Gobbett & Hutchison 1969; Rafek *et al.* 1989).

The sedimentary rocks, which dominate the eastern portion of the road as far as Banding, occur as variably interbedded sequences of fine sandstones, siltstones and shales with local occurrences of tuffaceous material. The arenaceous rocks are dominantly quartzitic whilst the finer materials, in addition to clay minerals illite and mica, can contain significant amounts of carbonaceous material. These rocks, which can be variably and lightly metamorphosed, are considered to belong to the Baling Group of the Lower Palaeozoic.

Metamorphic sequences become more dominant beyond Banding, with the occurrence of phyllites, quartzites and schists, variable between chlorite mica, mica and quartz mica in composition. Up to the occurrence of the first granite, these rocks have been mapped as metamorphosed sequences of the Baling Group. West of the Baling Group the metamorphic rocks are mapped as members of the Sungei Tiang and Sungei Mangga Schists. The nature of these foliated rocks is variable and, particularly in the western sections of the road, are intruded by dioritic material associated with the main granites.

The characterization of tropically weathered *in situ* materials

Existing standard approaches to investigation, particularly with respect to sampling and testing, are, in many tropical material environments, incapable of adequately dealing with TWIM problems, either through the inadequacy of procedures in dealing with fabric sensitive materials or through an inability to represent a complex and non-homogenous soil–rock mass. It is commonly suggested, for example by Brand (1985) and De Mello (1972), that geotechnical design in TWIM environments should be by modified precedent or semi-empirical methods rather than by traditional classical methods. A key element in the modified precedent approach is the accurate collection of relatively easily obtainable and inexpensive site information which may be correlated with similar sites or otherwise be adapted for design purposes. There was therefore in the development of the EWH characterization procedure an emphasis on logical and accurate data collection and description techniques.

The characterization methodology proposed for TWIMs seeks to integrate general investigation principles into a project-related programme utilizing, where relevant, those ground investigation procedures most suited to their nature. This approach by itself is not specific to TWIMs, but the particular applicability is in the use of relevant data gathering and data interpretative techniques ranging from desk study data collection through to sampling and laboratory testing; many of which require adaptation from the norm for use in tropical environments. A fundamental aspect of the current research into TWIMs is the utilization of an integrated earth science approach incorporating aspects of geology, geomorphology, soil mechanics, mineralogy and rock mechanics (McGown & Cook 1994). Key factors emphazised as regards TWIMs within the overall investigation procedures are as follows:

- definition of terrain/geological environment (terrain classification)
- rigorous description technique allied to behaviour tests
- mineralogical and fabric analysis
- sampling techniques related to the sensitivity of material fabric
- emphasis on initial soil-rock laboratory index testing
- *in situ* behaviour assessment

Particular attention is placed on the material mineralogy, texture and fabric in addition to the mass structure. *In situ* geotechnical testing and observation of mass behaviour also have major roles to play in the establishing of the mass character (Ramlan *et al.* 1994).

Typical TWIMs characterization procedures

A number of geotechnical procedures have been integrated into the characterization process. For the purposes of this paper a number of these are summarized below.

Description. A field description system has been developed based on the systematic observation and description of masses, profiles and their constituent materials by means of standard proformas allied to a coded system of options easily transferable to commercial database software. One of the objectives of the material research has been to identify key index tests or observations, either individually or in groups, relevant to particular rock types and their weathering patterns. Standard approaches to the field collection of discontinuity data can be incorporated into the overall system (Brown 1988).

Classification. A general approach to TWIM profile classification has been adopted that allows specific systems to be developed to deal with individual project requirements. The use of overall standard classification systems is fraught with problems and it is unrealistic to expect them to function effectively in the highly variable parent materials within the TWIMs environments. Figure 4 outlines the basic classification process which can use appropriate general or genetic systems in an overall approach to TWIMs. Two basic criteria underlie the classification; firstly weathering, and secondly engineering behaviour. The development of specific classification systems may be based on a combination of engineering, genetic and behavioural grounds, the exact make-up being a practical function of project requirements.

Sampling. TWIM profiles are likely to contain materials that will be difficult to sample in an undisturbed and representative manner. Particular attention has been paid to the class of sample that can be achieved as compared to that required for testing. The use of simple 80 mm diameter plastic tubes cut into soil-rock exposures materials has proved effective.

Index testing. A TWIM testing programme needs to consider the use of both soil mechanics and rock mechanics type procedures, with particular attention being paid to areas of overlap between soil and rock testing to ensure effective correlation. Bearing in mind the

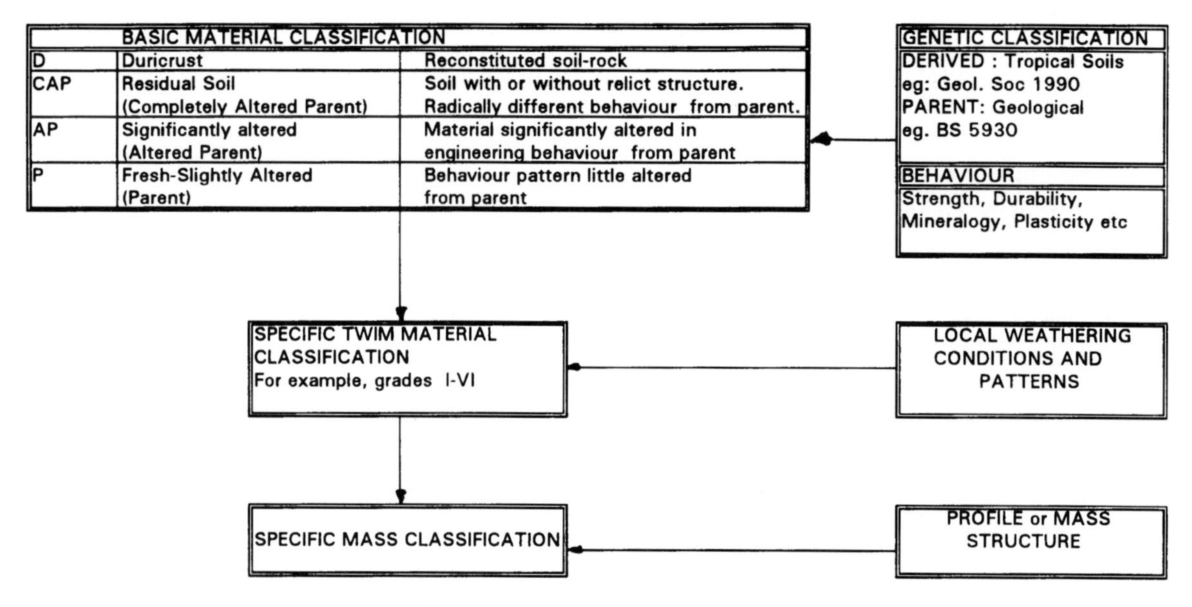

Fig. 4. Basic weathering classification.

difficulties of sample recovery, the laboratory programmes were be based around large numbers of relevant index tests allied to a limited number of high quality, more sophisticated tests and specialist mineralogical and fabric examination.,

Characterization applied to the EWH

The characterization of typical profiles along the EWH as well as being a basic objective of the engineering geological research study, also underpinned the assessment of *in situ* mass and material for hazard and risk identification. At a primary level the organization of the route into geological and landform units gave a sound and logical basis for the collation assessment of recovered data. The relationship of the mass and profile character to the highway was seen as having a fundamental influence on its engineering performance. This involved not only *in situ* geotechnical nature but the physical association of the highway to the topography and the underlying weathering pattern. Research into the nature, extent and variability of the weathering patterns for the various rock types was of direct practical use to the hazard and risk project.

Detailed material behaviour patterns gave support information as to engineering performance. The analysis of identified high-risk areas required the input of detailed geotechnical information both at a material and mass level. Research undertaken on the detailed characterization of the materials and masses for the representative profiles was able to supply such information.

Conclusions

The EWH hazard and risk assessment project has indicated the important role of engineering geology in such projects at several levels: in the development of terrain systems, in the classification of weathering patterns and in the characterization of materials. It has contributed significantly to evaluation of statistical data and hence to the production of hazard maps and to the assessment of risk. The assessment of weathering pattern, mass character and their relationship to the highway formed a significant part of the engineering judgement filter to the statistical outputs.

Acknowledgements. Grateful acknowledgement is made to Jabatan Kerja Raya (JKR) and the EWH Long Term Stability Team for their agreement to produce this paper. A particular debt is acknowledged to the IKRAM, SRE and ZAABA engineers and geologists for their dedicated efforts both in the field and in the analysis.

References

BRAND, E. W. 1985. Geotechnical engineering in tropical residual soils. *In*: *Proceedings of the 1st Conference on Geomechanics in Tropical Lateritic & Saprolitic Soils*, Brasilia, 3.

BROWN, E. T. (ed.) 1988. *Rock Characterisation, Testing and Monitoring Suggested Methods*. International Society of Rock Mechanics. Pergamon Press,

De Mello, V. F. B. 1972. Thoughts on soil engineering applicable to residual soils. *In*: *Proceedings of the 3rd SE Asian Conference*, SMFE, Hong Kong.

Gobbett, D. J. & Hutchinson, C. S. 1973. *Geology of the Malay Peninsula*. Wiley, New York,

McGown, A. & Cook, J. R. 1994. Origin and nature of Malaysian Weathered Rocks and Soils. Keynote Paper, Geotropika '94, Regional Conference in Geotechnical Engineering, Malacca, Malaysia.

Rafek, A. G., Komoo, I. & Tan, T. T. 1989. Influence of geological factors on slope stability along the East–West Highway, Malaysia. *In*: *Proceedings of the International Conference on Engineering in Tropical Terrains*, Malaysia, 79–93.

Ramlan, N., Cook, J. R. & Lee Eng Choy 1994. An approach to the characteristics of tropically weathered soil-rock profiles. Geotropika '94, Regional Conference in Geotechnical Engineering, Malacca, Malaysia.

Mitigating geohazards affecting mountain roads in northeast Somaliland

Paul Nathanail & Judith Nathanail

Centre for Research into the Built Environment, Nottingham Trent University, Nottingham NG1 4BU, UK (e-mail: crbe@crbe.co.uk)

Abstract. Roads linking the Gollis plateau with sea ports on the Gulf of Aden descend a steep escarpment. The roads are continually affected by erosion and degradation. These usually gentle processes constitute significant geohazards to the communities on the plateau which rely on the roads to export their agricultural produce as they attempt to rebuild the local economy following years of civil war. Engineering geological advice was provided to local communities involved in maintaining and repairing three roads, in areas occupied by different Somali clans. The absence of advice at an earlier stage had resulted in much wasted effort in inappropriate realignments and gradients.

By tailoring the advice to the materials, tools and skills available locally, one road could be re-opened and two could be significantly improved, reducing travel times for laden lorries by up to half.

Introduction

The Somaliland Republic, a breakaway state from the rest of Somalia, is unrecognized by the international community. Its boundaries coincide with the former British Somaliland Protectorate. Civil war culminated in the overthrow of the then ruler, Siad Barre in 1991 and gave rise to a state of national chaos. Central government has no finances and little control over the guerrilla forces that brought it to power. The national economy is in a state of ruin: the entire manufacturing base and almost all engineering plant having been destroyed or looted, the economy being sustained by donations sent by Somalis working abroad.

The country is dominated by hills capped by Mesozoic and Cenozoic marine sediments. The cores of these hills are Precambrian rocks of metamorphic and igneous origin which have long been reputed to bear precious and semi-precious minerals (Macfadyen 1933). The main population centres have grown on the 1500 m high plateau and are served by ports along the Gulf of Aden coast. A major fault scarp steps down from sedimentary formations at plateau level down to the Gulf of Aden.

Following the end of the civil war, aid agencies began the task of helping with the reconstruction of basic infrastructure (Action Aid Somaliland 1992*a*, *b*; PAI HIZ 1991). At the invitation of Action Aid Erigavo, the authors inspected three roads in the Sanaag Region that were either in the process of being repaired or were being considered for funding of repairs. The roads had suffered from the results of erosion, flooding, poor initial construction and lack of maintenance.

The Sanaag Region traditionally relied upon roads to transport produce for export to the sea ports of Xiis and Maydh. Somali society is very strongly clan-oriented. Deep-seated mistrust between clans led to the need to work simultaneously on three roads from the Erigavo plateau to the coast to enable each clan to control their own link with the ports (Action Aid Somaliland 1994).

The available resources were manual labour, with raw materials limited to nearby naturally available aggregate. The only cement factory in Somaliland had been destroyed during the civil war. Wire mesh for use as gabions would have quickly been stolen.

Engineering geological advice was given to village elders and non-government organization (NGO) workers who would be co-ordinating and funding the work. The advice focused on the sections of the roads which descended the slopes of the east African Rift valley where inadequate drainage, poor road alignment and fast vehicle speeds had combined to make the routes all but impassable to the slow, poorly maintained lorries that were being used to transport produce to the ports.

Sanaag region

The Sanaag region (Fig. 1), is one of the largest in Somaliland and comprises the plateau of the Gollis mountains (Fig. 2), the Gollis escarpment and the coastal lowlands (Fig. 3). The regional capital is Erigavo. The economy is dominated by livestock farming (camels, goats and sheep) with irrigated agriculture in valleys with permanent streams.

NATHANAIL, P. & NATHANAIL, J. 1998. Mitigating geohazards affecting mountain roads in northeast Somaliland. *In*: MAUND, J. G. & EDDLESTON, M. (eds) *Geohazards in Engineering Geology*. Geological Society, London, Engineering Geology Special Publications, **15**, 231–237.

Fig. 1. Location of Sanaag Region in northern Somalia.

Livestock and other local produce, such as frankincense from trees growing on limestone outcrops and boulders, are exported from the seaports of Maydh and Xiis.

Geology

Erigavo is built on the Lower to Middle Eocene Taleh Evaporites; gypsum is well-exposed in the rocky outcrops which occur in the middle of most of the town's roads. The Upper Cretaceous to Lower Eocene Auradu Limestone formation makes up the spectacular cliffs of the escarpment. The lower reaches of the escarpment are formed of Cretaceous limestones and marls. Precambrian metamorphic rocks form the low hills around Maydh and Xiis (Fig. 4). The rocks range from granitic gneisses to quartzites. Pleistocene to Recent sediments of fluvial and colluvial origin fill most of the valleys. Beach, wind-blown and marine sediments are found along the coast (Macfadyen 1933; Beydoun 1970, Merla *et al.* 1979).

The Gollis escarpment is characterized by deeply incised valleys and torrential flows which cause considerable erosion.

Tabah road, Erigavo to Maydh

The Tabah road links Erigavo on the plateau with the port of Maydh (PAI HIZ 1991) (Fig. 5). The road traverses the gypsiferous Taleh Evaporites around Erigavo; the Auradu Limestone Formation at the top of the escarpment; the Precambrian metamorphic formations in the foothills around Maydh (Fig. 5) and Recent

Fig. 2. Deeply incised plateau west of Erigavo.

Fig. 3. Coastal lowlands and sand dunes.

river and beach deposits. The following comments are based on a traverse from Maydh to Erigavo.

At Damas Copse, a new alignment of the road follows the south bank of the river. Differential erosion at a limestone–phyllite boundary created a rough trafficking surface while limestone bedding planes were attacked by weathering, creating steps in the road surface. In order to dissipate some of the energy of flood waters, placing rip rap upstream of the rough part of the road was recommended, with local grouting to smooth the stepped limestone surface.

At Horuba, a proposed 1 km realignment of the road to avoid the river bed was evaluated. The proposed route generally follows phyllite or phyllite-derived colluvium, which give a good, if easily erodable, trafficking surface. It was recommended that local outcrops of coralline limestone, which were difficult to excavate and gave rise to an uneven running surface, be avoided. Areas

Fig. 4. Foothills of Gollis Mountains.

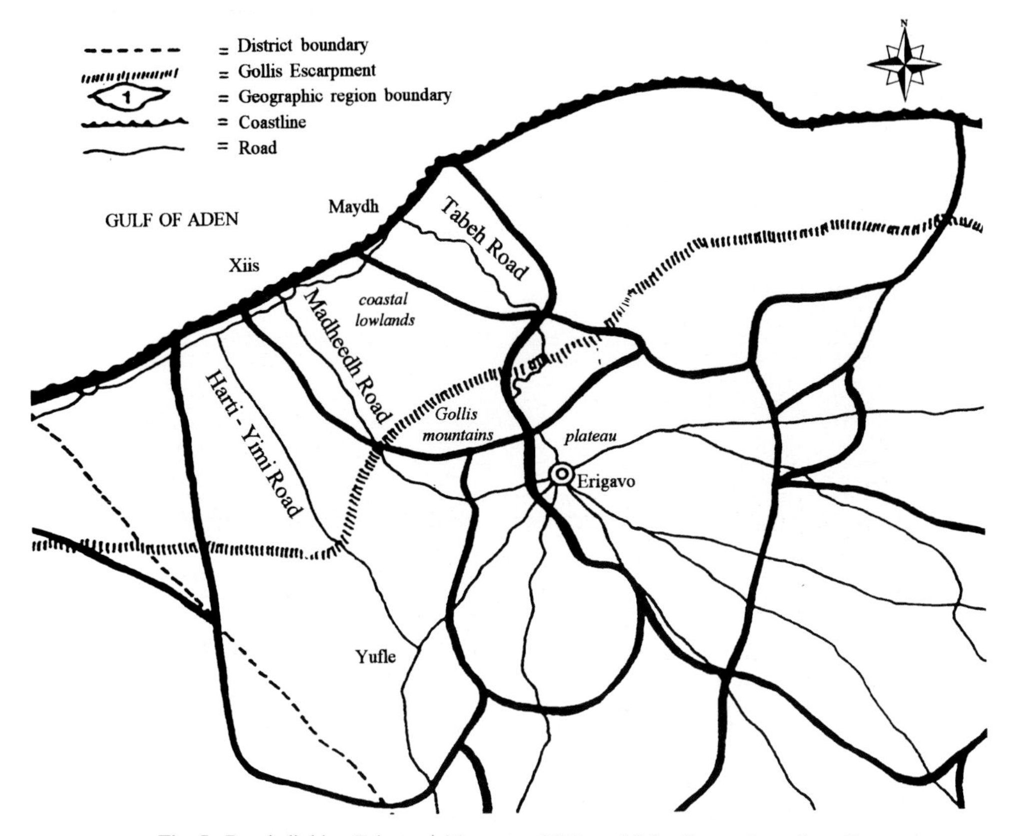

Fig. 5. Roads linking Erigavo with ports of Xiis and Maydh, northern Somalia.

requiring excavation of the slope were also avoided as this would undercut the slope and require significant maintenance.

Along the portion of the road leading up to the Erigavo plateau, recent repair work on drainage ditches involved excavation at the toe of cuttings and had resulted in an overhang in colluvial soils which would collapse, probably in the first storms, resulting in blocking of drainage ditches!

The road surface had been built of very coarse gravel and cobbles, resulting in a very uneven surface. Although well-graded sand and gravel would give a better running surface, there were no local supplies of fine aggregate and the only method available to reduce coarse aggregate to an appropriate size was hand-breaking.

Culverts and pole drains were recommended to divert surface waters away from the road surface, while ensuring that erosion of the slope beneath the road was prevented by making sure water was channelled to areas away from the road. The culverts or pole drains also served to control the speed of descending vehicles thereby reducing wear and tear on the road, especially above sharp corners where wear and tear was greatest.

Most retaining walls were in good condition, requiring only local repair to the mortar. However, one partially collapsed retaining wall required significant repairs that were beyond the current capabilities of the local workers. Fortunately the wall was above the road and did not therefore threaten the stability of the running surface itself.

Near the top of the escarpment the road is forced to follow a narrow gorge and consequently had to cross the river channel in several places. Much surface water flows along the road during and immediately after periods of heavy rain. Water is directed away from the river and along the road by thick vegetation that partially blocks the river bed. Clearing the vegetation downstream of the road reduced the amount of water diverted on to the road. Vegetation upstream of the road was left in place to dissipate some of the energy of the storm waters. The construction of raised embankments where the road approaches the river would further have reduced the amount of water diverted onto the road; but would require substantial effort in terms of earth moving and significant quantities of cement to prevent erosion of the embankment.

Overall the Tabah road was in generally good condition with locally poor stretches; usually the result of surface water damage. Problems have arisen out of poor maintenance:

- drainage ditches need reconstruction and clearance;
- river crossings need improvement and route realignment;
- unstable retaining walls need repair.

Remediation is required only in places and is deliberately 'low-tech' and low cost. Nevertheless these improvements will result in a significant reduction of travel times.

Madheedh road, Erigavo to Dayaha to Xiis

The Madheedh road is 85 km long and links Dayaha and the surrounding agricultural area with the port of Xiis (Action Aid Somaliland 1992*b*) (Fig. 2). The route from

Fig. 6. Section of road following the river bed.

Erigavo through Dayaha to Gudmo-Afafod was constructed by the British; the remainder was constructed by local communities and the complete route was opened in 1983. Clan elders reported that no traffic had used the road during the past two years. In several places the road had to be cleared of boulders and large pot-holes had to be infilled before progress could be made.

The geology is similar to that of the Tabah road but it follows many more river beds and the terrain is much more undulatory. The Madheedh road was in very poor condition along long stretches when visited and was probably not passable by lorries. The main problems may be summarized as follows:

- erosion where the road follows river beds (Fig. 6);
- large boulders blocking the road in river beds;
- very poor trafficking surfaces along long stretches.

Previous attempts to repair the road had only been partially successful and there was a clear need for close supervision of future work, to ensure the following:

- adequate depth of drainage ditches;
- gradients kept to less than 1 in 10;
- well-graded aggregate used for road surface.

Harti–Yimi road, Yufle to Xiis

The Harti–Yimi road links the village of Yufle, on the plateau, with Xiis (Fig. 2). It is a rough, dirt road whose construction in 1975 was funded from local taxes and contributions from inhabitants of the town of Boroma, whose clan ancestor is buried along the route (Action Aid Somaliland 1992*a*). The route was driven from Xiis along the coast then inland up to Yufle, generally going from northwest to southeast, and provided the main link from Erigavo to the coast during the civil war. Elders from the village of Gudmo Biyas have been co-ordinating community efforts to repair the road.

The lower stretches of the road were being repaired by local communities with assistance from Action Aid. The road was generally in good repair with the exception of approximately 10 km from the coastal lowlands up the escarpment. The escarpment is much lower here than further east and comprises limestone dipping gently south, out of the hillside. This section comprises several switchbacks and entails a climb of several hundred metres. Four-wheel-drive vehicles could achieve speeds of 10–15 $km\,h^{-1}$ over this steep section. Lorries, however, can rarely go faster than 5 $km\,h^{-1}$. In some cases, lorries were reported to have to unload half the payload, climb the escarpment, unload and return for the other half; a process taking some 5–6 h to cover a distance of no more than 10 km. The road surface follows bedding planes along much of the climb up the escarpment. Drainage ditches are almost entirely absent. Superficial deposits, comprising slope wash deposits, are generally less than 1 m in thickness. The geotechnical problems in this section may be summarized as as follows:

- poor trafficking surface;
- inadequate drainage;
- excess side slopes;
- gradients locally too steep;
- height lost through poor alignment of road.

The trafficking surface generally comprises either limestone bedding planes or limestone gravel and cobbles. Each of these had inherent problems (see below) and where half the road width is on superficials and the other half on bedrock, severe side slopes had developed.

The bedding planes provide a good surface where they have been etched and pitted by localized chemical weathering. Elsewhere, the bedding planes are planar, smooth and polish easily, forming an obstacle to traffic (extensive areas of burnt off rubber were observed along some switchbacks). It was advised that the running surface should be improved by roughening the smooth rock to provide a better grip, for instance by removing small chunks of rock using hammers and chisels.

The running surface formed by the gravel and cobbles was uneven – as a result of the erosion of fine material. It was suggested that the large cobbles and boulders be replaced with finer material. This could be achieved by breaking the cobbles *in situ* to provide a range of aggregate sizes.

In the absence of ditches, the road acted as the drainage path for surface water runoff. As a result, erosion of the superficial deposits and finer aggregate has led to large ruts and an uneven surface which causes even light vehicles such as Land Rovers to tilt at alarming angles. In places the superficials had been completely removed. Drainage channels on the upslope side of the road, and culverts to take water under the road and down the mountainside, were advised.

The gradient of the road was too steep at the corners of almost all switchbacks, reducing the speed of vehicles to almost nil. The long limbs of the switchbacks were too long and did not gain sufficient height – in some cases height was lost! The switchbacks required realigning to ensure the limbs climbed and the gradient at the corners did not exceed 1:10.

The worst section of the road, some 750 m long, was towards the top of the escarpment. The dipping limestone bedrock formed the surface of the road and had not been cut to provide a level road surface. This resulted in a severe side gradient and again caused light vehicles to tilt precariously. Small cuttings in the limestone were needed to provide a level road surface but were beyond the capability of the local workers. A realignment was being constructed but had encountered the problem of dipping limestone beds close to the surface. Since excavating rock was difficult with the available hand tools, much of this section had been built on embankments supported by dry stone retaining walls. The realigned part of the road had a number of problems:

- the road was too narrow;
- the retaining walls were vulnerable to erosion and to loading by traffic coming too close to the edge;
- the running surface had been formed from local soil which contained too many silt and clay sized fines and not enough coarse sand and gravel and would therefore erode very easily.

The road needed to be widened by about 2 m, by excavating into the limestone bedrock, and have a running surface formed of well-graded sand and gravel. Drainage channels and culverts were needed to prevent surface water running across the road. Large stones should be placed at 10–20 m intervals along the outer edge to prevent vehicles from overloading the retaining walls.

Conclusions

The dynamic geomorphological setting of the Sanaag region, coupled with social and economic disruption has resulted in rapid deterioration of three major road links between plateau and coast. Geomorphological processes which would not qualify as geohazards in most parts of the world are having a significant adverse impact on the attempts to regenerate the local economy. Scarce resources are having to be deployed to repair and maintain roads constantly being damaged by erosion, flooding and mass movements. The poor condition of all three roads, the long distances involved and the clan structure of Somali society meant that all three roads had a role in carrying traffic between the regional capital, Erigavo, the ports of Maydh and Xiis and the villages in between. A number of problems were observed repeatedly:

- road surface of very coarse gravel or cobbles;
- lack of drainage;
- steep gradient at approaches to river crossings.

Much work had already been done and two of the three roads are presently in use, but some of this effort had been misdirected due to the absence of engineering advice at an early stage. The local communities appear to have the will to carry out the work themselves but lack equipment and skills beyond the basic ones.

Engineering geological understanding of materials, processes and products could be applied to the conditions in northern Somaliland. Advice had to be appropriate to the capabilities of local populations working with small numbers of hand tools on a ‘food for work’ basis. Earlier involvement of engineering geologists could have prevented much wasted effort and ensured more efficient use of scarce resources and energy. Nevertheless, by giving advice directly to the local communities on the ground, better use of those resources in future work is likely.

Acknowledgement. The logistical support of Save the Children Fund during our time in Somaliland is gratefully acknowledged.

References

Action Aid Somaliland. 1992*a*. *Harti – Yimi road: project proposal*. Action Aid Somaliland, Erigavo (unpublished).

——1992*b*. *Rehabilitation of Madheedh Road*. Action Aid Somaliland, Erigavo, (unpublished).

——1994. *Programme review and evaluation*. Action Aid Somaliland, London (unpublished).

Beydoun, Z. R. 1970. Southern Arabia and Northern Somalia comparative geology. *Pholisophical Transactions of the Royal Society of London*, Series A, **267**, 267–292.

Macfadyen, W. A. 1933. *The Geology of British Somaliland*, with 1:100000 scale geological map, Crown Agents, London.

Merla, G., Abbate, E., Azzaroli, A., Bruni, P., Canuti, P., Fazzuoli, M., Sagri, M. & Tacconi, P. 1979. *A geological map of Ethiopia and Somalia (1973) and comment with map of major landforms*. CNR Italy, Centro Stampa Firenze.

PAI HIZ. 1991. *Rehabilitation of Tab'aa Road: Summary*. PAI HIZ, 91-PAI-SL-27 (unpublished letter report).

A hazard map of the Magnesian Limestone in County Durham

M. R. Green,[1] R. A. Forth[2] & D. Beaumont[3]

[1] Soil Mechanics Limited, Glossop House, Finchampstead, Wokingham, Berkshire RG40 4QW, UK
[2] Department of Civil Engineering, University of Newcastle, Newcastle upon Tyne NE1 7RU, UK
[3] Durham County Council, Civil and Geotechnical Laboratory, County Hall, Durham DH1 5UQ, UK

Abstract. A large part of County Durham is underlain by carbonate rocks of Permian age, principally Magnesian Limestone. In recent years problems have been encountered in constructing on the Permian carbonates, due to dissolution followed by subsidence and/or sink hole formation.

Features believed to be triggers for dissolution of limestone have been mapped and a weighted factor hazard map has been created. The map is based on an extensive review of existing site investigation data including a study of aerial photographs. Records of dissolution features noted by the County Council engineers have been incorporated into the hazard map. The map is intended as a guide for County Council Engineers who are planning site investigations within the area. The preparation of the map and its limitations are discussed.

Introduction

County Durham is situated in the northeast of England, and a large area of the east of the county is underlain by carbonate rocks of Permian age.

A review of factors thought to be affecting the dissolution of the carbonates was carried out. The results of this review were then rated according to their perceived importance to produce a map showing the degree to which problems with dissolution might be expected to be encountered when constructing pavements or roadways.

It is intended that the weighted factors hazard map will be of use to planners and engineers within the County Council.

Geology

The Permian Limestone forms a prominent west-facing escarpment in County Durham. The geological structure of the Magnesian Limestone is relatively simple, with the rocks being gently folded and having an eastward dip of 2° to 4° (Fig. 1). There is little major faulting, although a large number of minor faults and dislocations occur. These minor faults provide major pathways for migrating calcium sulphate rich groundwaters and they are often associated with narrow areas of dedolomite which is generally harder and more resistant than dolomite.

The Magnesian Limestone has three main units, the Lower, Middle and Upper Magnesian Limestone (Fig. 2). The base of the Upper Limestone is composed of a complex of breccias that foundered as a result of the removal by solution of underlying anhydrites and evaporites. The foundering of the rocks affected their character in a variety of ways, from gentle subsidence to multiphase severe fracturing that resulted in brecciation of the rocks (Smith 1993).

Sequences of glacial and periglacial drift (Fig. 3) attain maximum thickness in buried valleys, although the average thickness is about 15 m (Smith & Francis 1967). Where thin tills exist on limestone benches and on the dip slopes of limestone, cuestas are commonly punctured by lines of sinkholes a few metres in diameter which form where till has collapsed into solution features.

Methodology adopted

Several authors have dealt with the prediction of the occurrence of sinkholes and solution features as well as other natural hazards. A review of these methods was carried out to try to determine which particular method best suited the data from County Durham. 'Hazard' is defined as the probability that a particular area will be affected by a destructive collapse in a given time interval and is a complex function of the probability of collapse and of sinkhole type. The estimation of risk is a complex task. A realistic approach is to collect data on all parameters that are reasonably easy to measure and that could be reliable precursors of dangerous solution activity.

A number of quantitative approaches can be adopted in assessing the likelihood of sinkholes, including correlations with rainfall data, use of geomorphological information and historical data, and probability analyses. These should all provide an indication of the relative probability of failure.

Presentation of results should be done in a manner that allows people of all disciplines to understand the results easily. This can be facilitated by the use of broad

Green, M. R., Forth, R. A. & Beaumont, D. 1998. A hazard map of the Magnesian Limestone in County Durham. *In*: Maund, J. G. & Eddleston, M. (eds) *Geohazards in Engineering Geology*. Geological Society, London, Engineering Geology Special Publications, **15**, 239–246.

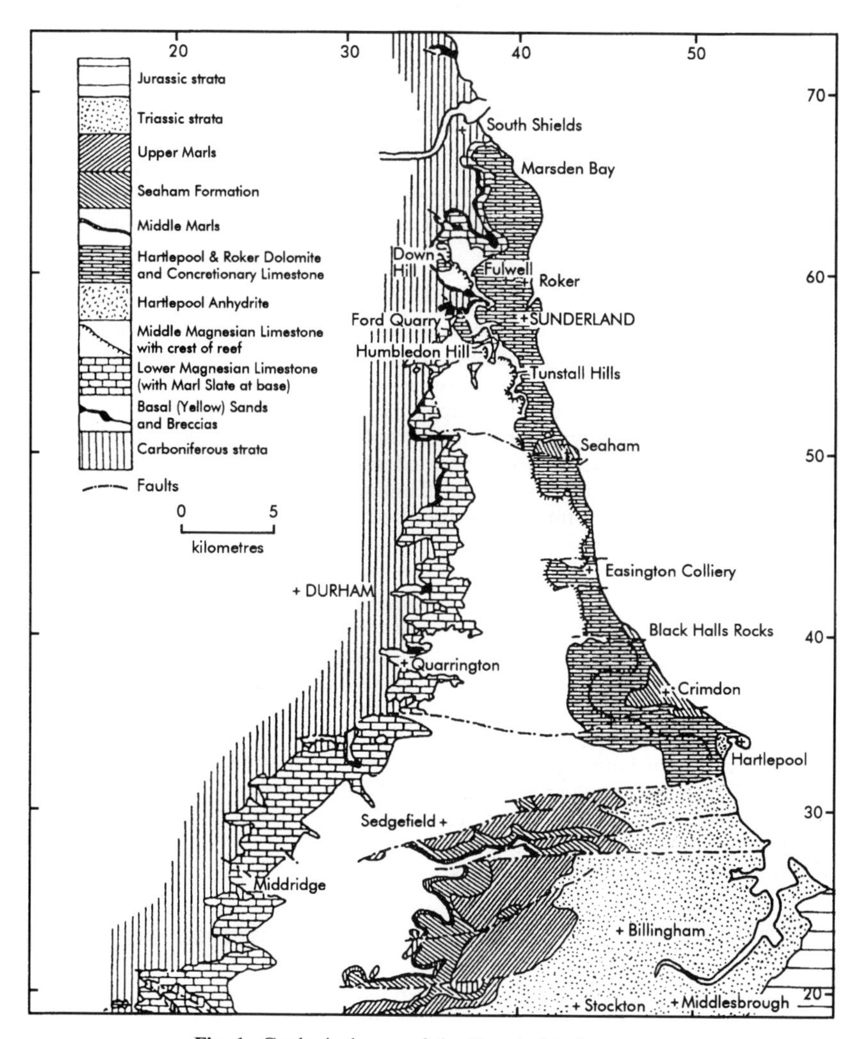

Fig. 1. Geological map of the County Durham area.

qualitative terms for the description of the relative degree of risk. A quantitative risk assessment represents a rational framework for the evaluation of the probability and the degree of damage. The results permit zoning of sites and facilitate comparisons between sites in terms of relative risks. When quantitative risk analysis is applied to site design an approach should be adopted that is based on a mixture of professional judgement and a probabilistic approach as opposed to the sole use of statistical techniques. The tolerable risk level should also be determined, which may relate to such things as the degree of tilt caused by subsidence that a structure can tolerate before failure.

Using all collected map and field data, a model can be developed in which each factor is weighted. The most important factors affecting collapse and subsidence can be statistically determined and then a series of maps combining two or more factors can be made to aid in predicting collapse and subsidence susceptibility (Ogden 1984). This was the method chosen for the East Durham data. Due to the size of the area (approximately 600 km^2) and the obvious time constraints, the method of weighting factors was chosen. In this method a whole series of factors that may be contributing to sinkhole/solution feature development are assessed and mapped onto individual maps.

A database was set up based on grid squares. In this case over 600 1 km^2 squares were used as the basis for the database. Each factor that affects or is thought to affect sinkhole development was scored or weighted according to its degree of importance. The degree of importance was decided upon by a combination of visual examination, professional experience and literature studies.

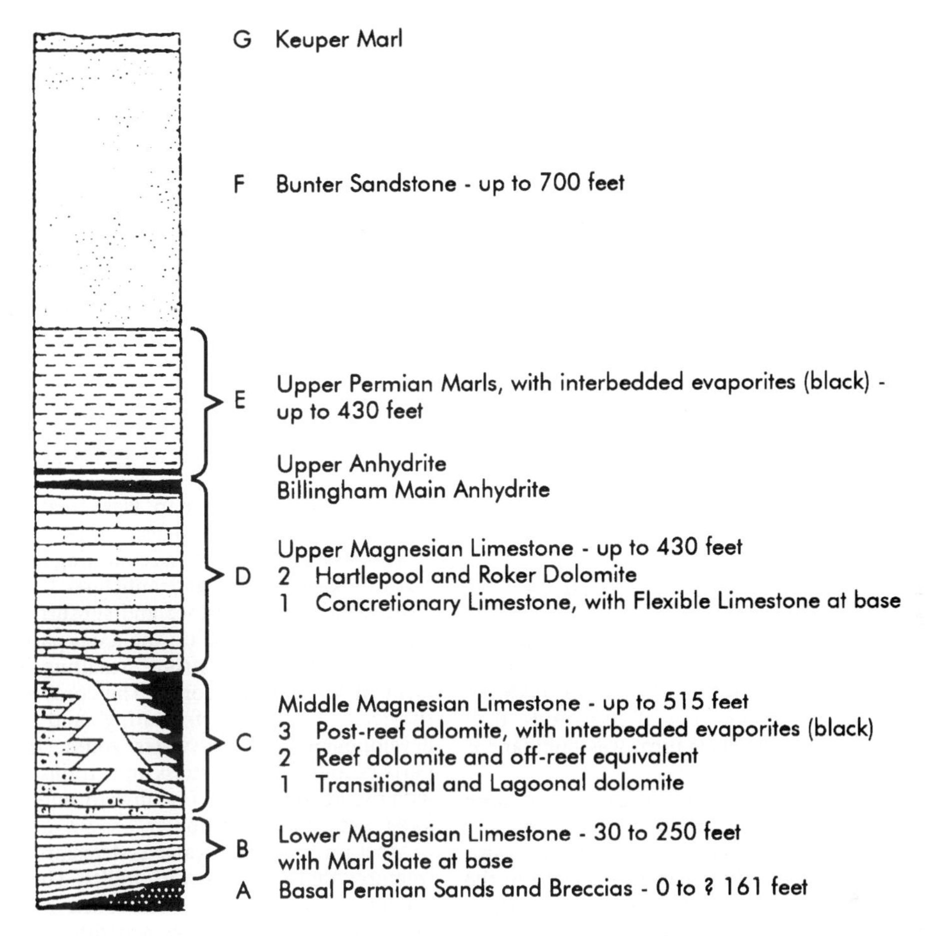

Fig. 2. Generalized section of the rocks of County Durham (Smith & Francis 1967).

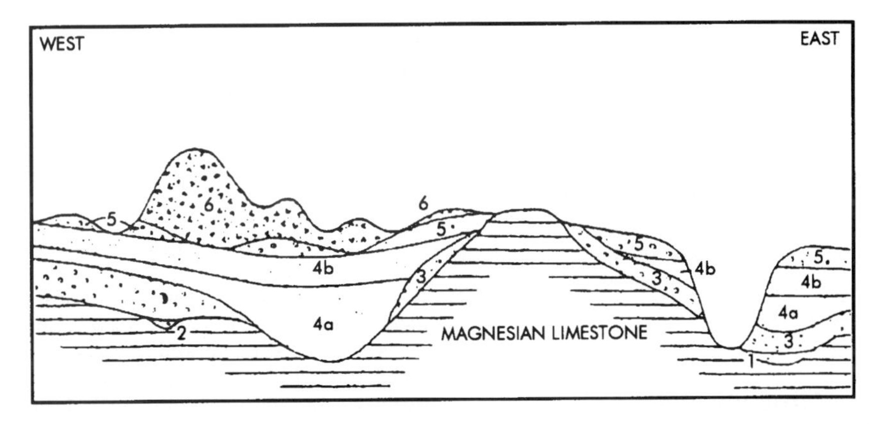

Fig. 3. Diagrammatic section showing the relationship of drift: 1, Scandinavian Drift; 2, Lower Gravels; 3, Lower Boulder Clay; 4, Middle Sands; 5, Upper Boulder Clay; 6, Morainine Drift and Upper Gravels (Smith & Francis 1967).

Once all factors affecting sinkhole development were weighted, the individual 1 km^2 squares were looked at individually to see if any of them were affected by those factors. For example, if intersecting faults lay within a square, 20 points were attributed to that square. Each factor was then looked at to see if it occurred in that square. This was carried out for all of the 634 squares.

A review of factors thought to influence solution of limestone was carried out to try and determine their relative importance in the solution process so that the various factors could be weighted to produce a hazard map.

A contour map of drift thickness was produced from details shown on British Geological Survey Maps, site investigation data held by Durham County Council and available geological memoirs. From the contour map, drift was seen to thin to the west and north of the area, although it should be noted that drift thickness can vary over a very short distance so a regional picture is gained. Solution failures appear to occur where the drift is thin and as a consequence was weighted accordingly.

Pre-existence of solution features

A map showing solution features was compiled from a variety of sources. Firstly, British Geological Survey maps at 1:10 000 scale were reviewed and any sinkholes shown were plotted. A proportion of these were then observed in the field so that their form was familiar for future aerial photographic study.

Some 375 linear km of aerial photographs were observed (750 photographs) with stereo pairs, and any suspicious features were plotted. Mottled fields were also noted as these indicate that the limestone is close to the surface. The photographs were taken in 1968 at various times of the year. It was found that features stood out much more clearly if the photographs had been taken in December or January. It is believed that this is due to better tonal variations occurring because the sinkhole solution features retain moisture. When photographs from the summer months were observed, features that had been seen in the winter photographs appeared to be completely absent.

The presence of solution features generally indicates that limestone is close to the surface and obviously the areas are prone to solution so are heavily weighted for the map. Wherever possible, solution features seen in aerial photographs were confirmed in the field. Part of the study area was not covered by aerial photographs, thus introducing bias into the study.

Groundwater

A depth to groundwater contour map was produced on a contouring package (SURFER) from data supplied by the National Rivers Authority. The readings of groundwater levels were taken in April 1994. The level of groundwater does not change considerably throughout the year, with a maximum of 4 m change occurring in the south of the county (P. Butler, NRA, pers. comm.). The presence of groundwater is important in the solution processes occurring in the subsurface although continued fluctuation of the groundwater is more important than the actual level with regard to the process of solution.

Without a probabilistic analysis of the effect of groundwater on solution processes, weightings were difficult to attribute. Hydrological surface features do not adversely affect the development of solution features and are consequently attributed a low weighting.

Structural geology

A structural geology map was created from British Geological Society geological maps at various scales. Synclines are commonly the collecting and storing structures of karst water, and anticlines form the dividing structures between surface and subsurface water. Folding concentrates stress on the axial parts of folds, and deformations can be strong, resulting in the development of karst water, zones. The cores of anticlines are subjected to tensile fracturing and this helps the development of karst water zones. In the plunging parts of anticlines the stress and strains are complicated with fissures and radial fractures forming the controlling karst water flow direction.

Most underground water systems are connected with fractures that correlate with faults. The distribution of sinkholes appears to be controlled by faults. At the intersection of faults large negative landforms such as basins and valleys occur. The deep fault is a fundamental condition of karst water deep circulation and deep karst development. It can be seen that faults are far more prevalent in the north and west of the study area. Not many folds are seen in this area. Thus areas of intersecting faults are weighted heavily, as are synclinal areas.

Roads and railways

Roads play a definite role in solution failure development. Trunk roads, motorways and 'A' roads are generally well drained and any problems occurring are rectified quickly. Smaller roads are less well drained and problems are not generally picked up quickly; however, problems on such roads are less important than those on main roads. Problems do tend to occur when a drainage system's capacity is exceeded. Cuttings for

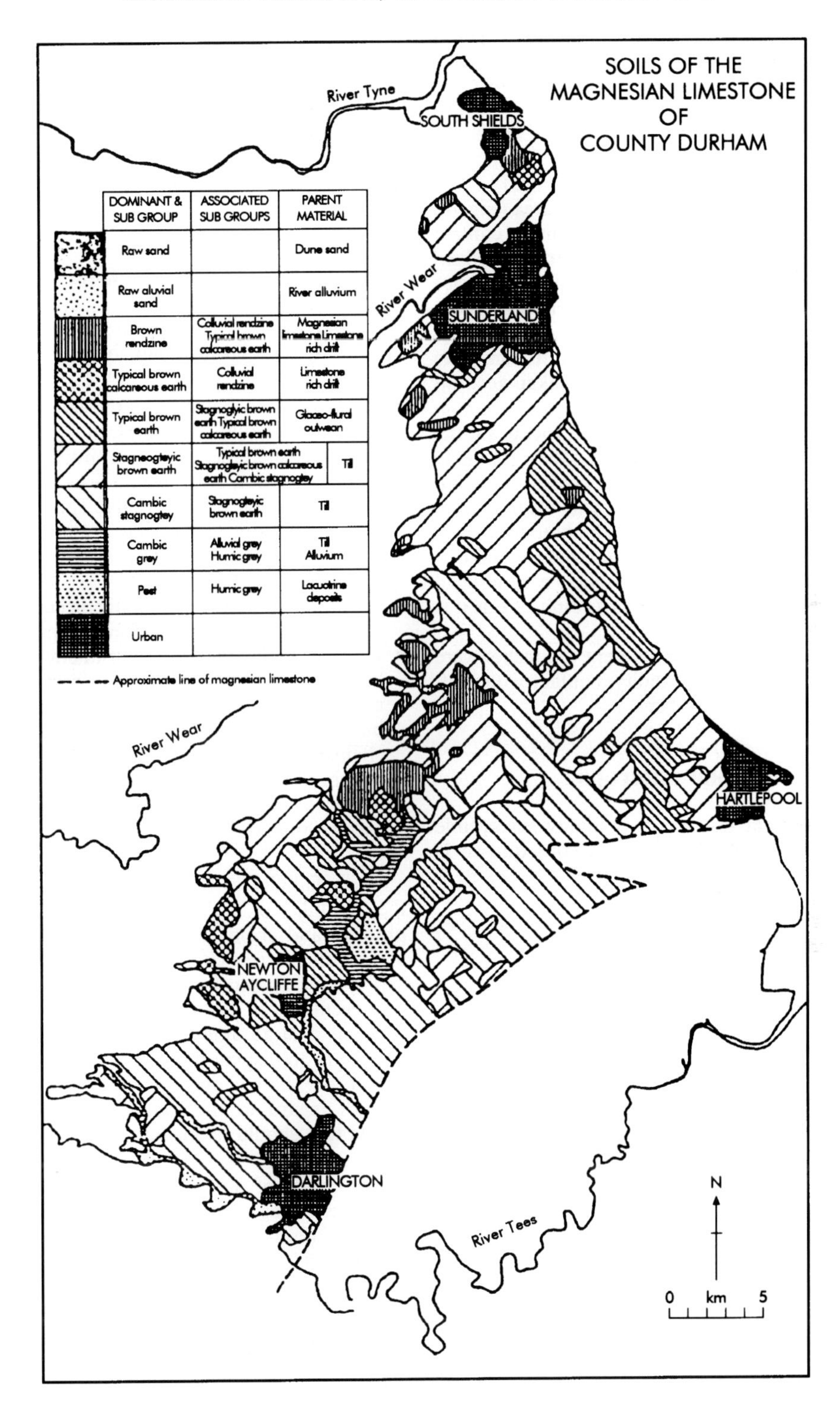

Fig. 4. Soils of the Magnesian Limestone of County Durham (Alexander, in Dunn 1980).

roads and railways increase the potential for water to enter the limestone and so constitute a risk and were weighted accordingly.

Drift composition

There are 16 types of drift that occur in the area (Fig. 4). Of these, only five or six are actually in direct contact with the underlying limestone. The rest are superficial deposits such as glacial lake sediments and river terrace deposits. However, if these deposits are close to rockhead and overlie a fissured boulder clay their permeabilities may be an important factor in transmitting surface water to rockhead. This has been taken account of in the weightings.

Of the deposits that are in contact with rockhead, the sands and gravels are most important as they are likely to be permeable and able to transmit surface water to the limestone. If the top layers are calcreted, permeability will be significantly reduced. The boulder clay is generally fairly impermeable although when fissured it can transmit water easily. Peat, although a superficial deposit, can supply acid water and as a result of this it is heavily weighted.

Weightings attributed to the different sediments were done so on the basis of permeability, contact with bedrock and likely degree of fissuring.

Limestone outcrops

Where limestone is at outcrop it is highly susceptible to weathering and solution, and as a consequence is heavily weighted.

Rock composition

A map was created from Geological Survey maps (Smith & Francis 1967; Smith 1971; Dearman & Coffey 1981). The weightings attributed to the various units were given on the basis the chemistry of the rock, the degree of bedding and laminations, the degree of faulting and brecciation, the weathering state of the material and the hardness of the rock. A brief summary for each rock type is given below.

- Sandstone: not soluble
- Permian Upper Marl: up to 55% non-carbonate material
- Seaham Formation: heavily faulted, variable composition
- Concretionary Limestone: bedded, laminated, calcitic and dolomitic, very soluble
- Permian Middle Marl: similar to Permian Upper Marl
- Hartlepool and Roker Dolomite: soft, bedded, granular, dolomite, brecciated
- Middle Magnesian Limestone: brecciated, dolomite, soft, variable in nature
- Reef: dolomite, shells and detritus
- Lower Magnesian Limestone: evenly bedded, dolomitic, brecciated cavities exist
- Basal Sand: not important
- Alston Group: not important

Conclusions

Whatever the methodology used for the estimation of hazard, a useful hazard map should be

- *Applicable*: the dynamics of the sinkhole must be clear so that adequate assumption about the type of expected sinkhole, the magnitude of the sinkhole and the location of the sinkhole can be used to predict the distribution of the events. This is achieved when the dynamics of the process are understood or when the main factors affecting it are identifiable (Pereschi & Bernstein 1989).
- *Reliable*: the database should accurately reflect the distribution of sinkholes from the past.
- *Flexible*: by changing the parameters when new data become available, the techniques of mapping should adequately provide new information. Computer-based hazard maps are flexible as it is possible to change the initial and boundary conditions of the system of equations.
- *Rapid*: hazard maps should be produced in a short enough time to be effective and usable.

The reliability of the results obtained depends upon the capability of the models to adequately describe the physical and chemical behaviour of the sinkholes and also on the correct choice of the input parameters. This can be investigated by testing and validating models of past events.

When a large amount of data are being used it is important to display the results in a visual or graphical way rather than as mathematical data or tables.

With regard to the hazard maps there are several points that should be made:

- The choice of weighting factors was subjective, with choices being made according to knowledge gained from literature study, discussion with professional engineers and visual analysis of the individual factor maps.
- The method of weighted factors would have been significantly improved if a probability analysis of the factors causing dissolution had been carried out. For example, with drift thickness it may be found that 80% of solution features occur where the drift is less than 10 m thick. Thus it can be seen that drift

thick-ness is an important factor and a suitable weighting could then be applied. This was not carried out simply because of time constraints associated with the very large study area (>600 km^2).

- The area chosen to be covered by the hazard map reflected the area of County Durham that is underlain by Magnesian Limestone and thus is relevant to County Council Engineers.
- Hazard maps are normally drawn for relatively small areas and they are particularly suited to site-specific investigations where the factors causing dissolution can be subjected to a rigorous statistical analysis, which cannot be carried out in this instance.
- The large number of solution features observed (approximately 400) makes it difficult to thoroughly investigate their causes. If the cause of each solution feature could be found then a map giving degrees of importance of features could be developed in conjunction with a hazard map.
- The effect of varying weighting factors on the final hazard map has not been investigated as the data set is so large as to deter this. There are 10 factors used and more than 600 1 km^2 squares to attribute these factors to, representing a database of over 6000 variables. Contouring of the map is also extremely time-consuming.
- The weighting method used is a non-time-dependent method. In County Durham it is not necessary to predict when sinkholes will occur as most problems are a result of ancient sinkholes being reactivated by construction works so a time-dependent method does not need to be adopted. It is far more important to predict the likely locations of sinkholes rather than when new ones will develop. If a time-dependent study was deemed necessary, aerial photographs from a series of years would need to be consulted to observe the development of new features and also their rate of development.

Bias has been introduced into the hazard map by the following factors:

- The presence of solution features was investigated using aerial photography. The photographs were not all taken during the same season and, as already explained, the resolution tends to be poorer during the summer months so features could be missed.
- Aerial photographs were not available for the south-east of the area and so any solution features occurring in this area would not be noted. Thus in this area on the hazard map risk may in reality be greater than it appears from the map.

Some factors have not been accounted for on the maps as the number of factors had to be restricted to keep the data within manageable limits. The aspect of slopes may be an important factor in controlling the soil CO_2 productivity and thus dissolution potential. The degree to which this affects the dissolution potential is difficult to quantify so it has been omitted from the hazard factors. The soil CO_2 levels throughout the region could not be added to the hazard map as very few results are available.

To improve the hazard mapping techniques, we would recommend the following:

- A full probabilistic analysis of all factors used for the hazard map should be carried out. This would be extremely time-consuming but would lead to the attribution of weightings to the hazard map on a scientific quantitative basis rather than it being subjective.
- We would have liked to have investigated the role of the aspect of slopes on the dissolution potential and would recommend this be carried out on a statistical basis as with all the other factors.
- The processes of formation of sinkholes and solution features in County Durham are not well understood and for this reason we would recommend a study of the profile of solution features and the character of the materials within them. This would enable a database to be developed of the type of sinkholes and their suitability for constructing upon. This could be achieved by drilling or probing with window sampling or dynamic probing and carrying out laboratory testing on samples.
- The level of CO_2 in soils obviously plays a major role in the dissolution potential of an area, and as such it is recommended that where risk of sinkholes is indicated to be high on the map, measurements of soil CO_2 are taken and a database created.
- The solubilities of the different limestone units in the area have not been determined to any great degree - of accuracy as information that is available seems to be conflicting. It is recommended that samples of all the different units of limestone are collected (from quarries possibly) and tested by such methods as slake durability to try and develop a ranking of solubilities.

The hazard maps produced give the user an indication of the likelihood of encountering a solution feature that already exists. The map does not indicate when new sinkholes will develop or the likely severity of the problem if a sinkhole is encountered.

It is likely that the hazard maps produced will be used by the Durham County Council planners as an aid to planning site investigations and desk studies so that problem areas can be anticipated and dealt with in an appropriate way.

Acknowledgements. The facilities and data provided by the Civil and Geotechnical Laboratory, Highways Department, Durham County Council, were invaluable in the production of the hazard map.

References

DEARMAN, W. R. & COFFEY, J. R. 1981. An engineering zoning map of the Permian Limestones of NE England. *Quarterly Journal of Engineering Geology*, London, **14**, 41–57.

DUNN, T. C. (ed.) 1980. *The Magnesian Limestone of Durham County*. Durham County Conservation Trust.

OGDEN, A. E. 1984. Methods for describing and predicting the occurrence of sinkholes. *In*: *Proceedings of the 1st Multidisciplinary Conference on Sinkholes*, Orlando, Florida, 15–17 October 1984., A. A. Balkema, Rotterdam.

PARESCHI, M. T. & BERNSTEIN, R. 1989. Modelling and image processing for visualisation of volcanic mapping. *IBM Journal of Research and Development*, **33**(4), July 1989.

SMITH, D. B. 1971. *The stratigraphy of the Upper Magnesian Limestone in Durham: a revision based on the Institutes Seaham borehole*. Natural Environment Research Council, Institute of Geological Sciences, Report No. 71/3, HMSO.

——1993. *Durham coast management plan. Report on relevant aspects of the coastal geology*. Posford Duvivier Environmental.

—— & FRANCIS, E. A. 1967. *Geology of the Country between Durham and West Hartlepool*. Natural Environment Research Council, Geological Survey of Great Britain. HMSO, London.

Landslide susceptibility mapping using the Matrix Assessment Approach: a Derbyshire case study

Martin Cross

Arcadis Geraghty & Miller International, Inc. Wharfedale House, 6 Feastfield, Horsforth, Leeds LS18 4TJ, UK

Abstract. Civil engineering schemes such as new highways and railway lines, regional planning and large-scale land-management projects in areas known to have a landslide problem require regional landslide susceptibility evaluation. The Matrix Assessment Approach (MAP) is introduced as a medium-scale landslide hazard mapping technique for establishing an index of slope stability over large areas. The method allows the relative landslide susceptibility to be computed over large areas using a discrete combination of geological/geomorphological parameters. MAP was applied to a region in the Peak District, Derbyshire. The model identified key geological/geomorphological parameters involved in deep-seated failures, provided an effective means of classifying the stability of slopes over a large area and successfully indicated sites of previously unmapped landslides. The resultant regional landslide susceptibility index provides useful preliminary information for use at the desk study and reconnaissance stages of large-scale civil engineering works such as highway construction.

Introduction

Regional landslide evaluation constitutes a major task for many regulating/planning authorities. Various techniques have been developed for the identification of landslide zones and the construction of maps showing different degrees of landslide hazard and risk (Degraff 1978; Lawrence 1981; Varnes 1982; Brabb 1984; Hansen 1984). This paper examines a computer-assisted medium-scale hazard mapping technique known as the Matrix Assessment Approach (MAP).

MAP is a quantitative method for establishing an index of instability over a large area. The method uses existing or readily obtained geological and geomorphological data to define a hillslope's susceptibility for landsliding to occur within a bounded region. In its simplest form MAP lacks the ability to predict landslide hazard risk in terms of probability or confidence intervals; however, it does allow the determination of landslide susceptibility to be evaluated over large areas using only a few key measurable factors. The landslide susceptibility classes are defined by discrete combinations of specific geological and geomorphological attributes. Most regional assessments of slope instability produce a map showing the location of existing and at times even fossil landslides. Much more difficult is the task of evaluating sites which are close to instability or even where instability may exist but has not been previously recognized (Anderson & Richards 1987). MAP reduces the subjective judgements involved in the evaluation process by using a simple objective procedure. Very large data sets can be generated by MAP, and because of the data manipulations that are required, a computer-based approach is necessary.

This paper describes a more advanced version of the Matrix Assessment Approach than that used by DeGraff (1978) by reference to a case study in the Derbyshire Peak District (Cross 1987).

Basic principles of MAP

MAP is based on a grid system which is created across the area or region of concern. The region for the landslide susceptibility assessment must be defined and boundaries clearly established. Each cell of the grid is classified on the basis of selected geological and geomorphological parameters that are believed to have a contributing influence on stability, and can be obtained relatively easily from existing sources (Johnson 1981). All discernible landslides are then mapped within the bounded area. This is accomplished by the use of existing geological maps and through a combined programme of aerial photograph interpretation and field checking. All types of landslides are mapped irrespective of their classification. Most of the landslides identified in the Peak District study were rotational slides comprising either single (72%) or multiple (13%) rotated forms, often associated with extensive debris aprons or mud-sliding (Cross 1987). The whole area affected by landsliding is mapped from the back-scar to the toe of the displaced mass. All grid squares which display morphological features associated with

Cross, M. 1998. Landslide susceptibility mapping using the Matrix Assessment Approach: a Derbyshire case study. *In*: Maund, J. G. & Eddleston, M. (eds) *Geohazards in Engineering Geology*. Geological Society, London, Engineering Geology Special Publications, **15**, 247–261

landsliding are identified. The map scale selected and accordingly the size of the grid squares used for the data collection must allow for the representation of the smallest significant detail of each type of matrix attribute used in MAP. The size of the grid chosen should be both small enough for the matrix attribute classes to be approximately constant within each unit in order to provide the optimum representation/coincidence of geological and geomorphological features in the region concerned.

A uniform grid overlay can be superimposed over parameter base maps, i.e. topographical, geological, soils etc., drawn/digitized to the same scale, to facilitate data collection. Typical variables collected include bedrock lithology, superficial deposits, slope steepness, slope aspect, relative relief, altitude, height above the valley floor, etc. The assigned matrix attribute can be derived either by noting the value of the mid-point of the grid square, by determining the largest area contained within the square or by counting the frequency of variables (i.e. the number of contours within the square as in the case of relative relief or between the centre of the grid square and a specific point on the base map such as the bottom of the valley as in the case of height above the valley floor). In such cases data relating to small (but perhaps critical) areas within each grid square may be lost, therefore the problem of grid size is important. Through Geographic Information Systems (GIS), once the key data sets are digitized (i.e. contours, geology, soils), all other derivatives can be obtained relatively simply using the appropriate software. The GIS approach also allows cell squares to be modified in size if required. The application of GIS to landslide hazard zonation is described by Carrara *et al.* (1991) and Van Western (1993). A smaller grid size will produce a more detailed map but will be more expensive to produce in terms of staff resource time, in data collection, computer time, and data storage space (McCullagh *et al.* 1985; Cross 1987). The choice of matrix attributes is discussed by reference to a Derbyshire Peak District case study.

The objective of the data collection is to provide a final classification for individual grid squares in terms of each attribute influencing landsliding. This will lead to the construction of separate grid maps (one for each landslide attribute) incorporating data to be compared and combined in order to provide a landslide susceptibility classification and final landslide susceptibility index (LSI) for each grid square. Each landslide attribute is subdivided into convenient sub-groups or classes. It is not necessary to seek an absolute value to represent a physical state. Therefore, in the case of bedrock geology, bedrock combination, soils and superficial deposits, it is only necessary to identify the lithological types within the grid square, to place these in suitable groups, and then assign a code number to each group. Slope aspect directions acquire their classes according to their compass direction (i.e. N, NE, E, etc.) and each can be classified with a specific code number. Relative relief, height or altitudinal variables can be grouped into convenient classes which cover the complete range of measured values identified. Slope steepness can be assigned to one of several slope classes devised to cover the full range of slope values recorded. At this stage, therefore, each grid square on each attribute map shows, by a numerical code, the appropriate slope classification for the hillslope area covered by that grid square.

When a map showing the aerial extent of landslides is compared with each grid map of parameter classes constructed at the same scale, it is possible to identify the combinations of parameters (and there may be more than one combination) which occur on those hillslopes which have already experienced failure. Having reached this stage it is possible to calculate the area of landsliding associated with each combination of attributes. An assessment of landslide susceptibility can be achieved by searching for all other grid squares with the same combination of attributes but which have no recorded landslides. The technique of MAP assumes that such sites are close to failure or manifest undetected, perhaps old or masked landslides.

The task of carrying out such a search manually is extremely time-consuming. Therefore, a computerized approach is advisable. The use of a computer enables the search operation for grid squares that possess critical combinations of attributes to be carried out quickly and efficiently. With the use of a computer it is relatively straightforward to calculate the landslide area factor and the corresponding regional area factor. A *landslide area factor* (LAF) is defined as the total area occupied by landslides possessing a discrete combination of hillslope attributes. A *regional area factor* (RAF) is the total area within the region occupied by a particular combination of hillslope attributes. Each unique set of attributes for each grid square therefore has its own corresponding RAF. For each set of attribute combinations, a *landslide susceptibility index* (LSI) can be calculated using the following expression:

$$\frac{\text{Landslide area factor (LAF)}}{\text{Regional area factor (RAF)}} = \text{Landslide susceptibility index (LSI)}$$

The closer the LSI is to unity, the more susceptible that particular combination of attributes is to slope failure. The index values can then be used to produce a new map showing both the existing landslides and the LSI value by an appropriate shade class in each grid square for the particular combination of attributes found in that grid square. Figure 1 shows a schematic diagram of MAP for a case using seven attributes, where:

$$\text{LSI (0.695)} = \frac{\text{LAF (98 ha)}}{\text{RAF (141 ha)}}$$

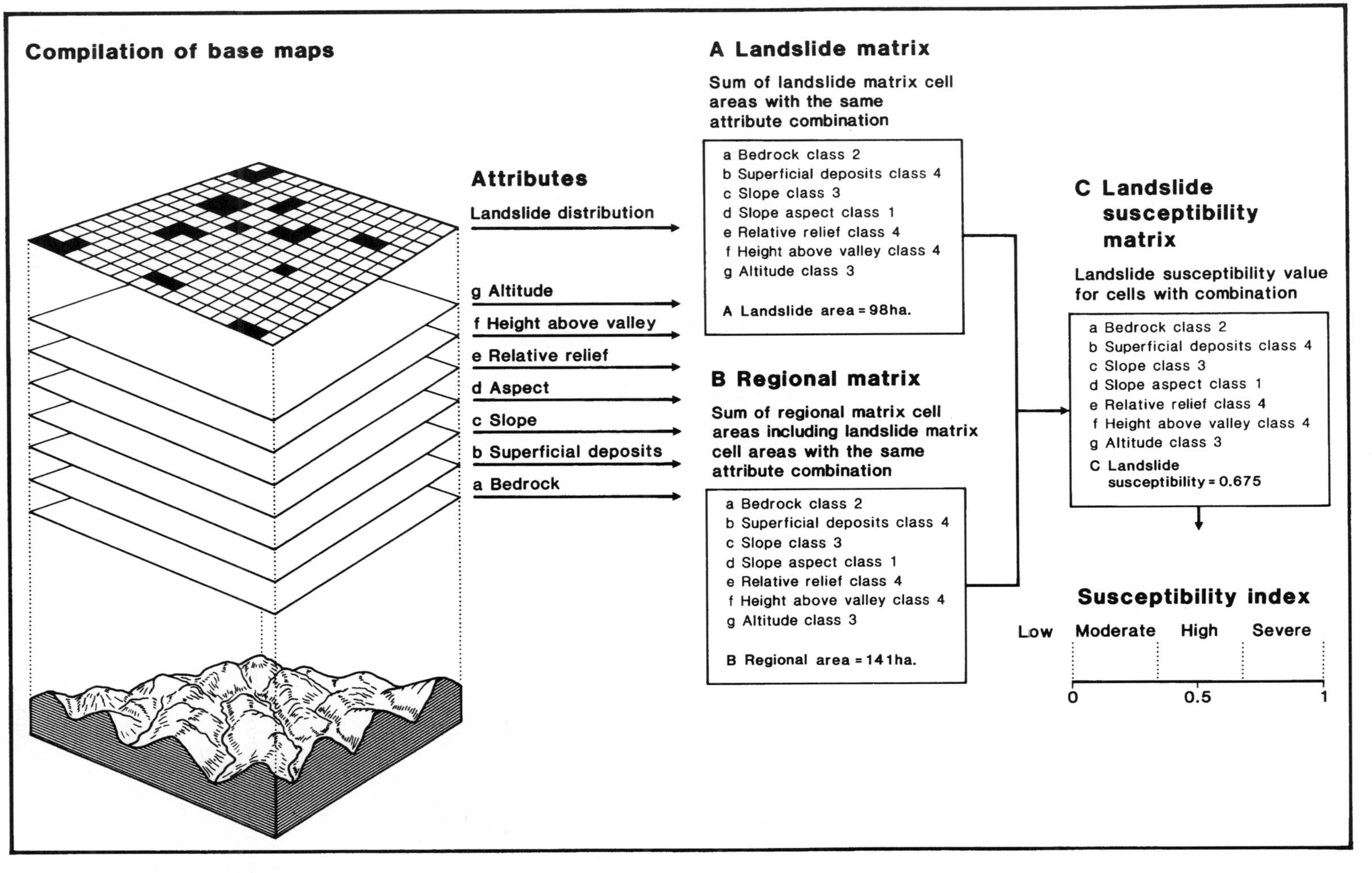

Fig. 1. Schematic diagram of the Matrix Assessment Approach for a case using seven geomorphological/geological attributes.

This particular combination can be regarded as having a high susceptibility to landsliding.

MAP landslide susceptibility attributes

An awareness of the local conditions which may lead to a situation of instability is required in order to select the slope susceptibility attributes for a particular area. Recognition of the potential for instability can be achieved from a preliminary assessment of the expected landforms and the materials, together with the processes and the timescales at which they are evolving (Johnson 1980; Crozier 1984; Cooke & Doornkamp 1974).

A practical limit on the number of attributes/parameters used is set by the ability of the compiler to acquire the appropriate data during a desk study. Costs tend to increase when more parameters are used, so in regional investigations the number of variables should be constrained to avoid prohibitive costs. The number of possible combinations of variables produced by combining a set of different parameters will be equal to the product of the number of variable classes of each parameter used in the assessment. Some possible outcomes will not exist because of autocorrelation between variables. In the case of the Peak District case study, MAP uses a maximum of nine geological and geomorphological attributes for slope susceptibility assessment. These are described below.

Bedrock geology. Particular bedrock units may be more susceptible to landsliding than others; for example, massive bedrock strata are not prone to landsliding in the same way as weaker, fractured bedrock strata or clay formations. Tendency towards slope failure in the UK, by strata type, is described by Jones & Lee (1994). Bedrock used as a matrix attribute is assumed to incorporate the lithological, stratigraphic, mineral composition and characteristics associated with the state of rock mass stress into the landslide susceptibility assessment.

Superficial deposits. The susceptibility of a slope towards a state of failure may be due to the presence of landslide-prone superficial deposits, i.e. regolith or head derived from a particular bedrock unit (Carson & Petley 1970; Taylor & Spears 1970; Carson 1971). Head deposits, created by the downslope soil movement in periglacial environments, have often been reduced to their residual shear strength and often come to rest on slopes at their angle of limiting stability (Harris 1972; Johnson 1987).

The presence of a permeable superficial material overlying a more impermeable bedrock unit can result in the formation of high pore-water pressures (perched water-table conditions) which may trigger shallow instability. Therefore, the type of superficial material taken in conjunction with the type of bedrock unit it overlies are important factors for landslide susceptibility assessment.

Slope steepness. The hillslope angle supplies the potential energy gradient on which a landslide moves to attain a more stable lowered energy state. The steeper the slope, the more liable it is to be unstable. Slope angle is employed as a measurable matrix attribute regardless of the geological/geomorphological factors responsible for the measured inclination (Freeze 1987; Kirkby 1987).

Slope aspect. Slope aspect is the compass direction in which a slope faces. Aspect is used to identify any significant slope orientation and steepness of the ground surface that may coincide with structural conditions (i.e. joint, bedding plane and fault plane directions and inclinations) that may initiate slope failure.

Relative relief. Relative relief, defined as the difference in height between the bottom and top of slopes recorded within each grid square, provides a further measure of the gravitational forces which exist within the unit. In their discussions of hard rock slopes, and drawing largely on the work of Terzaghi (1962), Carson & Kirkby (1972) provide a useful analysis of the factors controlling the critical height above which failure will occur.

Height above the valley floor. The height of a particular grid unit above the valley floor taken in conjunction with the relative relief of that particular unit provides information on the potential energy available for landsliding. In general terms the height of a stable slope is controlled by the cohesive and frictional strength along the bedding planes, the dip of the bedding planes in relation to the hillslope angle, and the unit weight of the rock (Richards & Lorriman 1987).

Height above sea level. The height of a point above sea level may provide information about elevational effects on landsliding, such as the altitudinal position of weaker strata, structural weaknesses associated with bedding planes, or seepage areas (e.g. spring lines) present on the valley sides

Soils. The classification of soil type (i.e. soil associations) indirectly provides information on the soil water regime (Avery 1980; Clayden & Hollis 1984). The soil water regime relates to the cyclical seasonal variation of wet, moist or dry states of a soil. The duration and degree of soil waterlogging can be described according to the system of wetness classes, grading from wetness class I (well drained) to wetness class VI (almost permanently waterlogged within 40 cm depth), (see Tables 1 and 2). The incidence of waterlogging on a slope

Table 1. *Soil wetness classification*

Wetness class	*Duration of waterlogging*
I	The soil profile is not waterlogged within 70 cm depth for more than 30 days[1] in most years.[2]
II	The soil profile is waterlogged within 70 cm depth for 30–90 days in most years.
III	The soil profile is waterlogged within 70 cm depth for 90–180 days in most years.
IV	The soil profile is waterlogged within 70 cm depth for more than 180 days, but not waterlogged within 40 cm depth for more than 180 days in most years.
V	The soil profile is waterlogged within 40 cm depth for more than 335 days in most years.
VI	The soil profile is waterlogged within 40 cm depth for more than 355 days in most years.

[1] The number of days specified is not necessarily a continuous period.
[2] 'In most years' is defined as more than 10 out of 20 years.

depends on the soil and site properties, underdrainage and climate. The presence of a particular soil association provides an indication of groundwater levels and particularly the presence of perched water-tables on the hillside. Such areas are often associated with processes involving active weathering of particular strata, poor drainage, high pore-water pressures, and are commonly associated with areas of instability (Anderson & Richards 1987; Fredlund 1987).

Table 2. *Wetness classification according to soil type*

Soil type	*Wetness class*
Typical brown earth	I
Typical brown podzolic soils	I
Brown rankers	I
Humo ferric podzols	I
Typical brown alluvial soils	I
Iron pan stagnopodzols	III/IV
Cambic stagnogley soils	IV
Cambic stagnohumic gley soils	V
Raw oligo fibrous peat soils	VI

Bedrock combination. Critical areas of instability may occur where a particular stratigraphical succession or combination of bedrock types occur on the hillslope. For example, where a massive well-jointed sandstone overlies a relatively impermeable shale or mudstone stratum, slope instability may result because of the development of strong hydrostatic pressures in the sandstones (Johnson & Vaughan 1983; Johnson 1987).

Derbyshire case study

Study area

Figure 2 shows the study region (32.2 km × 29.0 km) used for the application of the Matrix Assessment Approach (MAP). The region (932 km^2) is divided into northern and southern sectors. MAP was initially applied to the southern sector and tested for its predictive accuracy in the northern sector.

Landslide susceptibility map compilation

Ordnance Survey 1:25 000 scale sheets were used to construct the base maps. A grid was placed over the base maps, with each grid square covering 1.56 ha (125 m × 125 m) of land. Corresponding solid geology, drift deposit, pedological and other topographical derivative maps were digitized at the same scale. Information was then transferred digitally from the base maps onto the common grid format. Landslides were identified initially from 1:10 000 geological maps, and then boundaries were modified and previously unmapped landslides were added to the base map through aerial photograph interpretation (1:12 000 panchromatic aerial photographs) and field mapping. All classes of instability were included as previously described.

Each susceptibility attribute along with its class subdivision is shown in Tables 3–11. The attributes were classified in two ways: firstly a fine classification to obtain maximum information by subdivision into a large number of classes, and secondly a coarse classification using few subdivisions. The greater the subdivision for each attribute, the greater the possible number of combinations of attributes that have to be examined. Because of the potentially large number of combinations that need to be examined, the use of computing facilities is an essential part of the landslide susceptibility assessment, particularly for the rapid production of high quality mapping. The computer software systems, computer analysis and mapping procedures used for MAP enabled large data sets to be generated, and allowed the manipulation of different classifications and combinations of landslide susceptibility variables. An important part of the study was the testing, discarding and refinement of hypotheses regarding the interrelationships between the landslide susceptibility attributes being studied. The final susceptibility maps were printed using a colour inkjet plotter.

The software used to perform the various operations can be divided into four classes: (i) topographic model creation, (ii) derivative mapping, (iii) image processing,

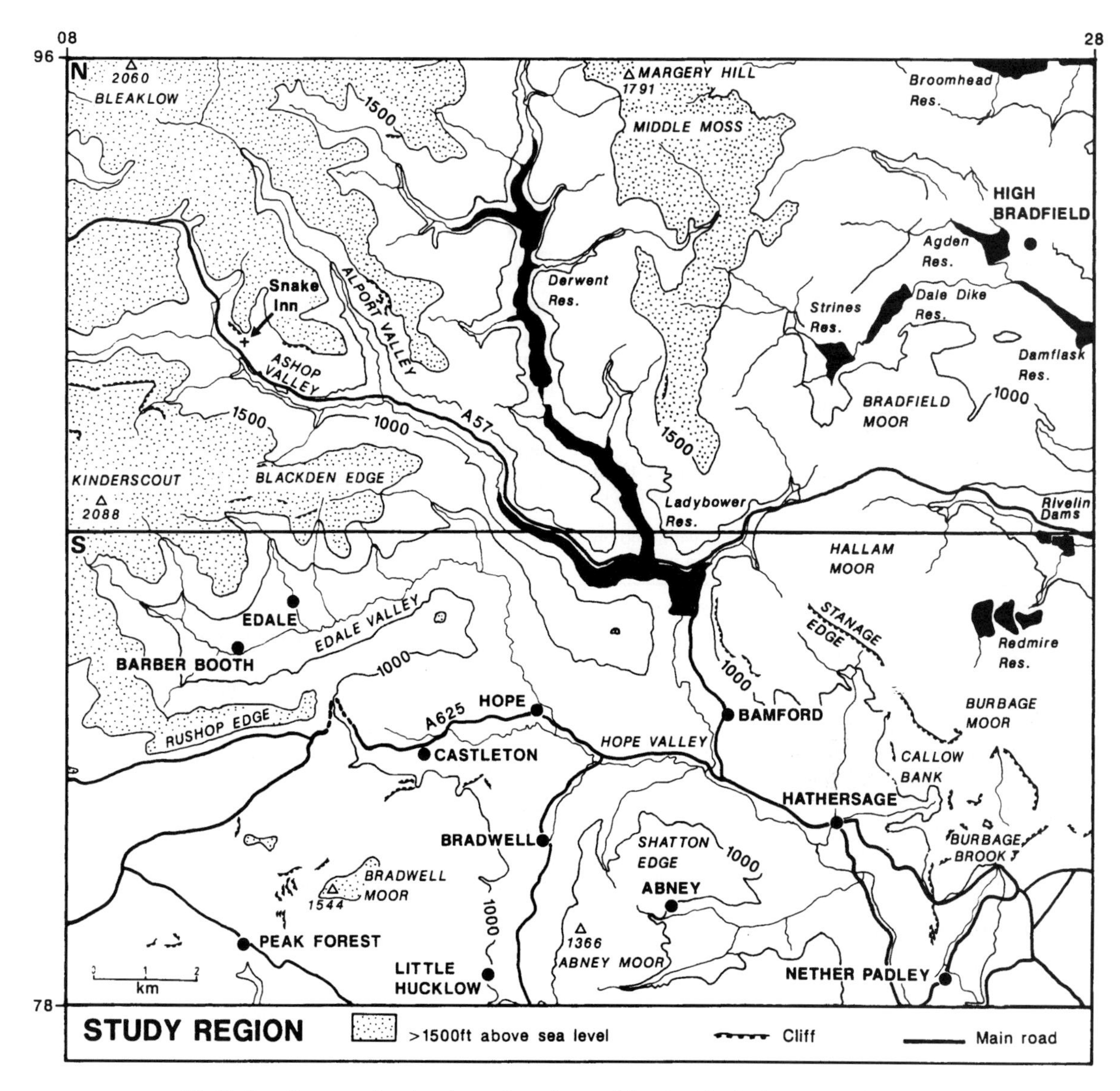

Fig. 2. Location map showing the study region used for the Matrix Assessment Approach.

and (iv) final map output. Three computer software systems were used: PANACEA, DERIVATIVE and MIRAGE. These have been developed for use on super-micros (i.e. Unix 6800) and written in Fortran 77 (McCullagh *et al.* 1985).

Computer analysis and mapping

The process of creating the landslide susceptibility grid from the nine terrain attributes has been described theoretically above. The practical problem lies in handling the total possible combinations when using the fine subdivision of all nine attributes. The maximum possible number of combinations for the Derbyshire case study area is 11 520. The number of zero entries in a table of all possible combinations will therefore be high. Because of this fact the data storage and retrieval system is designed so that only those combinations actually in use are retained for the rest of the analysis. This means that data storage for all combinations is unnecessary, and when a susceptibility ratio has been calculated for each

Table 3. *Slope aspect classification*

Fine classification	*Coarse classification*
1. N (1)	1. N, NE
2. NE (1)	2. E, SE
3. E (2)	3. S, SW
4. SE (2)	4. W, NW
5. S (3)	
6. SW (3)	
7. W (4)	
8. NW (4)	

Table 4. *Relative relief classification* (ft)

Fine classification	*Coarse classification*
1. 0–25 (1)	1. 0–25
2. 25–50 (2)	2. 25–75
3. 50–75 (2)	3. 75–125
4. 75–100 (3)	4. 125+
5. 100–125 (3)	
6. 125–150 (4)	
7. 150+ (4)	

Table 5. *Height above valley floor classification* (ft)

Fine classification	*Coarse classification*
1. 0–100 (1)	1. 0–200
2. 100–200 (1)	2. 200–400
3. 200–300 (2)	3. 400–600
4. 300–400 (2)	4. 600–700
5. 400–500 (3)	5. 700+
6. 500–600 (3)	
7. 600–700 (4)	
8. 700–800 (5)	
9. 800+ (5)	

Table 6. *Height above sea level clasification* (ft)

Fine classification	*Coarse classification*
1. 0–250 (1)	1. 0–500
2. 250–500 (1)	2. 500–1000
3. 500–750 (2)	3. 1000–1500
4. 750–1000 (2)	4. 1500+
5. 1000–1250 (3)	
6. 1250–1500 (3)	
7. 1500–1750 (4)	
8. 1750+ (4)	

Table 7. *Superficial deposits classification*

Fine classification
1. Head (1)
2. Hill peat (2)
3. River terraces (3)
4. Brown earths (4)
5. Iron pan stagnopodzols (5)
6. Brown podzolic soils (5)
7. Cambic stagnohumic gley soils (6)
8. Humic rankers (7)
9. Brown rankers (7)
10. Cambic stagnogley soils (6)
11. Humic ferric podzols

Coarse classification
1. Head
2. Hill peat
3. River terrace
4. Brown earths
5. Brown podzolic soils
6. Stagnogley soils
7. Rankers

Table 8. *Soils classification*

1. Typical brown earths (5.14) (Trusham, Bearstead 1, Malham 1 & 2, Rivington 2, Crediton) (4)
2. Raw oligo-fibrous peat soils (10.11) (Winter Hill) (1)
3. Cambic stagnogley soils (7.13) (Brickfield 3, Bardsley, Dale) (3)
4. Iron pan stagnopodzols (6.51) (Newport 1) (2)
5. Cambic stagnohumic gley soils (7.21) (Wilcocks 1) (3)
6. Humo ferric podzols (6.31) (Angelzark) (2)
7. Typical brown podzolic soils (Withnell 1) (2)
8. Rankers (3.13) (Crwbin, Wetton 1) (5)
9. Typical brown alluvial soils (5.61) (6)

Coarse Classification
1. Raw peat soils (10.1)
2. Brown podzolic soils (6.1)
3. Stagnogley soils (7.1)
4. Brown earths (5.4)
5. Rankers (3.1)
6. Brown alluvial soils (5.5)

cell a vast search is not required to find any given combination.

Rather than creating an actual nine-dimensional matrix in the computer memory, a database structure was used to index only those combinations that existed. The approach used was designed to maintain rapid access to the data and determine when a given combination was not accessed. The input/output overheads allowed the use of a memory-held hash-table approach: this was considered to be appropriate for a super-micro system, which had in the region of 1 Mb of memory available for any given process.

Table 9. *Bedrock geology classification*

Fine classification
1. Edale Shale (d_4) (Namurian Millstone Grit Series). (1)
2. Kinderscout Grit (KG) (Kinderscoutian R_1). (2)
3. Shale Grit (SG) (Kinderscoutian R_1). (2)
4. Mam Tor Beds (MT) (Kinderscoutian R_1). (3)
5. Monsal Dale Beds (Mo), Knoll reefs (K), Flat reefs (Kf), Litton Tuff (Z), Lower Millers Dale Lava (B_{11}), (Monsal Dale Group d_{3b}). (4)
6. Bee Low Limestone (BL), Apron reefs (Rap), Woo Dale (W), Dolerite (igneous intrusives) (D), Volcanic tuff (V), (Bee Low Group d_{3b}). (4)
7. Rivelin Grit (RG), Redmire Flags (RE), Heydon Rock (HR), Marsdenian (R_2). (5)
8. Crawshaw Sandstone (CRS) (Lower Coal Measures d_{5a}). (5)
9. Rough Rock (R) (Rough Rock Group). (6)
10. Lower Coal Measures (d_{5a}) (Lower Coal Measures d_{5a}). (6)
11. Eyam Limestones (EM), Longstone Mudstone (LSM) (Eyam Group). (4)

Coarse classification
1. Edale Shales (d_4) (Namurian Millstone Grit Series)
2. Kinderscout Grit (KG), Shale Grit (SG), (Kinderscoutian R_1)
3. Mam Tor Beds (MT), (Kinderscoutian R_1)
4. Dinantian (Carboniferous Limestone Series d^{2-3})
5. Marsdenian R_2 and Yeadonian R_2
6. Lower Coal Measures (d_{5a})

Table 10. *Slope steepness classification (percentage)*

Fine classification	*Coarse classification*
1. 0–10	1. 0–15
2. 10–20	2. 15–35
3. 20–40	3. 35–70
4. 40–50	4. 70+
5. 50–75	
6. 75+	

Derbyshire landslide susceptibility mapping

The landslide susceptibility index (LSI) was computed for the detailed nine-attribute full class subdivision, and also for a limited number of attributes and reduced class subdivisions in order to compare the LSI ratios obtained and to study the sensitivity of the LSI ratios to altered classification systems. If it is possible ultimately to obtain an acceptable estimate of landslide susceptibility by using either fewer (than nine) attributes and/or a coarser subdivision (of the selected attributes) then the amount of data acquisition and handling is greatly reduced. This, in turn, has advantageous implications regarding the computer memory size required for the susceptibility assessment.

The PANTONE halftone mapping output system (one of the units of the PANACEA digital terrain model system) was used to produce variable maps of slope steepness, relative relief, height above sea level and height above valley floor. The grids were produced using an inkjet printer, with each grid cell being a single 3 × 3 halftone unit. The MIRAGE mapping system was used to produce coloured maps of bedrock geology, bedrock combination, slope aspect, superficial deposits and soils.

One of the reasons for varying attribute combinations and attribute classification was to determine whether a decrease in the data used in the case study reduced the validity of the LSI to an unacceptable point. This was tested by comparing the effects of decreasing the number of attributes used in the Matrix Assessment Approach and the coincidence of high susceptibility LSI ratios within grid cells known to be affected by landsliding. Figure 3 shows the spatial distribution of landslides for the southern and northern areas of the case study.

Table 11. *Bedrock combination classification*

Fine classification
1. Edale Shale (d_4). (1)
2. Shale Grit (SG), Kinderscout Grit (KG), (Kinderscoutian R_1). (2)
3. Mam Tor Beds (MT). (Kinderscoutian R_1). (3)
4. Dinantian. (Carboniferous Limestone Series d^{2-3}). (4)
5. Rivelin Grit (RG), Redmire Flags (RE), Heydon Rock (HR) (Kinderscoutian R_1), Crawshaw Sandstone (CRS) (Lower Coal Measures). (5).
6. Lower Coal Measures (d_{5a}). (7)
7. Edale Shale/Kinder Scout Grit Group combination (SG/d_4, KG/d_4) (Kinderscoutian R_1) (8)
8. Mam Tor Beds/Edale Shale (MT/d_4) (Kinderscoutian R_1) (2)
9. Mam Tor Beds/Kinderscout Grit Group combinations (d_4/HR, d_4/RG,d_4/RE) and Edale Shale/Rough Rock Group combinations (d_4/R) (Marsdeian R_2, Yeadonian G_1) (5)

Coarse classification
1. Edale Shale (d_4).
2. Kinderscout Grit, Shale Grit, Mam Tor Beds/Kinderscout Grit, Mam Tor Beds/Shale Grit.
3. Mam Tor Beds, Edale Shales/Mam Tor Beds
4. Dinantian (Carboniferous Limestone Series d^{2-3}).
5. Middle Grit Group, Rough Rock Group, Edale Shale/Middle Grit Group, Edale Shale/Rough Rock Group.
6. Lower Coal Measures.
7. Edale Shale/Kinderscout Grit, Edale Shale/Shale Grit.

Results

Table 12 shows the frequency of landslide susceptibility values for each landslide susceptibility class using seven, four and three attributes for both fine and reduced class sets. The first class represents water areas and contained 144 cells. The landslide susceptibility classes then increase by steps of 0.1 to a top class covering the range of 0.8 to

(a)

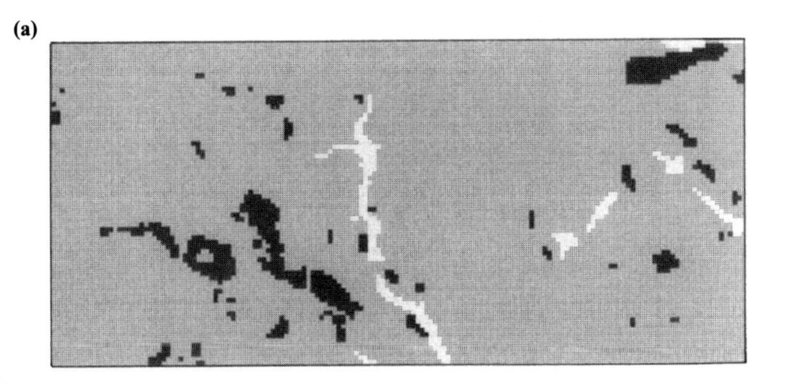

(b)

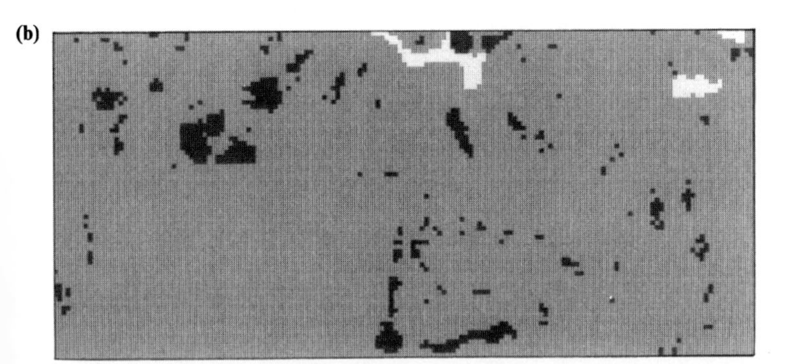

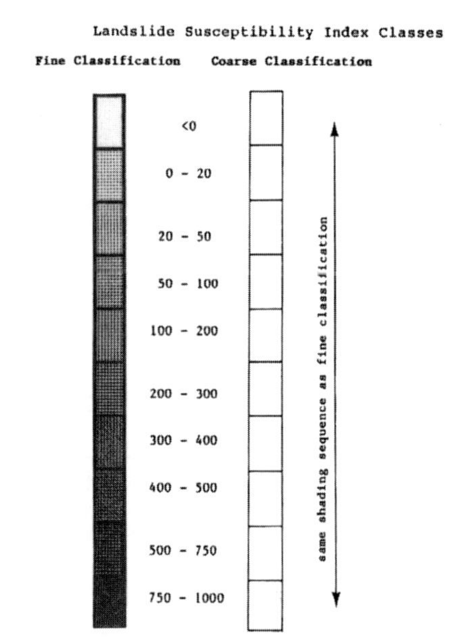

Fig. 3. (**a**) Landslide distribution for the northern sector of the study region. (**b**) Landslide distribution for the southern sector of the study region.

1.0. In the case of the seven-variable full class set, the number of combinations actively used was 6560 out of a maximum of 11 520. This indicated that most combinations only contained a few cells. Thus, using the seven-variable fine classification, cells tended to have either a high LSI ratio or a low one with few intervening values. This is demonstrated in Fig. 4a which shows the spatial distribution of susceptibility indices for the nine-variable full class set case. Figure 4a compares favourably with Fig. 3b which shows the actual landslide distribution in the southern section of the case study.

As the number of variables was reduced, the LSI ratios tended to decrease in magnitude, but covered a larger area. A comparison of LSI plots between the four-variable situation (bedrock geology, slope steepness, relative relief and height above valley floor) and the three-variable case shown in Fig. 4b (bedrock geology, slope steepness and slope aspect) shows that locationally the spatial distribution of the susceptibility index values are very similar, although the number of combinations observed was 1228 for the four-variable case, and 516 for the three-variable case. There was almost a 13:1 drop in the number of combinations between the seven-variable and three-variable full class sets. The ratio level for the three-variable case had fallen owing to averaging effects but the values covered a wider area of the study region than the LSI ratios in the seven-variable case. A comparison of LSI plots for the seven-variable full class set and the seven-variable coarse class set shows that the effects of a reduced class set are minimal despite the lower number of combinations involved; this pattern is also apparent for the three- and four-variable cases as shown in Table 12.

In order to assess the LSI values in more detail, the LSI values were printed out for each grid cell in the range between 0 and 1000 rather than 0 and 1.0 as previously described. A value of −1 was used to represent Ladybower, Derwent and other reservoirs. LSI values were highlighted within known landslide areas. As expected, the LSI ratios produced by the various attribute combinations tried, were greater in those grid cells corresponding to mapped landslide areas compared to non-landslide cells. The LSI ratios for mapped landslide areas were summed and the mean LSI ratio was calculated for four different attribute combination cases. Table 13 shows the mean LSI ratio values for mapped landslide areas for the four different attribute combination cases and the

Table 12. *Frequency of landslide susceptibility index values for each landslide susceptibility class using seven, four and three variables for both fine and reduced class sets*

Landslide susceptibility classification (LSI)	*Seven variables (fine class set)*	*Seven variables (reduced class set)*	*Four variables (fine class set)*	*Four variables (reduced class set)*	*Three variables (fine class set)*	*Three variables (reduced class set)*
Reservoirs	144	144	144	144	144	144
0–0.1	10 533	9522	9200	8702	9326	8601
0.1–0.2	70	841	1240	2068	1234	2449
0.2–0.3	63	435	499	537	450	316
0.3–0.4	63	281	249	43	176	6
0.4–0.5	5	77	52	10	140	0
0.5–0.6	174	87	84	10	36	4
0.6–0.7	36	53	6	3	6	0
0.7–0.8	0	26	9	0	0	0
0.8–1.0	432	54	46	3	8	
No. of combinations	6 560	1617	1228	192	516	74

Variable combinations used in Table 12. Seven-variable case: bedrock geology, slope steepness, slope aspect, height above sea level, height above valley floor, relative relief, superficial deposits. Four-variable case: bedrock geology, slope steepness, height above valley floor, relative relief. Three-variable case: bedrock geology, slope steepness, slope aspect.

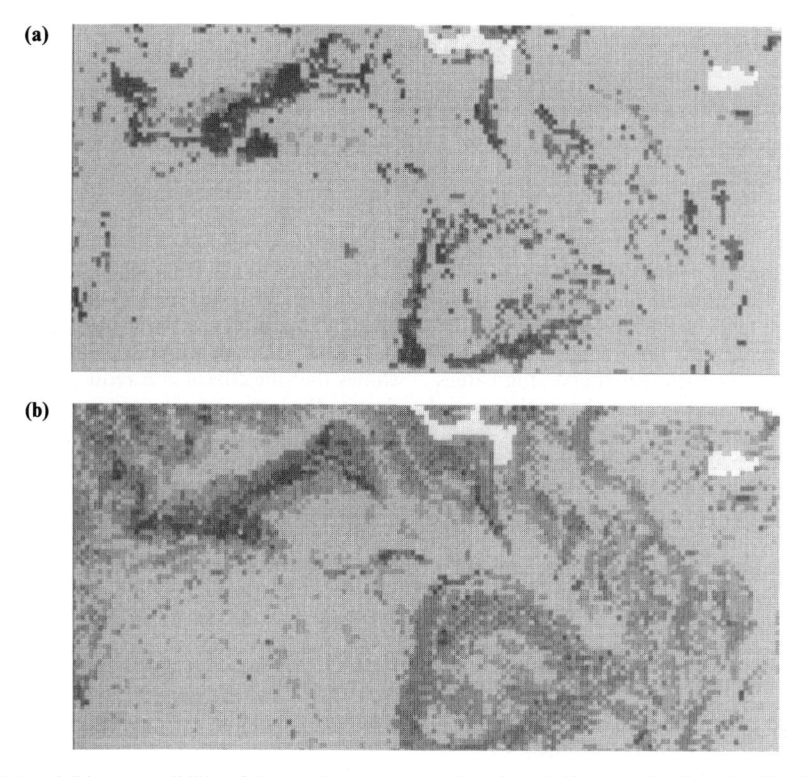

Fig. 4. (**a**) Landslide susceptibility of the southern sector, using nine attributes and a fine classification (slope aspect, superficial deposits, bedrock geology, relative relief, height above valley floor, slope steepness, height above sea level, bedrock combination, soils). (**b**) Landslide susceptibility of the southern sector, using four attributes and a fine classification (bedrock geology, relative relief, slope steepness, bedrock combination).

Table 13. *The mean LSI ratio values calculated for mapped landslide areas for four different attribute combinations and the frequency of none landslide cells possessing an LSI value greater than the calculated mean*

Attribute combination	*Mean LSI ratio for landslide areas*	*Number of none landslide cells > mean LSI ratio value*
Nine-variable, fine class set[1]	583	51
Seven-variable, fine class set[2]	479	136
Four-variable, fine class set[3]	208	449
Four-variable, fine class set[4]	127	1317

[1] Nine-variable, southern sector (fine class set): slope aspect, superficial deposits, bedrock geology, relative relief, height above the valley floor, slope steepness, height above sea level, bedrock combination, soils.
[2] Seven-variable, southern sector (fine class set): slope aspect, superficial deposits, bedrock geology, relative relief, height above the valley floor, slope steepness, height above sea level.
[3] Four-variable, southern sector (fine class set): slope aspect, relative relief, slope steepness, bedrock geology.
[4] Four-variable, southern sector (fine class set): slope aspect, relative relief, slope steepness, bedrock combination.

frequency of non-landslide cells possessing an LSI value greater than the calculated mean. The intention of calculating this mean LSI value was to indicate those non-landslide grid cells that possessed LSI ratios greater than the mean LSI value; these grid cell locations could then be considered to have a high degree of susceptibility to landsliding. The frequency of LSI ratios greater than the mean calculated for each attribute combination 1–4 is shown in Table 13. For the nine-variable, fine class case, 51 grid cells contained LSI values greater than the mean and were located in areas not mapped as containing landslides. The same 51 grid cells were also identified as having LSI values greater than the mean for attribute combinations given as 2, 3 and 4. Since the LSI ratios for the nine-variable full class subdivision tended to give either a high LSI ratio or a low one with few intervening values, the 51 grid cells identified were considered to represent the most susceptible areas for instability in the southern sector of the study region.

Using a seven-variable fine class set, a further 85 grid cells (discounting the 51 already determined by the nine-variable, fine class case) were identified as having a high susceptibility to failure. A comparison of the LSI ratio values for the nine-variable, fine class set and the seven-variable fine class set showed the spatial distribution of the LSI to be similar. Thus, it seems that the addition of the two extra variables (bedrock combination and soils) in the nine-variable full class set has had little effect on the overall distribution of LSI ratio values. The seven-variable fine class set included the variables, bedrock geology and superficial deposits; these are very similar with respect to their spatial distribution to the variables bedrock combination and soil respectively, therefore, one might expect little change in the LSI ratio distribution, with the omission of bedrock combination and soil variables in the nine-variable case.

A comparison of LSI ratios for the nine-variable reduced class set and seven-variable reduced class set showed the distributions to be very similar and not very different to the fine class set distributions. Thus, the seven-variable reduced classification appears to provide a reasonably accurate and acceptable level for landslide susceptibility classification.

The five-variable combination (bedrock geology, slope steepness, slope aspect, relative relief, and height above the valley floor) is very similar to the four variable case (slope aspect, bedrock geology, relative relief and slope steepness). Therefore, it appears that the variable height above valley floor used in the five-variable case has a minimal effect on the final LSI ratios produced.

Table 14 shows there are 449 cells greater than the mean LSI ratio for landslide areas for attribute combination 3 as opposed to 1317 for attribute combination 4. Table 14 also shows that the mean LSI ratio for attribute combination 3 is 208, this is much greater than the corresponding mean LSI ratio for attribute combination 4 (i.e. 127). This appears to confirm that bedrock combination is not as effective as bedrock geology, as a predictive attribute.

A comparison of LSI ratio plots where combinations of three variables had been used, shows that the variable combination (bedrock geology, slope steepness and slope aspect) is the most effective variable combination had been used. This particular attribute combination was used by DeGraff & Romesburg (1980) in their susceptibility mapping for the USDA Forest Service. A comparison of the three-variable case plots showed that slope aspect as a combinatorial variable is important and that bedrock combination is a poor choice of attribute for landslide susceptibility assessment. By using only three variables, one is still able to observe a general pattern of landslide susceptibility in

Table 14. *Field investigation in the northern sector*

No.	*Grid ref.*	*Name*	*Degree of instability*
1.	(130 925)	Hope Woodlands	Rotational slipping
2.	(135 897)	Hayridge Farm	Rotational slip
3.	(145 907)	Hucklow Lees Barn	Rotational slip and mudflow
4.	(153 889)	Woodlands	Rotational slip
5.	(166 923)	Marebottom	Small rotational slip
6.	(170 897)	Nabs Wood	Small rotational slip stabilized by retaining wall
7.	(175 897)	Pike Low	Small rotational slip
8.	(179 885)	Hag Side	No evidence of instability, but much of the lower section of slope had been removed for a car park
9.	(182 891)	Old House	Rotational slip
10.	(245 881)	Hollow Meadows	Rotational slip
11.	(262 891)	Load Brook	Small translational debris slide
12.	(273 931)	High Bradfield	Rotational slip
13.	(280 919)	Cliffe House	Rotational slip
14.	(280 939)	Bent Hills	Rotational slip
15.	(285 904)	Stacey Bank	Earth embankment stabilized above Damflask Reservoir outflow

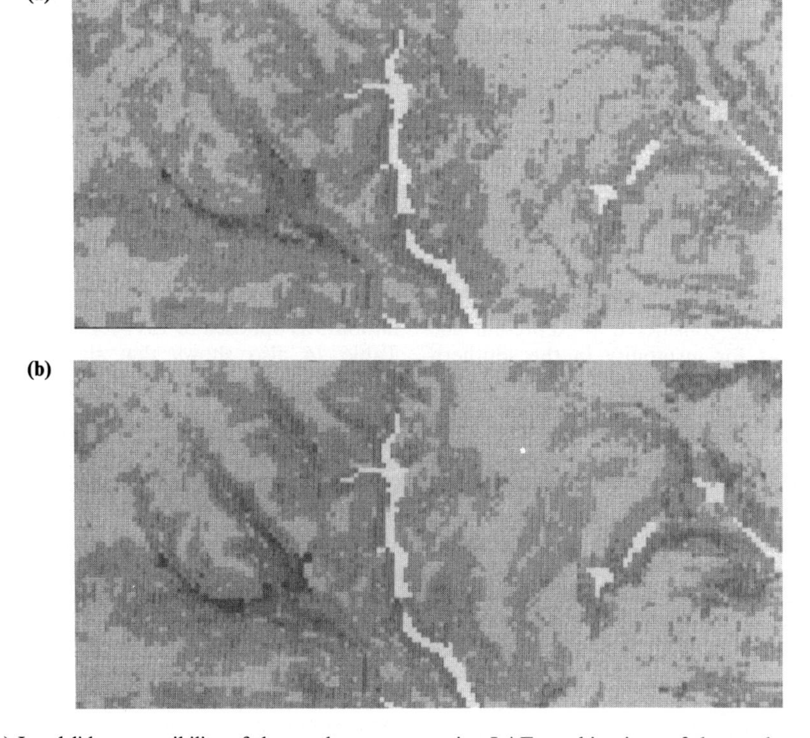

Fig. 5. (**a**) Landslide susceptibility of the northern sector, using LAF combinations of the southern sector to predict landslide susceptibility in the northern sector. Four attributes, reduced classification (bedrock geology, relative relief, slope steepness, slope aspect). (**b**) Landslide susceptibility of the northern sector, using LAF combinations of the northern sector to predict landslide susceptibility in the northern sector. Four attributes, reduced classification (bedrock geology, relative relief, slope steepness, height above sea level).

the study area. As slope steepness and bedrock geology were the only variables not changed in the various trials (three-variable cases), these two variables were considered to have the most dominating effect on the landslide susceptibility assessment model.

By systematically comparing the resultant changes caused by altering the number and combination of slope susceptibility variables as shown, one can begin to establish which variables appear to be the most important for landslide susceptibility assessment. The evidence gained from comparing the LSI value in each figure suggests that the following variables appear to be the most important: bedrock geology, slope steepness, slope aspect, relative relief and soil.

The reason why the use of the soil variable is considered here to be better than the superficial deposits data can be ascribed to two main factors:

- The classification used and the scale of mapping (1:25 000) carried out by the Soil Survey of England and Wales was such that it provided far more accurate and detailed information on the distribution of different soil groups.
- The soil classification provided a useful source of information regarding ground moisture conditions, i.e. through the soil wetness classification (see Tables 1 and 2). The superficial deposits classification could not take into account ground moisture conditions.

In order to determine whether the technique successfully identified landslide-sensitive slopes, a test was conducted to establish whether the LAF (landslide area factor combinations) in the southern sector could be used to predict the position of known landslides in the area immediately north of the southern sector. Because of the limited time and available computer storage space, only four variables (bedrock geology, slope steepness, relative relief and slope aspect) were collected for the northern area (RAF cells), using exactly the same procedure as that used for the southern sector landslide susceptibility mapping. Figure 5a shows the result of using the four variables and a reduced class set to predict landslide areas using the southern sector LAF in the northern section. This demonstrated that the technique was successful in identifying a classification of slope susceptibility to landsliding. Figure 5b shows the landslide susceptibility classification produced when the LAF combinations obtained from the northern sector are used in MAP rather than the southern sector attributes. A comparison between the distribution of landslide susceptibility index (LSI) values of Figs 5a and b shows that they are very similar although the LSI ratios in Fig. 5b have consistently slightly greater magnitude, particularly in known landslide areas. This demonstrates that the four variables (bedrock geology, slope steepness, relative relieve and slope aspect) used in MAP had similar importance in relation to landsliding for both the northern and southern sectors of the study area.

Table 15. *Field investigation in the southern sector*

No.	*Grid ref.*	*Name*	*Degree of instability*
1.	(086 869)	Edale Head	Shallow debris slide
2.	(093 855)	Lee House	Rotational slip
3.	(105 839)	Chapel Gate	Rotational slip
4.	(131 853)	Hollins (1)	Rotational slip
5.	(135 851)	Hollins (2)	Rotational slip
6.	(193 819)	Shatton Edge	Rotational slip
7.	(200 820)	Westfield	Rotational slip
8.	(195 860)	Ashop Farm	Rotational slip
9.	(196 853)	Win Hill	Rotational slip

The actual LSI ratios used to produce Fig. 5a in the northern sector were plotted for each grid cell. The highest LSI ratios were identified and actual landslide areas were delineated on the same plot. Those cells with high LSI values tended to correlate with areas of large multiple rotational landslide complexes along Snake Pass and Alport Dale. However, the technique was not very successful in locating small landslides in other areas; this may be attributable to the size of the grid used.

A test to determine whether the technique had successfully identified a classification of slope susceptibility to landsliding involved a quick walk-over survey in those areas represented by high LSI ratio cells and not containing mapped landslide areas. A number of cells containing high LSI ratio values from both northern (15 cells) and southern sectors (nine cells) were selected at random. These areas were visited and their degree of instability was noted (see Tables 14 and 15).

The field survey provided satisfactory evidence that the matrix assessment model provided a useful technique for classifying slope stability in both northern and southern sectors.

None of the landslide sites given in Tables 14 and 15 had previously been mapped on the 1:50 000 geological map sheets or had been recognized through aerial photograph interpretation as landslide areas. The landslide sites identified can then be added to the landslide distribution maps used in MAP and the landslide area factor (LAF) can be amended so that a more precise landslide susceptibility assessment of the study region can be undertaken.

Conclusions

The landslide combination matrix technique coupled with an on-line interactive graphics facility enabled the user to test hypotheses concerning:

- the optimum set of attributes to help explain slope failure, and
- the spatial pattern of slope susceptilbity for large areas.

The study has also shown that the matrix combination model is:

- sensitive to changes in the combination and number of variables used;
- less sensitive to changes in the classification; and
- that bedrock geology and slope steepness were the most important variables used but relative relief, slope aspect and soils also appear to be important susceptibility variables.

The Matrix Assessment Approach provides a useful preliminary medium-scale mapping technique which may be implemented during the reconnaissance stage of a proposed large-scale project that affects a region, e.g a highway scheme, a forest management scheme or a regional development planning scheme. The landslide susceptibility map produced using this technique should be considered primarily as a guide to slope instability and not an absolute indicator of existing or potential landslides. At the engineering design stage more detailed geotechnical investigations are required in the indicated areas (i.e. those cells containing high LSI values) to determine precise local variation. The basic data collection for the landslide susceptibility assessment could also be used by other specialists to carry out a variety of environmental investigations as part of a single multidisciplinary and multipurpose project for land resource evaluation and planning. With increasing technological advances in the field of Geographical Information Systems (GIS), map digitization and large-scale satellite imagery, the preparation of suitable base maps for MAP will become more cost-effective. The application of the Matrix Assessment Approach in the study area has provided a useful example of how techniques in automated cartography and numerical analysis can be integrated not only to provide a better understanding of the landslide susceptibility problem but also in the development of interactive display software systems of use to researchers, planners and engineers.

References

ANDERSON, M. G. & RICHARDS, K. S. 1987. Modelling slope stability: the complimentary nature of geotechnical and geomorphological approaches. *In*: ANDERSON, M. G. & RICHARDS, K. S. *Slope Stability*. Wiley, Chichester, 1–9.

AVERY, B. W. 1980. *Soil classification for England and Wales (higher categories)*. Soil Survey Technical Monograph, No. 14, Soil Survey, Harpenden.

BRABB, E. E. 1984. Innovative approaches to landslide hazard and risk mapping. *In*: *Proceedings of the 4th International Symposium on Landslides*, Vol, 1, Toronto, 307–324.

CARRARA, A., CARDINALI, M., DETTI, R., GUZETTI, F., PASQUI, V. & REICHENBACH, P. 1991. GIS techniques and statistical models in evaluating landslide hazard. *Earth Surface and Processes and Landforms*, **16**(5), 427–445.

CARSON, M. A. 1971. *Application of the Concept of Threshold Slopes to the Laramie Mountains, Wyoming*. Institute of British Geographers Special Publication, **3**, 31–48.

—— & KIRKBY, M. J. 1972. *Hillslope Form and Process*. Cambridge University Press.

—— & PETLEY, D. 1970. The existence of threshold hillslopes in the denudation of the landscape. *Transactions of the Institute of British Geographers*, **49**, 71–96.

CLAYDEN, B. & HOLLIS, J. M. 1984. *Criteria for Differentiating Soil Series*. Soil Survey Technical Monograph No. 17, Soil Survey, Harpenden.

COOKE, R. U. & DOORNKAMP, J. C. 1990 *Geomorphology in Environmental Management*, second edition. Clarendon, Oxford.

CROSS, M. 1987. *An Engineering Geomorphological Investigation of Hillslope Stability in the Peak District of Derbyshire*. PhD Thesis, University of Nottingham.

CROZIER, M. J. 1984. Field assessment of slope instability. *In*: BRUNSDEN, D. & PRIOR, D. B. (eds) *Slope Instability*. Wiley, Chichester, 1103–1142.

DEGRAFF, J. V. 1978. Regional landslide evaluation: two Utah examples. *Environmental Geology*, **2**, 203–214.

—— & ROMESBURG, H. C. 1980. Regional landslide susceptibility assessment for wildland management, a matrix approach. *In*: COATES, D. R. & VITEK, J. D. (eds) *Thresholds in Geomorphology*. Allen & Unwin, London, 401–414.

HANSEN, A. 1984. Landslide hazard analysis. *In*: BRUNSDEN, D. & PRIOR, D. B. (eds) *Slope Instability*. Wiley, Chichester, 523–602.

HARRIS, C. 1972. *Processes of Soil Movement in Turf-Banked Solifluction Lobes, Okstindan, Northern Norway*. Institute of British Geographers Special Publication, **5**, 155–174.

FREDLUND, D. G. 1987. Slope stability analysis incorporating the effect of soil suction. *In*: ANDERSON, M. G. & RICHARDS, K. S. (eds) *Slope Stability*. Wiley, Chichester, 113–143.

FREEZE, R. A. 1987. Modelling inter-relationships between climate, hydrology, and hydrogeology and the development of slopes. *In*: ANDERSON, M. G. & RICHARDS, K. S. (eds) *Slope Stability*. Wiley, Chichester, 381–403.

JOHNSON, R. H. 1980. Hillslope stability and landslide hazard: a case study from Longendale, North Derbyshire, England. *Proceedings of the Geologists Association*, London, **91**, 315–325.

——1987. Dating of ancient, deep-seated landslides in Temporate regions. *In*: ANDERSON, M. G. & RICHARDS, K. S. (eds) *Slope Stability*. Wiley, Chichester, 561–595.

—— & VAUGHAN, R. D. 1983. The Alport Castles, Derbyshire: a South Pennine slope and its geomorphic history. *East Midland Geographer*, **8**, 79–88.

JONES, D. K. C. & LEE, E. M. 1994. *Landsliding in Great Britain*. HMSO, London.

KIRKBY, M. J. 1987. General models of long-term slope evolution through mass movement. *In*: ANDERSON, M. G. & RICHARDS, K. S. (eds) *Slope Stability*. Wiley, Chichester, 359–379.

LAWRENCE, J. H. 1981. Urban capability as part of land use planning. *In*: *Geomechanics in Urban Planning*. Institution of Professional Engineers, New Zealand, **9** (2G), 328–336.

McCullagh, M. J., Cross, M. & Trigg, A. D. 1985. New technology and super-micros in hazard map production. *In*: *Proceedings of the Second UK National Land Surveying and Mapping Conference and Exhibition*. University of Reading, 1–16.

Richards, K. S. & Lorriman, N. R. 1987. Basal erosion and mass movement. *In*: Anderson, M. G. & Richards, K. S. (eds) *Slope Stability*. Wiley, Chichester, 331–355.

Taylor, R. K. & Spears, D. A. 1970. The breakdown of British Coal Measure rocks. *International Journal of Rock Mechanics and Mining Science*, **7**, 481–501.

Terzaghi, K. 1962. Stability of steep slopes on hard weathered rock. *Géotechnique*, **12**, 251–70.

Van Western, C. J. 1993. *Application of Geographic Information Systems to Landslide Hazard Zonation*. ITC Publication No. 15, ITC, Enschede, the Netherlands.

Varnes, D. J. 1982. Methods of making landslide hazard maps. *In*: *Landslides and Mudflows: Reports of Alma-Ata International Seminar*, October 1981. UNESCO/UNEP, Centre of International Projects, GKNT, Moscow, 388–406.

SECTION 5

GEOHAZARDS ASSOCIATED WITH UNDERGROUND SUBSIDENCE AND CAVITIES

Subsidence hazards caused by the dissolution of Permian gypsum in England: geology, investigation and remediation

A. H. Cooper

British Geological Survey, Kingsley Dunham Centre, Keyworth, Nottingham NG12 5GG, UK

Abstract. About every three years natural catastrophic subsidence, caused by gypsum dissolution, occurs in the vicinity of Ripon, North Yorkshire, England. Holes up to 35 m across and 20 m deep have appeared without warning. In the past 150 years, 30 major collapses have occurred, and in the last ten years the resulting damage to property is estimated at about £1 000 000. Subsidence, associated with the collapse of caves resulting from gypsum dissolution in the Permian rocks of eastern England, occurs in a belt about 3 km wide and over 100 km long. Gypsum ($CaSO_4.2H_2O$) dissolves rapidly in flowing water and the cave systems responsible for the subsidence are constantly enlarging, causing a continuing subsidence problem.

Difficult ground conditions are associated with caves, subsidence breccia pipes (collapsed areas of brecciated and foundered material), crown holes and post-subsidence fill deposits. Site investigation methods that have been used to define and examine the subsidence features include microgravity and resistivity geophysical techniques, plus more conventional investigation by drilling and probing. Remedial measures are difficult, and both grouting and deep piling are not generally practical. In more recent times careful attention has been paid to the location for development and the construction of low-weight structures with spread foundations designed to span any subsidence features that may potentially develop.

Introduction

Gypsum ($CaSO_4.2H_2O$) is widespread in the Permian and Triassic strata of England, but because of its very soluble nature it generally occurs only in low ground with an extensive cover of Quaternary superficial deposits; natural exposures of gypsum are rare. Its presence, especially in the Permian strata, is often indicated by natural catastrophic subsidence and existing subsidence features. In the urban area of Ripon, North Yorkshire, the catastrophic subsidence has caused about £1 000 000 worth of damage in the last 10 years and has generated problems for both planners and engineers. An understanding of the gypsum dissolution mechanisms and the geological controls of the subsidence are crucial for effective planning control and engineering practice. The problem is also of interest to property owners and insurance organizations.

Gypsum dissolution and subsidence history

Gypsum dissolves rapidly; a process illustrated by the dissolution of a gypsum cliff adjacent to the River Ure [NGR SE 307 753], 5 km N of Ripon (James *et al.* 1981). At this locality a block of gypsum, amounting to approximately 9 m^3, was dissolved, in only 18 months, by a river water flow velocity of about 1 m s^{-1}. Over the next 10 years the new gypsum face was undercut by a further 6 m collapsing again in 1989 (James 1992). Subterranean gypsum dissolution will form caves (Ryder & Cooper 1993) and fast flowing water in them may remove up to 0.5–1 m of gypsum per annum from the cave walls.

In England, Ripon suffers the worst subsidence caused by gypsum dissolution. Here at least 30 major collapses have occurred in the past 150 years and further collapses might be expected every two to three years (Cooper 1986, 1988, 1989, 1995). The subsidence hollows are commonly 10–30 m in diameter and reach 20 m in depth; east of the city they attain dimensions of up to 80 m diameter and 30 m deep. Numerous sags and small collapses on farmland go undetected, but careful analysis of historical levelling data suggests some areas of general ground lowering (Hogbin 1994). Many of the subsidence features at Ripon are arranged in a reticulate pattern (Fig. 1) related to the intersections of the joint systems in the rock (Cooper 1986).

Geology

The strata of the subsidence belt (Fig. 2) are mainly of Permian age. They comprise a lower carbonate (Cadeby Formation, 60 m thick) overlain by the Edlington Formation which consists of up to 40 m of gypsum overlain

COOPER, A. H. 1998. Subsidence hazards caused by the dissolution of Permian gypsum in England: geology, investigation and remediation. *In*: MAUND, J. G. & EDDLESTON, M. (eds) *Geohazards in Engineering Geology*. Geological Society, London, Engineering Geology Special Publications, **15**, 265–275.

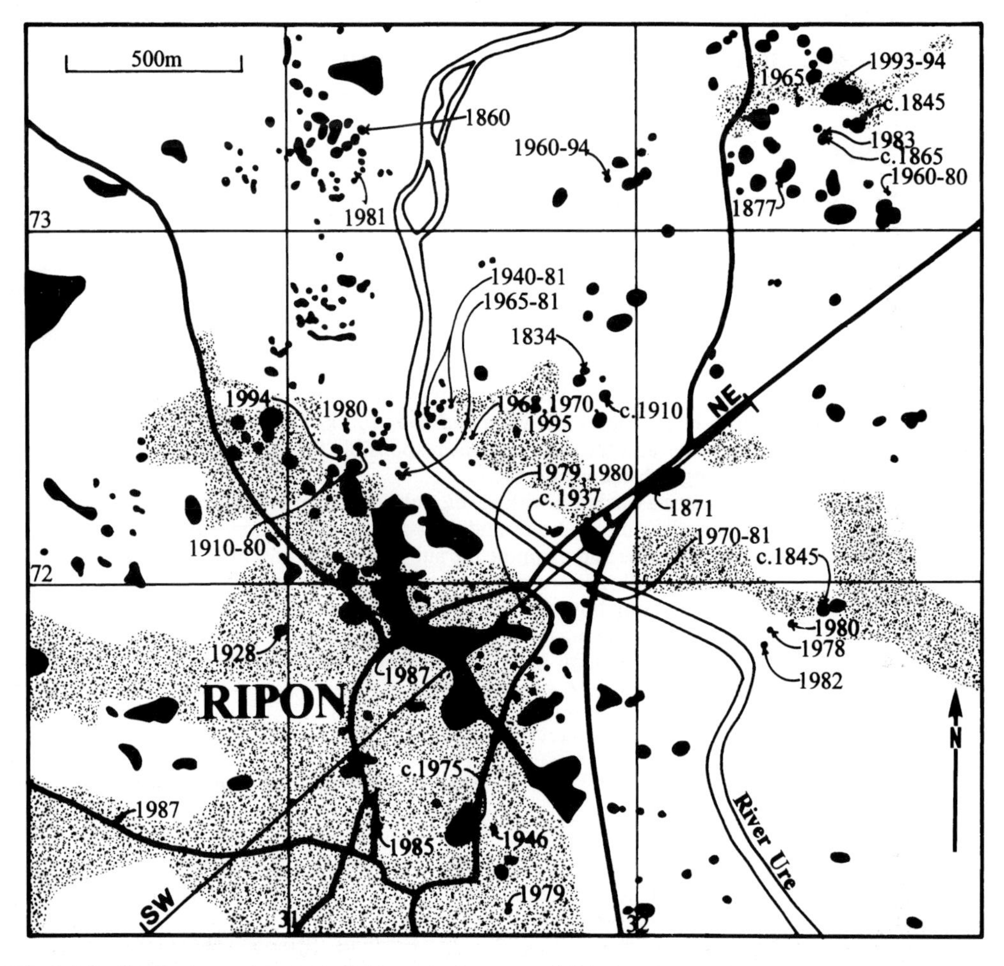

Fig. 1. The distribution and ages of subsidence hollows recorded in the vicinity of Ripon, North Yorkshire. The built-up area is shaded; subsidence hollows are shown in black. The approximate line of the section (SW–NE) illustrated in Fig. 2 is shown.

by 10 m of gypsiferous mudstone and mudstone. This is followed by another carbonate (Brotherton Formation, 12 m thick) overlain by up to 10 m of gypsum, then about 12 m of mudstone belonging to the Roxby Formation. The Cadeby and Brotherton Formations are both aquifers. The succession is capped by the red sandstone of the Triassic Sherwood Sandstone Group, the major regional aquifer, which attains 300 m in thickness. The rocks dip eastwards at about 2°–3° degrees and are partially concealed by Quaternary superficial deposits.

At Ripon the bedrock is cut through by a deep, largely buried, valley approximately following the course of the River Ure, and partially filled with up to 22 m of Devensian glacial and post-glacial deposits (Powell *et al.* 1992; Cooper & Burgess 1993). The buried valley intersects the carbonate and gypsum units creating a hydrological pathway from the bedrock to the river. Considerable groundwater flow occurs along this pathway and artesian water emanates from the Permian strata as springs which issue along the valley sides and up through the Ure Valley gravels. Artesian water, with a head above that of the river, has been encountered in some boreholes. Much of this water is nearly saturated with, or rich in dissolved calcium sulphate resulting from the dissolution of gypsum.

The greatest concentration of active subsidence hollows coincides with the areas marginal to the buried valley. Much of the sand and gravel partially filling the valley is cemented with calcareous tufa deposited from

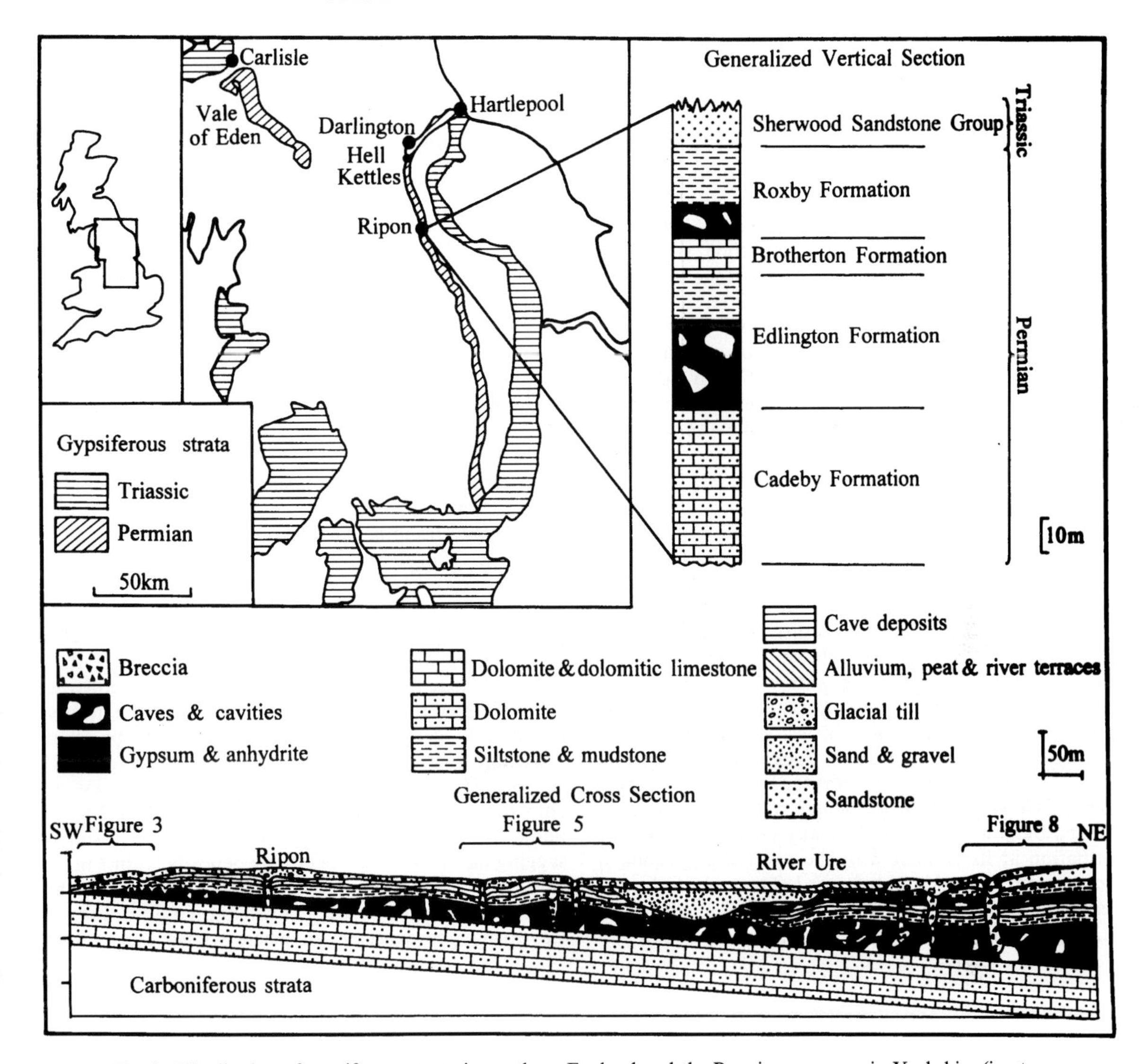

Fig. 2. Distribution of gypsiferous strata in northern England and the Permian sequence in Yorkshire (inset map). The sketch geological cross-section (approximately along line SW–NE in Fig. 1) illustrates the geology around Ripon. It shows caves in the gypsum sequence and their upwards propagation as breccia pipes. The legend is also that for Figs 3, 5 and 8, whose approximate positions are shown on the cross-section.

the groundwater which is also rich in dissolved carbonate (Cooper 1988). Subsidence is more extensive than it appears on the floodplain of the River Ure, because many of the subsidence features are infilled by overbank deposits.

Two factors delineate the subsidence belt. Westwards, the limit is the base of the lowest gypsum unit, which coincides with the base of the Edlington Formation; eastwards it is bounded by the down-dip transition from gypsum to anhydrite (Cooper 1986). The gypsum belt extends from Hartlepool, through Darlington and Ripon to near Doncaster. It has a maximum width of about 3 km and extends to a depth of around 100 m. From west to east across the belt the gypsum units show a progression from complete dissolution of gypsum, through buried pinnacled gypsum karst, to gypsum with caves and subsidence features, then finally to anhydrite (Fig. 2).

Variation of the subsidence features at Ripon

The localized character of the subsidence features is dependent on the underlying geological sequence and the superficial deposits at the surface. In the west of Ripon, only one sequence of gypsum is dissolving; in the east two are dissolving (Figs 2, 3, 5 and 8). In some places competent sandstone or limestone strata are present at the surface, in others soft mudstone is present. The situation is further complicated by differences in the Quaternary superficial deposits, which vary from thick valley-fill deposits to a blanket covering of glacial till (Fig. 2). Because of these variables, and as an aid to site investigation and subsequent design work, the nature of the geology and subsidence features from west to east across the area are described below.

Cadeby Formation

In the west of the area the Cadeby Formation, up to 60 m thick, forms the major escarpment of the Permian sequence. It is a local aquifer and yields water under artesian conditions which causes dissolution of the gypsum that lies immediately above it. Although the Cadeby Formation mainly consists of only moderately soluble dolomite, it does have caves and karst features in a few areas, such as near Doncaster, Wetherby, Knaresborough and the equivalent rocks in Sunderland. Generally it does not constitute a great geological hazard. However, in places where it has been dedolomitized by flowing water, especially near the top of the formation at its contact with the overlying gypsum, it may present a considerable hazard. The end product of dedolomitization may be a porous calcitic mush of low bearing strength that disintegrates to a 'sand' and this localized potential lack of strength should be considered if the rock is intended for founding piles.

Edlington Formation and associated gypsum

The main gypsum sequence in the Edlington Formation of Yorkshire rests immediately above the dolomite of the Cadeby Formation (Figs 2 3, 5 and 8). At Ripon it comprises up to 40 m of massive gypsum overlain by about 10 m of red-brown calcareous and gypsiferous mudstone. In the west at its feather edge the gypsum dissolves to form a pinnacled gypsum karst into which the overlying mudstones and Quaternary deposits founder (A in Fig. 3). From borehole information, and by comparison with the pinnacled gypsum karst seen in the Vale of Eden (Ryder and Cooper 1993), it can be expected that within this belt the gypsum pinnacles may have vertical sides, or be undercut to become cavernous (E in Fig. 3). The pinnacles and dissolution furrows may be many metres in height. The lower part of the gypsum in contact with the underlying Cadeby Formation may also be considerably dissolved and cavernous (C in Fig. 3). Throughout much of the Ripon area the pinnacled gypsum karst is concealed beneath 10–20 m of collapsed mudstone and superficial deposits; these will not bridge large cavities and the size of the collapses will be limited by the bridging capabilities of the massive gypsum. In general, individual subsidence depressions less than 10 m across occur within this belt of rock (C and G in Fig. 3), but in some areas they are amalgamated into larger subsidence areas (B in Fig. 3). Some of the subsidence hollows will be the result of sagging and flow of the overlying deposits (C in Fig. 3), while others will be underlain by breccia pipes, possibly penetrating to the Cadeby Formation (G in Fig. 3). Caves may be expected anywhere within the gypsum,

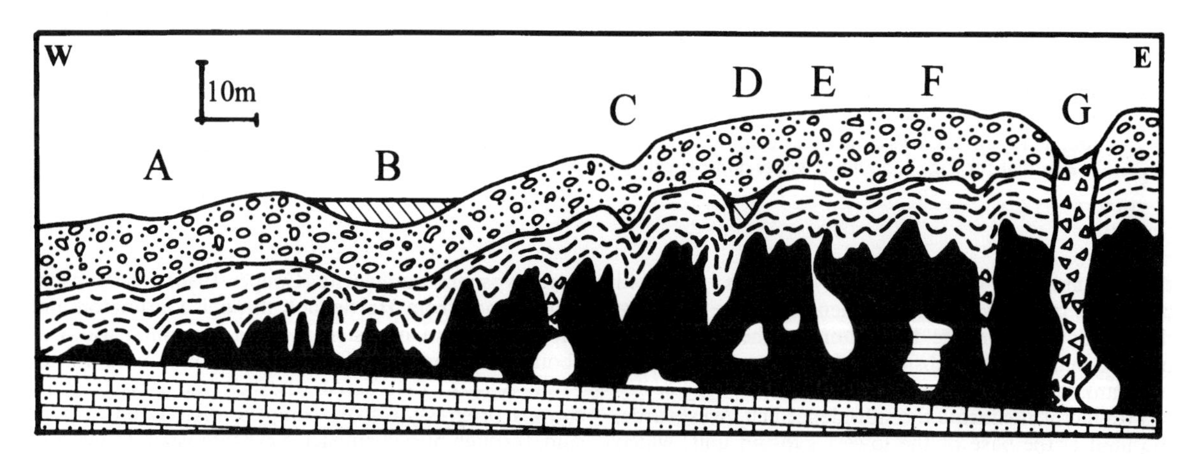

Fig. 3. Stylized cross-section through gypsum dissolution subsidence features in the west of the Ripon subsidence belt. See Fig. 2 for legend and the text for an explanation.

Fig. 4. Phreatic cave in the Permian, B-Bed gypsum at Houtsay Quarry [NY 624 276], Cumbria. The phreatic tube shows some preferential dissolution along some of the gypsum beds; the surface of the cave is also scalloped by dissolution. Some of the banding near the roof of the cave is water staining. The floor of the conduit is covered by thick glutinous mud to a depth of about 0.5 m (Ryder & Cooper 1993).

and caves partially filled in with laminated clays (F in Fig. 3; Fig. 4), deposited from material held in suspension by cave waters, and collapse breccias have been recorded. Within this area and in many of the areas to the east, fossil subsidence features predating the Quaternary glaciation may also be present. Thus subsidence hollows completely filled with glacial till, or laminated clay glacial deposits beneath the till, may occur (D in Fig. 3).

Brotherton Formation

Above the Edlington Formation, the Brotherton Formation comprises thin-bedded dolomitic limestone 8–14 m thick (Figs 2 and 5). It is an aquifer that feeds water down-dip to dissolve the overlying gypsum in the Roxby Formation. The Brotherton Formation can bridge considerable cavities that emanate upwards from the underlying gypsum (E in Fig. 5). However, when it fails it can do so catastrophically, producing hollows up to 20 m across such as those at Magdalens Road (in 1979 and 1980), Ripon [SE 3170 7192], (Fig. 6), and Nunwick [SE 3184 7473] (Cooper 1986; C in Fig. 5). Since up to 40 m of gypsum could be dissolved from beneath it (B in Fig. 5), the whole of the Brotherton Formation outcrop has a potential for subsidence.

At Ripon Golf Course [SE 312 732], the closely spaced subsidence hollows form a reticulate pattern of conical depressions probably formed over joint intersections in the gypsum. In the topographically lower parts of Ripon, such as Princess Road [SE 315 718] and Dallamires Lane [SE 318 703], similar groupings of intersecting conical subsidence hollows have become infilled with thick peat and clay deposits (A in Fig. 5). These areas have caused severe subsidence to many properties (Fig. 7). Attempts to overcome such severe foundation conditions have been made by piling. However, the piles terminate in the glacial till deposits and only partially cope with the problems of compressible peat and soft deposits which commonly suffer considerable further consolidation after construction. This type of foundation design does not address the problem of further gypsum dissolution and collapse at depth.

The Brotherton Formation limestones can bridge fairly large cavities and some of the gypsum caves below it have not collapsed. Instead they have become partially choked with cave deposits including laminated clay with subordinate peat and gravel washed into the cave system (D in Fig. 5). Thus it is possible to have a complex Quaternary sequence, at depth, underground beneath undisturbed surface geology. Breccia pipes penetrate the Brotherton Formation and include unconsolidated foundered material (Patterson *et al.* 1995), either Quaternary deposits or solid strata. In these pipes foundered areas of brecciated red-brown mudstone (derived from the Roxby Formation) or red sandstone (derived from the Sherwood Sandstone Group) commonly occur. Some of these collapse structures must be ancient because the stratigraphically younger formations, preserved in the pipes, have now been removed by erosion from the local area. Similar features penetrate the Brotherton Formation near Leeds at Sherburn in Elmet (Smith 1972). On the floodplain of the River Ure, subsidence hollows filled with unconsolidated organic-rich sand and silt occur. Some of these are further complicated by the presence of tufa-cemented deposits precipitated from artesian groundwater rich in dissolved carbonate.

Roxby Formation and associated gypsum

Above the Brotherton Formation, the Roxby Formation includes up to 10 m of gypsum overlain by about

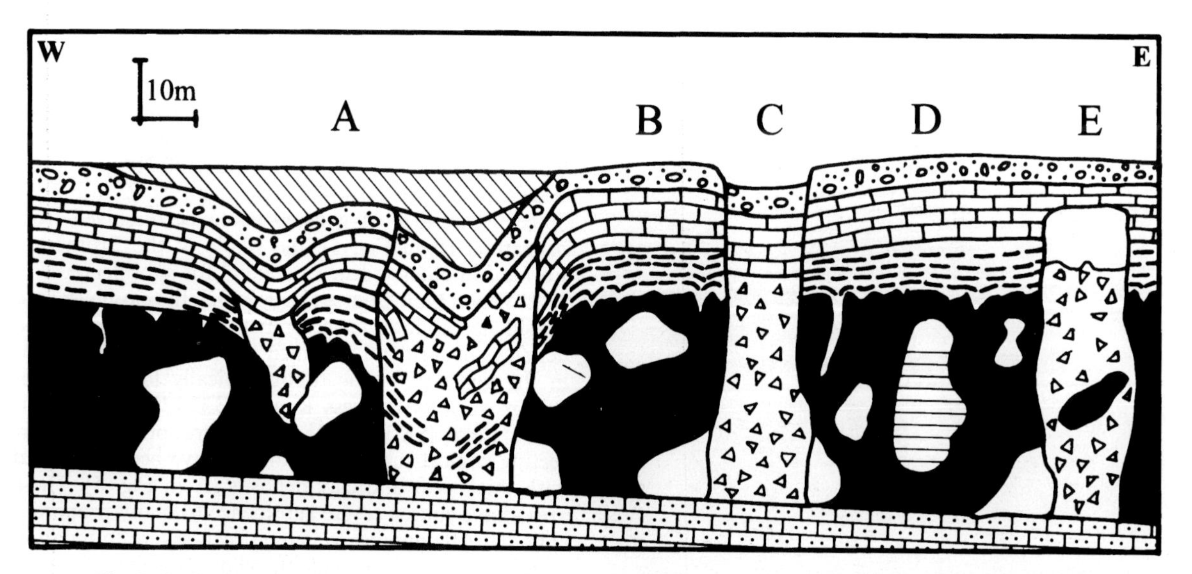

Fig. 5. Stylized cross-section through gypsum dissolution subsidence features in the centre of the Ripon subsidence belt. See Fig. 2 for legend and the text for an explanation.

Fig. 6. Subsidence hollow that formed on 28 July 1979 behind houses on Magdalens Road, Ripon [SE 3170 7192]. This hole was filled in and subsided a similar amount in 1980. (Photo: Acrill Newspapers)

Fig. 7. Subsidence-damaged houses on Princess Road, Ripon [SE 3156 7266] showing a sagging mode of failure. These properties have suffered differential subsidence caused by settlement of peat within a large amalgamated subsidence hollow; active gypsum dissolution may also have contributed to the problem. This property has recently been demolished and has been rebuilt in a similar style on a piled foundation.

10 m of gypsiferous and calcareous mudstone. Areas where the Roxby Formation is at rockhead have two potentially unstable sequences of gypsum. The complicated gypsum cave systems are linked by near-vertical subsidence and breccia pipes which unite the hydrological regimes of both the two dolomitic formations and the two gypsum sequences. Subsidence hollows developed on the Roxby Formation may reach about 20 m across, and are concentrated along the margins of the River Ure buried valley where artesian water escapes from the Permian sequence. Some hollows formed over this formation, such as the one developed at Sharow [SE 3238 7182] in 1982 (Cooper 1986, Fig. 3(b)), formed catastrophically and are deep-seated. The gypsum sequence in the Roxby Formation will be karstified, at rockhead, with pinnacles and caves similar to those in the Edlington Formation gypsum.

Sherwood Sandstone Group

The lower part of the Sherwood Sandstone Group is affected by subsidence. Up to 50 m total thickness of gypsum can be removed from beneath this Group up to 1–1.5 km from its western boundary. Spectacular subsidence hollows and amalgamated hollows up to 80 m across and 30 m deep occur near Hutton Conyers [SE 3259 7305]. Active subsidence hollows in this belt reach 40 m across, such as the one at Hutton Conyers that severely damaged a house in 1993 and 1994 [SE 3257 7339] (A in Fig. 8). The Sherwood Sandstone can bridge large cavities, but may fail catastrophically, resulting in large cylindrical subsidence pipes such as the one that developed near Ripon Railway Station in 1834 (C in Fig. 8; Fig. 9). Where thick drift deposits overlie the sandstone, the subsidence hollows form close groupings of conical-shaped depressions; good examples being Corkscrew Pits [SE 3200 7315] and near Hutton Conyers [SE 3273 7305] (D in Fig. 8). The breccia pipes associated with the subsidence can propagate up from the lower parts of the Edlington Formation which may lie as much as 100 m below surface. Like the more western areas, great variations in caves and cave deposits (B in Fig. 8) can occur beneath the area. The great depth and complexity of the subsidence problem makes detailed site investigation expensive and difficult.

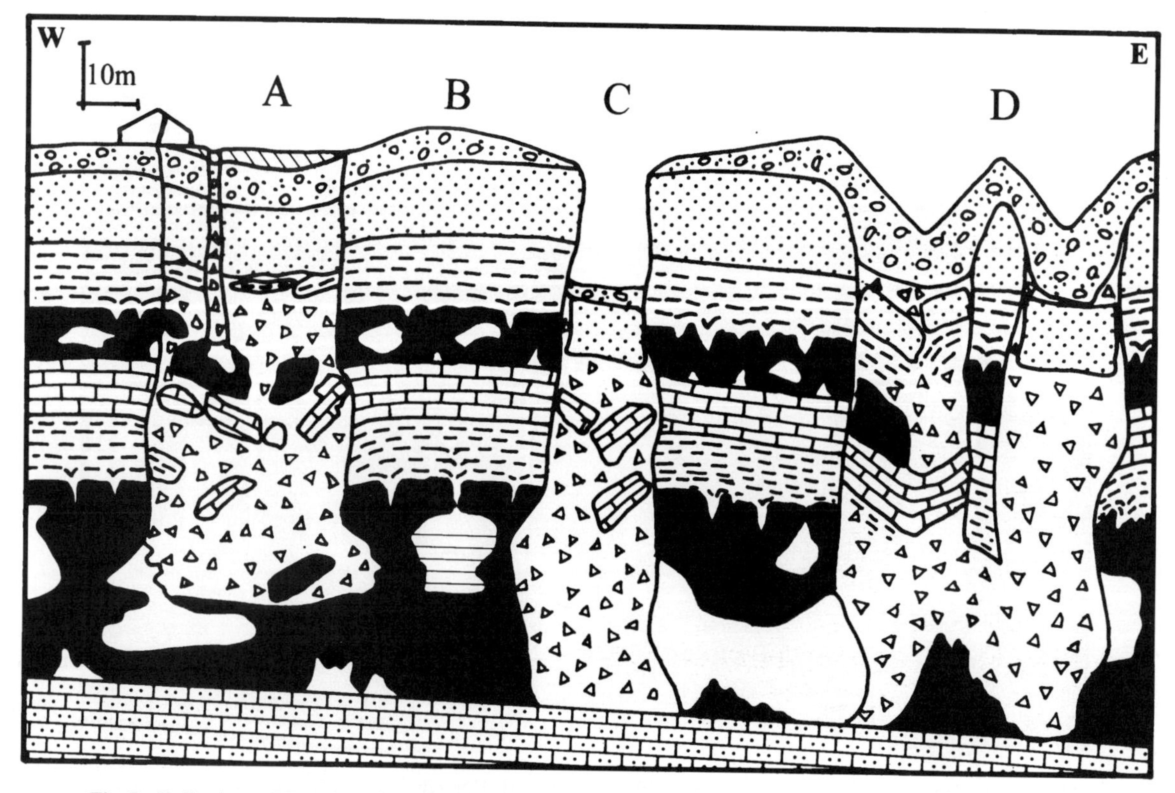

Fig. 8. Stylized cross-section through gypsum dissolution subsidence features in the east of the Ripon subsidence belt. See Fig. 2 for legend and the text for an explanation.

Gypsum dissolution and subsidence at Darlington

The subsidence-prone belt extends from Ripon northwards to Darlington and Hartlepool (Fig. 2), where the sequence is similar to that at Ripon, but the carbonate formations have different names. Around Darlington up to 40 m of gypsum is present in the Edlington Formation and up to 7 m in the Roxby Formation. Two types of subsidence occur in this northern area: (a) catastrophic collapse similar to that at Ripon and (b) more widespread settlement; the distinction between the two types is controlled by the hydrogeology and the thickness and lithology of the overlying glacial deposits.

South of Darlington at Hells Kettles, catastrophic collapse occurred in 1179. Four subsidence hollows up to 35 m in diameter and 6 m deep were formerly present (Longstaffe 1854), but one of these is now filled in. These hollows are very similar to those at Ripon (Cooper 1986, 1989). Artesian water emanates from Hells Kettles and from sulphate-rich springs nearby at Croft. Like Ripon, the sequence here dips gently eastwards and the outcrop of the carbonate formations is a groundwater recharge area. The groundwater moves down-dip to the low ground of the wide, partly-buried, valley of the River Tees. The subsidence appears to be associated with the margins of the buried valley, as at Ripon.

The southern half of Darlington has suffered subsidence related to gypsum dissolution. Here the subsidence has been prolonged, less severe (generally less than 0.3 m), and spread over subsidence depressions up to several hundred metres in diameter. Local boreholes have proved thick gypsiferous strata similar to those at Ripon, and cavities were encountered in one borehole. The bedrock surface forms a very broad valley filled with around 50 m of glacial and post-glacial deposits which include water-saturated sand and plastic laminated clays. As the gypsum dissolution proceeds, it appears that the overlying water-saturated sand flows into the gypsum cavities. Support is removed over wide areas causing broad subsidence depressions at the surface. The subsidence belt continues to the northeast of Darlington, extending to the coast at Hartlepool where thick deposits of anhydrite underlie part of the town.

Fig. 9. Subsidence hollow formed in July 1834 near Ripon Railway Station [SE 3186 7260]. The cylindrical hollow is 14 m in diameter and 15 m deep with red Sherwood Sandstone exposed in its sides.

Investigation of gypsum subsidence-prone areas for development

The depth of the gypsum sequences over many of the areas in question, and the evolving nature of the subsidence phenomenon, make site investigation and remediation difficult. Generally, only shallow site investigations have been undertaken over the gypsum subsidence belt. Detailed investigations, for modern developments, are now demanded by the planning authorities. If sites are investigated by boreholes alone, the size of the subsidence features demand closely spaced boreholes (at around 10 m intervals or less) drilled to the base of the gypsum; commonly this is 40–60 m deep under the city of Ripon. For such investigations it is important to core the solid strata and to have it logged by a geologist competent in identifying gypsum in all its forms. In the bulk of archival site investigation data, gypsum (except for satin spar) is usually identified as limestone. A potential problem with site investigation by drilling is the likelihood of triggering a subsidence event in unstable ground, either by vibration or circulation of drilling fluids; this has not happened yet, but should be considered when planning site investigations, the safety of the drill crew, and the associated insurance cover.

An alternative investigation technique is to use geophysics as part of a phased drilling and probing investigation. At Ripon, microgravity has successfully been used to delineate anomalies that have subsequently been drilled (Patterson *et al.* 1995). Computer-based modelling and field investigations (D. M. McCann pers. comm., 1992) show that microgravity can delineate breccia pipes and large cavities that breach, or come near to, the surface. However, even large caves, at depth, are difficult to image and edge effects of superficial deposits can partially conceal anomalies. At Ripon, subsidence features have also been investigated using ground conductivity electromagnetic and galvanic resistivity methods (gradient array; BGS RESCAN technique) (J. P. Busby pers. comm., 1992). These have proved a faster survey method than microgravity and have shown many anomalies, but few have yet been drilled to check the information, and electrical methods are difficult to use in built-up areas.

Evaluation of possible remedial measures

The dates and locations recorded for the historically recent subsidence events suggest concentration of water flow in the cave systems along certain specific paths (Fig. 1). The close grouping of subsidence hollows suggests that once a collapse has occurred the cave partially chokes and the dissolution continues in the adjacent strata. This commonly produces linear belts of subsidence related to the joint pattern. It also means that localities adjacent to, or in line with, existing subsidence hollows are probably more at risk from future subsidence (Fig. 10).

Engineers have suggested that grouting can be used to stabilize gypsum caves; this technique has been used in the Palaeogene gypsum of the Paris area (Toulemont 1984), but the long-term outcome of the work is not reported. In general, grouting of a gypsum cave system is not advisable. Unless the caves are small, proved to be abandoned and completely dry, filling them with grout could alter the groundwater regime. This could cause dissolution in the adjacent ground in the same way that natural collapse may block a cave system and through diversion of water channels cause dissolution nearby. There could also be problems caused by locally raising the local water-table which could trigger off subsidence. If a dry abandoned cave system was to be grouted, sulphate-resistant cement would have to be used.

Conventional piling, as already practised at Ripon, is also problematical; piles through disturbed and unconsolidated deposits may achieve the required bearing strength on the base of the pile in either the glacial deposits or the bedrock below. However, since the

Fig. 10. Subsidence hollow that appeared in March 1987 near the Clock Tower [SE 3132 7281] in Ripon. This subsidence feature (5 m long, 3.2 m wide and 2.2 m deep) is in an area of glacial till adjacent to a large area of amalgamated peat-filled subsidence hollows (extending along Princess Road; see Fig. 7).

bedrock contains gypsum beds, caverns might be present and these might propagate upwards thereby destabilizing the piled structures. It might be feasible to pile through the gypsum sequences, using bored piles, to the carbonate formation below, if the latter is not dedolomitized, but this could involve piling to depths of about 80 m to the east of the city. The use of sulphate-resistant cement would add to the cost and there is a danger that dissolution and collapse of the strata could place additional loads on the piles. This would necessitate the use of piles with a negative skin friction. Because of he prohibitive costs and likely difficulties associated with piling, it is largely impractical except in the west of the subsidence belt, or for all but the most expensive and sensitive structures.

An alternative approach devised for Ripon (Patterson *et al.* 1995) has been to delineate and avoid any subsidence hollows and breccia pipes. The constructions have then been placed within the site over the best ground conditions and designed to have minimal impact on the subsurface. They have also been designed to span any subsidence features that may potentially develop. This sort of approach can only be undertaken after extensive site investigation by engineering geologists working in close liaison with foundation engineers.

Acknowledgements. The help of numerous colleagues is acknowledged, especially D. McCann and J. Busby for information about geophysical techniques and M. Culshaw for engineering geology. P. Allen, T. Charsley, M. Culshaw and J. Powell are thanked for critically reviewing the manuscript. The geological survey of the Ripon area was supported by the Department of the Environment. A recent resurvey updating the information was supported by the Department of the Environment in association with Travers Morgan Ltd; A. Thomson of Travers Morgan Ltd is thanked for useful discussion. This paper is published with permission of the Director, British Geological Survey (NERC).

References

Cooper, A. H. 1986. Foundered strata and subsidence resulting from the dissolution of Permian gypsum in the Ripon and Bedale areas, North Yorkshire. *In*: Harwood, G. M. & Smith, D. B. (eds) *The English Zechstein and Related Topics*. Geological Society of London, Special Publication No. **22**, 127–139.

——1988. Subsidence resulting from the dissolution of Permian gypsum in the Ripon area; its relevance to mining and water abstraction. *In*: Bell, F. G., Culshaw, M. G., Cripps, J. C. & Lovell, M. A. (eds) *Engineering Geology of Underground Movements*. Geological Society of London, Engineering Geology Special Publication No. **5**, 387–390.

——1989. Airborne multispectral scanning of subsidence caused by Permian gypsum dissolution at Ripon, North Yorkshire. *Quarterly Journal of Engineering Geology (London)*, **22**, 219–229.

——1995. Subsidence hazards due to the dissolution of Permian gypsum in England: investigation and remediation. *In*: Beck, F. B. (ed.) *Karst Geohazards: Engineering and Environmental Problems in Karst Terrane*. Proceedings of the 5th Multidisciplinary Conference on Sinkholes and the Engineering and Environmental Impacts of Karst, Gatlinburg, Tennessee, 2–5 April 1995. A. A. Balkema, Rotterdam, 23–29.

—— & Burgess, I. C. 1993. *Geology of the Country around Harrogate. Memoir of the British Geological Survey*, Sheet 62 (England and Wales).

Hogbin, P. R. 1994. The uses of Ordnance Survey maps and data in engineering geology. MSc Thesis, Imperial College, University of London.

James, A. N. 1992. *Soluble Materials in Civil Engineering*. Ellis Horwood, Chichester.

——, Cooper, A. H. & Holliday, D. W. 1981. Solution of the gypsum cliff (Permian Middle Marl) by the River Ure at Ripon Parks, North Yorkshire. *Proceedings of the Yorkshire Geological Society*, **43**, 433–450.

LONGSTAFFE, W. H. D. 1854. *The History and Antiquities of the Parish of Darlington.* The proprietors of the Darlington and Stockton Times, London: republished by Patrick & Shotton.

PATTERSON, D. DAVEY, J. C. & COOPER, A. H. 1995. The application of microgravity geophysics in a phased investigation of dissolution subsidence at Ripon, Yorkshire. *Quarterly Journal of Engineering Geology (London)*, **28**, 83–94.

POWELL, J. H., COOPER, A. H. & BENFIELD, A. C. 1992. *Geology of the Country around Thirsk.* Memoir of the British Geological Survey, Sheet 52 (England and Wales).

RYDER, P. F. & COOPER, A. H. 1993. A cave system in Permian gypsum at Houtsay Quarry, Newbiggin, Cumbria, England. *Cave Science*, **20**, 23–28.

SMITH, D. B. 1972. Foundered strata, collapse breccias and subsidence features of the English Zechstein. *In*: RICHTER-BERNBURG, G. (ed.) *Geology of Saline Deposits.* Proceedings of the Hanover Symposium 1968. (Earth Sciences, 7). (Paris: UNESCO).

TOULEMONT, M. 1984. Le karst gypseux du Lutétien supérieur de la région parisienne. Charactéristiques et impact sur le milieu urbain. *Revue de Géologie Dynamique et de Géographie Physique*, **25**, 213–228.

Chalk solution features at three sites in southeast England: their formation and treatment

S. J. Rhodes[1] & I. M. Marychurch[2]

[1] WSP Environmental Limited, Intec 4 Wade Road, Basingstoke RG24 8NE, UK
[2] Card Geotechnics Limited, Alexander House, 50 Station Road, Aldershot GU11 1BG, UK

Abstract. Solution features are common phenomena within Chalk areas and the classical theory states that solution features are formed entirely by dissolution of the Chalk as a result of chemical weathering, probably during the Quaternary period. These solution features pose a risk to development and infrastructure works in these areas due to the presence of soft or loose infill materials. As a result a number of treatment methods have been developed to overcome the problems of construction in these areas.

Observations of solution features encountered at three sites in southeast England, all underlain by Upper Chalk, are reviewed and the treatment methods designed to overcome the hazards posed are discussed.

Introduction

Large areas of southeast England are underlain by Chalk, covered by variable thicknesses of Tertiary and/or recent drift deposits. Where the overlying deposits are thin, the Chalk interface is generally irregular and frequently punctuated by solution features, comprising near-vertical columns or pipes into the Chalk, and cone-shaped depressions. These features, according to classical methods of formation, are formed entirely by chemical weathering of the Chalk. According to West & Dumbleton (1972), the features are infilled with material derived from the overlying deposits. This infill is typically soft or loose in nature, and may contain voids.

The presence of solution features can be a significant problem for civil engineering works as the infill material is generally more compressible than the surrounding Chalk and liable to progressive or sudden collapse. This has resulted in problems with settlement of structural foundations, road pavements and services.

Chalk: its origin and engineering description

Chalk in the UK may generally be described as a moderately weak fine-grained white carbonaceous rock containing nodules and bands of flint in certain horizons. The Upper Chalk is the most recent of the late Cretaceous deposits and is unconformably overlain by the Tertiary deposits. The two periods are separated by a marine regression and a period of uplift and erosion. It is estimated that approximately 150 m of Chalk was removed during this period (Bennison & Wright 1972). The near-surface Chalk in southeast England generally weathers to a matrix-dominant material comprising white/grey sand and silt size fragments with lumps of intact Chalk and flints.

Several classifications have been developed to grade the various weathering states of Chalk, including those by Ward *et al.* (1968), Wakeling (1966), Jenner & Burfitt (1974), Spink & Norbury (1989) and CIRIA (1994). The system developed by Spink and Norbury has been used to classify the weathering grades of Chalk within this review.

This system comprises six weathering grades, with Grades V and VI Chalk considered to be highly weathered and reworked due to periglacial disturbance, either *in situ* or involving some solifluxion. During the periglacial period the ground was frozen, with the permafrost extending to depths of hundreds of metres. When in a permanently frozen state, little weathering of the Chalk occurred due to its impermeable nature. However, seasonal thawing and re-freezing of the upper 7 m or so, within the active layer, resulted in the Chalk being broken down into a silt and sand sized matrix which typifies the Grade VI and V Chalk. Chalk of Grade IV and less is generally considered not to have undergone reworking, showing structure in the form of bedding. Fractures within the Chalk of Grade IV and below are considered to be the result of stress relief due to weathering and erosion during the Tertiary unconformity and possible tectonic disturbance.

RHODES, S. J. & MARYCHURCH, I. M. 1998. Chalk solution features at three sites in southeast England: their formation and treatment. *In*: MAUND, J. G. & EDDLESTON, M. (eds) *Geohazards in Engineering Geology*. Geological Society, London, Engineering Geology Special Publications, **15**, 277–289.

Description of solution features and the classical theory of formation

The presence of solution features in the form of pipes was identified as early as 1906 by Osborne-White (1906) in the geological memoir for Hungerford and Newbury. A number of papers have been published discussing the occurrence and formation of solution features and their engineering significance, most notably by West & Dumbleton (1972), Higginbottom (1965) and Fookes & Higginbottom (1971).

Solution features are typically present where the Chalk is overlain by a thin, typically between 1 m and 10 m thick, layer of granular Tertiary or superficial drift deposits (West & Dumbleton 1972) and where groundwater is present at depth within the Chalk. They have also been identified in areas where the Chalk is not overlain by superficial deposits but are close to the margin. This is thought to be the result of erosion of the overlying deposits leaving the Chalk exposed (Higginbottom 1965).

The formation of solution features is widely considered to result from the dissolution of the Chalk by surface water which has been concentrated into a localized area, perhaps due to topographic features such as a surface depression. The inflowing water is concentrated into a fissure or a group of fissures within the Chalk giving rise to localized deep weathering. Chemical weathering of the Chalk occurs vertically along the full length of the fissure. Chalk adjacent to the fissure is weathered in a horizontal direction producing greater weathering at the top of the Chalk profile, where the Chalk is less massive, and the Chalk becomes less weathered with depth due to the water becoming saturated with calcium carbonate. This gives the classical solution features their stylized inverted cone shape, although many solution features observed in the field do not show a well-developed cone shape and may be vertical pipes.

Solution features are typically circular or elliptical in plan and may vary in size from a few centimetres to tens of metres in diameter (Culshaw & Waltham 1987). The typical inferred formation of a solution feature is shown in Fig. 1. It is generally thought that solution features were formed during periglacial periods (Gibbard 1985). A higher volume of water was available as meltwater from the regressing glaciers during this time, and the lower temperatures allowed dissolution of more carbon dioxide producing a more corrosive solution of carbonic acid.

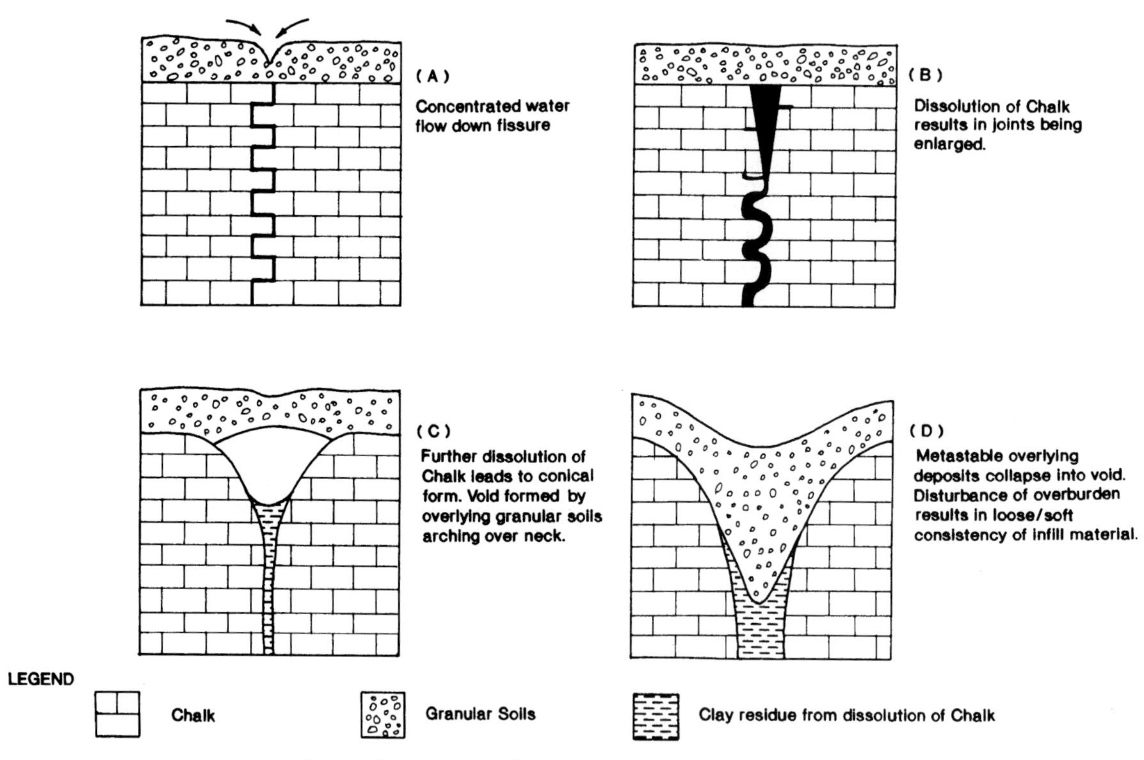

Fig. 1. Classical theory for development of solution features.

Limitations of the classical theory of formation

There are problems with assuming that solution features are formed purely by chemical dissolution of the Chalk during periglacial periods. Under this environment only the upper 7 m or so of the ground would thaw thus preventing deeper weathering of the frozen Chalk. Solution features, however, have been identified at depths of 20 m or even greater. Repeated freezing and thawing of the active layer would rapidly break the Chalk down into a silt-sized matrix-dominant material with a low permeability, restricting continued chemical weathering.

Chemical weathering of carbonate rocks may occur during a periglacial climate within sub-permafrost taliks, the permanently thawed zone which separates the permafrost from the active layer (Gibbard 1985). However, this does not explain their formation within the reworked Chalk. The theory of post-Quaternary dissolution of the Chalk requires surface water to flow through a mantle of low-permeability Grade VI/V Chalk.

Alternative method of formation

Observations from sites in southeast England show that not all of the solution features identified conform to the classical theory of formation. In particular, many of the features identified did not contain soft infill materials and lacked either the voids or surface depressions associated with their collapse, which are associated with the classical theory of formation. It is considered that some of these features may be the result of both periglacial reworking and post-glacial dissolution. The continued seasonal freezing and thawing cycles which occurred during the periglacial climate may have led to the development of ice wedge pseudomorphs, which may act as preferential flow paths concentrating surface water flow and allowing the deeper weathering of the *in situ* Chalk.

The principal difference between a solution feature formed this way, as opposed to the classical theory, is that deep weathering may not have occurred and therefore these features may not have the deep neck infilled with soft/loose debris. This mode of formation does not necessitate the presence of significant voids into which the infill material falls due to void collapse, typically associated with the archetypal feature, and therefore the infill material may be more dense. In addition, there is an absence of surface depressions as the infill material has not collapsed into the feature. The possible method of formation for these features is shown in Fig. 2.

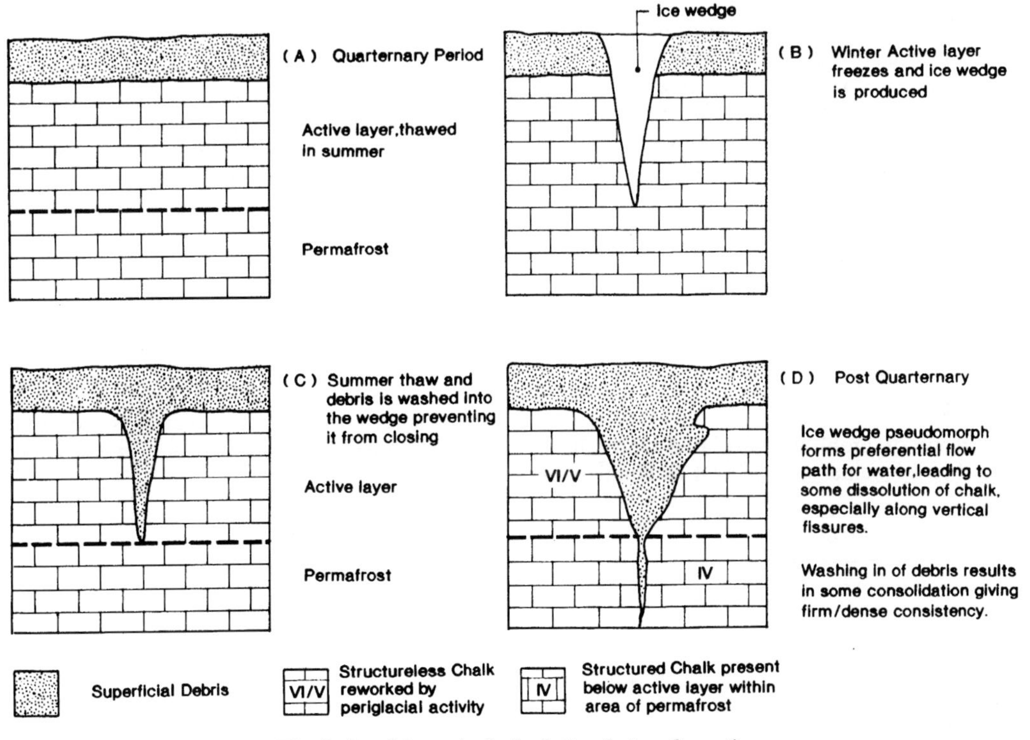

Fig. 2. Possible method of solution feature formation.

Infill materials

Where chemical dissolution of the Chalk has occurred an infill material comprising a clayey residue together with an accumulation of flints may be present. Observations by West & Dumbleton (1972), Higginbottom (1965) and Fookes & Higginbottom (1971) show the infill material to be soft clay or loose sand material. It is considered that the nature of the infill is loose/soft due to the formation of voids at the base of the feature caused by dissolution of the Chalk with collapse causing void migration and subsequent loosening of the infill material. Voids have been identified in a number of solution features due to the overlying soil arching and so producing a meta-stable cavity.

Engineering hazards

The solution feature infill represents a hazard to building structures, civil engineering and highway works. Primarily solution features are infilled with loose materials which may not be able to support the applied loadings. The materials are generally more compressible than the surrounding intact Chalk and long term settlement may occur. Solution features may contain voids within the infill material which can migrate to the surface by progressive collapse of the meta-stable overburden deposits. Collapse can be initiated by the following:

- Stresses induced by foundation loadings or live loadings on carriageways.
- Inflow of water into the solution features. Typically solution features manifest themselves after periods of heavy rainfall. Collapse may be initiated by human-made sources of water such as soakaways. Similarly partial collapse of a feature may cause rupturing of drains or water mains allowing a concentrated inflow of water into the feature and so initiating further collapse.
- Lowering of the groundwater by de-watering may initiate collapse of features due to the increase in effective stress.
- Vibrations from adjacent site activities, such as compaction plant.

The occurrence of problems associated with Chalk solution has been documented by West & Dumbleton (1972) in relation to the M4 and M40, by Culshaw & Waltham (1987) in relation to the M25 at South Mimms, and by McDowell (1989) at several sites in West Sussex.

Investigation of solution features

Any ground investigation, whether targeted or based on a random approach, can only expect to confirm the presence of solution features at a site and characterize their size, nature and likely extent. In addition, where there is surface evidence of potential solution features these can be investigated. The probability of identifying all of the solution features at a site from the investigation is small and therefore it is necessary to confirm their presence or absence during the site preparation works, and to have in place methods which will overcome the hazards based on an assessment of the associated risks.

Engineering treatment

A number of engineering methods have been used in the UK to treat sites underlain by solution features below structures, roads and infrastructure. Table 1 summarizes the economically feasible and commonly adopted treatment methods.

Excavation and replacement

The simplest method of treatment for shallow infilled solution features is to excavate the infill material and replace it with engineered fill or concrete. Partial removal of the infill can be undertaken and a concrete plug placed within the feature, providing that the plug has sufficient bearing on intact Chalk or can develop sufficient skin friction from the concrete/Chalk interface.

Ground treatment

Granular infill may be suitable for compaction-based ground treatment methods to enhance the bearing capacity and stiffness of the material to accept loading. Compaction methods may include dynamic compaction and vibroflotation.

Table 1. *Remedial engineering options*

Method of treatment	*Structures*	*Roads*	*Infrastructure*	*Relative cost*
Excavation/replacement	•	•	•	Low cost
Ground treatment (compaction)	•	•	•	Moderate
Ground treatment (grouting)	•	•	•	High
Reinforced capping		•	•	Low/moderate
Modified foundations	•			Moderate/high

Dynamic compaction. This involves the repeated impact of a weight onto the formation to densify the underlying soils and may be used to improve the engineering properties of loose granular infill materials by

- introducing collapse settlements of voids within the infill material;
- compacting the skeletal structure of the infill material.

The application of this method reduces long-term settlement of the infill material and improves its bearing capacity and increases its density, so reducing the soil's permeability. Sands, silts and some clays can be improved using dynamic compaction techniques. Dynamic compaction alone may not provide permanent treatment for solution features as future reactivation of the feature can occur. Surface depressions formed by the collapse of the solution feature may be infilled with granular material incorporating a membrane to prevent ingress of water (Koch 1984).

There are restrictions on the depth of treatment by dynamic compaction and in its effectiveness below the water-table. Dynamic compaction can only be undertaken away from existing structures as the energy generated may be transmitted within the soils and could initiate settlement of solution features below buildings. The vibrations may also induce structural distress of buildings.

Vibroflotation. This involves the inserting of a vibrating poker to densify granular soils (vibro displacement) or to replace cohesive soils with stone columns (vibro replacement), improving the overall engineering properties of the formation. Vibroflotation techniques can be used closer to existing structures than dynamic compaction as the vibrations are limited to a small radius around the columns, and can improve the ground to a greater depth than dynamic compaction. However, the stone columns may act as a preferential drainage path for surface water, initiating further development of the solution feature.

Grouting

Grouting involves the injection of a fluid cementitious or chemical grout into the ground under gravity or low pressure to infill voids and improve the properties of granular soils. Initially coarse-grained grout is used to infill cavities. Following this, secondary and tertiary grout points may be installed and a less viscose grout used to infill smaller cavities and voids.

Cohesive infill can be improved by hydrofracturing the clay and injecting columns of low viscosity grout to improve the overall properties of the infill material. This method, referred to as compaction grouting, is considered the most appropriate method for treating solution features below existing structures (Henry 1989).

Grouting can be expensive due to the volumes of grout frequently required, especially with fine-grained soils where an expensive low viscosity grout is needed. In addition, Chalk forms a major aquifer, from which groundwater is abstracted, therefore the Environmental Agency may prohibit the injection of grout down solution features.

Geogrid-reinforced capping

In situations where ground movement may not be critical, or where maintenance may be accommodated, the option of using geogrid-reinforced capping may be considered. This method comprises a granular blanket incorporating a geogrid-reinforcing element (Fig. 3), which is placed over an infilled feature with a sufficient length of geogrid extending beyond the limits of the feature to provide anchorage. Should settlement or collapse of the infill occur, the fill/geogrid blanket will induce tensile forces to develop within the anchorage zone thus preventing immediate collapse of the overlying construction. Long term creep of the geogrid may occur, although the system can be designed to minimize this using additional reinforcement, or alternatively, future maintenance work may be accepted.

Modified foundations

Where structures are unable to avoid solution features a number of foundation options are available:

- piled foundations
- raft foundations
- cruciform foundations
- capping slabs

Piled foundations may be used where the ground investigation shows that the assessed risk of widespread and variable solution features is not great and the piles can extend to competent Chalk. Bored cast *in situ* piling systems allow the integrity of the Chalk to be confirmed below the level of the solution feature prior to installation. Where piles are constructed through infill material, account should be taken of negative skin friction on the section of the pile shaft located in the infill.

Raft foundations may be designed to span or cantilever over potential voids. It is common practice to accommodate a span over a potential void of 3 m width and a cantilever of 1.5 m–2 m. With wide foundations, the feature may be capped using a reinforced concrete capping slab over which the foundation is constructed.

Cruciform foundation systems comprise shallow reinforced strip foundations designed as beams which will span over potential voids. The foundation beams are extended beyond the footprint of the structure to permit the spanning action to develop, should a feature be encountered at an existing corner. This system is frequently applied to housing developments, etc., in areas

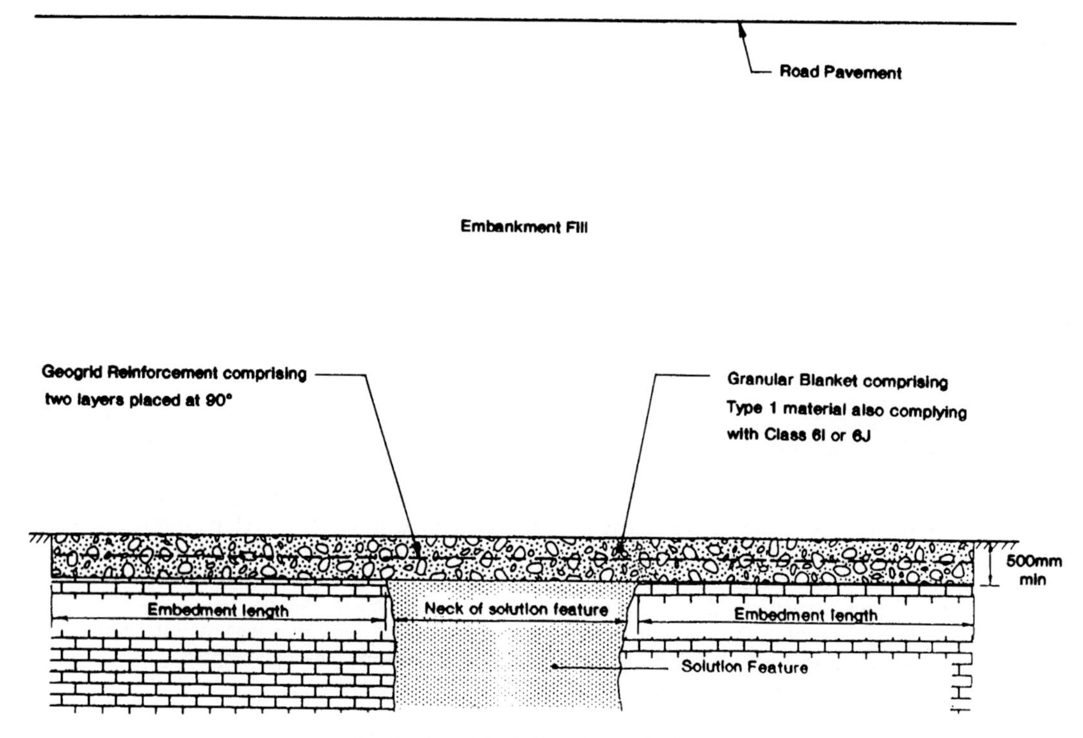

Fig. 3. Geogrid reinforced granular blanket.

prone to Chalk solution features where the profitability of the site may be marginal. Local authorities in these areas generally require that the foundations can span an unsupported length of 3 m and cantilever a distance of 1.5 m–2 m.

The methods described above have been derived from different approaches (e.g. from procedures adopted in mining areas) and require close understanding and co-operation between geotechnical and structural engineers. In particular, it is necessary to obtain a full understanding of the likely subsurface configuration of the features, which can only be gained from adequate site investigation. Strict control of construction activities needs to be exercised, particularly in respect of identifying the extent of the features. In addition it is important that the remedial measures do not act as sumps or conduits for surface water drainage as this may precipitate further movement.

Case studies

Solution features have been observed by the authors at three sites in southeast England, all of which are underlain by Upper Chalk. Specific details of the ground conditions and the configuration of the solution features are discussed below for each site and details are given regarding the engineering solutions designed.

Residential development, Dartford, Kent

A site in Dartford, Kent was being redeveloped for residential properties with strip foundations extending through a mantle of Brickearth and founding on to the underlying Chalk. During excavation for one of the foundations a solution feature was encountered.

The feature was investigated by excavating a series of trenches parallel with, and perpendicular to, the foundation trench, thereby allowing its full horizontal extent to be determined. This investigation showed the feature to be approximately 12 m in diameter. The depth of the feature was not proven during the excavation but it is known to exceed a depth of 5 m.

Surrounding Chalk was noted to be a structureless matrix-dominant material of Grade VI, grading into a Grade V with depth. Infill material was noted as comprising medium-dense sands and silts consistent with the overlying Brickearth deposits. No voids were

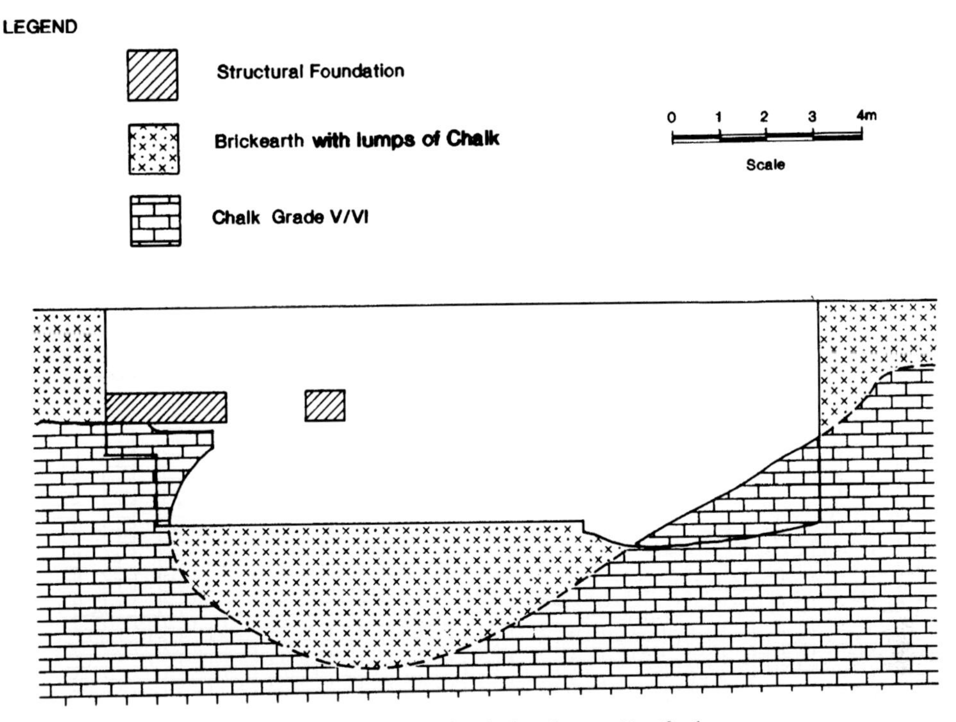

Fig. 4. Section through solution feature, Dartford.

noted within the infill and as it was present in a medium-dense condition it was concluded that void migration had not occurred. Some undercutting of the Chalk near to the neck of the feature was noted, giving the impression that apparently intact Chalk was underlain by Brickearth. A drawing showing the form of the feature is presented in Fig. 4. During excavation of the feature a number of columns of Chalk, Grade VI and V, were encountered together with amorphous lumps of Chalk within the infill material. This would suggest that the feature was made up from a number of interconnecting pipes of irregular form, infilled with a medium dense material, possibly originating as ice wedges, which were then connected together via weathering of the intervening Chalk. This would give a flattened base to the feature, consistent with the depth of the active layer.

Based on observations from the site, it was considered that the infill material would be more compressible than the surrounding Chalk, and settlement of an adjacent block had occurred resulting in a movement joint being incorporated into the structure. It was therefore considered that a remediation solution was required in order to maintain the long term integrity of the structures.

The foundations to the block were modified to span over the feature. This involved construction of a 12 m beam supported at either end by piled foundations so transferring wall loads to the adjacent intact Chalk. A section through the modified foundation design is shown in Fig. 5. Remaining units at the site were piled using the Continuous Flight Auger (CFA) system as this allowed the junction between the Chalk and overburden deposit to be confirmed during boring. The design of the piling system is shown in Fig. 6.

Site at Chieveley, Berkshire

The ground investigation for a proposed road improvement scheme identified solution features where the Chalk was overlain by Reading Beds and superficial deposits, and also where the Chalk outcropped at surface. Where present, the overlying Reading Beds materials comprised both sands and clays, often interbedded. The ground investigation utilized a range of exploratory methods including trial pits, boreholes, window samplers and dynamic probing. These exploratory holes were positioned at the location of every structure and at 50 m intervals along the proposed carriageways and slip roads. Where solutions features were encountered their lateral and vertical extent was investigated. It was considered that this frequency of

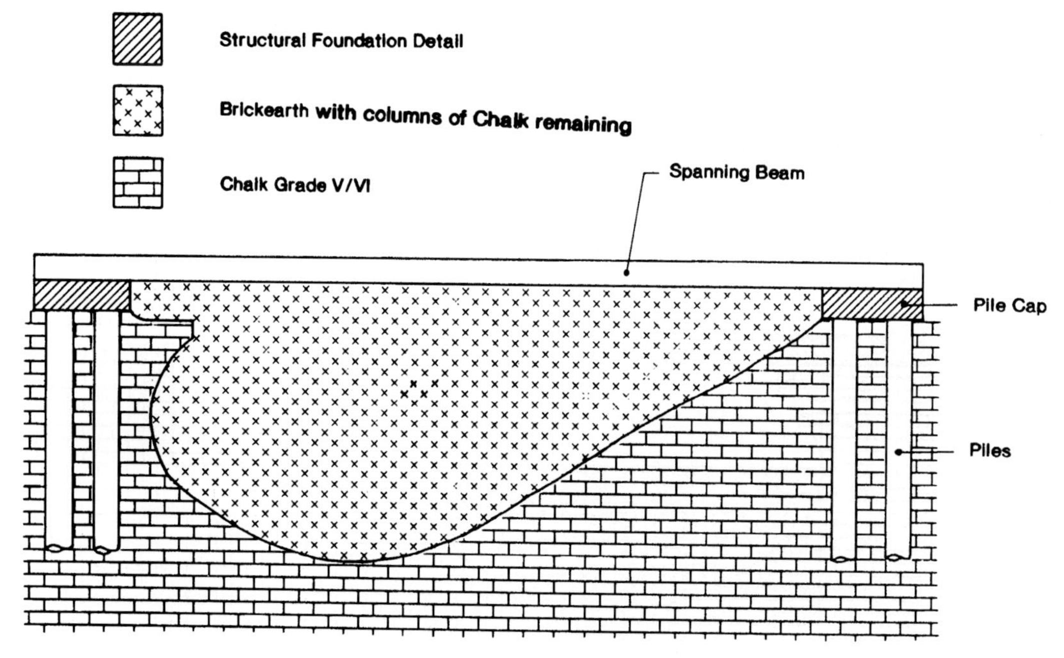

Fig. 5. Section through spanning beam detail.

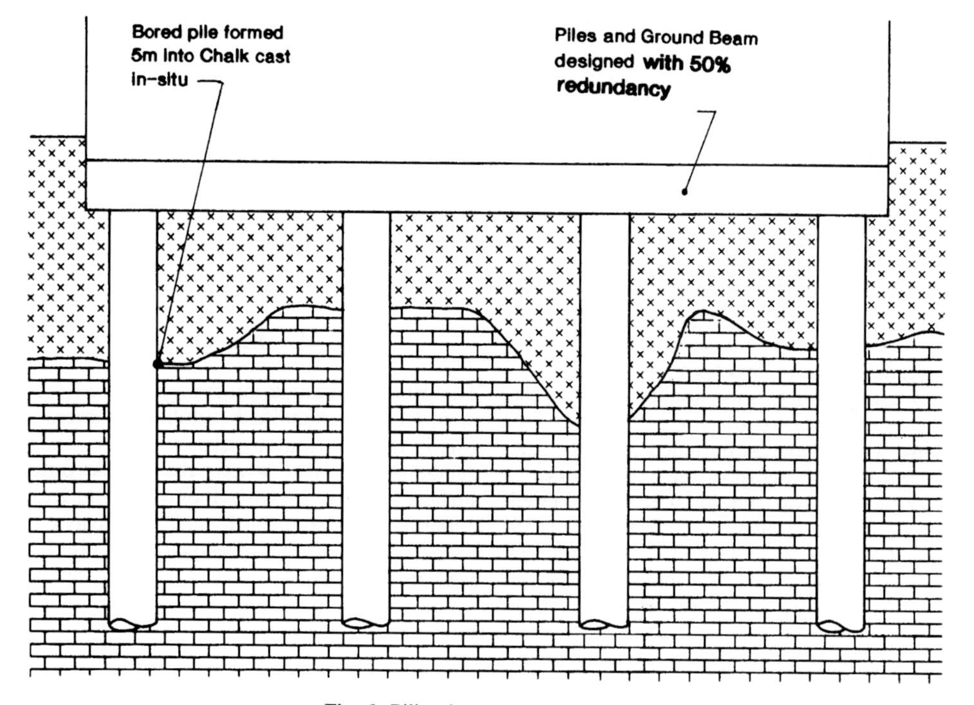

Fig. 6. Piling layout, Dartford, Kent.

investigation would give a good probability of identifying the type, nature and frequency of solution features so that mitigation measures could be designed prior to the construction works commencing.

The features ranged in size from pipes as narrow as 500 mm to solution features of up to 20 m in diameter. The solution features were identified within the Grade VI/V Chalk and within Grade IV Chalk. A diagram showing such a feature is presented in Fig. 7. Where this was encountered the interface between the solution feature and the Chalk was highly variable, with some features showing the infill material adjacent to intact Chalk and some with a weathering zone of Grade V/VI Chalk.

In almost all of the exploratory holes where solution features were encountered, the infill material was generally noted as being highly variable and comprising firm to stiff clays or medium-dense sands and gravels. Frequently both cohesive and granular soils were present as pockets and lenses. A void was noted in only one solution feature, which measured 500 mm by 300 mm in size. Columns of highly weathered Chalk were also noted, suggesting features being formed from more than one pipe joining together. A few of the solution features did contain a soft infill material.

Solution features were encountered to depths of up to 7 m below the surrounding Chalk level and up to 11 m below existing ground level. In general, the solution features identified were not associated with any surface depressions. Based on the depth of the features, the nature of their infill material and their general shape, it was considered that many of the features encountered were the result of development by weathering of ice wedge pseudomorphs.

A number of deeper solution features were encountered terminating at depths well below the active layer and containing a soft infill material. These features were present in areas where a significant thickness of granular Tertiary deposits directly overly intact Chalk. It is considered that these may have formed as more traditional features, with the Tertiary deposits providing protection, from periglacial reworking, to the Chalk.

It was considered that the presence of solution features at the site, which have the potential to give rise to long term settlements and reduced bearing capacity, represented a hazard. Therefore a number of remedial solutions were designed pertaining to structural foundations, highway earthworks and pavements at grade. Different design options were prepared for features of varying sizes and infill. These options are outlined below.

Where relatively small solution features, such as pipes of less than 1 m diameter, were present below the proposed bridge structures it was considered that excavating out the infill and replacing with structural fill or mass concrete was the most appropriate option as this allowed rapid treatment without any delays to the contract. Where larger features were present, reinforced concrete rafts were designed to span over the feature and transfer the load onto the adjacent Chalk. A number of designs were prepared in advance of construction for features of varying sizes so that, when a feature was identified during the foundation excavations, designs

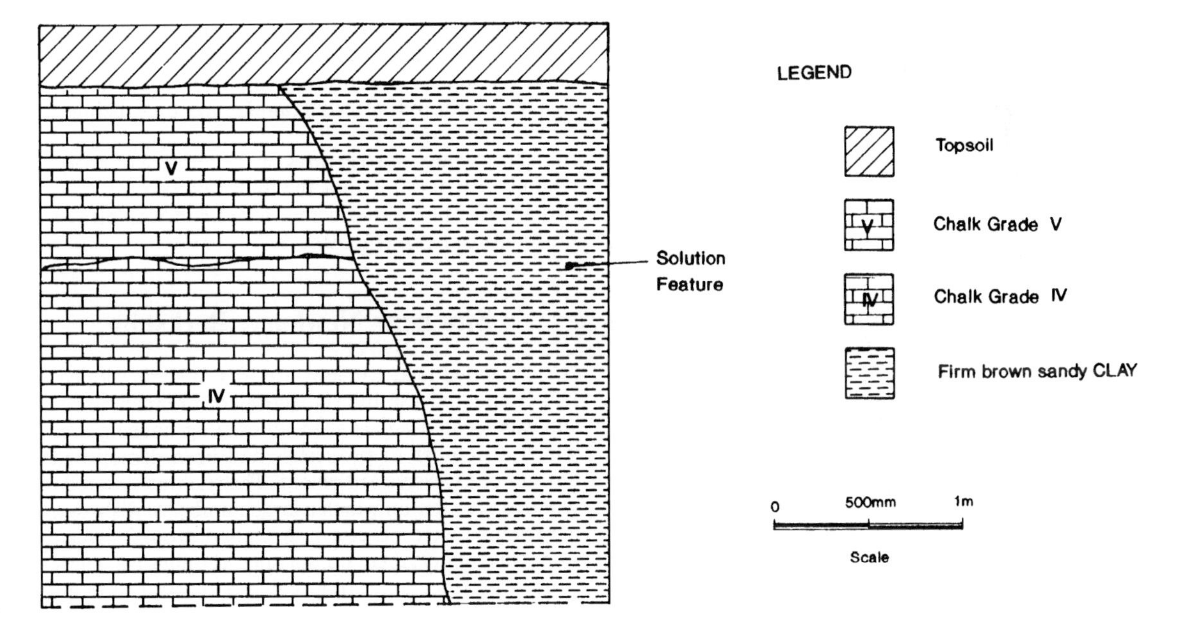

Fig. 7. Section through a typical solution feature at Chevely.

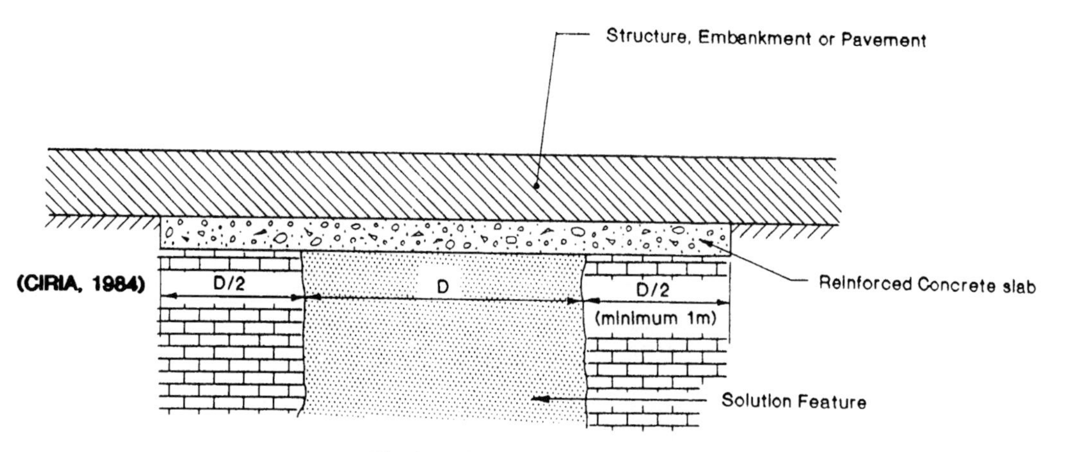

Fig. 8. Reinforced concrete slab.

were readily available, so minimizing delays. A drawing of a typical spanning slab is shown in Fig. 8.

Where particularly wide solution features, i.e. greater than about 8 m, were present below the bridge structures the use of a capping slab alone was precluded by the width of slab required to attain the bearing capacity necessary from the surrounding intact Chalk. Therefore an option to pile the structure or raft was available, with piles placed no closer than two pile diameters from the feature to prevent the pile interacting with the feature. This allows the necessary bearing capacity to be achieved without extending the footprint of the structure too far beyond the solution feature.

Solution features identified along the line of the embankments comprised shallow features infilled with medium-dense sand and gravels. It is considered that settlements would not be significantly greater than for the surrounding Chalk from the low loadings applied by the embankment. Therefore a geogrid-reinforced granular blanket similar to that shown in Fig. 3 was designed for use in areas where features are identified during the site preparation works. The geogrid reinforcement was designed, using published charts, to limit deflection of the carriageway to less than 1%, should collapse of the feature occur. Where solution features with soft cohesive infill were present, remedial measures were designed as follows:

- dynamic compaction or vibro replacement to consolidate the infill and capping the neck with a geogrid-reinforced granular blanket;
- excavation and filling of smaller features;
- concrete cap to span the feature (Fig. 8).

Solution features below the proposed roads were generally small in plan area and depth and were infilled with medium-dense sands and gravels. Many were located in cuttings and would be dug out during construction. Elsewhere treatment of the features was dependent on the size and the nature of their infill.

Small features, such as pipes of less than 1 m diameter, were to be treated using one of the following methods:

- placing a concrete cap over the neck of the feature with the conventional road construction placed on top;
- placing a geogrid-reinforced granular blanket to span the features should collapse of the infill occur.

Larger features with granular infill were to be compacted using dynamic compaction or vibro displacement and the subsequent depression infilled with granular fill incorporating geogrid-reinforcement.

Large features with soft cohesive fill were to be treated using vibro replacement forming stone columns within the infill and penetrating underlying Chalk. The stone columns improve the bearing capacity of the infill prior to the geogrid-reinforced granular mattress being constructed. Where vibro columns were not appropriate due to the proximity of structures or the absence of adequate confining material within the infill, the features were to be capped using either a reinforced concrete raft or an un-reinforced cap (Fig. 9).

Site near Newbury, Berkshire

A site located near to Newbury was to be developed for tied residential use by a government organization. The site was located adjacent to an existing housing area in which a number of properties had been demolished as a result of excessive settlement. The site was situated in a wooded area within which a large number of dish-shaped depressions were observed. A phased ground investigation was undertaken comprising geophysical surveys, cone penetrometer testing, cable percussion boreholes, continuous dynamic probing and trial pits. The investigation revealed approximately 10 m of Reading Beds

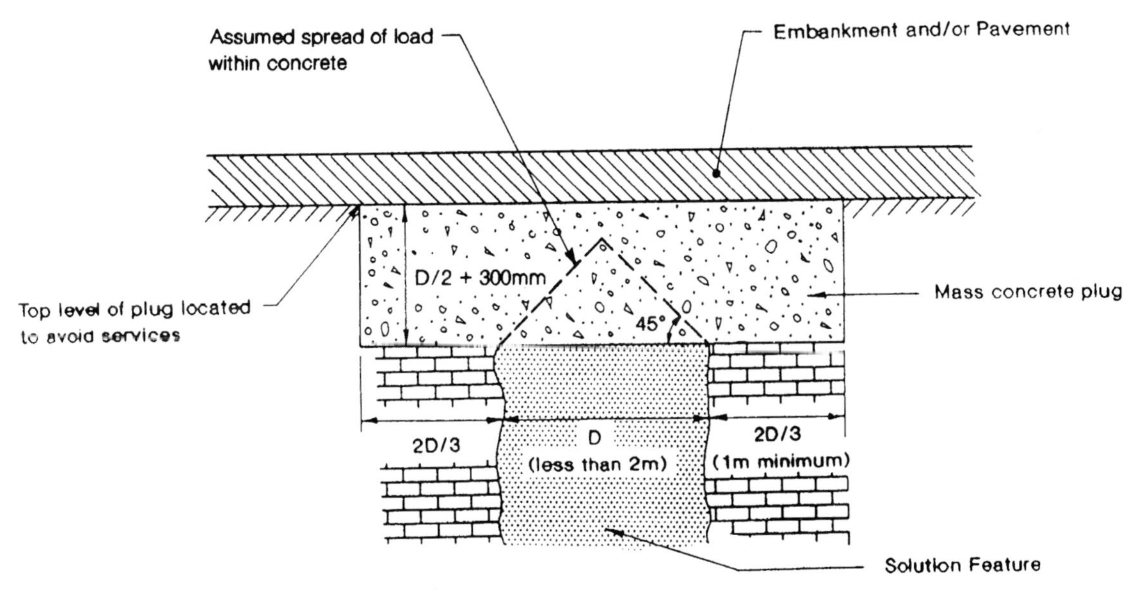

Fig. 9. Typical mass concrete cap.

sands underlain by Upper Chalk. At the locations of the depressions there was a marked change in density and colour of the sands, with the intact Reading Beds being generally pale yellow and dense, and the core of the depression containing very loose brown-grey sands with occasional gravel fragments, which in some areas contained voids.

The configuration of the features was determined by extensive profiling and excavation in the vicinity of known features. This revealed the infill deposits within the surrounding soils to be present in a generally wide and straight-sided form, not exhibiting the characteristic tapering shape that would normally be anticipated near to the surface. This configuration suggests a post-glacial mode of formation, with meltwaters gaining a lower pH as a result of increased solution of carbon dioxide and inclusion of weak organic acids from the sandy soils. This weak acid solution would have then attacked the joints in the Chalk. A progressive collapse of the overlying Reading Beds would then have taken place, assisted by continued ingress of meltwater through the preferential drainage path to give rise to the present form.

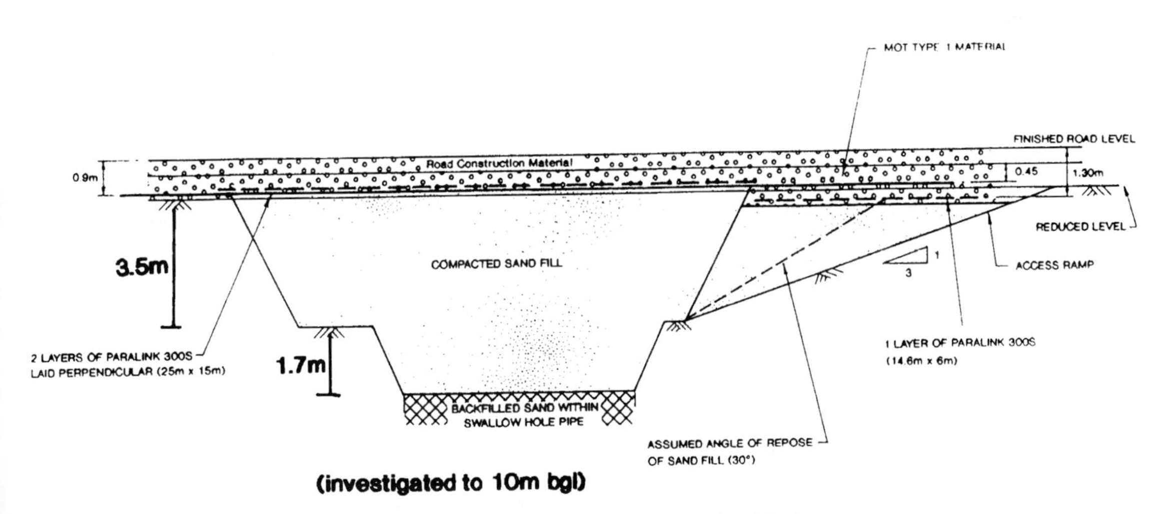

Fig. 10. Solution feature treatment used at Newbury.

The distribution of the features was not uniform across the site and the layout of the development could not avoid all the features, hence the structures were located in areas of potentially lower risk and the associated roads and infrastructure were routed across higher risk areas.

A number of options were examined in order to mitigate the potential risk to the development:

- grouting
- ground treatment
- capping larger features at the Reading Beds/Chalk interface
- near-surface capping

Grouting and ground treatment were rejected on the grounds of excessive cost, and the possibility that stone columns formed by vibroflotation could act as preferential drainage paths thus potentially reactivating the features. A trial was carried out on one large feature to establish the viability of placing a cap at the top of the Chalk, although the diameter of the feature did not decrease with depth as predicted by classical methods of formation and the top of the Chalk was not revealed in the 10 m deep trial excavation.

Accordingly it was decided that as the development was to be managed, the client was prepared to accept the risk that some maintenance would have to be carried out. The following solutions were adopted.

Foundations to the houses were to be stiff reinforced concrete rafts, designed to span 3 m in the event of a collapse within the footprint of the structure and to cantilever 1.5 m at the edges. Each housing plot was inspected by a geotechnical engineer on excavation for the presence of features and, where these were encountered, a judgement was made as to the potential risk of future subsidence occurring which could compromise the integrity of the house. In general terms plots were rejected where solution features exceeding 2 m were encountered, and where the soils surrounding the features were in a loose condition. As part of this strategy a number of alternative locations were available in the event that a particular plot was not considered suitable.

Excavations for drainage works and road pavements were also carefully inspected and where solution features were encountered they were capped. A number of capping types were designed and selection of the appropriate design was based on the observed dimensions of the features at the particular invert/formation level. Concrete caps, both mass concrete and reinforced concrete were designed for small features, generally less than 1.5 m in diameter. In landscaping areas a geogrid solution was used for small features.

In the case of larger features capping solutions utilizing geogrid reinforcement were designed. The configuration of the capping system depended on the dimensions of the feature and comprised the use of two perpendicular layers of high tensile unidirectional geogrid sandwiched within a Type 1 Subbase material which gave a satisfactory coefficient of friction with the geogrid (Fig. 10). Where this system was located below the invert of drainage runs sufficient material was replaced above the geogrid to develop the anchorage force required without requiring an excessively long length of geogrid.

The geogrid systems were designed such that, in the event of a catastrophic collapse of the solution feature, immediate surface subsidence would be limited to an extent whereby no damage could be sustained by vehicles passing over the area. Geogrids were selected on the basis of their strain characteristics such that the slope between the edge of the collapsed feature and the point of maximum deflection would be sufficiently gentle to accommodate traffic in the event of a collapse. The system was also designed, based on the time-dependent strain characteristics of the geogrids, to maintain its integrity for a minimum period of four weeks, during which time emergency maintenance/repairs could be undertaken. On handover of the development, the client had developed a programme of regular inspection and monitoring to identify any areas requiring attention.

Conclusions

The sites at Dartford and Chieveley showed solution features, many of which did not conform to the classical method of formation and it was therefore considered that these may have been formed from the weathering of relict ice wedges. Several deeper solution features infilled with soft materials were identified at the Chieveley site, generally where a significant thickness of granular soils overly the Chalk. The solution features at the site near Newbury were also in areas where the overlying deposits are thicker and where the Chalk would have been below the permafrost layer, thus these features are probably post-glacial. In all of the sites discussed the Chalk has been dissolved to a greater degree than the classical method of formation predicts.

Regardless of their method of formation, the occurrence of solution features poses a geotechnical hazard to the future development of sites underlain by Chalk. However, with careful investigation and design, these hazards can be overcome using simple and often cost-effective engineering solutions.

References

BENNISON, G. M. & WRIGHT, A. E. 1972. *The Geological History of the British Isles*. Arnold, London.

CIRIA 1984. *Construction Over Abandoned Mine Workings*. Construction Industry Research and Information Association, *Special Publication 32*.

——1994. *Foundations in Chalk*. Funders Report CP/13, Construction Industry Research and Information Association.

CULSHAW, M. G. & WALTHAM, A. C. 1987. Natural and artificial cavities in ground engineering hazards. *Quarterly Journal of Engineering Geology*, **20**, 139–151.

FOOKES, P. & HIGGINBOTTOM, I. E. 1971. Engineering aspects of periglacial features. *Quarterly Journal of Engineering Geology*, **3**, 85–115.

GIBBARD, P. L. 1985. *The Pleistocene History of the Middle Thames Valley*. Cambridge University Press, Cambridge.

HENRY, J. F. 1989. Ground modification techniques applied to sinkhole remediation, *In*: *Proceedings of the 3rd Multidisciplinary Conference on Sinkholes*, 327–332.

HIGGINBOTTOM, I. H. 1965. The engineering geology of Chalk. *In*: *Proceedings of a Symposium on Chalk in Earthworks and Foundations*. London, Institution of Civil Engineers, 1–13.

JENNER, H. N. & BURFITT, R. H. 1974. *Chalk: an engineered material*. Paper presented to the Southern Area ICE.

KOCH, H. F. 1984. Sinkholes in South-eastern North Carolina – a geologic phenomenon and related engineering problems. *In*: *Proceedings of the 1st Multidisciplinary Conference on Sinkholes*, 243–248

MCDOWELL, P. 1989. Ground subsidence associated with doline formation in chalk areas of southern England. *In*: *Proceedings of the 1st Multidisciplinary Conference on Sinkholes*, 129–143.

OSBORNE-WHITE, H. J. 1907. *The Geology of the Country around Hungerford and Newbury*. Memoirs of the Geological Survey of England and Wales, Explanation of Sheet 267, HMSO, London.

SPINK, T. W. & NORBURY, D. R. 1989. The engineering geological description of chalk. *Chalk symposium*. Thomas Telford, London.

WAKELING, T. R. M. 1966. Foundations on chalk. *In*: *Proceedings of a Symposium on Chalk in Earthworks and Foundations*. ICE, London, 153–161.

WARD, W. H., BURLAND, J. B. & GOLTIS, R. W. 1968. Geotechnical assessment of a site at Mundeford, Norfolk for a large proton accelerator. *Geotechnique*, **18**, 339–431.

WEST, G. & DUMBLETON, M. J. 1972. Some observations on swallow holes and mines in chalk. *Quarterly Journal of Engineering Geology*, **5**, 171–177.

A basic downhole geophysical approach to the investigation of shallow mineworkings

Stephen Weston

Johnson, Poole & Bloomer, Copthall House, New Road, Stourbridge, West Midlands DY8 1PH, UK (Present address: Mine Investigation & Stabilisation Ltd., 88 Shakespeare Gardens, Rugby CV22 6EZ, UK)

Abstract: Present investigations for shallow mineworkings generally consist of a desk study followed by rotary openhole or corehole drilling. The former, although relatively cheap, produces relatively low quality and quantity data. Investigations can maintain a high data retrieval at a low cost by combining openhole drilling with the downhole measurement of natural gamma radiation.

Detailed analysis of these data will allow a more confident interpretation to be made. This includes the correlation of solid mineral seams, collapsed workings and voids, and therefore more precise recommendations can be made. Examples of the benefits of natural gamma logs, rotary openholes and detailed analysis are shown.

Introduction

Shallow mining in the Black Country, i.e. the South Staffordshire Coalfield area around Dudley, Cradley, Tipton and Bilston (Fig. 1), is known to have taken place from Roman and Saxon times and reached its peak during the early 19th century. The local Carboniferous strata contain many seams of coal, fireclay and ironstone at shallow depths. These seams, the best known being the Staffordshire Thick (or Ten Yard) Coal, were extensively worked (Beete-Jukes 1853; Whitehead & Eastwood 1927). The Thick Coal is commonly at a depth of less than 40 m. These seams, combined with the underlying Silurian limestones, formed the backbone of the Victorian mining industry in the Black Country.

There were no statutory requirements for the owners or their agents to maintain and deposit plans of workings until 1872, by which time the majority of the mining had already taken place. The mining industry in South Staffordshire typically consisted of numerous small pits, and of the plans produced, only a small proportion has survived to the present day. It is fair to assume that there are no records for most of the old mineworkings.

Many existing properties and a significant proportion of development sites in the Black Country have to contend with the results of the industrial past of the area. These properties are underlain by old, shallow mineworkings that are still capable of causing subsidence, and this is of concern to both owners and insurance companies. Littlejohn (1979) and Healy & Head (1984) gave general recommendations for investigating and accommodating the problems caused by shallow mineworkings.

Although the majority of the old workings have collapsed, residual subsidence may still occur. Workings that have remained open, or partly open, pose a bigger threat to the structural stability of nearby buildings. There have been numerous cases involving damage to buildings, underground services and the disruption of transport arising from workings collapsing. The local authorities' planning departments are aware of this problem and frequently demand that the geology and mining history of any site that is potentially affected is researched. Any suspected shallow mineral seams within influencing depth should be revealed by the research and these would then have to be either proven stable or made stable.

Rotary drilling is used to investigate for shallow workings. Although it apparently fulfils the present requirements of those concerned, it has a number of shortcomings that involve poor quality and quantity data retrieval. It is, however, possible to solve this dilemma.

The author suggests that the measurement of natural gamma emissions in a rotary openhole, followed by detailed analysis, is the most time- and cost-effective method of obtaining sufficiently high quality data to enable a non-ambiguous interpretation to be made of the condition of the mineral seams at shallow depth, in addition to their identity and the geological structure. It follows that any recommendations for the stabilization will then be made with greater confidence.

The natural gamma log dates from the early days of geophysical downhole logging and was first used in 1939. It is widely used in the oil, water and mineral exploration industries. It has even been used before on site-specific investigations for shallow mineworkings.

WESTON, S. 1998. A basic downhole geophysical approach to the investigation of shallow mineworkings. *In*: MAUND, J. G. & EDDLESTON, M. (eds) *Geohazards in Engineering Geology*. Geological Society, London, Engineering Geology Special Publications, **15**, 291–296.

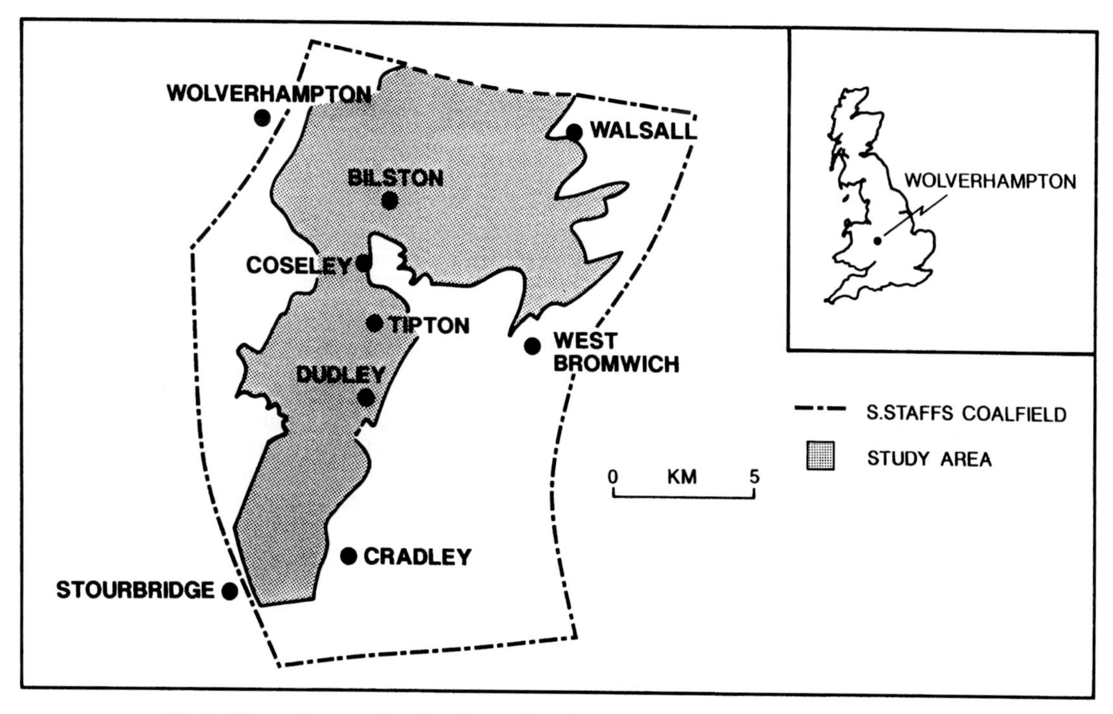

Fig. 1. The study area, showing the outline of the former South Staffordshire Coalfield.

The work discussed in this paper has an immediate benefit to any individual site but it is also part of a more widespread and exhaustive project over a wider area to quantify the usefulness of this method.

Following an appraisal of present methods of investigation, the basic theory behind natural radiation and the radioactivity encountered within the strata encountered will be followed by examples which demonstrate the usefulness of the gamma log and rotary openhole.

Present methods of investigation

Before it is possible to explain the potential benefits of natural gamma logging, it is necessary to understand the current standard method of investigation. An initial research phase will involve the examination of geological maps and existing mine plans from a number of archives. This will establish an anticipated stratigraphical succession and structure, and highlight any possible shallow seams which may have been of economic importance in the past. If no recorded evidence of shallow mining is revealed, its presence may still be suggested by the geological interpretation or the mining history of the neighbourhood. Even if there is conclusive evidence for shallow workings, a physical investigation may be beneficial to determine their current state or extent and thereby allow a firmer estimate of the cost of their stabilization.

Shallow mineral seams are presently physically investigated by rotary drilling. This can be by the openhole method, where the hole is deepened by either a rock roller or a percussive down-the-hole hammer. In both cases, the bottom of the hole is broken up and the rock chippings carried to the surface by the air or water drilling flush. The alternative to openhole drilling is the coring method, where a solid stick of rock is extracted from the hole by a circular hollow drilling bit.

On smaller investigations, openholes are generally used, with an eye on both engineering requirements and cost. On larger, more sensitive schemes, coreholes are often preferred, as there is usually a corresponding large budget available for the investigation to provide the greater degree of confidence that is often required. It is the usual practice on larger projects to use a combination of both methods to keep costs and time in the field to a minimum but still permit sufficient data retrieval of a sufficiently 'high' quality. The reliance on rotary drilling appears to fulfil present requirements, but there is always room for improvement.

Openhole drilling has the advantage of being approximately one-fifth to one-tenth the cost of coring and can take less than one-quarter of the time. It relies, however,

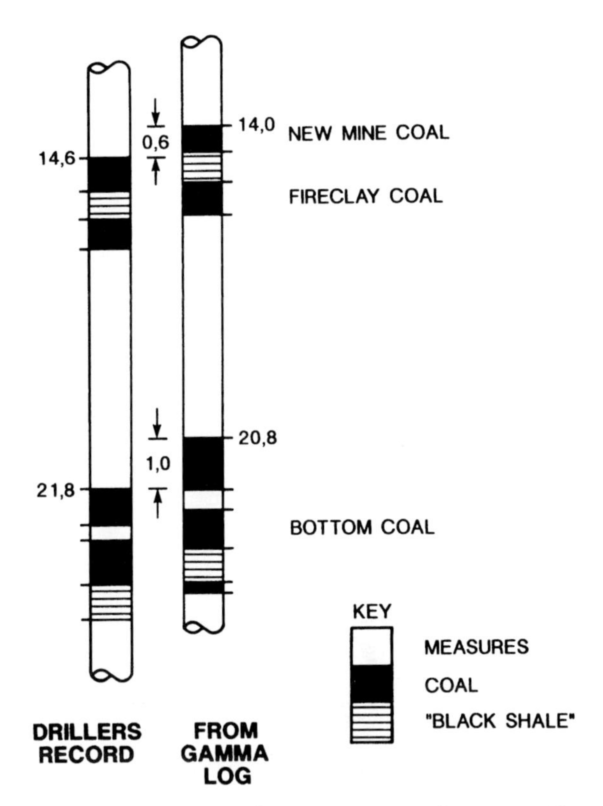

Fig. 2. The 'pull-down effect' on strata sometimes seen with rotary openhole drilling.

on the skill of the driller to identify strata changes and thicknesses. Thinner strata may be missed and depths are somewhat uncertain due to the length of time taken for the flush returns to appear at the surface as the hole becomes deeper and overall the hole becomes 'stretched'. This pull-down effect has sometimes been noted during this project and an example is given in Fig. 2. This effect can be significant in borderline cases where a seam was judged to be just too deep to require stabilization from the drillhole information alone. Flush samples may be contaminated by chippings from higher up the hole and 'ghost strata' may appear in the driller's log. Also, should the drilling flush be lost down the hole, perhaps due to encountering broken ground or a worked seam, it becomes impossible to identify any strata below the flush loss, only whether the strata are hard, soft or broken. Openhole drilling is generally considered to be inferior to coring, as the data retrieved are poorer in quality and quantity.

The results from cored drillholes are dependent on good core recovery, and thus again on the driller's skill. It is not possible to obtain core of broken ground, packed waste in the worked seams or, obviously, any voids. Yet these three indicators are the most important ones to the interpretation and subsequent recommendations, and their presence can be determined by relatively inexpensive but less precise openholes.

Although much useful information can be obtained from either drilling method, each has its 'blindspots'. Both methods may leave vital gaps in the information obtained and this may make any interpretation less confident than desired. These gaps can be filled by natural gamma logging, which is used to complement and enhance the information from openholes, not to replace it.

Based on contractor's rates from recent projects, five openholes drilled to 40 m, taking five or six rig days, would cost approximately 20% of that for five coreholes taken to the same depth that would take 15–20 rig days. Natural gamma logging would only increase the cost of the openhole investigation to approximately 30% of that for all coring. Where knowledge of the local geology and mining is uncertain, four openholes and one corehole would cost approximately 45% of the five coreholes.

Natural gamma logs through the coal measures: theory and background

The natural gamma emissions observed from any formation are a function of the radioactive isotopes present, these usually being potassium-40, uranium and thorium. Mudstone tends to have a relatively higher content of the radioactive elements than sandstone, which in turn has more than coal. Although exceptions can be demonstrated, this general principle holds firm. A number of factors contribute to the differences seen in the natural gamma logs. It is well known that a natural gamma log is not as repeatable as other downhole geophysical logs, and the same hole logged twice will produce a slightly different log the second time. This is commonly thought to be due to the non-uniform nature of radioactive decay. Other measurement factors, such as logging speed and counting (sampling) rate and physical factors such as drillhole diameter, casing and groundwater level have to be considered and allowances for these can be made during logging and analysis. Its use as a method of correlation is well documented (Barker *et al.* 1984), and an example from this project is shown in Fig. 3.

In addition, when comparing logs from adjacent drillholes, the vertical and lateral variability of the strata must contribute to some of the differences noted. A sound knowledge of this variability, from local mining history, sedimentology and the detailed analysis of the information obtained, is most important to make full use of these logs.

Any particular coal seam shows a relatively uniform gamma log signature, due mainly to its stable depositional environment. Coal finds its origin in a peat swamp. A swamp whose peat will form a high quality coal will have very little detrital input. Poor quality coals

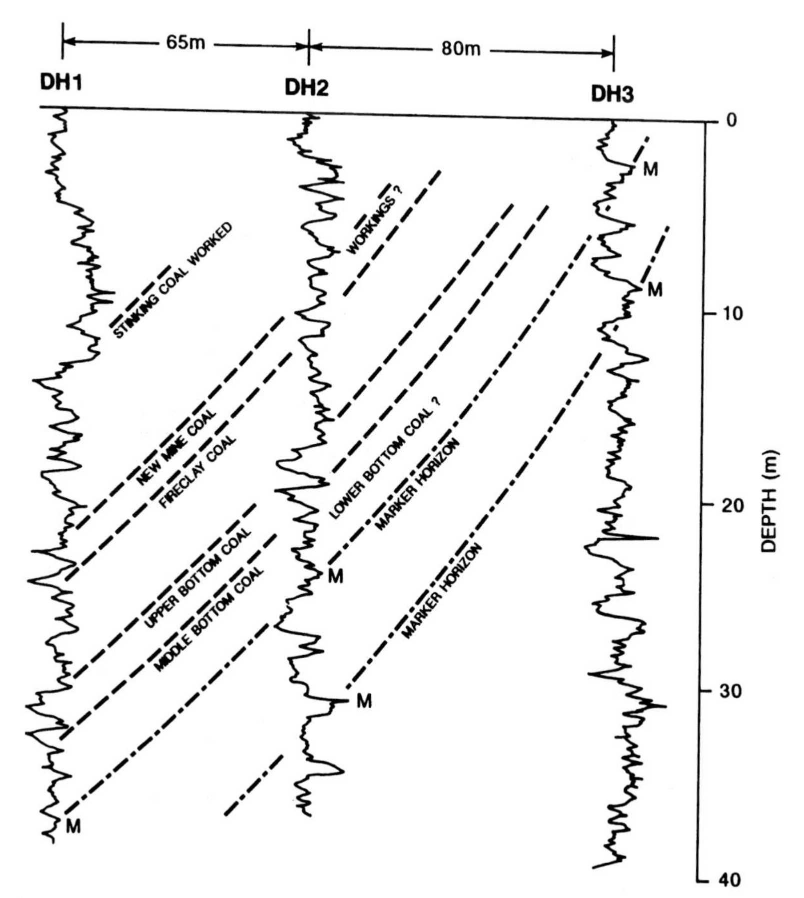

Fig. 3. Use of gamma logs to correlate between drillholes. Note the coal and non-coal marker horizons. Peaks marked 'M' are marker horizons.

or carbonaceous mudstones (black shales) and mudstones with carbonaceous bands indicate increasing detrital input.

This suggests that a good quality coal should be indicative of very stable conditions. As previously mentioned, coal has a very low natural radioactivity. Figure 4 shows seven intersections of the Bottom Coal within one site, with the logs arranged along strike. A fair degree of similarity can be seen in this case, perhaps due more to the local differences in depositional environment than to the nature of radioactive decay. Reference would have to be made to seam exposures in an opencast excavation to support this finding.

Whereas coal seams have formed in a stable environment, the intervening measures have been deposited in a relatively dynamic one: might we not expect the section of the logs representing the intervening measures to show more lateral differences?

It is vitally important to the correlation of solid mineral seams, collapsed or backfilled workings and voids, that as much attention is paid to the intervening measures as to the seams. Are there sedimentological reasons to explain the frequent changes in the natural gamma logs? Mining archives and corehole records, together with comparisons to modern coal-forming environments, can suggest geological, rather than radioactivity based, causes for some of the different features in the gamma logs.

There are 10 major coal seams and various ironstones and fireclays in approximately 200–300 m of strata, all of which have been worked in many places throughout the coalfield. With many mineshaft sections and coreholes giving detailed geological control, the South Staffordshire Coalfield is an ideal study area.

Field technique

A combination of investigation techniques and a detailed analysis has been used to determine the depth

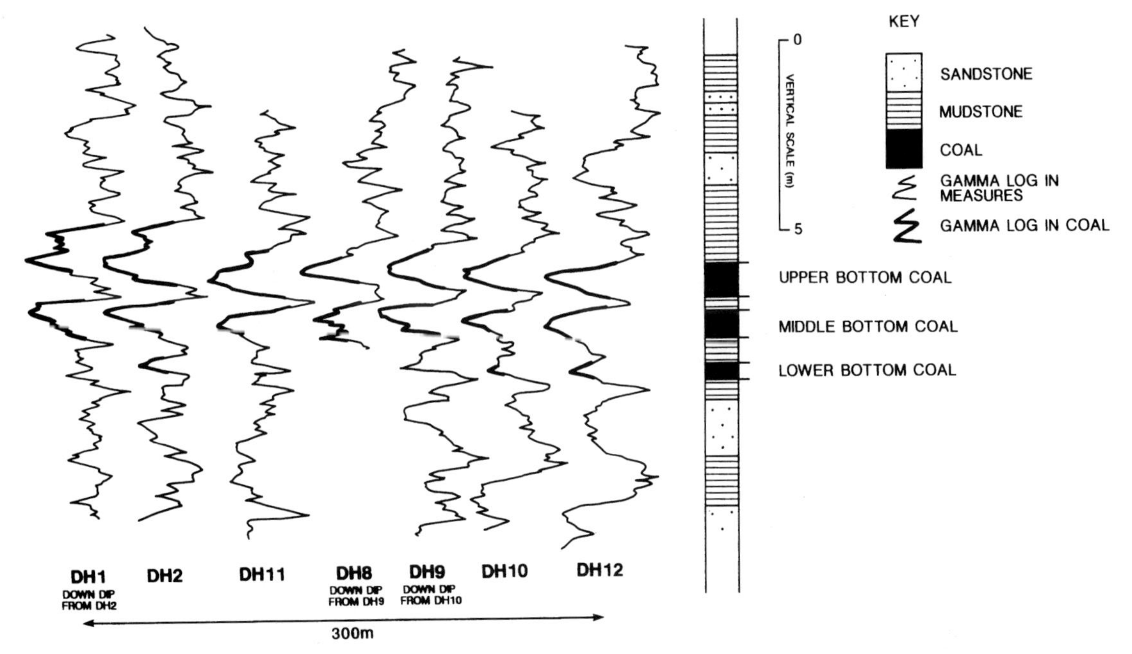

Fig. 4. Variations in gamma logs seen in seven intersections of the Bottom Coal.

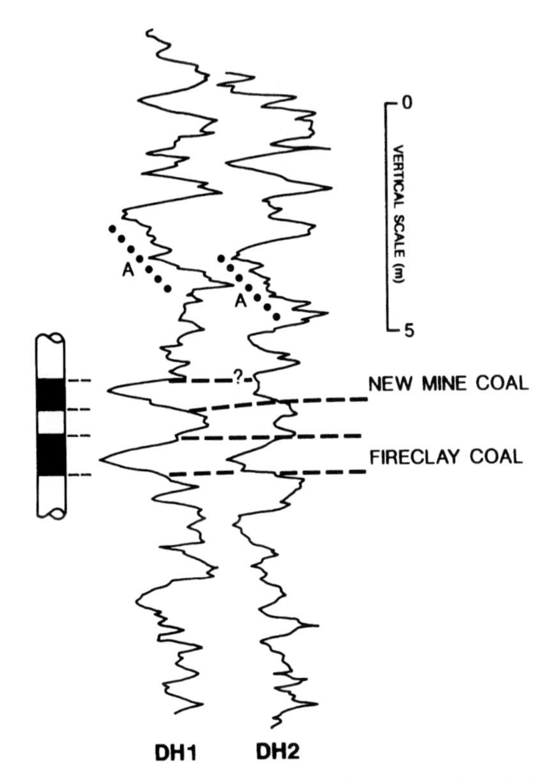

Fig. 5. Example of gamma log through a suspected worked coal seam.

and condition of the coal present, and sometimes other mineral seams. The strata between the mineral seams are subject to the same degree of analysis. The natural gamma log is used in conjunction with the geological information from the drillhole and information obtained from research.

The logging of the hole is undertaken using a Mount Sopris logger which has a pen plotter. The trade-off between cost and data quality has so far not justified more sophisticated equipment.

The natural gamma logging tool can be run in both dry and wet holes, and in cased or uncased holes. It is normal practice to use plastic casing to reduce the chance of loosing a tool down the hole; something that is important in any hole encountering broken ground or old workings. These conditions dramatically increase the chances of the hole collapsing. Although the casing decreases the amount of radiation reaching the tool, the reduction is relatively insignificant in comparison with the amount of radiation in an uncased hole. It is possible to use thin metal casing and thin walled steel NW size casing, as in the majority of logs given in this paper. Results with thicker casing are usually disappointing.

Figure 5 indicates the New Mine Coal to have been worked. On this occasion, the drillers' daily records gave no indication of workings in the New Mine Coal. The underlying Fireclay Coal acts as a clear reference point. The anticipated trough for the New Mine Coal seen in DH1 is missing in DH2. The gradual increase in both logs above the New Mine Horizon (feature A) has 'dropped' in the second log, perhaps due to collapsed workings. These two coals were encountered in six drill-holes, all of which showed similar roof strata, suggesting a non-geological cause. With the drillers' records alone, the impression given would have been that there was no problem with mining. The gamma log, on the other hand, suggests that such a conclusion is premature, and there may still be open or only partially collapsed workings nearby. Only trial grouting of the ground would confirm the suspicions of workings. On this occasion, the depth of the seam was within that given in the guidelines for stabilization (Healy & Head 1984). A 'misdiagnosis' might have proven costly to the future property owner.

Conclusions

The initial results of gamma logging to enhance and complement simple openhole drilling have been encouraging. The work has been shown to be cost-effective and more detailed interpretations have been possible. The initial work has indicated that the next stage of research work should include the following:

- the use of an electrical conductivity sonde to obtain a more accurate depth for coal seams;
- the determination of the optimum data collection settings;
- the use of data filtering and analysis;
- the comparison of investigation findings with later drilling and grouting works;
- obtaining representative rock samples to determine amounts of radioactive elements in the laboratory;
- to build up a database of natural gamma logs from the South Staffordshire Coalfield for both study and reference.

Acknowledgements. My thanks go to Johnson, Poole & Bloomer for encouragement, financial assistance and help, especially C. Knipe, the Senior Partner, for supporting this research project and acting as my external supervisor, and to G. Worton for his geological assistance. R. Oldnall kindly prepared the figures. Also thanks to R. Barker, my internal supervisor at the University of Birmingham, for his guidance, and to P. Fenning of Earth Science Systems for his assistance.

The following JPB clients were kind enough to allow the publication of this information: Bryant Homes Mercia Limited and The West Midlands Regional Health Authority.

Correspondence can be sent to the author by E-mail at stephen@mine.woden.com

References

Barker, R. D., Lloyd, J. W. & Peach, D. W. 1984. The use of resistivity and gamma logging in lithostratigraphical studies in the Chalk in Lincolnshire and South Humberside. *Quarterly Journal of Engineering Geology*, **17**, 71–80.

Beete-Jukes, B. 1853. On the geology of the South Staffordshire Coalfield. *In: Records of Sciences Applied to the Arts*, **1**(11), 149–335.

Healy, P. R. & Head, J. M. 1984. *Construction Over Abandoned Mineworkings*. Construction Industry Research and Information Association, Special Publication 32.

Littlejohn, G. S. 1979. Surface stability in areas underlain by old coal workings. *Ground Engineering*, March 1979, 23–30.

Whitehead, T. H. & Eastwood, T. 1927. *The Geology of the Southern Part of the South Staffordshire Coalfield.* Memoir of the Geological Survey of England and Wales.

Digital image texture analysis for landslide hazard mapping

P. J. Mason,[1] M. S. Rosenbaum[2] & J. McM. Moore[1]

[1] T. H. Huxley School of Environment, Earth Science and Engineering, Imperial College of Science, Technology and Medicine, Prince Consort Road, London SW7 2BP, UK
[2] Faculty of Environmental Studies, The Nottingham Trent University, Burton Street, Nottingham NG1 4BU, UK

Abstract. Textural enhancements using co-registered, multi-temporal, digital remotely sensed imagery have been used to study slope instability at three landslide localities. The remotely sensed imagery consists of Landsat TM, SPOT Pan and Daedalus 1268 Airborne Thematic Mapper (ATM); stereo, black and white aerial photographs have also been used. Image processing provides spatial and textural information concerning landform change, landslide location and morphology. The enhanced imagery can be integrated with digital elevation models (DEMs), as can other map and geotechnical data, into a GIS to analyse quantitatively the effect of slope geometries on the observed landsliding. The suitability of available imagery depends upon the type and scale of the study and the landslide dimensions. The case histories have shown that spatial resolution is the most important parameter for successful landslide identification. The optimum imagery is provided by a combination of textural and spectral enhancements using different types of imagery.

Introduction

Textural characteristics of remotely sensed imagery and image processing techniques have been used to enhance and interpret landslide morphology with the aim of establishing a landslide hazard mapping methodology.

Landsliding affects many terrain types and is a widespread problem in many climatic conditions, in many parts of the world. In order to study slope instability and to prevent or lessen its effects, landslides need to be identified, mapped and their stability assessed. The problem is commonly one of identification in cases where occurrence is believed to be widespread and surface expressions may be both complex and subtle. Satellite remote sensing can help solve this problem and a series of landslide case histories, in northwest Italy, southeast Spain and southeast England, have been investigated to this end. These landslides occur in different settings and are caused by a variety of factors, e.g. climatic (torrential storm) events and the combined effects of river erosion and human activity.

A digital elevation model (DEM) is appropriate for the analysis of topographic aspects of landslides using a GIS. Remotely sensed imagery can also be 'draped' over a DEM to create a three-dimensional visualization of spectral and textural information, greatly improving the understanding of spatial relationships between image texture and topography. Analysis can also be achieved in a quantitative manner by considering parameters which are relevant to further spatial analysis, e.g. slope angle, geotechnical data, proximity to drainage, rainfall intensity and land use, for which a relational database and GIS are ideally suited. Previous hazard assessment and slope stability analyses have been completed using GIS and probabilistic models, e.g. Rosenbaum & Jarvis (1985), Brass *et al.* (1991) and van Westen *et al.* (1994), but these studies did not consider temporal change or texture analysis. Our aim is ultimately to develop an image-based hazard assessment methodology incorporating site-specific information with established analytical techniques.

Landslide characterization

The type and timing of image data are very important for a slope stability study. Relevant considerations for data selection are landslide dimensions, time of year, state of vegetation and sun angle/direction. The primary data requirements for this type of study are sensor resolution, spectral information and multi-temporal coverage. For optimum enhancement of specific landslide features, image parameters need to be considered with reference to sensor specification and the details of each landslide locality.

Certain terrain features are characteristic of most landslides. These have been covered in detail by Schuster & Krizek (1978), Varnes (1984), IAEG (1990), WP/WLI (1990), Cronin (1992) and others. A selection of landslide terrain features which are most relevant to image/photo interpretation are summarized in Table 1. Such features all have potential textural expressions in

Mason, P. J., Rosenbaum, M. S. & Moore, J. McM. 1998. Digital image texture analysis for landslide hazard mapping. *In*: Maund, J. G. & Eddleston, M. (eds) *Geohazards in Engineering Geology*. Geological Society, London, Engineering Geology Special Publications, **15**, 297–305.

Table 1. *Landslide characteristics useful for photo-interpretation. Modified after Schuster & Krizek (1978)*

- Land masses undercut by rivers or waves
- Sharp break of slope at the main scarp
- Arcuate shape of crown wall
- Deep fissures or tension cracks in turf, soil or rock
- Hummocky surface of the slipped mass
- Spoon-shaped troughs in the terrain
- Elongate undrained depressions
- Disrupted drainage channels
- Vegetation changes indicating changes in ground moisture
- Displaced roads, fences and walls
- Accumulation of debris in drainage channels and valleys
- Stripped regolith, soil and vegetation revealing fresh rock surfaces
- Exposed slip plane
- 'Levées' between debris flows and slides

imagery. Only some are large enough to be detected in satellite imagery, e.g. an arcuate crown wall. Others may only be detectable using air photography, e.g. tension fissures and subtle undulations of hummocky slipped ground.

The main advantages of using multi-temporal remote sensing for a landslide study are the usefulness of imagery for visual identification and mapping, and the speed of analysis. In locations where there are numerous landslides, the aerial coverage provided by remote sensing leads to a very efficient method for preliminary mapping. In cases where landslides are isolated and relatively large, identification may be straightforward but remote sensing can nevertheless provide textural and spectral information concerning internal structures, vegetation growth and water content. The combination of human eye and brain is both efficient and intuitive for texture discrimination whereas computers alone (i.e. automated textural discrimination) are not as efficient. The best method for digital discrimination of natural features, on the basis of texture, therefore relies on the enhancement of high frequency information, the detection of temporal change and human interpretation.

Image data requirements

Resolution

The size and form of landslides vary considerably dependent upon their causes, type and setting. Landsat TM spatial resolution is set at 30 m for visible and infrared channels, SPOT XS at 20 m and Panchromatic at 10 m, and ATM imagery at 7.5 m (in the examples used here). In the study of large landslides (>500 m in length or width) the resolution of Landsat TM may be sufficient to identify internal structures, whereas small landslides (<100 m in length or width) may be too small to be positively identified using SPOT Panchromatic imagery. In such cases, the use of scanned air photography or airborne imagery will be necessary.

Image texture

Image texture is a combination of the magnitude and frequency of tonal change (Drury 1993). It is a product of the combined reflectance/emittance of all surface features within a particular area and is dependent on sensor type, climatic conditions, spatial resolution and spectral properties. The enhancement of spatial variations in image texture assists interpretation of the aerial limits of a landslide, internal structures, landslide propagation, deformation of the toe area, disturbances of vegetation distribution and human-made structures. Vegetated surfaces are often darker than exposed soils and may also produce image textures which are slightly different from those of adjacent areas. The same is true of areas subject to differential weathering.

Spectral properties of soils and rocks

Landslides modify the spectral properties of vegetated slopes and weathered rock surfaces, exposing bright soils and bare rock surfaces. Weathering variations and changes in vegetation may also indicate a history of movement over a considerable period of time. The enhancement of such variations can thereby assist the identification of landslides from such imagery. Multi-spectral information enables the enhancement and mapping of such spectral variations within and surrounding a landslide.

Multi-temporal coverage

High resolution images commonly reveal very complex tone and texture and so landslide identification may not be straightforward. Regular satellite coverage provides the opportunity for imagery acquired before and after landslide movement to be compared. Multi-temporal coverage also provides the opportunity to monitor stages of movement where landslides are intermittently active. There is evidence to suggest that landslides occur repeatedly in vulnerable areas (Govi & Sorzana 1982).

Seasonal variations in vegetation produce significant textural and spectral changes. Ideally 'before' and 'after' images should have been acquired at similar times of year so that illumination and sensor geometries are comparable. Interpretation of winter images is hampered by long topographic shadows and high contrast. The low illumination angle at such times means greater sensitivity to surface discontinuities and landslides on illuminated slopes but not on shadowed slopes (Stephens *et al.* 1988). The time period following a landslide event should also be considered. Regions characterized by warm, humid climates experience rapid growth, and re-vegetation begins quite quickly after the landslide event, inhibiting detection by remote sensing. Imagery should therefore be acquired at a known and reasonably short period after the landslide event. Comparison of imagery taken before and after the landslide enables the identification of changes in vegetation distribution and density of growth during the intervening time period. Some of the changes will be attributable to landsliding and some will be produced by other factors such as human activity. Image processing techniques, involving algebraic operations and spatial enhancements, can assist in the differentiation of such changes.

Digital image processing

Image integration and contrast enhancement

The image data sets were digitally processed using the raster based processing software called ERMapper (Earth Resource Mapping) run on SUN Sparc stations at Imperial College. SPOT Panchromatic, Landsat TM and Airborne Thematic Mapper (ATM) images of the three landslide case studies were used. Where multiple images were obtained for one location, they were co-registered to common coordinates. Prior to textural enhancements, the multi-spectral imagery used was stretched using the balanced contrast enhancement technique (BCET; Liu 1991). This technique stretches each band to the same mean and value range, using a parabolic function, without altering the basic shape of the histogram. Following textural enhancement, linear and piece-wise linear stretch functions were applied to restore the image information to a full eight-bit range.

Textural enhancement

Digital convolution filters and moving window operators provide efficient methods for enhancing spatial brightness variations at different frequencies and have been documented elsewhere (e.g. Niblack 1986; Drury 1993). This study has revealed that simple filters (illustrated in Fig. 1) have proved to be the most effective. Complex textural filtering techniques such as those developed by Haralick *et al.* (1973) and by Pietikainen *et al.* (1983) produce results which are rather confusing and difficult to interpret for landslide studies. The filters which have been found to be the most useful for landslide terrain study are listed below. Kernels 'a' and 'b' should be used with caution since they alter the form of the image histogram and, in doing so, degrade the image information.

- Sobel gradient filters are sensitive to vertical, horizontal or diagonal gradients. They are individually passed over an image and the products are added together to enhance multi-directional linear features and textural regions.
- Texture gradient filters (centrally weighted): are useful for enhancing and separating areas of high frequency information (rough textures) from areas of relatively uniform texture. This kernel produces a boundary value between regions of different digital texture whilst forcing the values in the regions themselves to similar levels, thus suppressing topographic variation.
- Laplacian (3 × 3), centrally weighted, filters are useful for general edge sharpening. The effect is an exaggeration of points and linear features in the image by creating a 'double' edge to areas having sharp gradients. The general form of the image histogram is maintained and therefore the image information is not altered.

Image subtraction

The simple subtraction of image from different dates, to create a temporal-change detection image, is a useful tool for the identification of terrain features produced during a known time period, such as those produced by slope movements. The subtraction of two images acquired at different times, assuming similar illumination geometries and conditions, also suppresses constant features such as field boundaries and buildings. Ideally the difference image should show terrain features which are caused only by landsliding, but this is unlikely to be perfect. Large landslides (>500 m in width or length) may be very distinct and relatively easy to identify by texture and morphology. Where

$$a)\ \text{filter}_x \begin{bmatrix} -1 & 0 & 1 \\ -2 & 0 & 2 \\ -1 & 0 & 1 \end{bmatrix} + \text{filter}_y \begin{bmatrix} -1 & -2 & -1 \\ 0 & 0 & 0 \\ 1 & 2 & 1 \end{bmatrix} \quad b) \begin{bmatrix} -1 & -1 & 0 \\ -1 & 3 & 0 \\ 0 & 0 & 0 \end{bmatrix} \quad c) \begin{bmatrix} -1 & -1 & -1 \\ -1 & 14 & -1 \\ -1 & -1 & -1 \end{bmatrix}$$

Fig. 1. (**a**) Sobel gradient filters (can be used separately or added together); (**b**) texture filter; (**c**) Laplacian gradient filter.

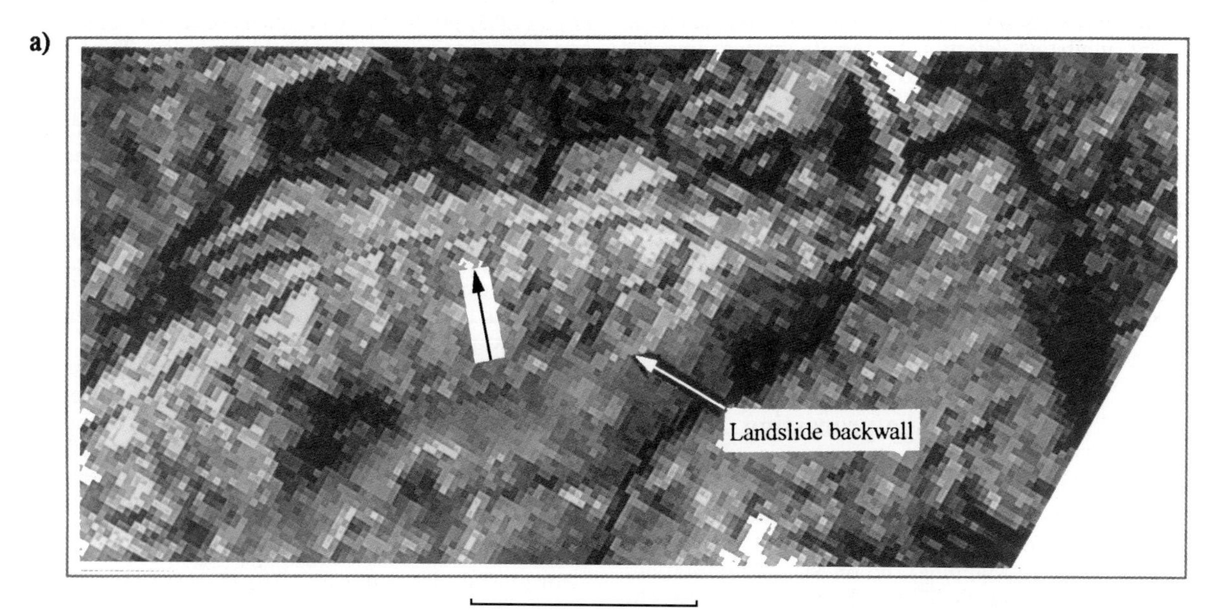

Fig. 2. Images of a landslide in southeast Spain illustrating the improved spatial detail provided by airborne imagery: (**a**) TM band 5; (**b**) ATM band 9; both have been Laplacian filtered. Arrow indicates direction of landslide movement.

landslides are small or produce subtle surface expressions, both textural and spectral information may be required to identify the landslides.

Illustrations of landslide morphology

The following examples illustrate some of the characteristic features of landslides (as listed in Table 1) that can be enhanced by image processing methods and remote sensing.

Crown and main scarp

Landsat TM (band 5) and ATM (band 9) sub-scenes (both edge enhanced using the Laplacian filter) from southeast Spain are shown in Figs 2(a) and (b) and illustrate the arcuate crown and back scarp of the landslide. Figures 2(a) and (b) demonstrate the increase in textural detail provided by the 7.5 m resolution of ATM over the 30 m pixels of TM. The distinct textural boundary between a very gently dipping relatively resistant plateau (upper part of the images) and the intensely denuded soft sediments beneath is best illustrated in the ATM image. Hybrid images combining TM and SPOT Pan show an improvement over the TM image alone and reveal the detail provided by ATM (Eyers 1994). Spectral analysis of this area using ATM has allowed detailed soil and vegetation mapping as well as identification of less obvious landslides and debris flows in the area (Wright 1995). Further detail concerning tension cracks and fracturing of the plateau can be extracted using black and white stereo air photographs for this area. The locality has a history of movement and instability and the landsliding has recently been reactivated by road-building along the river valley.

Hummocky textures of slipped material

An ATM sub-scene from southeast England is shown in Fig. 3. The image in Fig. 3(a) has been texturally enhanced using Sobel filters added together. Figure 3(b) shows a sketch interpretation of the image in Fig. 3(a). The area has a long history of movement (many movements have been documented over the last 100 years) produced by erosion of the foreshore, due to longshore drift, and resulting in the progressive degradation of the cliff line (Hutchinson *et al.* 1980). The hummocky textures are produced in the 'main body' of the landslide, and the 'accumulation' or 'foot' portion (IAEG 1990) by the many movements which have occurred on different parts of the landslide.

Temporal brightness variations

The 'difference' image from northwest Italy shown in Figs 4(a) and (b) illustrates the advantage of using multi-temporal data to identify distinct local brightness

a)

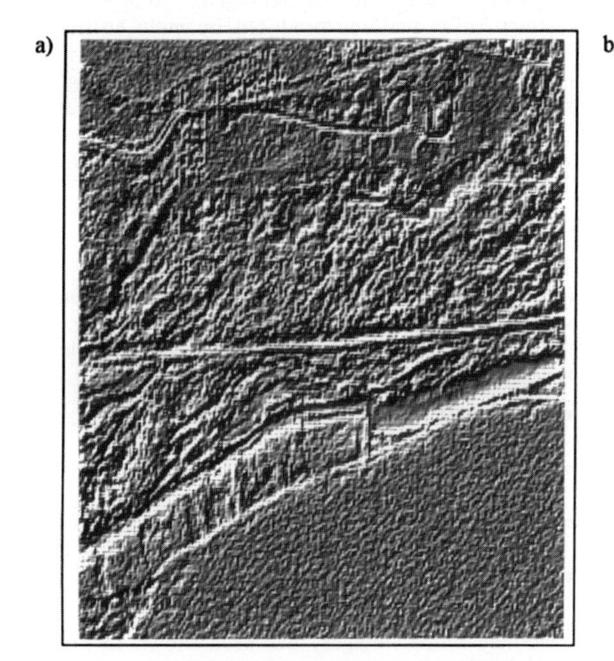

b)

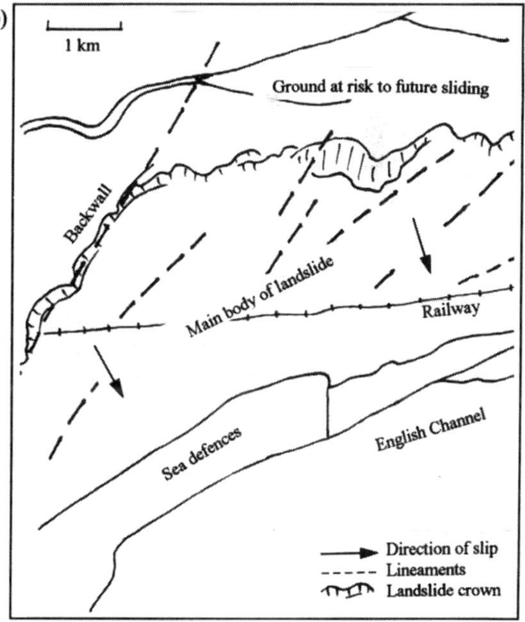

Fig. 3. Images of southeast England: (**a**) ATM band 9, Sobel filtered; (**b**) sketch interpretation. Note the distinct hummocky surface textures and lineaments in the main body of the landslide.

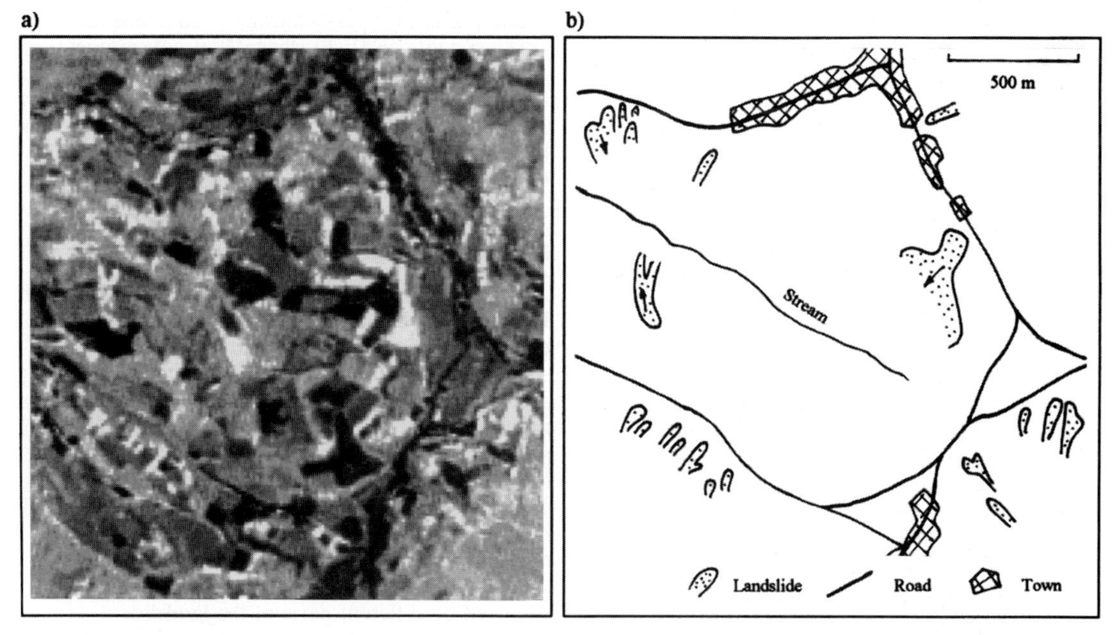

Fig. 4. Images of northwest Italy: (**a**) change detection image (SPOT Pan post- minus pre-landslide image) illustrating temporal brightness variations caused by landsliding (white = greatest change between the two images). The image has been Laplacian filtered to improve edges. (**b**) Sketch interpretation of landslides.

changes produced by a landslide event. Pre- and post-landslide images have been subtracted and filtered for general edge sharpening. Brighter tones indicate greater change. In this study a single post-landslide image provided insufficient information to identify all the landslides because of their very limited sizes and the complex image textures. The landslides have disrupted the ground surface and revealed rock surfaces with a high albedo, which are normally blanketed by soils and vegetation.

Drainage patterns and gullying

Sub-scenes from northwest Italy (SPOT Pan) and southeast Spain (ATM) are shown in Figs 5(a)–(f). Figures 5(a)–(d) illustrate the close relationship between debris slides and natural drainage channels in narrow valleys on steep hill slopes. As in Fig. 4, the landslides are evident by the very distinct brightness changes between the wooded hill slopes and the very bright rock surfaces. Figures 5(a) and (b) are texture filtered images acquired 5(a) pre- and 5(b) post-landsliding. Figure 5(c) shows the post-landslide image unfiltered, for reference. Figure 5(d) shows the distribution of landslides and drainage and indicates areas of landslide hazard. Another known example of landsliding of this type is North Island, New Zealand (Stephens & Trotter 1988).

Figures 5(e) and (f) illustrate the gully badland erosion of soft sediments in the 'main body' and 'accumulation' zone (IAEG 1990 and WP/WLI 1990) of the landslipped ground. Figure 5(e) shows ATM band 9 with 'Sobel' filters imposed and then added together to illustrate the distinct linear features produced by gully erosion. A sketch landslide hazard map is shown in Fig. 5(f). The distinctive drainage patterns produced highlight a sudden rock type change and a break of slope at the crown of the landslides.

Discussion

The combined use of remote sensing and field work (ground truth data collection) reveals the importance of spatial information to the study of landsliding. The following are considered important features revealed by the case studies examined in this work:

- the complex interplay between image textures, rock type and slope angle
- the influence of strata dip angle and topography on landslide morphology
- the influence of vegetation, drainage and differing slope form on landslide morphology
- the textural expression of high energy drainage on rocks of differing resistance

Remotely sensed imagery has been shown to provide spatial information concerning landslide locations and surface textures relating to slope instability. Using

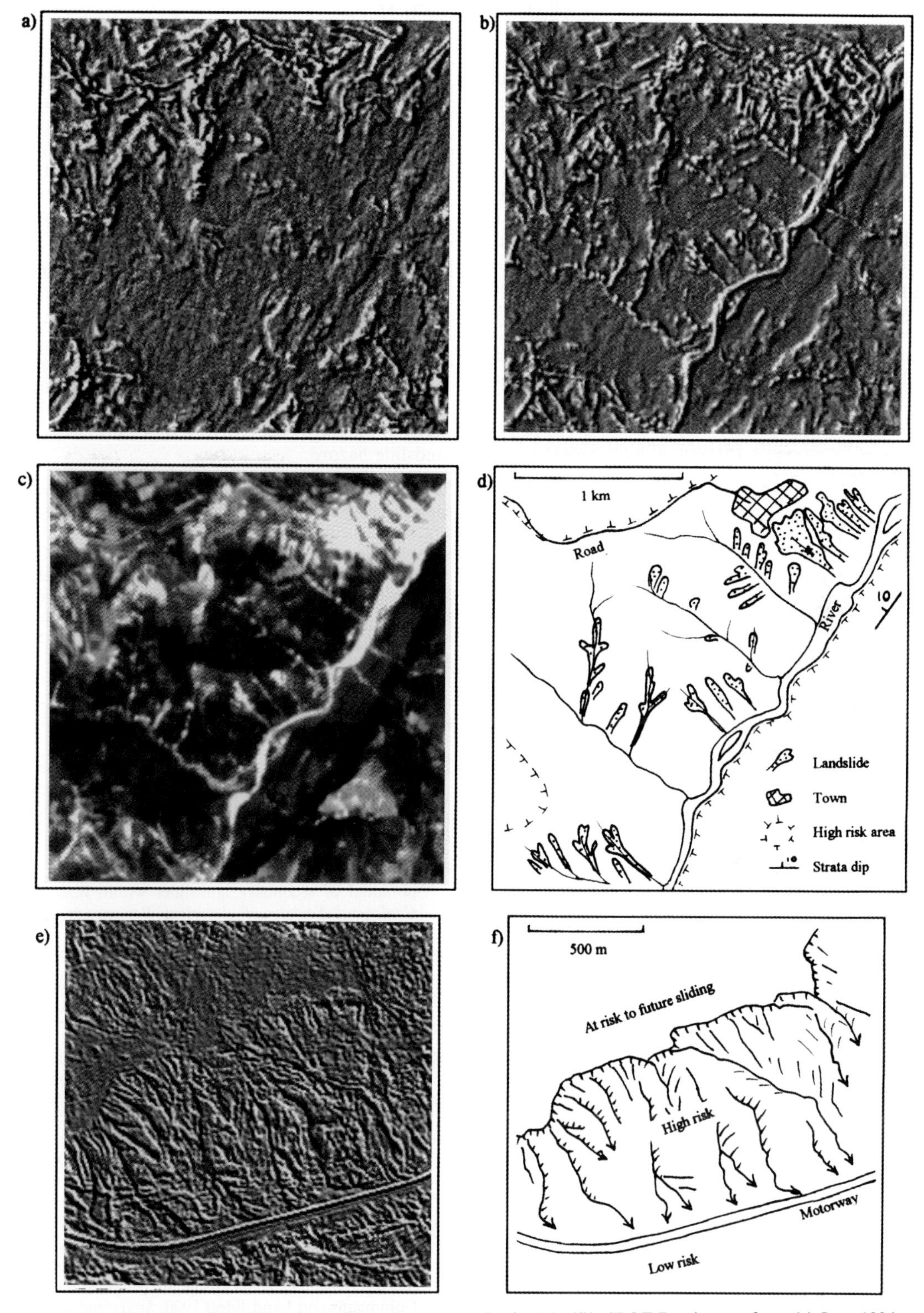

Fig. 5. Images of northwest Italy ((**a**)–(**d**)) and southeast Spain ((**e**)–(**f**)). SPOT Pan images from (**a**) June 1994 and (**b**) April 1995, both texture filtered; (**c**) April 1995 unfiltered; (**d**) sketch of drainge and landslide hazard; (**e**) ATM band 9 Sobel filtered; and (**f**) sketch map of landslide hazard.

texturally and spatially enhanced imagery, and or thematic information extracted from imagery, information relating to landslide type and occurrences has been derived. van Westen et al. (1994), identified 'National' (<1:1,000,000), 'regional' (1:1,000,000), 'medium' (1:25,000 to 1:50,000) and 'large' (>1:10,000) levels of scale for landslide hazard zonation using GIS. These categories appear to be useful but they apparently restrict the use of satellite imagery to the 'regional' scale of study, whereas both SPOT and TM imagery are also well suited to studies in the 'medium' scale range. The ongoing development of new satellite sensors with improved spatial and spectral capabilities, e.g. IRS-1D (Indian Space Research Organization), SPOT 5 and ASTER (Advanced Spaceborne Thermal Emission and Reflection Radiometer) will increase this range of scales even further.

GIS can subsequently perform spatial analysis tasks by using map algebra on the information gained from remote sensing and ground truth data. Combined with related geotechnical information, such as has been described in the GIS-based hazard assessment models described by Brass *et al.* (1991) and van Westen *et al.* (1994), areas susceptible to future slope stability can be targeted. Integration of such data using the GIS enables data assessment from which weighting of the relative influences of each variable on landslide occurrence can be established. The weighted factors can then be combined to provide a model which is flexible and sensitive to knowledge of the locality.

Conclusions

Digital imagery from a variety of sensors has been applied to the study of slope instability. The suitability of a particular type of imagery depends mainly on the dimensions, type and setting of the landslides. The following conclusions can be drawn:

- The three case studies, in Italy, Spain and southeast England, have shown that information concerning landslide location, type and morphology can be extracted from remotely sensed imagery.
- The most important factor governing image texture in landslide studies is spatial resolution. The lower limit of resolution for remotely sensed imagery will govern the size of the smallest landslide that can be detected. For example, the lower limit of known landslides in the northwest Italy case study, which can be resolved by the 10 m pixels of SPOT Pan, is approximately 100 m by 40 m. The resolution of airborne imagery and photography significantly improves on this detection, enabling detailed mapping and interpretation, but cannot provide the temporal coverage.
- Enhanced imagery can reveal spatial patterns and enable relationships to be established linking landslides to terrain type, climate, land use, geology, slope angle, aspect, rock type and human intervention.
- The information required for inclusion in an image-based database, suitable for defining the degree of instability using GIS, depends on the temporal factors such as rainfall and vegetation growth. Situations particularly suited to GIS analysis include, for example, cases where widespread landslides have been produced by a single event.

Such spatial information should then be combined with information more traditionally associated with landslide studies (such as a DEM, lithology, slope angle, groundwater levels and material strengths) in a GIS to construct the landslide hazard assessment model. The identification of landslides in differing situations using image analysis forms a valuable contribution to the decision-making process involved in a GIS assessment of landslide hazard.

Acknowledgements. The following are gratefully acknowledged for facilitating the research described in this paper: the MURST/British Council Agreement for the provision of travel subsistence money for fieldwork in Italy; P. Boccardo, G. Bottino and colleagues at the Politecnico di Torino for sharing data and knowledge; Imperial College Department of Geology for providing computing facilities and the Remote Sensing data archive; NERC for the provision of ATM imagery in Spain and southeast England.

References

BRASS, A., WADGE, G. & READING, A. J. 1991. Designing a Geographical Information System for the prediction of landsliding potential in the West Indies. *In*: JONES, M. & COSGROVE, J. (eds) *Neotectonics and Resources.* Belhaven, London.

CRONIN, V. S. 1992. Compound landslides: nature and hazard potential of secondary landslides within host landslides. *Geological Society of America, Reviews in Engineering Geology*, **IX**, 1–8.

DRURY, S. A. 1993. *Image Interpretation in Geology.* Allen & Unwin, London.

EYERS, R. 1994. *The use of satellite imagery for the detection of landslides in Southeast Spain.* MSc Thesis, University of London.

GOVI, M. & SORZANA, P. F. 1982. Frana di scivolamento nelle Langhe Cuneesi Febbraio–Marzo 1972, Febbraio 1974. *Bolletina della Associazione Mineraria Subalpina*, Anno **XIX**(1–2), 231–263.

HARALICK, R. M., SHANMUGAN, K. S. & DINSTEIN, I. 1973. Textural features for image classification. *IEEE Transactions on Systems, Man and Cybernetics*, **SMC-3**, 610–622.

HUTCHINSON, J. N., BROMHEAD, E. N. & LUPINI, J. F. 1980. Additional observations on the Folkestone Warren landslides. *Quarterly Journal of Engineering Geology*, **13**, 1–32.

IAEG (International Association of Engineering Geology, Commission on Landslides) 1990. Suggested nomenclature for landslides. *Bulletin of the International Association of Engineering Geologists*, **41**,13–16.

LIU, J. G. 1991. Balanced contrast enhancement technique and its application in image colour composition. *International Journal of Remote Sensing*, **12**(10), 2133–2151.

NIBLACK, W. 1986. *An Introduction to Digital Image Processing*. Prentice-Hall, London.

PIETIKAINEN, M., ROSENFELD, A. & DAVIES, L. 1983. Experiments with texture classification using averages of local pattern matches. *Institution of Electrical and Electronic Engineering, Transactions on Systems, Man and Cybernetics*, **MSC-13**(3), 421–426.

ROSENBAUM, M. S. & JARVIS, J. 1985. Probabilistic stability analysis using a microcomputer. *Quarterly Journal of Engineering Geology*, **18**(4), 353–356.

SCHUSTER, R. L. & KRIZEK, R. J. (eds) 1978. *Landslide Analysis and Control*. Special Report 176, Transportational Research Board Commission on Sociotechnical Systems, National Research Council, National Academy of Science, Washington, DC.

STEPHENS, P. R. & TROTTER, C. M. 1988. Landslides spotted by SPOT. *Streamland*. Department of Scientific and Industrial Research, Palmerston North, New Zealand, 75.

——, ——, DE ROSE, R. C., NEWSOME, P. F. & CARR, K. S. 1988. Use of SPOT satellite data to map landslides. *In*: *The 9th Asian Conference on Remote Sensing*, Bangkok, Thailand, 23–29 November, J-11-1 to J-11-6.

VAN WESTEN, C. J., SOETERS, R. & RENGERS, N. 1994. GISSIZ: training package for the use of geographical information systems in slope instability zonation. *In*: *Proceedings of the 10th Thematic Conference on Geologic Remote Sensing*, San Antonio, Texas, 9–12 May, I-387 to I-397.

VARNES, D. 1984. *Landslide Hazard Zonation: A Review of Principles and Practice*. Commission on Landslides of the International Association of Engineering Geology, United Nations Educational Social and Cultural Organisation, Natural Hazards, No. 3.

WP/WLI (International Geotechnical Society's UNESCO Working Party on World Landslide Inventory) 1990. A suggested method for reporting a landslide. *Bulletin of the International Association of Engineering Geologists*, **41**, 5–12.

WRIGHT, R. 1995. *An analysis of slope movements and associated lithologies using ATM data*. MSc Project, University of London.

SECTION 6

URBAN GEOHAZARDS IN DEVELOPING COUNTRIES

Geohazards and the urban environment

G. J. H. McCall

44 Robert Franklin Way, South Cerney, Gloucestershire, GL7 5UD, UK

Abstract. Recent escalation of urbanization throughout the world has increased the attention to geoscience in the urban setting. In the urban context, earlier attention to routine engineering geology and hydrogeology investigations was later augmented by the introduction of thematic mapping and the erection of urban databases to support environmental assessments and land use planning. More recently, the role of the urban environment in exporting pollution regionally and globally, and the problems of degradation of the urban environment itself, together with the immediately surrounding area on which the urban community depends for resources, have attracted attention; as has the mushrooming of megacities which threatens disastrous environmental problems. In this paper, the aim is to review the geohazards that threaten the urban setting (using the term 'geohazard' in the broadest sense, i.e. both natural and human-made, and both intensive rapid-onset and slow-onset pervasive geohazards).

To illustrate the extent of urban problems, the principal natural geohazards which threaten towns and cities, most of them of the intensive, rapid-onset type, are briefly considered. However, the emphasis is on the various human-made geohazards, most of them pervasive and of slow onset. Examples are cited of these human-made problems, all of which involve the interaction between human activity and natural processes. Details are given of some of the acute problems facing cities and towns, as well as urbanized regions.

Introduction

Urbanization has increased throughout the world during the last three decades, especially in coastal zones where an estimated 75% of the population now reside in cities and towns. The percentage of the world population that are urban dwellers is approaching 50%. Percentage statistics for 1960 and 1990 are given in Table 1 with a regional breakdown.

Geoscience has long been practised in the urban setting, but initially such studies were mainly concerned with engineering geology and hydrogeology investigations, and studies of urban geohazards involving instability such as earthquakes, subsidence and landslides. With the pressure for environmental control and stricter land use planning arising from environmental awareness, increased population densities, urban decay and the need for renewal of many long-established towns and cities in the Western world, came the development of urban thematic mapping and urban database preparation. The establishment of databases covering all existing subsurface information is a critical requirement for the support of urban land use planners, whether working in established towns and cities or new urban developments. Old cities and towns seldom possess such databases in an integrated form and they have to be laboriously gleaned from a variety of sources (governmental, commercial, archival, utilities). New towns and cities need to commence with a systematic compilation of such records. In both cases, the database needs to be continuously updated. Such studies usually involve specialized geological mapping as well as data collection, and where the existing geological maps are inadequate there may even be a need for geological remapping. Examples of such geotechnical and environmental coverage of established urban areas are studies by Gilsanz *et al.* (1985) in Madrid; Gozzard (1986) in

Table 1. *The change in urban populations by percentage, regionally and global: 1960 to 1990*

Region	*Urban population as % of total*	
	1960	*1990*
Africa	18.3	33.9
North & Central America	63.2	71.4
South America	51.7	75.1
Asia	21.5	34.4
Europe	61.1	73.4
Former USSR	48.8	65.8
Oceania	66.3	70.6
World	34.2	45.2

Source: World Resources Institute (1992), quoted in Pickering & Owen (1994).

McCall, G. J. H. 1998. Geohazards and the urban environment. *In*: Maund, J. G. & Eddleston, M. (eds) *Geohazards in Engineering Geology*. Geological Society, London, Engineering Geology Special Publications, **15**, 309–318.

Perth, Western Australia; Loudon & Mennim (1987) in Southampton; McCall & Chandler (1992) in Plymouth; and of a proposed new city by Cratchley & Denness (1972) in Milton Keynes.

Recently, however, the role of the urban setting in environmental change and damage on a broader scale has become a major concern. The urban setting is a major exporter of environmental damage on a regional and even a global scale; for example, acid rain and pollution of surface and groundwater systems on a regional scale (effects stemming largely from industry, power stations, smelting and domestic effluents), and pollution from transport, energy installations and industry contributing to the greenhouse effect and ozone layer depletion on a global scale. Besides these wider considerations, the urban setting is also damaging to itself and the area surrounding it, upon which it depends for the resources. It is not inconceivable that excessive urban growth of a megacity may in the end destroy the resources on which its very existence depends and bring it to a state of disaster. In addition to being prone to the many mainly human-made, slow-onset, pervasive geohazards, the urban concentrations of population concentrate the risk from mainly natural intensive, rapid-onset geohazards.

Reduction of the adverse environmental effects within and immediately adjacent to towns and cities will reduce the regional and global impact therefrom and so the problem of urban environmental degradation, if tackled on the local scale (individual cities or towns), will also reduce the regional and global impact of urban degradation.

This paper aims to enumerate the geoscience-related problems facing urban communities throughout the world and to provide some examples that demonstrate the widespread occurrence of severe problems of this nature in urban settings.

Table 2. *The principal urban geohazards*

Intensive, rapid-onset geohazards
Earthquakes and active faults
Volcanic eruption (lava flows, ash falls, núees ardentes, lahars)
Tsunamis
Landslides, avalanches, debris flows, rapid subsidence
Flooding, storms, cyclones, tornados
Wildfire related to earthquakes, combustion related to hazardous gases such as methane
Pervasive, slow-onset geohazards and other usually non-intensive processes
Foundation problems: peatlands and other compressible soils, low areas subject to flooding
Permafrost conditions requiring special and expensive building and construction techniques
Subsidence related to excessive groundwater abstraction, mine workings, hydrocarbon extraction or karstic conditions
Ground fissures, including those related to expansion and shrinkage
Neotectonic uplift
Groundwater pollution, by sewage, leachates from waste dumps, nitrates from agriculture or other sources, benzene, toxic metals and chlorinated organic solvents, toxic metals
Surface water pollution
Depletion of the groundwater resource, including loss of recharge due to urbanization
Rising groundwater levels due to cessation of or diminished abstraction, irrigation returns, etc.
Rising sea level (possibly due to global warming), affecting coastal cities and towns
Sea water intrusion in coastal situations
Contaminated soils from urban industrial sources
Saline soils
Compressibility, shrinkage, heave of soils
Radon emanation
Erosion and deposition
Soil loss due to deforestation and urban development
River bank failure and silting
Coastal changes due to urbanization

The principal geohazards facing urban communities

The term 'geohazard' is used in the broadest sense following the usage in McCall *et al.* (1992) to embrace both the intensive, rapid-onset and the pervasive, slow-onset types. The principal urban geohazards are listed in Table 2. The usage adopted here also follows that of Doornkamp (1989): 'a hazard of a geological, hydrological or geomorphological nature which poses a threat to man and his activities'; and that preferred by the present author (McCall 1992): 'a geohazard is one that involves the adverse interaction of man and any natural process of the Planet'.

Natural geohazards in the urban setting

Earthquake

Amongst the natural geohazards, earthquakes take the greatest toll of human life in single events, particularly in the less developed countries where buildings and constructions are commonly not specially designed. Examples are 240 000 deaths in Tangshan, China, in 1976, mainly in the city (Chen Yong *et al.* 1988); 30 000 in Killari, Maharashtra, India, in 1993 (Seeber 1994); and 5000 in Kobe, Japan, in 1995 (Esper & Tachibana 1995). Low-rise, poorly constructed buildings suffered most in Maharashtra, whereas in Mexico City, in 1985, much of the damage was due to the interaction of medium- to high-rise buildings (Degg 1992). Magnification of shock effects where buildings and constructions overlie thick superficial material including liquifiable sands and clays was an important factor in the cases of Anchorage in 1964 (Dubrovolny & Schmoll 1968), Mexico City in 1985 (Degg 1992), and San Francisco in 1989 (Gore 1995).

In developed countries the financial toll may be extremely high, though the casualties are relatively few; for example, in Los Angeles in 1994, there were 60 fatalities but the damage was estimated at more than $10 billion (Gore 1995). For the Kobe earthquake the cost is predicted to be immense, early estimates being about US$100 billion (Reid 1995) but a recent estimate exceeding US$200 billion (Esper & Tachibana 1995). A major earthquake in the nearby Tokyo conurbation of *c.* 30 million people would affect the entire global economy.

Volcanic hazards

Volcanic hazards also provide a major threat to many cities and towns: Armero, Colombia, was destroyed by a lahar (volcanic mud flow) in 1985, with 23 000 fatalities (Hall 1992); volcanic ash from Irazu, Costa Rica, caused vast financial loss in 1963–65; and Rabaul, New Britain, had to be entirely evacuated in 1994 because of catastrophic ash falls (McKee 1994). Quito, Ecuador, is under constant threat from lahars emanating from Cotopaxi, most of the outlying suburbs being set on historic lahar deposits (Mothes 1992).

Tsunamis

Destructive tidal waves may follow earthquakes or explosive submarine volcanic eruptions, and also large-scale submarine landslides. Towns situated on the Pacific coastlines are particularly prone to tsunamis: in the Sanriku district of Japan, 28 000 people drowned in a tsunami event in 1896 and 3000 in a 1933 event (Menard 1974). Oahu and the big island in Hawaii suffered 159 deaths and damage valued at US$25 million from a tsunami sourced south of Unimak Island, Alaska, thousands of kilometres away, in 1946 (Menard 1974). The Chilean earthquake of 1960 caused 61 fatalities in the city of Hilo, Hawaii, and 180 on the Japanese coast where damage amounted to US$500 million (Menard 1974).

Landslides

Landslides, which may be due to natural causes or triggered by human activities, do not usually cause a great number of fatalities in each event, but taken overall they cause immense financial damage to urban settings throughout the world, annually (Jones 1992). There are, however, exceptional very large-scale and catastrophic events. For example, an earthquake-triggered ice and rock avalanche from Huascaran Norte in Peru in 1987, caused 50–100 m^3 to descend at speeds of up to 400 $km\ h^{-1}$, obliterating the towns of Yungay and Ranrahirca with the loss of 25 000 lives (Reynolds 1992). Also, in the Vaiont Dam disaster in the Italian Alps in 1963, 250 million m^3 of rock slid into an impounded lake, causing waves 100 m high to overtop the retaining dam, drowning 2600 people in the valley below (Jones 1992). Hong Kong, where slopes are steep and rainfall is high, is prone to very costly slope failures and landslides. Los Angeles has suffered spectacular and costly landslide damage recently after heavy rainfall. Most Calabrian hill towns in southern Italy are subject to landslides: in Basilicata, 115 towns out of 131 are affected and the financial loss thereby is estimated at US$200 million in a single year (Jones 1992).

Natural subsidence

Natural subsidence causes extensive and costly damage in cities and towns with karstic limestone foundations: examples are Wuzhan City in China, Orlando in Florida and Ribeirao in Brazil (Landplan IV 1993).

Storms, flooding and wildfire

Great loss of life and damage to property can be caused by floods: figures for fatalities in 1993 are 7903 for earthquakes and 6807 for floods (Ayala 1994). Riverine floods affecting the Indus and Mississippi systems have been among the most recent catastrophic flooding events. Financial damage was to the fore in the Mississippi floods of 1993 (Melcher & Parrett 1993), due to a 1 in 100 year freak climatic event, and much of the estimated $10–15 billion damage was due rupture of levees and inundation of urban developments. Human casualties were relatively few in that case, but were greater in the Indus floods, where the city of Hyderabad was threatened. Storm surges associated with cyclonic storms are also a cause of great loss of life and property. In Bangladesh, in 1970, the densely populated low-lying region of the Ganges delta suffered a storm surge from the Bay of Bengal during a hurricane and between 250 000 and 500 000 people were killed (Cobb 1993). A similar disaster occurred in the same area in 1991 following another cyclonic storm surge, and fatalities were estimated at 139 000 (Cobb 1993). Much of the damage was to urban settlements, including cities of Chandpur and Kazipur, which were in danger of washing away. The combination of very high wind velocity and rainfall was experienced in the case of Hurricane Andrew in 1992 (Gore 1993), which is estimated to have caused the largest amount of property damage in any recent disaster in the United States, estimated at $30 billion, much of it in urban areas of the southeast. The cities of Darwin and Townsville in Australia were flattened by cyclones in the 1970s.

Inundation and flooding as a result of rising sea levels threaten many coastal cities: this may be a combination of natural sea transgression and rise due to global warming, related to anthropogenic causes. The industrial cities of Shanghai, Tianjin and Guangzhou on the coast of China are also threatened (Wang Sijing & Zhao Xitao 1992).

Fire

Fire is a common consequence of earthquakes, as in San Francisco in 1906 (Andrews 1963) and Kobe in 1995 (Reid 1995). It may also occur during drought conditions, sparked naturally, by human carelessness or arson; and such events have caused much property damage and loss of life recently in the outer suburbs of Sydney and Melbourne, Australia, and in the Los Angeles conurbation.

Radon emanation

Of the pervasive geohazards listed in Table 2 radon emanation is regarded by some as the most important of the natural geohazards (Akerblom 1994; Sutherland 1994). In Gegui City, Yunnan Province, China, with a population of 1.9 million people, lung cancer is reported to be 4–40 times the natural incidence (Landplan IV 1993) and this is attributed to radon emanation, which shows a marked geological control in its distribution, occurring over placer deposits and areas where houses are built abutting hills of dolomite or monzonite.

Permafrost

Permafrost in northern Canada and Siberia necessitates very costly building and construction practices.

Such natural hazards are always with us, their effects being concentrated and accentuated in the urban setting. They are the subject of the ongoing International Decade for Natural Disaster Reduction (IDNDR) programme (Housner 1989). They cannot be removed and must be tackled by mitigation procedures (often of limited potential), such as avoidance of urbanization in hazard-prone sites, and emergency warning and preparedness systems. Education of threatened urban populations as to the nature of the risks and involving them in warning systems is also important (Delos Reyes 1992). Evacuation of urban populations when severe risk arises is a viable option, but involves a difficult decision for socio-economic reasons: repeated evacuation of homes and property on account of threatened events which prove to be groundless means that eventually there will be little or no response from the population when an event takes place. Evacuation was effected successfully at Rabaul, but at Armero, though there was adequate warning (Hall 1992), it was not implemented – with disastrous consequences. Alerts in 1958 and 1959 for threatened tsunamis in Hawaii, when nothing happened, resulted in widespread disregard of the warning in 1960, when a tsunami event was actually experienced (Menard 1974).

Human-made geohazards

It is not the natural geohazards that are at present causing most concern amongst environmentalists. It is the human-made group, which is to a large extent resultant on ill-managed urbanization and industrialization, that causes most concern. Strictly no geohazard is entirely man-made, for these geohazards involve the interaction of human activities with natural processes: these geohazards are, however, resultant on the actions of humans.

To illustrate the scope of these problems in urban settings, a number of examples of acute urban problems are described below.

Urban problems in China

China, with its immense and growing population, has immense problems of urban degradation (Table 3). There are many ancient towns and cities which have been industrialized within a very short period and are rapidly expanding, and until recently development was not accompanied by careful land-use planning with reference to environmental damage, though much is now being done in the way of prevention and rehabilitation. A number of case studies were presented at a conference held in Beijing in 1993 (Landplan IV 1993; Wang Sijing & Wang Cunyu 1994). Fifty cities in China are reported to have problems of industrial pollution of groundwater, 33 phenol–cyanogens–mercury–chromium related and 17 arsenic–fluorine related. Saline intrusion is a major problem rife in many coastal cities, as well as rising sea level (Wang Sijing & Zhao Xitau 1992). Pollution of karst aquifers under cities is very common and land stability problems are widespread. Active faulting and fissuring, related to neotectonics, is a major natural problem in some cities, and is exacerbated by increased groundwater extraction from a deep confined aquifer in the case of Xian City (Forster 1995).

Urban problems in Eastern and Central Europe

Many industrial cities and towns of the former Soviet Union have severe environmental problems related to neglect of environmental safeguards and planning, but detailed information from these is lacking. Industrialization in the urban setting along the coast of the Baltic Sea has caused extreme environmental damage in the Baltic States, and this pollution, coupled with that from fertilizers, pesticides and sewage, is extending throughout the Baltic (Vesilind 1989). Equally alarming is pollution in

Table 3. *Some urban problems affecting cities in China*

City	*Earthquake*	*Landslide*	*Ground subsidence*	*Neotectonic, ground fissuring*	*Water supply deficiency*	*High water-table*	*Pollution of surface and ground water*	*Saline intrusion*	*River-bank failure and silting*	*Waste disposal problems*	*Karstic ground-water source problems*	*Karstic subsidence*	*Rising sea level problems*	*Radon emanation*
Beijing	×		×											
Changbo[1]			×											
Chengdu						×								
Fuzhou			×											
Gegui														×
Handan[2]							×							
Lanzhou	×	×	×	×										
Longku[1]								×						
Lhaizon[1]								×						
Nanjing		×			×		×		×	×				
Nanlong[1,3]	×		×						×	×				
Nunchang[4]					×		×							
Shanghai[1]			×										×	
Tangshan	×													
Tianjin[1]			×											
Tong-chuan		×		×					×					
Wuzhan												×		
Xi'an[1,5]			×	×			×			×				
Zibo											×			

Source: Landplan IV (1993).
[1] Due to overabstraction of groundwater.
[2] Pollution from industry (chlorine sulphate, cyanogens, mercury).
[3] Problems of shock liquefaction of superficial deposits.
[4] Pollution due industrial sources (benzene), agrichemicals and sewage.
[5] Pollution from industrial waste stacks and domestic waste.

Table 4. *Environmental geo-problems in cities of Eastern and Central Europe*

City	*Ground-water pollution*	*Surface water pollution*	*Ground-water resource depletion*	*Soil contamination*	*Flooding*	*Land subsidence*	*Sea-water contamination*
Belgrade					×		
Bratislava	×						
Bucharest		×					
Budapest	×			×	×		
Łodz	×	×					
Moscow	×	×	×				
Prague	×						
Riga							×
Sofia	×				×		
Tallinn	×		×	×			
Tirana		×	×	×		×	
Vilnius				×			
Warsaw		×					

Polish Silesia along the Oder, Vistula, Nysa Luzycka and Bug river catchments (Helios Rybicka 1996). There has been extensive exploitation here of hard and brown coals, copper, zinc and lead ores, native sulphur and rock salt, with little or no environmental safeguards. Weathering of spoil heaps is the main source of the pollution, but sewage pollution is also widespread.

There has been some systematic collection of data from the cities of central and eastern Europe (E. F. J. De Mulder pers. comm., 1994) and this indicates that groundwater pollution, surface water pollution and soil contamination are the main human-made problems. Flooding and land subsidence are also important, the latter problem also affecting Polish Silesia as a result of underground mining. The scale and intensity of such problems is generally greater than in Western Europe, but documentation and mapping of surface and subsurface conditions in the urban areas of Western Europe is generally remarkably good. Prague is particularly well provided with detailed maps and an up-to-date database of geotechnical information. Acute problems threatening these cities are summarized in Table 4.

Foundation and other problems in Bangladesh

The problems of cyclone/storm surge related flooding in Bangladesh have been mentioned above. Dakha and other conurbations also have acute problems and the urban population is expected to double there by AD 2000. The urban developments of Dakha City spread over both high level and low level areas, much of the recent expansion being onto lower ground with no proper planning control. Such areas are artificially raised above the normal flood level by landfill for residential and commercial development. Domestic refuse is often used for landfill material. Peat fields and other organic soils lie 20 m below the good foundation layer. Hydraulic fill techniques are also in use. All this produces severe foundation problems, and subsidence is widespread after building has been completed. All towns in Bangladesh are at risk from catastrophic monsoonal flooding and most of the Ganges–Brahamaputra outflow plain would be submerged with any substantial rise in sea level due to global warming: Dakha is less than 8 m above sea level. The bulk of the available crop and pastoral lands, which supply food to this exploding urban population, would be submerged as well as much of the low-lying urban developments.

Urban subsidence in Asia and Europe

Thailand, like Bangladesh, is liable to severe flooding after heavy rainfall: in 1988, 55 800 houses were destroyed, with 375 fatalities, and 1.35 million people were affected. The cities also experience severe problems from land subsidence due to excessive extraction of groundwater (Nutalaya 1989). Remedial measures have been initiated but none has proved completely successful. The result is severe inundation of urban areas including Bangkok and cities in Samut Prakan, Samut Sakhon, Nonthaburi and Pathum Thani provinces. Nearby in Viet-Nam increasing problems of groundwater abstraction are being encountered in four cities, as well as related land subsidence, sea water incursion, flooding, foundation failure and pollution of groundwater and surface water. Ho Chi Minh City (Pham *et al.* 1994), Haiphong, Da Nang and Hue are all similarly affected.

The same problem affects other cities in eastern Asia. Tokyo, Osaka, Niigata, Saga, Nagoya, Saitama, Aomori and Nobi in Japan are affected by this problem as well as foundation problems on peatlands and compressible soils. Shanghai and three other cities in China, Manila,

Jakarta and Surabaya, are affected by groundwater-extraction-related subsidence. San Carlos City, Brazil, in a sabkha setting, is experiencing both subsidence and groundwater rise.

The most famous example of this problem in Europe is Venice, where subsidence in the lagoonal setting is accompanied by large-scale industrial and domestic pollution. London is reported to be experiencing a combination of subsidence and rising water-tables, producing a relative change two to three times as fast as in the case of Venice.

Hydrocarbon extraction can produce similar effects of urban subsidence as is occurring above the Wilmington oilfield in the Los Angeles conurbation, California.

Problems of rising water-tables under cities of the Middle East

The cities of Kuwait, Doha, Cairo, Riyadh, Jizzan, Tabrik, Buraidh, Madinah and Jubail in Arabia and Egypt all suffer from the effects of rising groundwater levels (George 1992). The problem relates to the implementation of major water supply systems drawing on sources outside the city limits and the abandonment of extraction from wells situated within those limits. The rise of water is also partly due to irrigation returns from nearby agriculture. The result is the invasion of water into foundations and basements, with the damage and destruction of buildings and roadworks; and coupled with this there is spillage from septic tanks into the storm water system causing it to be heavily polluted, with the spread of disease. In Europe, the problem of rising groundwater is reported from Paris (and London, see above), whereas in Luneberg, Germany, the groundwater level appears to be rising but is in fact stationary, the apparent rise being counteracted by salt-mining-related land subsidence. In Hamburg, the groundwater level fluctuates, but there are problems of groundwater rise and related pollution.

Exhaustion of surface and groundwater resources supplying cities of the Middle East

Cities are expanding in the arid countries of the Middle East beyond the capacity of the available groundwater and surface water resources that support them. In the case of both Amman, Jordan, and San'a, Yemen, available groundwater resources are stretched to the limit and the cities continue to expand.

In the case of Islamabad, Pakistan, a new city of 1.3 million people (and the twin city of Rawalpindi), there has been an immense demand for building and construction materials and their extraction has severely damaged the aquifer, and depleted water supplies. A similar problem may be experienced in the Mendip Hills near the English cities of Bath and Bristol, where it has been suggested that intensive limestone quarrying may, if continued, eventually threaten groundwater resources.

Reliance on 'fossil' water sources by cities in North Africa

A number of the cities of North Africa obtain their supplies from 'fossil' groundwater, i.e. groundwater stored in aquifers in a period of much wetter climate long ago. Such supplies are finite and there is an acute problem in these arid cities of finding adequate alternative water sources to replace them.

Groundwater pollution

Many Indian cities have acute problems of groundwater pollution. In Madras, which relies mainly on supplies pumped from wells outside the city or surface water, there is industrial contamination of a shallow aquifer under the city (Somasundaram *et al.* 1993). Out of 93 wells tested, TDS were high in 43, sulphate in 14%, nitrate in 70%, nitrate values being up to $1040\,mg^{-1}$. Heavy metal pollution is also evident (e.g. Hg, As, Pb and Cd). Thus the shallow groundwater is grossly polluted throughout the city area. Bhopal and Ludhiana suffer from industrial pollution of groundwater, Lucknow from nitrate pollution, Uttar Pradesh and Faridabad from chromium pollution, and Jaipur from groundwater pollution from refuse dumps.

The Lima–Callao conurbation in Peru (Rojas *et al.* 1994) has a population of 6.5 million and it relies largely on an unconfined aquifer in alluvial sediments. This is severely contaminated by bacterial pollutants, as well as by chemical contaminants from industry, mining and agriculture. Similar problems are faced by Greater Bandung, Indonesia, with a population of 2.5 million (Wagner & Sukrisno 1994): intensive land-use and explosive population and economic growth have led to the infiltration of domestic sewage and industrial contaminants as well as leachate from waste disposal into the aquifers exploited for the central urban water supply.

Bacterial contamination of groundwater is a problem in the European cities of Orvieto, Italy, and Tilberg, the Netherlands, as well as in Merida, Mexico and Tirupathi, India, and nitrate contamination of groundwater related to sewage contamination is acute on Long Island, New York, and in Stockholm, Tirupathi, and Cairo, Egypt, which also suffers from toxic metal contamination. In Kumasi, Ghana, there is severe contamination of the groundwater supply by leachates from waste disposal sites. São Paulo, Brazil, has similar problems of groundwater pollution.

Pollution spreads downwards particularly rapidly in cities set in karstic limestone terrains: as in Bermuda and Wuzhan, China (Landplan IV 1993).

In Taiwan (Zich 1993), heavy industrialization has not been matched by environmental protection and than 90 000 factories emit cadmium, chromium, zinc and other toxic substances. Less than 4% of the sewage is treated. Forty-four rivers are contaminated and half the drinking water reportedly comes from polluted sources. Problems of industrial pollution have also been encountered in Perth, Western Australia (Landplan IV 1993), which is underlain by a shallow unconfined aquifer; there has also been loss of recharge due to sealing of much of the surface as the city of more than one million people has expanded.

Cities affected by solvent pollution

Pollution of aquifers by industrial solvents is widespread. In England, the problem is particularly acute in Coventry (Lerner 1994), where chlorinated solvents have been extensively used, particularly in the motor industry. These penetrate to the water-table and spread out as a plume. Copenhagen is also severely affected (Markussen & Møller 1994), a particularly acute problem because Denmark relies almost entirely on groundwater; removal of this form of pollution is expensive and difficult and wells may have to be taken out of use permanently on account of it. New Jersey, Long Island, Tilberg and Milan also have similar problems; Tilberg, in addition, has a problem with cyanide pollution (Somasunderam *et al.* 1993).

An unusual problem at Halifax, Nova Scotia, is chlorine pollution from de-icing salts, and Markussen & Møller (1994) have mentioned a similar problem in Copenhagen, derived from salting of roads in winter.

Coastal problems

Many of the problems of the coastal cities have been mentioned above, but the problem of saltwater intrusion into the freshwater bodies due to overpumping of groundwater is one of the most acute. Besides the cities of China already cited, this problem is experienced by the Atlantic coast cities of North America and by Egypt's heavily populated Mediterranean coast (Attia 1995).

In Florida, there are severe problems of coastal change caused by coastal urbanization.

Mining-related subsidence

Many cities and towns where a mining industry has been developed and underground mines are still active or worked out have problems of subsidence due to collapse of underground workings which may reach the surface as collapse or crown holes. Mining-related subsidence is particularly common in Polish Silesia (Helios Rybicka 1996) and in the Ruhr Valley cities of Germany, such as Essen and Duisberg-Bielefeld; at the latter, coal mining related subsidence has affected the largest river port in the world. Salt mining in Luneberg has caused a similar problem and in the town of Malartic, Quebec, a street of frame houses had to be moved on account of subsidence into old gold workings.

Discussion

Geohazards affecting urban areas are of four types:

(a) rapid-onset, very destructive, large-scale hazards (such as earthquakes, volcanic eruptions, lahars and other major debris flows) which cause great losses in property and lives and are difficult to mitigate against except through some planning and emergency responses;
(b) hazards which may be quick but are less extensive (e.g. pollution incidents, subsidence events, slower landslides) and may cause risks to people but more often result in economic losses: these can be addressed through planning and preventative action;
(c) slow, cumulative hazards which may affect areas of varying sizes (e.g. some types of pollution) which may be relatively easy to deal with in the early stages but which may build up to major problems that are difficult to address;
(d) pervasive hazards which have small-scale localized effects but which over the years constitute a major source of economic loss (e.g. shrinkage and heave of soils: the mean annual loss in the USA due to expansive soils was estimated at US$2200 million; Robinson & Speiker 1978).

There has been a tendency for both the public and politicians to notice and react to types (a) and (b) although the community memory of such events is alarmingly short. Recently type (c) has been appreciated much more, mainly due to problems of contaminated land and polluted ground and surface water, and has in fact become a major issue of the late 20th century. Type (d) tends to be overlooked even though major losses are involved.

Conclusion

It seems apparent from the above set of examples that the most significant human-made urban geo-problems are those related to ground and surface water and that of these pollution (bacteria and nitrate related to sewage; toxic metals, nitrate and organic solvents related to industry) is the most widespread. Pollution of surface water, including pollution of the seas from industrial and sewage discharges from fringing cities and towns, as in the case of the North Sea, Baltic and Mediterranean, is much more apparent and offensive to the general public and tends to be corrected, if only in the long term, but pollution of groundwater is more insidious, not being obvious, yet it is immensely damaging to this critical urban resource. Groundwater rise and land subsidence related to water extraction cannot be separated from pollution of the water resource which accompanies both conditions, and loss or decrease of the surface and groundwater resource, another major problem, may be related to its pollution. Cities and towns in the coastal setting are particularly prone to problems of sea-water intrusion into groundwater bodies, rising sea levels, and pollution of the sea water by sewage effluent which may carry substantial toxic metals.

Mineral, industrial and radioactive waste disposal is a problem with regard to both choice of method and finding adequate space for disposal: waste disposal can produce pollution of ground and surface water from leachates, and also air-borne dispersal of dust, which often carries toxic components that cause damage to health if inhaled and may eventually pass to the water system. Collapse of solid waste stacks in urban settings can cause loss of life (as at Aberfan, South Wales in 1966) and produce health risk and urban disruption.

The emphasis on water pollution problems in the literature studied may to some extent be subjective in view of the 'major issue status' at the present time mentioned above. Other equally important problems may be understated in the literature. Instability problems such as subsidence related to mining, or landslides related to unsafe engineering practices (e.g. excavating the toe of steep slopes) are widespread urban problems, as are foundation problems related to building on unsuitable soils and geological formations or on sites affected by instability. Degradation of soils and land by mineral extraction and industry requires much input by geoscientists into rehabilitation at the present time. There is a major problem in finding space in overcrowded cities for new constructions and transportation systems and quite novel techniques of engineering are being developed including much more use of underground space. The processes and problems of atmospheric urban pollution and its secondary effects are also very important and perhaps at present under-investigated.

It is apparent that geoscientists are going to be heavily involved with urban problems for many years. It is also clear that all cities and towns must couple their land-use planning activities with a sound basis of geoscience investigation, data compilation, interpretation and advice.

References

Akerblom, G. 1994. Ground radon monitoring procedures in Sweden. *Geoscientist*, **4**(4), 21–27.

Andrews, A. 1963. *Earthquake*. Angus and Robertson, London.

Attia, F. A. R. 1995. Groundwater in the Nile aquifer system and desert fringes of Egypt. *In*: Nash, H. & McCall, G. J. H. (eds) *Groundwater Quality*, AGID Special Publication No. **17**, Chapman & Hall, London, 123–129.

Ayala, F. 1994. Natural disasters in the World: update 1993. *Cogeoenvironment Newsletter*, **6**, 6.

Chen Yong, Tsoi-Kam-Ing, Chen Feibi, Gao Zhenzuan, Zou Qijia & Chen Zhangli 1988. *The Great Tangshan Earthquake of 1976 – An Anatomy of Diaster*, Pergamon Press, Oxford.

Cobb, C. E. 1993. Bangladesh – when the water comes. *National Geographic Magazine*, **183**(6), 118–134

Cratchley, C. R. & Denness, B. 1972. Engineering geology in urban planning with an example from the new town of Milton Keynes. *In*: *Proceedings of the 24th International Geological Congress*, Montreal, **13**, 14–23.

Degg, M. 1992. Some implications of the Mexican earthquake of 1985 for hazard assessment. *In*: McCall, G. J. H., Laming, D. J. C & Scott, S. C. (eds) *Geohazards – Natural & Man-Made*. AGID Special Publication No. **15**, Chapman & Hall, London, 105–114.

Delos Reyes, P. J. 1992. Volunteer observers program: a tool for monitoring volcanic and seismic events in the Philippines. *In*: McCall, G. J. H., Laming, D. J. C. & Scott, S. C. (eds) *Geohazards – Natural & Man-made*. AGID Special Publication No. **15**, Chapman & Hall, London, 13–24.

Doornkamp, J. C. 1989. Hazards. *In*: McCall, G. J. H. & Marker, B. R. (eds) *Earth Science Mapping for Planning, Development & Conservation*. Graham & Trotman, London, 157–175.

Dubrovolny, E. & Schmoll, H. R. 1968. Geology as applied to urban planning: an example from the Greater Anchorage Borough. *In*: *Proceedings of the 23rd International Geological Congress*, **12**, 39–56.

Esper, P. & Tachibana, E. 1995. The lesson of the Kobe earthquake. *In*: Maund, J. G., Penn, S. & Culshaw, M. G. (eds) *Geohazards and Engineering Geology*. Abstract Volume, 31st Annual Conference of the Engineering Group of the Geological Society, Coventry University, 10–14 September 1995, 137–151.

Forster, A. 1995. Active ground fissures in Xi'an, China. *Quarterly Journal of Engineering Geology*, **28**, 1–4.

George, D. J. 1992. Rising groundwater: a problem of development in some urban areas of the Middle East. *In*: McCall, G. J. H., Laming, D. J. C. & Scott, S. C. (eds) *Geohazards – Natural & Man-made*', AGID Special Publication No. **15**, Chapman & Hall, London, 171–182.

Gilsanz, J. de P., Carvillo, J. de D. C., Ruiz, L. I. O., Alonso, S. G., Deltell, E. A., Bombin, R. E. & Martinez, C. P. 1985. *Mapa fisiografico de Madrid*. Escale 1:2000000. Communidad de Madrid, Consejeria de Agricultura y Granaderia.

Gore, R. 1993. Andrew aftermath. *National Geographic Magazine*, **183**(4), 2–37.

——1995. Living with California's faults. *National Geographic Magazine*, **187**(4), 2–35.

Gozzard, J. R. 1986. Perth Sheet 2034 II and parts of Sheets 2034 III and 2134 III, *Perth Metropolitan Region Environmental Geology Series*, Geological Survey of Western Australia, Perth, Western Australia.

HALL, M. L. 1992. The 1992 Nevado del Ruiz eruption: scientific, social and governmental response and interaction before the event. *In*: MCCALL, G. J. H., LAMING, D. J. C. & SCOTT, S. C. (eds) *Geohazards – Natural & Man-made*. AGID Special Publication No. **15**, Chapman & Hall, London, 43–52.

HELIOS RYBICKA, E. 1996. Environmental impact of mining and smelting industries in Poland. *In*: APPLETON, D., FUGE, R. & MCCALL, G. J. H. (eds) *Environmental Geochemistry and Health*. Geological Society, London, Special Publications, **113**, 183–193.

HOUSNER, G. W. 1989. An International Decade for Natural Disaster Reduction: 1990–2000. *Natural Hazards*, **2**, 45–75.

JONES, D. K. C. 1992. Landslide hazard assessment in the context of development. *In*: MCCALL, G. J. H., LAMING, D. J. C. & SCOTT, S. C. (eds) *Geohazards – Natural & Man-made*. AGID Special Publication No. **15**, Chapman & Hall, London, 117–141.

LANDPLAN IV 1993. *International Conference on Geoscience and Urban Development*, Beijing, China, Abstract Volume.

LERNER, D. 1994. Chlorinated solvent pollution of an industrialised urban area: summmary of findings and implications for groundwater protection and clean up. *In*: NASH, H. & MCCALL, G. J. H. (eds) *Groundwater Quality*. AGID Special Publication No. **17**, Chapman & Hall, London, 185–190.

LOUDON, T. V. & MENNIM, K. C. 1987. *Applied geology mapping Southampton Area. Mapping techniques*. BGS Research Report ICSO/87/3.

MCCALL, G. J. H. 1992. Natural and man-made hazards: their increasing importance in the end-20th century World. *In*: MCCALL, G. J. H., LAMING, D. J. C. & SCOTT, S. C. (eds) *Geohazards – Natural & Man-made*. AGID Special Publication No. **15**, Chapman & Hall, London, 1–4.

—— & CHANDLER, P. 1992. *Applied Geological Mapping Study of the Plymouth-Plymstock Areas, South Devon*. Report by GAPS Geological Consultants for the Department of the Environment, London, MI Press, Long Eaton, Nottingham.

——, LAMING, D. J. C. & SCOTT, S. C. (eds) 1992. *Geohazards – Natural and Man-made*. AGID Special Publication **No. 15**, Chapman & Hall, London.

MCKEE, C. 1994. The Rabaul eruption of 1994. *Australian Geologist*, **93**, 24–26.

MARKUSSEN, L. M. & MØLLER, H.-M. F. 1994. Groundwater quality under Copenhagen. *In*: NASH, H. & MCCALL, G. J. H. (eds) *Groundwater Quality*. AGID Special Publication No. **17**, Chapman & Hall, London, 175–184.

MELCHER, N. B. & PARRETT, C. 1993. 1993 Upper Mississippi River Floods. *Geotimes*, **38**(12), 15–17.

MENARD, H. W. 1974. *Geology, Resources & Society*. Freeman, San Francisco.

MOTHES, P. 1992. Lahars of Cotopaxi Volcano, Ecuador. *In*: MCCALL, G. J. H., LAMING, D. J. C. & SCOTT, S. C. (eds) *Geohazards – Natural & Man-made*. AGID Special Publication No. **15**, Chapman & Hall, London, 53–63.

NUTALAYA, P. 1989. Flooding and land subsidence in Asia. *Episodes*, **12**(4), 239–248.

PHAM, VAN NGOC, BOYER, D., NGUYEN, THI KIM THOA & NGUYEN, VAN GIANG 1994. Deep ground-water investigation by combined VES/MTS methods near Ho Chi Minh City, Viet Nam. *Ground water*, **32**(4), 675.

PICKERING, K. T. & OWEN, L. A. 1994. *An Introduction to Global Environmental Issues*. Routledge, London and New York.

REID, T. R. 1995. Kobe wakes up to a nightmare. *National Geographic Magazine*, **188**(1), 112–136.

REYNOLDS, J. M. 1992. The identification and mitigation of glacier-related hazards: examples from the Cordillera Blanca, Peru. *In*: MCCALL, G. J. H., LAMING, D. J. C. & SCOTT, S. C. (eds) *Geohazards – Natural and Man-made*. AGID Special Publication No. **15**, Chapman & Hall, London, 143–157.

ROBINSON, G. D. & SPEIKER, A. M. 1978. *Nature to be Commanded*. US Government Geological Survey Professional Paper No. **950**, US Government Printing Office, Washington, DC.

ROJAS, R., HOWARD, G. & BARTRAM, J. 1994. Groundwater quality and water supply in Lima, Peru. *In*: NASH, H. & MCCALL, G. J. H. (eds) *Groundwater Quality*. AGID Special Publication No. **17**, Chapman & Hall, London, 159–167.

SEEBER, L. 1994. Killari – the quake that shook the world. *New Scientist*, **142**(1919), 25–29.

SOMASUNDARAM, M. V., RAVINDRAN, G. & TELLAM, J. H. 1993. Groundwater pollution of the Madras aquifer, India. *Ground water*, **31**(1), 4–11.

SUTHERLAND, D. S. 1994. Radon workshop – geology, environment, techniques. *Geoscientist*, **4**(2), 27–29.

VESILIND, P. J. 1989. The Baltic: arena of power, *National Geographic Magazine*, **175**(5) 602–635.

WAGNER, W. & SUKRISNO 1994. Natural groundwater quality and groundwater contamination in the Bandung Basin, Indonesia. *In*: NASH, H. & MCCALL, G. J. H. (eds) *Groundwater Quality*. AGID Special Publication No. **17**, Chapman & Hall, London, 168–174.

WANG SIJING & WANG CUNYU 1994. *Proceedings of LandPlan IV International Conference on Geoscience in Urban Development*, August 1993, Beijing, China. China Ocean Press.

—— & ZHAO XITAO 1992. Sea-level changes in China – past and future: their impact and countermeasures. *In*: MCCALL, G. J. H., LAMING, D. J. C. & SCOTT, S. C. (eds) *Geohazards – Natural & Man-made*. AGID Special Publication No. **15**, Chapman & Hall, London, 161–169.

ZICH, A. 1993. Taiwan – the other China changes course. *National Geographic Magazine*, **184**(5), 2–33.

Hazards induced by groundwater recharge under rapid urbanization

A. R. Lawrence,[1] B. L. Morris[1] & S. S. D. Foster[2]

[1] British Geological Survey, Hydrogeology Group, Wallingford OX10 8BB, UK
[2] British Geological Survey, Groundwater & Geotechnical Surveys Division, Nottingham NG12 5GG, UK

Abstract. Urban population growth across most of Latin America and Asia is relentless. Rapid urbanization has profound impacts on the hydrological cycle, including major changes in groundwater recharge. Existing infiltration mechanisms are radically modified, and new ones introduced, with evidence of an overall increase in recharge rates. There is also widespread diffuse pollution of groundwater by nitrogen compounds (normally nitrate, but sometimes ammonium), salinity (especially sodium chloride) and dissolved organic carbon. Groundwater contamination by petroleum and chlorinated hydrocarbons, related synthetic organic compounds, and, on a more localized basis, pathogenic bacteria and viruses, is also encountered. An analysis of the processes involved is made through citing detailed case histories.

Introduction

Urban hydrogeological cycle

The provision of water-supply, sanitation and drainage are key elements of the urbanization process. Major differences in the development sequence exist between higher-income areas, where the process is normally planned in advance, and lower-income areas, where informal settlements are progressively consolidated into urban areas, but common factors to most urbanization are impermeabilization of a significant proportion of the land surface and major importation of water from outside the urban limits.

Sanitation and drainage arrangements, which are fundamental to a consideration of the hydrological cycle (Lindh 1983), generally evolve with time and vary widely with differing patterns of urban development. In most developing towns and cities installation of mains sewerage systems lags considerably behind population growth and water-supply provision.

Urbanization causes radical changes in the frequency and rate of groundwater recharge, with a general tendency for volume to increase significantly and for quality to deteriorate substantially (Foster 1990). These changes cannot be measured directly and are thus difficult to quantify. The changes in recharge caused by urbanization in turn influence groundwater levels and flow regimes in underlying aquifers. This can take considerable time, because the response constants of groundwater systems are normally the largest of all components of the urban hydrological cycle (Ven 1990). It will usually be decades before aquifers reach equilibrium with the process of urbanization.

Abstraction frequently masks changes in recharge regime, especially when groundwater levels are falling due to local resource overexploitation. The most troublesome groundwater mounds are likely to develop under low-lying urban areas on relatively low-permeability unconfined aquifers, especially if they are not exploited for water-supply, as a result of poor water quality. Rising groundwater levels can affect the stability, integrity and operation of subsurface engineering structures and installations. Thus, while the hydrogeologist tends to speak of 'the impact of urbanization on groundwater' in terms of large-scale, long-term, temporal changes, the geotechnologist tends to think in terms of 'the impact of groundwater on urban structures' in a site-specific, steady-state sense.

Analysis of impacts on groundwater

A general summary of the principal urban processes affecting groundwater, and their relative impact, is given in Table 1. The impact of any given process will vary considerably with climatic regime, hydrogeological environment, and with the pattern of urban development itself.

Modifications to natural systems

Surface impermeabilization and drainage. Urbanization results in land-surface impermeabilization, which reduces direct infiltration of excess rainfall, but also tends to lower evaporation and thus to increase, as well as accelerate, surface runoff. Pluvial drainage arrangements

Lawrence, A. R., Morris, B. L. & Foster, S. S. D. 1998. Hazards induced by groundwater recharge under rapid urbanization. *In*: Maund, J. G. & Eddleston, M. (eds) *Geohazards in Engineering Geology*. Geological Society, London, Engineering Geology Special Publications, **15**, 319–328.

Table 1. *Summary of the impacts of urbanization processes on groundwater*

Process	*Effect on subsurface infiltration*			*Implications for quality*	*Principal contaminants*[1]
	Rates	*Area*	*Time base*		
(A) *Modifications to natural system*					
Surface impermeabilization and drainage					
roofs, paved areas	reduction	extensive (but see mains drainage)	permanent	minimal	none
stormwater soakaways	increase	extensive	intermittent	marginally negative	Cl, HC, CHC
mains drainage	reduction	extensive	intermittent to continuous	none	none
surface water canalisation	marginal reduction	linear	variable	none	none
Irrigation of amenity areas	increase	restricted	seasonal	variable	N, Cl
(B) *Introduction of water-service network*					
Water-supply system	increase	extensive	continuous	positive	none
Sanitation measures					
unsewered	increase	extensive	continuous	negative	N, FP
mains sewerage	marginal increase	extensive	continuous	marginally negative	N, FP, DOC

[1] Cl, chloride and other major ions; DOC, dissolved organic carbon; N, nitrogen compounds (nitrate or ammonium); HC, hydrocarbon fuels; CHC, chlorinated hydrocarbons; FP, faecal pathogens.

determine whether there will be a net change in overall groundwater recharge rate, but anything from a major reduction to a modest increase is possible.

Surface impermeabilization processes include the construction of roofs, and of paved areas, such as major highways, minor roads, parking lots, industrial patios, airport aprons, etc. While the proportion of the land area covered is a key factor, it should be noted that some types of urban pavement, such as tile, brick and porous asphalt, are, in fact, quite permeable and, conversely, that some unpaved surfaces become highly compacted with reduced infiltration capacity (Bouvier 1990).

If no pluvial drainage system is installed, runoff will either infiltrate at the edge of impermeable surfaces or via soakaways, enter river channels, or accumulate in land-surface depressions. If pluvial (so-called stormwater) drainage is installed, the mode of drainage-water disposal will exercise an important influence on urban groundwater recharge rates. The reduction in direct groundwater recharge, as a result of land-surface impermeabilization, is quite commonly offset by increases in indirect recharge, when drainage is routed to soil soakaways or infiltration basins.

Although this is excellent water conservation practice, where drainage from major highways and industrial patios is included, it will represent a significant increase in the risk of groundwater pollution, due to spillages of hydrocarbon fuels and industrial chemicals in particular. Soakaways are also all too often used for the casual disposal of liquid waste from residential areas, such as used motor oils or for the illegal connection of septic-tank overflows.

Urbanization also involves radical modification to the condition of surface water courses, which in turn can affect either the recharge and quality, or the discharge, of groundwater. The effects vary widely with hydrogeological environment and climatic type.

Irrigation of amenity areas. In climates where irrigation has to be continuously or intermittently practised, excessive application is commonplace in parks and gardens, especially if water is applied by flooding from irrigation channels or hose pipes, as opposed to more efficient methods, such as sprinklers, which are more often used on sports fields.

In urban areas with permeable soils, over-irrigation can result in locally very high rates of groundwater recharge, although these are normally limited to a relatively small proportion of the total area urbanized. While irrigation return water is normally of relatively good quality, this is not the case where urban wastewater is used, which tends to overload the soil with nitrogen, and in some situations, cause microbiological and/or organic groundwater contamination.

Introduction of water-service network

Water-supply system. Most urbanization involves large water imports from beyond urban limits, except where local groundwater resources are adequate, in both quantity and quality, to provide the major contribution.

When expressed in hydrological terms, the amount of water in circulation in distribution systems is very significant in relation to excess rainfall, even in relatively humid climates (Foster *et al.* 1993). Since water mains for the most part are constantly pressurized, they are highly prone to leakage and in permeable soils most of this leakage occurs without surface manifestation as infiltration to the ground. This leakage becomes an important component of groundwater recharge.

Quantifying this recharge is, however, difficult as no direct measurements are feasible. Moreover, a significant proportion of subsurface leakage may in fact be intercepted by tree roots or enter sewers (or other subsurface ducts and trenches situated at greater depths than water mains), and ultimately form part of the overall surface runoff from the urban area.

However, the proportion of water put into distribution that apparently does not reach the consumer can vary from 20 to 50%. For one district of Lima, Peru, mains water leakage was estimated to be equivalent to 360 mm a^{-1} (Geake *et al.* 1986).

While this source of recharge is often extremely important in the groundwater balance of unconfined urban aquifers, the existence of excessive leakage represents a major loss of revenue to water-supply undertakings, since even if most of the lost water is recuperated from local production boreholes, the additional energy costs for pumping are considerable.

Sanitation measures. In urban areas of developing nations in particular, the provision of mains sewerage lags greatly behind pollution growth and considerably behind the provision of mains water-supply. In many cases only small areas in the centres of cities are, in fact, sewered.

Although unsewered (*in situ*) sanitation units, such as latrines, cesspits and septic tanks, can provide an adequate service level for excreta disposal in urban areas under favourable site conditions, such on-site sanitation measures exert a major influence on groundwater recharge and are the predominant control on groundwater quality (Morris *et al.* 1994).

Unsewered sanitation greatly increases the rate of urban groundwater recharge. Since consumptive use of water on domestic premises is usually taken to be 5–10%, more than 90% of the water-supply provided will end up as recharge to groundwater where sullage water is also disposed via sanitation units. Rates of recharge will generally be large and very significant in the urban groundwater balance, especially in areas of higher population density. If a significant proportion of the water-supply is derived from local groundwater, large-scale recycling will occur, whereas if a significant proportion is obtained from external sources, there will be a substantial increase in net groundwater recharge.

In some hydrogeological conditions, notably those with fractured bedrock close to the surface and/or with a very shallow water-table, the standard design of an *in situ* sanitation unit results in a high risk of penetration of pathogenic bacteria and viruses to nearby groundwater sources. This has been the proven vector of pathogen transmission in numerous disease outbreaks. It most often happens in densely populated settlements, but can also occur in more prosperous urbanizations, where individual houses have both private shallow wells and septic tanks.

The deployment of *in situ* sanitation measures to serve areas of moderate/high density population, will often result in an excessive load of nitrogen to the subsurface, and can cause widespread groundwater pollution problems by nitrate or, more rarely, by ammonium. The main factors determining the severity of such pollution are the non-consumptive per capita water use, the natural infiltration rate and the proportion of the gross nitrogen load that will be leached to groundwater as nitrate (Foster & Hirata 1988). The latter is very variable with type and operation of *in situ* sanitation unit and local soil conditions, but in many documented cases exceeds 50%.

Sullage waters will also increase the risk of groundwater contamination, because of the widespread use of household products and community chemicals, which contain persistent halogenated synthetic organic compounds.

The installation of a mains sewerage system greatly reduces the rate of urban groundwater recharge, but does not eliminate the risk of groundwater contamination since there is increasing evidence of significant rates of sewer leakage.

Land surface storage and disposal of industrial effluents. In many developing nations, extensive fringe urban areas remain without sewerage cover. Increasing numbers of industries, such as textiles, metal processing, vehicle maintenance, laundries, printing, tanneries, and photo processing, tend to be located in such areas. Most of these industries generate liquid effluents, such as spent lubricants, solvents and disinfectants, which are often discharged directly to the soil and can also represent a serious long-term threat to groundwater quality.

Bigger industrial plants often use large volumes of process water and commonly utilize lagoons for the handling and concentration of liquid effluents. Moreover, urban wastewaters are increasingly being treated by retention in shallow oxidation lagoons, prior to discharge in rivers, to the ground or for reuse in irrigation. These lagoons are often unlined, with high rates of seepage loss, and have a considerable impact on local groundwater quality.

Urban field-survey areas

The hydrogeological conditions and urbanization situation in each of in the four survey areas are indicated in

Table 2. *Summary of geological conditions and service arrangements in urban survey areas*

Urbanization	*Lima, Peru La Molina Alta*	*Santa Cruz, Bolivia*	*Merida, Mexico*	*Hat Yai, Thailand*
Year of survey	1984	1989	1990	1990
Survey area (ha)	650[a]	14 500[b]	16 000[b]	1 000[b]
Population	21 000	650 000	525 000	140 000
Overall population density (people ha^{-1})	30	45	35	140
Maximum population density (people ha^{-1})[c]	50	120	175	210
Industrial zones	none	major	major	major
Geological environment	intermontane alluvium	alluvial outwash plain	karst limestone	coastal alluvium
Altitude (mASL)	180–320	390–420	8–10	5–10
Aquifer type	phreatic	semi-confined	phreatic	semi-confined
Depth to groundwater (m)	25–45	1–6[d]	6–8	1–5[d]
Mean rainfall ($mm\,a^{-1}$)	20	1200	1000	1900
Temporal distribution	n/a	seasonal	seasonal	seasonal
Mains water-supply coverage (%)	100	70	80	40
Water-supply ($l\,day^{-1}\,person^{-1}$)[e]	880	210	460	200
Proportion urban groundwater (%)	100	100	35	60
Mains sewerage coverage (%)	70	20	0	0[f]
Stormwater drainage system	none	10% cover, some to ditches	none, to soakaways	extensive, to canals
Unsewered sanitation units	septic tanks	septic tanks, pit latrines	septic tanks, cesspits	septic tanks, direct to drains

[a] Outer high-income residential suburb.
[b] Entire city within outer ring road or municipal limit.
[c] As recorded in part of survey area.
[d] Shallowest aquifer perched above regional water-table.
[e] Gross value including distribution loses.
[f] Majority connected to open drain/canal system, only small proportion direct to ground.

Table 2 and the most significant features are briefly summarized below.

It must be recognized that all hydrological data collection in urban areas is difficult and problems occur when estimating processes characterized by diffuse flowpaths for which direct measurement is impossible. A further complication arises in defining the survey limit and this affects results directly, since recharge rates are presented as averages over the entire survey area.

Various methods can be employed to make, and to corroborate, survey estimates, including statistical analysis of related parameters, hydrometeorological analysis, aquifer mathematical modelling and groundwater chemical indicators, but in most instances estimates will be subject to a considerable degree of uncertainty.

Lima, Peru

The Lima survey area comprised a relatively new, high-income residential suburb (La Molina Alta) located on an intermontane, valley-fill, alluvial aquifer. It is of relatively small area with extremely arid climate, and prior to urbanization was under irrigated agriculture with water being provided by a system of surface canals. These canals continue to operate, delivering water for landscape lagoons and amenity irrigation. This provides incidental recharge to an aquifer which is overexploited for public water-supply. The area has high coverage of mains water and sewerage systems, and extremely high per capita water usage, largely as a result of excessive irrigation of extensive private gardens.

Santa Cruz, Bolivia

The entire city within the outer ring road comprised the survey area in this case. Santa Cruz is a low-rise, relatively low-density city and one of the fastest growing in the Americas. It is unusual in that, up to the present, all of its water-supply is derived from wellfields within city limits, abstracting from a semi-confined, outwash-plain, alluvial aquifer. The city has relatively high coverage of mains water-supply, considering its very rapid growth, but only the older central area has mains sewerage. It has few parks, but all houses, including those close to the city centre, have substantial gardens and thus the proportion of the land surface impermeabilized is relatively low. Despite seasonal high-intensity rainfall, stormwater drainage to lined canals is only just being developed to overcome local flooding. For the most part, surface runoff infiltrates around the margins of impermeable areas into the sandy subsoil.

Merida, Mexico

Merida is similar in overall areal extension, population density and climatic conditions to Santa Cruz, but although a much higher proportion of the land area is paved, municipal parks and squares are more common. The city is situated on an extremely flat peninsula of highly permeable karstic limestone, from which it derives all of its water-supply, although the greater part (almost 70%) is imported from wellfields outside urban limits. There is no mains sewerage nor stormwater drainage, all wastewater being returned to the ground via unsewered sanitation units and all surface drainage via soakaways.

Hat Yai, Thailand

Although much smaller in total population than Santa Cruz and Merida, Hat Yai has a higher overall population density and a very high density (more than 200 people per hectare) in the city centre. As a result, a high proportion of the city area is impermeabilized. The water-supply situation is complex with most mains water-supply being imported from outside the urban limits and derived from surface sources. However, a minor proportion is obtained from groundwater within the urban area itself and local groundwater resources

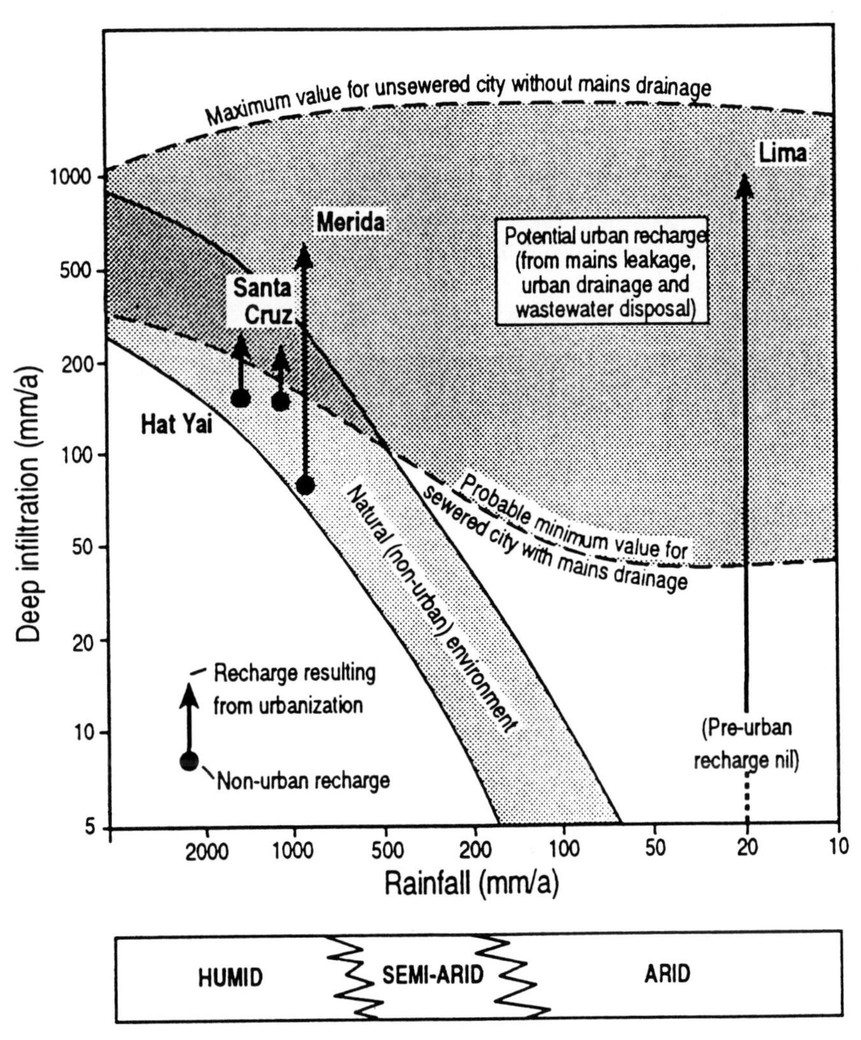

Fig. 1. Potential range of increase of subsurface infiltration due to urbanization.

also provide industrial and commercial users. As a result, groundwater abstraction represents perhaps as much as 60% of the total supply.

The city is situated on low-lying, coastal-alluvial deposits and in consequence experiences problems with drainage. It is estimated that about 20% of wastewater disposal is directly to the ground via unsewered sanitation units, the remainder being connected to open drains which discharge into larger canals, which also receive stormwater runoff. However, as a result of overexploitation of groundwater and piezometric lowering in the semi-confined aquifer within the urban area, canal seepage now represents the single most important component of groundwater recharge.

Integrated effects of urbanization

Groundwater recharge

An attempt has been made to assess the net effect of urbanization on groundwater recharge (Fig. 1). This diagram indicates very approximately the normal range of rainfall–infiltration relations for natural (non-urban) conditions and the potential recharge resulting from mains leakage, wastewater percolation and urban drainage, recognizing that the former two will vary widely with population density and development level. Mains water leakage is the most consistent source of urban recharge, since it will normally comprise more than 20% of gross water production and will thus generally exceed 100 mm a^{-1}.

For largely unsewered cities it is possible for 90% of gross water production to find its way by various routes into the ground, since consumptive use (normally not exceeding 10%) will be the only loss. This is most likely to be the case in arid regions underlain by permeable strata, where irrigation of amenity areas is likely to be another important factor.

In the more humid regions, a larger proportion of urban drainage and wastewater will normally be directed to surface watercourses, because of the larger volumes of the former and reduced infiltration and storage capacity of the ground resulting from more frequent occurrence of clay strata and a shallow water-table. As a consequence the increase of deep infiltration due to urbanization will be much less spectacular, but often still highly significant.

The shallow geology and hydrogeological regime is of crucial importance when trying to predict the effect of urbanization on groundwater. This is well illustrated by the contrasting response in the cases of Merida and Hat Yai. In the former case, the highly permeable karstic limestone formation readily accepts all water discharged

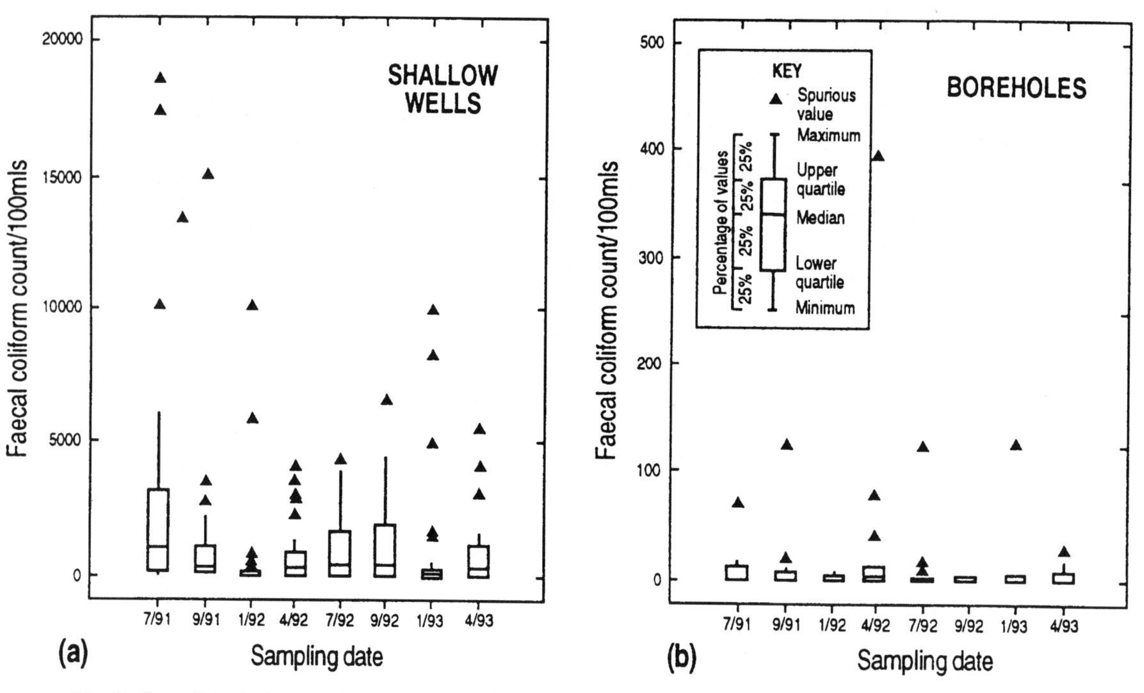

Fig. 2. Bacteriological contamination of the karst limestone aquifer beneath Merida: (**a**) shallow wells (<12 m) (**b**) boreholes (>18 m).

to the ground, albeit at the expense of groundwater quality. Deep infiltration to groundwater is estimated to have increased from 180 to 600 mm a^{-1} as a result. In the case of Hat Yai, shallow semi-confining beds are of insufficient permeability to accept urban drainage and wastewater and this effect is compounded by the existence of a shallow water-table. In consequence most is discharged to surface watercourses, but even so, groundwater recharge is estimated to have increased from 170 to 370 mm a^{-1} as a result of urbanization.

Groundwater quality

In urban residential districts without, or with incomplete, coverage by mains sewerage, seepage from unsewered sanitation systems such as septic tanks, cesspits and latrines, probably represents the most widespread and serious diffuse pollution source. For groundwater, the immediate concern from on-site sanitation systems is a risk of direct migration of pathogenic bacteria and viruses to underlying aquifers and neighbouring groundwater sources (Lewis *et al.* 1982).

The karstic limestone aquifer beneath Merida is especially vulnerable in this respect and widespread and gross bacteriological contamination of the urban groundwater was observed; faecal coliform counts in excess of 1000 per 100 ml were not uncommon in samples obtained from shallow wells (Fig. 2). In Merida, the septic tanks discharge to soakaways whose bases are in fissured limestone and only 1–3 m above the water-table, thus opportunities for pathogen attenuation within the unsaturated zone are limited.

Whilst bacterial contamination of shallow wells, in all types of geological formations, is believed to be widespread, it is likely, in the case of the unconsolidated deposits, to be more frequently a feature of improper design and construction of the well rather than of aquifer contamination. Migration of pathogens through unconsolidated strata to deep water-supply wells is extremely unlikely and any bacteriological contamination almost certainly reflects poor design and construction of the borehole.

The nitrogen compounds in excreta do not represent as immediate a hazard to groundwater but can cause much more widespread and persistent problems. The resulting concentrations of nitrate in groundwater are normally high, although quite variable. In Merida, despite the exceptionally high percentage of nitrogen that is leached to the water-table (up to 100%; Morris *et al.* 1994), the mean groundwater nitrate concentration is

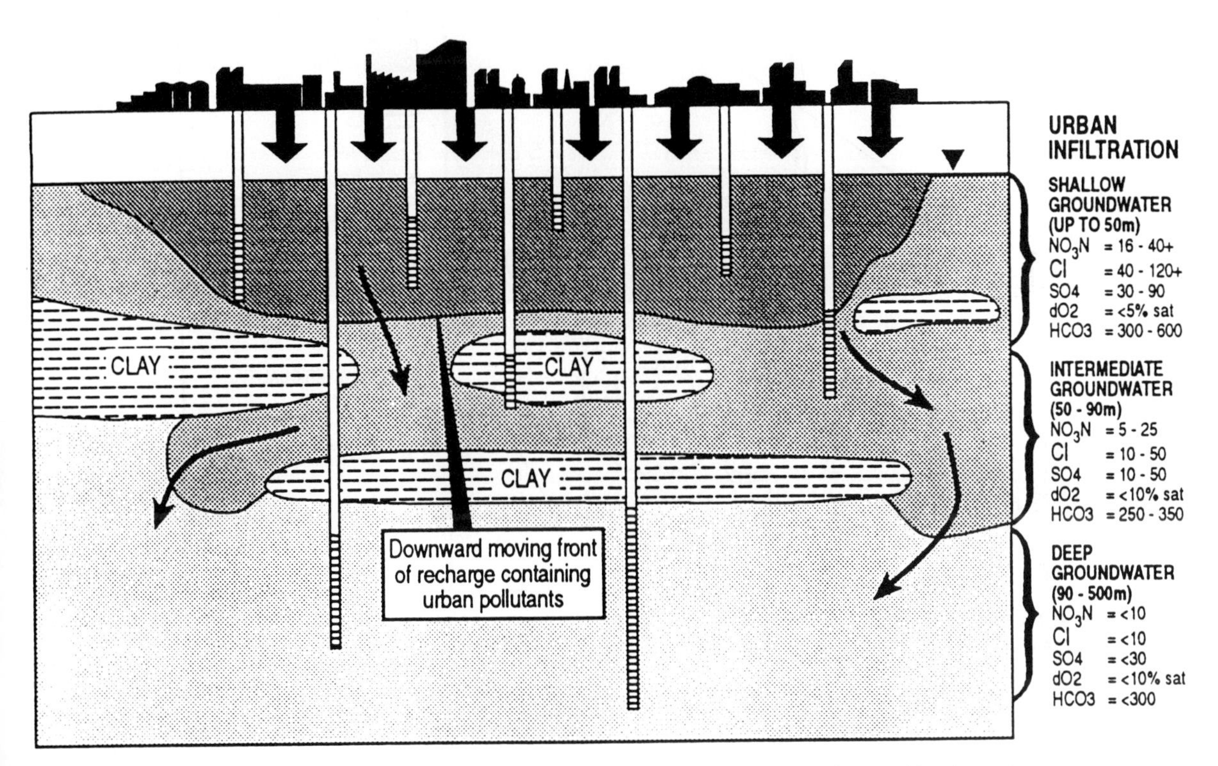

Fig. 3. Downward migration of contaminated water induced by pumping from deep aquifers beneath Santa Cruz.

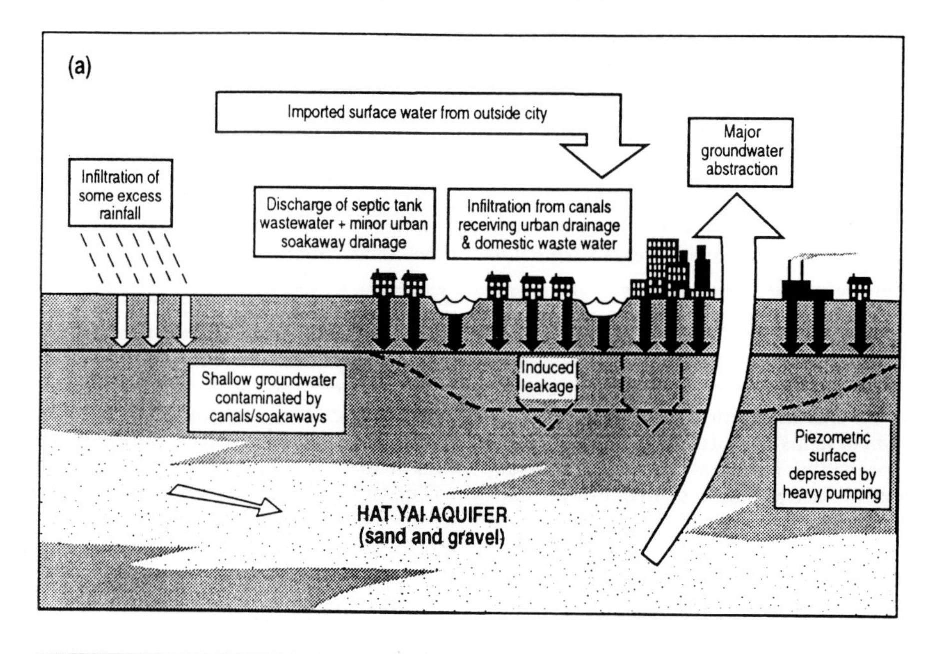

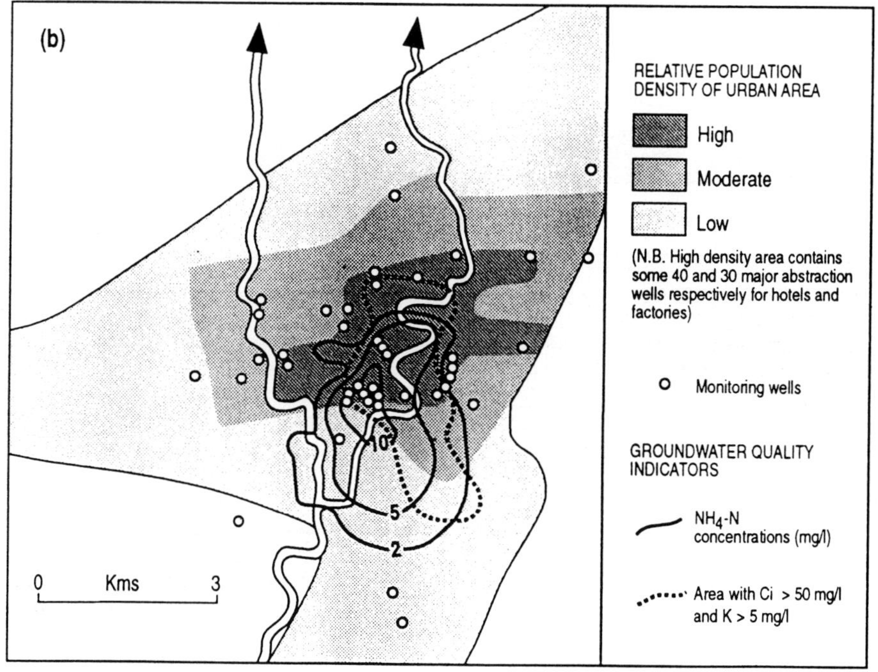

Fig. 4. Incipient effects of urban development on the Hat Yai coastal alluvial aquifer system of Thailand: (**a**) schematic and (**b**) general plan.

only 4 mg N l^{-1}, largely as a result of the relatively low urban population density and the considerable dilution afforded by both aquifer throughflow and high urban water use.

Conversely, in Santa Cruz probably only 10–20% of the nitrogen deposited is leached to the underlying alluvial aquifers; however, the greater population density and lower dilution (less throughflow and lower

Table 3. *Urban impact on groundwater quality*

City	*Aquifer*	*Nitrate* $(mg\,N\,l^{-1})$		*Chloride* $(mg\,l^{-1})$		*Bicarbonate* $(mg\,l^{-1})$		*Sulphate* $(mg\,l^{-1})$	
		Back-ground	*Urban range*	*Back-ground*	*Urban range*	*Back-ground*	*Urban range*	*Back-ground*	*Urban range*
Merida	karst limestone (0–40 m)	<5	5–30[a]	<30	30–200[c]	<350	350–550[d]	<20	20–80
Santa Cruz	Alluvium outwash plain								
	Shallow aquifer (0–45 m)	<10	10–40[a]	<40	40–>120[a]	<300	300–600[e]	<30	30–90
	Deep aquifer (45–200 m)	<5	5–25[a]	<10	10–>50[a]	<250	250–350[e]	<10	10–30
Hat Yai	Shallow alluvium (20–50 m)	<2	2–22[b] includes NH_4-N	5–10	10–80[b]	0–20	20–90[e]	0–10	10–40

Sources: [a] Unsewered sanitation; [b] seepage from canals; [c] saline water at depth; [d] natural; [e] degradation of organic wastes; [f] detergents, road runoff.

water consumption) ensures that average nitrate concentrations in the shallow aquifer are in the range 10–40 mg N l^{-1}. The lower percentage of deposited nitrogen that is oxidized and leached to groundwater beneath Santa Cruz as compared to Merida reflects in part the lower dissolved oxygen status of the groundwaters in the alluvial aquifers. Groundwater abstraction from the deep alluvial aquifers, beneath Santa Cruz, has induced the downward movement of shallow groundwaters, and elevated nitrate and chloride concentrations, are observed even at depths approaching 90 m (Fig. 3).

In Hat Yai, where the disposal of excreta to the ground by on-site sanitation systems is not always possible (because of surfacing of the water-table during the rainy system), human faeces and other wastes are discharged into surface water courses. As a consequence, they receive heavy loads of untreated effluent. Elevated groundwater nitrogen concentrations (mostly as ammonium) occur close to, and as a direct result of leakage from, these canals (Fig. 4). The presence of ammonium (as opposed to nitrate) reflects the low dissolved oxygen status of the groundwaters and the absence of a significant unsaturated zone.

In addition to elevated nitrogen (as nitrate or ammonia) in urban groundwaters, increased concentrations of chloride (mostly from on-site sanitation systems), sulphate (from detergents and road runoff) and bicarbonate (derived from degradation of organic wastes) are frequently observed (Table 3).

The accidental spillage, leakage or improper disposal of industrial effluent can also cause serious pollution. In Merida, a survey of water-supply boreholes revealed widespread contamination by chlorinated solvents of the limestone aquifer. Although concentrations were generally low at less than 10 ppb, it was considered that these values were underestimates of the true concentration due to the volatility of the solvents and the inherent difficulties in the collection and analysis of samples (Gooddy *et al.* 1993).

Principal conclusions

- Urbanization generally results in significant increases in groundwater recharge, when cities are built upon unconfined or semi-unconfined aquifers, and the effect is particularly pronounced in more arid regions.
- The main processes responsible for this increase include water mains leakage, *in situ* wastewater disposal, urban drainage soakaways and over-irrigation of amenity areas.
- The potential recharge contribution from these processes varies considerably with level of economic development, population density, climatic type and hydrogeological conditions, but almost always exceeds the reduction in direct rainfall infiltration due to surface impermeabilization.
- The presence of less permeable strata at shallow depths beneath urban areas may result in mounding or perching of groundwater significantly above the regional water-table and cause serious negative impacts on subsurface engineering structures and installations. In most cases, however, it does not have this result, because aquifers are sufficiently permeable to prevent water-table mounding and/or because heavy abstraction of groundwater within the urban area depresses aquifer water-levels.
- This paper shows that urbanization has a profound effect on groundwater recharge, not to mention

quality. Considering the scale of implications for water management and engineering structures, it is important that more urban hydrogeological studies are undertaken. It could be argued that these should be an essential component of all city master plans, since the information they provide is at the heart of efficient engineering of the urban infrastructure and environment.

References

BOUVIER, C. 1990. Concerning experimental measurements of infiltration for runoff modelling of urban watersheds in western Africa. *IAHS Publication*, **198**, 43–49.

FOSTER, S. S. D. 1990. Impacts of urbanization on groundwater. *IAHS Publication*, **198**, 187–207.

—— & HIRATA, R. 1988. *Groundwater Pollution Risk Assessment: A Methodology using Available Data.* Pan American Centre for Sanitary Engineering and Environmental Sciences (CEPIS), Lima, Peru.

——, MORRIS, B. L. & LAWRENCE, A. R. 1993. Effects of urbanization on groundwater recharge. *In*: *Groundwater Problems in Urban Areas.* ICE International Conference, June 1993, London.

GEAKE, A. K., FOSTER, S. S. D., NAKAMATSU, N., VALENZUELA, C. F. & VALVERDE, M. L. 1986. *Groundwater Recharge and Pollution Mechanisms in Urban Aquifers of Arid Regions.* BGS Hydrogeology Research Report 86/11.

GOODDY, D. C., MORRIS, B. L., VASQUEZ, J. & PACHECO, J. 1993. *Organic Contamination of the Karstic Limestone Aquifer underlying the City of Merida, Yucatan, Mexico.* BGS Technical Report WD/93/8.

LEWIS, W. J., FOSTER, S. S. D. & DRASAR, B. 1982. *The Risk of Groundwater Pollution by On-Site Sanitation in Developing Countries.* WHO-IRCWD. Report 01-82, Duvendorf, Switzerland.

LINDH, G. 1983. *Water and the City*. UNESCO, Paris.

MORRIS, B. L. LAWRENCE, A. R. & STUART, M. E. 1994. *The Impact of Urbanization on Groundwater Quality (Project Summary Report)*. BGS Technical Report WC/94/56.

VAN DE VEN, F. H. M. 1990. *Water balances of urban areas.* IAHS Publication, **198**, 21–32.

Natural hazards in the urban environment: the need for a more sustainable approach to mitigation

Martin Degg

University College Chester, Parkgate Road, Chester CH1 4BJ, UK

Abstract. In a variety of environmental and socio-economic circumstances, the medium to long-term effectiveness and sustainability of traditional approaches to managing natural hazards and their impact on urban areas are increasingly being questioned. This is reflected in two trends of particular relevance to the geosciences and engineering, namely (a) changing attitudes towards the use of 'hard' engineering to manage environmental processes; and (b) calls for greater involvement and responsibility on the part of the community and individual in hazard mitigation. The latter, in particular, is increasingly seen as crucial to reducing vulnerability to loss in both developed and developing societies. It requires a conceptual change in how the issue of human vulnerability is viewed and portrayed, and greater public accessibility to information about environmental processes, the potential dangers they pose and the ways in which these can be mitigated. In this respect the geoscientist has a vital role to play in providing the non-specialist with essential earth science information, in formats that are environmentally, economically and culturally appropriate.

Introduction

The declaration of the 1990s as the United Nations 'International Decade for Natural Disaster Reduction' (IDNDR) has helped to focus the attention of the international community on natural hazards and their potentially disastrous impacts upon built environments. Most significantly, perhaps, it has produced meetings to which specialists from many diverse fields have felt the need to contribute. Physical scientists, social scientists, engineers, insurers, environmentalists, development agencies and politicians have found themselves attending the same conferences and workshops, to mutual benefit.

There is an increasing realization that traditional approaches to managing the impact of natural hazards upon urban areas cannot be sustained in the medium to long term, and that the dual pressures of global population growth and rapid urbanization require a fundamental change in how societies perceive and manage human vulnerability. Spiralling economic losses from hazard impacts in the West and unacceptably high human losses in Third World countries, coupled with concerns about global climate change, necessitate that the dialogue should continue apace.

Recent historical context

Since the Industrial Revolution the dominant approach to tackling environmental hazards in the West has been what Hewitt (1983) described as 'technocratic'. This ideology considers that disasters such as those stemming from a flood or landslide in an urban area are caused by nature, and therefore that the most effective way to prevent such occurrences is for society to control nature. This control is deemed possible through the application of science, technology and engineering to monitor and 'fix' the environment, often by way of major projects intended to combat the areal extent and physical magnitude of the natural processes at work, e.g. in the construction of levees and flood barriers to control river flooding, and the development of advanced technologies to forecast extreme environmental events such as storms and earthquakes. The scale of the societal response is such that the individual urban dweller is often largely uninvolved in the hazard mitigation process, as there is a strong sense that 'everyday or "ordinary" human activity can do little except make the problem worse' (Hewitt 1983: 6).

In some respects this approach to dealing with natural hazards in the urban environment has been successful, most notably in reducing the vulnerability of human life to such hazards in Western societies. For example, during the period 1969–89 the average loss of life per hazard impact in the West was 19, compared to 38 fatalities per impact during the period 1947–67. By way of contrast, the average fatality per disaster ratio for the Third World during 1947–67 was almost 1000:1, and this increased to over 2000:1 during 1969–89 (Degg 1992). In other respects, however, the technocratic approach has important shortcomings, most notably:

- it is high cost and requires long-term planning and a high degree of social organization. As such it tends to be organized at the institutional level, often with little local input and awareness of the problem on the part of the lay-person (i.e. it is 'top-down');

DEGG, M. 1998. Natural hazards in the urban environment: the need for a more sustainable approach to mitigation. *In*: MAUND, J. G. & EDDLESTON, M. (eds) *Geohazards in Engineering Geology*. Geological Society, London, Engineering Geology Special Publications, **15**, 329–337.

- it is relatively inflexible with emphasis on using 'hard' engineering to *control* the environment, rather than on working with it. Badly thought out control mechanisms can backfire and increase the intensity of the hazard process (e.g. the rate of coastal erosion) or transfer the problem elsewhere;
- it encourages the development of marginal land, thereby multiplying risk and increasing the potential for catastrophic loss should the protective technology fail, or a larger than expected event occur.

In the West these shortcomings can be linked to a trend towards increasingly frequent large-scale economic losses from natural hazards. For example, winter storms in Western Europe during early 1990 cost US$10 billion, floods along the Missouri and Mississippi rivers during July–August 1993 cost at least US$4 billion, and earthquakes in southern California in 1989 and 1994 (the latter followed by mudslides) cost at least US$11 billion. To date these losses have largely been absorbed through the provision of government relief funds and through the widespread use of insurance as a loss-sharing mode of adjustment, but the scale of losses during the last two decades raises serious questions about how long this relatively fatalistic and *laissez-faire* approach to protection can be sustained. The Swiss Reinsurance Company (1993) has shown that whilst insurance losses due to human-made catastrophes were relatively constant in real terms during 1970–92, losses due to natural catastrophes increased approximately ten-fold. The losses are expected to continue to spiral into the next century (Fig. 1) as, for example, Californians come to terms with the fact that the San Andreas fault zone appears to have entered its active cycle following a relative lull in activity lasting 70–80 years, and Tokyo steadies itself for an overdue major earthquake which some government observers predict will cost hundreds of billions of dollars (*The Economist* 1991). The ramifications of such an impact on Tokyo for the *global* economy are discussed at length by Hadfield (1995). In addition, the global warming debate has heightened concerns about the possibility of intensified European and tropical windstorm activity, with additional implications for the magnitude and frequency of surface processes such as flooding and ground movement (Berz & Conrad 1993; Doornkamp 1993; Jones 1993).

In developing countries the extent and rate of increase of urban vulnerability to natural hazards are such that technocratic responses do not offer a viable mechanism for bringing about widespread reduction in human losses. Since the 1970s some development agencies and social scientists have been arguing that technocratic ideology, transferred to developing countries, is actually increasing vulnerability to loss because it is socially, culturally and economically 'inappropriate'. Economically, the implementation of high cost schemes is often beyond the means of developing countries other than through aid or transfer arrangements with the West. Susman *et al.* (1983) have questioned the role of this sort of arrangement in promoting indebtedness and dependency of recipient country upon donor, and have linked this directly to processes of impoverishment and increased vulnerability in Third World countries. Culturally, problems arise when overtly scientific approaches to hazard mitigation are applied to regions where non-scientific interpretations of natural hazards predominate (e.g. in societies where such occurrences are viewed as part of 'God's Will').

Despite these misgivings the technological approach remains attractive because of its emphasis on environmental control. It can be used to provide an environmental scapegoat that diverts attention away from politically sensitive socio-economic issues, such as poor quality housing, that are the real cause of many so-called 'natural' disasters. For example, although of

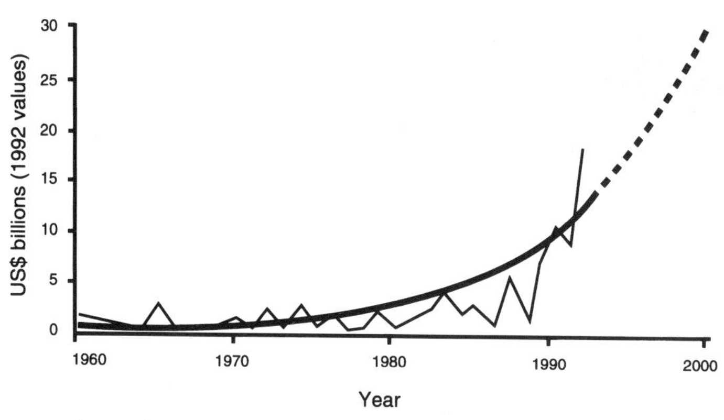

Fig. 1. Real insurance losses due to natural catastrophes, 1960–92, and extrapolated trend to the year 2000. Source: The Munich Reinsurance Company (in Berz & Conrad 1993).

only moderate magnitude (5.6), the northern Egypt earthquake of 12 October 1992 caused considerable damage and resulted in over 550 deaths and almost 10 000 injuries (Castellano & Milano 1993). Some 3500 adobe-type buildings in villages of the northern Nile valley and Faiyûm depression collapsed, whilst the failure of a 14-storey concrete apartment block in an affluent Cairo suburb, and closure of over 1000 state schools due to earthquake damage (a further 3500 required extensive renovation), revealed some woefully inadequate building practices (Khater 1992; Degg 1993). Investigations showed that the 14-storey apartment block had originally been only eight storeys high and that the extra floors had been added to it illegally (Fig. 2). Despite this, in the immediate aftermath of the earthquake, with over 40 000 people officially homeless, political and media attention in Egypt was focused not so much upon building standards and regulation enforcement, as upon the 'failure' of scientists at the country's national seismological observatory to predict the event. Overseas specialists were subsequently invited to discuss ways of improving Egypt's seismic network and earthquake 'prediction' capabilities. It is unfortunate that the concept of reliable earthquake prediction remains a fallacy to which many societies cling for peace of mind. The truth of the matter is that it is cheaper and easier to invest in earthquake prediction research than to devise and enforce policies, many of which may initially be unpopular with an electorate, aimed at bringing about genuine reductions in vulnerability to earthquake impact (e.g. through land use and construction regulation).

Fig. 2. Extended medium-rise building in the affluent northern Cairo suburb of Heliopolis. Such practise, often carried out without licence, was relatively commonplace prior to the 1992 earthquake.

The need for change

The recognition of a need to move away from excessive technocratic ideology is spreading to many different aspects and levels of hazard and disaster work. There are two trends of particular relevance to the geosciences and engineering:

- changing attitudes towards the use of 'hard' engineering (structural controls) to manage environmental processes;
- calls for greater involvement and responsibility on the part of the community and individual in hazard mitigation.

The former has become the focus of considerable political and media attention, particularly in the context of river and coastal management. A growing number of scientists are expressing the view that technological attempts to prevent rivers flooding onto floodplains are misguided and doomed to failure (Press 1994). The 1993 floods in the Mississippi Basin, which contains the most heavily engineered water courses in the world, heightened concern about whether or not such control mechanisms ultimately intensify flood hazard. The 15 m high wall that protects the city of St Louis (population of 900 000), on the confluence of the Mississippi and Missouri rivers, held out, although the waters came within 1 m of topping it. The bottleneck effect caused by this confinement of the river intensified the flooding upstream (Fig. 3), where many levees failed and an area the size of England was flooded, in places to depths of 12 m. Furthermore, the catastrophic nature of the inundation that accompanied sudden levee failure increased the destructive powers of the floodwater, which caused severe damage in valley towns such as Des Moines, Quincy and Davenport. Downstream of St Louis, high discharges within the confined channel

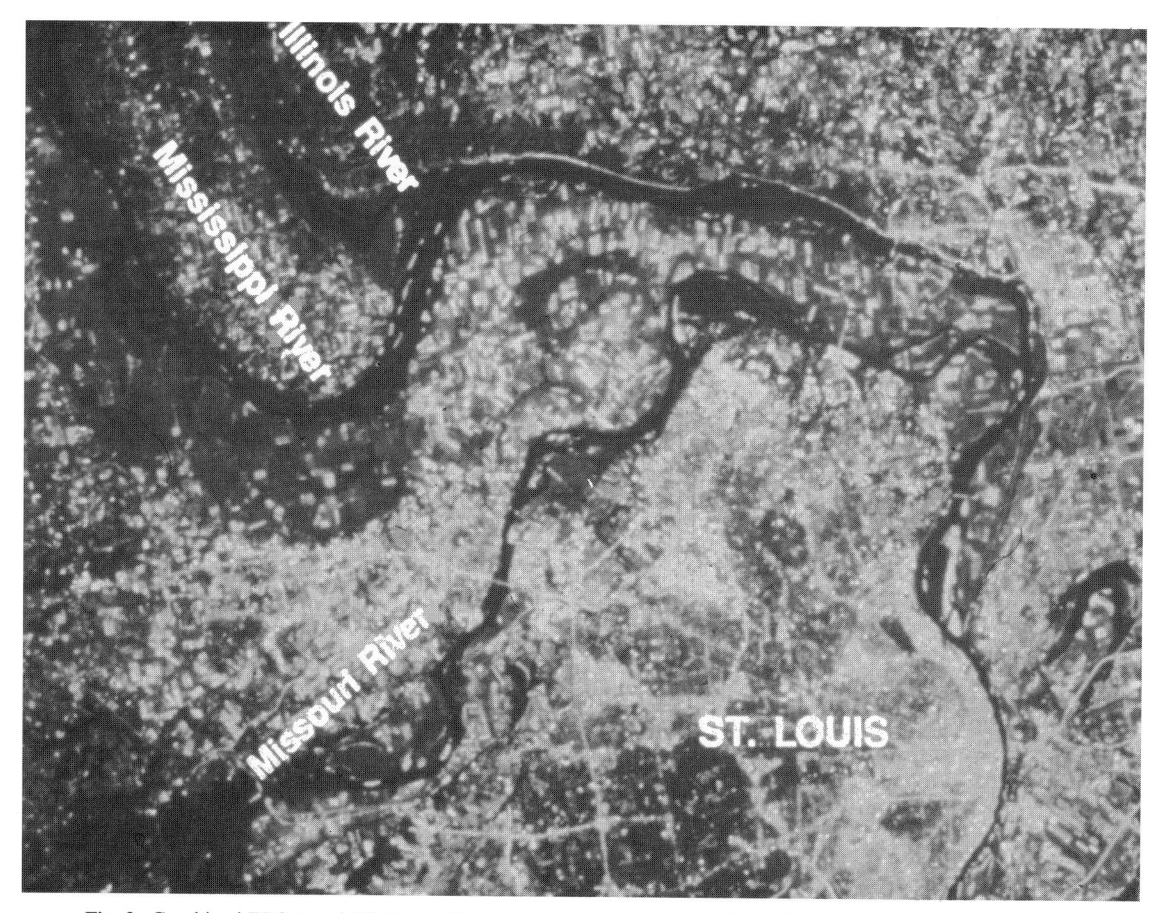

Fig. 3. Combined ERS-1 and SPOT satellite image revealing the extent of flooding around the confluence of the Mississippi, Missouri and Illinois rivers upstream of St Louis on 14 July 1993. Image processed by ITD/SRSC. SPOT image courtesy of SPOT Image Corp. © CNES 1993. © ESA 1993.

caused severe scouring and bank erosion, undermining defence mechanisms. A heated debate ensued in the United States concerning whether the damaged defences should be rebuilt, or whether the river should be allowed to return to a more natural state in which large parts of the floodplain are preserved as washlands to help dissipate peak flows (Mairson 1994).

Similar changing attitudes are apparent in Europe where, for example, on the Rhine Delta, Dutch authorities have started an ambitious plan to remove defensive dykes and encourage the river to reclaim its floodplain. The ultimate aim is to convert 15% of the Netherlands to its original state as washland which, during most years, would rarely be flooded for more than a few days (Evans 1994). Dykes will be built around existing settlements rather than along water courses, and the floodplains will remain agriculturally productive through the use of hardy crops. In Britain, traditional approaches to coastal engineering are increasingly being questioned on the basis of both economics and effectiveness. Engineered structures intended to prevent coastal erosion are extremely expensive to emplace (e.g. approximately £3000 per square metre of sea wall) and maintain, and can have a detrimental effect on coastal sediment supply and movement, leading to increased erosion further along the coast. The emphasis increasingly is upon managing coastal *systems* rather than on protecting individual *sites*, and on recognizing that some sections of coastline may have to be sacrificed to benefit the majority. The concept of *managed retreat* is slowly replacing that of coastal *defence*, and the importance of land use planning to avoid hazardous areas from the outset is being stressed (Brooke 1992; Carter 1993; Lee 1993; Hooke & Bray 1995). For example, the Department of the Environment Planning Policy Guidance Note on coastal planning (PPG20), of September 1992, stresses that 'new development should not generally be permitted in areas which would need expensive engineering works, either to protect developments on land subject to erosion by the sea or to defend land which might be inundated by the sea. There is also the need to consider the possibility of such works causing a transfer of risk to other areas' (DoE 1992: 2.13).

Calls for greater community involvement in hazard mitigation reflect a growing conviction of the benefits to be gained from switching more of the emphasis in hazard protection from the national and regional levels to the local level. Action at the local level, and on the part of the individual citizen, is increasingly seen as crucial to reducing vulnerability to loss in both developed and developing societies. As Nuhfer (1994: 17) has commented, 'like knowledge about health and hygiene, geologic knowledge held by specialists can only produce widespread benefits when it becomes applied through the actions of an informed public'. Some scientists have joined social scientists in calling for fundamental changes in how the issue of human vulnerability to natural hazards is viewed and portrayed. Coburn & Spence (1992) suggest that the conceptual change required is analogous to that which occurred in popular attitude to disease in Western societies from the mid-19th century onwards. Prior to this so-called 'Sanitary Revolution' outbreaks of disease in Western society were largely regarded in a fatalistic manner as unpredictable, unlucky events over which the individual had no control. The 'revolution' gradually changed this attitude by convincing people of the need for *personal* hygiene – of the fact that there are small but significant steps that every individual can take in day-to-day living to reduce the chances of being affected by disease. It is argued that people must now be encouraged to change their predominantly fatalistic attitudes to natural hazards, and to view the fundamental requirements of hazard mitigation as factors that are within their control. To quote Coburn & Spence (1992: 324):

> Just as public health depends on personal hygiene, so public protection depends on personal safety. The type of cooking stove a family uses, and their awareness that a sudden impact could tip it over is more important in reducing the risk of conflagration than the community maintaining a large fire brigade. The type of house a family builds and where they consider a suitable place to live affects the potential for disaster in a community more than sophisticated earthquake warning systems or large engineering projects to prevent rockfalls or stabilize landslides.

In Western societies in particular, the prospect of changing insurance practices is likely to force individual property owners to become more aware of natural hazards and, ultimately, to accept greater responsibility for hazard mitigation. As far as windstorm, flooding and subsidence hazard in the United Kingdom are concerned, escalating losses and greater market competition have led insurers to seek more accurate information about hazard exposure. The aim is to introduce 'fairer' premiums through differential ratings and excesses based on more precise assessments of hazard exposure (Ellis 1994). Premiums are currently defined on a relatively coarse basis using averaged hazard exposure values for postcode districts (e.g. NG7), containing up to 9000 households, and postcode sectors (e.g. NG7 2) containing about 1700 households. Although the long-established working practices of the insurance industry are difficult to change (Doornkamp 1995), plans are being advanced to use the whole postcode to locate specific addresses and charge premiums commensurate with the risk to which individual properties are exposed. One trial system currently being developed uses Ordnance Survey address information to give every home in the country a unique reference code, and hazard exposure ratings derived from large-scale digital geological, topographical and hydrological maps and insurance company loss statistics (Emrich *et al.* 1993; Anon 1994; Glaskin 1994). The data are integrated within a Geographical Information System or GIS (a spatially referenced computer database system), ensuring instant access to information at a much greater level of locational detail than has hitherto been attempted. Some regions of the National Rivers Authority (NRA) are now also using GIS to provide almost instantaneous replies to telephone enquiries about flood risk at particular sites (Farthing 1994).

An eventual outcome of this may well be that property owners in low risk areas will experience falling insurance premiums, whereas those in high risk areas will experience rises. It is conceivable that, ultimately, the seeking of information on insurance rating (and thereby hazard exposure) will become an integral part of the property buying process. A benefit of this may be that it will discourage new development in hazardous areas, although existing developments in areas of high hazard may prove very expensive to insure and difficult to sell. Owners of property in hazardous areas might be encouraged to take positive steps to reduce vulnerability to loss in return for premium reductions, rather as some insurance companies now offer reductions in household premiums for owners who install alarms to discourage burglars. For example, the owner of a building on a river flood plain who has taken recognized measures to flood-proof the property could be rewarded with a discounted premium. What seems certain is that the days of the insured property owner living in ignorance of potential threats posed by the local natural environment are set to end.

Prerequisites of change

A fundamental requirement of the types of change discussed above is greater public accessibility to information about environmental processes, the potential dangers they pose and the ways in which these can be mitigated. This applies both to calls for more flexible approaches to managing hazard, in that an informed public is more likely to be receptive to policy changes if it understands the reasoning behind them, and to calls for greater emphasis on hazard management at the local level. In this respect, a major challenge facing policy-makers and geoscientists in the West is how to provide such information and to encourage people to act upon it.

In Britain, the need to increase public awareness of the geological and geomorphological environment has led the Department of the Environment (DoE) to commission a number of national reviews. The work completed so far includes reviews of the distribution and nature of hazards caused by landsliding, mining instability, natural underground cavities, earthquakes, natural contamination, and erosion, deposition and flooding. The DoE has also initiated a programme of applied earth science mapping (Anon 1991), aimed at presenting important earth science information in a format comprehensible to the non-geologist (e.g. Edwards *et al.* 1987; Doornkamp 1988; GAPS 1992). These products are aimed initially at the professional concerned with land development (e.g. the local government planner, engineer, property developer, lawyer and landowner), but important findings of some of the work have been published in formats more accessible to the lay-person (e.g. Jones *et al.* 1989; Jones & Lee 1995). A very different trial initiative in 1991 was the opening of a Geological Information Centre, or 'slip shop', at Ventnor on the southern coast of the Isle of White, to increase local understanding of landsliding in the town (Fig. 4). Ventnor is situated within a landslide complex that continues to experience localized episodic movement, causing considerable concern within the community. A DoE-commissioned study of the landslide showed that conventional engineering approaches were unlikely to prevent all further movement and that greater emphasis ought to be placed on hazard management within the community (Lee *et al.* 1991) by preventing inappropriate construction, through regular inspection of drainage systems, soakways, septic tanks and retaining walls, and by introducing a voluntary hose-pipe ban. The information centre, which was funded by South Wight Borough Council, sought to put this message across to the general public, and was open for a 10-week period during which it received some 2000 visitors (Rendel Geotechnics 1991).

In the United States, it is becoming widely accepted that the public is paying dearly for its uninformed approach to dealing with natural hazards, and applied geoscience education is increasingly seen as fundamental to reducing losses. Some educational institutions and government and community organizations have introduced evening classes that teach how to access information on, and respond to, natural hazards. The University of Colorado has introduced an evening course entitled 'Geology for Homeowners', whilst the American Institute of Professional Geologists (AIPG) has published the *Citizens' Guide to Geologic Hazards* (Nuhfer *et al.* 1993). In addition to describing hazards such as those associated with swelling soils, landsliding, radon, asbestos, subsidence and floods, the book summarizes ways of recognizing and mitigating or minimizing the hazards. It also covers insurance aspects and shows where to get specialist geoscience information and assistance (e.g. the names, addresses and phone numbers of pertinent government agencies). As such the book, which has been described as possibly 'the single-most beneficial publication by geologists to the general public' (Baars 1994), is aimed not at the geoscientist but at people who are largely unaware of natural processes and their potential impact upon the built environment.

The challenges facing policy-makers, geoscientists and development agencies working in Third World countries are compounded by socio-economic and political problems that are often seen to present a more immediate threat to survival than natural hazards (e.g. lack of access to clean drinking water and secure housing, and unemployment and underemployment). With reference to the comments made above concerning the 'Sanitary Revolution' in Western societies, it is unfortunately the case that the majority of Third World residents live in societies that have yet successfully to pass through such a revolution. In addition, religious and cultural differences may mean that alternative (non-scientific) interpretations of hazards and disasters prevail. For example, a disastrous earthquake may be perceived not so much as the consequence of inadequate construction or poor location in the face of an environmental threat, but as the product of divine retribution against a 'corrupted' society. The latter interpretation implies that hazard mitigation should focus on social reform rather than on improving construction practices (which might be seen as pointless, because if divine authority has willed an earthquake disaster then humans are powerless to prevent it). Indeed, by trying to prevent a repeat occurrence in this way rather than by tackling the social issues that are thought to be incurring divine wrath, the individual and/or society might be seen as trying to go against divine will, thereby increasing the likelihood of further retribution.

Under these circumstances the process of increasing the role of the individual and community in hazard mitigation often needs to be enveloped in broader development strategies aimed at bringing about general improvements in living standards through self- and community-based help in Third World countries (e.g. see Maskrey 1989; Aysan *et al.* 1995). Culturally sensitive educational programmes have a key role to play in this respect (Degg & Ibrahim 1995). These should be of both a general nature, aimed at increasing awareness of how natural processes work (whilst not necessarily challenging sensitive interpretations of cause), and of a

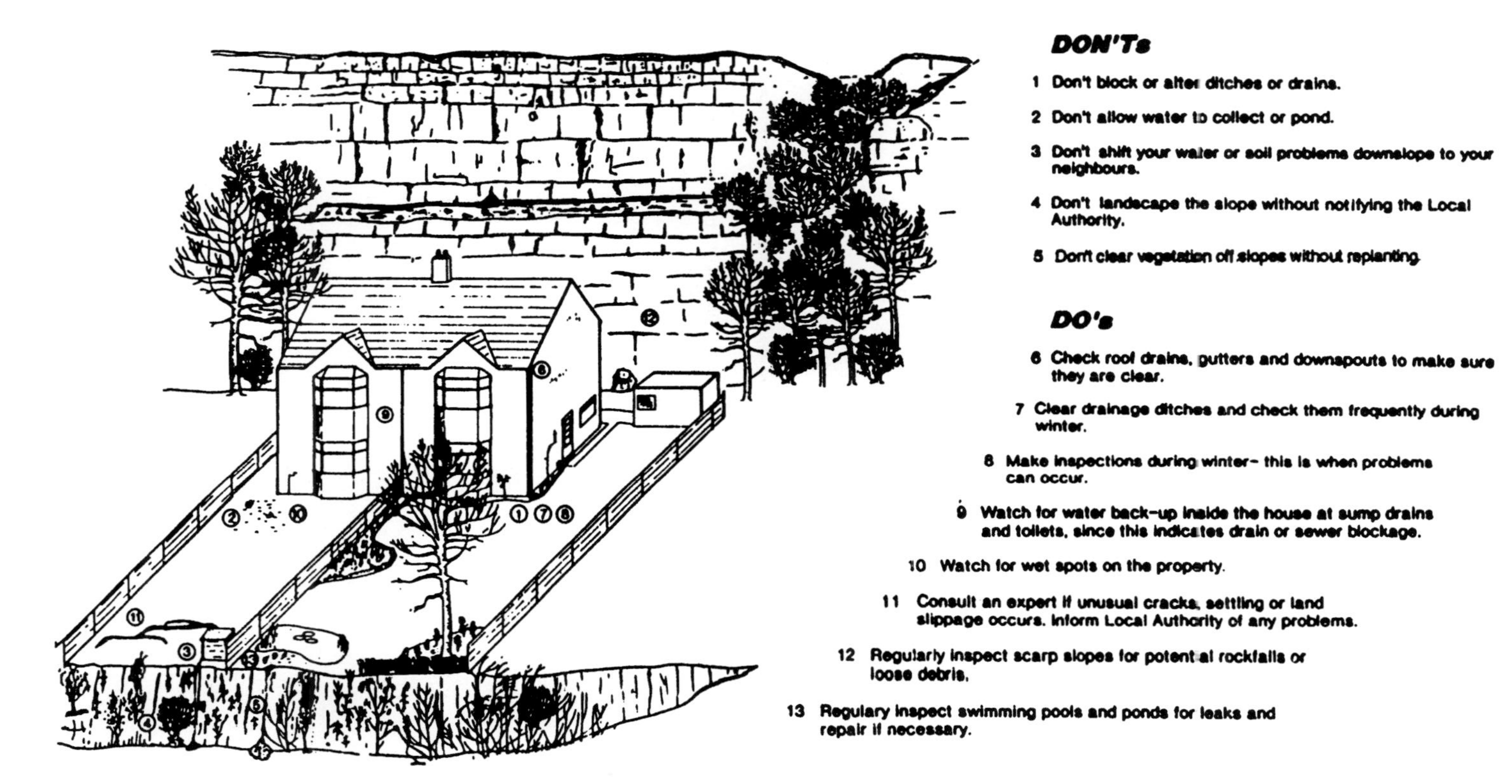

Fig. 4. Advice sheet for residents of Ventnor, Isle of White, aimed at encouraging greater community involvement in landslide hazard mitigation. After Geomorphological Services Ltd (1991) and Lee *et al.* (1991).

more specific nature aimed at putting across specialist knowledge related to hazard mitigation in a development context. The importance of the former is emphasized by the September 1992, northern Egypt earthquake during which a considerable number of schoolchildren died as a result of panic reaction and crushing in schools that were never in danger of collapse. The importance of the latter, in societies where the vast majority of dwellings are erected by their inhabitants, has been emphasized by numerous recent disasters in which inadequate local building practises have largely been responsible for high fatality and injury rates. In both respects the geoscientist clearly has a key role to play in helping to devise education and training programmes aimed at bringing about genuine and lasting reductions in vulnerability by helping people to help themselves.

Summary

To date, the dominant approach to tackling the issue of urban vulnerability to natural hazards in the West has been a strongly technological one with emphasis on fixing the environment through 'hard engineering' (e.g. the construction of coastal and river defences, and the development of advanced technologies to forecast episodic events like earthquakes). This approach has been transferred to developing countries with varying degrees of success/failure. Spiralling economic losses in developed countries and continuing high human losses in developing countries have led to increasing calls for more flexible approaches to hazard management, with greater emphasis on non-structural ('soft engineering') measures and more effective response at the local (community and individual) level. There is, in particular, a recognized need to switch more of the emphasis in hazard management from centrally planned ('top-down') control and prediction strategies, to programmes that accept and promote the limitations of the natural environment and place greater emphasis on effective, flexible response at the local level. In defining the function of the geoscientist in this respect, analogies are being drawn between the historically successful role of the medical profession in promoting personal hygiene and health to reduce the incidence of disease within society, and that yet to be played by geoscientists in encouraging people to protect themselves and their investments from natural hazards, to the collective benefit of society.

Acknowledgements. The author is indebted to E. M. Lee, Rendel Geotechnics and South Wight Borough Council for information regarding their work at Ventnor, and to J. C. Doornkamp and E. J. Shufflebotham for comments on the text. Figure 3 is reproduced with the permission of Radar Sat International Inc., Canada, and Fig. 4 by permission of the Department of the Environment.

References

ANON 1991. *Applied Earth Science Mapping*. Proceedings of a seminar held at the Geological Society of London on 21 May 1990. HMSO, London.

——1994. Address-Point finds a home in the insurance industry. *Ordnance Survey Data News*, Summer, 6.

AYSAN, Y., CLAYTON, A., CORY, A., DAVIS, I. & SANDERSON, D. 1995. *Developing Building for Safety Programmes: Guidelines for Organizing Safe Building Improvement Programmes in Disaster-Prone Areas*. Intermediate Technology Publications, London.

BAARS, D. L. 1994. Book review: 'The citizen's guide to geologic hazards', by NUHFER, E. B., PROCTOR, R. J. & MOSER, H. *Geotimes*, **39**(1), 25.

BERZ, G. & CONRAD, K. 1993. Winds of change. *The Review*, June, 32–35.

BROOKE, J. S. 1992. Coastal defence: the retreat option. *Journal of the Institution of Water and Environmental Management*, **6**, 151–156.

CARTER, R. W. G. 1993. Retreat to the future: a new approach to coastal defence. *National Trust Magazine*, **70**, 28–36.

CASTELLANO, M. & MILANO, G. 1993. The Egyptian earthquake, 12 October 1992. *STOP Disasters*, **12**, 15.

COBURN, A. & SPENCE, R. 1992. *Earthquake Protection*. Wiley, Chichester.

DEGG, M. R. 1992. Natural disasters: recent trends and future prospects. *Geography*, **77**(3) 198–209.

——1993. The 1992 'Cairo earthquake': cause, effect and response. *Disasters*, **17**(3), 226–238.

—— & IBRAHIM, H. A. M. 1995. Geoeducation and the International Decade for Natural Disaster Reduction. *Geoscientist*, **5**(4), 7–8.

DEPARTMENT OF THE ENVIRONMENT 1992. *Coastal Planning*. PPG20, HMSO, London.

DOORNKAMP, J. C. (ed.) 1988. *Planning and Development: Applied Earth Science Background, Torbay*. Geomorphological Services Ltd (for the Department of the Environment), Newport Pagnell.

——1993. Clay shrinkage induced subsidence. *The Geographical Journal*, **159**(2) 196–202.

——1995. Perception and reality in the provision of insurance against natural perils in the UK. *Transactions of the Institute of British Geographers*, **20**(1), 68–80.

ECONOMIST, THE 1991. Waiting for the big one. *The Economist*, 7 December, 131–133.

EDWARDS, R. A., SCRIVENER, R. C. & FORSTER, A. 1987. *Applied Geological Mapping: Southampton area*. BGS Research Report IC80/87/2, British Geological Survey, Keyworth.

ELLIS, S. 1994. Floods bring a torrent of rises in insurance. *Sunday Times*, 13 February, 3.

EMRICH, J., WADGE, T. & KELK, B. 1993. The cost of subsidence. *Intercity Magazine*, July–August, 38–39.

EVANS, R. 1994. Run river run. *Geographical*, **LXVI**(7), 17–20.

FARTHING, K. 1994. The water is wide: customising GIS to manage flood enquiries. *Mapping Awareness*, **8**(1), 30–32.

GAPS Geological Consultants 1992. *Applied Geological Mapping Study of the Plymouth–Plymstock Area, South Devon*. M1 Press, Nottingham.

Geomorphological Services Ltd 1991. *Coastal Landslip Potential Assessment: Isle of White Undercliff, Ventnor*. Department of the Environment, London.

GLASKIN, M. 1994. Insurers address the risk factor. *The Guardian*, 12 February, 33.

HADFIELD, P. 1995. *Sixty Seconds that will Change the World: The Coming Tokyo Earthquake*. Pan Books, London.

HEWITT, K. (ed.). 1983. *Interpretations of Calamity*. Allen & Unwin, London.

HOOKE, J. M. & BRAY, M. J. 1995. Coastal groups, littoral cells, policies and plans in the UK. *Area*, **27**, 358–368.

JONES, D. K. C. 1993. Slope instability in a warmer Britain. *The Geographical Journal*, **159**(2), 184–195.

—— & LEE, E. M. 1995. *Landsliding in Great Britain*. HMSO, London.

——, BROOK, D. & BRUNSDEN, D. 1989. Grounds for improvement. *Geographical Magazine*, **LXI**(8), 38–43.

KHATER, M. 1992. *Reconnaissance report on the Cairo, Egypt earthquake of October 12 1992*. Technical Report NCEER-92-0033, National Center for Earthquake Engineering Research, Buffalo.

LEE, E. M. 1993. The political ecology of coastal planning and management in England and Wales: policy responses to the implications of sea-level rise. *The Geographical Journal*, **159**(2), 169–178.

——, DOORNKAMP, J. C., BRUNSDEN, D. & NOTON, N. H. 1991. *Ground Movement in Ventnor, Isle of White*. Department of the Environment, London.

MAIRSON, A. 1994. Great flood of '93. *National Geographic*, **185**(1), 42–81.

MASKREY, A. 1989. *Disaster Mitigation: A Community Based Approach*. Oxfam, Oxford.

NUHFER, E. B. 1994. Geologic hazards and educational outreach. *Geotimes*, **39**(1), 16–18.

——, PROCTOR, R. J. & MOSER, P. H. 1993. *The Citizen's Guide to Geologic Hazards*. American Institute of Professional Geologists, Colorado.

PRESS, F. 1994. Humankind and Earth's natural system. *Geotimes*, **39**(1), 4.

RENDEL GEOTECHNICS 1991. *Getting the message across: ground movement and public perception*. Report prepared by Rendel Geotechnics for South Wight Borough Council.

SUSMAN, P., O'KEEFE, P. & WISNER, B. 1983. Global disasters, a radical interpretation. *In*: HEWITT, K. (ed.), *Interpretations of Calamity*. Allen & Unwin, London, 263–283.

SWISS REINSURANCE COMPANY 1993. Natural catastrophes and major losses in 1992: insured damage reaches new record level. *Sigma*, **2/93**.

Landfill disposal of urban wastes in developing countries: balancing environmental protection and cost

A. R. Griffin[1] & J. D. Mather[2]

[1] Brown and Root Environmental, Thorncroft Manor, Dorking Road, Leatherhead, Surrey KT22 8JB, UK
[2] Department of Geology, Royal Holloway, University of London, Egham, Surrey TW20 0EX, UK

Abstract. Controlled landfilling with minimal engineering is appropriate for urban waste disposal in most developing countries. Case histories from Tanzania, the Gambia and Mauritius show how unsanitary dumpsites can be replaced by controlled landfills using simple technology at a sustainable and realistic cost commensurate with Gross Domestic Product (GDP). In each country care has been taken to tailor schemes to local conditions.

Introduction

In the United Kingdom landfill is the most important disposal route for both domestic wastes and for the more hazardous wastes produced by industry. Despite increasing regulatory controls, economic initiatives to encourage waste reduction and recycling, and environmental objections, it seems likely to remain the main route for some time to come.

The bulk of the waste is generated within urban environments but as there is little spare land available disposal takes place on the rural margins. For example, in London, collected waste is transferred to barges or trains for shipment to marshes on the Thames Estuary, old brickpits in Bedfordshire or worked out sand and gravel excavations in Oxfordshire.

The industry is changing rapidly in response to recent legislation. Prior to 1972 the only controls were under planning law and public health legislation. Most landfills operated on the principal of dilute and disperse, although this was not as a result of any coherent policy. Some legitimacy was given to this management strategy by Department of the Environment (DoE) funded research in the 1970s. However, the EC Groundwater Directive and more recently the National Rivers Authority Groundwater Protection Policy have meant that such landfills can no longer be licensed. They have been replaced by the engineered containment landfill where the management strategy is to minimize leachate formation, contain it within the site and when it does form collect it, control it and dispose of it. In many schemes the objective appears to be to create a dry tomb for the waste. Such a strategy has its own problems, not least of which are the considerable costs of installing complex lining and capping systems and the length of time for which management will be required once the site has been filled and income ceased to be generated (Mather 1994). Fortunately, these issues are now beginning to be addressed with emphasis placed on encouraging waste degradation so that landfill licenses can be surrendered within a reasonable time-scale.

The situation in many developing countries contrasts markedly with that in the United Kingdom. In urban areas there is a bias towards city centres, ringroads, airports and other status symbols. Waste disposal takes place not so much on landfill sites as on 'dumpsites' or 'tips'. Wastes are dumped in a poorly or uncontrolled manner on the fringe of urban areas. Aerobic degradation takes place and much of the waste is burnt. Little groundwater pollution occurs but sites are environmental eyesores with smoke, smells, birds and rats endemic. Major disasters can occur. For example, in Turkey a hillside dumpsite had a refuse face about 50 m high. It was covered by the municipality but this had little overall effect except to turn parts of the inside of the tip anaerobic. On 28 April 1993 local residents reported an explosion and a rumbling sound. Methane ignited and at 10.15 am there was a major avalanche of uncompacted refuse. Some 1 000 000 m^3 of refuse flowed down the valley, burying some 11 houses and killing 39 people (Wilson 1993).

Clearly there is a need to bring some semblance of order to the appalling dumpsites which exist in some developing countries. This paper considers the options which are available, ranging from the introduction of engineered containment to simple changes involving little cost or advanced technology. The principles suggested are illustrated by their application to urban waste management programmes in three developing countries with which one of the authors (A.R.G.) has recently been involved.

Principles

The waste management industry in the United Kingdom and many of the developed nations of the world is at last

GRIFFIN, A. R. & MATHER, J. D. 1998. Landfill disposal of urban wastes in developing countries: balancing environmental protection and cost. *In*: MAUND, J. G. & EDDLESTON, M. (eds) *Geohazards in Engineering Geology*. Geological Society, London, Engineering Geology Special Publications, **15**, 339–348.

beginning to shed its uncontrolled image with improved training for operatives and the use of engineering techniques to provide for waste containment and site monitoring. The increasing use of geomembrane liners at containment sites requires skilled personnel and specialized equipment to lay and join individual geomembrane sheets to strict quality control standards. The cost of waste disposal is also increasing markedly as the demand for more sophisticated containment systems increases. Multi-barrier systems using both mineral liners and geomembranes are becoming commonplace as regulatory controls are tightened. The overall picture is one of increasing costs and an increasing need for specialized trained staff.

Unfortunately it is the combination of a lack of technical knowledge and a lack of finance which is the cause of many of the problems experienced by developing countries. There is a need to balance the benefits to be derived from improved waste disposal practices with the benefits which might accrue from alternative uses of those funds. It is difficult to justify any expenditure on improving dumpsites if a significant proportion of the population is close to starvation. There must be a willingness to pay for any improvements and this will only happen if the introduction of an effective waste management service provides visible benefits to the community as a whole. The perception that waste accumulation may be related to disease is often lacking or overlooked, although the nuisance which it creates is not. If garbage is removed from residential areas and drains (a regular site for uncontrolled dumping and an attraction for vermin), it is possible to demonstrate that the local environment has been improved and the quality of life enhanced. It follows that the first requirement is for effective storage of the waste at source and an efficient waste collection service using appropriate plant and equipment. Provision of such a service can immediately reduce adverse environmental impacts and demonstrate the value of investment in waste management.

As indicated in the introduction, collected waste is commonly transported to the fringes of urban areas or into surrounding rural areas for disposal. The disposal site may be an old quarry, excavation, or valley where infilling takes place or waste can be built up into a hill or mound on almost any piece of land. Enormous improvements can be made to open dumpsites by the simple process of compacting the waste as it is deposited and covering it at the end of each working day. If done properly this eliminates rodents, blowing paper and plastic, and reduces odour. The waste needs to be deposited in layers a few metres thick rather than being tipped over the quarry sides or down hillslopes to form steep faces.

The advanced lining and sealing technology employed in the developed countries is not necessary to produce major benefits in health and hygiene. Such technology was introduced to protect groundwater resources where strict regulatory controls apply and in countries where adequate expertise and financial resources exist. In developing countries there is a need to balance the requirements for water supply and the requirements of waste disposal. Such a balancing of interests may conclude that it is unnecessary to protect a small minor aquifer from localized pollution if it means that waste can be disposed of in a controlled and hygienic manner.

If dilute and disperse sites are to be accepted a requirement exists to choose sites with care and to carry out a risk assessment as a precursor to any development. Although some localized minor groundwater contamination might be acceptable, significant migration of either leachate or landfill gas is not, particularly if extensive use of groundwater is made for public water supply. Geological and hydrogeological assessments are essential to characterize potential sites and it is necessary to construct an outline water balance based on the best available information to predict leachate volumes. It needs to be emphasized that if sophisticated engineering is to be avoided more attention needs to be paid to the siting of landfill sites as the underlying strata and groundwater conditions become more important. It is suggested that controlled landfilling with minimal engineering is likely to be appropriate for waste disposal in most developing countries and will yield major benefits in terms of health and hygiene. The introduction of advanced technology in the form of lining and capping systems, accompanied by leachate collection and treatment, is generally inappropriate and should be resisted. It will lead to long-term problems of leachate control and management which have not yet been adequately thought through in the United Kingdom. The objective must be to replace unsanitary dumpsites with controlled landfills using simple technology at a sustainable and realistic cost commensurate with Gross Domestic Product.

Case histories

The principles outlined above can be illustrated with respect to urban waste management programmes that have recently been carried out in Tanzania, the Gambia and Mauritius. Each of these countries has different environmental problems but all require controls and improvements in waste disposal practices. Both Tanzania and the Gambia have low GDPs at US$260 and US$342 per capita per year respectively (1989 estimates). The latter has a significant tourist industry concentrated along the coastal zone (termed the Tourist Development Area), which, while providing benefits to the economy, creates an added burden to the waste management services. Mauritius, while relatively affluent, with a GDP of US$3500, has been lax on certain areas of environmental management, particularly waste disposal, although waste collection is efficient and has improved further through recent privatization.

Tanzania

Waste management studies have been carried out on five towns in the north of Tanzania. The population of the towns is increasing annually at an average rate of about 4.5%, increasing the burden on the currently under-resourced services. While the per capita waste generation rate is not high, probably less than 0.3 kg $person^{-1}$ day^{-1}, the collection rate by the cleansing services is poor due to a lack of suitable collection vehicles or plant. A waste survey completed in two of the towns indicated that the bulk of waste generated is vegetable or kitchen waste with a significant percentage of yard wastes (leaves, branches, etc.). Within the commercial areas paper and packaging increases (Fig. 1). The towns are mostly agriculturally based and industrial wastes are limited, but some special wastes such as hospital and surgical waste

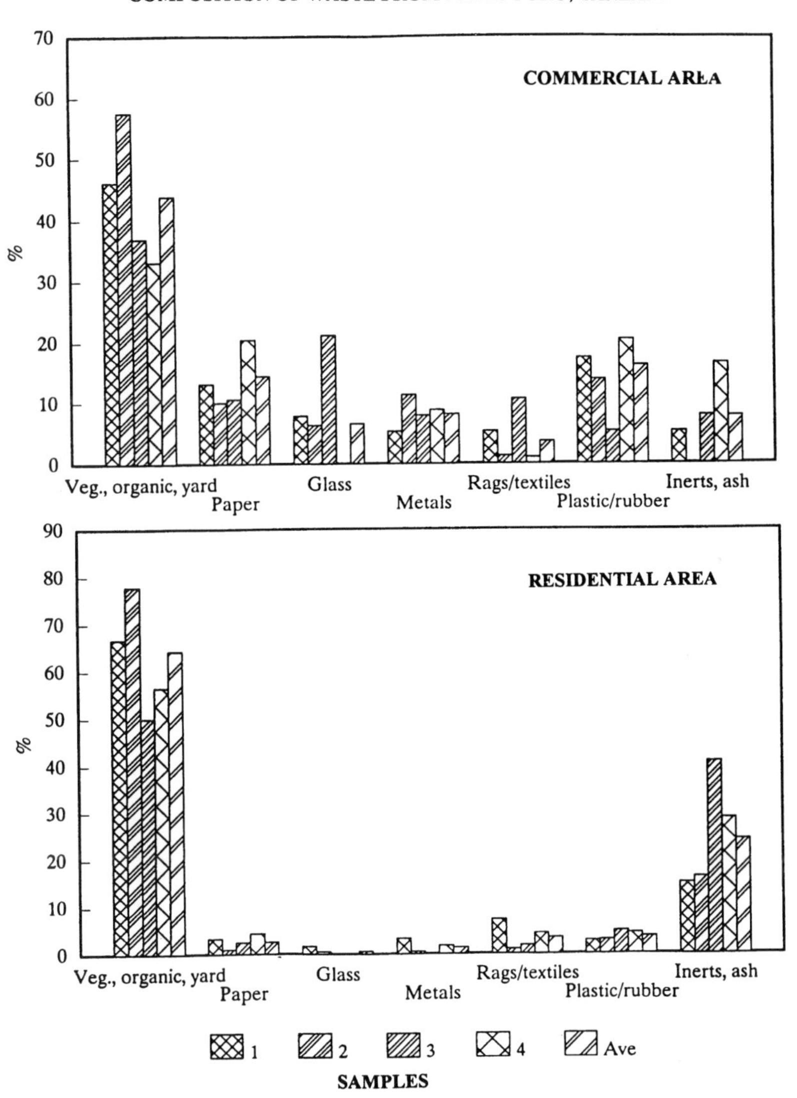

Fig. 1. Compositions of collected waste at Morogoro, Tanzania. Top, commercial area; bottom, residential area. Four separate samples of waste were analysed from each area.

are collected as there are no working incinerators. The high putrescible content combined with a high rainfall, can lead to the generation of leachates although not continuously because of seasonal variations.

Existing dumps are in or close to the towns. In Arusha, for example, the dumpsite is located some 2–3 km from the town centre. This site covers an area of about 4 ha on the edge of a shallow valley and the waste is emptied and spread in an uncontrolled manner. Much of the waste is burnt. Due to urban expansion the boundaries of the dump are now built on, mainly by squatters, and planting of corn has actually occurred on the waste itself, which has no soil cover. An effective waste management system for Arusha requires improvements to existing waste storage, waste collection and transportation as well as to improved disposal. Studies on affordability and willingness to pay indicated that to sustain the system an annual charge to householders would have to be set at a level of some US$1–2 per person per year. This precluded the development of lined systems with leachate treatment facilities. A controlled landfill was accepted as an appropriate standard and a number of potential sites examined. Two sites were considered: a basalt quarry 2–3 km away from the town and a roadstone quarry some 14 km from the town but adjacent to the main highway.

A risk assessment eliminated the first site on grounds of geology and hydrogeology. The site, although of adequate void capacity, was in heavily jointed rock and close to a stream, serving as a water source, and to habitation. It was considered that there was a significant risk from the migration of leachates and from possible future methane generation, and that extensive lining or ground treatment processes would be required.

The second site was located at the base of a parasitic volcanic cone. The cone, according to the published geological survey sheet (1:125 000 Sheet 55 Arusha, 1983), consists of basic-ankaramite basalt lava and scoria with an estimated age of 1.7 Ma. The cone is surrounded by a flat plain of alluvium and black/grey shrinkable soils. The cone has been quarried for road aggregates and two abandoned quarries at the base of the cone, selected for the landfill site, showed exposures of medium to coarse weakly cemented volcanic ash and scoria. Both quarries, without additional preparatory excavations, had a void capacity of about 180 000 m^3 and good road access was available from the main highway into the quarry floors. The deeper of the two quarries had near-vertical faces up to 15 m in height.

The main water sources in the area were from surface runoff. Two adjacent streams some 200–300 m from the site are usually dry for most of the year. Groundwater resources are not used in this area. The volcanic scoria has a moderate permeability, but has a high absorption capacity and was believed to have beneficial properties enhancing biological breakdown of effluent akin to slags used in sewage filter beds. While leachate generation is expected during the rainy season it was not considered that this would represent a hazard to the surrounding environment and no receptors under a risk were identified. Minimal engineering is required to develop the site and will be restricted to provision of a perimeter drain (to channel upslope runoff away from the site), some excavation and stockpiling of the ash (for final cover), and infrastructure works including a reception office and sanitary facilities, a small maintenance workshop and a fuel storage tank. No leachate treatment system will be provided. It is expected from field permeability tests that any leachates generated will disperse through the underlying strata, initially through the landfill floor and then through the sidewalls as the base becomes clogged with fines. The principle of dilute and disperse is considered acceptable from the perception of likely risk.

The estimated capital cost for development is expected to be less than US$60 000 with an annual operation and maintenance charge for the site (including capital depreciation of facilities and landfill plant) of US$65 000. However, because of the distance from the town an intermediate transfer station will be required. In Arusha and several other towns under study (Moshi and Morogoro), the waste management plan was adapted primarily to suit the location of the waste disposal site.

The Gambia

The majority of the Gambia's population is centred on the capital Banjul, which is also the main commercial area, and to the south of the town in an area called the Kombo Peninsular (Fig. 2). Due to the natural expansion and in-country migration the urban population in this area is growing at an average annual rate of 8%. The populations and population densities of the capital Banjul and the Kanifing district of the Kombo Peninsular, taken from the 1993 census, are as follows:

	Population	*Density (per km^2)*
Banjul	42 407	3468
Kanifing	228 945	3030

This population is swollen by tourism which is concentrated on the narrow coastal belt southwards from Banjul. The Gambia has some 100 000 visitors per year (concentrated in some 22 hotels), the majority of which visit in the winter season. The tourist population adds a significant burden to existing under-resourced waste collection services as it produces a disproportionally large amount of waste.

Currently, solid waste is disposed of from Banjul at an uncontrolled dumpsite to the west of the town on the margins of mangrove swamps. Waste from the remaining areas has been dumped at a number of sites including a large quarry (Bakoteh Quarry). The waste collection service, which is carried out by the Cleansing Services (a Department of the Ministry of Health and Social Welfare) is under-resourced and can only collect a

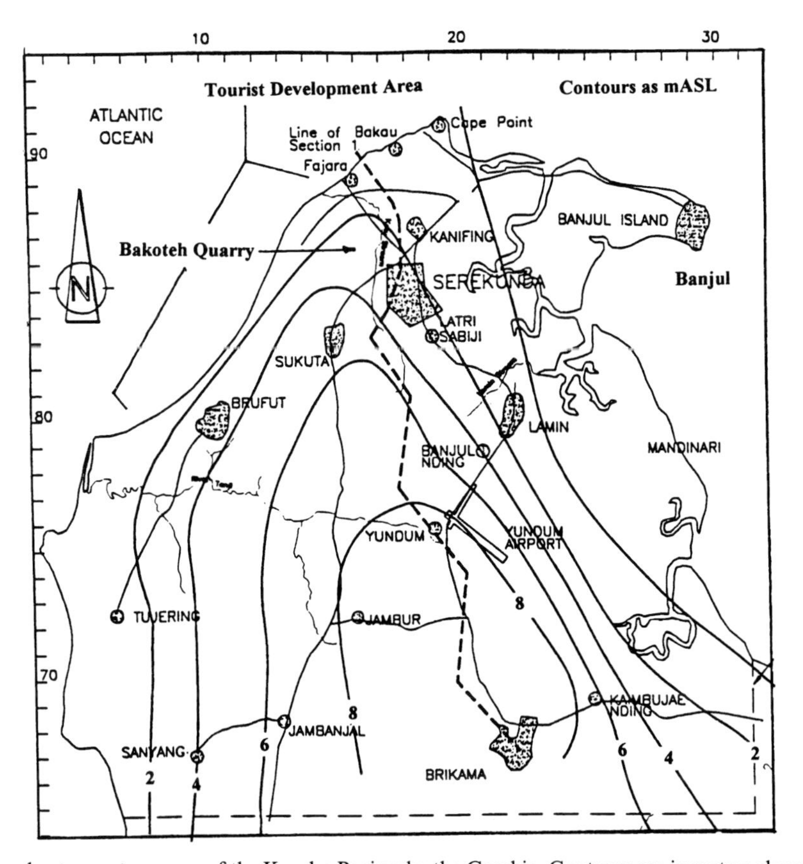

Fig. 2. Groundwater contour map of the Kombo Peninsula, the Gambia. Contours are in metres above sea level.

small amount of waste generated. This leads to a large amount of dumping by private individuals and hotels, much of which is uncontrolled.

The existing dump sites are unsanitary. Burning is common, either deliberately by scavengers or by accident, and large volumes of noxious smoke are regularly visible. While the Gambia has a limited industrial base, some wastes that are produced are of concern. These include oil sludges from the power station (which are currently being stored in drums) and sanitary wastes from the hospitals, the latter being disposed of at the dumps.

In common with Tanzania, any improvements to waste management must be sustainable. While affordability studies have yet to be carried out, the general impression is that residents would be willing to contribute to a visibly effective system that would have to concentrate on improving storage systems and introducing a regular and reliable collection system. It would seem likely that, as in Tanzania, affordability may be as little as US$1 or US$2 per person per year. This precludes extensive capital expenditure and limits development of dumps to the level of controlled landfill sites.

Banjul and the Kombo Peninsula's sole water supply is groundwater. The Gambia is underlain by alluvial deposits which contain a number of aquifer units (Fig. 3). The upper sandy and lateritic unit varies in thickness up to 11 m and is drawn on by many hand-dug wells. Where public water supplies are available this water is not used for drinking due to the risk of pollution in some areas from wastewater derived from a high density of pit latrines. The lower sand aquifer which is 5–30 m thick is separated from the upper zone by silts and clays (0–50 m thick) and produces all the pumped water supplies. The silts and clays form an aquitard and the sequence acts as a leaky aquifer system. Pumping from the lower aquifer has led to a lowering and locally drying out of the upper aquifer. There is also a potential risk of saline intrusion from unlicensed pumping of the aquifer in the coastal region. The adoption of a controlled landfill site with limited infrastructure development precludes the use of liners and associated leachate drains and leachate treatment systems.

The existing quarry site (Bakoteh) was considered as a potential landfill site. The quarry had been excavated for lateritic gravel which is used for aggregates and enclosed

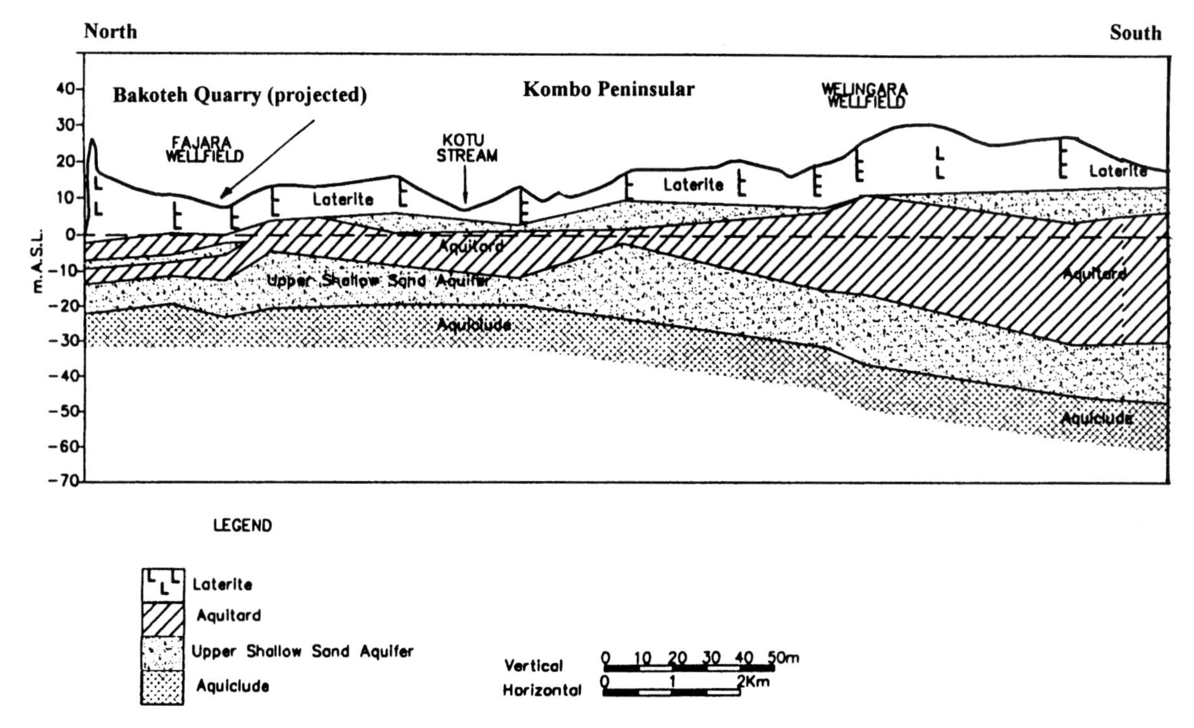

Fig. 3. Geological/hydrogeological cross-section through the aquifer, Kombo Peninsula, the Gambia. The line of cross-section is marked on Fig. 2.

an area of some 500 m by 300 m. The geology consists of 2 m of dense weakly cemented sandstone above at least 2 m of sandstone with abundant gravel size nodules of laterite. The quarry is up to 6 m deep in places and the floor is probably close to the groundwater level of the upper aquifer. Dumping is currently occurring and some ground pollution is present from waste oil dumping. The site is close to the coastline. The underlying aquifer is partially polluted and to the north further pollution may be occurring from unlined sewage lagoons. The area between the dump and the coast is mostly the Tourist Development Area (TDA) which is the planned area for hotels. There water supplies are from the public services or from deep wells to the lower aquifer. The potential risk from the site, considering the potential migration routes and the absence of any major receptors, was considered negligible. However, improvements to the site through proper management including controlled codisposal of certain wastes, such as oils, could reduce potential pollution further.

Despite the acceptable technical conditions at the site it was considered that the site should have a limited life due to encroachment by urban development, but should be the main site in the short term to take all the waste from the area enabling faster infilling and a quicker return to use by early restoration. An alternative site some distance from the urban area and away from the main aquifer recharge areas will have to be identified in the near future.

Mauritius

The island of Mauritius produces some 600 tonnes per day of household waste and probably a further 150 tonnes per day of commercial and industrial waste. Before 1992 this waste was dumped at eight or nine sites around the island in an uncontrolled manner. Much of the waste was burnt, usually by scavengers, and smoke and fumes created a hazard. This was a particular hazard at the Roche Bois site close to the capital Port Louis.

Mauritius (area 1860 km^2, population 1.06 million) is a relatively affluent island (1992 GDP US$3250 per capita) primarily reliant on sugar cane, textiles and the tourist industry. Some light industry including electronics is now developing. The waste collection system is well organized and partially privatized but the municipal services do suffer from under-resourcing of plant, much of which is old and inappropriate. However, while the collection system is well established, disposal (until recently) is not.

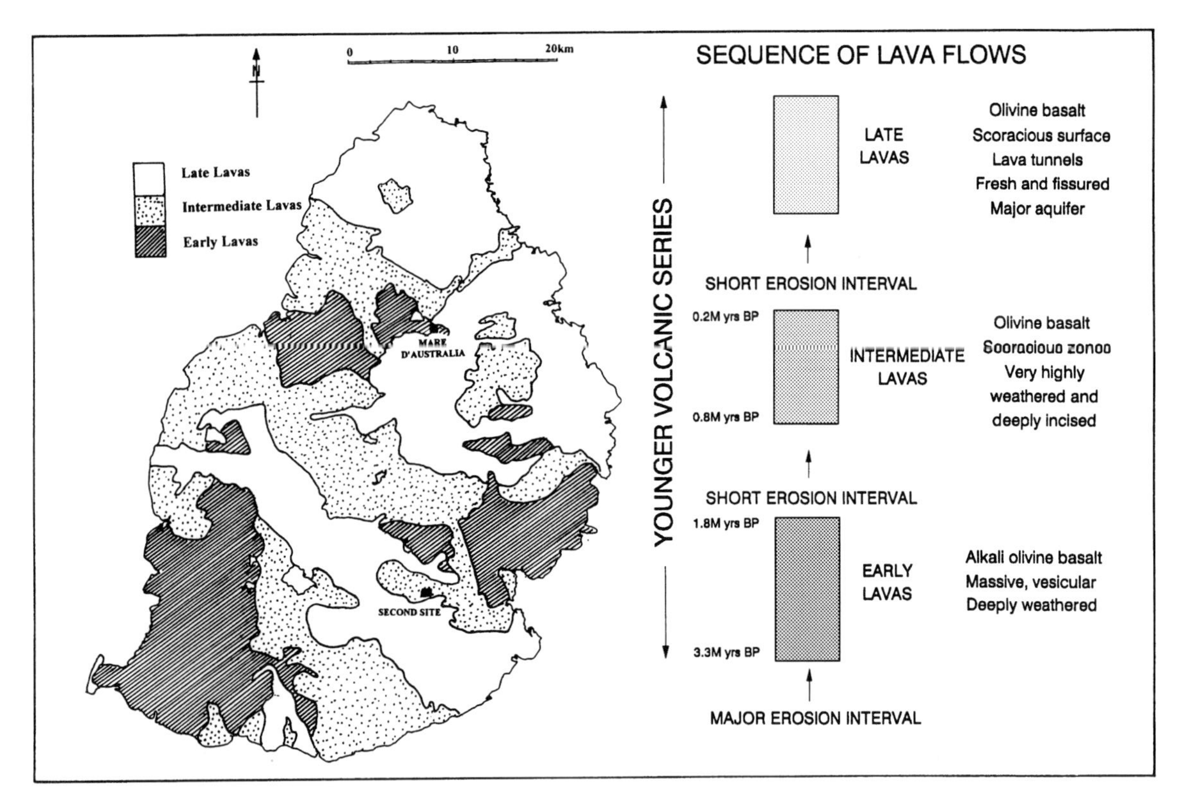

Fig. 4. The younger volcanic series in Mauritius and outline geology.

The government responded swiftly to criticisms of its poor environmental record with the introduction of extensive legislation and the creation of a Ministry of the Environment. One issue to be tackled is waste disposal with the phasing out of dumpsites and the transfer of waste to one or two engineered landfill sites. Several potential sites have been identified and subsequently investigated and detailed environmental impact assessments carried out on two of the most promising sites (Servansingh Jadav & Partners and Binnie & Partners 1992). One site is to be established and engineered to a standard for the safe disposal of 'hazardous' waste, which in Mauritius can include chemical wastes, especially agrochemicals.

A large percentage of Mauritius's water supply is from groundwater resources and a key issue in the siting and engineering of the landfills was the protection of these sources. Mauritius is entirely volcanic but formed by two major periods of activity. Within the younger eruptive period three phases are evident: early, intermediate and late lavas (Fig. 4). The latter are fresh and highly permeable and represent the main groundwater resources. The intermediate lavas are usually highly weathered and less permeable and were therefore considered suitable locations for development of the landfills. Mauritius has a high rainfall ranging from 800 to 3600 mm per annum across the island. Unfortunately the optimum sites for landfill development, both for geological and logistical reasons, tended to be in areas of higher rainfall and it is therefore anticipated that leachate generation will be high. Control of leachate through drainage, collection and treatment is a major issue to avoid potential flooding or spillages of contaminated waters onto adjacent agricultural areas

One of the potential sites is located close to the village of Mare d'Australia and was located to serve most of the more populated and the commercial/industrial areas of north, central and western parts of Mauritius (Fig. 5). The site is located in a broad semicircular valley and underlain by impermeable intermediate lavas and away from major surface abstraction points. However, because of the topography, the site needs to be developed as a series of benches and terraces and phased to allow early restoration. Restoration and rapid return to agricultural use is a major issue at this location and dictates the outline methodology for development.

Development of the site will include excavation of the weathered rock as a source for lining and restoration,

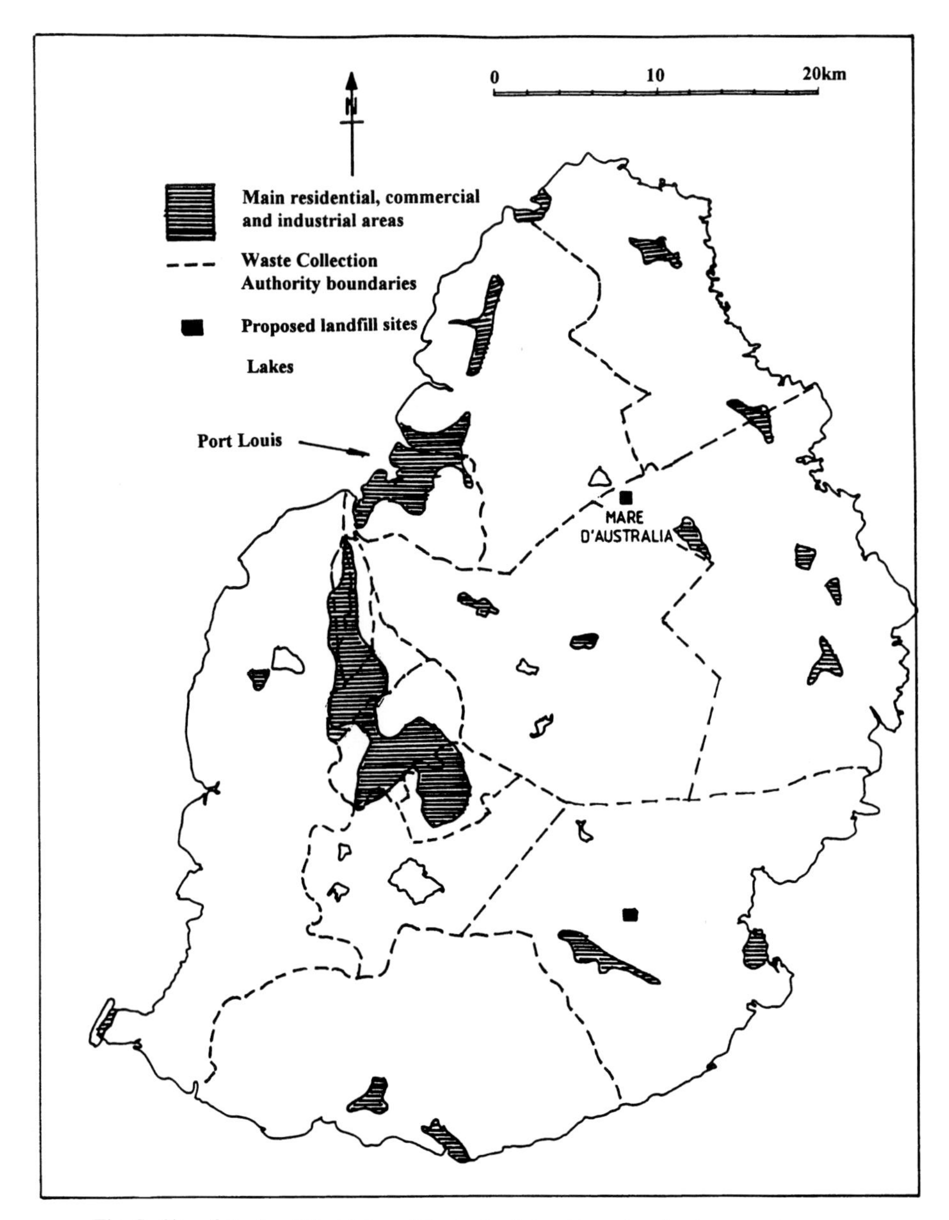

Fig. 5. Site of the landfill at Mare d'Australia, Mauritius, showing its central location.

compaction of the weathered material as a basal liner, construction of basal leachate drainage systems, leachate collection and treatment system (by lagoons) and site control offices (Fig. 6 shows a provisional layout prepared for the EIA). Aggregates for drains are to be obtained from basalt boulders which litter the adjacent fields. This debouldering programme will be an additional benefit by improving the quality of the agricultural areas. While the estimated capital costs of the development are not available, it would be expected that this could be in excess of US$1–2 million. Recurrent operation and management costs are also expected to be high. It is unlikely that more than one or possibly two such sites will be sustainable or affordable. An integrated waste management plan for the island has been developed to enable the bulk of the urban and rural waste, including the commercial and industrial waste, to be taken to the engineered site(s) through a series of

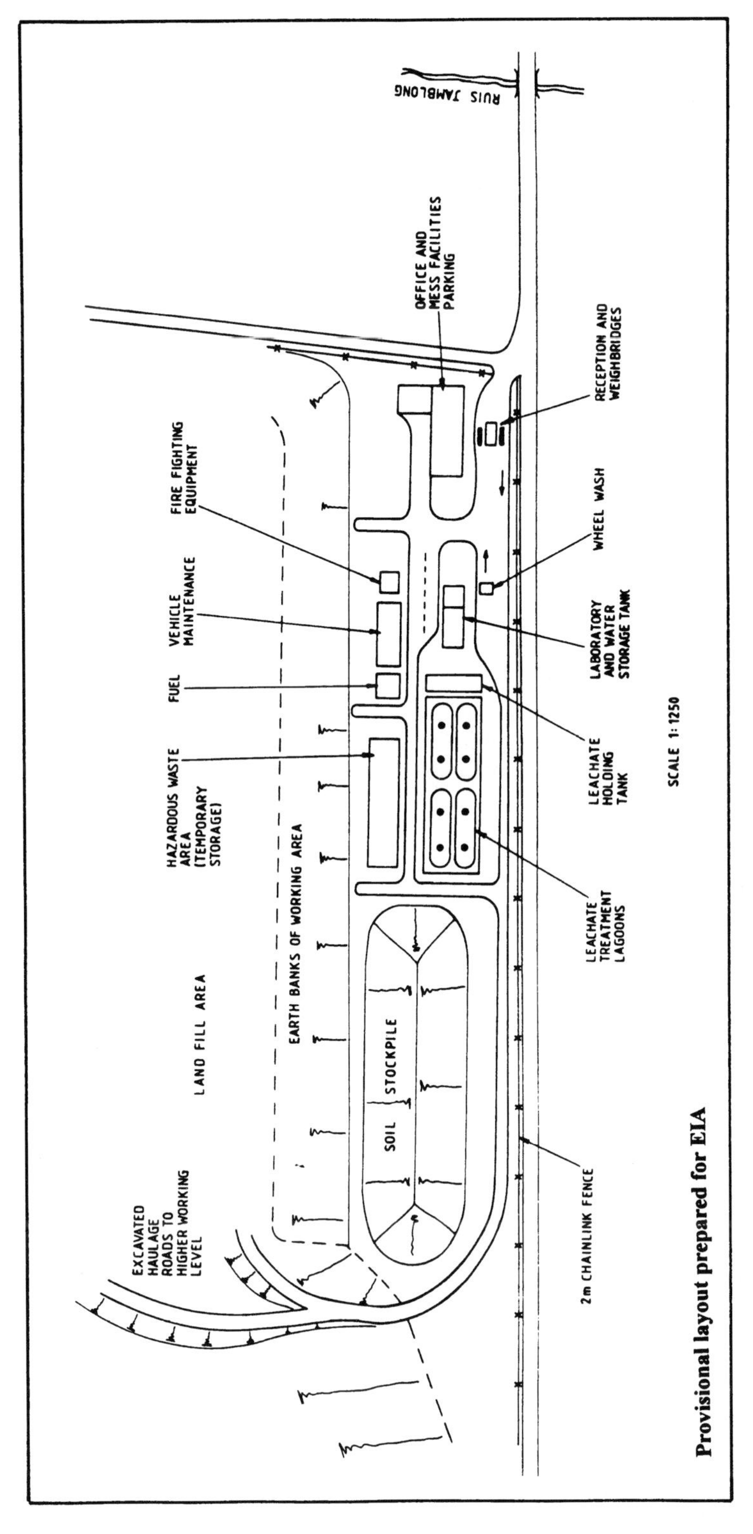

Fig. 6. Provisional layout of the works at the Mare d'Australia site, Mauritius.

small strategically placed waste transfer stations. With expected growth in waste generation and improved efficiency of collection, waste levels may rise as high as 1000 tonnes day^{-1} by the year 2002. At these levels it is estimated that the proposed site has a capacity of at least 35 years. Depreciation of capital over an extended period will reduce the required revenues to a level that may be sustainable.

Conclusions

The objective of replacing unsanitary dumpsites, which are the general method of waste disposal in many developing countries, by controlled landfills using simple technology at a sustainable and realistic cost commensurate with GDP, has been illustrated by schemes recently undertaken in Tanzania, the Gambia and Mauritius. In the former two countries, with GDPs of US$260 and US$342 respectively, an annual charge of some US$1–2 per person is all that householders are likely to be willing to pay for improved waste management.

Controlled landfills at existing quarry sites have been proposed for both these countries. Sites will operate on a dilute and disperse basis but risk assessments demonstrate that local groundwater supplies are not at risk. In both countries some local groundwater contamination is acceptable in return for major improvements in health and hygiene resulting from the elimination of unsanitary dumpsites.

In Mauritius, with a GDP of US$3250 per capita per annum, more funding is available to support improved waste disposal facilities. A large percentage of the island's water supply is derived from groundwater and its protection is a key issue. At the site selected, to serve the most populated area of the island, it is proposed to provide a liner, constructed from local weathered rock, and leachate collection and treatment facilities. However, even on a relatively prosperous island such as Mauritius it is unlikely that more than one or two such sites will be sustainable or affordable.

None of the projected schemes uses the highly engineered systems now the norm in North America and Western Europe. In each case care has been taken to tailor schemes to local conditions. Simple technology has been used to provide solutions to the problem of waste disposal at a sustainable and affordable cost. This must be the objective of any waste management project in developing countries.

References

MATHER, J. D. 1994. Preventing groundwater pollution from landfilled waste – is engineered containment an acceptable solution? *In*: NASH H. & MCCALL, G. J. H. (eds) *Groundwater Quality*. Chapman & Hall, London, 191–195.

SERVANSINGH JADAV & PARTNERS AND BINNIE & PARTNERS 1992. *The development of a sanitary landfill site for domestic and commercial/industrial waste and return to agricultural use*. Environmental Statement for the Ministry of Environmental and Quality of Life, Mauritius.

WILSON, L. 1993. A dumpsite guilty of the greatest crime. *Waste Management*, October 1993, 44–46.

SECTION 7

GEOHAZARDS ASSOCIATED WITH CONTAMINATED LAND

Acid mine drainage and pollution: two case histories from South Africa

F. G. Bell,[1] S. E. T. Bullock[1] & S. Geldenhuis[2]

[1] Department of Geology and Applied Geology, University of Natal, Durban, 4041, South Africa
[2] Steffan, Robertson and Kirsten Inc., Oxford Road, Illovo, Johannesburg, South Africa

Abstract. Different types of waste waters and effluents may be produced as a result of mining. They arise due to the extraction or preparation of the mineral deposit or from the disposal of associated spoil. Generally the major pollutants are suspended solids, dissolved salts or acidity. In the latter case acid mine drainage refers to the oxidation of sulphide minerals, notably pyrite, which are exposed in the mine or are present in the spoil. The primary oxidation products of pyrite are ferric and ferrous sulphates, and sulphuric acid. Two case histories are provided which deal with the problem of acid mine drainage associated with mines in South Africa. The first refers to a tin mine in the Transvaal where acidic waters were produced by heaps of pyrite discard. As the mine was due to close, the problem of possible groundwater pollution due to acid mine drainage had to be investigated as part of the environmental management programme. The second case history involves a coal mine in the Eastern Transvaal which was seeping an appreciable load of acidic effluent and salts into the headwaters of the Vaal River, which is a major source of water to the main industrial conurbation of the country. Investigations which included a study of the history of the site, geology, geophysics, hydrology, hydrogeology and geochemistry succeeded in clarifying the sources of the acidic seepage and the remedial measures that would have the greatest chance of success.

Introduction

The term acid mine drainage is used to describe natural oxidation of sulphide minerals that occur in mine rock or waste which are exposed to air and water. This is a consequence of the oxidation of sulphur in the mineral to a higher oxidation state and, if aqueous iron is present and unstable, the precipitation of ferric iron with hydroxide. It can be associated with underground workings, with spoil heaps, with tailings ponds or with mineral stockpiles. Acid mine drainage is responsible for problems of water pollution in major coal and metal mining areas around the world. However, it will not occur if the sulphide minerals are non-reactive or if the rock contains sufficient alkaline material to neutralize the acidity. The character and rate of release of acid mine drainage is influenced by various chemical and biological reactions at the source of acid generation. If acid mine drainage is not controlled it can pose a serious threat to the aquatic environment since acid generation can lead to elevated levels of heavy metals and sulphate in the water which obviously has a detrimental effect on its quality. The development of acid mine drainage is time-dependent and may evolve over a period of years.

Certain conditions including the right combination of mineralogy, water and oxygen are necessary for the development of acid mine drainage. Such conditions do not always exist. Consequently acid mine drainage is not found at all mines with sulphide-bearing material. The ability of a particular mine rock or waste to generate net acidity depends on the relative content of acid-generating minerals and acid-consuming or neutralizing minerals. Acid waters produced by sulphide oxidation of mine rock or waste may be neutralized by contact with acid-consuming minerals. As a result the water draining from the parent material may have a neutral pH value and negligible acidity despite ongoing sulphide oxidation. However, if the acid-consuming minerals are dissolved, washed out or surrounded by other minerals, then acid generation continues. Where neutralizing carbonate minerals are present, metal hydroxide sludges, such as iron hydroxides and oxyhydroxide, are formed.

The primary chemical factors which determine the rate of acid generation include pH value; temperature; oxygen content of the gas phase if saturation is less than 100%; concentration of oxygen in the water phase; degree of saturation with water; chemical activity of Fe^{3+}; surface area of exposed metal sulphide; and chemical activation energy required to initiate acid generation. In addition, the chemolithotropic bacteria *Thiobacillus ferrooxidans* may accelerate reaction by its enhancement of the rate of ferrous iron oxidation. It also may accelerate reaction through its enchancement of the rate of reduced sulphur oxidation. *Thiobacillus*

BELL, F. G., BULLOCK, S. E. T. & GELDENHUIS, S. 1998. Acid mine drainage and pollution: two case histories from South Africa. *In*: MAUND, J. G. & EDDLESTON, M. (eds) *Geohazards in Engineering Geology*. Geological Society, London, Engineering Geology Special Publications, **15**, 351–364.

ferrooxidans is most active in waters with a pH value around 3.2. If conditions are not favourable, the bacterial influence on acid generation will be minimal.

The initial reaction for direct oxidation of pyrite, either abiotically or by bacterial action, according to Lundgren & Silver (1980) is

$$2FeS_2 + 2H_2O + 7O_2 - 2FeSO_4 + 2H_2SO_4 \quad (1)$$

Subsequent biotic and abiotic reactions which lead to the final oxidation of pyrite by ferric ions (indirect oxidation mechanism), can be represented as follows:

$$4FeSO_4 + 2O_2 + 2H_2SO_4 - 2Fe_2(SO_4)_3 + 2H_2O \quad (2)$$

$$Fe_2(SO_4)_3 + 6H_2O - 2Fe(OH)_3 + 3H_2SO_4 \quad (3)$$

$$4Fe^{2+} + O_2 + 4H - 4Fe^{3+} + 2H_2O \quad (4)$$

$$FeS_2 + 14Fe^3 + 8H_2O - 15Fe^{2+} + 2SO_4 + 16H \quad (5)$$

Reaction (1) shows the initiation of pyrite oxidation, either abiotically (auto-oxidation) or biotically. *Thiobacillus ferrooxidans* converts the ferrous iron of pyrite to its ferric form. The formation of sulphuric acid in the initial oxidation reaction and concomitant decrease in the pH make conditions more favourable for the biotic oxidation of the pyrite by *Thiobacillus ferrooxidans*. The biotic oxidation of pyrite is four times faster than the abiotic reaction at pH 3.0 (Pugh *et al.* 1984).

The development of acid mine drainage is a complex combination of inorganic and sometimes organic processes and reactions. In order to generate severe acid mine drainage ($pH < 3$) sulphide minerals must create an optimum micro-environment for rapid oxidation and must continue to oxidize long enough to exhaust the neutralization potential of the rock.

Accurate prediction of acid mine drainage is required in order to determine its level of control. The objective of acid mine drainage control is to satisfy environmental requirements using the most cost-effective techniques. Due to the potential impact on the environment, regular monitoring is required of acid mine drainage. The major objective of a monitoring programme involved with acid mine drainage is to monitor the effectiveness of the prevention–control–treatment techniques and to detect whether the techniques are unsuccessful at the earliest possible time.

Prediction of the potential for acid generation involves the collection of available data and carrying out static tests and kinetic tests. A static test determines the balance between potentially acid-generating and acid-neutralizing minerals in representative samples. One of the frequently used static tests is acid–base accounting. However, static tests cannot be used to predict the quality of drainage waters and when acid generation will occur. If potential problems are indicated, the more complex kinetic tests should be used to obtain a better insight of the rate of acid generation. Kinetic tests involve weathering of samples under laboratory or on-site conditions in order to confirm the potential to generate net acidity, determine the rates of acid formation, sulphide oxidation, neutralization, metal dissolution and to test control and treatment techniques. The static and kinetic tests provide data which may be used in various models to predict the effect of acid generation and control processes beyond the time-frame of kinetic tests.

There are three key strategies in acid mine drainage management: firstly, control of the acid generation process; secondly, control of acid migration; and, thirdly, the collection and treatment of acid mine drainage (Connelly *et al.* 1995). Source control of acid mine drainage involves measures to prevent or inhibit oxidation, acid generation or containment leaching. If acid generation is prevented, then there is no risk of the resultant contaminants entering the environment. Such control methods involve the removal or isolation of sulphide material, or the exclusion of water or air. The latter is much more practical and can be achieved by the placement of a cover over acid-generating material such as waste or air-sealing adits in mines. Migration control is considered when acid generation is occurring and cannot be inhibited. Since water is the transport medium, control relies on the prevention of water entry to the source of acid mine drainage. Water entry may be controlled by diversion of surface water flowing towards the source by drainage ditches; prevention of groundwater flow into the source by interception and isolation of groundwater (this is very difficult to maintain over the long term); and prevention of infiltration of precipitation into the source by covers, but again their long-term integrity is difficult to ensure. Release control is based on measures to collect and treat acid mine drainage. Collection requires the collection of both ground and surface water polluted by acid mine drainage, involving the installation of collection trenches and wells. Treatment processes have concentrated on neutralization to raise the pH and precipitate metals. Lime or limestone are commonly used, although offering only a partial solution to the problem.

Case history 1

A tin mine at Rooiberg in the northwest Transvaal was forced to cease production because of the decline in the price of the metal. However, before a mine in South Africa can close officially, a certificate of closure must be granted to the mine by the Department of Mineral and Energy Affairs. Section 2.11 of the Minerals Act (August 1991) also states that mines which are to close, must implement an environmental management programme. The programme should include remedial measures to ensure that no pollution will occur when the mine is closed. Accordingly a ground and surface water investigation was undertaken at the mine as part of the environmental management programme. The objectives of the investigation were, firstly, to assess whether pollution of ground and surface water had occurred as a result of mining, secondly, to identify any sources of pollution and, thirdly, to offer solutions to any problem of pollution which existed.

Hydrogeology

Almost all the groundwater in the region occurs in secondary aquifers associated with the Boschoffsberg Quartzite Member (Fig. 1). These secondary aquifers consist of weathered and fractured rock, and lie directly

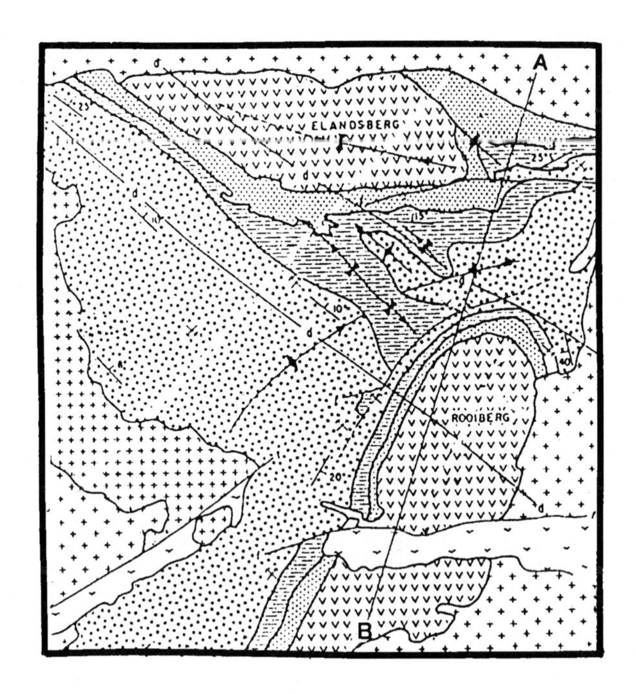

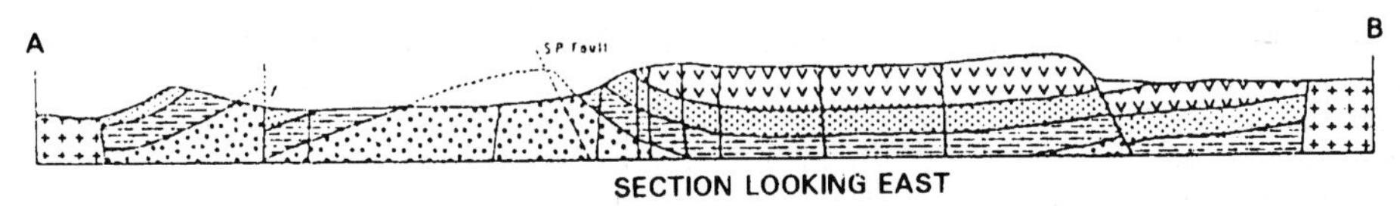

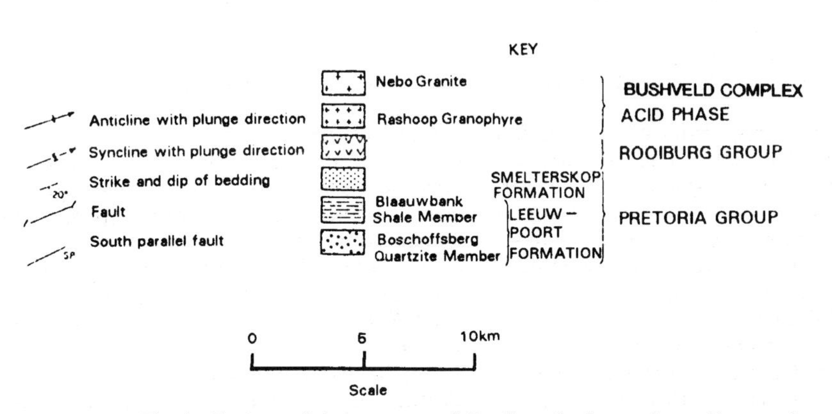

Fig. 1. Geology of the area around Rooiberg in the northwest Transvaal.

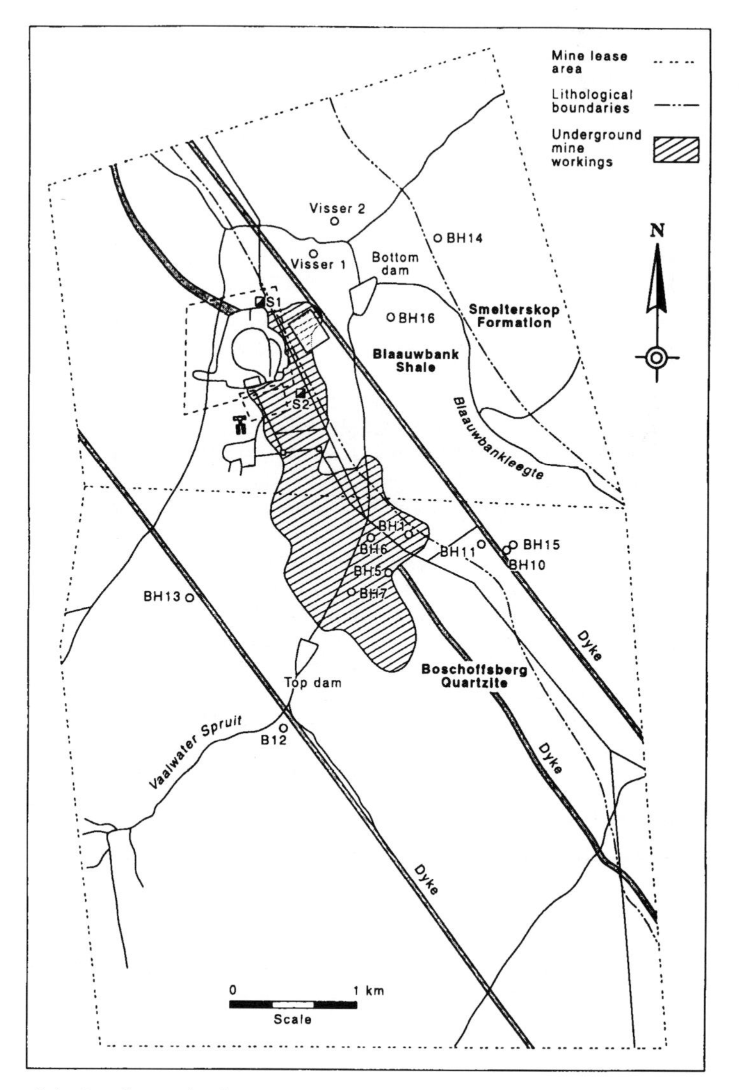

Fig. 2. Geology of the Rooiberg mine lease area showing the location of major dumps, boreholes, sampling points and trenches.

beneath the soil surface. The quantity of groundwater stored in the secondary aquifers is limited, while the permeability of such aquifers is generally low. Nevertheless groundwater represents the primary supply of potable water for both domestic purposes and stock watering in the area. Any possibility of pollution affecting these limited groundwater resources represents a significant threat to the community.

A series of dykes compartmentalizes the groundwater. Therefore water moves along fractures until it intersects a dyke where is it impounded. Lateral movement along the dykes is from the southeast to the northwest.

Extensive underground mine workings (hatched area in Fig. 2) complicate the prediction of groundwater movement in the area. These workings at present are being supplied by natural groundwater recharge. The underground tunnels override any structural control on groundwater movement in the vicinity of the mine. As water continues to flow into the mine workings, it will exert some influence on the groundwater regime until the workings are flooded.

A perched water-table exists around the major dumps at the mine and is due to the presence of a layer of ferrocrete in the soil horizon. Water movement in the perched water-table was determined by digging a series of trenches around the mine and by observing from which direction water seeped into the trenches. Movement of this water is topographically controlled and eventually finds its way into the Boschoffsberg Quartzite aquifer system. Although the quality of the water in the perched water-table is poor, attenuation processes improve the quality as it percolates into the aquifer system proper.

Investigation

In order to ascertain whether pollution of ground and surface water was occurring, samples of water and soil were taken for analysis. Water from all the boreholes on the mine lease area was sampled, as well as from the boreholes of local farmers. Water samples were taken after a borehole had been purged for a period so as to ensure that the water sampled was actually from the aquifer and not water that had been lying stagnant in the borehole. Further groundwater samples were obtained from the underground workings. A number of the trenches which contained groundwater were also sampled. Surface water samples were taken from the two reservoirs at the mine site and streams. In the case of the latter, samples could only be taken after thunder showers as water flow in the streams generally only lasted a short while after a rain storm.

Chemical analyses of water and soil samples

Table 1 lists the results obtained from analyses carried out on surface water in the mine lease area. The high sulphate concentration and low pH in some of the surface samples presumably was attributable to acid mine drainage.

It can be seen from Table 2 that some of the groundwater samples also have extremely high concentrations of sulphate. As with the surface water, these high values in the groundwater are associated with the acid mine drainage. Acidic waters, however, do not only originate at the surface. Oxidation of sulphides in old underground workings also results in the formation of acidic water. The pH of this water, however, shows no evidence of acid mine drainage (see S1 and S2 in Table 2), as any acid waters that do form are quickly neutralized by ankerite ($Ca\{Mg, Fe^{+2}, Mn\}\{C_3\}_2$), which is present in relative abundance in the host rock. Therefore the signs of mine water degradation are higher sulphate concentrations, higher electrical conductivity and higher total dissolved solids (TDS) values.

The results of groundwater analysis of samples were plotted on trilinear diagrams. Figure 3(a) shows the trilinear diagrams for water sampled from boreholes BH10 to BH16. The major anion diagram indicates that bicarbonate is the dominant ion, with minor amounts of sulphate and chloride, whereas the cation diagram suggests that calcium, magnesium and sodium are not dominant. Water taken from farm boreholes has a lower concentration of chloride anion and magnesium cation than in the boreholes mentioned previously, but the calcium and sodium concentrations are greater (Fig. 3(b)). Trilinear diagrams of water sampled from trenches and subsurface workings are shown in Fig. 3(c). The anion triangle shows a distinct skewness towards the sulphate ion. The lower bicarbonate concentration

Table 1. *Analysis of surface water*

	Stagnant water	*River water*	*Bottom dam*	*Vaalwater*	*Top dam*	*South African limits*[a]	
						MLNR[b]	*MPLIR*[c]
pH	2.4	4.1	7.4	7.1	7.3	6.0–9.0	5.5–9.5
COD	1410	38	–	83	–	[d]	[d]
EC ($mS\,m^{-1}$)	865	57.2	65	82	138	70	300
Total hardness	32409	219	249	523	663	300	650
Total N	38.4	2.5	<0.2	0.2	<0.2	6	10
Ca	163	64	82	66	93	150	200
Mg	74	12	32	30	22	70	100
Na	4	11	65	51	97	100	400
K	<1	9	8	23	10	200	400
SO_4	11189	236	174	119	57	250	600
Cl	18	6	42	34	42	200	600
F	<0.1	<0.1	1.5	0.7	8.37	1.0	1.5
Fe	3580	2.4	–	0.2	–	0.1	1

All units expressed as milligrams per litre, except pH and where stated.
[a] Source: Anon (1993).
[b] Maximum limit for no risk.
[c] Maximum permissible limit for insignificant risk.
[d] Chemical oxygen demand (COD): general standards should not exceed $75\,mg\,l^{-1}$ after applying the chloride correction.

Table 2. *Results of groundwater analysis*

Sample	pH	EC	NO_3	Ca	Mg	Na	K	SO_4	Cl	HCO_3	CO_3	TDS	Hardness	Alkalinity
T2	7.43	187	3.9	17.1	14.3	364	30.6	427	121.0	372.1	3	1 167	102	310
T3	6.30	444	11.8	401	289	418	18.5	2 937	26.7	24.4	0	4 114	2 171	20
T4	5.9	421	4.9	539	373	202	23.8	3 275	735	18.3	0	4 404	2 866	15
T11	3.01	3130	29.4	493	3376	228	12.7	47 772	86.9	0.0	0	51 946	15 123	0
T12	2.9	2370	64.8	627	621	102	6.9	14 580	33.4	0.0	0	16 039	4 121	0
T13	2.8	1680	19.6	509	561	70	0.7	27 070	53.5	0.0	0	28 284	2 579	0
T15	3.2	857	7.4	614	309	103	12.5	2 857	173.8	0.0	0	4 078	2 807	0
S1	7.54	497	4.7	475	430.8	367	20.1	3 463	25.4	280.6	0	4 933	2 729	240
S2	7.74	190	1.0	153	123.5	131	5.2	805.5	62.2	353.8	9	1 467	601	305
BH1	7.99	129	42.9	79	74.2	54	19.1	405.6	63.6	85.4	0	781	434	70
BH5	8.15	165	61.4	106	85.7	92	8.8	581.2	48.6	112.0	0	1 057	520	103
BH6	7.73	182	57.5	129	99.5	85	10.0	683.4	49.4	122.0	0	1 175	632	100
BH7	8.07	172	87.3	102	88.9	101	7.4	522.7	59.4	201.3	0	1 070	457	165
BH10	7.2	69	0.2	70	3.1	69	9.0	12.0	4.3	374.4	0	448	317	375
BH11	7.3	71	0.4	64	3.2	68	6.0	15.0	12.6	374.3	0	461	334	375
BH12	7.8	112	<0.2	76	3.7	138	4.0	263.0	12.3	296.2	0	728	452	298
BH13	7.2	58	<0.2	53	4.5	54	3.0	62.0	17.4	263.4	0	377	226	265
BH14	7.1	51	0.4	59	3.4	20	3.0	7.0	10.3	308.6	0	332	286	309
BH15	7.5	75	0.2	75	3.3	91	9.0	7.0	4.6	407.0	0	488	314	409
BH16	7.3	87	0.3	102	3.7	54	17.0	148.0	24.5	297.4	0	566	452	298
Visser 1	7.65	81	2.6	22.8	57	71.7	2.1	23.0	6.4	536.8	0	435	439	440
Visser 2	7.58	78	1.1	55.7	85.5	19.5	3.3	4.1	4.6	610	0	478	490	500
Sleepwa	8.57	51	2.3	44.9	34.4	19.7	2.0	76.5	31.9	122.0	6	279	154	110
Knoppies-kraal	7.74	45	20.8	32	23.6	37.6	2.4	5.6	18.5	189.1	0	235	22	155
Strydom	7.18	68	2.5	62.9	46.4	33.3	1.7	3.7	5.1	536.8	0	424	349	440
Blockdrift	7.32	77	0.0	76.5	54.0	30.0	1.4	2.0	6.8	494.1	0	439	405	–
Nieuwpoort	7.16	107	13.5	74.3	56.9	76.7	0.3	15.4	80.8	530.7	0	583	420	435
Blaauwbank	8.06	36	63.4	12.3	7.9	18.3	11.4	18.8	10.2	36.6	0	161	23	30
South African limits[a]														
MLNR[b]	6–9	70	6	150	70	100	200	200	250			450	300	
MPLIR[c]	6.5–9.5	300	10	200	100	400	400	600	600			2 000	650	

[a] Except for pH and EC ($mS\,m^{-1}$), all units are expressed in milligrams per litre.
[b] Maximum limit for no risk.
[c] Maximum permissible limit for insignificant risk.

and higher sulphate ion concentration is attributed to acid mine drainage. In the cation triangle some samples show a tendency towards the magnesium ion. This also is attributed to acid mine drainage, magnesium being more soluble in waters with low pH and so is concentrated in the polluted water. Calcium values are relatively high due to cation exchange processes.

A number of trenches were excavated around the dumps, and soil samples were taken to determine the extent of the pollution. They all were taken from a depth of 300 mm from the top of the trench. These samples were scanned for a number of elements using an inductively coupled plasma mass spectrometer (I.C.P-MS). The results are given in Table 3, which shows elevated levels of heavy metals in many samples, presumably associated with acid mine drainage. In addition, an attempt was made to ascertain the depth to which acidic waters had penetrated the substrate below the major pyrite dump (see below). This was done by digging a trench to a depth of 2.5 m with a back-actor. Soil samples were taken at different depths within the trench. The results of the analysis are given in Table 4 and suggest that acidic waters have penetrated to a depth greater than 2.5 m. There is a relationship between the pH value and the iron percentage, i.e. as the iron content increases the pH decreases.

Metal ions and toxic salts are brought into solution more readily in acid conditions ($pH < 5.5$). Therefore as acidic surface waters move through the pyrite dumps they absorb heavy metals and toxic salts. Once in the soil, however, the pH of the water increases as it comes in contact with cations sorbed on to clay minerals. For example, a predominance of cations such as Ca, Mg, Na and K tends to raise the pH when they are released from clay. As the water looses acidity, heavy metals and toxic salts may begin to precipitate in the soil. The relatively high heavy metal concentrations (Table 3) in the soil accumulated in this fashion.

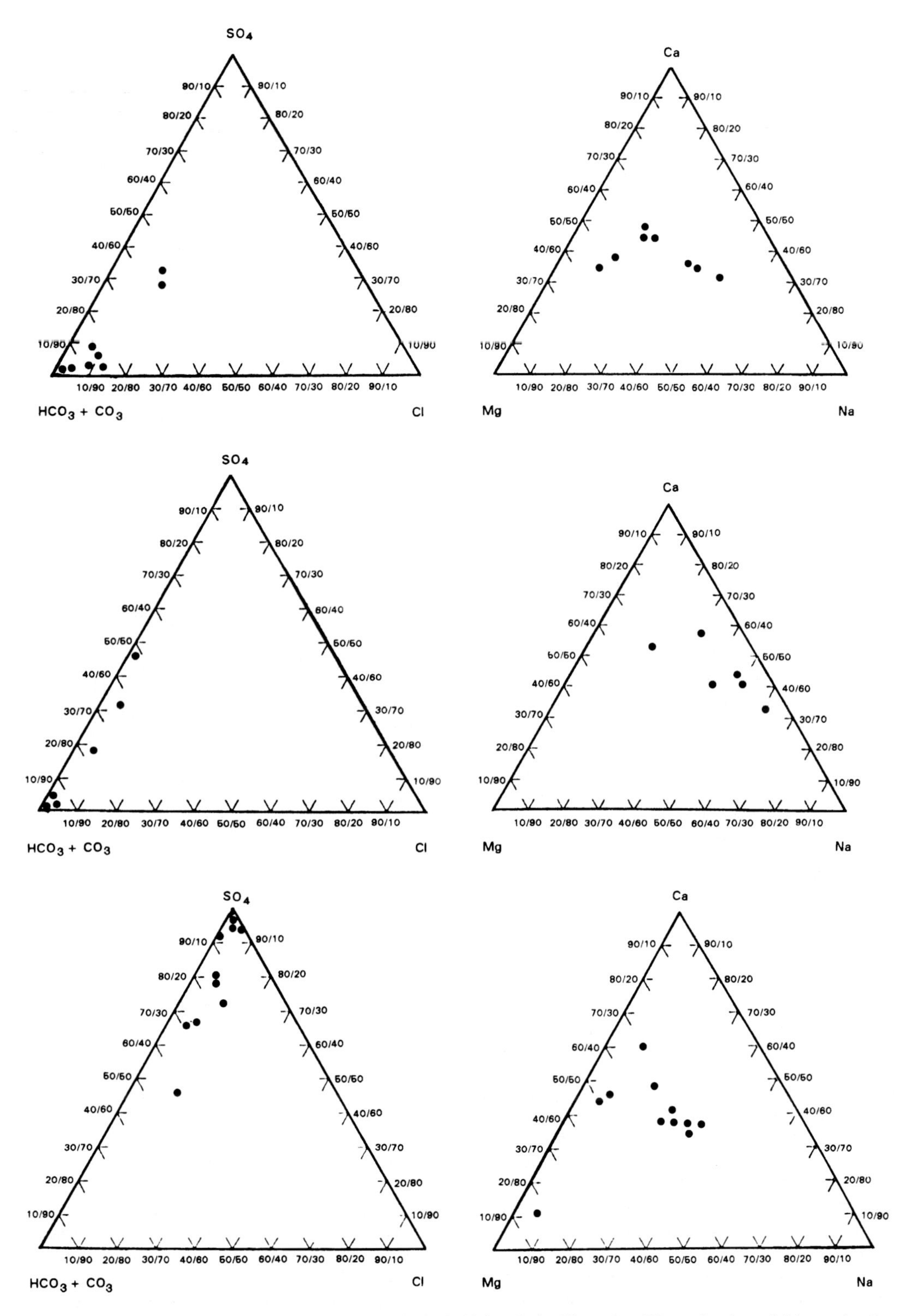

Fig. 3. Trilinear diagrams for (**a**) boreholes B10 to B16, (**b**) boreholes Visser 1 to Blauuwbank, and (**c**) trenches 2 to 15 and underground workings.

Table 3. *Semi-quantitative analysis of soil samples*

Element	*T1*	*T2*	*T3*	*T4*	*T5*	*T6*	*T7*	*T15*	*Sand*	*Slag*	*R1*	*R2*	*R3*	*R4*	*R5*
pH	4.36	7.78	5.04	4.62	2.9	7.1	3.73	3.04	8.24	8.05	6.54	6.25	2.84	4.37	6.67
EC	152	120	182	360	469	342	314	31 000	47	2 000	35	2210	1 228	279	47
Ca	871	95	8011	4424	3227	3982	4792	147	428	3 836	141	1666	496	3997	374
Mg	378	25	434	1846	473	1577	894	47 296	130	19 419	50	2176	9 949	643	122
Al	113	46	0	85	1399	2	860	30 782	0	8	8	17	15 829	167	8
Mn	98	0	91	421	345	46	262	46 202	6	14	4	380	759	89	8
Fe	15	3	2	10	454	9	10	24 731	1	16	1	19	3 790	24	2
Co	20	0	3	43	91	1	67	3 916	0	1	0	285	823	39	0
Ni	9	0	1	8	11	1	28	484	0	0	0	37	23	11	0
Cu^{2+}	178	1	0	35	450	0	225	14 820	0	5	0	12	3 924	34	2
Cu^{3+}	179	1	0	35	440	0	231	15 192	0	5	0	12	3 944	35	1
Zn	4	0	0	7	4	0	3	226	0	0	0	3	16	2	0

[a]All values except pH and electrical conductivity ($mS\,m^{-1}$) are in $mg\,kg^{-1}$. T1, etc. = trench samples; R1, etc. = stream samples.

Table 4. *Results of soil analysis from a trench dug below the major pyrite dump*

Depth	*pH*	*Fe (%)*	*S (%)*
0.11	5.4	4.41	0.19
0.22	5.2	4.47	1.30
0.36	6.0	3.58	0.35
1.26	4.1	6.65	0.66
2.2	4.6	6.31	0.49

Samples were also taken of the sediment in the stream (Table 3) running alongside the dumps (Fig. 2). Sediment sample R5 was taken from the river bed just as it exits the mine lease area. Analysis of this sample suggests that pollution in the stream has not spread beyond the mine lease area as yet.

Source of acid mine drainage and pollution

Consequently an attempt was made to determine the source of the pollution by examining the waste dumps at the tin mine. The dumps included pyrite dumps, slimes dumps, a slag dump, a sand dump, and a rock dump (Fig. 2). Samples from the major dumps were taken from a depth greater than 500 mm below the surface. This was considered a suitable depth below which oxidation and leaching had not occurred. All the samples were taken from the upper parts of the dumps as both leaching and oxidation are more significant on the sides.

An estimated 8125 tonnes of pyrite material had been deposited in dumps. An elemental analysis of material from the pyrite dump showed that sulphur accounted for some 45% and iron for nearly 36%. According to Halbert *et al.* (1983), the generation of sulphate with bacteria, at a pH of 3.0 and a temperature of 21°C is 1.17 moles kg^{-1} tailings per month. Taking into consideration the fine particle size of this waste material, that the pH value of the paste is 3.2, that the surface temperature of the pyrite dumps exceeds 40°C in the summer months and 21°C in the winter, and the fact that *Thiobacillus ferrooxidans* cultures are present in the crust on the dumps, production rates of sulphate are likely to exceed 1.17 moles kg^{-1} tailings per month. In fact, Halbert *et al.* (1983) maintained that the rate of generation of sulphate increased approximately threefold for every 10°C increase in temperature. Therefore assuming an average annual surface temperature of 30°C, the sulphate generation rate on the pyrite dumps would be around 3.68 mole kg^{-1} tailings per month. This implies that in one month 353.5 mg of sulphate will form from 1 kg of waste. This rate of sulphate generation, however, is probably never reached because the yellow coating on the dumps inhibits the access of oxygen and so reduces the rate of oxidation. Nevertheless, after rain has fallen the yellow coating partly dissolves, although within two or so days the crust can reform, paying testimony to the initial rapid oxidation rates.

Further proof of the high generation rates of sulphate was obtained from the results of acid base account tests on the pyrite material. Acid base accounting allows determination of the proportions of acid-generating and neutralizing minerals present. Initially the pH of a sample paste is measured to assess the natural pH value of the material, as well as to determine if acid generation has occurred prior to analysis. Generally if the pH of the paste is less than 5, then there probably has been acid generation in the sample. Next the total sulphur content of the sample is measured and the maximum potential acidity as sulphuric acid is calculated from the sulphur content. Lastly, the neutralization potential is found by using a base titration procedure of a pre-acidified sample. The results gave an acid potential of 1410.625 $gCaCO_3\,kg^{-1}$, a neutralizing potential of $-40.6\,gCaCO_3\,kg^{-1}$, and a net neutralizing potential of $-1451.7\,gCaCO^3\,kg^{-1}$. According to Brodie *et al.* (1989), samples with a negative net neutralization potential and a ratio of neutralizing potential to acid potential

Table 5. *Composition of slimes material*

Major elements (>1%)	Silica Aluminium Sodium Potassium	Tin Iron Calcium Magnesium
Minor elements (1%–>0.1%)	Rubidium Titanium	Zirconium Phosphorus
Trace elements (<0.1 %)	Manganese Sulphur Copper Cadmium	Chromium Barium Lead Strontium

of less than 1:1 have a high potential for acid generation. The acid base accounting results therefore clearly demonstrate that the pyrite dump material had an extremely high potential to produce acidic waters.

A number of old slimes dumps are associated with the tin mine. Samples of this material were analysed by X-ray fluorescence (XRF) to determine their composition. The results are summarized in Table 5.

The results of acid–base accounting carried out on slimes material showed an acid potential of 5.0 $gCaCO_3$ kg^{-1}, a neutralizing potential of 40 $gCaCO_3\,kg^{-1}$ and a net neutralizing potential of 35 $gCaCO_3\,kg^{-1}$. Hence the slimes material has a potentially acid-consuming character (Brodie *et al.* 1989) and will not contribute to the acid mine drainage problem at the tin field.

The slag dump contains an estimated 5000 t of material. The paste pH of the material was 8.1. The results of acid–base accounting tests done on the slag material showed an acid potential of 10 $gCaCO_3\,kg^{-1}$, a neutralizing potential of 40 $gCaCO_3\,kg^{-1}$ and a net neutralizing potential of 30 $gCaCO_3\,kg^{-1}$. Consequently the slag material has a potentially acid-consuming character. This is not necessarily surprising as sulphides had been extracted prior to dumping the slag.

The sand dump had an estimated volume of some 6000 t. This material had a paste pH of 8.2. As the name suggests this material has a dominant sand size fraction and consequently a high permeability. The moisture content of the sand dump material is low. Acid–base accounting tests which were carried out on the sand material indicated that the acid potential was 30 $gCaCO_3\,kg^{-1}$, the neutralizing potential was 38 $gCaCO_3\,kg^{-1}$ and the net neutralizing potential was 8 $gCaCO_3\,kg^{-1}$. Accordingly the sand material falls into the zone of 'uncertainty', i.e. the material can either be acid-consuming or acid-generating, depending on the local conditions.

The fragments in the rock dump varied in size from 20 mm up to 300 mm. A mine rock classification developed by Brodie *et al.* (1989) was used to determine whether acidic waters seeped from this dump. This classification uses visual, physical and geochemical characteristics to classify mine rock, in particular to identify relatively homogeneous rock units with respect to the quality of drainage water. The classification is based on six key properties, namely, particle size, sulphide type, sulphide surface exposure, alkali type, alkali surface exposure, and slaking characteristics. The relative importance of each of these properties is taken into account by a weighting system. The highest acid rock drainage (ARD) potential value obtainable in the classification is 68 and the lowest ARD potential is −20. The rock dump material had a value of 2 and therefore probably did not produce any acidic seepage.

Thc investigation showed that both ground and surface waters in the immediate vicinity of the mine had been affected to a varying degree. Acidified water was the main culprit and this was produced primarily by breakdown of material in the pyrite dumps. The increased acidity of such water meant that metal ions and toxic salts were taken into solution. When the pH rises these metal ions and toxic salts are precipitated in the soil. Accordingly there are concentrations of metal ions in the soil surrounding the major dumps.

In a regional context, however, it would seem that these heavy metals do not pose a major pollution threat at the present, as the high concentrations only occur very close to the main disposal site. In addition, the groundwater gradients around the major dumps are low and groundwater movement is generally towards the north and northeast. If these low groundwater gradients are taken into consideration, together with the dyke system in the vicinity of the major dumps (a dyke occurs 500 m to the north of the dumps and another occurs 1 km to the southwest) and the impermeable nature of the Boschoffsberg rocks, it is unlikely that affected groundwater will migrate further than 1000 m to the north and northeast of the major dumps. None the less, it was decided to remove the pyrite dumps and fortunately a buyer was found who wanted the pyrite for the manufacture of sulphuric acid.

Case history 2

The Loubert Coal mine is situated some 10 km southeast of Ermelo in the Eastern Transvaal in the catchment of the Vaal River. The mine dates back to 1928. The coal mined belongs to the Vryheid Formation of the Karoo Supergroup and is particularly high in sulphide. Therefore the discards and slurry contain even higher levels of sulphide. Consequently at present there is concern about the discharge of polluted water from the mine into the nearby Human Spruit.

Both underground and opencast mining are carried out at the site (Fig. 4). Some underground workings were abandoned in 1943 and are situated on a topographic high which drains towards the Human Spruit in the south and

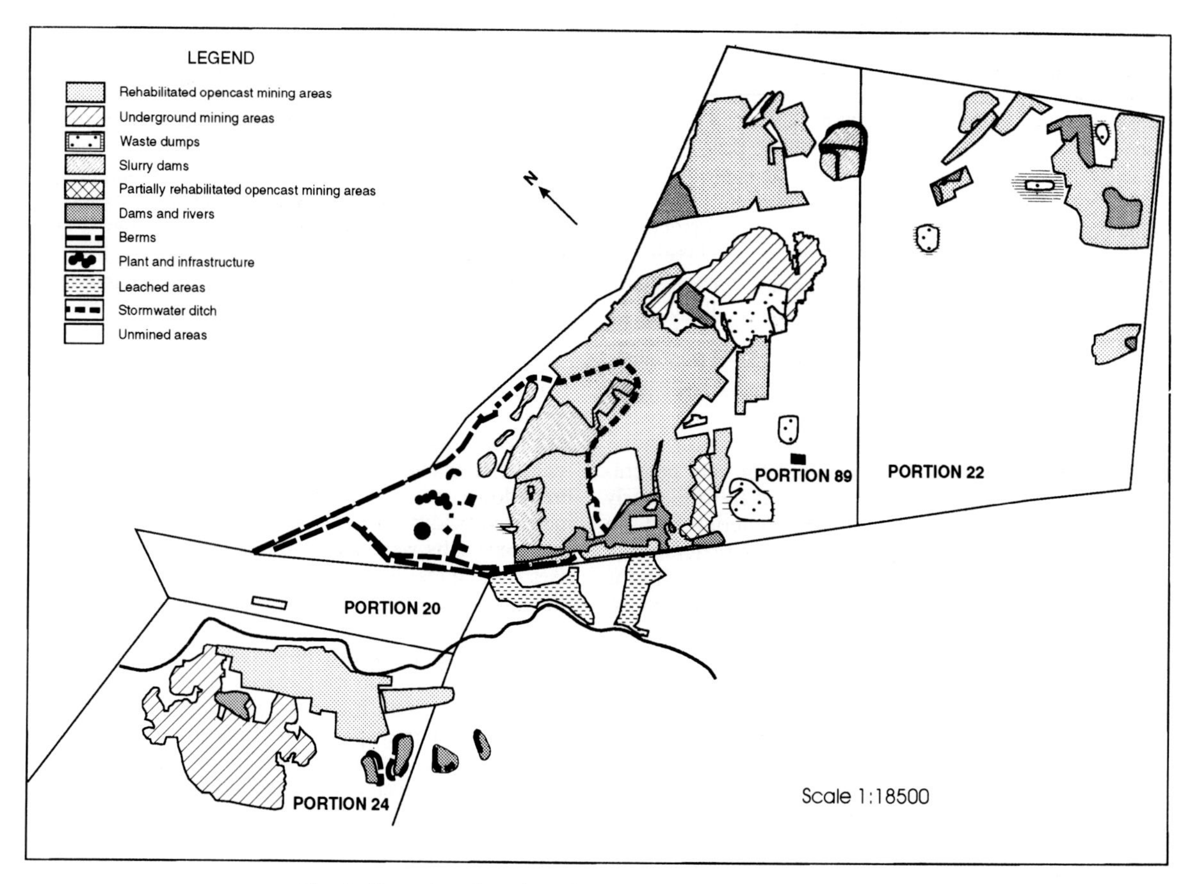

Fig. 4. General surface plan of Loubert Mine, eastern Transvaal.

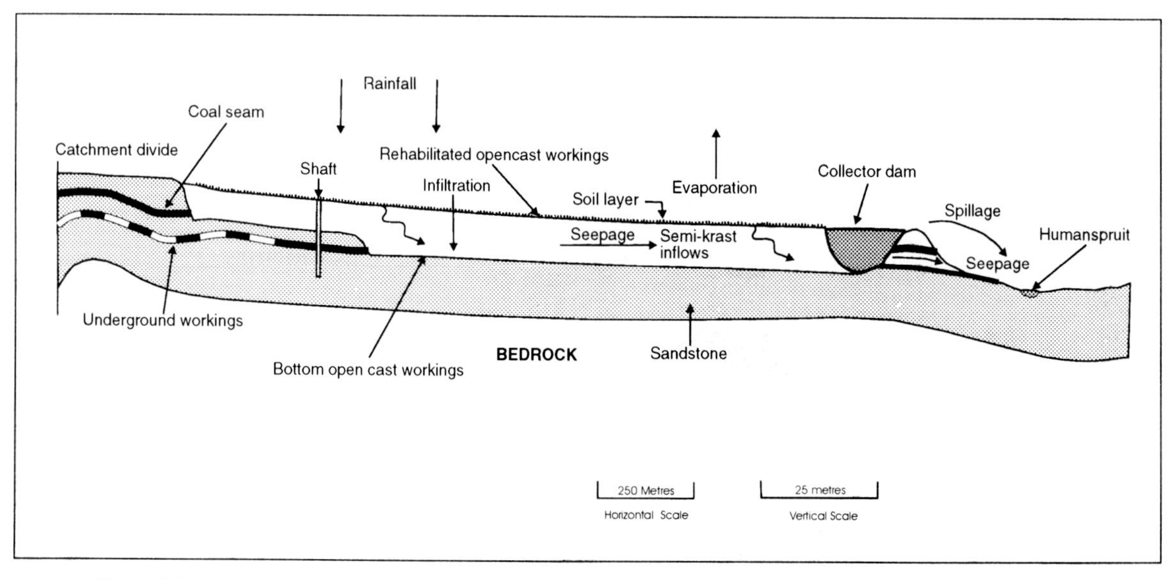

Fig. 5. Schematic section of surface runoff and groundwater movement across the southern part of the site towards the collector dams.

towards the Witpunt Spruit in the north. Most of the opencast operations were carried out to the south of the water divide in the area and it is here where the most obvious pollution problems occur. Water draining to the south collects in three pollution control reservoirs (note that in South Africa, reservoirs are referred to as dams) which are situated just above a small sandstone escarpment, and spillage and seepage from these reservoirs flow into the Human Spruit (Fig. 5), and have a significant effect on the soils around and the water quality in the stream.

A limited amount of opencast mining was done to the north of the divide and the area has been largely restored, with the exception of a few unfilled pits which now contain water. There was also a very small section of underground mining in this area but it was not connected with the other underground workings. The restoration in this area seems to be satisfactory and pollution problems are not immediately apparent.

A major part of the pollution problem on the southern slope of Portion 89, to the immediate southeast of the mine, emanates from the old underground workings. The lower (C_l) seam was mined in the underground workings, while the upper (C_u) seam was opencasted over the old workings and both the seams were opencasted away from the underground workings. The pillar between the underground and opencast workings is about 15 m, and the opencasting did not break into the underground workings. Seepage through the C_l seam in the highwall was not significant, but acid seepage into the footwall of the opencast workings and in the adjacent valley was a problem during the opencast operation.

Investigation

The desk study that preceded the investigation proper involved the examination of maps, drawings, previous water quality analyses and copies of relevant correspondence. It showed that the information on the surface and underground activities of the mine was not well correlated and it was accordingly decided to place the information from the different sources onto a GIS database in order to facilitate interpretation and later planning of remedial measures.

The surface water hydrological analysis concentrated on the area comprising the plant and the restored

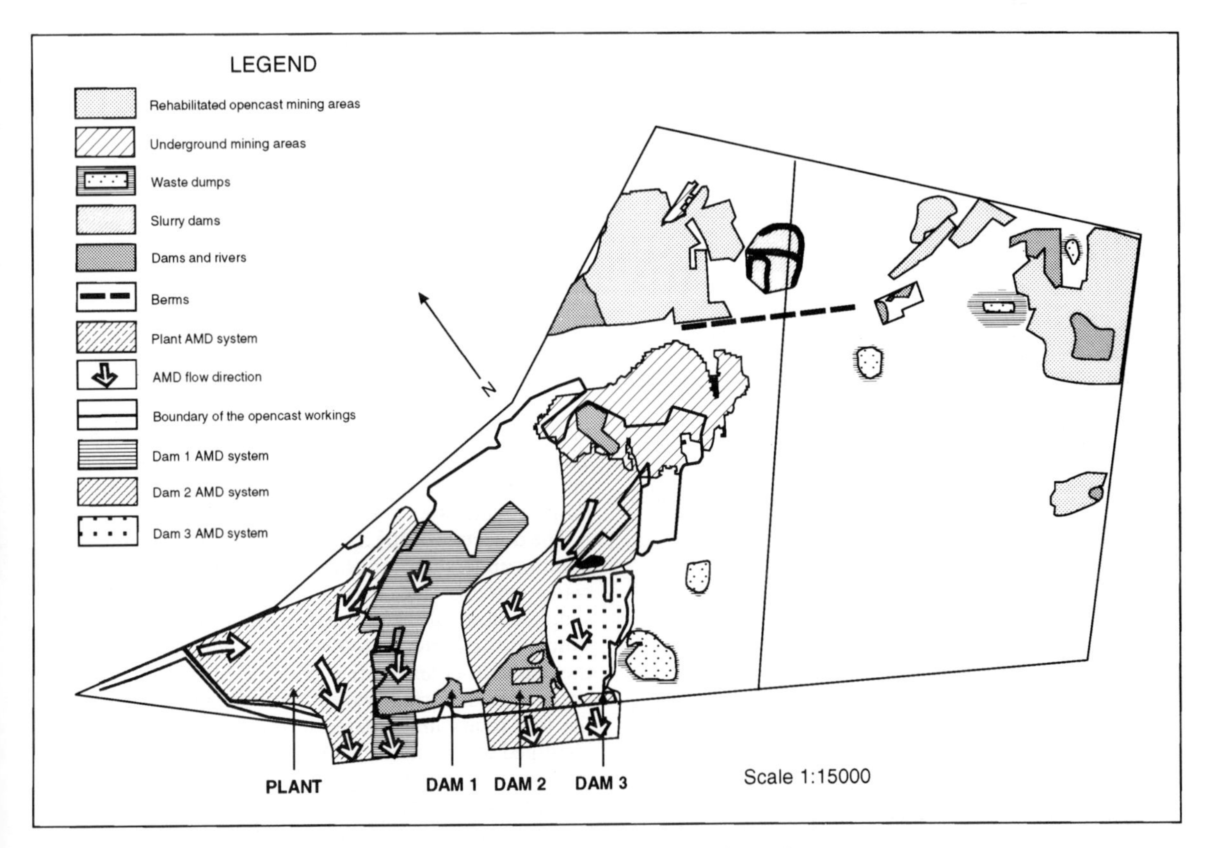

Fig. 6. Seepage flow systems at Loubert Mine.

Table 6. *Analysis of water from four sampling sites*

Date	*Rain (mm)*	*pH*	*EC*	SO_4	*Zn*
1. Human Spruit upstream of mine					
13/3/91		7.28	33	25	
25/6/91		8.28	52	95	
13/11/91		5.88	39	146	0.10
22/11/91		7.78	40	18	0.01
26/11/91		6.58	35	55	0.11
4/12/91	38.88	8.88	24	36	0.04
5/12/91	4.38				
9/12/91	3.58				
10/12/91	30.00				
11/11/91	22.88				
19/12/91	9.88	6.58	27	46	0.01
17/12/91	10.88				
19/12/91		6.68	35	39	0.02
3/1/92		5.78	42	17	0.03
6/1/92	35.00				
13/1/92	12.00				
16/1/92	22.00	7.68	42	162	0.02
21/1/92	2.00				
25/1/92	6.00				
28/1/92	2.50				
2. Human Spruit downstream of the reservoir					
23/3/88		3.40	150	900	
7/2/91		3.50	39	70	
13/3/91		4.30	65	275	
25/6/91		3.20	176	968	0.59
13/11/91		2.38	398	9582	1.05
13/11/91		2.00	216	1200	0.68
20/11/91		2.58	288		
22/11/91		2.48	233	1269	0.66
26/11/91		2.18	400	2859	0.74
4/12/91	30.88	3.98	48	154	0.06
5/12/91	4.58				
9/12/91	3.58				
10/12/91	38.00				
11/12/91	22.00				
12/12/91	9.80	3.68	72	374	0.11
17/12/91	1800				
19/12/91		2.00	237	1952	0.97
3/1/92		2.00	437	3232	1.92
6/1/92	35.00				
7/1/92		2.06	482	2879	0.68
13/1/92	12.00				
16/1/92	22.00	2.00	112	553	0.24
21/1/92	2.08				
25/1/92	6.00				
28/1/92	2.58				
3. Overflow from central reservoir (Dam 2)					
23/3/90		2.68	397	3000	
7/2/91		3.38	75	300	
13/2/91		2.50	462	275	
25/6/91		2.78	520	4600	4.68
4/9/91		2.00	567	5002	3.87
13/11/91		2.00	464	4850	3.18
20/11/91		2.10	468		
22/11/91		2.00	498	4413	4.00
26/11/91		2.00	584	4299	4.47
4/12/91	38.00	2.00	372	4336	3.33

Table 6. (*continued*)

Date	*Rain (mm)*	*pH*	*EC*	SO_4	*Zn*
3. Overflow from central reservoir (Dam 2) (continued)					
5/12/91	4.58				
9/12/91	3.58				
10/12/91	30.66				
11/12/91	22.00				
12/12/91	9.00	2.38	452	4387	2.77
17/12/91	10.00				
19/12/91		2.00	416	4671	3.81
3/1/92		2.18	362	4373	3.38
6/1/92	35.00				
7/1/92		2.00	368	4509	3.49
13/1/92	12.00				
16/1/92	22.00	2.00	493	4428	3.44
21/1/92	2.00				
25/1/92	6.00				
28/1/92	2.50				
4. Overflow from mine return water reservoir (Dam 1)					
4/9/91		2.38	327	2196	0.35
13/11/91		2.28	263	1537	0.61
22/11/91		2.38	288	1006	0.66
26/11/91		2.28	283	1758	0.74
4/12/91	30.00	2.18	268	1978	0.55
5/12/91	4.58				
9/12/91	3.58				
10/12/91	30.88				
11/12/91	22.00				
12/12/91	9.00	2.58	257	1786	0.45
17/12/91	10.00				
19/12/91		2.10	228	1948	0.42
3/1/92		2.20	344	1999	0.62
6/1/92	35.00				
7/1/92		2.10	371	2183	0.68
13/1/92	17.00				
16/1/92	22.00	2.20	363	2363	0.72
21/1/92	2.00				
15/1/92	6.08				
28/1/92	2.58				

opencast workings to the north of the Human Spruit. This area, of approximately 187 ha, was identified as the major source of pollution to the Human Spruit from both groundwater and surface water flows.

The objective of the hydrological analysis was to determine an understanding of the existing surface and groundwater systems and their interaction. Consequently, a simple empirical water balance model was developed to simulate the current mine water system and to quantify the effluent entering the Human Spruit, taking into account both surface runoff and seepage. Spillage from the two reservoirs situated in the south of the site was identified as the major source of polluted water discharging into the Human Spruit. The hydrological analysis therefore concentrated on the development of a model to simulate the variation in the monthly spillage from these reservoirs. The model based on monthly water balance took into account the climatic conditions, topography, mine infrastructure, surface water runoff and groundwater seepage flow direction. Empirical methods were used to determine the runoff, infiltration, seepage and evaporation from the catchments and mine infrastructure.

Electromagnetic (EM) geophysical surveys were employed to map the likely positions of near-surface groundwater pollution plumes and seepage pathways in the backfilled opencast mine areas to the east of the plant. The objective of the mapping was to provide additional information on the possible nature, depths and directions of polluted near-surface flow in the backfill and to provide possible target sites for location of monitoring boreholes. The intention was to use EM data to better define the hydrogeological model for acid mine drainage and to incorporate the results into the rehabilitation plan. The EM survey allowed the identification of the boundaries of the opencast and underground mining areas, as well as identifying the major polluted groundwater flow directions (Fig. 6). It also helped confirm that the old mining plans were essentially correct.

Because of the very coarse character of the backfill in the backfilled opencast areas, voids are present in the backfill. Where these voids are interconnected, they form open flow channels where the flow is very rapid. The major flow during times when the system is recharged by rainfall or stormwater runoff, then occurs through these conduits. Pollution tends to seep out of the backfill slowly and degrades the quality of the water present in the conduits through dispersion, diffusion and the dissolution of pollutants. Therefore the quality of water exiting in springs and seepage points shows an improvement during high infiltration periods but the quality deteriorates when flows decrease. This has a significant impact on the options available for rehabilitation of such systems.

Four sample sites were chosen to monitor water quality. These were on the Human Spruit upstream of Loubert mine; the Human Spruit downstream of Loubert mine; overflow from the central reservoir; and overflow from the mine return water reservoir (i.e. Dam 1 on Fig. 6). A summary of the results is given in Table 6. The results confirm that the site as a whole is strongly acid-generating and that there is little difference in the quality of water from different surface sources.

When the rainfall over the period of the study is examined in conjunction with the other data it can be seen that it has no effect on the quality of the upstream water in the Human Spruit. This sampling point was upstream of all the coal workings. On the other hand, rainfall had a significant effect on the downstream water samples, especially on the conductivity and dissolved salts, while the pH is not as significantly improved during rainfall events. In other words, an inverse relationship exists between rainfall and conductivity. Rainfall has an apparently negligible effect on the quality of the water spilling from the central reservoir. This indicates that the rinsing effect of the flow through the backfilled opencast areas is not effective in promoting good quality runoff during moderate rainfall events. In the same way, rainfall has an apparently minimal effect on the water spilling from the plant return water reservoir (Dam 1).

It was concluded that the site is acid-generating except for the restored area to the southeast of the mine (Portion 22). At present, the major load of pollutants is generated by the restored and semi-restored opencast mining areas to the immediate southeast of the mine where most of the underground and opencast mining was carried out by a previous mine owner (Portion 89). Pollution emanating from the partially restored opencast and underground mining areas of Portion 24 is a threat to the Human Spruit, although the load to the stream appears to be minimal at present.

The water which infiltrates the backfill in the rehabilitated opencast area percolates through rock material containing sulphides; water also infiltrates through coarse discard placed on site and through the fine discard of the slurry ponds. It emerges mainly in the reservoirs and as seepage in the stream. Quality is poor, with conductivity ranging from 350 to 570 $mS\,m^{-1}$ for the central reservoir (Dam 2; TDS approximately 3500–5700 $mg\,l^{-1}$) and 230–370 $mS\,m^{-1}$ for the plant return reservoir (Dam 1; TDS approximately 2300–3700 $mg\,l^{-1}$. With annual spillage from the two reservoirs projected to be in the region of 300 000 m^3, this converts to a load of approximately 1200 $t\,a^{-1}$. These figures do not include the potential contribution to the system by inflow from the underground workings.

Pollution control options for acid generation

Options for controlling acid generation are limited since the site is no longer in operation. It would not be economical to attempt to remove the sulphide-bearing material from the site. In addition, exclusion of oxygen by the use of a water cover is not practical, taking the nature of the topography and low rainfall into account. Exclusion by the use of an oxygen-impermeable liner for the whole site or the worst parts of the site is feasible but would be very expensive and subject to practical limitations in terms of making such a cover permanently effective. In the same way the exclusion of water from the system would be costly and difficult. Accordingly it was recommended that acid generation control should not be considered.

Hence attempts would have to be made to minimize the movement of the acid drainage once it has formed. Measures to do this are essentially based on controlling the water inflow into the system and in so doing constraining the transport medium. Options include the following:

- Diverting clean surface runoff from the veld. The stormwater drainage system installed at the mine

- could be upgraded relatively easily. An important aspect of the upgrade would be to prevent the water from entering the two reservoirs (Dams 1 and 2), as it does at present.
- The mine should use the water from both these reservoirs for plant operation. At present, only the water from one reservoir (Dam 1) is used. This would have the effect of reducing the volume of water in the reservoirs and therefore reducing the spillage and seepage.
- Inflow points to the backfill should be sealed, as should be the opening in the bed of the clean stormwater channel (Fig. 4). Sealing could be carried out with a number of materials. There is a possibility that locally available pulverised fly-ash (PEA) could be used for this purpose because of its alkalinity.
- Ponding of water on the backfilled areas should be eliminated. This water becomes acid and infiltrates into the groundwater system, leading to an increased pollution load and better transport of the acid mine drainage. The amount of earthworks required to achieve this would be relatively small.
- Runoff from the restored areas should be promoted. If the quality of the runoff, or of runoff from identified areas is acceptable, it should be diverted out of the system. Implementation would require monitoring to determine the success of the measures.
- Impermeable liners should be placed on part of the restored area if the previous measures do not give an acceptable reduction in pollution load.
- Attempts should be made to cut off or isolate portions of the groundwater system if the load reduction is still not acceptable. It may be that the inflow from the underground workings would need to be isolated in this way.

Collection and treatment of the remaining seepage and spillage load may be considered as part of an interim measure, or less preferably as a long-term option. In the latter case attention would have to be given to financial guarantees to sustain the treatment after closure of the mining operation.

Conclusions

Acid mine drainage is associated with coal and metalliferous mines, and is due to the oxidation of sulphide minerals in mine rock or waste which are exposed to air and water. It can lead to both ground and surface water pollution with elevated levels of heavy metals and sulphate in the water. Accurate prediction of acid mine drainage is needed if control measures are going to be successful. Hence any investigation which is carried out to determine the source(s) of acid mine drainage and its character obviously involve water sampling and analysis, together with determination of the direction of water movement, notably that of groundwater. Prediction of the potential for the generation of acidic waters not only involves data collection but is likely to involve static, and may be kinetic tests.

The investigation allows an acid mine drainage strategy to be developed. This can involve an attempt to control the process of acid generation. Alternatively an attempt can be made to control the migration of acidified waters. Thirdly, acid mine drainage can be collected and treated. At the tin mine, the primary source of acid mine drainage was not difficult to recognize, it being the material in the pyrite dumps. However, the general impermeable character of the subsurface rocks, the groundwater gradients and compartmentalization by the dyke system suggested that it would be unlikely for the affected water to migrate beyond the site. Be that as it may, the pyrite dumps (i.e. the sources of the pollution) were removed. Similarly, the sources of acid mine drainage were identified at the coal mine. This time it was concluded that the remedial measures should involve an attempt to control the movement of acid mine drainage since its generation could not be prevented.

References

ANON 1993. *South African Water Quality Guidelines, Volume 1: Domestic Uses*. Department of Water Affairs and Forestry, Pretoria,.

BRODIE, M. J., BROUGHTON, L. M. & ROBERTSON, A. 1989. A conceptional rock classification system for waste management and a laboratory method for ARD prediction from rock piles. *British Columbia Acid Mine Drainage Task Force, Draft Technical Guide*, **1**, 13–135.

CONNELLY, R. J., HARCOURT, K. J., CHAPMAN, J. & WILLIAMS, D. 1995. Approach to remediation of ferruginous discharge in the South Wales coalfield and its application to closure planning. *Minerals Industry International, Bulletin Institution Mining and Metallurgy*, **1024**, 43–48.

HALBERT, B. E., SCHARER, J. M., KNAPP, R. A. & GORBER, D. M. 1983. Determination of acid generation rates in pyritic mine tailings. *In: Proceedings 56th Annual Conference of Water Pollution and Control Federation, Atlanta*. Offprint.

LUNGREN, D. G. & SILVER, D. 1980. Ore leaching by bacteria. *Annual Review of Microbiology*, **34**, 263–283.

PUGH, C. E., HOSSNER, L. R. & DIXON, J. B. 1984. Oxidation rates of iron sulphides as affected by surface area, morphology, oxygen concentration and autotrophic bacteria. *Soil Science*, **137**, 309–314.

GIS techniques for mapping and evaluating sources and distribution of heavy metal contaminants

C. Fragkos,[1] M. S. Rosenbaum,[2] M. H. Ramsey[3] & K. L. Goodyear[3]

[1] Department of Geography, King's College London, Strand, London WC2R 2LS, UK
[2] Faculty of Environmental Studies, The Nottingham Trent University, Burton Street, Nottingham NG1 4BU, UK
[3] Environmental Geochemistry Research Group, T H Huxley School of the Environment, Earth Science & Engineering, Imperial College of Science Technology and Medicine, Prince Consort Road, London SW7 2BP, UK

Abstract. Spatially dependent phenomena related to heavy metal concentration in a drainage system have been evaluated using GIS methodology, so providing an indication of areas which may contain contaminated land. The geochemical distribution is used to identify the heavy metal sources. The GIS facilitates dispersion modelling defined by a digital elevation model. This can be used to delineate drainage catchments and to examine its interaction with stream sediment geochemistry. The resulting drainage basin-segment images can then be used to identify heavy metal sources. Such modelling of geochemical dispersion of metals within drainage basins allows predictions to be made as to which areas of contamination originate from which point source. The accuracy of such maps depends fundamentally on the accuracy of the dispersion model. This model includes both descriptive and deductive components concerning the influential criteria for describing the distribution of heavy metals. Each criterion is defined using an interactive discussion process that leads to the generation of a set of weighting coefficients. The integration of such criteria is achieved using fuzzy set methodology to produce images ranked according to how well they meet the selected site assessment criteria.

Introduction

A Geographical Information System (GIS) is designed to manage spatial information so that it can support decision-making processes based on spatial data (Maguire *et al.* 1991; Bonham-Carter 1994). Many kinds of environmental phenomena are spatially dependent, and so the application of GIS is an attractive tool for their analysis (Townshead 1991).

Mathematical modelling, using map algebra, can be performed within such systems. The simplest such model is based on Boolean operators (Varnes 1974; Robinove 1989) whereby only two possible conditions are recognized, true or false, i.e. a model based on Boolean logic will either include or exclude an element from a set of information. Within a GIS it is also possible to develop models that incorporate uncertainty by the use of the fuzzy set theory. This can be achieved by ascribing a permissible outcome to a set of observations, the membership of the outcome being based on subjective judgement (Zadeh 1965; Zimmerman 1985).

Objectives

The identification of areas containing high heavy metal concentrations, including knowledge of their sources, is of great importance for the assessment of land contamination (Ramsey *et al.* 1994). It is therefore necessary to identify the location and origin of contamination sources within such areas of elevated concentrations.

The distribution of heavy metal concentration within a drainage system can be estimated using stream sediment samples since these represent the material carried by water downstream within the catchment area. The subsequent identification of heavy metal sources then enables areas to be identified where contamination is coming from natural or undocumented anthropogenic sources. In order to apply a GIS to this process, three objectives were addressed:

- Determine the general distribution of heavy metals within segments of a drainage system.
- Map and classify segments where heavy metal concentrations exceed a threshold which indicates the possibility of that area containing contaminated land.
- Identify the various possible sources of the heavy metals and thereby establish their origin as being either natural or anthropogenic.

This work develops the approach of Ramsey *et al.* (1994) by the development of more realistic models for geochemical dispersion aimed at enabling them to identify contamination sources and the possible extent

Fragkos, C., Rosenbaum, M. S, Ramsey, M. H. & Goodyear, K. L. 1998. GIS techniques for mapping and evaluating sources and distribution of heavy metal contaminants. *In*: Maund, J. G. & Eddleston, M. (eds) *Geohazards in Engineering Geology*. Geological Society, London, Engineering Geology Special Publications, **15**, 365–372.

of land contamination. These objectives have been tackled by developing a geochemical dispersion model using GIS modelling techniques and by evaluating the available sets of information which could influence the ground contamination by applying fuzzy logic methodology.

Case study: the Allen Basin in southwest Cornwall

The Allen Basin is located north of Truro in southwest Cornwall. This has provided a study area suitable for establishing a geochemical dispersion methodology. The reasons for selecting this case study were as follows:

- the availability of detailed geochemical information at high sampling density;
- the existence of both natural and anthropogenic sources in the same drainage basin, creating areas of contaminated land;
- the availability of detailed descriptions of the conditions that control the distribution of heavy metals and the location of several contamination sources within the particular area (Jones & Tombs 1976; Gianni Gali 1985).

The data set is available from a geochemical survey undertaken with the purpose of identifying the location and origin of heavy metal contamination sources (Goodyear *et al.* 1996). Information regarding heavy metal concentrations, notably of lead, zinc and copper, in the fine ($<170\,\mu$m) fraction of stream sediments is available, together with information on heavy metal concentrations from water samples taken at the same sampling sites.

The approach that has been developed to tackle the three objectives listed previously has been achieved using the following steps:

1. Create the GIS database incorporating coverages of the geochemically related information (stream sediment samples, water samples, geology, land use, vegetation, known sources of heavy metal contamination, topography, slope).
2. Develop a digital elevation model (DEM) for the study area necessary to compute the individual segments within the drainage basin above each sample point, and then assign the heavy metal concentrations to each segment. The DEM has been computed from digitized topographic contours.
3. (a) Develop a dispersion and contamination source identification model, based on the theoretical model proposed by Hawkes (1976).
   (b) Define influential criteria from the available sets of information, i.e. identify the degree of influence of each set of information to the dispersion model.
   (c) Generate decision rules that can be translated into mappable factors, i.e. quantify the predetermined influential criteria to reflect the perceived influence of each factor on the geochemical dispersion.
   (d) Merge the factor images together, using fuzzy set methodology to incorporate uncertainty regarding their degree of influence.
   (e) Assign known sources to observed areas of contamination, where possible.
   (f) Generate maps of areas which could contain potentially contaminated land, ranked according to how well they meet the predetermined criteria.

The dispersion model

The identification of sources of contamination (step 3(a) above) depends upon realistically modelling the geochemical dispersion of contaminants from their sources. A theoretical downstream decay model for stream sediment dispersion has been proposed by Hawkes (1976). This has been adapted for use within a GIS. The model was initially designed for the localization of geochemical anomalies in mineral exploration. It is based on the concept that as erosion removes material from within the drainage basin, the stream sediment will yield a representative sample of all the surface material coming from the basin provided that erosion is occurring at an equal rate. At specific sample points, the metal content of a sediment sample (Me_s) may be found by weighting the metal contents of the mineralised and background regions according to their respective areas:

$$Me_s A_t = Me_m A_m + Me_b(A_t - A_m) \quad (1)$$

where:

Me_s = predicted metal concentration at the sample point at site 's' ($\mu g\, g^{-1}$);
A_t = total area of the basin draining into the sample point (km^2);
Me_m = concentration at a known source of contamination within the drainage basin ($\mu g\, g^{-1}$).;
A_m = area of the known source of contamination (km^2);
Me_b = background metal concentration, away from any contamination ($\mu g\, g^{-1}$).

Equation (1) may be rearranged to give:

$$Me_s = (Me_m + Me_b)\frac{A_m}{A_t} + Me_b \quad (2)$$

The development of the GIS model, based on equation (2) is presented in outline in Fig. 1. This first

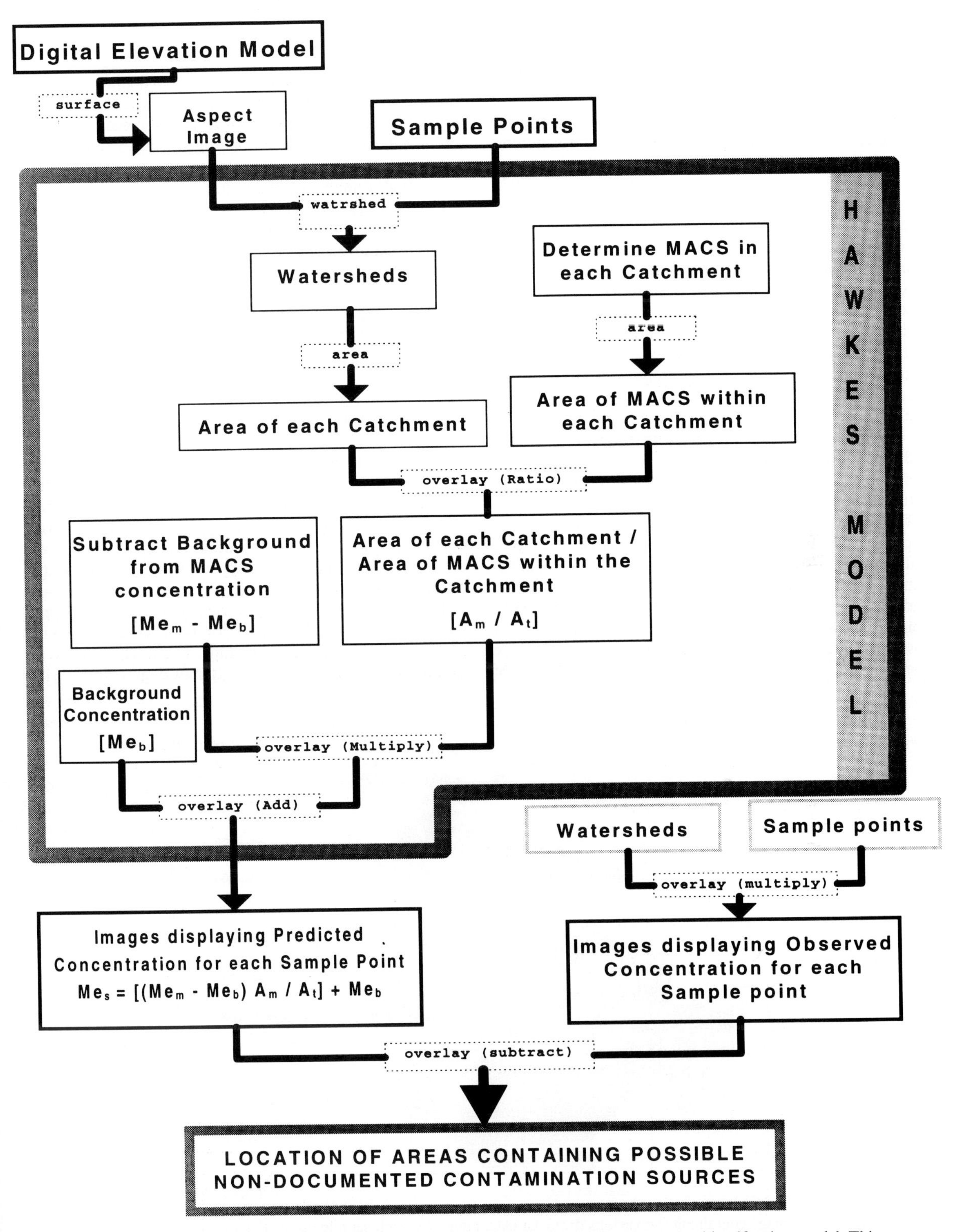

Fig. 1. Flow diagram illustrating the GIS methodology used for developing the source identification model. This is based on Hawkes' (1976) hypothesis for describing downstream decay of geochemical sediment anomalies. The various steps of the process are contained in solid boxes, and the GIS commands using Idrisi (Eastman 1993) are contained in the dotted boxes.

requires the generation of a digital elevation model (DEM), which can be computed from topographic contours. The DEM can then be used to compute the surface drainage and so define the drainage catchments of each sample point using 'watershed' algorithms (Peucker & Douglas 1985; Merkel & Sperling 1993). This leads to the calculation of the area of each catchment. The next stage assigns the heavy metal concentration values to each known contamination source ('Mining Activity Contamination Site', MACS).

A comparison of the model's prediction with the observed values of heavy metal concentration at each drainage segment can provide an important indication of the existence of non-documented, sources of heavy metal within the catchment. Discrepancies between the predicted and the observed concentrations may also be due to errors in the field measurements or limitations of the dispersion model. Work is in progress to quantify and reduce such sources of errors. The aim is to evaluate their origin, i.e. discriminate between anthropogenic and natural contamination sources.

The metal initially used as the basis for developing the dispersion model was zinc, selected because it is generally the most soluble and mobile of the three chemical elements for which adequate data were available, i.e. lead, zinc and copper (Rose *et al.* 1979: 555, 560, 580). Zinc (as shown in Fig. 2) is here used as an indicator for possible contamination sources contributing to drainage systems and not as a direct indicator of the degree of heavy metal contamination at any particular site. It is important therefore to emphasize that the information regarding the concentration of zinc within the study area has just been used for modelling purposes and not as evidence of how contaminated a particular sampling point or drainage catchment happens to be.

Later stages of this work will consider the role and contribution of heavy metal sources to the possibility of creating ground contamination.

Incorporating uncertainty using fuzzy set theory

The methodology used hitherto for dispersion modelling has not incorporated uncertainty or any measure regarding possible error propagation. The incorporation of uncertainty is believed necessary for decision-making methodologies to be successfully applied in practice (Spiegelhalter 1986). The fuzzy logic approach offers a possible framework appropriate for incorporating uncertainty, enabling ground to be classified according to the degree of importance of factors to the dispersion problem.

The approach using fuzzy logic is based on defining variable boundary thresholds for a set which represents a possible state in the ground. The inclusion of an element of information is a matter of degree and partial

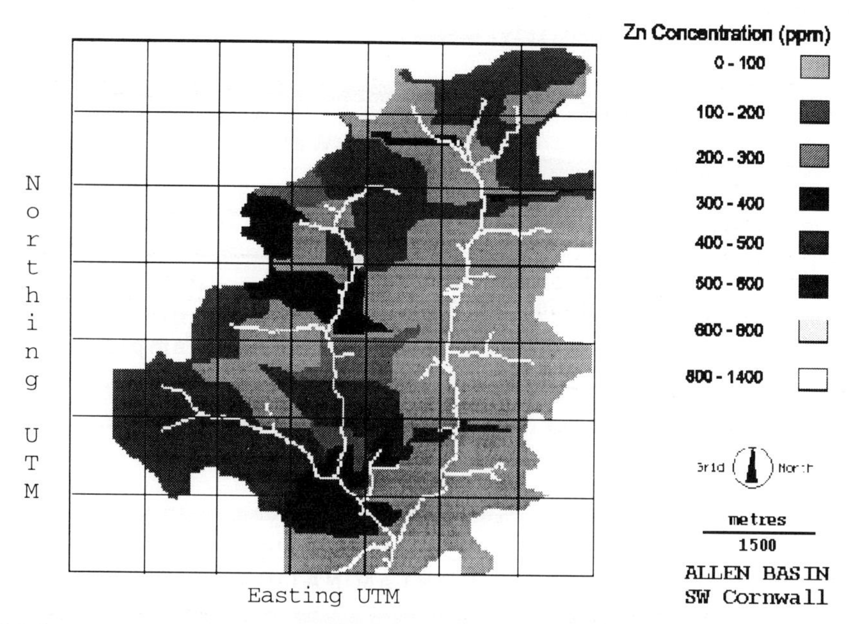

Fig. 2. GIS image containing the various drainage catchments for each sample point. Each drainage catchment has been assigned the zinc concentration value at its respective sample point and classified accordingly (the grid has 1 km spacing).

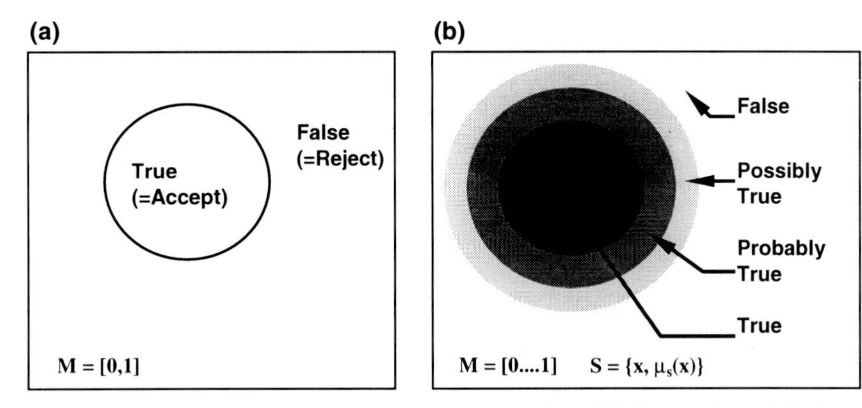

Fig. 3. Diagram illustrating the differences between (**a**) Boolean logic and (**b**) fuzzy set logic. M is the membership of the fuzzy set S where x is the value of the parameter and $\mu_s(x)$ is the fuzzy membership value of the set.

membership of an element is permitted, as shown in Fig. 3. The concept can be described by comparing Boolean logic with fuzzy logic. For Boolean logic the membership (M) of a set is defined as true or false (i.e. 1 or 0) and deals only with definite statements. For fuzzy logic, membership values $\{\mu_s(x) \in M = [0 \dots 1]\}$ of a fuzzy set recognize a range between, and including, non-membership ($\mu_s = 0$), full membership ($\mu_s = 1$), and possible membership ($0 < \mu_s < 1$, where μ_s is the fuzzy membership value of a set), and can therefore deal with statements of possibility. Every value of x is associated with a value $\mu_s(x)$, and the ordered pairs $[x, \mu_s(x)]$ are known collectively as a 'fuzzy set'.

This approach then continues by using the empirically chosen weighting coefficients, derived from the analysis of the observations. This process enables the membership of a specific class of behaviour to be assigned on a systematic basis, using a knowledge of the processes involved. Consequently, it enables the calculation and combination of estimates of the relative importance for each item of information.

The shape of a fuzzy set membership function is shown in Fig. 4. This can take on a variety of shapes such as 'linear' (Fig. 4(a)) and 'stepped' (Fig. 4(b)). Fuzzy membership functions may also be S-shaped ('sigmoidal') or J-shaped (Schmucker 1982; Burrough 1989). In addition, fuzzy membership functions can be expressed as numerical tables whereby classes for any layer of information can be associated with a corresponding fuzzy membership value; GIS can incorporate such information as tables of attribute values.

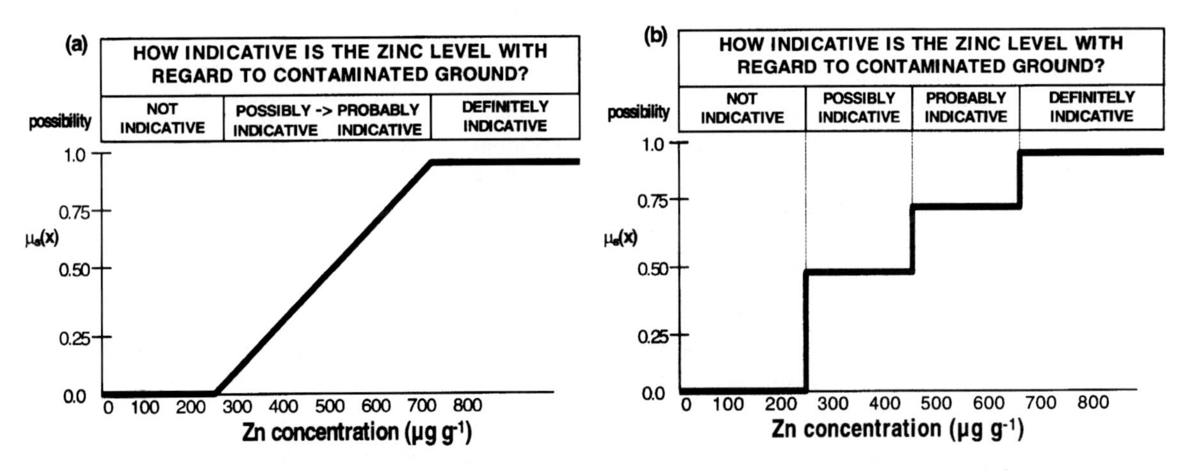

Fig. 4. Conceptual diagrams illustrating (**a**) a 'linear' fuzzy membership function and (**b**) a 'stepped' fuzzy membership function.

Fuzzy membership values must lie in the range from 0 to 1 and are chosen to reflect the degree of membership of each set. The choice is based on subjective judgement, using a group selection process as described elsewhere (Rosenbaum 1995).

The presence of a possible state existing within the mapped area can be expressed in terms of fuzzy membership, and thus be stored as a specific field within the map attribute table for the GIS. For example, the values of zinc in the catchment can be represented as fuzzy numbers in terms of their membership of two fuzzy sets: the first fuzzy set is a 'favourable indicator for identifying heavy metal contamination sources', and the second set is a 'possible indicator for identifying heavy metal contamination sources'. The membership functions for these two sets need not look the same.

Each fuzzy set is weighted according to the value of its perceived contribution to the calculation of the membership function for the combined model. In the simple case of zinc given above, the two fuzzy sets representing the indicators of heavy metal concentration would provide a combined model. This can be achieved by applying the following sequence of operations:

1. Decide on the fuzzy set membership functions and values appropriate for classifying the risk of each area containing a high heavy metal concentration.
2. Assign an appropriate weighting for each fuzzy set membership function for every feasible class. This requires decisions regarding the possibility of occurrence of heavy metal ground contamination on the basis of the worth of each contributory criterion, i.e. set of information.
3. Establish the relative importance of each criterion to the assessment of possibility of heavy metal contamination on a scale from 9 to 1/9, from which a weighting coefficient describing relative importance can be calculated. The relative importance is optimized and normalized so that the coefficients sum to 1.0. These coefficients can then be applied on each criterion by multiplication.
4. Identify the maximum value at each pixel regarding the degree of possibility of heavy metal ground contamination being contained within the area, based on a ranking of all the feasible fuzzy sets. The closer to 1.0, the greater will be our belief that the ground at the location of the pixel contains an area of high heavy metal concentration.

After establishing these fuzzy relationships, the fuzzy sets can then be combined using one or more of the following fuzzy operators (Eastman *et al.* 1993; Bonham-Carter 1994). The first fuzzy operator is the 'fuzzy algebraic product', defined as:

$$\mu_s(x) = \prod_{i=1}^{n} \mu_i \tag{3}$$

where μ_i is the fuzzy membership function for the $i = 1 \ldots n$ sets of information. Given that numbers with values less than 1 have been multiplied, the result will be smaller than, or equal to, the smallest contributing membership value.

The second fuzzy operator is the 'fuzzy algebraic sum', defined as:

$$\mu_s(x) = 1 - \prod_{i=1}^{n} (1 - \mu_i) \tag{4}$$

In this case the result is always larger than, or equal to, the largest contributing fuzzy membership value.

The third fuzzy operator is the 'gamma' operator, which is a combination of the fuzzy algebraic product and fuzzy algebraic sum operators previously described, given as:

$$\mu_s(x) = (\text{fuzzy algebraic sum})^{\gamma} * (\text{fuzzy algebraic product})^{1-\gamma} \tag{5}$$

where γ is a parameter chosen in the range 0 to 1 (Zimmerman & Zysno 1980). When γ is 1 the combination is the same as the fuzzy algebraic sum, and when it is 0 the combination equals the fuzzy algebraic product. The 'gamma' fuzzy operator allows a more flexible approach than does either the fuzzy algebraic sum or the fuzzy algebraic product because the sensitivity of the contribution can be adjusted by judicious choice of γ.

Discussion

Uncertainty existing within the particular data set may concern the accuracy of the measurements, the extent of the drainage catchments, the distribution of the metals in each particular catchment, and the degree of sediment contribution of a point within a catchment that affects the metal concentration value measured at the sample point corresponding to the catchment (Kern & Stednick 1993).

Such uncertainties can be quantified with respect to the degree of importance of each factor concerning the possibility of an area containing high heavy metal concentrations, i.e. possible areas of land contamination. The different sets of information used for generating the classification map to identify the possibility of heavy metal contamination being present can be summarized as follows:

- A fuzzy set containing the zinc concentration based on the stream sediment samples, classified in relation to the median of figures cited in Wedepohl (1978) for sandstones and quartzites (the local bedrock is predominantly shales and sandstones).
- A fuzzy set concerning measurements of zinc concentration based on the water samples taken at the same sampling sites as the stream sediments. This fuzzy set has been assigned a moderate weighting

coefficient taking into account the fact that metal concentration values in stream waters can be subject be seasonal changes.

- A fuzzy set concerning the mapped anthropogenic sources of contamination (MACS) for which measurements on zinc concentrations are available. This set has been classified according to the measured concentration values, giving a higher weighting coefficient to sites with very high values (e.g. 2040 $\mu g\,g^{-1}$ Zn) than to sites with low values (e.g. 105 $\mu g\,g^{-1}$ Zn); the upstream areas are excluded from this buffering process.
- A similar fuzzy set to that of the MACS concerning the distance from mineralized (Pb–Zn) veins; the upstream areas are again excluded from this buffering process.

Further fuzzy sets can be used to indicate the possible influence of the presence of high heavy metal concentrations, based on (a) distance from water courses, (b) distance from farm buildings, (c) distance from residential areas and (d) distance from woodland.

Conclusions

The methodology has demonstrated that GIS can be a powerful tool for the description and evaluation of the dispersion of heavy metals in the Allen Basin drainage system. The model developed here provides primarily a means of identifying the location and origin of possible sources of heavy metal contamination. The GIS analysis of the available data has suggested the presence of areas containing non-recorded sources of heavy metals dispersing into the stream sediments.

Subsequently the results of such a process can prove very useful in the overall assessment of the possibility of the occurrence of areas with elevated concentrations of metals. It is necessary to emphasize that the process needs further refinement. Nevertheless, problems of uncertainty in such an assessment process can be tackled using fuzzy set methodology.

A methodology based on fuzzy set logic has been applied to incorporate uncertainties relating to the use of zinc as an indicator for areas containing high heavy metal concentrations and thus of areas potentially contaminated by heavy metals. Such a methodology allows the incorporation of diverse influential criteria in the decision-making process and thereby reinforces any known indications of high metal concentrations being present, suggested by the application of a source identification model.

In future stages of the work, a testing and feedback methodology needs to be developed by the comparison of different geochemical modelling techniques and integrating additional necessary information into the already assembled data set for the particular study area.

References

BONHAM-CARTER, G. F. 1994. *Geographic Information Systems for Geoscientists: Modelling with GIS*. Pergamon, New York.

BURROUGH, P. A. 1989. Fuzzy mathematical methods for soil survey and land evaluation. *Journal of Soil Science*, **40**, 477–492.

EASTMAN, J. R. 1993. *IDRISI Version 4. 1 Update Manual.* Graduate School of Geography, Clark University, Massachusetts.

——, KYEM, P. A. K., TOLEDANO, J. & JIN, W. 1993. GIS and decision making. *Exploration in Geographic Information System Technology*, 4. UNITAR, Geneva.

GIANNI GALI, N. 1985. Sample acquisition and interpretation of analytical results in a fine fraction stream sediment survey, River St Allen, Truro, Cornwall. *Bolletino, Societa Geologica Italiana*, **104**, 217–221.

GOODYEAR, K. L., RAMSEY, M. H., THORTON, I. & ROSENBAUM, M. S. 1996. Source identification of Pb–Zn contamination in the Allen Basin, Cornwall, S.W. England. *Applied Geochemistry*, **11**, 61–68.

HAWKES, H. E. 1976. The downstream dilution of stream sediment anomalies. *Journal of Geochemical Exploration*, **6**, 345–358.

JONES, R. C. & TOMBS, J. M. C. 1976. Geochemical and geophysical investigations around Garras mine, Cornwall. *Transactions of the Institute of Mining and Metallurgy*, Section B, **85**, 158.

KERN, T. J. & STEDNICK, J. D. 1993. Identification of heavy metal concentrations in surface water through coupling of GIS and hydrochemical models. *In*: KOVAR, H. & NACHTNEBEL, H. P. (eds) *Applications of GIS in Hydrology and Water Resources Management*. IAHS Publication No. **211**, 559–570.

MAGUIRE, D. J., GOODCHILD, M. F. & RHIND, D. W. 1991. An overview and definition of GIS. *In*: MAGUIRE, D. J., GOODCHILD, M. F. & RHIND, D. W. (eds) *Geographical Information Systems: Principles and Applications*. Longman Scientific & Technical, Harlow.

MERKEL, B. & SPERLING, B. 1993. Raster based modelling of watersheds and flow accumulation. *In*: KOVAR, H. & NACHTNEBEL, H. P. (eds) *Application of GIS in Hydrology and Water Resources Management*. IAHS Publication No. **211**, 193–200.

PEUCKER, T. K. & DOUGLAS, D. H. 1985. Detection of surface specific points by local parallel processing of discrete terrain elevation data. *Computer Graphics and Image Processing*, **4**(3), 375–378.

RAMSEY, M. H., HARTLEY, G. H. & ROSENBAUM, M. S. 1994. Interpretation and source identification of heavy metal contamination of land using Geographical Information Systems (GIS). *In*: COTHERN, C. R. (ed.) *Trace Substances, Environment and Health*. Science Reviews, Northwood, 95–104

ROBINOVE, C. J. 1989. Principles of logic and the use of digital geographic information systems. *In*: RIPLE, W. J. (ed.) *Fundamentals of Geographical Information Systems: A Compendium*. American Society of Photogrammetry and Remote Sensing, 61–80.

ROSE, A. W., HAWKES, H. E. & WEBB, J. S. 1979. *Geochemistry in Mineral Exploration*. Academic, London.

ROSENBAUM, M. S. 1995. Coping with uncertainty in ground engineering. *Proceedings for the Engineering Geology and the Channel Tunnel Symposium*. University of Brighton.

SCHMUCKER, K. J. 1982. *Fuzzy Sets: Natural Language Computations and Risk Analysis*. Computer Science Press, Rockville, MD.

SPIEGELHALTER, D. J. 1986. Uncertainty in expert systems. *In*: GALE, W. A. (ed.) *Artificial Intelligence and Statistics*. Addison-Wesley, Reading, MA, 17–55.

TOWNSHEAD, J. R. G. 1991. Environmental databases and GIS. *In*: MAGUIRE, D. J., GOODCHILD, M. F. & RHIND, D. W. (eds) *Geographical Information Systems: Principles and Applications*. Longman Scientific & Technical, Harrow.

VARNES, D. J. 1974. *The logic of maps with reference to their interpretation and use for engineering purposes*. United States Geological Survey Professional Paper 837.

WEDEPOHL, K. H. 1978. *Handbook of Geochemistry*. Springer, Berlin.

ZADEH, L. A. 1965. Fuzzy sets. *Information and Control*, **8**, 338–353.

ZIMMERMANN, H. J. 1985. *Fuzzy Set Theory – and Its Applications*. Kluwer-Nijhoff, Boston.

—— & ZYSNO, P. 1980. Latent connectives in human decision making. *Fuzzy Sets and Systems*, **4**, 37–51.

Improved methods for developing and visualizing remediation strategies

W. Hatton,[1] G. Hunter,[1] D. Hall[2] & D. Haigh[2]

[1] KRJA Systems Limited, Springfield House, Hucknall Lane, Nottingham NG6 8AJ, UK
[2] Golder Associates, Landmere Lane, Edwalton, Nottingham NG12 4DG, UK

Abstract. Remediation of contaminated sites can be both costly and time-consuming. In order to ensure that a cost-effective remediation method is selected, several options are usually considered. VULCAN, a 3D modelling package, has been used to demonstrate the development and visualization of remediation strategies using site assessment data.

At a former chemical manufacturing facility, 71 samples were collected from 31 boreholes. Concentrations of the contaminant were measured in all samples and visualized using the 3D package.

Remedial options were considered and a benefit analysis performed. This considered positive benefits, such as attainment of clean-up levels, and negative benefits, such as excavation and associated costs. Optimization of the most cost-effective method was then performed using the Lerchs-Grossman algorithm.

Introduction

Interpretation of data collected during site characterization of contaminated sites requires the analysis of highly variable contaminant data. The variability of the data is often modelled in the × and y plane, with little attention paid to visualizing the data in the third, z, dimension (Eddy & Looney 1993). The use of block modelling techniques to model contaminant concentration provides not only a powerful 3D appreciation of the contaminant distribution, but will easily allow cost, benefit and volumetric analysis of the contaminated site using block modelling scripting techniques. Scripting techniques allow rapid re-calculations and instantaneous visualization of 'what-if' conditions. A block model is a matrix of blocks, held in 3D space. Each block can hold data from physical measurements or data which have been interpolated or calculated from other block variables. It is possible to process, generate and interrogate these blocks in the manner of a 3D spreadsheet. Physical and economic factors can be combined and interrogated with visual tools, making model validation rapid.

This paper describes a new application of the Lerchs-Grossman algorithm, which is in common usage by open-pit mining engineers to find the optimum open-pit design. In 1965, Lerchs & Grossman developed an algorithm using mathematical graph theory capable of producing an pit design from a block model of an orebody. VULCAN's software was used to display the results of the 'Whittle 3D' optimization package, using the algorithm to calculate and visualize the best remediation design, by maximizing a benefit variable in the block model. CAD and mine design tools were used to calculate the volume of contaminants needed to be remediated. The major benefits that this technique gives are summarized as follows:

- A full 3D appreciation of the contaminants distribution is easily achieved.
- A block modelling approach allows rapid validation of the interpolated contaminant field and the associated clean-up costs.
- The Lersch-Grossman algorithm ensures maximized benefits are returned from any planned remediation design.

The paper concentrates on the technique rather than reporting a specific case history.

Data preparation

The contaminant sample data were collected by Golder Associates and supplied to the VULCAN package via an ASCII file. The data set consisted of 71%age concentration samples of a single contaminant variable collected from 31 borehole locations (Fig. 1). The data were read into a database, from which the values for contaminant concentration were interpolated into a 3D block structure. The rockhead interface was recorded in the 31 boreholes, supplied in ASCII format, and the surface was modelled using a standard Delauney triangulation routine. The contaminant data and the geology were combined into one data structure using block modelling.

Block modelling and contaminant estimation

Block modelling is a technique which originated in mining geology to estimate the grade of an orebody

HATTON, W. A., HUNTER, G., HALL, D. & HAIGH, D. 1998. Improved methods for developing and visualizing remediation strategies. *In*: MAUND, J. G. & EDDLESTON, M. (eds) *Geohazards in Engineering Geology*. Geological Society, London, Engineering Geology Special Publications, **15**, 373–382.

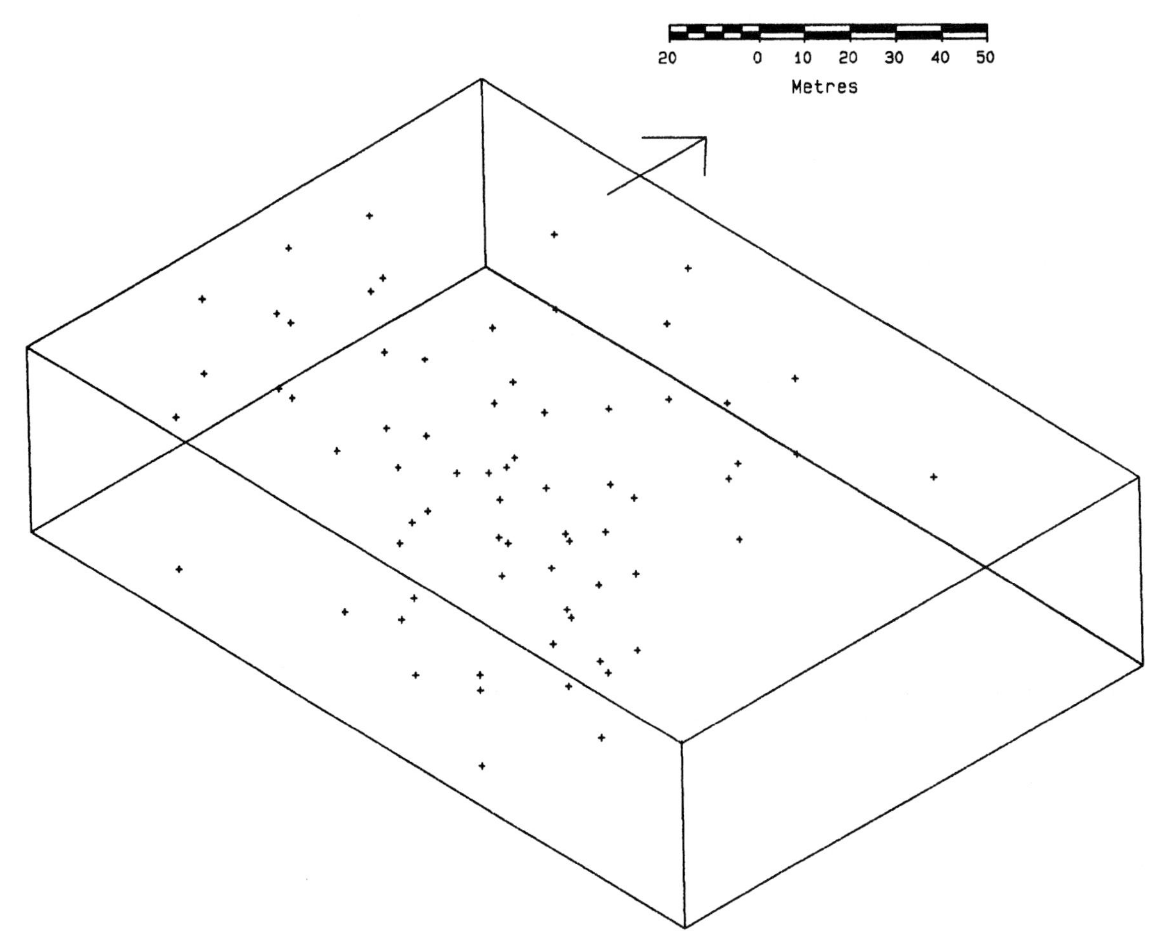

Fig. 1. Sample locations.

prior to mining. The technique holds the estimated grade or contamination in a matrix of blocks, each indexed to their position in 3D space. The block size can itself be made up of a smaller sub-block size, allowing more accurate modelling of locally irregular geological surfaces. The measured contaminant values are then interpolated into a variable limited to the individual block size, using a standard inverse distance interpolation method. The inverse distance method takes all the sample values lying within a user-defined neighbourhood to interpolate a single value at the centre of the individual block.

Naively giving an equal weight to each of the sample values which fall within the estimation neighbourhood may not produce an accurate estimate. An alternative is to give more weight to the closest samples to the estimation location and less weight to those which are farthest away. One obvious way to do this is to make the weight for each sample inversely proportional to its distance from the point being estimated (equation (1)):

$$V = \frac{\sum_{i=1}^{n} V_i * (1/d_i)^p}{\sum_{i=1}^{n} (1/d_i)^p} (1)$$

where V is the estimated value; n is the number of samples to estimate V with; $d_i \ldots d_n$ are the distances from each of the n sample locations to the point being estimated; $V_i \ldots V_n$ are the sample values; and p is an exponent.

Note the exponent p, whereby the weights can be made inversely proportional to the power of the distance. Traditionally the most common choice for the power is 2. Alternatives to inverse distance are geostatistical methods such as kriging. Englund *et al.* 1992 discuss at

Table 1. *Inverse distance estimation parameters*

Printout of Inverse Distance Estimation Run:

```
The number of discretization points per block are X:   1  Y:   1  Z: 1

Sample points are limited by a true ellipsoid.
       Search radius in the x direction is    20.00
       Search radius in the y direction is    20.00
       Search radius in the z direction is     2.00

No high-yield limit is applied.
No drill hole sample limit.

Power        2.000000
Aniso        1.000000     1.000000     1.000000
Octant search is requested with:
The maximum number of samples per octant :          20
The minimum number of samples for estimation is:    1

Inverse distance is used to interpolate grades.

Sub-cells in block model are interpolated at parent block size.
That is, all sub-cells in a parent cell of common geological characteristics
will receive the parent cell block value.

Azimuth:      0.000 Plunge:      0.000 Dip:   0.000

Estimation Variable :  conc
Default :      0.000
Number of samples :  nosamps
Average Distance :  avdist
selected 71 data points

Beginning estimation ...0
Transforming samples
Sample statistics after transformation

x_min:         996.00        y_min:        -1165.00      z_min:         65.10
x_max:        1118.00        y_max:        -1000.00      z_max:         69.10

nx     :       7             ny     :       9            nz     :        3

The above min, max and n (No of cells in each direction) are used to divide the
area into blocks of equal size and record samples belonging to each block for
quick access.  The block size's are equal to the search distances.

sorting the samples
indexing the samples
       71 composites have been rotated and indexed
       1000 blocks estimated      block number  :          1626
       2000 blocks estimated      block number  :          2960
completed  : 3359/3359  blocks
```

Table 2. *Block model details*

```
Model name            :     covs04_conc
Number of blocks      :     3359
Number of variables   :     8
Number of schemas     :     2
Origin                :     0.000000   0.000000   0.000000
Bearing/Plunge/Dip    :     90.000000   0.000000   0.000000
Created on            :     12:04:09    8-Jun-95
Last accessed on      :     08:08:32   23-Jun-95
Model is unindexed.
```

Variables	Default	Type	Description
Conc	-999.00	float	Contaminant concentration
nosamps	0	Integer	
avdist	0.00	Integer	
geology	drift	name	Solid geology drift interface
cost	999	float	Excavation cost (pounds/t)
benefit	0	float	Benefit (pounds/t)
dollar	-999	float	Dollar value (benefit-cost)
pit	0	integer	Shell excavation phase no)
volume	-	predefined	
xlength	-	predefined	
ylength	-	predefined	
zlength	-	predefined	
xcentre	-	predefined	
ycentre	-	predefined	
zcentre	-	predefined	

```
Translation Tables  :    geology  :
                               drift  = 0
                               rock   = 1
schema <parent>
Offset   minimum :    990.000000   990.000000   64.600000
         maximum :   1200.000000  1130.000000   69.6000000
Blocks   minimum :     10.000000    10.000000    1.000000
         maximum :     10.000000    10.000000    1.000000
  No of blocks     :   21   14   5

Schema <subblocking_2>
Offset   minimum :  990.000000   990.000000   64.600000
         maximum :  1200.000000   1130.000000   69.6000000
Blocks   minimum :  2.000000   2.000000   0.500000
         maximum :  10.000000 10.000000   1.000000
  No of blocks     :   105   70   10
```

length the merits of such techniques. Kriging is a geostatistical method used to estimate the value of a point or block, using a linear combination of the available samples in, or near, that block. The principles and practical application of kriging are discussed in more detail by Isaacks & Srivastava (1989). VULCAN's grade estimation module can interpolate the contaminant value using inverse distance, nearest neighbour, ordinary kriging, probability indicator and indicator kriging.

The parameters used in the contaminant estimation and the block model creation are given in Tables 1 and 2. Table 1 indicates the search radius in the X, Y and Z directions, the inverse distance power for calculation, the estimation variable and sample statistics. Table 2 lists the variables stored in the block model, the number of blocks and the dimensions of parent and sub-blocks. The main advantages of using block modelling to estimate contaminant concentrations are as follows

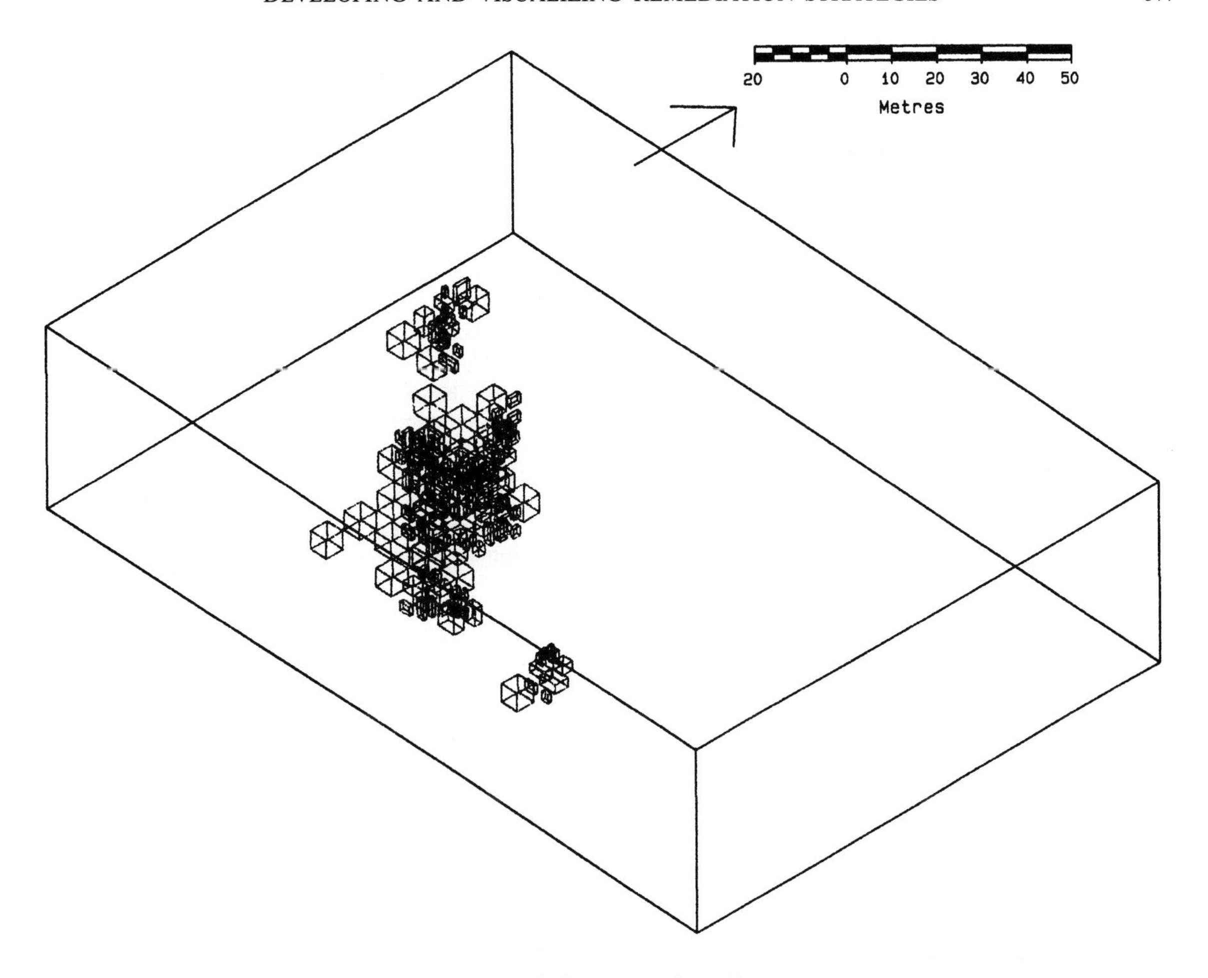

Fig. 2. Block concentration >10.

- The block model is fully integrated within VULCAN's 3D visualization tool. The contaminant concentration can be viewed in a number of ways using coloured slices on any plane through the model to display concentration, and can be contoured in any plane. Zonal contours of concentration can be used as 3D shells to construct solid 3D triangulations, enabling volumetric calculations to be performed at any contaminant concentration. In Fig. 2 the blocks with a concentration above a defined 10% threshold level are highlighted.
- The block model allows rapid manipulation to derive new variables by calculation. This is achievable by writing simple scripts, combining 'IF' statements to develop scenario indicators. An example is shown in Table 3 which calculates a cost to clean up each block in the model. The variables in the script are either block model variables or as defined in the script.
- A clean-up cost variable can then be passed to the 'Lersch-Grossman' optimization algorithm, (Whittle 3D). The optimal remediation scheme can then be calculated, maximizing benefits, based upon clean-up costs, excavation rates and contamination values.
- Rapid 'What-if' scenarios can be modelled and re-run quickly, allowing more rapid model validation.

Cost–benefit analysis

For the purposes of this work we are drawing parallels between open-pit mining and remediation design. In the mining situation, each block in the block model has a positive revenue value (derived from the final value of the product) and a negative cost value (which includes excavation, transport, processing and dependent overhead costs). These costs can be calculated using the known or interpreted data held in the block model along

Table 3. *Block model reports*

Criteria	*Volume*	*Average concentration*	*Total concentration × volume*
Total block model:			
Conc. <5.0	136 000	0.472	1580
Conc. 5.0–10.0	7 300	6.681	1433
Conc. >10.0	3 700	19.976	3277
Total	147 000	1.272	6290
Optimum shell:			
Conc. <5.0	5 652	2.527	183
Conc. 5.0–10.0	5 750	6.770	851
Conc. >10.0	3 270	20.934	2755
Total	14 672	8.292	3789
Excavation design:			
Conc. <5.0	14 798	1.930	686
Conc. 5.0–10.0	6 216	6.665	1132
Conc. >10.0	3 330	20.934	3083
Total	24 344	5.739	4901

with each block's position in space and the maximum allowable slope angles. Using this information, the status of each block can be calculated, either lying inside or outside the optimum excavation 'shell'.

To apply this technique to remediation design we need to assign a positive benefit value to each block, along with a negative cost value. The benefit value could be obtained by considering the following factors:

- contaminant concentrations
- clean-up level (regulatory or via risk assessment)
- vicinity of environmental or human receptors
- migration pathways

A simplistic benefit value may be obtained by considering the contaminant concentration relative to the clean-up level, for example. Having assigned this value the negative costs can be assigned easily by considering treatment/excavation costs, transport costs, disposal costs, and dependent overhead costs. The optimum excavation shell can now be calculated.

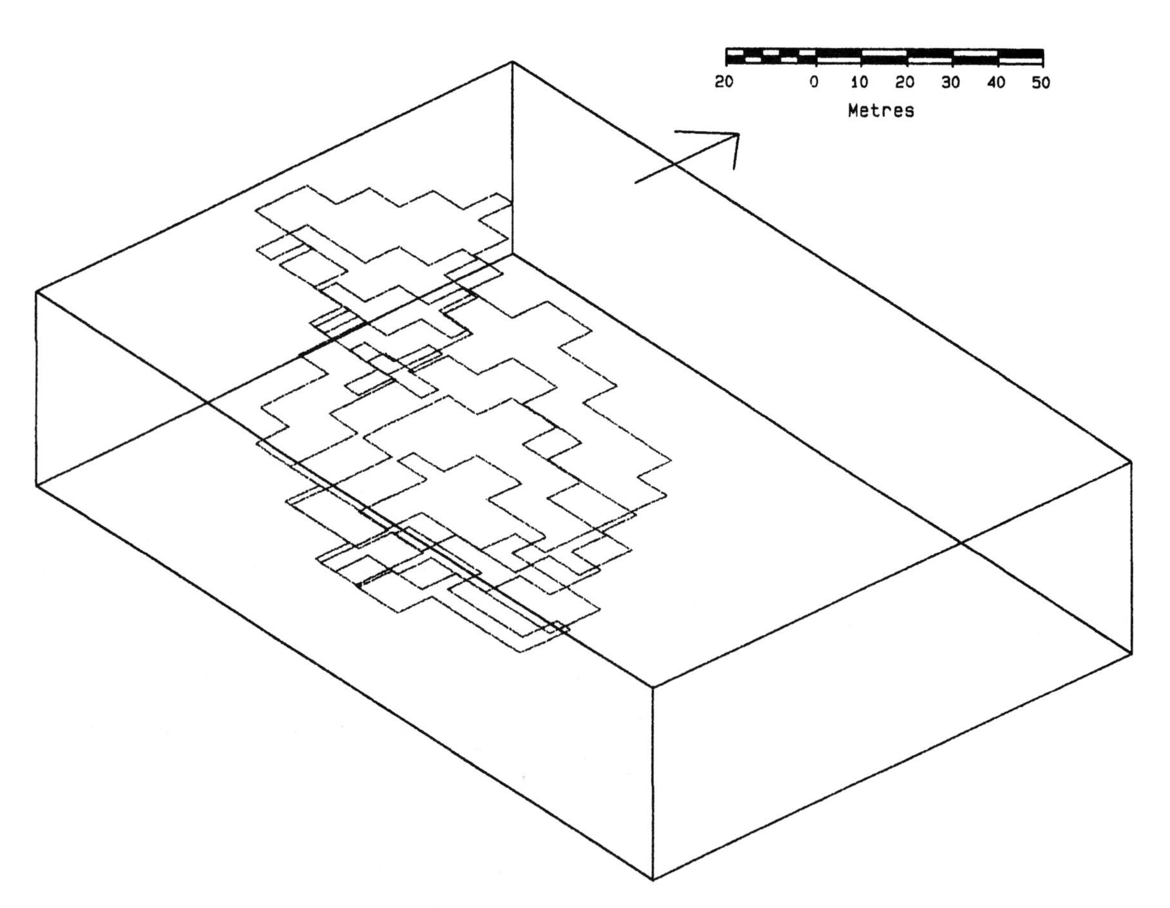

Fig. 3. Optimum shell contours.

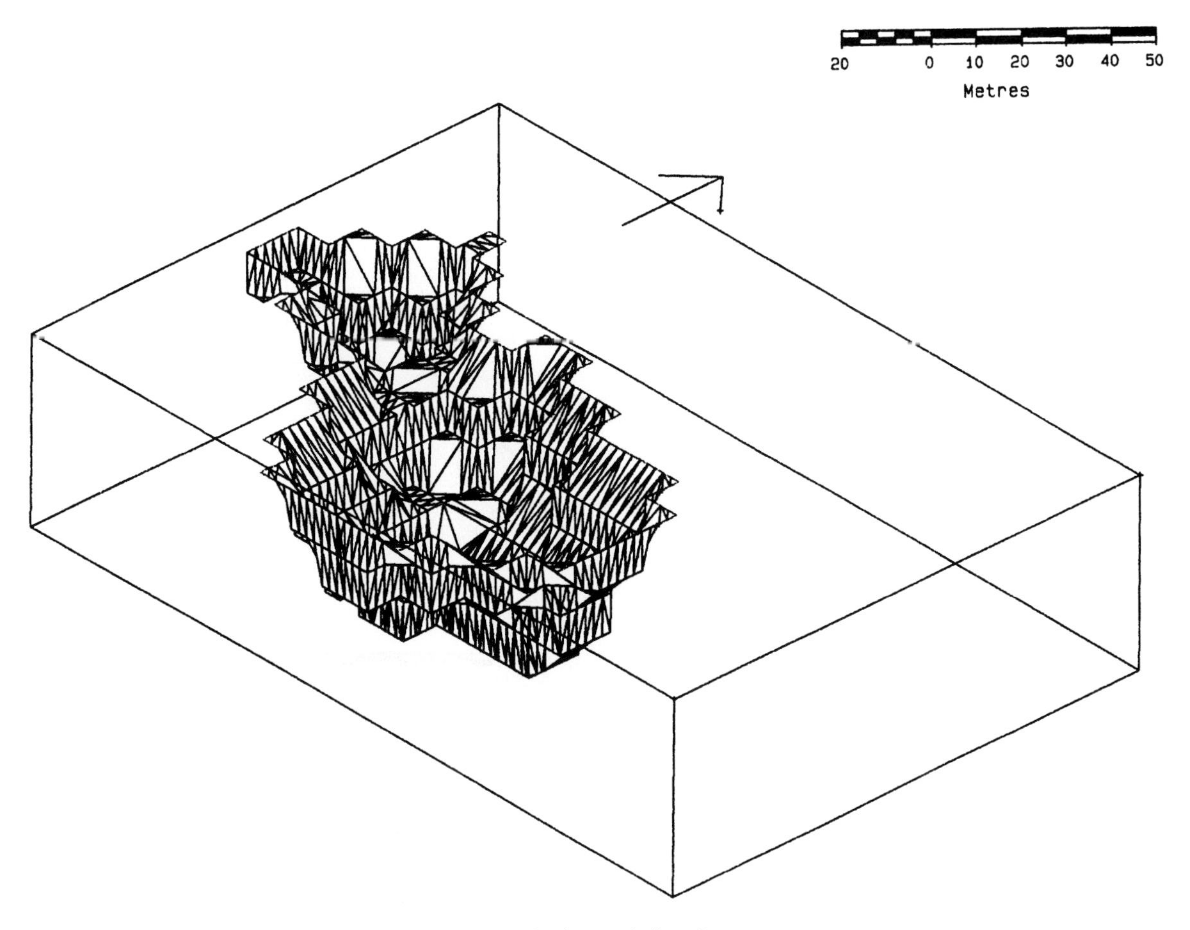

Fig. 4. Optimum shell surface.

The use of block model scripts in conjunction with the 3D CAD features allows easy calculation of the economic parameters described above. For example, excavation/treatment costs can be varied with depth (using the elevation value attached to each block) or varied by zones reflecting transport distance (maybe applicable over a large and long site). Alternatively, costs of disposal can be calculated from contaminant concentration and geological zone.

Optimization of remediation excavation

Having considered the economics of the remediation site, as outlined in the previous section, we have enough information to perform an optimization. For this case study, we have employed the Lerchs-Grossman method (Lerchs & Grossman 1965) which is the most common method used in open-pit mine planning world-wide. This takes a block model and determines which blocks should be mined to obtain the excavation design with the highest total value. This set of blocks defines the optimum pit design. The algorithm considers two types of information: maximum allowable slope angles and a 'dollar' value.

The first type of information is related to the required excavation slopes. For each block in the model, the Lersch-Grossman algorithm needs details of what other blocks must be removed to uncover it. This information is presented as a list of pairs of block identification numbers, known as structure arcs, which specify that if the first block in the pair is to be mined, then the second must be mined to uncover it. The graph method repeatedly applied calculates an inverted cone of arcs for any desired slope design.

The second type of information consists of the economic value of each block, or 'dollar' value, which

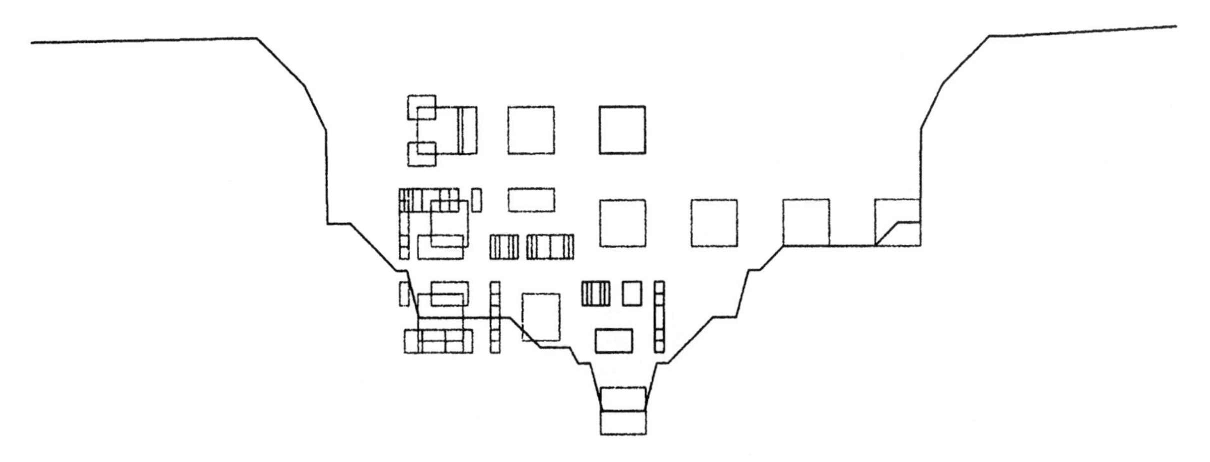

Fig. 5. Section view – blocks and shell.

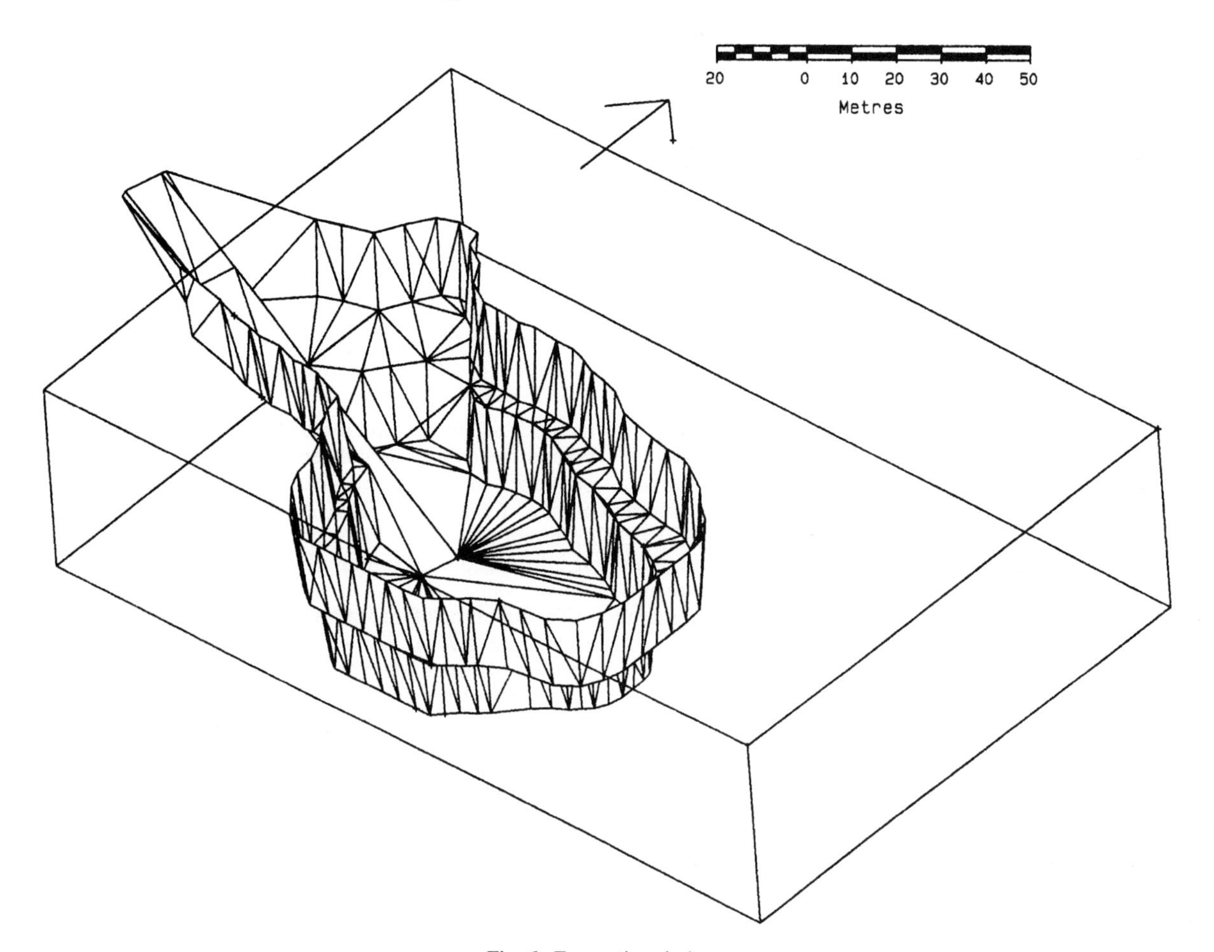

Fig. 6. Excavation design.

is the sum of the positive benefit value and the negative remediation cost. Blocks with a positive 'dollar' value (generally those with contaminant concentration) will contribute to the total value of the remediation, whereas blocks with negative values (generally those without any contamination such as overburden) will reduce the total value.

Table 3 illustrates the method used to calculate the 'dollar' value. When calculating the remediation costs it should be noted that the transport and excavation costs are related to the tonnes of material to be moved, whereas any treatment or disposal costs will be related to the tonnes of material and contaminant concentration.

Whittle 3D is a set of programs which efficiently find the Lerchs-Grossman optimum pit shell and is interfaced to the VULCAN design and vizualisation package. Whittle 3D takes input from the block model along with slope angle parameters, then outputs the result back into the block model. This program features a number of extensions to the basic optimization described above, which include handling multiple contaminants and varying slope angle.

Optimization results

The result of a Whittle 3D optimization is a list of blocks which should be excavated, which can be contoured to give a rather angular excavation shell (Fig. 3). Output in this format is useful for comparing shells produced with different economic or concentration estimation parameters. Figure 4 illustrates the contours as a triangulation, which will allow accurate volume calculations to be made. The volume between this shell and the original topography was calculated to be 14 700 m^3; however, this shell does not represent a practical excavation design.

Figure 5 shows a section view through the shell with the blocks of concentration greater than 10%. This allows the results to be checked to identify any high concentration blocks that have been excluded from the optimum shell, and to ascertain if the optimization criteria are acceptable. The block model has been interrogated to give the total tonnes of material contaminated above 10% and the respective amount that is contained within the optimum shell.

Further design work is required to produce a practical design incorporating minimum excavation widths and roadways with correct gradients and widths. Figure 6 shows one of many possible excavation designs based on the optimum shell, which incorporates practical gradients for roadways.

When incorporating a practical ramp design into an optimum pit shell, the ramp can either be placed inside the shell (thus reducing the total benefit of the shell) or it can be placed outside the shell (thus increasing the excavated volume and total cost of the shell). In this instance, it was chosen to place the ramp outside the shell, but in a location where some contaminated material outside the shell could be excavated. This material is sub-optimum, meaning the cost of uncovering and excavating the material exceeds the benefit; however, if we excavate this material to build a ramp we still receive this benefit.

Table 4 illustrates how the volume of the excavation exceeds the shell, for each of the concentration cut-offs. This table can be used to compare differing practical excavation designs with the optimum shell. The 'total concentration × volume' gives a measure of total contaminants moved, which can be seen to have increased from the optimum shell to the excavation design.

Summary and possible applications of this technique

The described technique has numerous applications in the assessment of remedial options. Various treatment/excavation methods can be visualized rapidly using this

Table 4. *Benefit equation script*

- Density fixed for whole model (could use block model variable instead)

  density = 1.5

- Elevation of top of model

  top = 70.1

- Define benefit–cost factor per tonne-concentration

  f1 = 10

- Define mining cost per tonne-metre depth

  f2 = 10

- Define processing cost per tonne-% concentration

  f3 = 0.1

- Define tonnes

  tonnes = volume × density

- Calculate benefit of each block based on concentration

  benefit = concentration × f1 × tonnes

- Calculate cost of excavating and processing each block

  cost =((top-zcentre) × f2 tonnes)
  + (f3 × tonnes × concentration/100)

- Calculate total cost per block

  dollar = benefit – cost

technique which will aid identification of the most appropriate scenario. Using this visualization, the environmental impact of each option may be more easily assimilated.

The technique also allows several different remedial options to be assessed against each other. The most cost-effective or most appropriate method can therefore be easily identified.

The VULCAN 3D package also allows optimization of combined treatment methods at sites where a number of treatment techniques are considered.

Acknowledgements. The authors would like to acknowledge Golder Associates for permission to use the data presented in this paper.

References

EDDY, C. A. & LOONEY, B. 1993 Three-dimensional digital imaging of environmental data: selection of gridding parameters. *International Journal of Geographic Information Systems*, **7**(2), 165–172

ENGLUND, E. J., WEBER, D. & LEVIANT, N. 1992. The effects of sampling design parameters on block selection. *Mathematical Geology*, **24** (3), 329–343.

ISAAKS, E. H. & SRIVASTAVA, R. M. 1989. *Applied Geostatistics.* Oxford University Press, Oxford.

LERCHS, H. & GROSSMAN, I. F. 1965. Optimum design of open pit mines. *Trans CIM*, **LXVIII**, 17–24.

WHITTLE, J. 1988. *Beyond Optimisation in Open Pit Design.* 1st Canadian Conference on Computer Applications in the Mineral Industry, Quebec City, March 1988.

SECTION 8

PLANNING AND GEOHAZARDS

Incorporation of information on geohazards into the planning process

Brian R. Marker

Minerals Division, Department of the Environment, 2 Marsham Street, London SW1P 3BE, UK

Abstract: Urban areas require geological resources for construction and maintenance. Most of these come from rural areas and are often considered separately from the planning of the urban area. Uncontrolled urban growth may sterilize resources or lead to contaminative activities occurring close to vulnerable aquifers. Foundation problems may influence both rate and costs of urban development and geological hazards may threaten property or public safety. Urban areas may grow across potential problems of their own making, such as waste disposal sites. Such problems can be minimized if (a) development takes place in the right places and is properly designed and executed; and (b) the implications of urban development are considered in the context of the availability and supply of minerals, water and land, and the environmental effects of exploiting these.

To achieve balanced consideration of planning and development options, sound geological and geomorphological information is needed at (a) the site level, for specific developments; and (b) the regional level, to guide general allocations of land for particular uses and alert those designing site investigations to the range of problems to be considered.

Geological and related factors relevant to decision making are often neglected. There have been increasing efforts by geologists to secure adequate basic information for selected urban areas but results often have been presented in a form which does not meet the requirements of planners and other decision makers, who may have had no geological training. Whilst planning of land use in urban areas will always be led by economic, social and conservation pressures, it is important to involve planners in identification of relevant issues and the design and presentation of results.

Introduction

The word 'planning' conveys different messages to different people. For instance, it may mean planning of investment by developers and financial institutions; planning of responses to civil emergencies; or the planning of land use.

This paper is concerned with the last of these, but will need to touch on the others. The discussion is based on direct experience of the planning system in England and Wales. However, most of the points made are likely to be more generally applicable.

Land use planning

Systems for planning of land use vary greatly but most have features which are broadly similar to the approach taken in England and Wales. The overall aim is to regulate the physical development and use of land in the public interest (DoE 1993). Within this aim, the objective is to facilitate development and to strike the right balance between development and the need for conservation and protection of the environment – the balance between wealth creation and the quality of life.

The key principle is that development should not be undertaken without planning permission obtained from a responsible authority – usually a local authority. *Development* may be defined as the carrying out of building, engineering, mining or other operations in, on, over or under the land, or the making of any material change in the use of any buildings or other land.

In the British system, there are exceptions to this. For instance, the use of land for the purpose of agriculture or forestry falls outside the land use planning system and a General Development Order gives general permission in advance for certain defined classes of development such as small extensions and alterations to dwellings. That is not necessarily the case elsewhere.

The main elements of land use planning in England and Wales are

- the development plan, and
- the control of development.

The development plan

The development plan sets out strategic policies for land use in the area. In the British system, the provisions are not prescriptive but are intended to provide a firm basis for rational and consistent planning decisions and co-ordinating the needs of development, including provision of infrastructure and the interests of conservation.

MARKER, B. R. 1998. Incorporation of information on geohazards into the planning process. *In*: MAUND, J. G. & EDDLESTON, M. (eds) *Geohazards in Engineering Geology*. Geological Society, London, Engineering Geology Special Publications, **15**, 385–389.

In many other countries the development plan has a zoning function which is much more directive. However, in both cases, there may be one or more tiers of plans which

- set the broad policies for land use in the selected area, which is usually part or all of a local authority area;
- set out criteria which may be used in deciding planning applications;
- give general indications of specific allocations of land where particular types of development may be considered favourably or in which certain types of development will not normally be acceptable.

Key topics covered in development plans usually include the following:

- housing
- conservation and green belts
- major industrial, business, retail and other employment-generating development
- facilities for health and education
- strategic transport and highway facilities
- land reclamation and re-use
- tourism, leisure and recreation
- mineral working
- waste disposal

With the exception of minerals and, to some extent, waste disposal, most of these topics are not associated with geological or geomorphological issues by most non-geologists. Indeed, minerals and waste disposal planning are widely regarded as specialist topics for which separate staff are employed, thus isolating these from the mainstream of planning. The steps required in preparing and maintaining a development plan are as follows:

- to undertake a survey of the principal physical characteristics of the area;
- to formulate draft policies for land use and criteria for making decisions on planning applications;
- to undertake public consultation and to amend the draft plan in the light of the findings;
- to secure approval for the plan;
- to monitor the effectiveness of policies and changes in the plan area as a basis for regular review of the plan and, if necessary, amendment of it.

Such surveys commonly include consideration, for example, of mineral resources, suitable locations for waste disposal, and geological or geomorphological Sites of Special Scientific Interest. Few other geologically related factors were taken into account as a matter of course, even in areas with relatively extensive ground problems. For instance, the document setting out Government guidance on land for housing currently makes no reference to suitability of land in terms of natural hazards such as subsidence, landslides, erosion or flooding. Equally, few development plan policies were propounded in respect of, for example, unstable land or flooding until relatively recently. As a result, land allocations in some plans may be suitable for specific uses in socio-economic terms but may be subject to natural or human-made geohazards which may act counter to the purpose of the allocation.

Control of development

Control of development involves the making of decisions on planning applications for specific developments. In Britain, such decisions are based on planning grounds and with regard to the provisions of the development plan and to any other material considerations. This gives a very wide range of potentially relevant factors, including geological and related conditions, but these may not be recognized by non-geologists. It is stated unambiguously that it is the responsibility of the developer to satisfy the planning authority as to the suitability of the site for the proposed development for instance through site investigation. This was sometimes interpreted as meaning, for example, that instability of the ground, whether due to mining, landslides or other causes, was not a matter which should be taken into account when reaching a planning decision. Because of this uncertainty, advice was issued in 1990 to make it quite clear that land instability can be a material planning consideration.

Planning permission may be granted subject to such conditions as the local planning authority or the Secretary of State may think fit, provided such conditions are necessary, relevant to planning and to the proposed development, enforceable, precise and reasonable. An adequate understanding of the interactions between the site and the proposed development and of any implications for adjacent land and property is needed in order to propound sound planning conditions. The geological dimension has sometimes been neglected. Instances have occurred, for example, of development cutting into a slope and destabilizing old landslides which had not been detected during site investigation.

Attitudes to the urban area

A *developer* sees an urban area as an opportunity for investment and wealth creation and considers, in particular, potential market trends, the labour force, the site location (especially in relation to transport routes), and the likely costs of the development in relation to expected profits. The ground conditions at possible development sites play a small part in this planning of investment, particularly in the early stages, partly because of a generally low level of appreciation of physical constraints arising from ground conditions but also because, for many types of development, the costs of preventive and remedial works are small compared with the total outlay. The costs of site investigation may often be a very minor element of the budget but this does not prevent some developers skimping the cash at this stage despite the risk of more expensive problems later due to redesign, remedial works and delays. However, it is usually only in the case of the smallest developments, such as a small group of houses, that the cost of proper investigation and treatment of land is likely to preclude the development from going ahead.

The *planner*, except for specialists in minerals and waste disposal, tends to see the area in socio-economic terms, guided by the need to secure development and prosperity at the least cost to the quality of life. In many urban areas, environmental improvement, reduction of the legacy of past industrial damage, and conservation of the cultural heritage are major concerns. But earth sciences are not seen as being relevant. This should not surprise us because few planning courses contain any coverage of the role of earth sciences in decision making. There are, of course, exceptions. In many areas with major natural hazards, such as cities on the Pacific Rim, it is well understood that planning has an important role, for instance, in placing infrastructure, potentially hazardous installations and emergency response facilities in the least vulnerable settings.

The academic *geologist* traditionally saw urban areas as a mosaic of interesting rocks, some of which contained natural resources, but with buildings getting in the way of mapping and scientific interpretation. This attitude has changed greatly over the last 20 years not least because the provision of advice on urban development and on the rising demand for natural resources is a major source of employment for the profession. But the change has not been communicated adequately to non-geologists. This may be because professional advice is normally given in professional language, and may not be very comprehensible to the uninitiated.

The nature of the urban area

An urban area usually depends on the surrounding countryside for mineral resources for use in construction, industry and production of energy. The sources may fall under the remit of other authorities, leading to them being considered separately from urban trends. The impact of an urban area, therefore, extends far beyond its boundaries and may raise problems of pressure for resources of sand and gravel from good agricultural land or of crushed rock from scenic upland areas. The demand may be partly offset by measures such as recycling of materials or energy saving but, in the main, the demand needs to be met. The penalty may be damage to valuable environments far from the urban area in which the demand arises. The implication is that there needs to be a good appreciation of the nature, extent and quality of minerals and water resources, and of the likely demand in order to safeguard and to make the best possible use of the resource. In a relatively small and heavily urbanized area like Britain this catchment is most of the country. The production of a wide-ranging inventory of natural resources, therefore, is fundamental to the process of preparation of development plans and to sustainable development of the resources.

Even the maintenance of the historical and cultural fabric may require a geological element. For example, there is a need for compatible sources of stone to be found for restoration of old buildings, and mineralogy is of key importance in developing approaches to protecting stone damaged by atmospheric pollution.

The pattern of growth of the urban area influences the balance of issues that may be relevant today and may store up problems for the future, as the following examples illustrate:

- A high rate of growth causes the urban area to extend progressively onto less suitable land, which may be subject to geohazards; local minerals and agricultural land may be sterilized; resources are tapped from an increasingly wide area; and transport links have to be augmented to carry these and the population. The provision of infrastructure may be outstripped and this, and industry, may lead to contamination of surface and underground water. The urban area may build across former quarries, now backfilled, waste tips and mines. If not recorded, the previous existence of these may be forgotten until land is redeveloped.
- Containment of the urban area, for instance by a green belt, tends to cause urbanization to leap frog the safeguarded area, particularly on key roads, with potential sterilization of resources in the previously rural area. Within the city the emphasis is placed on recycling of land and thus on confronting a legacy of geohazards from past land uses such as contaminated land, and made ground of uncertain properties.
- Dispersal or decentralization away from the urban core leads to suburban sprawl or, ultimately, to conurbation. The density of development is often lower but the land take higher thus giving the potential to sterilize more resources. This may also lead to problems of overstretched infrastructure, for instance areas may not be on mains drainage. Whilst this may be acceptable in some areas, the use of soakaways may exacerbate subsidence into artificial or natural cavities, or may contribute to slope instability where there is a potential for this to occur.
- The rarer option of creating a new urban area gives the opportunity for a well planned start, including a full assessment of the potential for geohazards but, often, the pressures to develop land quickly may

negate the opportunity. In addition, this approach can cause major resource and conservation losses.

The whole process of urbanization – of concentration of population, industry and infrastructure – is one that increases the risks that may arise from geohazards such as subsidence, flooding, earthquakes, volcanic activity, and the demand for, but sterilization of, resources. It becomes increasingly important that earth science information, in a broad sense, is taken into account.

Reducing losses and risks

It is obvious to the earth scientist that losses and risks can be reduced if sound information on the following is readily available:

- the nature and uses of ground materials
- the availability and quality of water and the susceptibility of aquifers to pollution
- the nature and extent of potential natural and human-made hazards

at a general level for use in broad-brush planning and at a detailed level to assist site-specific work.

However, it is often far less obvious to the planners or developers that this is necessary. The planner may leave the matter to the developer. The developer may appreciate the need for work on the development site. There may not, however, be a willingness to pay for studies of the wider context of the site except where, for instance, environmental assessment of a major development is required. Both may see any proposal for wider geological study as an attempt by geologists to secure more work rather than something which has real benefits to them, particularly if the results of previous work were incomprehensible to the lay-person.

A key step is to establish a general awareness of the consequences of not taking earth science information into account and that a sound geological and geomorphological understanding can

- reduce costs, delays and risks
- safeguard valuable materials
- provide environmental benefits

To date, few attempts have been made to assess the costs and benefits of geological and related research in respect of land use planning (e.g. de Mulder 1988).

It further needs to be made clear that sound information is needed at the general level. It is necessary to alert planners and developers to the following:

- the possible extent of resources and the relationships of these to environmental constraints; and
- to potential problems and hazards so that it can be ensured that these are considered when allocations of land are made or applications to develop land are submitted.

It is also important that adequate site investigations are undertaken. Generalized information helps to set a context for the design of site investigations and may also help in making a case to the prospective developer for adequate funding of investigations.

Any discussion of problems – only some of which are hazards – needs to be accompanied by the understanding that most can be dealt with for a modest cost compared with the cost of the development, particularly if problems are recognized early on. Cost may preclude small-scale development in some cases but if it is not funded properly it would be unwise to go ahead. In particular, there is a need to demonstrate clearly what the costs and risks to the community of not taking these matters into account may be.

There is a growing international trend towards the preparation of 'Environmental Geology Maps' (EGMs) which identify resources and constraints and can be used as an earth science input into the planning base. Many of these are backed up by databases which draw together existing information in an easily accessible form and allow more efficient retrieval of information which is relevant to site-specific work. However, the coverage of such maps and databases is very sporadic in most countries and seems likely to remain so. Part of the reason for this may be that EGMs are said to be designed for use by planners but are often presented in a form that is much better suited to the needs of geologists and engineers (DoE 1991; McCall & Marker 1989).

The principal requirements of the planner are to know what issues may be relevant to decision making in particular parts of a plan area, how important these issues are, and when professional advice is needed from earth scientists. For instance, an indication that past mining may have taken place, accompanied by a rider that any application for development in the area should be accompanied by a mining stability report prepared by a competent person, is enough. The planner can easily check that a report is submitted, that the credentials of the person who prepares it are adequate, and can take note of the findings. Some geologists are concerned that maps may be misused by non-specialists. The risk of this happening is greatly reduced if the planning map emphasizes *when and what advice is needed and where it can be obtained.* This is best undertaken in a two-way process involving both planners and earth scientists.

There is a need also to provide more general advice which explains the significance of earth science factors to practical planning and guidance on how to take these matters into account. Such guidance can lock maps and databases into the planning process and help to ensure that these are used (e.g. DoE 1990, 1996).

Another important point is that information on ground materials and conditions accrues continually, especially in urban areas, through site investigations. The results of these can be used to improve the available information for planning if these are made available.

A problem in the past has been that paper maps have become outdated rapidly. Digital cartography now gives the opportunity for frequent updating. The use of Geographic Information Systems may extend this flexibility in the future and there are signs already that planning authorities are showing more interest in incorporating geological information into GIS than they have been in using paper maps in planning. There is a danger in this in that simply digitizing the geology will not give the information that is needed.

There are other difficulties. The safeguarding of mineral resources, for example, is not always a popular action since there may be concern that the possibility of future quarrying may affect land values adjacent to the site. Similarly, there may be a reluctance to admit to potential instability or contamination of land on the grounds that recognition may affect property values, loans, insurance and investment confidence. The need for sensible decision making has to be weighed against such possible effects.

These issues need to be addressed and solved.

Conclusion

Comments in this paper have been based largely on circumstances in England and Wales but discussions at international conferences have demonstrated that the position is basically similar in most countries. There may be differences in the perceived balance between wealth creation and environmental protection in planning and in the legislation under which planning operates but there is a general under-use of earth science information which needs to be addressed.

There is a need for general guidance to planners on the significance of earth science information – information at the general level to guide development plan policies and to provide a better context for design of site investigations. Recording of site specific data will help to make site desk studies more thorough. More flexible means of mapping will allow information to be updated frequently. However, it is essential that results for planners are presented in such a way that they fit into the planning process and can be understood by non-specialists. In the case of urban areas, it is important to remember that the population centre feeds on the surrounding countryside. There is a need, therefore, to consider the wider implications of urban trends, for instance through regional planning, in order to secure and safeguard natural resources.

It may not be possible to translate approaches developed in one country directly to another due to differences in language, ideas and culture. In some cases, especially where literacy is low, it may be necessary to involve local people, as well as planners, in the process of preparing advice.

However, in general, steps of the sort outlined here may help to convince decision makers that there is a need to consult the earth scientist early on and that research in this field is providing valuable information for planners and not just work for geologists.

Acknowledgements. The content of this paper has developed through discussions with many individuals over the years who are too numerous to list, but special mention should be made of R. C. Mabey and D. Brook (DOE), M. G. Culshaw (British Geological Survey), E. F. J. de Mulder (Rijks Geologische Dienst) and G. J. H. McCall.

References

De Mulder, E. F. J. 1988. Thematic applied Quaternary maps – a profitable investment or expensive wallpaper? *In*: De Mulder, E. F. J. & Hageman, B. P. (eds) *Applied Quaternary Geology*. Balkema, Rotterdam, 105–117.

Department of the Environment 1990. *Planning Policy Guidance: Development on Unstable Land*. PPG14, HMSO, London.

——1991. *Applied Earth Science Mapping – Proceedings of a Seminar*. HMSO, London.

——1993. *Planning Policy Guidance: General Policy and Principles*. PPG1 HMSO, London.

——1996. *Development on Unstable land: Landslides and Planning*. PPG 14 (Annex 1), HMSO, London.

McCall, G. J. H. & Marker, B. R. (eds) 1989. *Earth Science Mapping for Planning, Development and Conservation*. Graham & Trotman, London.

Development advice maps: mining subsidence

M. J. Scott & I. Statham

Arup Geotechnics, Cambrian Buildings, Mount Stuart Square, Cardiff CF1 6QP, UK

Abstract. In 1985 the Department of the Environment/Welsh Office commissioned the 'South Wales Desk Study – Mining Subsidence'. This project was to develop a method for preparing Development Advice Maps giving advice to planners and developers on abandoned shallow mining. The pilot area for this study was centred in Ebbw Vale and supplemented by a detailed review of mining subsidence throughout the whole of South Wales. Subsequent to this study a follow on project was undertaken in 1991 in the Borough of Islwyn. The purpose of the study was to confirm the methodology developed during the first study and produce Development Advice Maps which could be introduced into the planning process at Islwyn Borough Council. Maps were introduced into the planning system with planning guidelines and were monitored for 15 months. The monitoring period has indicated that the Development Advice Maps, drafted in accordance with the procedure developed from the South Wales Mining Desk Study, are generally reliable.

This paper discusses the method of preparation of Development Advice Maps in South Wales and proposes a strategy of risk management with which to appraise potential mining subsidence.

Introduction

During the post-war period, and particularly since the 1960s, mining in the South Wales Coalfield has progressively declined and now has virtually ceased apart from opencast mining and small licensed mines. The Department of Environment/Welsh Office for many years has recognized the unwelcome legacy of potential ground collapse from past mining and other processes. In the early 1980s, they initiated a programme of research to develop methods for assessing the potential for ground instability and for bringing this information into the planning process.

In 1990, the DoE/WO published Planning Policy Guidance No. 14, *Development on Unstable Land*. This guidance document summarizes the causes of instability, the responsibilities of various parties when considering development of unstable ground, and the role of planning control.

Further, the latest edition of the *Building Regulations*, published in 1991, recognizes that instability of ground is a material consideration in development control. Finally, in 1994, Mineral Planning Guidance, MPG 12, *Treatment of Disused Mine Openings and Availability of Information on Mined Ground*, was published.

Consequently, recent planning legislation means that for the first time, planners have to consider ground instability at local plan and development control level.

Mining subsidence

Subsidence mechanisms

As mine cavities collapse, the effects of the collapse are transmitted towards the surface and may result in subsidence. This may be almost simultaneous with the mining or it may be delayed, resulting from the slow deterioration and eventual collapse of partial extraction workings.

The following major categories of mining subsidence are identified (Fig. 1):

- *Crownholes*: crater-like holes appearing at the surface following collapse of strata into a mine cavity. The collapse into the mine may rapidly affect all the overlying strata, and cause a crownhole; but usually the progressive upward 'migration' of a void from mineworkings to reach the surface takes many years after abandonment. Crownholes are normally associated with partial extraction mining, but may result from collapse of roadways remaining after total extraction has taken place.
- *General subsidence*: settlement of the ground surface over a wide area resulting from the collapse of part of a mine. This results from the collapse of either partial or total extraction mines and may occur while the mine is being worked or at any time afterwards.
- *Subsidence due to longwall mining*: settlement of the ground surface over a longwall mining panel, usually

Scott, M. J. & Statham, I. 1998. Development advice maps: mining subsidence. *In*: Maund, J. G. & Eddleston, M. (eds) *Geohazards in Engineering Geology*. Geological Society, London, Engineering Geology Special Publications, **15**, 391–400.

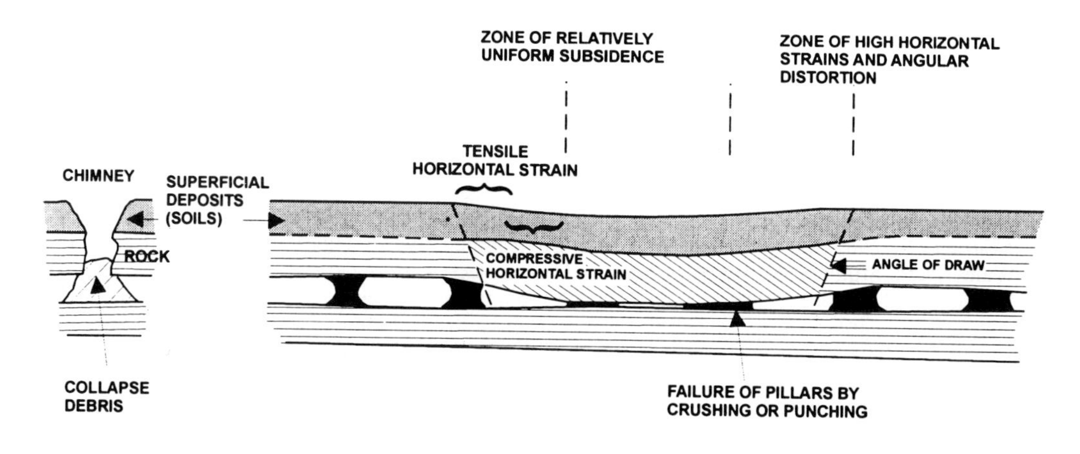

Fig. 1. Typical subsidence mechanisms into abandoned coal workings.

occurring within a short period of time after mining has passed beneath the location affected. This mechanism did not apply to the study area, deep mining having ceased in 1989.

The operation of the first two mechanisms is usually spatially and temporally indeterminate. It is seldom that the mining methods and geology are so regular that the general condition, permitting ready analysis, exists.

Subsidence mechanisms in general are discussed in detail in Ove Arup and Partners (1991), Whittaker & Reddish (1990) and Healy & Head (1984).

Islwyn study

The study area is part of Islwyn Borough and is defined as 'that part of the Borough underlain by the Mynyddislwyn Coal Seam'. It is irregular in shape and almost 30 km^2 in area, centred on the town of Blackwood in the Sirhowy Valley. In the north it reaches to Mynydd Manmoel, between the Ebbw and Sirhowy valleys, and to the upland area between Argoed and Aberbargoed, on the interfluve between the Sirhowy and Rhymney rivers. To the south is the ancient hamlet of Mynyddislwyn. On the eastern side, the area reaches almost to Newbridge, in the Ebbw Valley, whilst to the west it reaches almost to Ystrad Mynach and Bargoed in the Rhymney Valley. The location of the study area is shown in Fig. 2.

The general topography of the study area is a gently undulating plateau ranging from 200 m AOD in the south to 300 m AOD in the north, dissected by the rivers Rhymney, Sirhowy and Ebbw. The main river valleys trend almost N–S and are typically 150 m deep. The valley side slopes are steep in the north, often over 30°, but become much gentler around Blackwood and Pontllanfraith. In the south, the valleys once again become more sharply incised.

The surface bedrock geology consists of the Grovesend Beds of the Upper Coal Measures. These are mainly sandstones with subordinate mudstones, and several coal seams. The base of the Grovesend Beds is defined by the Mynyddislwyn Coal Seam.

In the north the beds dip gently at about 5°SSE. The south of the area is more heavily faulted and so the bedding dip is variable, although generally it remains gentle, about 5°, typically to the north or northwest. Faults follow two main trends, one NE–SW the second NE–SE. The latter is typical of faulting in the eastern part of the South Wales Coalfield.

The main coal seam exploited at shallow depth is the Mynyddislwyn, which is generally about 1.70 m thick. It is usually divided into two leaves, the Upper and Lower, which are generally separated by less than 1 m of strata in the north but can be separated by as much as 10 m in the south. Two smaller seams lie above the Mynyddislwyn: the Small Rider and Big Rider, which are 250 mm and 900 mm thick respectively. A number of impersistent, unnamed seams have been worked locally.

Details of the geology may be found in Squirrel & Downing (1969) and Strahan (1899).

Mining and seam reputation

Records indicate that the Mynyddislwyn Seam has a reputation for extensive workings almost everywhere in the study area. This was especially so in the north, where the Upper and Lower leaves are close together and the strata are little disturbed by faulting. To the south, the

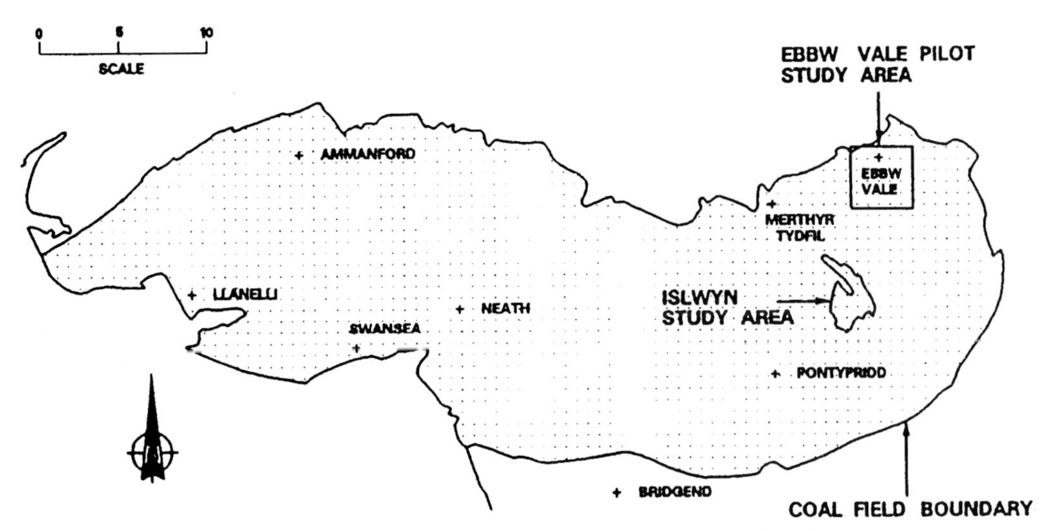

Fig. 2. The South Wales Coalfield showing study area locations.

two leaves are up to 10 m apart and have been exploited separately. In this area, the poorer quality, lower leaf is often unworked, although the upper leaf was still widely exploited. Many mine plans show secondary reworking and make reference to large areas of 'old workings' or 'coal worked out', with no details of the mine layout.

The associated rider seams and other mineral horizons have not been widely exploited and only occasionally present a potential for subsidence.

Methods of working

Some evidence for isolated areas of bell pit workings have been found. However, the majority of workings in the Mynyddislwyn Seam were by pillar and stall methods (Fig. 3). At first, extraction was only partial, the pillars remaining in place to support the roof. Later, almost total extraction was achieved by pillar and stall methods in a single operation.

Manual longwall methods were carried out in the late 19th and early 20th century, particularly on the Lower Mynyddislwyn and Big Rider seams.

Present condition of abandoned workings

No opencast mine excavations exist in the study area, where the condition of old workings might be examined directly. Site investigation and mineral exploration boreholes, however, provide an indirect method of assessment.

The findings of over 700 borehole records showed that significant voidage was only associated with the Upper Mynyddislwyn. On all other seams, few voids were revealed, probably confined to abandoned roadways. These findings agree with the South Wales Desk Study, (Ove Arup & Partners 1985), where inspection of opencast sites showed little voidage, apart from on thicker seams where the amount of backstowing was small because little waste was excavated to win the coal.

During this study, excavations into the Mynyddislwyn Seam for the Newbridge to Maesycymmer Bypass were inspected. Very few examples of significant upward migration of voids were observed during the excavations. However, some collapses were induced by site traffic where the depth to the roof of workings was very shallow, typically 2–3 m. The observed examples of upward void migration usually had reached no more than 2–3 m above the roof of the seam. The main voids observed in the excavations were on open roadways.

Preparation of development advice maps

Method

The method of preparing the maps was originally devised for the South Wales Desk Study (Ove Arup & Partners 1985). It was revised during the Islwyn Study to take account of recent changes in availability of data. The modified method is shown in Table 1.

Zones represented on maps

The maps show the following zones and features, an extract from one of the Development Advice Maps is shown on Fig. 4.

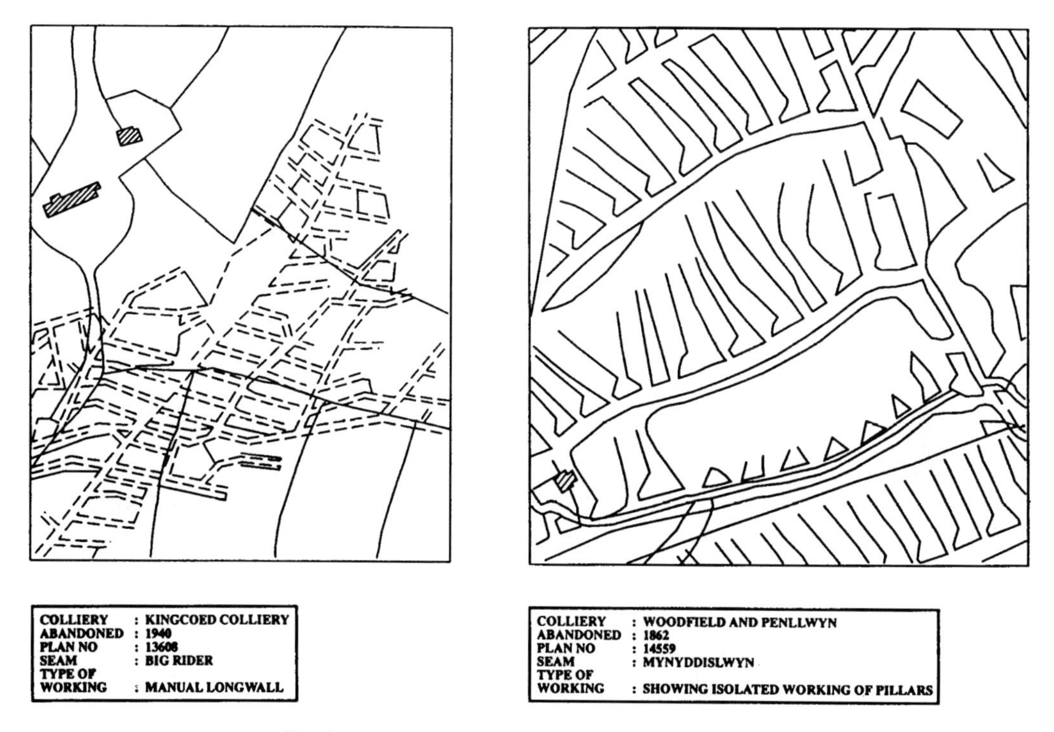

Fig. 3. Mine plans showing typical working methods.

- *Mine entrances*: shafts are shown as solid circles and adits are represented as arrows whose direction indicates the trend of the entrance into the ground. They have been derived from a 1:10000 scale plot from the British Coal Authority computer database, supplemented by historical plan information and observations from BGS maps and field slips.
- *Corrected seam outcrops*: where data gathered during the study are of sufficient quality, the outcrop positions can be altered.
- *Faults*: faults are taken largely from the geological map although some corrections have been made, such as those encountered on mine plans, or inferred from site investigations.
- *Mining subsidence consideration zones*: these are the zones where abandoned shallow mining should be taken into consideration in the planning process. The zones correspond to areas where 90% and 100% respectively of all subsidence incidents would be expected to occur, above the seam in question.
- *Uncertainty zones*: are shown along the positions of faults and seam outcrops on the plan.
  - *Faults*: fault uncertainty zones are taken as 50 m. Here the 'red' and 'yellow' zones extend for an arbitrary distance of 50 m beyond the recorded fault position.
  - *Outcrops*: outcrop uncertainty zones are typically based on topography, 30 m on valley slopes and up to 100 m on valley floors. Some adjustment has taken place where borehole information allows.

Selection of rock cover limits

The rock cover limits for the 'mining consideration' zones are based on back-analysis of subsidence incidents which have occurred on the relevant seams, and their equivalents throughout the coalfield. H/T ratios have been determined where H is rock cover thickness over the worked seam and T is the void thickness, taken as 2 m (the thickness of a typical roadway).

For the Mynyddislwyn Seam and its equivalents elsewhere in the coalfield, there are records for 33 incidents, 24 of which were collapses into workings or adits below rockhead. A histogram of H/T versus the number of incidents is given in Fig. 5. The values of H/T corresponding to 90% and 100% respectively of all incidents are 12 and 15, i.e. rock cover thicknesses of 24 m and 30 m.

However, subsidence incidents on the Mynyddislwyn Seam in the study area with H/T ratios above 10 relate to incidents where the seam thickness is close to 2 m, i.e. in the northern part of the study area where the separation

Table 1. *Modified method for preparing development advice maps in South Wales*

	Activities	*Source of data*
Step 1	*Prepare best available geological map.* BGS published map is used as starting point, modified where additional information is available and checked against coalfield-wide system for seam correlation. Estimate of drift thicknesses included.	BGS: geology maps (all available editions), file information and field slips. Opencast Executive: prospecting records, development maps. Readily available site investigation reports: e.g. from local authorities, British Geological Survey, etc.
Step 2	*Establish 'workable seams' and positions of recorded mine entries*	Study of local mining history: readily available archival material, local historians. Coal Authority Abandoned Mines Department: review mine plan catalogue and view mine plans likely to contain geological information and seam elevations. Study mine plans to assist in dates, methods, seams worked. Obtain computer printout of mine entrances. Mine plans, historical plans and geological maps: a strategic area review of all of the above should enable most mine entrances to be located and checked against the computer database for reliability. Aerial photographs: typically available from 1946 through to the present, useful for showing areas of ground instability from bell pitting and shallow unrecorded workings.
Step 3	*Establish 'seam reputations'* for the area, modified as necessary from the previous whole coalfield study.	South Wales Mining Subsidence Desk Study Report: particularly local subsidence incidents (brought up to date). Opencast Executive: experience in nearby pits, if available. Shallow Workings: exposed in existing mines, if possible. Aerial Photographs: will help identify those areas that have experienced ground instability and can be used with the subsidence record to determine mining consideration zones.
Step 4	*Draw Development Advice Map*	South Wales Mining Subsidence Desk Study Report: procedure for choosing zones, see below.

between the two leaves is small. In the southern section of the study area no examples of subsidence incidents with H/T greater than 4 have been discovered. In the south the Mynyddislwyn Seam is separated into two leaves some 10 m apart, which have always been worked independently. Therefore the values of H/T in the southern part of the study area may be taken as 6 and 10 for limits of 90% and 100%, giving rock covers of 12 and 20 m. These figures are based on the general subsidence record for the South Wales Coalfield taken for gently dipping coal seams (Ove Arup and Partners 1985).

Mine entrances

It should be realized that the mine entrance data are a compilation of many sources with no interpretation as to whether different sources show the same entrance but in slightly different positions. This applies equally to the Coal Authority Database, which is itself a compilation. It is therefore certain that there are a greater number of records than of entrances they represent. Conversely, early unrecorded mineworkings may have been completely obliterated and may not be shown on any primary sources of data. Hence there is always a possibility of unrecorded entrances associated with the seams.

Discussion

The subsidence hazard in South Wales

The subsidence incident database collected in two studies (Ove Arup & Partners 1985, 1995) indicate the following:

- Almost all incidents are 'crownholes' in form; delayed general subsidence is rare, only three examples having been collected.

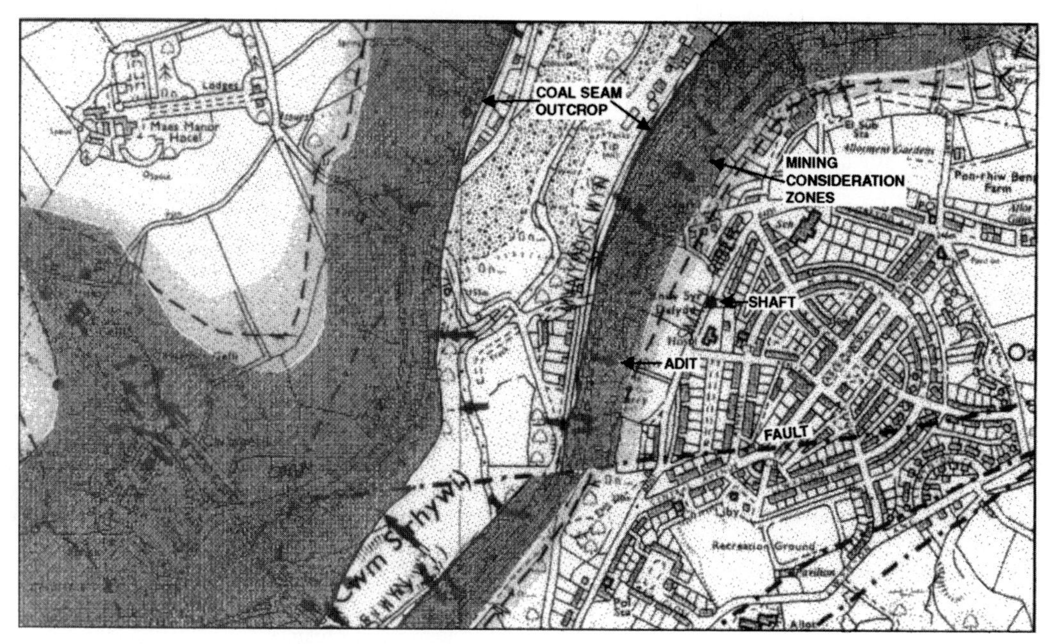

Fig. 4. Extract from Development Advice Map.

- About 75% of subsidence incidents are caused by the collapse of mine entrances or workings close to outcrop.
- Mine entrances pose the greatest risk and account for over 50% of recorded incidents.
- The recorded rate of occurrence of incidents is very low, on average between 10 and 15 per year.

For 90% of incidents, H/T was less than 6, and in 75% of all cases it was less than 2. It is important to realize the basis upon which the H/T ratio has been assessed. In almost all cases there was no detailed information on depth to seam, which was usually interpreted from geological considerations. Similarly, the original height of void (T) was rarely known and an

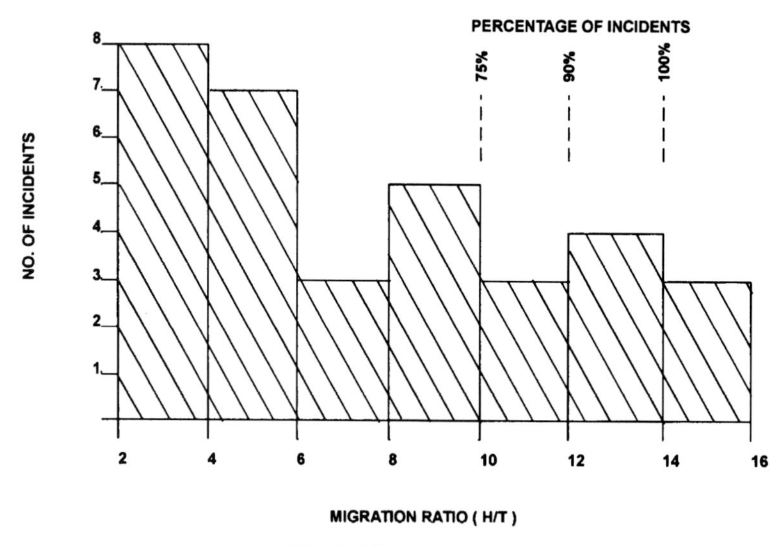

Fig. 5. Migration ratios.

assumed value of 2 m, a typical roadway height, has been taken. Hence, there is nothing absolute about the H/T values calculated; had a different assumption been made, the ratios would have changed. This presents no problem when applying the subsidence results to the production of the maps, as long as the assumptions remain consistent. However, comparisons with other published predictions of height of collapse migration above workings should be treated with caution.

Figure 6 shows a plot of D/T against H/T for some 40 crownholes where depth was recorded, D being the depth of the crownhole at the surface, and T and H as defined previously. Also plotted are the anticipated limits to stable arch formation (based on Garrard & Taylor 1988), assuming a roadway of 2 m width, and to void migration by collapse and bulking theory. The latter assumed cylindrical collapse and 25% bulking. Only 20% of the incidents occur within the limits expected from stable arch formation as defined by Garrard & Taylor (1988). For void migration and bulking of collapsed debris, nearly 60% of the incidents could be explained. However, the remaining 40% fall outside the limits of these deterministic models. This suggests a large element of non-deterministic behaviour for incidents at the upper limits of the crownhole process. These incidents are probably 'special circumstances', e.g. where collapsed material has continually been removed from the mine to form a larger void, or where groundwater movement may prevent debris accumulating beneath the collapse feature as it forms.

It is concluded that a deterministic model for assessing crownhole migration is misleading and that a statistical approach should be used based upon the subsidence record. This assumes that the past is a reasonable predictor of the future, which may not be true. However, studies of subsidence incidents over a period of 30 years have shown no increase in the occurrence of serious incidents that have presented a risk to property or life. Consequently, it is reasonable to assume that ongoing subsidence in the South Wales Coalfield is a result of residual voidage and is not likely to increase in the future.

There is strong evidence from data gathered from grouting contracts, opencast sites and site investigations that migration above workings generally extends to a limited height above the original void. Records of over 10 000 drill and grout boreholes have been collected from many sites around South Wales. In almost every case, the record of broken ground, workings or voids associated with a seam occurred at or immediately above the anticipated elevation of the worked horizon. The only boreholes which encountered voids migrating further were above major access roadways into mines; here collapses would be continually removed by miners to keep the roadway open and thus effectively increase the void height on abandonment. Observations made from opencast sites (Ove Arup & Partners 1985), site investigation boreholes and the Newbridge to Maesycymmer Bypass (Ove Arup & Partners 1995), also indicate that migration of voids rarely exceeds twice the original void height.

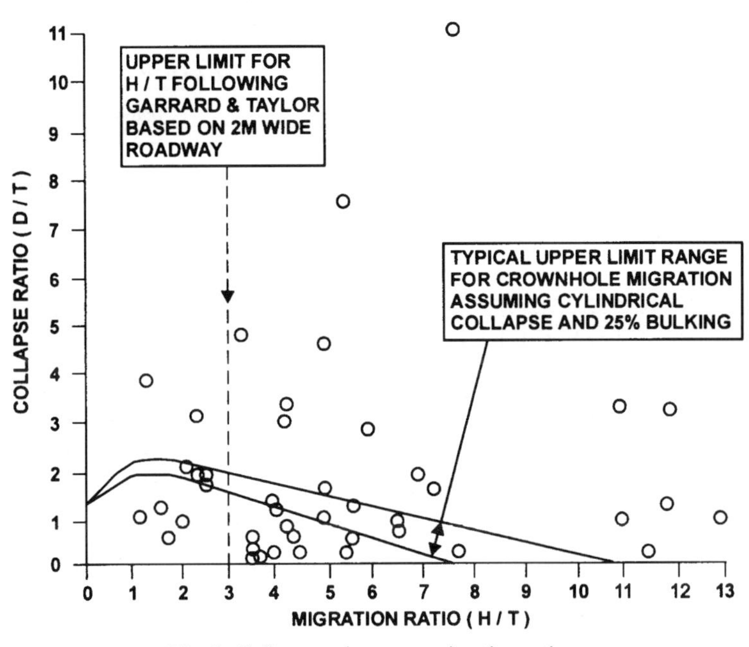

Fig. 6. Collapse ratio versus migration ratio.

Table 2. *Annual risk of subsidence due to abandoned mining in South Wales*

Location in coalfield	*Annual risk level*	*Risk classificaiton*		*Public expectation*
		To life	*To property*	
Whole coalfield	1×10^{-7}	Very unlikely	Negligible	About what would be expected in a mining area for structures
At mine entrances	1×10^{-4}	Some risk	Unlikely	Not acceptable in public buildings, workplaces or houses
Near-seam outcrops	5×10^{-6}	Slight chance	Unlikely	About what would be expected in open spaces in a mining area
Elsewhere (away from entrances and outcrops)	1×10^{-8}	Practically impossible	Negligible	Better than would be expected in a mining area for structures

From the above it may be concluded that the majority of residual voids have already collapsed to a stable profile. Otherwise, voids would be encountered at a wide range of depths as they progress to the surface.

Under these circumstances it seems unlikely that the rate of crownholing is about to increase significantly.

Risk assessment

Risk (R) may be defined as:

$$R = F \times C$$

where F is the frequency of an undesirable event and C is the consequence of it happening. Consequences may be financial loss as a result of damage or disruption of economic activity, or may mean injury or death. This definition of risk is accepted by the Health and Safety Executive, who publish guidelines of tolerable risks to life and limb (Health and Safety Executive 1989).

Cole (1988, 1993) has proposed a general strategy for making engineering decisions on the treatment of mine-workings based on risk assessment. He has used the notions of 'risk classification' and 'public expectation' to derive a matrix of risk acceptance by society. To this matrix, the 'annual risk level' calculated by Statham & Treharne (1991) has been added for various locations, e.g. 'whole coalfield', 'mine entrances'. These are shown in Table 2.

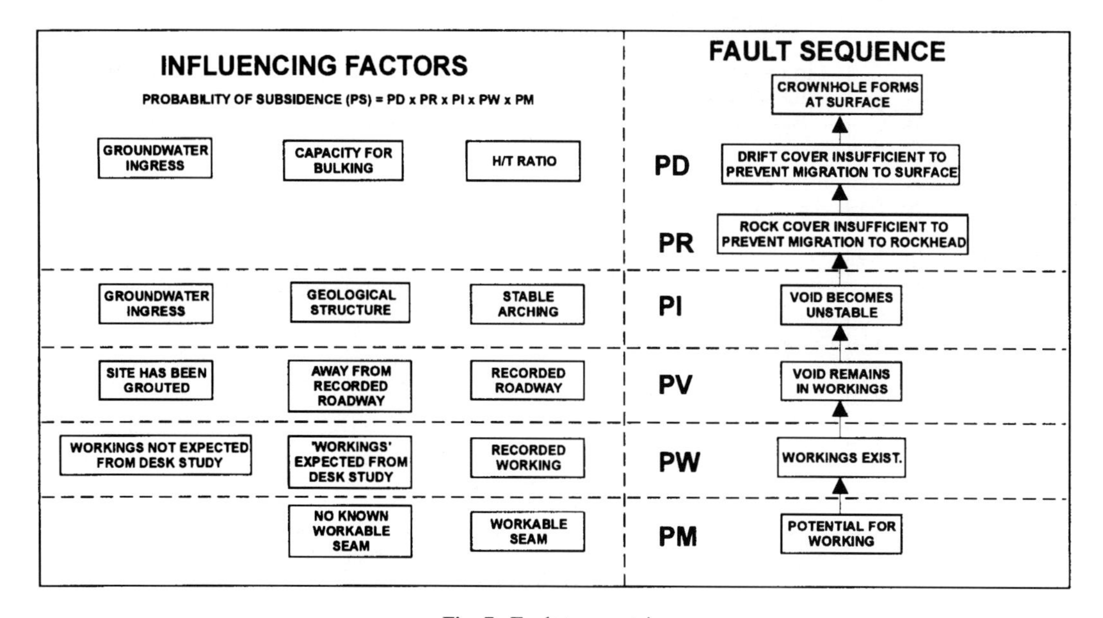

Fig. 7. Fault tree matrix.

It must be realized that 'acceptance of risk' is an emotive and subjective issue. It varies considerably between individuals and with different circumstances for the same individual.

The assessment of risk due to mining subsidence is a twofold process, involving first the assessment of the probability of a subsidence event occurring, i.e. the subsidence potential; and secondly, the consequences resulting from that event.

For a crownhole to occur, the following must coincide at any point:

- there must be a potential for mining (PM);
- there must be workings under the site (PW);
- a void must remain in the workings (PV);
- the void must be unstable and able to migrate to the surface (PI);
- the thickness of the rock overburden must be insufficient to choke the collapse feature, before reaching rockhead (PR);
- the thickness of drift must be insufficient to prevent the chimney reaching the surface (PD).

A typical 'fault tree matrix' for assessing crownhole subsidence risk at a given site is shown in Fig. 7. Associated risk assessment is discussed in more detail in Statham & Scott (1992).

Potential for wider application in the UK

There is no reason why the principle of Development Advice Mapping should not be extended to other coal mining areas the UK. The general methods used in South Wales are considered applicable, subject to modification depending on local conditions and data availability.

However, age and extent of workings, methods of mining and geological characteristics (seam stratigraphy, geological structure) all influence the subsidence process and it would be dangerous to assume that the findings from this study and the previous study (Ove Arup and Partners 1985) in connection with subsidence are directly transferable elsewhere in the UK.

Accordingly it would be necessary to carry out a first-stage study of each mining area to establish the general historical and mining framework, the reputations of the various seams and the mechanisms of subsidence which predominate. From this, the general principles for establishing risk zones may be determined, as in the previous study.

Future trends

The technology of data input, handling, manipulation and interrogation is developing very rapidly. GIS linked to databases are now widely available and provide a powerful tool for routine handling of any area-based data. There is no reason why Development Advice Maps for specific planning consultations cannot be mounted in such a system and used as a planning and development control mechanism, or in the wider application of local and structure plan compilation.

Acknowledgements. This research project was promoted by the Department of the Environment and managed by various officers; in particular, assistance from H. Payne and C. Morgan of the Welsh Office and D. Brook of the DoE was much appreciated. Thanks also go to members of the Steering Committee Group who gave much helpful criticism throughout the study. The views presented in this paper are those of the authors and not necessarily those of the Welsh Office or Department of the Environment.

References

COLE, K. W. 1988. *Ground Engineering. Building Over Abandoned Shallow Mines: A Strategy for the Engineering Decisions on Treatment.*

——1993. Building over abandoned shallow mines. Paper 1: Considerations of risk and reliability. *Ground Engineering.*

—— & STATHAM, I. 1992. General (areal) subsidence above partial extraction mines. *Ground Engineering* March 1991 Part 1 p. 44–55; April 1991 Part 2 p. 36–40. Parts I & II.

DEPARTMENT OF ENVIRONMENT/WELSH OFFICE 1990. *Development on Unstable Land.* Planning Policy Guidance No. 14.

GARRARD, G. F. G. & TAYLOR, R. K. 1988. Collapse mechanisms of shallow coal workings. *In*: BELL, F. G., CULSHAW, M .G., CRIPPS, J. C. & LOVELL. M. A. (eds) *Engineering Geology of Underground Movements.* Geological Society, London, Engineering Geology Special Publications, **5**, 3–32.

GEOMORPHOLOGICAL SERVICES LTD 1987. *Review of Research into Landsliding in Great Britain.*

HEALTH AND SAFETY EXECUTIVE 1989. *Quantified risk assessment: Its input to decision making.* HMSO, London.

HEALY, P. R. & HEAD, J. M. 1984. *Construction over Abandoned Mineworkings.* CIRA Special Publication No. 32.

HMSO 1991. *Building Regulations.* HMSO, London.

MINERAL PLANNING GUIDANCE 1994. *Treatment of Disused Mine Openings and Availability of Information on Mined Ground.* MPG 12.

OVE ARUP & PARTNERS 1985. *South Wales Mining Subsidence Desk Study.* Report No. 85/1100, Open File Report at Welsh Office/DOE.

——1991. *Review of Mining Instability in Great Britain.*

——1995. *Islwyn Shallow Mining, Development Advice Maps.* Department of the Environment/Welsh Office.

SQUIRRELL, H. C. & DOWNING, R. A. 1969. *Geology of the South Wales Coalfield, Part 1: The Country around Newport*, 3rd edition HMSO, London.

STRAHAN, A. 1899. *The Geology of the South Wales Coalfield, Part 1: The Country around Newport*, 1st edition, HMSO.

STATHAM, I. & SCOTT, M. J. 1992. *Grouting in the Ground Conference, An Analysis of Drill and Grout Records from the South Wales Coalfield.*

—— & TREHARNE, G. 1991. Subsidence due to abandoned mining in the South Wales Coalfield, UK: causes, mechanisms and environmental risk assessment. *In*: JOHNSON, A. (ed.) *Land Subsidence*. IAS Publication No. **200**, 143–152.

WHITTAKER, B. N. & REDDISH, D. J. 1989. Subsidence: occurrence, prediction and control. *Developments in Geotechnical Engineeering*, **56**. Elsevier, Amsterdam.

Some issues associated with the preparation of a review of foundation conditions in Great Britain

David Earle,[1] Brian Marker,[2] Paul Nathanail[3] & Judith Nathanail[3]

[1] ENSR International Limited, 16 Frogmore Road, Hemel Hempstead, Herts HP3 9RW, UK
[2] Department of the Environment Transport and the Regions, Eland House, Bressenden Place, London SW1E 5DU, UK
[3] Centre for Research into the Built Environment, Nottingham Trent University, Burton Street, Nottingham NG1 4BU, UK

Abstract. The Department of the Environment commissioned Wimpey Environmental Limited, with the National House-Building Council, to undertake research to establish whether planners should take greater account of foundation conditions in preparing development plans and in considering planning applications.

Existing information on factors such as compressible ground, shrinkage and swelling, saturated and loose granular deposits, frost susceptibility, groundwater conditions, durability of construction materials, and gas emissions, was reviewed. It was concluded that some of these factors may be material planning considerations whilst others are likely to remain only a matter for site development and Building Control.

Soil survey digital data were used as a basis for 1:250 000 and 1:625 000 scale maps showing the general geographical extent of selected potential foundation problems in Great Britain. These, together with reports, provide information on ground conditions to planners and developers which will help to increase awareness of the problems and of the purposes, procedures and benefits of thorough site investigation. The preparation of the maps was guided by a database of site investigation information which should be of use to those who are undertaking more detailed consideration of sites.

Background

In the mid-1980s, the Department of the Environment set out a strategy for research with the aim of securing better consideration of land instability and related geological issues in land use planning and control of development. The research was intended to form the basis for preparing and implementing planning policy guidance.

The research consisted of

- a series of national reviews of specific stability and safety problems; and
- investigations of selected areas in order to give practical demonstrations of how ground problems should be taken into account in planning and development.

The national reviews commenced with consideration of landsliding (Jones & Lee 1994) and went on to examine mining instability (Arup Geotechnics 1992), natural underground cavities (Applied Geology Ltd 1994), seismic risk (Ove Arup & Partners 1993), natural contamination (Appleton 1995), other foundation conditions (Wimpey Environmental Ltd and National House-Building Council 1995) (the subject of this paper), and erosion, deposition and flooding (Rendel Geotechnics 1995). The results, together, comprise a comprehensive overview of ground problems in England, Scotland and Wales.

The planning policy guidance commenced with general guidance on *Development on Unstable Land* (DoE 1990) and has been amplified by an annex on landslides and planning. The need for an additional annex on subsidence and planning is under consideration and this will draw upon results from the reviews of mining instability, natural underground cavities, and of other foundation conditions.

The 'Review of the significance of foundation conditions for planning and development in Great Britain' was commissioned by the Department of the Environment in 1991. The work was undertaken by Wimpey Environmental Ltd (now ENSR International Limited) in association with the National House-Building Council (NHBC) and was completed in 1995. The main purposes of the research were as follows:

- to establish the nature and extent of selected foundation conditions in Great Britain;
- to review the administrative and technical responses available for dealing with the ground problems;

EARLE, D., MARKER, B., NATHANAIL, P. & NATHANAIL, J. 1998. Some issues associated with the preparation of a review of foundation conditions in Great Britain. *In*: MAUND, J. G. & EDDLESTON, M. (eds) *Geohazards in Engineering Geology*. Geological Society, London, Engineering Geology Special Publications, **15**, 401–407.

- to assess which, if any, of the foundation conditions required a response through the planning system; and
- to indicate the nature of information on foundation conditions that is required for use in the planning system and the best means for presenting such information.

The selected foundation conditions were limited, in the contract, to those which had not been dealt with fully, or at all, in the other reviews in the series or within other departmental research programmes. These were compressible ground, shrinkage and swelling, saturated and loose granular deposits, frost susceptibility, groundwater conditions, durability of construction materials, previous development, fill (but excluding contaminated land), and gas emissions. All mentions of 'foundation conditions' in this paper refer only to those in this list.

This paper outlines briefly the reasons for undertaking the work and the approaches used, and considers the significance of foundation conditions in planning.

The planning system

The planning system in England and Wales aims to regulate the development of land in the public interest. There are two main aspects:

- local planning authorities prepare development plans which set out policies for the use of land and indicate criteria which may be relevant to the decision of planning applications; and
- planning applications are submitted by developers and are decided by planning committees on the basis of information that is material to the decision.

The physical condition of the ground is commonly taken into account in the planning process in some specific settings. In coalfield areas, for example, there are often development plan policies pertaining to the use and development of mined ground and there is a general awareness that, in some parts of such areas at least, the stability of the ground may be material to whether development goes ahead or not, or the form which it takes. There are many other ground factors which are less familiar to planners, however, and which are rarely taken into account in the planning process even though some of these give rise to widespread problems. These include most of the issues identified for study in the review of foundation conditions.

Some planners have felt that there is no need to consider such factors because these are taken into account under the Building Regulations (DoE & The Welsh Office 1991). Whilst it is a basic principle that there should be no duplication between planning provisions and the building regulations, the regulations do have a limited scope. They apply only within the site and only to the construction and occupancy of buildings and not to other parts of the site or other site works. In addition, not all development is built development and subsequent changes to the use of land do not come within the building regulations but are covered by the planning system. Because of this, there was a need to consider to what extent the selected foundation conditions were relevant to planning.

Significance of foundation conditions in planning

A useful way of considering foundation conditions is in terms of the scale of their effects. Foundation conditions vary from minor difficulties which can be readily dealt with by simple preventive or remedial measures, to major problems which may present a significant threat to property or people. The first step was to establish the level of significance for each of the selected foundation conditions in respect of the land use planning process. The following criteria may be used:

- problems which may pose a hazard to life and limb require special consideration, for instance emissions of explosive or asphyxiant gases; less immediately obvious are contaminated fill or unsafe structures left from previous development;
- the cost of remedial measures which may be extremely high in respect of, for instance, gas emissions, contaminated land, or unexpected groundwater conditions;
- the potential difficulty of identifying the nature and scale of a problem such as gases originating outside the boundary of a proposed development site, some types of contamination where prior uses of the site are not known, and some adverse groundwater conditions;
- the extent of a particular problem and the ease with which it can be delineated.

On the basis of such considerations it is possible to place the selected foundation conditions into a crude relative ranking (Table 1). In general terms, those ranked towards the top of the table are more significant to development and use of land and are, therefore, more likely to be material planning considerations in a given set of circumstances. Thus, it was concluded that groups A to C may constitute material planning considerations but groups D and E were unlikely to be so. A distinction can also be made between those ground conditions which may be relatively difficult to predict (group C) and those which are more easy to assess (groups D and E) and thus are more readily accommodated within the planning process. Groups A to C (conditions 1 to 6), therefore, may be considered, where appropriate, as material to the formulation of development plans or to planning decisions. In general, those conditions which are higher in the sequence shown in the table are more likely to be of interest to *planners* than those which are

Table 1. *A Classification of the selected foundation conditions in terms of relative significance to planning and development*

Group (A, B)	Condition	Rank	Group (C, D, E)
A, B	Methane	1	C
A, B	Carbon Dioxide	2	C
A, B	Fill	3	C
A, B	Previous Development	4	C
B	Groundwater	5	C
B	Saturated loose granular deposits	6	C
B	Shrinkability	7	D
B	Compressibility	8	D
	Sulphate	9	E
	pH	10	E
	Frost-susceptibility	11	E

A: Potentially hazardous to life
B: Possibly involving significant costs to overcome or mitigate the condition
C: Difficult to predict (may also be costly and/or hazardous)
D: Predictable from site investigations (may be costly to overcome or mitigate the condition)
E: Predictable from site investigations (low cost implications)

lower down. In some localities, however, the sequence may need to be modified.

The potential future impacts of gound problems, and the associated costs, may be a factor in selection of sites for development and thus of interest to prospective developers. Local planning authorities may also need to take this into account when allocating land for specific purposes in local plans or when preparing briefs in respect to the sale of land. Table 1 can be recast in terms of future impacts (Table 2).

The classification in Table 2 is set out with the planning system in mind. It does not mean, of course, that those placed near the bottom of the table are less important in other contexts, such as site investigation and evaluation, where all relevant conditions must be taken into account when appropriate.

Having established that some foundation conditions are more likely to be material planning considerations than are others, it is necessary to consider how these might be taken into account by planners in such a way that the provisions of the Building Regulations are not duplicated.

The planning response

It seems, therefore, that some of the foundation conditions considered in this work may be material planning

Table 2. *Future impact of foundation conditions on planning and development*

Rank	*Condition*	*Future impact on planning and development and associated costs*
1	Methane	Increasing significance, impact and cost implications because of a backlog of problems, regeneration of derelict land and increased legislation
2	Carbon dioxide	
3	Fill	
4	Previous development	
5	Groundwater	Steady significance, impact and cost implications (or possibly falling because of improved methods, better information, etc.)
6	Saturated loose granular deposits	
7	Shrinkability	Falling significance, impact and cost implications because of better awareness, improved ground treatment and more knowledge of the problems.
8	Compressibility	
9	Sulphate	
10	pH	
11	Frost susceptibility	

considerations in some areas. The limitations to the scope of the Building Regulations are such that some ground problems may not be considered at all in relation to proposed new development unless they are taken into account in the planning process. There is, thus, some role for the planning system without duplicating the functions of building control.

The handling of any individual application for development will need to take account of any potential hazard both to the development itself and to the surrounding area. Whilst each application should be considered on its merits, the local planning authority will need to be satisfied by the developer that any significant hazard has been taken into account. Where there are good reasons to suppose that there may be problems which could make the ground unsuitable for development a specialist investigation and assessment should be secured. It is the responsibility of the developer to investigate the ground conditions at any specific site. The local planning authority will need to satisfy itself, however, that adequate information is provided with the planning application for a decision to be taken. If information is inadequate the authority may request additional information or may refuse the application.

This presupposes that both the developer and the planning authority are alert to the possibility of specific ground problems occurring in a particular development site. Such awareness may be provided by statutory and other consultees who comment on a planning application, or by other departments within a local authority such as building control which may be consulted. However, there may be a risk that potential problems may be overlooked unless there is a readily available source of information.

The best way of indicating that particular issues may need to be taken into account in the planning process is by mentioning these in the relevant development plan either through policies, within lists of criteria for considering planning applications, or in supplementary planning guidance. All local planning authorities are charged with undertaking surveys of the principal physical characteristics of their area as a basis for preparation of, or amending, a development plan. Unless there is a means of alerting the relevant authority to issues which might not otherwise be obvious, there is a danger that some of these may be overlooked and thus may not be obvious to planning applicants either.

As a first step, therefore, it is important to have a general national statement of the nature and extent of various ground problems so that the relevant authorities can consider whether these are relevant to their area and are significant enough (a) to warrant more detailed examination within the development plan; and (b) to be drawn to the attention of prospective developers as matters which may be taken into account in deciding a planning application. Part of the purpose of the present research has been to prepare such information.

Preparation of national maps

Three types of data, each with particular advantages and failings, could be used for the preparation of national maps showing the selected foundation conditions. These data types are

- site investigation data
- geological survey data
- soil survey data

Site investigation data

Site investigation data provide specific and factual information about particular locations. For example, it might be possible to say, on the basis of trial pits, boreholes and *in situ* tests, that underlying a particular site is loose granular material associated with a high water-table. Thinking in terms of a localized site or a restricted area, this would be extremely valuable information in providing forewarning of possible adverse ground conditions. The database complied as part of this review is information of this type. The clear advantage of these data lies in the site-specific application; the value generally becomes diluted if used in an attempt to delineate a regional distribution, because of the sparsity of data points. Obviously, the regional mapping possibilities improve with increasing numbers of data points and for some foundation conditions it may then be possible to obtain useful indications of distributions. Using this type of data, it should be remembered that the data points may not be of equal value. Some site investigations will be more comprehensive than others and some may have simply failed to detect, or to search for, particular ground condition problems.

Geological survey data

An engineer or geologist concerned with the ground conditions at a site will usually refer first to Geological Survey data, generally in the form of hard copy maps and associated memoirs. Considering regional foundation condition mapping, it would be possible to allocate certain potential foundation conditions to particular geological formations. In order to undertake this process most efficiently, digital data would be preferred because of their ease of manipulation compared with paper maps. For Great Britain, comprehensive digital geological maps are not available, although this situation will undoubtedly change. Geological mapping of Great Britain is available at a variety of scales although complete coverage for all scales is not yet available. Geological mapping, in some cases, dates back many years. Revision and remapping are constantly undertaken but there are inevitable differences in approach to mapping in different periods and this is especially true in the mapping of drift deposits.

Soil survey data

Soil Survey data deal essentially with surface and near-surface superficial materials to a maximum depth of consideration of about 1.5 m and are concerned with pedological rather than engineering soils. Foundations are often constructed at fairly shallow depths and in many cases are placed in drift rather than on bedrock. Soil Survey data are often a reflection of the underlying 'solid' geology and a case can be made for soils data representing both 'solid' and 'drift' geology.

A complete coverage of soils data in digital format exists for Great Britain, albeit that the data have been produced by two separate Soil Surveys (for England and Wales and for Scotland) using different soil classification schemes. Nevertheless, there is a relatively evenly based Soil Survey for the whole of Great Britain compared with the Geological Survey where mapping originates from different dates. Soil survey data are produced by the Soil Survey and Land Research Centre (SSLRC) for England and Wales and by the Macaulay Institute for Scotland.

Table 3 compares the three sources of information for constructing foundation condition maps.

Because of the complete digital coverage and the close correspondence of these data to surface geology, soils data were used in preparing a set of summary maps at 1:250 000 and 1:625 000 scale. The soils data were used to classify areas in terms of general ground conditions and the results were moderated using site investigation records held on file by Wimpey Environmental Limited and the NHBC.

Use of the maps

The maps should not be used as more than a general guide to the ground conditions in any specific area. They certainly must never be used in place of adequate site investigation. Their purpose is to alert planners and developers to the range of foundation conditions which may occur so that consideration can be given to what form of more detailed investigation is required and the nature of the professional advice that should be secured. The maps are accompanied by a report which describes the background to the study, explains how the results should be used, and provides an introduction to aspects

Table 3. *Comparison of methods of constructing foundation conditions maps*

Using site investigation data	*Using Geological Survey data*	*Using Soil Survey data*[a]
Used to produce inventorial foundation condition maps	Used to produce interpretative or predictive foundation condition maps	Used to produce interpretative or predictive foundation condition maps
Consists of site-specific ground information which requires a large number of data points to be useful on a regional scale, with consequent cost implications	Consists of regional information on 'solid' and 'drift' geology	Consists of regional information on surface and near-surface materials but often related to underlying geology
–	Available at a variety of map scales but not full coverage of Great Britain at all scales	Available at a variety of map scales and full coverage of Great Britain available at 1:250 000
Site investigation information can be compiled onto databases	Complete digital coverage not available	Complete digital coverage of Great Britain at 1:250 000
Site investigations are not all of equal value	Mapping carried out in different parts of the country at different dates with different emphasis (e.g. variation in detail of drift mapping)	Evenly based mapping of Great Britain at 1:250 000 scale although different soil classification schemes employed in England and Wales and in Scotland
Original site investigation reports may be available for consultation	Memoirs and regional guides available for detailed and background information	Bulletins and engineering databases provide ancillary information
Coverage is often concentrated in urban and near-urban areas with density of site investigations lower outside these areas	Complete coverage of geological mapping including urban areas	No coverage in urban areas for England and Wales
–	Paper maps inexpensive	Paper maps inexpensive. Costs of leasing digital data high

[a] Sources: Hartnup & Jarvis (1979), Olson (1981), Brink *et al.* (1982), Lee & Griffiths (1987), Reeve (1989), Hodgson & Whitfield (1990).

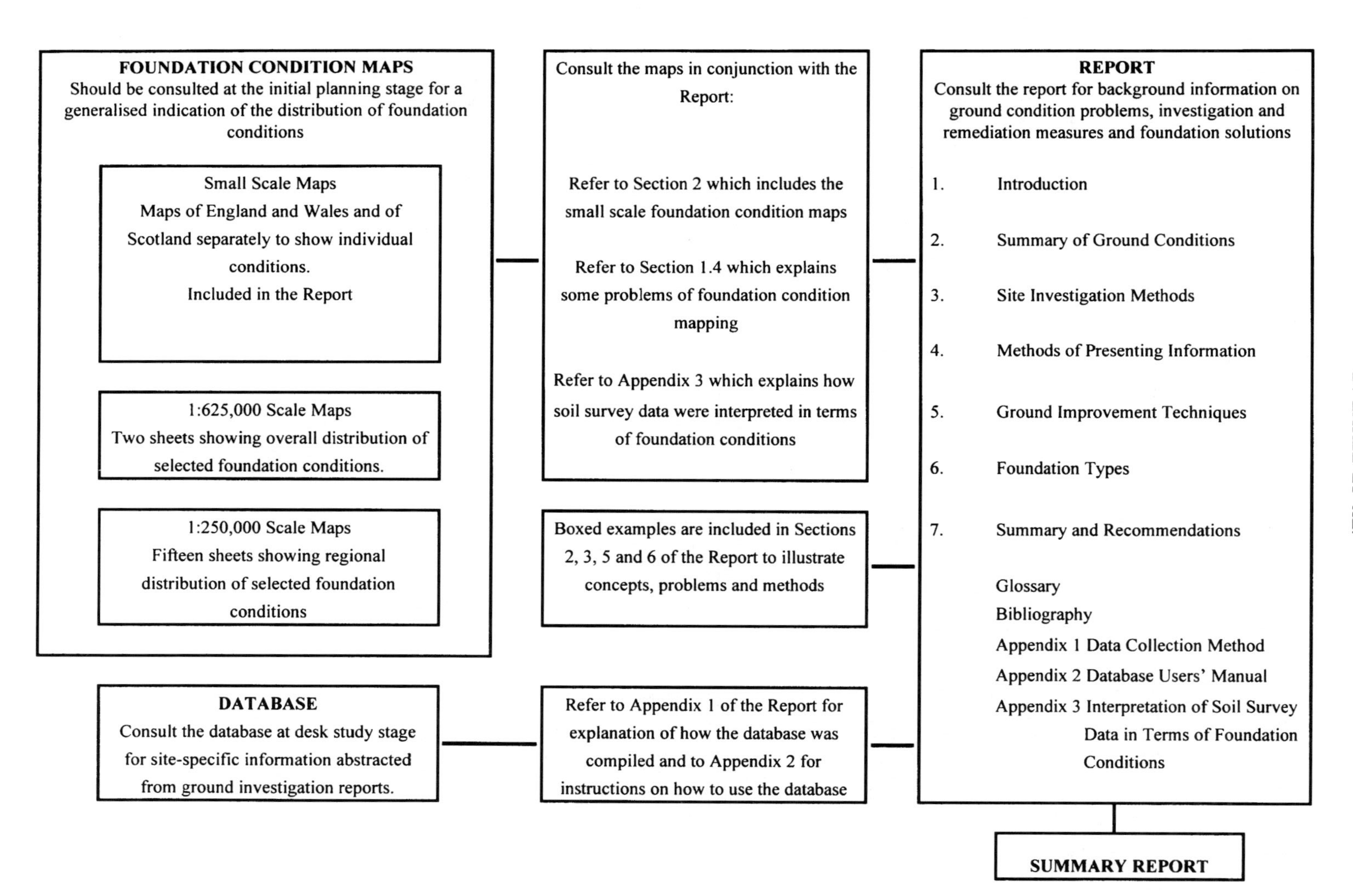

Fig. 1. Relationship between elements of the review.

of ground conditions, site investigation and foundations. The maps should be used in conjunction with the report. Figure 1 explains the relationship between elements of the review.

Conclusions

A brief outline has been given of some issues associated with a Review of Foundation Conditions in Great Britain undertaken for the Department of the Environment. The use of soil survey data to construct foundation condition maps and the significance of foundation conditions to planning and development have been touched on. The underlying purpose of the research has been to increase awareness amongst planners and developers of the range of potential ground condition problems and to emphasize the benefits of reliable and timely site investigation.

Acknowledgements. Thanks are due to the members of the Steering Committee who guided the project. The work was undertaken by Wimpey Environmental Limited and the National House-Building Council in collaboration. Many other staff of Wimpey Environmental (and previously Wimpey Laboratories) provided research input, advice and assistance. Digital soils data were leased from the Soil Survey of England and Wales and the Macaulay Land Use Research Institute. Any views expressed are those of the authors and not necessarily those of the Department of the Environment.

References

APPLETON, J. D. 1995. *Radon, methane, carbon dioxide, oil seeps and potentially harmful elements from natural sources and mining areas: relevance to planning and development in Great Britain.* BGS Technical Report WP/95/4, British Geological Survey, Keyworth.

APPLIED GEOLOGY LTD 1994. *A review of instability due to natural underground cavities in Great Britain.* Applied Geology Ltd, Royal Leamington Spa.

ARUP GEOTECHNICS 1992. *Review of Mining Instability in Great Britain: Summary Report.* HMSO, London.

BRINK, A. B. A., PARTRIDGE, T. C. & WILLIAMS, A. A. B. 1982. *Soil Survey for Engineering.* Clarendon Press, Oxford.

DEPARTMENT OF THE ENVIRONMENT 1990. *Planning Policy Guidance: Development on Unstable Land.* PPG14 and PPG14 (Annex 1) March 1996, Department of the Environment, London.

—— & THE WELSH OFFICE 1991. *The Building Regulations.* Fourth Impression (with amendments) 1994. Department of the Environment, London.

HARTNUP, R. & JARVIS, M. G. 1979. Soils in civil engineering and planning. *In*: JARVIS, M. G. & MACKNEY, D. (eds) *Soil Survey Applications*, Soil Survey Technical Monograph No. 13.

HODGSON, J. M. & WHITFIELD, W. A. D. 1990. *Applied Soil Mapping for Planning, Development and Conservation: A Pilot Study.* HMSO, London.

JONES, D. K. C. & LEE, E. M. 1994. *Landsliding in Great Britain.* HMSO, London.

LEE, E. M. & GRIFFITHS, J. S. 1987. The importance of pedological soil survey in land use planning, resource assessment and site investigation. *In*: CULSHAW, M. G., BELL, F. G., CRIPPS, J. C. & O'HARA, M. (eds) *Planning and Engineering Geology.* Geological Society Engineering Geology Special Publication No. 4, 453–465.

OLSON, G. W. 1981. *Soils and the Environment. A Guide to Soil Surveys and Their Applications.* Chapman & Hall, London.

OVE ARUP & PARTNERS 1993. *Earthquake Hazard and Risk in the UK.* Ove Arup & Partners, London.

REEVE, M. J. 1989. Soils. *In*: MCCALL, G. J. H. &, MARKER, B. R. (eds) *Earth Science Mapping for Planning Development and Conservation.* Graham & Trotman, London, 119–156.

RENDEL GEOTECHNICS 1995. *Erosion, deposition and flooding in Great Britain – a summary report.* Rendel Geotechnics, Birmingham.

WIMPEY ENVIRONMENTAL LTD AND NATIONAL HOUSE BUILDING COUNCIL 1995. Foundation conditions in Great Britain: a guide for planners and developers. Wimpey Environmental Ltd, Hayes, Middlesex.

Development of 'rockhead' computer-generated geological models to assist geohazard prediction in London

P. J. Strange, S. J. Booth & R. A. Ellison

British Geological Survey, Kingsley Dunham Centre, Keyworth, Nottinghamshire NG12 5GG, UK

Abstract. Much of London is founded on Quaternary river terrace deposits, consisting mainly of sand and gravel associated with the course of the River Thames. The base of these Quaternary river terrace deposits, the 'rockhead' surface, is generally planar and rests on London Clay. Locally, however, the terraces may include steep-sided, gravel-filled hollows, some of which penetrate other solid formations below the London Clay, such as the Lambeth Group (formerly known as the Woolwich and Reading Beds).

Unforeseen engineering problems involving deep foundation construction or tunnelling beneath London have resulted from failure to predict irregularities in the base of Quaternary deposits. Detailed knowledge of the three-dimensional shape of the Quaternary deposits can assist in predicting ground conditions. In order to provide such a model more than 22 000 borehole records from London have been encoded into a digital, relational database. Contour plots of specific geological surfaces, including the top of the Chalk, the base of the London Clay and the base of the Quaternary deposits have been generated from the database using computer-modelling software.

Quality assurance is an important part of the databasing procedure and suspect borehole information is discarded at an early stage. As computer modelling proceeds, further data may be eliminated if considered anomalous. The computing infrastructure provides graphical presentations of borehole logs, contour plots, isopachyte (thickness) plots of the major geological units together with cross-sections, fence diagrams, block diagrams and perspective views, giving an enhanced, three-dimensional understanding of the geology of London.

Introduction

The solid geology of London is dominated by Cretaceous Chalk strata and overlying Palaeogene clays and sands (mainly the London Clay Formation and the Lambeth Group). The superficial deposits consist mainly of Quaternary river terrace sand and gravel, and alluvium along the River Thames and its tributaries (Fig. 1).

In early to middle Quaternary time, the ancestral Thames flowed to the north of London, along the alignment of the present day Vale of St Albans (Gibbard 1977). With the southward advance of the Anglian ice sheet, the course of the ancestral Thames was blocked and the river diverted to the approximate line of its present course. Fluvial sands and gravels, assigned to seven river terraces, were subsequently deposited in a 15–20 km wide valley. These terrace deposits average 5 m in thickness and occupy benches cut largely into London Clay, with each terrace base at a well-defined topographic level (Fig. 1).

During the last glacial (Devensian) episode in Britain, the sea-level dropped to about −100 m OD. The Thames cut down into the solid geology during this period and localized deep drift-filled hollows were formed (Berry 1979). These features consist of 'funnel shaped' depressions up to 500 m in diameter and 70 m depth into bedrock. Their mechanism of formation and infilling is speculative, but the presence of ground ice and the effect of hydrostatic pressure in the Chalk have been suggested as possible causes (Berry 1979; Hutchinson 1980). In the subsequent rise in sea level during the Holocene, in the last 8000 years, the valley was drowned and the river terrace gravels overlain by alluvial silt and clay and peat.

'Rockhead' geohazards and engineering

There are a number of geohazard issues associated with the base of the superficial deposits in the London area. These include the shape of 'rockhead', depth to 'rockhead', lithology of the Quaternary deposits and their relationship to bearing capacity and shear strength. The term 'rockhead' as described in this paper refers to the upper surface of the solid geology, although it is recognized that many of the solid geology units of the London area cannot be classified as 'rock' in engineering terms. The presence of water as perched water tables in the drift deposits where they overlie the London Clay also presents a significant geohazard. Similarly, high hydrostatic head when superficial deposits are in hydraulic continuity with deposits below the London Clay can be a concern. Rising groundwater levels are a

STRANGE, P. J., BOOTH, S. J. & ELLISON, R. A. 1998. Development of 'rockhead' computer-generated geological models to assist geohazard prediction in London. *In*: MAUND, J. G. & EDDLESTON, M. (eds) *Geohazards in Engineering Geology*. Geological Society, London, Engineering Geology Special Publications, **15**, 409–414.

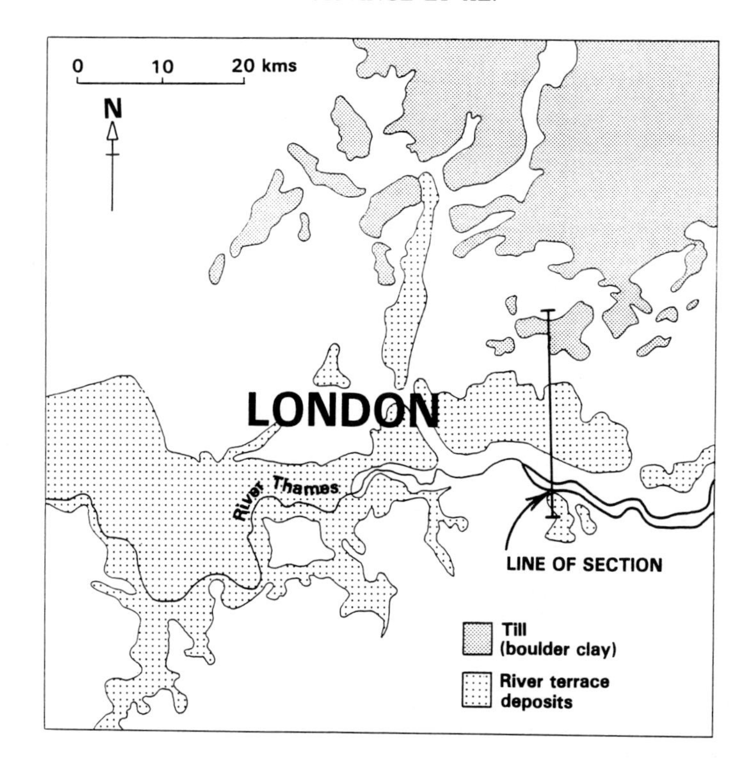

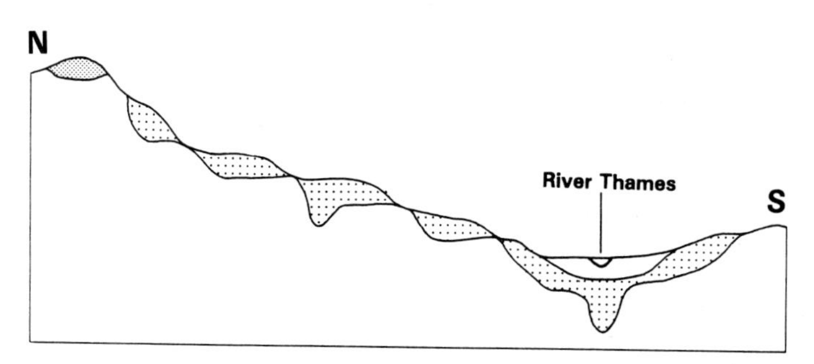

Fig. 1. Distribution of principal superficial deposits of the London region and generalized cross-section.

problem in this last situation, and it is widely recognized that groundwater levels are rising steadily beneath central London. When the groundwater level was depressed, the pyrite in the London Clay oxidized to produce sulphate cations (Kinniburgh *et al.* 1994). Now, however, where the groundwater table is rising through the basal beds of the London Clay there is a serious problem of concentration of acidic fluids which can corrode iron pipework and tunnel linings.

At least 700 km of tunnels have been excavated beneath London for a variety of purposes including underground railways, a complex sewer network, the enclosure of rivers, road and utilities (water supply, electricity, post office, telecommunications), and air raid shelters.

It is well known that the London Clay provides an excellent tunnelling medium, particularly for using tunnel-boring machines. The preference for tunnels to be excavated in London Clay is exemplified by the distribution of the London Underground railway system, mainly north of the Thames where London Clay is at 'rockhead'. However, due to a variety of human-made and geological factors, it was in some cases difficult to confine the tunnels to the London Clay strata and frequently problems, largely due to water ingress, ensued.

Many unforeseen engineering problems involving tunnelling beneath London have resulted from a failure to predict the 'rockhead' levels. In a number of cases, whilst tunnelling through London Clay, excavations

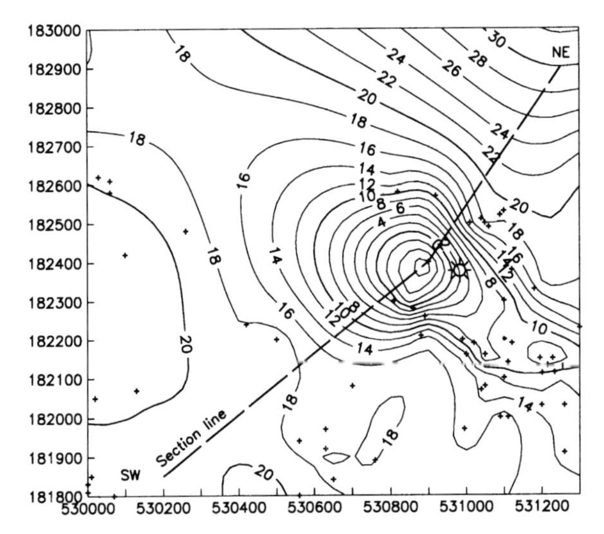

Fig. 2. Computer-generated contour plot of superficial deposits, Clerkenwell area. Symbol shows location of Mount Pleasant Post Office. Scale 1:10 000; contours at 2 m vertical interval relative to Ordnance Datum.

have encountered the rockhead – superficial deposits interface, resulting in a rapid flushing of hundreds of cubic metres of gravel and water into the tunnel, creating a serious subsidence cavity at surface. For example, during tunnel construction beneath Green Park, the tunnel face unexpectedly broke into the gravel-filled depression of the former Tyburn Valley, resulting in complete burial of the digger shield and the consequent delay to construction schedules (DoE 1975).

Besides the threat arising from perched groundwater tables at the base of the river terrace gravels, the prevention of river water inundation into tunnels has been a concern from the very early days of Brunel's first tunnel under the Thames. During the Blackwall Tunnel construction in the late 19th century, tunnelling came to within 1.5 m of the river bed of the Thames. It was necessary to temporarily dump layers of clay up to 3 m in thickness on the river bed to prevent excessive water entering the workings (Hay & Fitzmaurice 1897).

The presence of the deep drift-filled hollows at a number of localities along the Thames Valley and its associated tributaries was described by Berry (1979) and Hutchinson (1980). Such features, penetrating far below the predicted upper surface of the London Clay, have usually been recognized during detailed site investigations prior to tunnelling, but on occasion even a very dense borehole coverage has missed these localized depressions. Described by engineers as 'funnelling down', these hollows were encountered in the construction of the Post Office Railway in the vicinity of Mount Pleasant, Clerkenwell, in 1924 (Figs 2 and 3) (Dalrymple-Hay 1928). During excavation of the Victoria Line, similar features were penetrated with the associated tunnel inundation and surface collapse (Higginbottom & Fookes 1971). On the same line, between Vauxhall and Brixton, site investigation assisted in identifying steep-sided depressions filled with sandy gravel to below −15 m OD, enabling remedial measures to be taken in advance of the tunnelling (Wakeling & Jennings 1976).

Surface engineering operations have also been affected on many occasions by the irregular rockhead surface. Detailed site investigations assisted with the recognition of the problem prior to construction of the Battersea Power Station (Edmunds 1930). Here, a localized scour hollow extended to a depth of 33 m below the general rockhead level. However, many examples can be quoted of the failure to recognize such geohazards. A typical case occurred at a building development site in Victoria Street, where a drift-filled hollow was not recognized during the initial site investigation (Wakeling & Jennings 1976). The maximum excavation depth extended 19 m below street level and quickly became flooded. A major dewatering exercise was needed to lower the groundwater levels before construction could continue.

Borehole database and data processing

The majority of borehole records held in the BGS archive relate to site investigation holes, drilled in the last 50 years to depths generally between 5 m and 40 m. Older records (some dating back to 1870) are usually

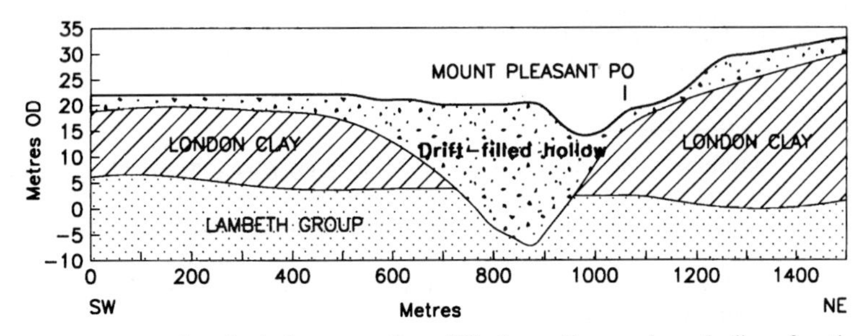

Fig. 3. Computer-generated geological cross-section of Clerkenwell area, along the line of section shown in Fig. 2.

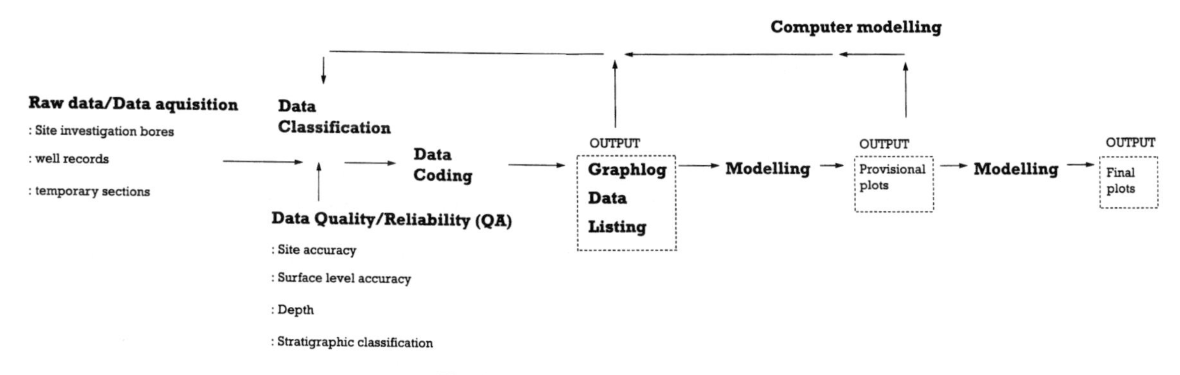

Fig. 4. Data processing flowchart.

from deeper boreholes for water supply. Major tunnelling projects such as the Thames Water Ring Main and the London Underground Jubilee Line have provided modern deep borehole information.

Over 22 000 borehole records, from the British Geological Survey's National Geoscience Records Centre, were selected for entry into a digital relational borehole database. The records were checked for accuracy using a BGS quality assurance (QA) procedure, and reliability values of the raw borehole data were assigned. These QA procedures imposed reliability codings on five parameters relating to the borehole record:

- accuracy of grid coordinates
- reliability of the borehole log
- accuracy of the Ordnance Datum
- reliability of the stratigraphical classification
- reliability of the recorded depths in the borehole log

Using any combination of these reliability factors, information of a specified quality can be selected from the relational database. The digital information was refined as the computer modelling proceeded, and anomalous values were reassessed and either deleted from the database or corrected. Since large numbers of borehole records were available, it has been preferable to discard questionable data to enable a geologically satisfactory result to be achieved. However, the data are retained in the database for reference purposes and may be used at a later date if required (Fig. 4).

The data are processed on the Natural Environment Research Council (NERC) Vax 8550 and 6510 mainframe computer cluster at the BGS Keyworth campus, using the following software packages (Fig. 5):

- ORACLE – a relational database management system used to store, retrieve and manipulate data (Anon 1990);
- GRAPHLOG – a BGS-developed programme (Flower 1987) used to produce graphical borehole logs for reference purposes (available on disk and as hard copy).

Subsequent modelling is carried out on a Unix-based, Silicon Graphics Indy workstation using commercial software (Fig. 5):

- EARTHVISION – workstation-based modelling software (Dynamic Graphics 1992). Generically derived from modelling software designed for the oil industry; it is used for creating gridded coverages, contouring, drawing cross-sections and perspective block diagrams.

Selected data are retrieved from the database using ORACLE into files suitable for use in the modelling package. Several file types are used: scattered data files for point reference; polygon files that include, for example, areas of outcrop of a particular geological formation; and annotation files that contain digital topographic information.

Although it is possible to compute contour displays and perform various analytical tasks directly from scattered data, the modelling software uses these data and a griding algorithm to calculate surface grids. For computational purposes, a surface is generated by using

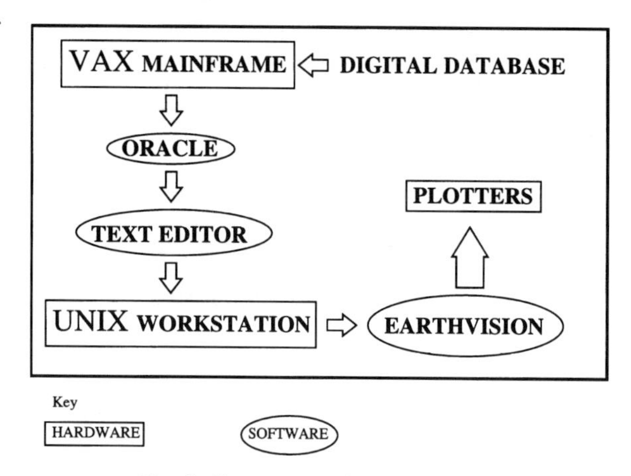

Fig. 5. Computer modelling flowchart.

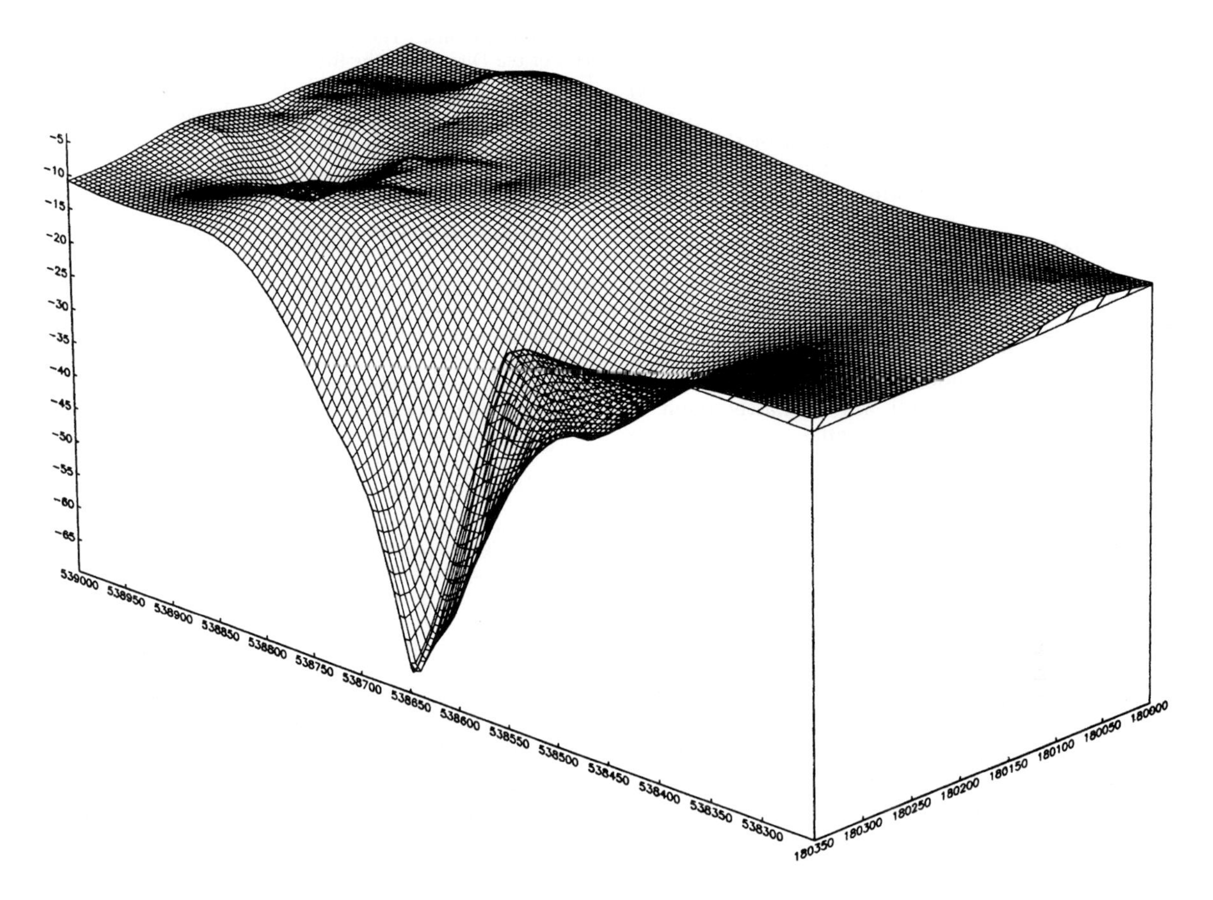

Fig. 6. Perspective view of the Blackwall scour hollow.

the algorithm to calculate the height of the surface (relative to Ordnance Datum) at a theoretically unlimited number of points. In practice, a grid with nodes at 10 m or 20 m spacing is used. These grids provide the basis for subsequent modelling and geological analysis.

By combining data from geological maps and the borehole database, the software allows the data to be interrogated, manipulated and corrected to produce geological models consistent with the known surface geology.

Computer-generated geological models

LOCUS has concentrated on modelling five main geological surfaces:

- base of made ground
- base of Quaternary ('rockhead')
- base of London Clay
- top of Thanet Sands
- top of Chalk

Of these, rockhead is proven by the highest borehole density. An example of a computer-generated rockhead contour plot for an area of central London, at Clerkenwell, is shown in Fig. 2. This figure illustrates rockhead contours at 2 m vertical intervals; it is possible to select specific intervals and to employ colour or hachure infill at yet different intervals in order to highlight geological features such as 'breaks of slope' associated with terrace margins. Computer-generated geological cross-sections can be produced with the minimum of effort from the grids (Fig. 3).

A perspective view of a deep drift-filled hollow in the vicinity of the Blackwall Tunnel (Fig. 6) clearly shows the 'funnelled' form of the feature, and illustrates a convenient method of visualizing the contoured data.

Practical use of geological models

The rockhead surface is of particular value in highlighting irregularities which may present serious foundation construction and tunnelling problems if not recognized prior to the design stage. It is known that such irregularities may also be associated with unconsolidated fine-grained superficial sediments rather than the usual sands and gravels of the river terrace deposits. Thus, knowledge of the form of this surface can be utilized in designing additional site investigation to provide more information on the sediment properties.

The knowledge that unforeseen geohazards, such as the drift-filled hollows described above, can exist beneath the surface in many parts of London, should encourage engineers and earth scientists to undertake a thorough study of the geology of any proposed major development site. Additional costs incurred by a more detailed geological appraisal will ultimately result in significant cost benefits if potential geohazards are identified in advance of construction.

The implementation of digital geological map production and the comprehensive LOCUS database and its 3D geological modelling capability, provides a major step towards a Geographical Information System (GIS) which will allow for ready visualization of the subsurface geology of London.

Geohazard prediction in London can be assisted by the preparation of contour plots depicting the base of the superficial deposits, but care must be taken to cover a wide area around a particular site or linear route, enabling the identification of features which may occur close to, but not necessarily on, the site. Such features may well affect groundwater movement, for example, and therefore exert an influence on the site itself. In the case of linear routes, the investigation of potential geohazards is best accomplished by the examination of the 3D geology of a strip of ground which extends at least 200 m (and preferably 500 m) on either side of the route.

Unusual features identified for example on the rockhead contour plots should be investigated by conventional site investigation methods. The presence of a major drift-filled hollow may result in a route realignment, if feasible, or at least forewarn the consultants and contractors of the existence of a geohazard.

Acknowledgements. This paper is presented with the permission of the Director of the British Geological Survey, Natural Environment Research Council. The authors are grateful to D. Bridge and A. Forster for reviewing the paper.

References

ANON 1990. *ORACLE Documentation.* Oracle Corporation, California,

BERRY, F G. 1979. Late Quaternary scour hollows and related features in central London. *Quarterly Journal of Engineering Geology*, **12**, 9–29.

DALRYMPLE-HAY, H. H. 1928. The Post Office tube railway, London. *Engineering*, January, 92–96.

DOE. 1975. *Probing ahead for tunnels: a review of present methods and recommendations for research.* Transport and Road Research Laboratory, Supplementary Report 171UC, Department of the Environment, London.

DYNAMIC GRAPHICS 1992. *EarthVision Documentation (Release7).* Dynamic Graphics Inc., USA.

EDMUNDS, F. H. 1930. *Some gravel-filled pipes in the London Clay at Battersea.* Summary of Progress Geological Survey. London, 12–24.

FLOWER, S. M. 1987. *GRAPHLOG Documentation.* Natural Environment Research Council, Computer Services, Edinburgh.

GIBBARD, P. L. 1977. Pleistocene history of the Vale of St Albans. *Philosophical Transactions of the Royal Society of London*, **B280**, 445–483.

HAY, D. & FITZMAURICE, M. 1897. The Blackwall Tunnel. *Min. Institution of Civil Engineers*, **130**, 50–79.

HIGGINBOTTOM, I. E. & FOOKES, P. G. 1971. Engineering aspects of periglacial features in Britain. *Quarterly Journal of Engineering Geology*, **3**, 85–117.

HUTCHINSON, J. N. 1980. Possible late Quaternary pingo remnants in central London. *Nature*, London. **284**, 253–255.

KINNIBURGH, D. K, GALE, I. N, SMEDLEY, P. L. *et al.* 1994. The effects of historic abstraction of groundwater from the London Basin aquifers on groundwater quality. *Applied Geochemistry*, **9**(2), 175–195.

WAKELING, T. R. M. & JENNINGS, R. A. J. 1976. Some unusual structures in the river gravels of the Thames Basin. *Quarterly Journal of Engineering Geology*, **9**, 255–263.

Subsidence hazard assessment as a basis for planning guidance in Ripon

Alan Thompson,[1] Peter Hine,[1] Denis Peach,[2] Lindsay Frost[3] & David Brook[4]

[1] Symonds Travers Morgan, Symonds House, Wood Street, East Grinstead, West Sussex RH19 1UU, UK
[2] Formerly Symonds Travers Morgan , now with the British Geological Survey, Maclean Building, Crowmarsh Gifford, Wallingford, Oxfordshire OX10 8BB, UK
[3] Formerly Harrogate Borough Council, now with Lewes District Council, Lewes House, 32 High Street, Lewes, East Sussex BN7 2LX, UK
[4] Department of the Environment, Transport and the Regions, Eland House, Bressenden Place, London SW1E 5DU, UK

Abstract. The Ripon area of North Yorkshire provides a well-documented example of subsidence problems caused by the natural dissolution of gypsum deposits. This paper summarizes the findings of a programme of research, commissioned by the former Department of the Environment, which has aimed to assess the degree of hazard associated with this particular form of subsidence and to develop appropriate planning and engineering responses which might be applicable both to Ripon and to other affected areas. The paper draws attention to the sensitive nature of the problem and to the conflict which exists between the benefits of using hazard assessment to minimize the risk to future development within an area, and the possible disbenefits of widely publicizing the existence of a land instability problem.

In Ripon, the potential consequences of gypsum-related subsidence, in terms of building damage and road closures, can be significant. In addition, there may be a small risk of personal injury. It is therefore important that the planning of future development in the area should be guided by the results of the hazard assessment carried out in this study and that such development should be subject to controls and mitigation measures, where appropriate. However, the probability of this form of subsidence occurring at any given site is very low and the risk to individual properties is therefore generally very small, especially from the statistical viewpoint normally adopted by insurers. Taking account of these facts, forward planning and development control procedures have been recommended for use by the local planning authority.

Introduction

Gypsum, the hydrated form of calcium sulphate, occurs naturally within the Permian rocks of northeastern England, extending from the Darlington area in County Durham, through the length of Yorkshire to north Nottinghamshire. The gypsum occurs as a secondary mineral, formed by the hydration of original anhydrite deposits that are present at greater depth and over a wider area.

Gypsum is susceptible to rapid dissolution wherever there is active circulation of groundwater that is undersaturated with respect to calcium sulphate, leading to the localized development of substantial underground voids and cave systems. The spontaneous collapse of individual caverns allows upward migration of voids, either by gradual caving of thinly bedded strata or (occasionally) by the sudden failure of more competent, thickly bedded rocks, leading ultimately to subsidence of the overlying ground surface.

The Ripon area of North Yorkshire is a well-documented example of these phenomena, having been studied by the British Geological Survey for more than 20 years (Smith 1972; Cooper 1986, 1988, 1989, 1995; Powell *et al.* 1992; Cooper & Burgess 1993). The area was identified, in a recent study of natural underground cavities (Applied Geology Limited 1993), as one which would merit more detailed investigation as a trial area for the development of hazard assessment techniques and corresponding planning and engineering responses that could be applied both to Ripon and to other affected areas.

This paper summarizes the findings of a research contract, funded by the former Department of the Environment (DoE), now the Department of the Environment, Transport of the Regions (DETR), as part of its Minerals, Land Instability and Waste Planning Research Programme, which has sought to address these issues.

THOMPSON, A., HINE, P., PEACH, D., FROST, L. & BROOK, D. 1998. Subsidence hazard assessment as a basis for planning guidance in Ripon. *In*: MAUND, J. G. & EDDLESTON, M. (eds) *Geohazards in Engineering Geology*. Geological Society, London, Engineering Geology Special Publications, **15**, 415–426.

The aims of the project were

- to assess the working of the planning system in reducing hazards due to subsidence related to gypsum dissolution and
- to prepare a draft framework of advice suitable for use by planners, developers, land and property owners, insurers and others.

In any study of this type, which draws public attention to the existence of a natural hazard, there is a conflict between the benefits to be gained from a better understanding of the problem, which enables the risk to future development to be minimized, and the possible adverse consequences of the publicity, in terms of public alarm, insurance costs, mortgage lending and property values. A very important consideration throughout the study therefore was to recognize and to allow for the highly sensitive nature of the issue.

Land instability and planning

In 1990 the Department of the Environment published a planning policy guidance note (PPG 14), concerning development on unstable land. This provides general guidance for local planning authorities and developers on the ways in which land instability issues can be taken into account in the planning of new development.

PPG 14 draws attention to the fact that, although the ultimate responsibilities for determining the suitability of land for any particular purpose, and for the safe development and use of a site, rest with the developer and landowner, both the planning and building control departments of local authorities can help to ensure that information on land instability is taken properly into account, through forward planning and development control procedures.

Forward planning

Structure plans, local plans and unitary development plans set out policies with respect to a wide range of issues which may influence future development, including land instability. The approach adopted in considering hazards due to land instability is a matter of preference for each local planning authority, depending on the availability of information and resources, the accuracy with which different categories of hazard can be identified and quantified, and the extent to which different hazard zones can be related to particular planning responses.

One approach, used by South Wight Borough Council to deal with problems of landslip hazard in the town of Ventnor, is that of a planning guidance map clearly relating categories of hazard to planning responses. This was possible because individual landslip units in Ventnor can be readily identified on the basis of their geomorphological expression within the landscape (Geomorphological Services Ltd 1991). In contrast, Norwich City Council have adopted a more narrative approach in dealing with the less clearly defined hazard of subsidence associated with old chalk mines and natural underground cavities (Howard Humphreys & Partners 1993). For further discussion of the options available for dealing with geological issues in the planning of new development, see the DETR's *Guide to Good Practice on Environmental Geology in Land Use Planning* (Thompson *et al.* 1998).

Development control

Whether a planning guidance map or a narrative approach is adopted, information on land instability can be used by local planning authorities and building inspectors in the control of specific development proposals.

At the planning stage a number of options exist, ranging from the issuing of advisory notices (which point out the possible existence of a problem and recommend that the developer seeks specialist advice) through conditions relating to investigation of the ground and the implementation of remedial engineering measures; to the refusal of planning permission, either on the grounds of proven instability or on the grounds of failure to provide sufficient information to satisfy the local planning authority as to the stability of the site.

The choice is not a straightforward one however, since at one extreme, the local authority may be accused of not making full use of the information at its disposal, whilst at the other, poor presentation of information may result in adverse consequences for prices of property and the availability of insurance cover in areas which are perceived as being most affected.

Once a proposed development has progressed beyond the planning stage, building control is concerned primarily with the application of minimum standards of construction. With respect to ground conditions, this may include consideration of foundation design in relation to potential land instability, insofar as this can reasonably be foreseen. However, given that in certain cases the building control function is provided by an approved inspector, such as the National House Builders Council (NHBC), the local authority may not be able to rely on this as a means of exerting full control on new development.

An important objective for this study was therefore to determine the most appropriate form of forward planning and development control responses to the problems of subsidence caused by gypsum dissolution within the Ripon area and, more widely, to similar problems elsewhere. To be able to do this, however, it was first necessary to gain a clear understanding of the physical processes involved.

Controls of gypsum dissolution and subsidence

Within the Ripon area both gypsum and anhydrite are present within the Edlington and Roxby Formations of the Permian sequence. The anhydrite is metastable at shallow depths and tends to rehydrate to gypsum as it comes into contact with circulating groundwater within the near-surface zone (Murray 1964; Holliday 1970; Mossop & Shearman 1973). In the Ripon area, this research project has shown that the transition from anhydrite to gypsum occurs mainly within 40 to 120 m of the ground surface. At greater depths the circulation of groundwater appears to be generally insufficient to cause rehydration and is thus unlikely to be capable of substantial dissolution.

The area which is potentially susceptible to gypsum dissolution activity is therefore broadly constrained by two basic geological controls: the limits of outcrop of gypsum-bearing strata (in the west) and the limits beyond which the easterly dipping gypsum beds reach depths of more than 120 m (in the east). Between these limits the gypsum may or may not be at risk of dissolution within a given period of time, depending upon the flowpaths, chemistry and magnitude of local groundwater movement.

Even in areas where active dissolution of gypsum is taking place, the nature and timescale of any associated subsidence activity will vary according to the size of individual cavities and also the thickness and geotechnical properties of overlying strata. Cooper (1988), however, has shown that cavities can eventually migrate to the surface, before becoming choked, within all parts of the area affected by active dissolution, and that none of the overlying strata is completely resistant to collapse.

Given the very long timescale over which dissolution is likely to have been operating, cavities in all stages of migration towards the surface are likely to be present, irrespective of their individual rates of propagation. The assessment of subsidence hazard can be restricted, therefore, to an analysis of those factors which control the rates and distribution of gypsum dissolution. These fall essentially into three main categories:

- factors which control the distribution of massive gypsum beds within the study area;
- factors which control the rates of flow of groundwater within the gypsum layers; and
- factors which control the hydrochemical characteristics of groundwater entering and flowing through the gypsum beds.

Table 1. *Summary of factors likely to control gypsum dissolution*

1. *Geological controls*
   (a) The limits of outcrop of gypsum-bearing strata
   (b) The areas in which massive gypsum beds are present within about 120 metres of the ground surface
   (c) The detailed stratigraphy of the solid geology, particularly with regard to the occurrence and thickness of individual gypsum beds and the lithologies of adjoining strata
   (d) The geological structure of the area, in particular the nature and distribution of major faults, and folds, and the spacing and orientation of joint patterns
   (e) The nature, thickness and distribution of Quaternary drift deposits, particularly with regard to their influence on the spatial distribution of groundwater recharge to and discharge from the underlying solid strata
2. *Geomorphological controls and indicators*
   (a) The general topography of the area, particularly with respect to its influence on the regional patterns of groundwater flow
   (b) The morphology, position and age of development of the drift-filled buried valley which exists beneath the modern course of the River Ure
   (c) The nature of weathering of near-surface strata and its development over time, including karstification (the formation of solution-widened joints and fissures) within the limestones and massive gypsum beds
   (d) Long-term fluctuations of groundwater levels, in association with climatic change and with episodes of river incision and aggradation, throughout the Quaternary period
   (e) The effects of periglacial processes, especially cambering and valley bulging, on the basic geological structure
   (f) Geomorphological evidence for the nature and distribution of gypsum-related subsidence, including long-term foundering of strata over wide areas, the development of individual subsidence hollows, and historical subsidence events
3. *Hydrogeological controls*
   (a) The regional patterns of groundwater flow
   (b) The nature and spatial distribution of groundwater recharge (including artificial recharge, e.g. mains leakage)
   (c) The hydrogeological parameters (geometry, transmissivity and storage) of each of the main aquifer units
   (d) The nature and distribution of groundwater discharge zones
   (e) The observed variations in groundwater heads and head gradients
   (f) The resulting flowpaths and velocities of groundwater entering, flowing through and emerging from the massive gypsum beds
   (g) The observed hydrochemical characteristics of groundwater in various parts of the system

The factors in each of these groups are complex and are often interrelated. For example, aspects of the geological structure and stratigraphy of an area, which help to determine the present distribution of gypsum, are also likely to have a significant bearing on hydrogeological characteristics and groundwater flow paths within the area, and will thereby influence both the rates and chemical characteristics of groundwater flows. Table 1 lists the primary factors involved in the control of gypsum dissolution, grouped, for ease of data collection, according to the type of information involved. The nature and limitations of the available data in each of these groups are outlined below.

Geological controls

Solid geology

In general, the sequence of Permian and Triassic rocks is encountered in order of decreasing age, from west to east across the study area, reflecting the predominant gentle easterly dip of the strata. This pattern is disrupted, however, by the presence of buried valleys beneath the modern courses of the River Ure and its tributary, the River Skell, which have incised deeply into the Permian sequence to produce a more complex outcrop pattern. Figure 1 illustrates the three-dimensional relationships between the outcrop pattern, the underlying structure and the form of the rockhead surface beneath the cover of Quaternary drift deposits. In detail, the outcrop pattern is complicated further by the presence of small, isolated outliers of younger strata in various parts of the study area resulting from the localized foundering of strata in relation to gypsum dissolution beneath.

The details depicted in Fig. 1 are based on the latest geological interpretation of the borehole information currently available, including data that were obtained specifically for this purpose in the course of the DoE study. The details are substantially different from those portrayed on previously published BGS maps and are subject to further revision in the light of any additional ground investigation data which may be obtained

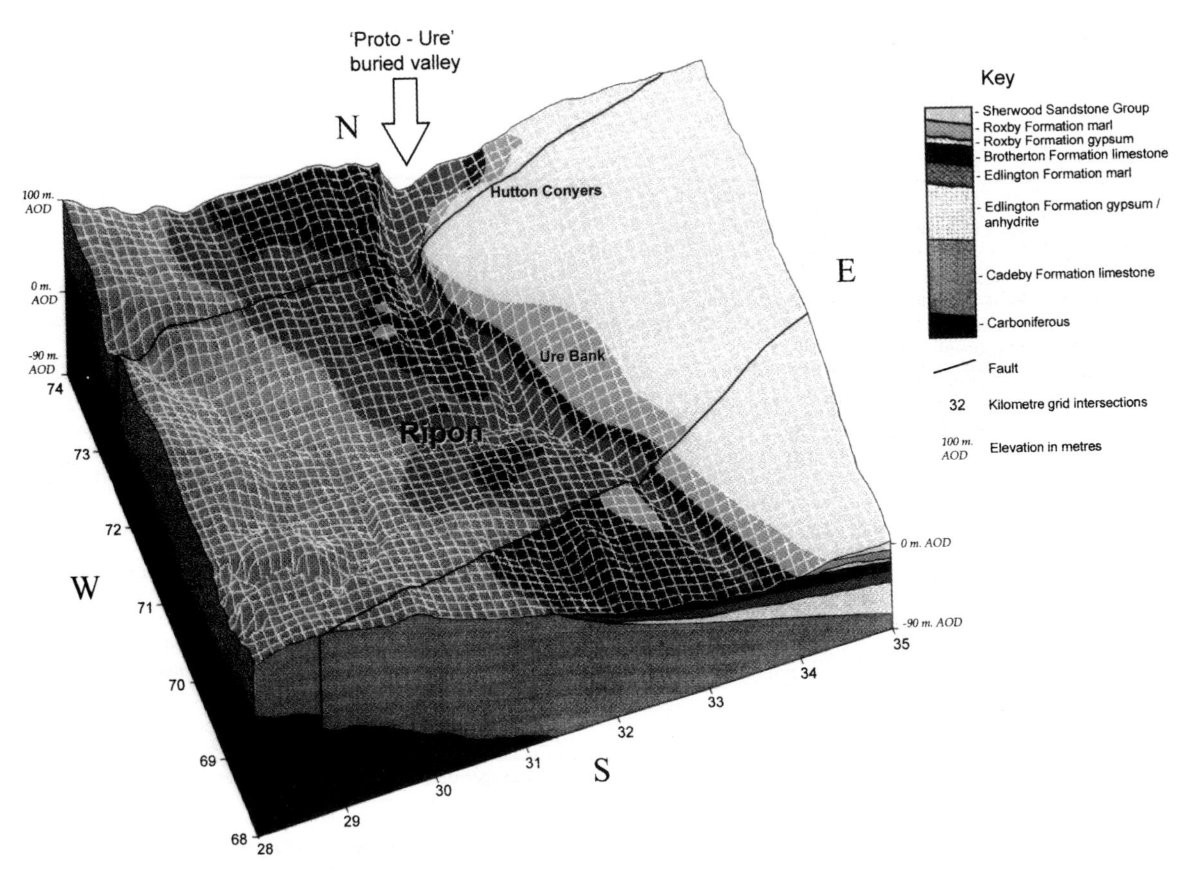

Fig. 1. Geological block diagram of the Ripon area. Geological boundaries as revised by A. H. Cooper (BGS) and A. Thompson (STM), 1996. Quaternary drift deposits not shown.

in future years. The current interpretation is shown in detail on the 1:10,000 scale BGS maps, revised by A. Cooper and A. Thompson in 1996.

Drift geology

Over the majority of the study area the various Permo-Triassic strata are concealed beneath a variety of Quaternary drift deposits ranging from glacial tills and glacio-fluvial sands and gravels to recent alluvium. The most widespread deposits are of glacial till. These are locally associated with lenses of glacial sands and gravels and have an important influence on the spatial distribution of precipitation recharge to the groundwater system.

Of greater importance to the overall patterns and rates of gypsum dissolution, however, are the Quaternary sands, gravels and cemented gravels (conglomerates) which infill a deep buried valley beneath the modern course of the River Ure. This extends to a proven depth of up to 53 m below the present flood plain, i.e. to elevations of up to 30 m below Ordnance Datum (Thompson *et al.* 1996). The buried valley gravels are largely of high transmissivity and provide a means by which groundwater emerging under artesian pressure from the Permian strata on both sides of the buried valley can escape towards the surface (see Fig. 2).

Geomorphological evidence of subsidence and foundered strata

Geological cross-sections compiled from the available borehole data illustrate the progressive thinning of the Edlington Formation, and in particular of the massive gypsum beds at its base, from east to west across the study area (Fig. 2). Whilst this might be partly due to variations in the original thicknesses of sediment, it is also considered likely to be at least partly a consequence of the progressive increase in gypsum dissolution as the strata are followed up-dip towards the ground surface.

The geological cross-sections also show a similar, but far less pronounced east–west thinning of the Roxby Formation, which again is regarded as being at least partially due to the progressive dissolution of the massive (but much thinner) beds of gypsum at its base.

In the case of the Edlington Formation, the geological cross-sections reveal that the general pattern of long-term subsidence is complicated by the effects of more

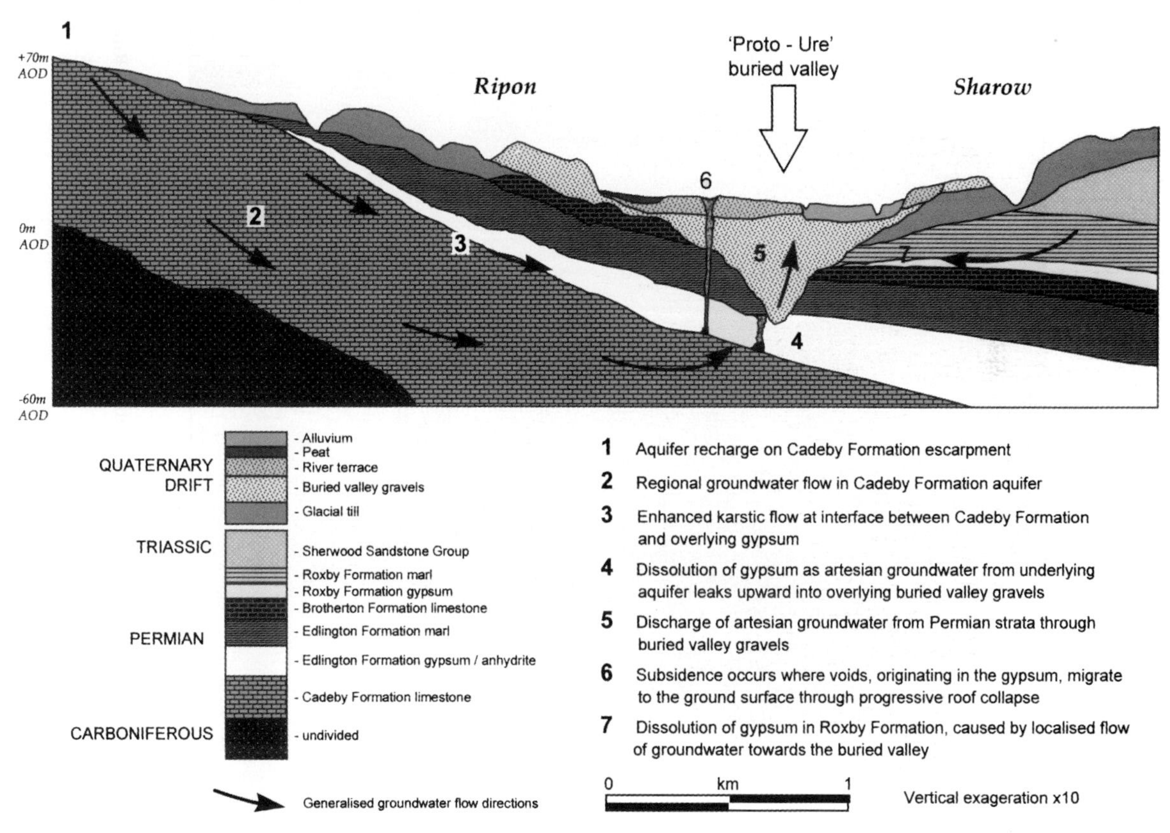

Fig. 2. Geological cross section through the central part of the Ripon study area, illustrating key aspects of the conceptual model of groundwater flows and gypsum dissolution.

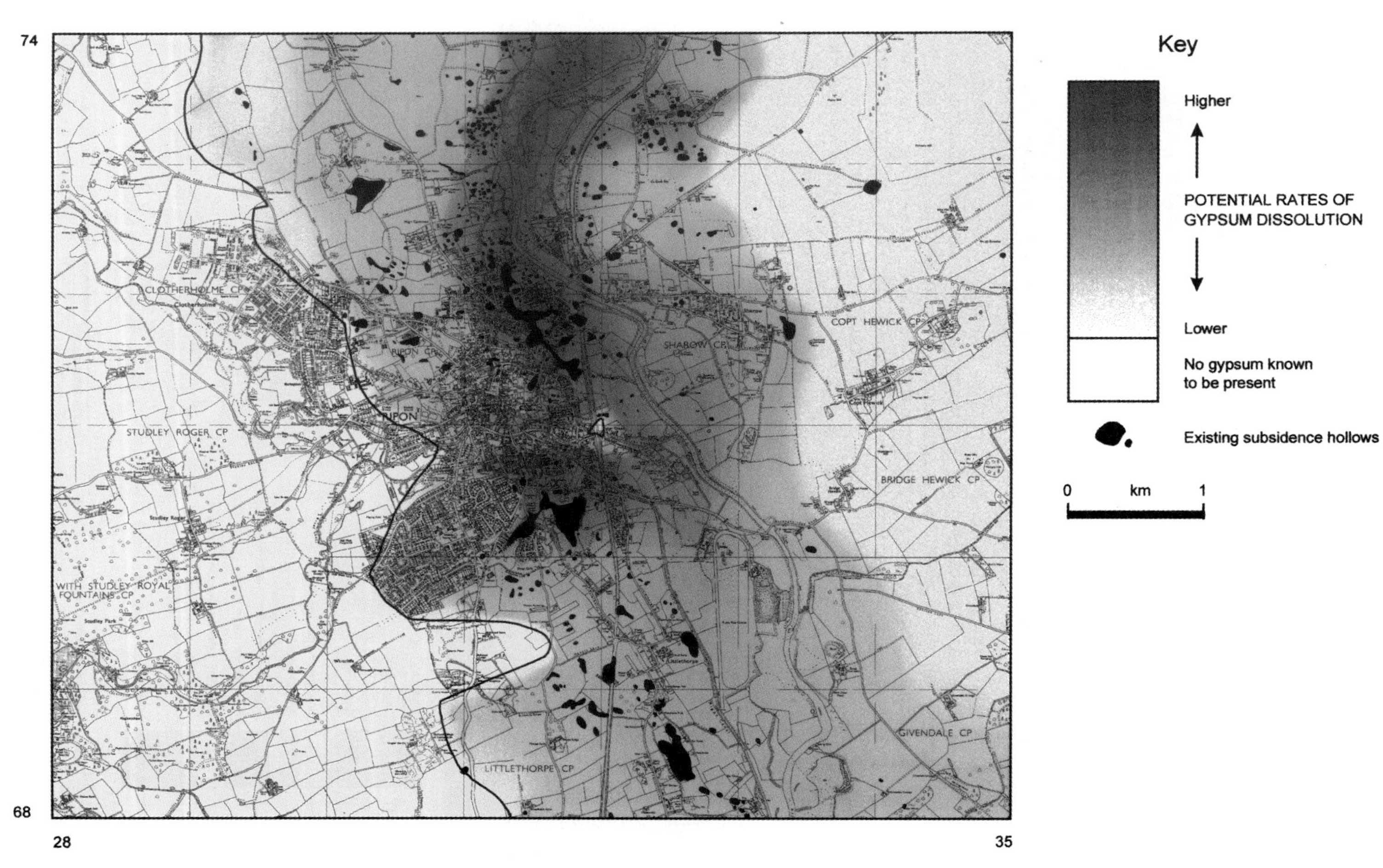

Fig. 3. Schematic representation of potential rates of gypsum dissolution in the Ripon area, based on the conceptual groundwater model developed by Thompson *et al.* (1996).

intensive, localized foundering at the western margin of the buried valley, especially within the northern part of the study area. This is manifested by a very marked thinning of the gypsum and by a steepening of the contact with the overlying Brotherton Formation on the western edge of the valley (see Fig. 2). It is also demonstrated by the patterns of postglacial subsidence hollows and recent subsidence events (Fig. 3). These show a marked concentration of subsidence activity along the western margin of the Ure Valley, especially upstream of the Skell confluence, and, to a lesser extent along parts of the eastern margin, particularly around Hutton Conyers. It should be noted that the relatively limited number of subsidence hollows within the intervening floodplain areas is a reflection of the limited age of the floodplain surface rather than of a lower frequency of subsidence activity.

The localized nature of foundering and subsidence activity at the margins of and beneath the buried valley suggests that this foundering is a relatively recent phenomenon, superimposed upon the more general, east–west thinning of the Edlington Formation since the formation of the valley. There appears to be no reason why such localized foundering should have occurred in these positions prior to the valley's incision. To understand the mechanisms involved requires an examination of the hydrogeological regime within the study area.

Hydrogeological conceptual model

The 'solid' strata of the Ripon district represent a complex, multi-aquifer system, comprising three main aquifer units (the Cadeby and Brotherton Formation limestones and the Sherwood Sandstone), separated by low permeability mudstones within the Edlington and Roxby Formations. Analysis of the limited data available on discrete groundwater heads within individual aquifers, and of the more widely available data on non-specific rest water levels in several hundred site investigation boreholes, has enabled the following generalized observations to be made concerning the hydrogeological regime within the study area.

Regional groundwater movement within the solid strata is predominantly from west to east across the district, controlled by the base levels provided by major river systems within the Vale of York and/or more distant sea levels. However, within the study area, a more local base level is provided by the River Ure which, by virtue of the hydraulic connections created by the gravel-filled buried valley beneath the modern floodplain, is able to intercept the regional groundwater flows within the Permian strata (see Fig. 2).

The principal aquifer unit within the Ripon area is the Cadeby Formation limestone. This is characterized by high transmissivities, with groundwater movement predominantly along joints, fissures and bedding planes within the rock, especially at the interface between the limestone and the overlying gypsum. Widespread dissolution of the gypsum might be expected to occur along this horizon, especially where low sulphate water (from the limestone) first encounters the gypsum and thus has a high potential for dissolution. This, however, could only be substantial in areas where the groundwater is moving rapidly through the system, i.e. where groundwater head gradients are high and where there are opportunities, in the form of pre-existing joints and fissures, for the water to discharge through the overlying gypsum deposits towards the surface.

Such conditions occur primarily beneath and along the sides of the 'Proto-Ure' buried valley, where the confining overburden of low permeability marl has previously been largely or entirely removed by erosion, and where fracturing of the massive gypsum has been induced by the resulting stress relief and subaerial weathering, prior to the infilling of the valley with Quaternary sediments (see Fig. 2). This effect is likely to be particulary noticeable on the western side of the valley, where the vertical head differences are greatest, and to diminish rapidly on the eastern side as the gypsum dips once again beneath a greater thickness of confining strata.

Analysis of the limited data available on discrete groundwater heads and groundwater chemistry suggests that, for a distance of up to 2 km to the east of the River Ure, groundwater within the Brotherton Formation limestone is likely to flow up-dip, towards the river, and that, within a few hundred metres of the valley it will tend to move obliquely upwards, through the overlying gypsum, causing locally enhanced dissolution of these deposits (see Fig. 2). From the evidence provided by the distribution of postglacial subsidence hollows and historical subsidence events, this mechanism seems to be operating most effectively in the vicinity of Hutton Conyers and Ure Bank.

Assessment of subsidence hazard

Whilst there are insufficient data to produce a reliable mathematical model of the hydrogeological conditions outlined above, not least because of the karstic flow mechanisms which are likely to be involved and the transient nature of individual groundwater flow paths within the highly soluble gypsum, there is sufficient evidence to confirm the general principles (Thompson *et al.* 1996). By combining the effects of the two main hydrogeological processes operating on the eastern and western sides of the Ure Valley, a generalized pattern of variations in the potential rates of gypsum dissolution can be envisaged, as shown schematically by the gradational shading in Fig. 3. (Note that the variations are shown more clearly and in greater detail in the original full-colour version of this map, copies of which can be obtained from the authors, on request.)

Figure 3 illustrates the expected concentration of dissolution activity beneath and at the margins of the buried valley, particulary in the northern part of the study area. Further south, the influence of water moving upwards from the Cadeby Formation is likely to be diminished, because of the greater thickness of marl which is thought to remain between the base of the buried valley and the underlying gypsum.

Figure 3 also shows a localized absence of predicted gypsum dissolution at the centre of the Ure Valley, at the point where it is joined by the River Skell. This corresponds to a deep scour hollow, proved by new borehole information, where all of the Edlington Formation has been eroded away and the buried valley gravels rest directly upon the underlying Cadeby Formation limestone.

The gradational shading on Fig. 3 may be thought of as a representation of the intensity of gypsum dissolution that is taking place, averaged over a long period of time. At any given time the actual distribution of gypsum dissolution activity will be far more localized, since the groundwater will flow predominantly along a karstic network of solution-widened joints, fissures and bedding planes within the gypsum. The precise locations of individual conduits cannot be determined, except by chance interception of voids within site investigation boreholes. Moreover, the patterns of flow will constantly change, as joints are widened by dissolution; as new joints are intersected and capture the flow; and as overlying strata collapse to divert the flow in other directions.

For these reasons it is both impossible and meaningless to portray the actual distribution of gypsum dissolution at any given moment in time, and in this respect, the time-averaged distribution, as portrayed in Fig. 3, provides a more useful representation of the degree of hazard associated with gypsum dissolution within the Ripon area. In other respects, however, this also has its problems, not least of which are the apparent anomalies of sites which show no signs of historical subsidence activity, even though they are located within an area of greater than average susceptibility.

A further problem lies in the difficulty of quantifying the relative degrees of hazard between one area and another, and of quantifying the difference between gypsum-related subsidence hazard and other insurance risks.

Suggested framework for planning guidance in Ripon

In view of these difficulties it would be inappropriate to use the schematic model outlined above for detailed planning or development control purposes. There is, however, a clear need to make use of the information now available to refine the existing framework of control, in order to avoid, or at least minimize, any future subsidence damage to new development within the Ripon area.

A simplified, three-fold zonation of the study area was therefore devised, based on development control areas which are more clearly distinguishable from one another and which are defined primarily by the differences in development control procedures which will apply in each case. The three areas, as shown in Fig. 4, comprise the following:

- *Area A*, where no gypsum is believed to be present and therefore no requirement for dealing with potential gypsum-related subsidence exists. This corresponds to the area which lies to the west of the outcrop of all gypsum-bearing strata, as shown on the new geological maps produced in the course of this study.
- *Area B*, where gypsum and/or anhydrite is known to be present, but only at depths which appear to be below the influence of groundwater moving towards the buried valley of the River Ure. This zone lies entirely to the east of the River Ure and to the east of the groundwater divide which is thought to lie approximately beneath the crest of the Sherwood Sandstone escarpment. Whilst gypsum-related subsidence might occur in this area, in association with existing cavities that have migrated near to the surface and are continuing to collapse, the likelihood of subsidence occurring at any given point is extremely slight and is likely to decrease further still towards the east.
- *Area C*, where gypsum and/or anhydrite is likely to be present at depths which are within reach of groundwater moving more rapidly towards the River Ure buried valley, and where there is, therefore, a greater likelihood of gypsum dissolution and associated subsidence activity. This zone comprises all parts of the study area in between Areas A and B and encompasses all areas of known subsidence activity as well as some intervening areas which appear to have remained undisturbed for many hundreds, or even thousands of years.

Notwithstanding some of the anomalies noted above, it follows that Area C should be the area in which greatest attention should be given to the control of new development. It also follows that, whilst the hazard from gypsum-related subsidence need not prohibit development in any part of the Ripon area (providing appropriate mitigation measures are employed), in order to minimize the risk of subsidence damage to new development, this should preferably be concentrated in those areas which are likely to be least affected (Areas A and B). This, of course, takes no account of other 'constraining' factors to development which might outweigh the hazard from gypsum dissolution. It must

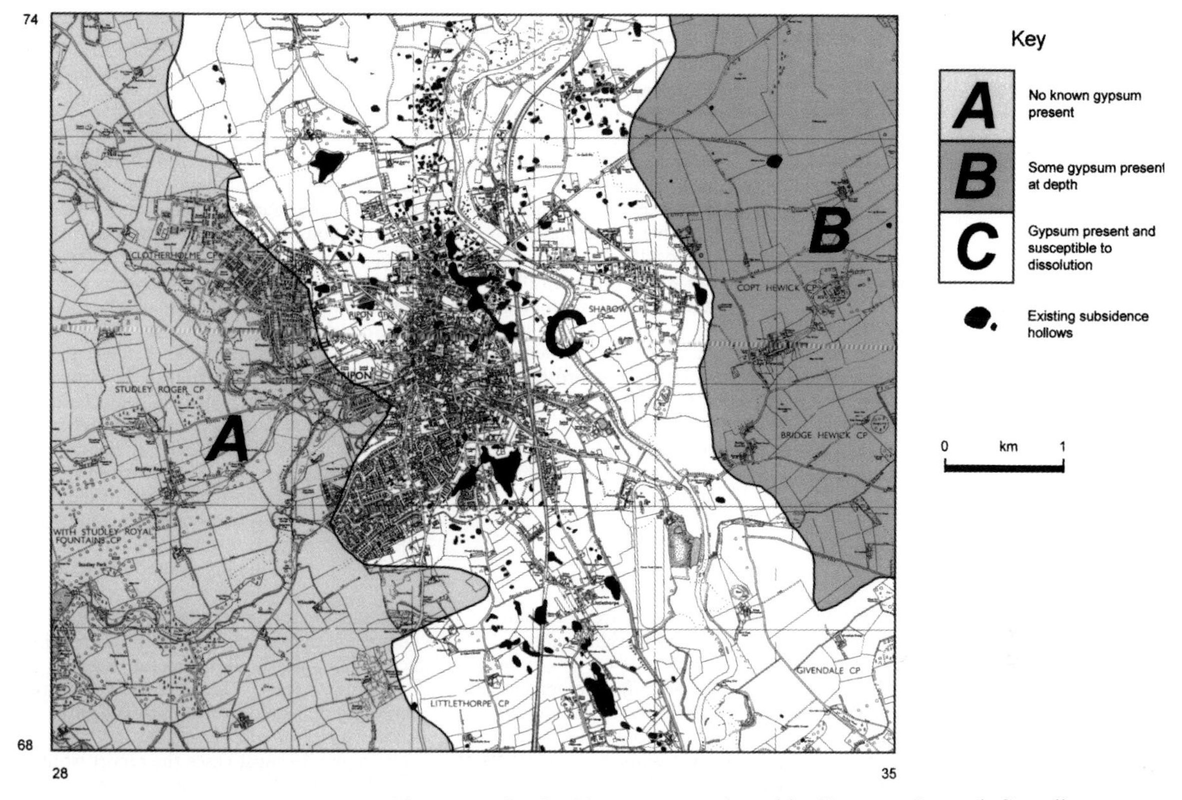

Fig. 4. Simplified development guidance map for the Ripon area, as adopted by Harrogate Borough Council Development control policies, recommended by STM, apply to Development Control Areas B and C.

also be emphasized that the boundaries between the different areas are based on the current understanding of geological and hydrogeological conditions. As new information is obtained there may be a need for these boundaries to be revised.

Implications for existing development

The identification of the geographical extent of ground problems can sometimes have implications for property values, property insurance, and mortgage lending. In order to avoid an unnecessarily adverse reaction to this information, it is important to emphasize that, for Ripon as a whole, the frequency of gypsum-related subsidence events is not particularly high, and that the risks to individual properties are generally very small, particularly from the statistical viewpoint normally taken by insurance companies.

From information compiled by the British Geological Survey, local engineering consultants and the insurance industry, it appears that gypsum-related subsidence events in Ripon occur about once per year on average, and that, in most cases, the subsidence takes place in open countryside, without causing damage to buildings. From an insurance perspective, this is a much lower level of risk than, for example, that associated with flooding of the River Ure, which occurs with a similar frequency but which can potentially affect a much greater number of properties at any one time.

Whilst the overall probability of a gypsum-related subsidence event occurring at any particular point is generally very low, the consequences of a major collapse could be extremely serious if it occurred, for example, beneath an occupied building. In view of this it is considered essential that the problem should be taken into account in both forward planning and development control procedures.

Development control procedures

The suggested development control procedures for each of the three areas identified on the development guidance map are summarized in Table 2. These aim to ensure that the risk of gypsum-related subsidence is minimized in all future building development carried out within the area, and that decisions regarding the assessment of potential instability on individual sites are taken by those who are best qualified to do so.

Table 2. *Summary of development guidance categories and suggested procedures*

Development control area	*Gypsum-related subsidence hazard*	*Forward planning implications*	*Suggested development control procedures*
A	No gypsum present according to current geological maps.	Areas suitable for development in accordance with the local plan. Gypsum problems impose no constraints on local plan development proposals.	No requirements with respect to gypsum at the planning stage. Building control measures may be needed if isolated outliers of gypsum are discovered during routine site investigations or construction work.
B	Slight subsidence hazard associated with very localized, existing, near-surface cavities, formed originally by the slow localized dissolution of deep-seated gypsum deposits.	Areas which are generally suitable for development in accordance with the local plan. Gypsum-related subsidence hazard may impose minor, localized constraints, which should be identified and taken account of in local plan development proposals.	A ground stability report, prepared by a competent person,[1] will normally[2] be required before planning applications for new building development in this area can be determined. In most cases it is likely that the report would need to be based only upon a geotechnical desk study and site appraisal, although site investigations to identify existing cavities may be required if problems are identified by the initial desk study. In recognition of the very limited degree of risk involved in this area, these requirements may often be imposed in the form of conditional planning consent.
C	Areas which may be potentially subject to localized on subsidence hazard, associated with both existing cavities and with the ongoing dissolution of gypsum deposits in areas affected by groundwater moving towards the Ure Valley.	Areas which are potentially subject to significant constraints on development. Local plan development proposals should identify and take account of these constraints, making use of the detailed hazard assessment contained within Symonds Travers Morgan's Technical Report.	A ground stability report, prepared by a competent person[1] will normally[2] be required before planning applications for new buildings, or those relating to change of use involving increased exposure of the public to a known risk of subsidence, can be determined. In most cases the report would need to be based on a geotechnical desk study and site appraisal followed up by a programme of ground investigation designed to provide information needed for detailed foundation design, unless adequate information from previous boreholes on the same site is available. Where planning consent is given, this may be conditional upon the implementation of approved foundation or other mitigation measures, designed to minimize the impact of any future subsidence activity.

[1] A 'competent person' in this context is a Geotechnical Specialist, as defined by the Site Investigation Steering Group of the Institution of Civil Engineers (Anon, 1993). Ideally, this person would also have appropriate experience in the investigation and remediation of gypsum-related subsidence problems within the Ripon area. Further guidance can be obtained from the current edition of the British Geotechnical Society's directory of Geotechnical Specialists.

[2] *Permitted development* under the Town and Country Planning (General Permitted Development) Order 1995 and other minor developments, including most *householder applications* such as modest extensions, will not normally be subject to the development control requirements set out above. In the case of householder applications, the council will issue an advice note, drawing the applicant's attention to the potential risk of subsidence, but it reserves the right to request ground stability reports in situations where there are particular reasons for greater concem, for example in locations which are close to sites of recent subsidence activity.

The suggested procedures include requirements for undertaking desk studies, walk-over surveys and site investigations, designed to assess the specific risk of subsidence for each proposed development, and to assess the most appropriate type of foundation design required to minimize such risk. A particular feature of the recommendations, is that site-specific studies undertaken for developers in support of planning applications should be accompanied by a **Ground Stability Declaration Form**, completed by a Geotechnical Specialist, as defined by the Institution of Civil Engineers' Site Investigation Steering Group (Anon 1993). The declaration form requires the Geotechnical Specialist to give a clear, professional opinion as to the adequacy of the available information, the stability of the site and its suitability for development. Where a risk of instability can be foreseen, the form also requires appropriate mitigation measures to be put forward. Further details of the declaration form are given in the STM's technical report on the Ripon study (Thompson *et al.* 1996) and in the *Guide to Good Practice on Environmental Geology in Land Use Planning* (Thompson *et al.* 1998).

In some areas, it might be technically feasible to overcome the risk of damage caused by gypsum-related subsidence to a building (but not necessarily to surrounding land or services) by founding the building on piles which transfer the load down to competent limestone strata beneath the lowest gypsum beds.

In many cases, however, this approach would require very deep, reinforced piles, which could withstand the possibility of lateral as well as vertical ground movement, and would therefore only rarely be economically viable.

In order to achieve an appropriate balance between the potentially serious consequences of subsidence and its very low probability of occurrence, the guidelines provided to aid the implementation of the suggested policies concentrate on the need to minimize the extent of any structural damage to buildings, and thus avoid injury to people, in the event of subsidence activity. This can generally be achieved by providing some measure of structural rigidity (e.g. through the use of reinforced rafts) without the need for using deep piled foundations. Further details are provided in the technical report produced for the DoE (Thompson *et al.* 1996).

Conclusions

The findings of the DoE-funded research, as outlined briefly in this paper, have shed considerable new light upon the mechanisms of gypsum dissolution and associated subsidence within the Ripon area, and have thereby enabled a broad assessment to be made of the spatial variations of potential subsidence hazard.

Due to the complexity of geological and hydrogeological conditions, however, especially within areas of previously foundered strata, only limited confidence can be placed in many of the boundaries that could be drawn between zones of differing subsidence hazard. For this reason it is considered inappropriate to use a detailed hazard zonation map for forward planning and development control purposes.

It is also considered that a hazard map, if used in isolation, could give a misleading impression of the severity of subsidence risk within the Ripon area. Whilst it is clear that gypsum-related subsidence is a significant land instability problem which must be taken into account in the planning of new development, it is also evident that, for Ripon as a whole, the frequency of gypsum-related subsidence events is low, and that the risk to individual properties is generally very low.

It is concluded, therefore, that the best way of dealing with the issue is through the use of a simplified development guidance map. Such a map has been prepared for three development control areas defined on the basis of geological, hydrogeological and geomorphological characteristics and records of subsidence events. The map and associated planning policies were adopted by Harrogate Borough Council for a two year trial period in 1996. Together, they provide a framework for the monitoring and control of new development which seeks to ensure that the worst consequences of gypsum-related subsidence are minimized, and that an acceptable balance between construction costs and hazard mitigation can be achieved.

The policies and procedures have been working well so far (July 1998), without creating additional difficulties for existing development, and without significantly deterring potential new developers. It is important, however, that the operation of policies and the information that becomes available as a result of their implementation should be monitored to enable future review of both the policies and the development guidance map as and when appropriate.

Acknowledgements. The authors are grateful to the former Department of the Environment and Harrogate Borough Council for permission to publish the paper. Thanks are also due to all Steering Group members, to numerous colleagues at Symonds Travers Morgan (STM) who assisted in this project (notably John Greig, Dave Williams, Kerry Tomkinson, Trevor Bishop & Keith Driscoll) and to the British Geological Survey, especially A. Cooper, who assisted STM with the revision of basic geological mapping and provided helpful liaison throughout the study.

References

APPLIED GEOLOGY LIMITED 1993. *Natural Underground Cavities in Great Britain.* Summary Report and Regional Reports for the Department of the Environment.

ANON 1993. *Without Site Investigation, Ground is a Hazard.* Institution of Civil Engineers' Site Investigation Steering Group, Thomas Telford, London.

COOPER, A. H. 1986. Subsidence and foundering of strata caused by the dissolution of Permian gypsum in the Ripon and Bedale areas, North Yorkshire. *In*: HARWOOD, G. M. & SMITH, D. B. (eds) *The English Zechstein and Related Topics.* Special Publication of the Geolocical Society of London No. **22**, 127–139

——1988. Subsidence resulting from the dissolution of the Permian gypsum in the Ripon area; its relevance to mining and water abstraction. *In*: BELL, F. G., CULSHAW, M. G., CRIPPS, J. C. LOVELL, M. A. (eds) *Engineering Geology of Underground Movements.* Engineering Geology Special Publication of the Geological Society of London No. **5**, 387–390.

——1989. Airborne multispectral scanning of subsidence caused by Permian gypsum dissolution at Ripon, North Yorkshire. *Quarterly Journal of Engineering Geology, London,* **22**, 219–229.

——1995. Subsidence hazards due to the dissolution of Permian gypsum in England: investigation and remediation. *In*: BECK (ed.) *Karst Geohazards*. Balkema, Rotterdam, 23–29.

—— & BURGESS, I. 1993. *Geology of the Country around Harrogate*. Memoir of the British Geological Survey, Sheet 62 (England and Wades).

DEPARTMENT OF THE ENVIRONMENT 1990. *Planning Policy Guidance: Development on Unstable Land*, PPG 14. HMSO, London.

GEOMORPHOLOGICAL SERVICES LTD 1991 *Ground Movement in Ventnor, Isle of Wight*. A Summary of the study of landslip problems in Ventnor, carried out for the Department of the Environment. Geomorphological Services Ltd.

HOLLIDAY, D. W. 1970. The petrology of secondary gypsum rocks: a review. *Journal of Sedimentary Petrology*, **40**(2), 734–744.

HOWARD HUMPHREYS & PARTNERS, 1993. *Subsidence in Norwich*. HMSO, London.

KLIMCHOUK, A. B. 1992. Large gypsum caves in the Western Ukraine and their genesis. *Cave Science*, **19**(1), 3–11.

MOSSOP, G. D. & SHEARMAN, D. J. 1973. Origins of secondary gypsum rocks. *Transactions of the Institution of Mining Metallurgy Section B: Appied Earth Sciences*, **82**, B146–B154.

MURRAY, R. C. 1964. Origin and diagenesis of gypsum and anhydrite. *Journal of Sedimentary Petrology*, **34**(3), 512–523.

POWELL, J. H., COOPER, A. H. & BENFIELD, A. C. 1992. *Geology of the Country around Thirsk*. Memoir of the British Geological Society, Sheet 52 (England and Wales).

THOMPSON, A., HINE, P. D., GREIG & PEACH, D. W. 1996. *Assessment of subsidence arising from gypsum dissolution*. Technical Report to the Department of the Environment, June 1996. Symonds Travers Morgan, East Grinstead.

——, ——, POOLE, J. S. & GREIG, J. R. 1998. *Environmental Geoogy in Land Use Planning: A Guide to Good Practice*. Report to the Department of the Environment, Transport and the Regions. Symonds Travers Morgan, East Grinstead.

Some geohazards caused by soil mineralogy, chemistry and microfabric: a review

F. G. Bell[1] & M. G. Culshaw[2]

[1] Department of Geology and Applied Geology, University of Natal, King George V Avenue, Durban, South Africa
[2] Coastal and Engineering Geology Group, British Geological Survey, Kingsley Dunham Centre, Keyworth, Nottingham NG12 5GG, UK

Abstract. The paper reviews three geohazards: swelling/shrinkage, dispersivity and collapsibility, that are not dramatic and do not usually cause disasters because they rarely result in loss of life, nor cause sudden huge financial losses. However, these unspectacular' geohazards result, over time, in the loss of millions of pounds and in disruption to people's lives due to damage to foundations or loss of land. The geological controls on the hazards are described; all result from the inherent microfabric, mineralogy and/or chemistry of the particular soils affected. From this understanding it is possible to design mitigation and remedial measures that will prevent, or reduce, the effects of these hazards; these measures are presented.

Introduction

There are several ways in which geohazards can be classified: a distinction can be made between natural geohazards (for example, earthquakes, landslides, swelling and shrinkage of soils) and man-made geohazards (for example, mine shafts, ground subsidence due to mining or fluid extraction and landfill gases). Geohazards also have been classified according to the natural environment in which they occur (for example, onshore, on the ground surface, in mountainous areas; or, offshore, on the seabed) (Hu Haitao 1993). An alternative approach is to look at process and to divide geohazards into primary and secondary ones. Primary geohazards are not triggered by other events and are often viewed as cyclical in their occurrence because attempts are made to predict them from the pattern of past events; they are controlled by global or regional geology and they affect regions. The geological processes that control these geohazards are often poorly understood. Secondary geo-hazards are triggered by other hazards, can be only partially predicted by analysis of past events, are controlled by local geology and affect sites or districts.

Many of the geological processes that control these secondary geohazards are quite well understood. For example, landslides may be triggered by an earthquake, by volcanic activity or by a period of prolonged and/or intensive rainfall. Secondary geohazards can result from the continuous erosive processes that affect the ground, including the effects of man. Landslides can be triggered by river erosion at the toe of the slide; ground collapse can be caused by the solution of more soluble rocks, or by the removal of minerals during mining. In most of these cases, the amount of ground movement is large enough, and often rapid enough, to be readily observable by the general public.

However, there is a further group of secondary geohazards that exists largely because of the geotechnical properties of the soils concerned. These properties are, in turn, related to the soils' microfabric and mineralogy. This group of geohazards includes the swelling and shrinkage of some clay soils, dispersivity in clay soils with particular chemical characteristics and the collapse potential of certain silty soils. The effects of these hazards, while being unspectacular and rarely causing loss of life, can result in considerable financial loss. Swell–shrink in clay soils, for example, has caused losses of well over ,2bn in recent years in the United Kingdom (Anon 1996). Collapsible loess covers large parts of central Europe, Russia, China and North America. The use of dispersive soils in southern Africa, Australia and the United States has led to the failure of dams and road embankments. This paper describes these 'unspectacular' geohazards, and how the geotechnical properties, microfabric, mineralogy and chemistry control the engineering behaviour of the soils and hence the engineering measures that are needed to overcome them.

Swelling and shrinkage in clay soils

The causes of swelling and shrinkage

Some clay soils undergo slow volume changes which occur independently of loading and are attributable to swelling or shrinkage; they are also often referred to as

BELL, F. G. & CULSHAW, M. G. 1998. Some geohazards caused by soil mineralogy, chemistry and microfabric: a review. *In*: MAUND, J. G. & EDDLESTON, M. (eds) *Geohazards in Engineering Geology*. Geological Society, London, Engineering Geology Special Publications, **15**, 427–441.

expansive soils.' These volume changes can give rise to ground movements which may result in damage to buildings. Low-rise buildings are particularly vulnerable to such ground movements since they generally do not have sufficient weight or strength to resist. In addition, shrinkage settlement of embankments can lead to cracking and break-up of the roads they support.

Problems caused by expansive soils in the United States were estimated to cost of over \$2bn each year in the 1970s (Robinson & Spieker 1978) and between \$6 and \$11bn annually by the 1990s (Nuhfer *et al.* 1993). This is estimated as twice the cost of flood damage, or of damage caused by landslides, and more than 20 times the cost of earthquake damage (Robinson & Spieker 1978). The principal cause of expansive clays is the presence of swelling clay minerals such as montmorillonite. For example, Popescu (1979) found that expansive clays in Romania contained between 40 and 80% montmorillonite. Construction damage is especially notable where expansive clay occurs at the surface in regions which experience alternating wet and dry seasons leading to swelling and shrinkage of these soils. Again, taking expansive clays in Romania as an example, Popescu (1979) noted that maximum seasonal changes in moisture content in these soils were around 20% at 0.4 m depth, 10% at 1.2 m depth and less than 5% at 1.8 m depth. The corresponding cyclic movements of the ground surface were between 100 and 200 mm.

Differences in the period and amount of precipitation and evapotranspiration are the principal factors influencing the swell–shrink response of a clay soil beneath a building. Poor surface drainage or leakage from underground pipes also can produce concentrations of moisture in clay. Trees with high water demand and uninsulated hot process foundations may dry out clay causing shrinkage. Cold stores also may cause desiccation of clay soil. The depth of the active zone in expansive clays (that is, the zone in which swelling and shrinkage occurs in wet and dry seasons respectively) varies. It may extend to over 6 m depth in some semi-arid regions of South Africa, Australia and Israel. Many soils in Britain, especially in southeast England, possess the potential for significant volume change due to changes in moisture content. However, owing to the damp climate in most years, volume changes are usually restricted to the upper 1.0 to 1.5 m in clay soils.

The potential for volume change of clay soil is governed by its initial moisture content, initial density or void ratio, its microstructure and the vertical stress, as well as the type and amount of clay minerals present. Cemented and undisturbed expansive clay soils often have a high resistance to deformation and may be able to absorb significant amounts of swelling pressure. Remoulded expansive clays, therefore, tend to swell more than their undisturbed counterparts. For example, Schmertmann (1969) maintained that some clays increase their swell behaviour when they undergo repeated large shear strains due to mechanical remoulding. He introduced the term 'well sensitivity' for the ratio of the remoulded swelling index to the undisturbed swelling index and suggested that such a phenomenon may occur in unweathered highly overconsolidated clay when the bonds which hold clay particles in bent positions have not been broken. When these bonds are broken by remoulding the clays exhibit significant swell sensitivity.

Expansive clay minerals take water into their lattice structure. In less dense soils they tend to expand initially into zones of looser soil before volume increase occurs. However, in densely packed soil with low void space, the soil mass has to swell more or less immediately to accommodate the volume change. Hence clay soils with a flocculated fabric swell more than those which possess a preferred orientation. In the latter the maximum swelling occurs normal to the direction of clay particle orientation.

Because expansive clays normally possess extremely low permeabilities, moisture movement is slow and a significant period of time may be involved in the swelling–shrinking process. Accordingly, moderately expansive clays with a smaller potential to swell but with higher permeabilities than clays having a greater swell potential may swell more during a single wet season than more expansive clays.

Expansive clays are often heavily fissured due to seasonal changes in volume which produce shrinkage cracks and shear surfaces. Consequently near-vertical fissures are found frequently at shallow depth with diagonal fissures at greater depth. Sometimes the soil is so desiccated that the fissures are wide open and the soil is shattered or micro-shattered. The soils normally are of medium to high or very high plasticity.

The surfaces of clay particles are negatively charged and so attract positively charged ions in the pore water of soil. However, the ability of clay minerals to attract positive ions varies significantly. When water is freely available, clay particles attract layers of bound water around them, referred to as 'diffuse double layers', and the spacing between the particles increases. If a suction is applied to the water, or the clay particles are compressed by an external force, the bound layers of water are reduced in thickness, some water is expelled, provided drainage conditions allow, and the clay particles move closer together. The suction, or externally applied pressures, required to bring the particles closer together can be very large (over 10 MPa). This is why clay soils harden when dried and why, when wetted, they can undergo large changes in volume or, if this is inhibited, exert high swelling pressures.

Therefore, the swell–shrink behaviour of a clay soil under a given state of applied stress in the ground is controlled by changes in soil suction. The relationship between soil suction and water content depends on the proportion and types of clay minerals present, their microstructural arrangement and the chemistry of the pore water. Changes in soil suction are brought about by moisture movement through the soil due to evaporation from its surface in dry weather, by transpiration from plants, or alternatively by recharge consequent upon precipitation. The climate governs the amount of moisture available to counteract that which is removed by evapotranspiration (the soil moisture deficit). In semi-arid climates there are long periods of high soil moisture deficit alternating with short periods when precipitation balances or exceeds evapotranspiration. The distribution of foundation problems caused by shrinkage in Britain during the period 1989 to 1991 was partly related to the geological distribution of swell-shrink susceptible clay soils but also to the level of soil moisture deficit; this was greatest in the southeast of England where such swell–shrink susceptible clay soils are also found.

The volume changes that occur due to evapotranspiration from clay soils can be conservatively predicted by assuming the lower limit of the soil moisture content to be the shrinkage limit. Desiccation beyond this value cannot bring about further volume change.

Transpiration from vegetative cover is a major cause of water loss from soils in semi-arid regions. Indeed, the distribution of soil suction in soil is primarily controlled by transpiration from vegetation and represents one of the most significant changes made in loading (that is, to the state of stress in a soil). The behaviour of root systems is exceedingly complex and is a major factor in the intractability of swelling and shrinking problems. The spread of root systems depends on the type of vegetation, the soil type and groundwater conditions. The suction induced by the withdrawal of water fluctuates with the seasons, reflecting the growth of vegetation and probably varies between 100 and 1000 kPa (equivalent to pF values 3 and 4 respectively). The complete depth of active clay profiles usually does not become fully saturated during the wet season in semi-arid regions. Nonetheless, changes in soil suction may occur over a depth of 2 m or so between the wet and dry seasons. The suction pressure associated with the onset of cracking is approximately pF 4.6. The presence of desiccation cracks enhances evaporation from soil.

The moisture characteristic (moisture content v. soil suction) of a soil provides valuable data concerning the moisture contents corresponding to the field capacity (defined in terms of soil suction; this is a pF value of about 2.0) and the permanent wilting point (the level at which moisture is no longer available to plants; this corresponds to a pF value of about 4.2), as well as the rate at which changes in soil suction take place with variations in moisture content. This enables an assessment to be made of the range of soil suction and moisture content which is likely to occur in the zone affected by seasonal changes in climate.

The extent to which the vegetation is able to increase the suction to the level associated with the shrinkage limit is obviously important. In fact, the moisture content at the wilting point exceeds that of the shrinkage limit in soils with a high content of clay and is less in those possessing low clay contents. This explains why settlement resulting from the desiccating effects of trees is more notable in low to moderately expansive soils than in expansive ones. When vegetation is cleared from a site, its desiccating effect is also removed. Hence the subsequent regain of moisture by clay soils leads to them swelling. Swelling movements on expansive clays in South Africa, associated with the removal of vegetation and subsequent erection of buildings, in many areas have amounted to about 150 mm, although movements over 350 mm have been recorded (Williams & Donaldson 1980).

Methods of predicting volume changes in soils can be grouped into empirical methods, soil suction methods and oedometer methods. Empirical methods make use of the swelling potential as determined from void ratio, natural moisture content, liquid and plastic limits, and activity. However, because the determination of plasticity is carried out on remoulded soil, it does not consider the influence of soil texture, moisture content, soil suction or pore water chemistry, which are important factors in relation to volume change potential. One of the properties which has been widely used to predict the potential expansiveness of clay soils is their activity. Expansive soils, however, can plot within the field of low expansion on the activity chart developed by Van der Merwe (1964), and vice versa. Over reliance on the results of this simple test, therefore, must be avoided. Consequently empirical methods should be regarded as simple swelling indicator methods and nothing more. As such, it is wise to carry out another type of test and to compare the results of the two before drawing any conclusions.

Soil suction methods use the change in suction from initial to final conditions to obtain the degree of volume change. Soil suction is the stress which, when removed allows the soil to swell. In other words, the value of soil suction in a saturated, fully swollen soil is zero. O'Neill & Poormoayed (1980) quoted the United States Army Engineers Waterways Experimental Station (USAEWES) classification of potential swell (Table 1) which is based on the liquid limit, plasticity index and initial (*in situ*) suction. The latter is measured in the field by a psychrometer. Soil suction is not easy to measure accurately. Filter paper has been used for this purpose (McQueen & Miller 1968). According to Chandler *et al.* (1992), measurements of soil suction obtained by the filter paper method compare favourably with measurements obtained using psychrometers or pressure plates. Nonetheless, there are a few factors which could affect the results significantly.

The oedometer methods of determining the potential expansiveness of clay soils are more direct methods. In the oedometer methods, undisturbed samples are placed

Table 1. *USAEWES classification of swell potential (from O'Neill & Poormoayed 1980)*

Liquid limit (%)	*Plastic limit (%)*	*Initial (in situ) suction (kPa)*	*Potential swell (%)*	*Classification*
Less than 50	Less than 25	Less than 145	Less than 0.5	Low
50–60	25–35	145–385	0.5–1.5	Marginal
Over 60	Over 35	Over 385	Over 1.5	High

in the oedometer and a wide range of testing procedures are used to estimate the likely vertical strain due to wetting under vertical applied pressures. The latter may be equated to overburden pressure plus that of the structure which is to be erected. The double oedometer test, however, has been shown to underestimate the amount of expansion likely to occur in a clay soil (Burland 1963). Conversely, the single oedometer method over-estimates the amount of potential swell. In oedometer testing, one-dimensional compression-swelling occurs. However, in reality most expansive clays are fissured which means that lateral and vertical strains develop locally within the ground. Even when the soil is intact, swelling or shrinkage is not truly one-dimensional. The effect of imposing zero lateral strain in the oedometer is likely to give rise to over-predictions of heave and the greater the degree of fissuring the greater the over-prediction. The values of heave predicted using oed-ometer methods correspond to specific values of natural moisture content and void ratio of the sample. Therefore, any change in these affects the amount of heave predicted.

Consequently, results obtained from the above mentioned tests can show a wide variation between one test method and another. This is illustrated in Table 2.

Table 2. *Prediction of swelling potential of expansive clay from Ladysmith, South Africa (after Bell & Maud 1995)*

Method	*Predicted Heave (mm)*	
	Maximum	*Minimum*
Empirical		
Van der Merwe (1964)	89	37
Brackley (1975)	215	18
Weston (1980)	59	8
Oedometer		
Swell under load	40	−23*
Frydman & Calabresi (1987)		
Double oedometer	127	33
Jennings & Knight (1957)		
Soil suction		
Brackley (1980)	57	16

* The negative sign indicates that compression rather than swelling of the soil sample was predicted.

Engineering control methods

Effective and economic foundations for low-rise buildings on swelling and shrinking soils have proved difficult to achieve. This is partly because the cost margins on individual buildings are low. Detailed site investigation and soil testing are out of the question for individual dwellings. Similarly, many foundation solutions which are appropriate for major structures are too costly for small buildings. Nonetheless, the choice of foundation is influenced by the subsoil and site conditions, estimates of the amount of ground movement and the cost of alternative designs. In addition, different building materials have different tolerances to deflections (Burland & Wroth 1975). Hence materials which are more flexible can be used to reduce potential damage due to differential movement of the structure.

Three methods can be adopted when choosing a design solution for building on expansive soils, namely, provide a foundation and structure which can tolerate movements without unacceptable damage; isolate the foundation and structure from the effects of the soil; or alter or control the ground conditions. The following precautions may be taken where differential movements up to 25 mm are likely to occur:

1. Deep strip foundations, preferably with nominal reinforcement.
2. Short, small diameter bored cast-in-place piles with suspended reinforced ground beams where swelling is anticipated.
3. A light stiffened raft. The design of the slab is dependent on assumed allowable relative deflections and it is usually necessary to incorporate certain anti-cracking features in the superstructure.
4. A light stiffened raft with beam support.
5. The provision of movement joints in the super-structure.
6. The use of soft cement mortar for all masonry.
7. The use of a floating floor slab with total separation from foundations and outer walls.
8. Flexible support of interior walls on the floor slab. Alternatively, support the interior walls on separate deep strip, beam or piled foundations.

Some of these methods are described in more detail by Anon (1980*a*, *b*). In addition to these construction details moisture control measures should be adopted as far as possible.

The isolation of foundation and structure has been widely adopted for 'severe' and 'very severe' ground conditions. Straight-shafted bored piles can be used in conjunction with suspended floors for severe conditions. The piles are sleeved over the upper part and provided with reinforcement. For severe conditions it may be necessary to place piles at appreciable depth (that is, below the level of fluctuation of natural moisture content) and/or use under reams to resist the pull-out forces.

The use of stiffened rafts is fairly commonplace (Bell & Maud 1995). The design of the slab is dependent on assumed allowable relative deflections and it is usually necessary to incorporate certain anti-cracking features in the superstructure such as flexible joints.

Moisture control is perhaps the most important single factor in the success of foundations on shrinking and swelling clays. The aim is to maintain stable moisture conditions with minimum moisture content or suction gradients. The loss of moisture around the edges of a building which leads to the moisture content of the soil under the centre of the building being higher gives rise to differential heave (the situation may be complicated by watering of gardens or where buildings are heated artificially). In order to control this, an attempt should be made to maintain the same moisture content beneath a building. This can be achieved by the use of horizontal and vertical moisture barriers around the perimeter of the building, drainage systems and control of vegetation coverage. Common sources of moisture are broken pipes and poor drainage. Hence, all water supply pipes and waste water pipes should have flexible connections and couplings; all rainwater pipes should be ducted well away from the foundations; storage tanks and septic tanks should be reinforced to minimize cracking and have flexible waterproofing; the ground should be sloped away from buildings to convey run-off away. Large trees and bushes should be situated at a distance of at least one and a half times their mature height away from foundations. Paving (approximately 2 m wide) around the perimeter of the building helps to reduce moisture fluctuations in the vicinity of the foundations. It also helps to reduce differential movements across the building. Ideally this paving should be laid on thick polythene or PVC.

A simple method of reducing or eliminating ground movements due to expansive soil is to replace or partially replace them with non-expansive soils. There is no requirement for the thickness of the replacement material but a minimum of 1 m has been suggested by Chen (1988). The material should be granular but it should not allow surface water to travel freely through the soil so that it wets any swelling soils in lower horizons. Hence the presence of a fine fraction is required to reduce permeability or a geomembrane can surround the granular material.

If expansive soil is allowed to swell by wetting prior to construction and if the soil moisture content is then maintained, the soil volume should remain relatively constant and no heave take place. Ponding is the most common method of wetting. This may take several months to increase the water content to the required depth, notably in areas with deep groundwater surfaces. Vertical wells can be installed to facilitate flooding and thus decrease the time necessary to adjust the moisture content of the soil. Williams (1980) described a case where severe damage due to swelling was corrected by controlled wetting.

The amount of heave of expansive soils is reduced significantly when the soil is compacted to low densities at high moisture contents. Expansive soils compacted above optimum moisture content undergo negligible swell for any degree of compaction. On the other hand, compaction below optimum results in excessive swell.

Many attempts have been made to reduce the expansiveness of clay soil by chemical stabilization. For example, lime stabilization of expansive soils, prior to construction, can minimize the amount of shrinkage and swelling they undergo. In the case of light structures, lime stabilization may be applied immediately below strip footings. However, significant SO_4 content (in excess of 5 000 mg/kg) in clay soils can mean that they react with CaO to form ettringite with resultant expansion (Forster *et al.* 1995). However, the treatment is better applied as a layer beneath a raft so as to overcome differential movement. The lime stabilized layer is formed by mixing 4 to 6% lime with the soil. A compacted layer, 150 mm in thickness, usually gives satisfactory performance. Furthermore the lime stabilized layer redistributes unequal moisture stresses in the subsoil so minimizing the risk of cracking in the structure above, as well as reducing water penetration beneath the raft. Premix or mix-in-place methods can be used (Bell 1988). Alternatively, lime treatment can be used to form a vertical cut-off wall at, or near, the footings in order to minimize movement of moisture.

The lime slurry pressure injection method also has been used to minimize differential movements beneath structures, although it is more expensive. The method involves pumping hydrated lime slurry under pressure into soil, the points of injection being spaced about 1.5m apart. The lime slurry forms a network of horizontal sheets, often interconnected by vertical veins. Injection can be used to form a seam line around the perimeter of a building to provide a barrier to moisture movement, the seams extending below the critical zone of change in moisture content.

Lime columns have been used instead of piles as foundations for light structures. The columns reduce total and differential movements and may be placed in a square pattern with a concentration beneath loaded walls. The load of the structure can be distributed to the lime columns by way of a concrete slab.

Cement stabilization has much the same effect on expansive soils as lime treatment, although the dosage of cement needs to be greater for heavy expansive clays (Bell 1995). Alternatively, they can be pretreated with lime, thereby reducing the amount of cement which needs to be used (Stamatopoulos *et al.* 1992).

Dispersive soils

The causes of dispersivity

Dispersion occurs in soils when the repulsive forces between clay particles exceed the attractive forces thus bringing about deflocculation so that in the presence of relatively pure water the particles repel each other. In non-dispersive soil there is a definite threshold velocity below which flowing water causes no erosion. The individual particles cling to each other and are only removed by water flowing with a certain erosive energy. By contrast, there is no threshold velocity for dispersive soil, the colloidal clay particles go into suspension even in quiet water and therefore are highly susceptible to erosion and piping. For a given eroding fluid the boundary between the flocculated and deflocculated states depends on the value of the sodium adsorption ratio (see below), the salt concentration, the pH value and the mineralogy.

Nonetheless, there are no significant differences in the clay fractions of dispersive and non-dispersive soils, except that soils with less than 10% clay particles may not have enough colloids to support dispersive piping. Dispersive soils contain a higher content of dissolved sodium (up to 12%) in their pore water than ordinary soils. The pH value of dispersive soils generally ranges between 6 and 8.

The sodium adsorption ratio (SAR) is used to quantify the role of sodium where free salts are present in the pore water and is defined as:

$$SAR = \frac{Na}{\sqrt{0.5(Ca+Mg)}} \quad (1)$$

with units expressed in meq/litre of the saturated extract. There is a relationship between the electrolyte concentration of the pore water and the exchangeable ions in the adsorbed layers of clay particles. This relationship is dependent upon pH value and may also be influenced by the type of clay minerals present. Hence it is not necessarily constant. An SAR of more than 6 suggests that the soil is sensitive to leaching. However, in Australia, Aitchison & Wood (1965) regarded soils in which the SAR exceeded 2 as dispersive. As can be seen from Table 3 the latter value seems to be the more appropriate.

The presence of exchangeable sodium is the main chemical factor contributing towards dispersive behaviour in soil. This is expressed in terms of the exchangeable sodium percentage (ESP):

$$ESP = \frac{\text{exchangeable sodium}}{\text{cation exchange capacity}} \times 100 \quad (2)$$

where the units are given in meq/100 g of dry clay. A threshold value of ESP of 10% has been recommended, above which soils that have their free salts leached by seepage of relatively pure water are prone to dispersion. Soils with ESP values above 15%, are highly dispersive. Those with low cation exchange values (15 meq/100 g of clay) have been found to be completely non-dispersive at ESP values of 6% or below. Similarly, soils with high cation exchange capacity values and a plasticity index greater than 35% swell to such an extent that dispersion is not significant. High ESP values and piping potential generally exist in soils in which the clay fraction is composed largely of smectitic and other 2:1 clays. Some illites are highly dispersive. On the other hand, high values of ESP and high dispersibility are rare in clays composed largely of kaolinites.

Another property which has been claimed to govern the susceptibility of clayey soils to dispersion is the total content of dissolved salts (TDS) in the pore water. In other words, the lower the content of dissolved salts in the pore water, the greater the susceptibility of sodium saturated clays to dispersion. Sherard *et al.* (1976) regarded the total dissolved salts for this specific purpose as the total content of calcium, magnesium, sodium and potassium in milliequivalents per litre. They designed a chart in which sodium content was expressed as a percentage of TDS and was plotted against TDS to determine the dispersivity of soils. However, Craft & Acciardi (1984) showed that this chart had poor overall agreement with the results of physical tests. Bell & Maud (1994) showed that the use of the dispersivity chart to distinguish dispersive soils had not proved reliable in Natal, South Africa. There the determination of dispersive potential frequently involves the use of a chart designed by Gerber & Harmse (1987) which plots ESP against cation exchange capacity (CEC). Dispersive soils occur in semi-arid regions; for example, in South Africa they are found in areas which have less than 850 mm of rain annually. More specifically, with few exceptions, they are present in areas which experience Weinert's (1980) climatic N values between 2 and 10. Dispersive soils tend to develop in low-lying areas with gently rolling topography and smooth relatively flat slopes where the rainfall is such that seepage water has a high SAR. In more arid regions where the N values exceed 10, the development of dispersive soils generally is inhibited by the presence of free salts, despite high SAR values.

Unfortunately, dispersive soils cannot be differentiated from non-dispersive soils by routine soil mechanics testing. Although a number of tests have been used to

Table 3. *Some physical and chemical properties of dispersive soils from Natal (after Bell & Maud 1994)*

Location	*Clay (%)*	w_L	I_p	A	LS	*pH*	*K*	*Ca*	*Mg*	*Na*	*TDS*	*ESP*	*SAR*	*CEC meq/100 g*	*EC*	*Dispersivity (%)*	*Dispersivity potential*
Makatini	30	28	14	0.47	6.0	8.6	2.6	9.0	8.0	97.6	117.2	39.4	33.4	97.9	900		HD
Makatini	29	43	24	0.83	8.0	8.6	2.6	181.9	78.9	115.6	379	30	10.1	132.86			HD
Paddock	40	31	12	0.3	5.3	8.6	1.6	95.9	147.2	23.5	268.2	89	2.1	66.3		44	HD
Paddock	44	33	14	0.32	6.0	8.6	0.02	0.3	0.4	4.4	5.1	16.6	7.5	60.2	52		HD
Winterton	31	27	13	0.42	4.0	8.6	0.03	0.8	0.4	7.7	8.9	24.4	9.9	50.9	95	71.9	HD
Winterton	44	35	10	0.23	5.3	8.95	0.04	0.8	0.5	8.3	9.6	16.1	10.2	36.6	110	53.2	HD
Makatini	30	28	14	0.47	6.0	8.6	14.7	293.4	201.5	21.3	530.9	30	1.1	132.86			HD
Makatini	12	40	14	1.17	6.0	8.85	0.1	0.6	0.5	2.7	3.9	17.3	3.6	137.3	38		HD
Winterton	37	28	13	0.35	5.3	7.5	0.04	1.9	1.5	4.8	8.2	6.3	3.7	53.5	81	35.4	D
Rietspruit	36	47	24	0.67	8.0	6.5	2.4	71.6	46.9	13.1	134	9.3	1.7	38.8			D
Rietspruit	60	66	34	0.57	10.0	7.55	1.8	160.4	139.1	17.1	318.4	5.3	1.4	53.8			D
Tala	22	45	6	0.27	0.7	5.05	5.3	33.7	4.8	4.8	48.6	8.8	1.1	25.9			D
Ramsgate	18	29	8	0.44	5	5.2	0.9	22.0	46.7	5.4	75	7	0.9	38	77.5		D
Ramsgate	21	30	10	0.48	5	5.1	1.2	15.8	31.7	4.6	53.5	8	0.9	30	64.8		D
Ramsgate	20	29	8	0.4	5	5.4	1.4	24.0	61.7	11.3	98.4	11	1.7	44	105.6		D
Ramsgate	23	28	4	0.17	5	5.4	1.4	15.6	63.3	14.6	94.9	15	2.3	43	98.6		D
Ramsgate	22	20	10	0.45	2	5.5	0.7	16.8	21.7	4.2	43.4	9	1.0	23	98.6		D
Ramsgate	21	32	8	0.38	6	5.6	0.9	18	55.0	9.9	83.8	12	1.7	40	70.4		D
Ramsgate	22	30	9	0.41	5	6.4	0.9	28.0	57.5	10.7	97.1	11	1.6	46	119.7		D
Underberg	30	34	16	0.53	5.3	5.95	1.1	33.2	18.4	4.0	57.2	6.7	0.8	20.0		35.8	M
Umzimkulwana	18	40	20	1.1	8.0	6.25	0.8	44.7	34.5	4.9	84.9	5.5	0.8	32.8			M
Umzimkulwana	48	42	23	0.48	8.0	7.00	1.5	66.9	67.4	7.9	143.7	5.3	1.0	30.8			M
Greytown	29	34	14	0.48	6.7	6.35	1.5	20.1	21.3	4.5	47.4	8.5	1.0	18.21			M
Rietspruit	40	17	16	0.4	6.7	6.85	3.5	74.5	60.8	5.6	144.4	3.8	0.7	37.1			M
Paddock	18	21	9	0.5	2.7	4.9	0.5	4.5	5.2	1.8	12	6.1	0.8	16.38			M
Paddock	17	18	7	0.41	2.3	4.75	0.3	2.8	1.4	2.0	6.5	8.5	1.4	13.88			M
Everton	16	38	16	1.0	5.0	5.5	5.5	18.1	26.0	2.4	52	4.3	0.5	0.5	34.69	39.4	M
Underberg	56	43	14	0.25	7.3	5.5	1.5	10.3	19.7	2.6	34.1	5.2	0.7	8.88		7.1	ND
Underberg	46	56	21	0.47	10.0	5.55	1.8	146.9	105.2	5.1	259	1.9	0.5	57.28		22.0	ND
Winterton	20	20	7	0.35	2.0	7.1	0.02	1.2	0.7	1.4	3.3	2.1	1.4	37.7	32	28.9	ND
Karkloof	51	44	13	0.25	6.0	5.55	0.9	4.3	1.7	1.1	8	9.2	0.6	3.24		4.9	ND
Mount West	38	37	15	0.39	6.7	4.65	2.3	18.3	16.8	2.3	39.7	2.9	0.5	20.79		3.0	ND
Claridge	39	33	13	0.33	6.0	5.5	0.9	25.7	26.2	2.2	55	3.0	0.4	18.59			ND
Greytown	50	36	14	0.28	6.7	5.95	3.4	35.7	26.4	3.1	68.6	4.2	0.6	14.6			ND
Paddock	25	24	9	0.36	3.4	4.85	0.3	1.8	2.3	2.5	6.9	8.1	1.9	12.36			ND
Claridge	45	40	18	0.4	7.3	4.8	3.8	40.3	93.8	4.1	142	2.4	0.5	37.6			ND
Claridge	54	49	22	0.41	8.8	5.15	1.9	34.9	23.1	3.0	62.9	2.1	0.6	14.5			ND
Paddock	41	57	27	0.66	10.7	6.1	2.5	18.5	27	3.0	51	5.6	0.6	13.7			ND
Howick	31	30	10	0.32	4.7	5.4	0.7	9.2	9.2	0.9	20	3.8	0.3	8.57		6.6	CND
Howick	41	38	13	0.32	6.0	5.5	0.7	8.8	8.1	1.3	18.9	5.7	0.4	5.61		4.7	CND
Claridge	70	54	30	0.43	12.3	5.0	1.6	23.9	15.1	1.3	41.9	1.8	0.3	7.4			CND
Everton	43	30	12	0.28	4.5	5.2	1.8	16.4	33.7	0.2	52.1	0.3	0.4	12.52		12.9	CND
Paddock	50	54	16	0.32	8.0	5.1	0.8	4.0	3.9	1.2	9.4	3.3	0.6	7.2			CND

w_L, liquid limit; I_p, plasticity index; *A*, activity; *LS*, linear shrinkage; K, potassium; Ca, calcium; Mg, magnesium; Na, socium; TDS, total dissolved solids; ESP, exchangeable sodium percentage; SAR, sodium adsorption ration; CEC, cation exchange capacity; EC, electrical conductivity (mS/m); H highly and very highly dispersive; D dispersive; M marginal; ND, non-dispersive; CND, completely non-dispersive (according to Gerber & Harmse 1987).

Table 4. *Suggested rating system for potentially dispersive soils (after Bell & Maud 1994)*

Crumb Test	Class Rating	Strong reaction 4	Moderate reaction 2	Slight reaction 1	No reaction 0
Dispersion Test	Class Rating	Highly dispersive 4	Moderately dispersive 2	Slightly dispersive 1	Non-dispersive 0
*ESP/CEC (meq/100 g clay)	Class Rating	Highly dispersive 5	Dispersive 3	Marginal 1	Non-dispersive 0
SAR	Class Rating	Over 10 3	2–10 1	Less than 2 0	
pH	Class Rating	Over 8 2	6–8 1	less than 6 0	

Highly dispersive = 18 and above; dispersive = 9–17; cautionary zone = 5–8; marginal = 1–4; non-dispersive = 0.
* Highly dispersive includes very highly dispersive. Non-dispersive includes completely non-dispersive of Gerber & Harmse (1987).

recognize dispersive soils, no single test can be relied on completely to identify them. For example, Craft & Acciardi (1983) found that the crumb test and the pinhole test at times yielded conflicting results from the same samples of soil. Subsequently, Gerber & Harmse (1987) showed that the crumb test, the double hydrometer test and the pinhole test were unable to identify dispersive soils when free salts were present in solution in the pore water, which is frequently the case with sodium-saturated soils. Atkinson *et al.* (1990) also concluded that because internal erosion is governed by true cohesion (that is, the strength at zero effective stress) and since true cohesion may be influenced by physicochemical effects in the soil grains, in the pore water and in the free water, the crumb test, dispersion test and pinhole test have shortcomings. They devised the cylinder dispersion test to examine erosion and dispersion of soils from the surfaces of cracks into water.

The various tests show that the boundary between deflocculated and flocculated states varies considerably among different soils so that the transition between dispersive and non-dispersive soils is wide. Because there is no clearly defined boundary between dispersive and non-dispersive soils it is wise to do several types of tests on at least a proportion of all samples.

As no one test can be relied upon to identify dispersive soil with absolute certainty, Bell & Maud (1994) suggested a tentative rating system (Table 4). This included some of the tests or properties which have been used to help recognize dispersive soils. The most important of these was the assessment based on ESP and CEC (meq/100g clay).

Engineering control methods

Serious piping damage to embankments and failures of earth dams have occurred when dispersive soils have been used in their construction. Severe erosion damage also can form deep gullies on earth embankments after rainfall. Indications of piping take the form of small leakages of muddy-coloured water from an earth dam after initial filling of the reservoir. The pipes become enlarged rapidly and this can lead to failure of a dam. Dispersive erosion may be caused by initial seepage through an earth dam in areas of higher soil permeability, especially areas where compaction may not be so effective such as around conduits, against concrete structures and at the foundation interface, through desiccation cracks or cracks due to differential settlement or those due to hydraulic fracturing (Wilson & Melis 1991).

In many areas where dispersive soils are found there is no economic alternative other than to use these soils for the construction of earth dams. However, experience indicates that if an earth dam is built with careful construction control and incorporates filters, then it should be safe enough even if it is constructed with dispersive clay. If no supply of clean pervious sand of the type used for a chimney drain is available, then a zone of silty sand or sandy silt can be considered as a line of defence against piping in dispersive soil used for a homogeneous earth dam. Such a chimney may not be pervious enough to act as a drain but it would be expected to prevent piping. Watermeyer *et al.* (1991) referred to the use of geotextiles as drainage filters in the Bloemhock dam in South Africa. Sherard *et al.* (1977) maintained that many homogeneous dams without filters, in which dispersive clay has been properly compacted, experience no leaks and so no failure has occurred. If a leak developed under unusual conditions such as cracking due to drying consequent upon a long period of low water level or earthquake shock, then failure could occur. In order to increase the security of such a dam, a blanket of sand filter can be placed on the downstream side covered with a weighted berm.

Alternatively, hydrated lime, pulverized fly ash, gypsum and aluminium sulphate have been used to treat dispersive clays used in earth dams. The type of stabilization undertaken depends on the properties of the soil, especially the ESP and the SAR. The quantity of hydrated lime used should be that which raises the shrinkage limit to a value near saturation moisture content based on the compaction density to be achieved in the embankment. Usually lime treatment is applied to the outer 0.3 m of the surface of the embankment. McDaniel & Decker (1979) found that the addition of 4% by weight of hydrated lime, converted dispersive soil to non-dispersive soil. However, homogeneous mixing of small quantities of lime may not be achieved, and mixing, besides introducing brittleness, disrupts work which can lead to shrinkage cracks developing in the dam.

Indraratna *et al.* (1991) showed that pulverized fly ash, derived from burning lignite, and with a relatively high content of CaO, can be used to stabilize dispersive soil. They claimed that such fly ash provides divalent and trivalent cations which promote flocculation of dispersant clay particles. Moreover, being a pozzolanic material, fly ash encourages self-hardening with time. Hence, if a small amount (around 6%) is mixed with dispersive soil, the soil may gain in strength and become resistant to erosion. However, it must be remembered that the composition and physical properties of fly ash vary depending on the coal used and the firing process, and therefore need to be determined in order to ascertain whether a particular fly ash will be suitable as a stabilizing agent.

Because of its relatively low cost and reasonable solubility in water, gypsum, when in a very finely divided powder form, is another stabilizing material which can be used. The rate of base exchange reaction is controlled by its solubility in water with a high pH value. The gypsum is mixed with the soil during construction. The quantity of gypsum added is equivalent to the excess sodium which it is required to replace in order to bring ESP values within the desired limits. The water in a reservoir can also be dosed with gypsum so that as seepage occurs, deflocculation in the soil is prevented.

Aluminium sulphate or alum also has been added to dispersive clays. Although alum is highly soluble in water, it only effects a cation exchange reaction (that is, it is not cementitious) and it is strongly acidic. Nonetheless Bourdeaux & Imaizumi (1977) stabilized dispersive clay at the Sobradinho dam site, Brazil, by using 0.6% aluminium sulphate, by dry weight of soil.

Dispersive soils also can present problems in earthworks, such as those required for roads on both the fill and cut slopes relating thereto. In the case of embankment fills, dispersive soil can be used provided it is covered by an adequate depth of better class material. Care has to be exercised in the placement and compaction of the fill layers so that no layer is left exposed during construction for such a period that it can shrink and crack, and thus weaken the fill. As in the case of the construction of earth dams, the dispersive soil material should be placed and compacted at 2% above its optimum moisture content to inhibit shrinkage and cracking therein. Where seepage areas or springs are located along the alignment of a road embankment that has to be constructed of dispersive material, special care has to be exercised in the provision of adequate subsoil drainage to such areas, otherwise the long-term stability of the embankment could be jeopardised by the development of piping in the dispersive soil. In some instances, to reduce erosion, it is prudent to stabilize the outer 0.3 m or so, of dispersive soil material in an embankment with lime, the soil material and lime being mixed in bulk prior to placement rather than being mixed *in situ*.

To minimize potential settlement and erosion problems when dispersive soil fill is placed against a structure such as a bridge abutment and wingwall or a culvert, such structures, where practically possible, are provided with sloping soil interfaces such that the soil settles on to the interface. In this way the possibility of cracks developing in the soil which could lead to piping erosion is reduced.

In the case of road and other earthwork cut slopes in dispersive soil, unless adequately protected from surface erosion, severe runnel and gulling erosion can develop thereon. To some extent, this problem can be reduced by providing a steeper than normal slope, for instance, 0.5 vertical in 1.5 horizontal, as the dispersive soil usually possesses adequate cohesion to stand in a stable condition up to a height of about 3 m (under natural conditions near-vertical slopes in erosion gullies and dongas in dispersive colluvium frequently stand satisfactorily to heights of 5 m). The steeper than normal slope means that comparatively less rain falls on the slope so that the amount of slope erosion is reduced. In the relatively few instances where cuts are located in more than 3 m of dispersive soil, flatter slopes of about 1 vertical in 2 horizontal have to be employed. Such slopes have to be adequately vegetated immediately on completion to limit erosion. Adequate open channel drainage (1 to 2 m in width) has to be provided at the toes of cut slopes. These require cleaning maintenance from time to time. Such channel drains should be concrete-lined to limit erosion along their length.

Dispersive soils have low natural fertility and frequently they are calcareous and have a relatively high pH value (8.0 to 8.5). Therefore, it can be difficult to establish and maintain suitable vegetation on fill constructed of, and cut slopes in, dispersive soils in order to inhibit surface erosion. Apart from adequate artificial fertilization, it is usually necessary to place topsoil on such slopes to ensure satisfactory vegetative growth. The steepest slope that can be satisfactorily topsoiled has a gradient of about 1 vertical by 1 horizontal, and even such a slope may have to have artificial anti-erosion measures installed on it to hold the topsoil and vegetation in place.

Collapsible soils

The causes of collapse

Soils such as loess, brickearth, certain wind-blown silts and some tropical residual clay soils may possess the potential to collapse. Such soils possess porous textures with high void ratios and relatively low densities. They often have sufficient void space in their natural state to hold their liquid limit moisture at saturation. At their natural low moisture content these soils possess high apparent strength but they are susceptible to large reductions in void ratio upon wetting. In other words, the metastable texture collapses as the bonds between the grains (or clay peds in the case of tropical residual clay soils) break down when the soil is wetted. Hence the collapse process represents a rearrangement of soil particles into a denser state of packing. Collapse on saturation normally only takes a short period of time although, for silty and sandy soils, the more clay such a soil contains, the longer the period tends to be.

The silt-sized aeolian deposits generally are characterized by a lack of stratification and uniform sorting. Quartz is the principal mineral present in these soils along with feldspar, mica and clay minerals. Illite is often the most abundant clay mineral but this may not be the case in brickearth. Mixed-layer clay minerals, montmorillonite, kaolinite, chlorite and vermiculite also have been reported, sometimes in appreciable amounts. Increasing clay mineral content decreases the likelihood of collapse. Calcium carbonate may occur as grains, as thin tube infillings and as concretions. In weathered soils the amount of calcium carbonate is significantly less than in unweathered soils and indeed it may be absent.

The fabric of collapsible soils generally takes the form of a loose skeleton of grains (generally quartz) and microaggregates (assemblages of clay or clay and silty clay particles). These tend to be separate from each other, being connected by bonds and bridges, with uniformly distributed pores. The bridges are formed of clay-sized minerals, consisting of clay minerals, fine quartz, feldspar or calcite. Surface coatings of clay minerals may be present on coarser grains. Silica and iron oxide may be concentrated as cement at grain contacts and amorphous overgrowths of silica occur on grains. As grains are not in contact, mechanical behaviour is governed by the structure and quality of bonds and bridges.

The structural stability of collapsible soils is not only related to the origin of the material, to its mode of transport and depositional environment but also to the amount of weathering undergone. For instance, Gao (1988) pointed out that the weakly weathered loess of the northwest of the loess plateau in China has a high potential for collapse whereas the weathered loess of the southeast of the plateau is relatively stable and that the features associated with collapsible loess are disappearing gradually. Moreover, in more finely textured loess deposits, high capillary potential plus high perched groundwater conditions have caused loess to collapse naturally through time, thereby reducing its porosity. This reduction in porosity, combined with high liquid limit, makes the possibility of collapse less likely. Gao concluded that usually highly collapsible loess occurs in regions near the source of the loess where its thickness is at a maximum, and where the landscape and/or the climatic conditions are not conducive to development of long-term saturated conditions within the soil. This view was supported by Grabowska-Olszewska (1988) who found that in Poland collapse is most frequent in the youngest loess and that it is almost exclusively restricted to loess which contains slightly more than 10% particles of clay size. Such soils are characterized by a random texture and a carbonate content of less than 5%. They are more or less unweathered and possess a pronounced pattern of vertical jointing. The size of the pores is all important, Grabowska-Olszewska maintaining that collapse occurs as a result of pore space reduction taking place in pores greater than 1 μm in size and more especially in those exceeding 10 μm in size. Phien-wej *et al.* (1992) reported pore size in loess in northeast Thailand frequently varying from 200 to 500 μm. A recent bibliographic review of collapse in loess soils has been provided by Rogers *et al.* (1994).

Popescu (1986) maintained that there is a limiting value of pressure, defined as the collapse pressure, beyond which deformation of soil increases appreciably. The collapse pressure varies with the degree of saturation. He defined truly collapsible soils as those in which the collapse pressure is less than the overburden pressure. In other words, such soils collapse when saturated since the soil fabric cannot support the weight of the overburden. When the saturation collapse pressure exceeds the overburden pressure soils are capable of supporting a certain level of stress on saturation and Popescu defined these soils as conditionally collapsible soils. The maximum load which such soils can support is the difference between the saturation collapse and overburden pressures. Phien-wej *et al.* (1992) concluded that the critical pressure at which collapse of the soil fabric begins was greater in soils with smaller moisture content. Nonetheless, under the lowest natural moisture content the soils investigated posed a severe problem on wetting (the collapse-potential was as high as 12.5% at 5% natural moisture content). During the wet season when the natural moisture content could rise to 12% there was a reduction in the collapse potential to around 4%.

Several collapse criteria have been proposed for predicting whether a soil is liable to collapse upon saturation and loading. For instance, Clevenger (1958) suggested a criterion for collapsibility based on dry density, that is, if the dry density is less than 1.28 Mg/m^3, then the soil is liable to significant settlement. On the other hand, if the dry density is greater than 1.44 Mg/m^3, then the amount of collapse should be small, while at intermediate densities the settlements are transitional. Gibbs & Bara (1962) suggested the use of dry density and liquid limit as criteria to distinguish between collapsible and non-collapsible soil types. Their method is based on the premise that a soil which has enough void space to hold its liquid limit moisture content at saturation is susceptible to collapse on wetting. This criterion only applies if the soil is uncemented and the liquid limit is above 20%. When the liquidity index in such soils approaches or exceeds 1 then collapse may be imminent. As the clay content of a collapsible soil increases, the saturation moisture content becomes less than the liquid limit so that such deposits are relatively stable. However, Northmore *et al.* (1996) concluded that this method did not provide a satisfactory means of identifying the potential metastability of brickearth. More simply, Handy (1973) suggested that collapsibility could be determined from the ratio of liquid limit to saturation moisture content. Soils in which this was less than 1 were collapsible, while if it was greater than 1 they were safe.

Collapse criteria have been proposed which depend upon the void ratio at the liquid limit (e_l) and plastic limit (e_p). According to Audric & Bouquier (1976), collapse is probable when the natural void ratio (e_o) is higher than a critical void ratio (e_c) which depends on e_l and e_p. They quoted the following criteria as providing fairly good estimates of the likelihood of collapse:

$$e_c = e_l \tag{3}$$

$$e_c = 0.85e_l + 15e_p. \tag{4}$$

Fookes & Best (1969) proposed a collapse index (i_c) which also involved these void ratios and was as follows:

$$i_c = \frac{e_o - e_p}{e_l - e_p}. \tag{5}$$

Previously Feda (1966) had proposed the following collapse index:

$$i_c = \frac{m/S_r - PL}{PI} \tag{6}$$

in which m is the natural moisture content, S_r is the degree of saturation, PL is the plastic limit and PI is the plasticity index. Feda also proposed that the soil must have a critical porosity of 40% or above and that an imposed load must be sufficiently high to cause structural collapse when the soil is wetted. He suggested that if the collapse index was greater than 0.85, then this was indicative of metastable soils. The Feda criterion is more conservative than that of Equation (5). However, Northmore *et al.* (1996) suggested that a lower critical value of collapse index, that is 0.22, was more appropriate for the brickearths of south Essex. Derbyshire & Mellors (1988) also had referred to a lower collapse index for the brickearths of Kent. This may be due to the greater degree of sorting of brickearths than loess.

The absolute collapse index (i_{ac}) also can be used to predict collapse, it being:

$$i_{ac} = m/S_r - PL. \tag{7}$$

According to Northmore *et al.* (1996) the critical value of absolute collapse index for the brickearths from south Essex would appear to be 6 above which they are metastable.

The oedometer test can be used to assess the degree of collapsibility. For example, Jennings & Knight (1975) developed the double oedometer test for assessing the response of soil to wetting and loading at different stress levels. The test involves loading an undisturbed specimen at natural moisture content in the oedometer up to a given load. At this point the specimen is flooded and the resulting collapse strain, if any, is recorded. Then the specimen is subjected to further loading. The test procedure subsequently was modified by Houston *et al.* (1988). The total consolidation upon flooding can be described in terms of the coefficient of collapsibility, (C_{col}) given by Feda (1988) as:

$$C_{col} = \Delta h/h = \frac{\Delta e}{1+e} \tag{8}$$

in which Δh is the change in height of the specimen after flooding, h is the height of the specimen before flooding, Δe is the change in void ratio of the specimen upon flooding and e is the void ratio of the specimen prior to flooding. Equation (8) can be written in terms of strain where

$$C_{col} = \frac{\Delta\epsilon}{1-\epsilon} \tag{9}$$

where $\Delta\epsilon$ represents the additional strain upon flooding and ϵ is the strain at the same stress before flooding. Table 5 provides an indication of the potential severity of collapse. This table indicates that those soils which undergo more than 1% collapse can be regarded as metastable. However, in China a figure of 1.5% is taken (Lin & Wang 1988) and in the USA values exceeding 2% are regarded as indicative of soils susceptible to collapse (Lutenegger & Hallberg 1988).

Audric & Bouquier (1976) described a series of consolidated undrained triaxial tests carried out, at natural moisture content and after wetting, on loess soil from Normandy. When collapsible loess was tested the

Table 5. *Collapse percentage as an indication of potential severity (after Jennings & Knight 1975)*

Collapse (%)	Severity of problem
0–1	No problem
1–5	Moderate trouble
5–10	Trouble
10–20	Severe trouble
Above 20	Very severe trouble

deviator stress reached a peak at rather small values of axial strain and then decreased with further strain. The pore water pressures continued to increase after the peak deviator stress had been reached. By contrast, in non-collapsible soils the deviator stress continued to increase and there was only a small increase in pore water pressure. The shear strength of the collapsible loess was always less than that of the non-collapsible type. Northmore *et al.* (1996) noted the same as far as the shear strength of collapsible and non-collapsible brick-earth from south Essex was concerned.

Engineering control methods

From the above it may be concluded that significant settlements can take place beneath structures in collapsible soils after they have been wetted (in some cases in the order of metres, Feda *et al.* 1993). These have led to foundation failures (Clevenger 1958; Phien-wej *et al.* 1992). Clemence & Finbarr (1981) recorded a number of techniques which could be used to stabilize collapsible soils. These are summarized in Table 6. Evstatiev (1988) also provided a survey of methods which can be used to improve the behaviour of collapsible soils. The methods suggested by Evstatiev include:

1. compaction by rollers, heavy tampers, soil piles, vibration, explosions;
2. addition of coarser material;
3. stabilization by injection or mixing with binders and chemical reagents;
4. replacement with soil 'cushions' to redistribute the loading;
5. jet grouting;
6. reinforcement;
7. use of geomembranes;
8. desiccation by drainage, electro-osmosis or hygroscopic materials;
9. slope regrading and the use of vegetation.

Prevention of saturation around house foundations by using flexible drains and pipes and ensuring that run-off is kept away by the use of concrete aprons will often be sufficient to avoid collapse taking place.

Another problem which may be associated with loess soils is the development of pipe systems. Extensive pipe systems, which have been referred to as loess karst, may run sub-parallel to a slope surface and the pipes may have diameters up to 2 m. Pipes tend to develop by weathering and widening taking place along the joint systems in loess. The depths to which pipes develop may be inhibited by changes in permeability associated with the occurrence of palaeosols.

Table 6. *Methods of treating collapsible foundations (based on Clemence & Finbarr 1981)*

Depth of subsoil treatment	Foundation treatment
	A. Current and past methods
0–1.4 m	Moistening and compaction (conventional extra heavy impact or vibratory rollers)
1.5–10 m	Over-excavation and recompaction (earth pads with or without stabilization by additives such as cement or lime). Vibro-flotation (free draining soils). Vibroreplacement (stone columns). Dynamic compaction. Compaction piles. Injection of lime. Lime piles and columns. Jet grouting. Ponding or flooding (if no impervious layer exists). Heat treatment to solidify the soils in place.
Over 10 m	Any of the aforementioned or combinations of the aforementioned, where applicable. Ponding and infiltration wells, or ponding and infiltration wells with the use of explosive.
	B. Possible future methods
	Ultrasonics to produce vibrations that will destroy the bonding mechanics of the soil. Electrochemical treatment. Grouting to fill pores.

Conclusions

Swelling, shrinkage, dispersion and collapse of engineering soils are all largely controlled by their microfabric, mineralogy and chemistry.

The main cause of swelling and shrinkage is the presence of swelling clay minerals such as montmorillonite. Swell–shrink response of a clay soil is also strongly influenced by differences in the period and amount of precipitation and evapotranspiration. This can be influenced by drainage, leakage and vegetation as well as climate. The potential for volume change of clay is governed by its initial moisture content, initial density or void ratio, its microfabric and the vertical stress, as well as the type and amount of clay mineral present. Remoulded expansive clays tend to swell more than undisturbed or cemented ones. The amount of swelling is also influenced by permeability during a single wet season. Swell–shrink behaviour of a clay soil under a given state of applied stress is controlled by changes in soil suction, the relationship between suction and water content being dependent on the clay mineralogy, the microfabric and the pore water chemistry. Volume changes can be predicted using empirical methods relating expansiveness to various index properties, by soil suction methods and by tests in the oedometer.

Engineering solutions to control swelling and shrinkage include provision of a foundation and structure that can tolerate movements without unacceptable damage, isolation of the foundation and structure from the effects of the soil and alteration or control of the ground conditions. Movement tolerant foundations are used where differential movements up to 25 mm are likely, while isolation of the foundation and structure is adopted for 'severe" and 'very severe" ground conditions. However, moisture control is probably the most important single factor. This can be done physically (by barriers, by the avoidance of leakages near the building, by the control of vegetation, by replacement of the soil) or by chemical stabilization.

Dispersion occurs in soils when the repulsive forces between clay particles exceed the attractive forces, hence causing deflocculation so that in the presence of relatively pure water particles repel each other. Erosion by flowing water usually requires a certain velocity of flow but in dispersive soils there is no threshold velocity because the soils contain a higher level of dissolved sodium in their pore water than ordinary soils. The presence of exchangeable sodium is the main chemical factor contributing towards dispersive soil behaviour. Routine geotechnical tests cannot distinguish dispersive from non-dispersive soils. An alternative method involves a rating system that utilizes a number of tests and properties that have been used to recognize dispersive soils.

In some areas, dispersive soils have to be used for constructing earth dams and embankments. With careful construction control and the incorporation of filters, successful construction is possible. Proper compaction (at 2% above optimum moisture content to inhibit shrinking and cracking) has proved successful in preventing piping. Alternatively, chemical treatment using hydrated lime, pulverized fly ash, gypsum or aluminium sulphate has also been applied. Erosion can also be reduced in road embankments by using steeper than normal slopes. The development of vegetation on the slopes also helps reduce erosion.

Collapse occurs in certain soils with high void ratios and relatively low densities. At their natural low moisture content these soils possess high apparent strength but are susceptible to large reductions in void ratio upon wetting. The metastable texture of the soils collapses as the bonds between silt-size particles break down on wetting. Such collapse usually takes place fairly quickly. Loess, brickearth and some wind-blown silts are all susceptible to collapse as well as some tropical residual clay soils. Several indices, based on geotechnical properties, have been proposed to assess collapsibility but probably none is applicable to all soils.

Current methods to treat collapsibility include wetting with heavy compaction, over-excavation and recompaction, vibroflotation, vibroreplacement, dynamic compaction, compaction piles, injection of lime, lime piles, jet grouting, reinforcement, geomembranes, drainage or drying of the ground and ponding or flooding, possibly with the use of vibration or explosives. Slopes have been treated by regrading and the use of vegetation. Prevention of saturation may be sufficient to avoid collapse in the vicinity of existing light structures such as houses.

Acknowledgements. This paper is published with the permission the Director of the British Geological Survey (NERC).

References

ANON 1980*a*. *Low-rise Buildings on Shrinkable Clay Soils, Part 2*. Building Research Establishment, Digest 241. Her Majesty's Stationery Office, London.

ANON 1980*b*. *Low-rise Buildings on Shrinkable Clay Soils, Part 3*. Building Research Establishment, Digest 242. Her Majesty's Stationery Office, London.

ANON 1996. Cost of domestic and commercial property claims. *Insurance Trends*, January 1996, p. 36.

AITCHISON, G. D. & WOOD, C. C. 1965. Some interactions of compaction, permeability and post-construction deflocculation affecting the probability of piping failures in small dams. *In*: *Proceedings of the 6th International Conference on Soil Mechanics and Foundation Engineering, Montreal*, **2**, 442–446.

ATKINSON, J. H., CHARLES, J. A. & MHACH, H. K. 1990. Examination of erosion resistance of clays in embankment dams. *Quarterly Journal of Engineering Geology*, **23**, 103–108.

AUDRIC, T. & BOUQUIER, L. 1976. Collapsing behaviour of some loess soils from Normandy. *Quarterly Journal of Engineering Geology*, 9 265–278.

BELL, F. G. 1988. Stabilization and treatment of clay soils with lime. Part II: Some applications. *Ground Engineering*, **21**, 2, 22–30.

——1995. Cement stabilization of clay soils with examples. *Environmental and Engineering Geoscience*, **1**, 139–151.

—— & MAUD, R. R. 1994. Dispersive soils: a review from a South African perspective. *Quarterly Journal of Engineering Geology*, **27**, 195–210.

—— & ——1995. Expansive clays and construction, especially of low rise structures: a view point from Natal, South Africa. *Environmental and Engineering Geoscience*, **1**, 41–59.

BOURDEAUX, G. & IMAIZUMI, H. 1977. Dispersive clay at Sobrandinho dam. *In*: SHERARD, J. L. & DECKER, R. S. (eds) *Proceedings of a Symposium on Dispersive Clays, Related Piping and Erosion in Geotechnical Projects*, ASTM Special Publication 623, Philadelphia, 13–24.

BRACKLEY, I. J. A. 1975. *Interrelationship of the Factors Affecting Heave of an Expansive, Unsaturated Clay Soil.* PhD Thesis, Department of Civil Engineering, University of Natal, Durban, South Africa.

——1980. Prediction of soil heave from soil suction measurements. *In: Proceedings of the 7th Regional Conference for Africa on Soil Mechanics and Foundation Engineering*, Accra, Ghana, 159–167.

BURLAND, J. B. 1963. The estimation of field effective stresses and the prediction of total heave using a revised method of analysing the double oedometer test. *Transactions of the South African Institution of Civil Engineers*, **5**, 19–24.

—— & WROTH, C. P. 1975. Allowable and differential settlement of structures including damage and soil-structure interaction. *In: Settlement of Structures.* British Geotechnical Society, Pentech Press, London, 611–654.

CHANDLER, R. J., CRILLY, M. S. & MONTGOMERY-SMITH, G. 1992. A low-cost method of assessing clay desiccation for low-rise buildings. *Proceedings of the Institution of Civil Engineers*, **92**, 82–89.

CHEN, F. H. 1988. *Foundations on Expansive Soils.* Elsevier, Amsterdam.

CLEMENCE, S. P & FINBARR, A. O. 1981. Design considerations for collapsible soils. *Proceedings of the American Society of Civil Engineers, Journal Geotechnical Engineering Division*, **107**, 305–317.

CLEVENGER, W. A. 1958. Experience with loess as foundation material. *Proceedings of the American Society of Civil Engineers, Journal of the Soil Mechanics and Foundations Division*, **85**, 151–180.

CRAFT, D. C. & ACCIARDI, R. G. 1984. Failure of pore water analyses for dispersion. *Proceedings of the American Society of Civil Engineers, Journal of the Geotechnical Engineering Division*, **110**, 459–472.

DERBYSHIRE, E. & MELLORS, T. W. 1988. Geological and geotechnical characteristics of some loess and loessic soils from China and Britain: a comparison. *Engineering Geology*, **25**, 135–175.

EVSTATIEV, D. 1988. Loess improvement methods. *Engineering Geology*, **25**, 341–366.

FEDA, J. 1966. Structural stability of subsidence loess from Praha-Dejvice. *Engineering Geology*, **1**, 201–219.

——1988. Collapse of loess on wetting. *Engineering Geology*, **25**, 263–269.

——, BOHAC, J. & HERLE, I. 1993. Compression of collapsed loess: studies on bonded and unbonded soils. *Engineering Geology*, **34**, 95–103.

FOOKES, P. G. & BEST, R. 1969. Consolidation characteristics of some late Pleistocene periglacial metastable soils of east Kent. *Quarterly Journal of Engineering Geology*, **2**, 103–128.

FORSTER, A., CULSHAW, M. G & BELL, F. G. 1995. The regional distribution of sulphate in rocks and soils of Britain. *In*: EDDLESTON, M., WALTHALL, S., CRIPPS, J. C. & CULSHAW, M. G. (eds) *Engineering Geology of Construction.* Geological Society, London, Engineering Geology Special Publication, **9**, 95–104.

FRYDMAN, S. & CALABRESI, F. G. 1987. Suggested standard for one-dimensional testing. *In: Proceedings of the 6th International Conference on Expansive Soils*, New Delhi, India, **1**, 91–100.

GAO, G. 1988. Formation and development of the structure of collapsing loess in China. *Engineering Geology*, **25**, 235–245.

GERBER, A. & HARMSE, H. J. VON M. 1987. Proposed procedure for identification of dispersive soils by chemical testing. *The Civil Engineer in South Africa*, **29**, 397–399.

GIBBS, H. H. & BARA, J. P. 1962. Predicting surface subsidence from basic soil tests. ASTM Special Technical Publication, **322**, 231–246.

GRABOWSKA-OLSZEWSKA, B. 1988. Engineering geological problems of loess in Poland. *Engineering Geology*, **25**, 177–199.

HANDY, R. L. 1973. Collapsible loess in Iowa. *Proceedings of the American Society of Soil Science*, **37**, 281–284.

HOUSTON, J. L., HOUSTON, W. L. & SPADOLA, D. J. 1988. Prediction of field collapse of soils due to wetting. *Proceedings of the American Society of Civil Engineers, Journal of the Geotechnical Engineering Division*, **114**, 40–58.

HU HAITAO 1993. The types and distribution of geohazards in China and some counter-measures. *Hydrogeology and Engineering Geology*, **20**, 2, 1–7.

INDRARATNA, B., NUTALAYA. P. & KUGANENTHIRA, N. 1991. Stabilization of a dispersive soil by blending with fly ash. *Quarterly Journal of Engineering Geology*, **24**, 275–290.

JENNINGS, J. E. & KNIGHT, K. 1957. The prediction of total heave from the double oedometer test. *Transactions of the South African Institution of Civil Engineers*, **7**, 285–291.

—— & ——1975. A guide to construction on or with materials exhibiting additional settlement due to collapse of grain structure. *In: Proceedings of the 6th African Conference on Soil Mechanics and Foundation Engineering*, Durban, 99–105.

LIN, Z. G. & WANG, S. J. 1988. Collapsibility and deformation characteristics of deep-seated loess in China. *Engineering Geology*, **25**, 271–282.

LUTENEGGER, A. J. & HALLBERG, G. R. 1988. Stability of loess. *Engineering Geology*, **25**, 247–261.

MCDANIEL, T. N. & DECKER, R. S. 1979. Dispersive soil problem at Los Esteros Dam. *Proceedings of the American Society of Civil Engineers, Journal of the Geotechnical Engineering Division*, **105**, 1017–1030.

MCQUEEN, I. S. & MILLER, R. F. 1968. Calibration and evaluation of wide ring gravimetric methods for measuring moisture stress. *Soil Science*, **106**, 225–231.

NORTHMORE, K. J., BELL, F. G. & CULSHAW, M. G. 1996. The engineering properties and behaviour of the brickearth of south Essex. *Quarterly Journal of Engineering Geology*, **29**, 147–161.

NUHFER, E. B., PROCTER, R. J. & MOSER, P. H. 1993. *The Citizens' Guide to Geologic Hazards.* The American Institute of Professional Geologists, Colorado.

O'NEILL, M. W. & POORMOAYED, A. M. 1980. Methodology for foundations on expansive clays. *Proceedings of the American Society of Civil Engineers, Journal of the Geotechnical Engineering Division*, **106**, 1345–1367.

PHIEN-WEJ, N., PIENTONG, T. & BALASUBRAMANIAM, A. S. 1992. Collapse and strength characteristics of loess in Thailand. *Engineering Geology*, **32**, 59–72.

POPESCU, M. E. 1979. Engineering problems associated with expansive clays from Romania. *Engineering Geology*, **24**, 43–53.

——1986. A comparison of the behaviour of swelling and collapsing soils. *Engineering Geology*, **23**, 145–163.

ROBINSON, G. D. & SPIEKER, A. M. (eds) 1978. *"Nature to be commanded..." Earth-science maps applied to land and water management*. United States Geological Survey Professional Paper 950.

ROGERS, C. D. F., DIJKSTRA, T. A. & SMALLEY, I. J. 1994. Hydroconsolidation and subsidence of loess: studies from China, Russia, North America and Europe. *Engineering Geology*, **37**, 83–113.

SCHMERTMANN, J. H. 1969. Swell sensitivity. *Geotechnique*, **19**, 530–533.

SHERARD, J. L., DUNNIGAN, L. P., DECKER, R. S. & STEELE, E. F. 1976. Pinhole test for identifying dispersive soils. *Proceedings of the American Society of Civil Engineers, Journal of the Geotechnical Engineering Division*, **102**, 69–85.

——, DUNNIGAN, L. P. & DECKER, R. S. 1977. Some engineering problems with dispersive clays. *In*: SHERARD, J. L. & DECKER, R. S. (eds) *Proceedings of a Symposium on Dispersive Clays, Related Piping and Erosion in Geotechnical Projects*, ASTM Special Publication, **623**, 3–12.

STAMATOPOULOS, A. C., CHRISTADOULIAS, J. C. & GIANNAROS, H. Ch. 1992. Treatment for expansive soils for reducing swell potential and increasing strength. *Quarterly Journal of Engineering Geology*, **25**, 301–312.

VAN DER MERWE, D. H. 1964. The prediction of heave from the plasticity index and the percentage clay fraction. *The Civil Engineer in South Africa*, **6**, 6, 103–107.

WATERMEYER, C. F., BOTHA, G. R. & HALL, B. E. 1991. Countering potential piping at an earth dam on dispersive soils. *In*: BLIGHT, G. E., FOURIE, A. B., LUKER, I., MOUTON, D. J. & SCHEURENBURG, R. J. (eds) *Geotechnics* in the African Environment. Balkema, Rotterdam, 321–328.

WEINERT, H. H. 1980. *The Natural Road Construction Materials of South Africa*. Academica, Cape Town.

WESTON, D. J. 1980. Expansive road treatment for southern Africa. *In*: *Proceedings of the 4th International Conference on Expansive Soils*, Denver, **1**, 339–360.

WILLIAMS, A. A. B. 1980. Severe heaving of a block of flats near Kimberley. *In*: *Proceedings of the 7th Regional Conference for Africa on Soil Mechanics and Foundation Engineering*, Accra, **1**, 301–309.

—— & DONALDSON, G. W. 1980. Building on expansive soils in South Africa: 1973–1980. *In*: *Proceedings of the 4th International Conference on Expansive Soils*, Denver, **1**, 834–844.

WILSON, C. & MELIS, L. 1991. Breaching of an earth dam in the Western Cape by piping. *In*: BLIGHT, G. E., FOURIE, A. B., LUKER, I., MOUTON, D. J. & SCHEURENBURG, R. J. (eds) *Geotechnics in the African Environment*. Balkema, Rotterdam, 301–312.

Index